INSTRUCTOR'S GUIDE FOR

BIOLOGY

FIFTH EDITION

CAMPBELL • REECE • MITCHELL

MARK SHERIDAN, Ph.D

North Dakota State University

with contributions from

NINA CARIS, Ph.D.
HAROLD UNDERWOOD, Ph.D.

Texas A&M University

An imprint of Addison Wesley Longman, Inc.

Menlo Park, California • Reading, Massachusetts • New York
Harlow, England • Don Mills, Ontario • Sydney
Mexico City • Madrid • Amsterdam

The Benjamin/Cummings Publishing Company, Inc.
2725 Sand Hill Road
Menlo Park, California 94025

CONTENTS

TO THE INSTRUCTOR

The idea for a comprehensive lecture outline came from the observation that the most arduous and time-consuming task for a new instructor is reading and outlining the textbook. We thought it would be helpful to provide extensive lecture notes on material covered in the text, so that more time could be devoted to lecture enhancement—a luxury usually afforded only to experienced teachers. This set of lecture notes is only intended to be a beginning. It is *not* recommended that you lecture on *all* material presented here or lecture *only* from material presented in the text. For experienced teachers who are first-time users of Neil Campbell's *Biology*, we hope these notes will ease the transition to a new book.

Each chapter in the *Instructor's Guide* has five major components. The lecture *Outline* is a complete outline of the textbook chapter, usually down to level two headings. *Objectives* are provided for all major topics covered in the text. These objectives are primarily at the knowledge and comprehension level, but whenever possible, higher-order categories are included. *Key Terms* for students to define or identify are listed at the end of the objectives. In most cases they are in order of their appearance in the chapter. The *Lecture Notes* are a summary of chapter content in outline form. New terms are defined and relevant points listed below in bulleted format for ease in lecturing. *References* listed at the end of each chapter are by no means an inclusive list of readings, but rather are the references actually used to prepare lecture notes. For practical reasons, we relied heavily on references already at our disposal; there are many excellent texts not cited here that would serve just as well.

How To Use These Notes

Read the Textbook Read the textbook chapter to become familiar with material your students will read and to note pedagogical aids that can be incorporated into a lecture. The artwork in Campbell's *Biology* is skillfully integrated with the text, and almost an entire lecture can be centered around one or two effective diagrams, charts, or tables. Additionally, the end-of-chapter challenge questions can be used to generate class discussion or to help students synthesize newly presented material.

Select Course Objectives Select the course objectives that suit your specific goals for the course and that are appropriate in the context of your department, college, or university. Once the objectives for your course are mapped out, they can be used as a guide for:

- *Lecture Preparation.* Preparing lectures with specific objectives in mind will help focus lecture material on important concepts and avoid overemphasis on the trivial – a common criticism of introductory science courses. Try to strike a balance between knowledge-based objectives and those requiring higher-order thought processes.

- *Advanced Organizers and Transitions for Lecturing.* An effective way to provide organizational cues is to change objectives from declarative statements to questions and use them in lecture as you make a transition from one topic to another. Prefacing a topic with a question creates anticipation. As students wait to hear the answer they attend better to the material.

- *Reading Assignments.* If students are not held responsible for all material in the text, you can use the course objectives to determine selected reading assignments.

- *Study Questions.* Transforming course objectives into study questions can inform students of your expectations and enhance learning. As students answer questions, they not only take a more active role in learning, but their attention is focused on the concepts you believe are important.

- *Exam Preparation.* The course objectives can be used as a guide for both the content and cognitive level of exam questions. We highly recommend writing or selecting test questions that address specific course objectives and that are at or below the cognitive level of the objective being tested. Our exams generally contain 65-70% knowledge level questions and 30-35% comprehension and application level questions. For our students, we have found that an exam with 100% simple recall questions is not challenging enough, encourages rote memorization, and does not test for understanding or complex learning outcomes such as the ability to solve problems or to apply knowledge in new situations. An exam with more than 35-40% application type questions is perceived by the students as too difficult, and some slower students have difficulty completing the exam within the allotted time.

Modify Lecture Outline and Lecture Notes Once you have selected course objectives, modify the lecture outline and lecture notes to include only the material you wish to cover in the course. One measure of success for these lecture notes is how well they are used. It would delight us to see dog-eared versions, rearranged with cut and paste, three-hole punched in binders, highlights and scribbles in the margins. Customize the notes into your own.

Integrate Visual Ancillaries After the basic lecture notes are assembled, integrate the color acetates, slides, or text figures from the Instructor's Presentation CD-ROM you will use during the lecture presentation. The new set of beautiful color transparencies for *BIOLOGY,* Fifth Edition, is well done and all-inclusive. The type has been enlarged for large lecture rooms and we have found them invaluable for making effective presentations. For those who prefer to use computer-based presentation programs, the Instructor's Presentation CD-ROM includes all the figures in the book as well as animation sequences of dynamic processes that are difficult for students to visualize, and thus difficult to teach.

CHAPTER 1
INTRODUCTION: THEMES IN THE STUDY OF LIFE

OUTLINE

I. Life's Hierarchical Order
   A. The living world is a hierarchy, with each level of biological structure building on the level below it
   B. Each level of biological structure has emergent properties
   C. Cells are an organism's basic units of structure and function
   D. The continuity of life is based on heritable information in the form of DNA
   E. Structure and function are correlated at all levels of biological organization
   F. Organisms are open systems that interact continuously with their environments
   G. Regulatory mechanisms ensure a dynamic balance in living systems
II. Evolution, Unity, and Diversity
   A. Diversity and unity are the dual faces of life on Earth
   B. Evolution is the core theme of biology
III. Science as a Process
   A. Testable hypotheses are the hallmarks of the scientific process
   B. Science and technology are functions of society
   C. Biology is a multidisciplinary adventure

OBJECTIVES

After reading this chapter and attending lecture, the student should be able to:
1. Briefly describe unifying themes that pervade the science of biology.
2. Diagram the hierarchy of structural levels in biology.
3. Explain how the properties of life emerge from complex organization.
4. Describe seven emergent properties associated with life.
5. Distinguish between holism and reductionism.
6. Explain how technological breakthroughs contributed to the formulation of the cell theory and our current knowledge of the cell.
7. Distinguish between prokaryotic and eukaryotic cells.
8. Explain, in their own words, what is meant by "form fits function."
9. List the five kingdoms of life and distinguish among them.
10. Briefly describe how Charles Darwin's ideas contributed to the conceptual framework of biology.
11. Outline the scientific method.
12. Distinguish between inductive and deductive reasoning.
13. Explain how science and technology are interdependent.

KEY TERMS

emergent property	holism	evolution	control group
population	reductionism	natural selection	variable
community	prokaryotic	scientific method	experimental group
ecosystem	eukaryotic	hypothesis	deductive reasoning
biome	taxonomy	inductive reasoning	scientific theory
biogenesis			

LECTURE NOTES

Biology, the study of life, is a human endeavor resulting from an innate attraction to life in its diverse forms (E.O. Wilson's *biophilia*).

The science of biology is enormous in scope.

- It reaches across size scales from submicroscopic molecules to the global distribution of biological communities.
- It encompasses life over huge spans of time from contemporary organisms to ancestral life forms stretching back nearly four billion years.

As a science, biology is an ongoing process.

- As a result of new research methods developed over the past few decades, there has been an information explosion.
- Technological advances yield new information that may change the conceptual framework accepted by the majority of biologists.

With rapid information flow and new discoveries, biology is in a continuous state of flux. There are, however, enduring unifying themes that pervade the science of biology:

- A hierarchy of organization
- The cellular basis of life
- Heritable information
- The correlation between structure and function
- The interaction of organisms with their environment
- Unity in diversity
- Evolution: the core theme
- Scientific process: the hypothetico-deductive method

I.   **Life's Hierarchical Order**

  A. **The living world is a hierarchy, with each level of biological structure building on the level below it**

   A characteristic of life is a high degree of order. Biological organization is based on a hierarchy of structural levels, with each level building on the levels below it.

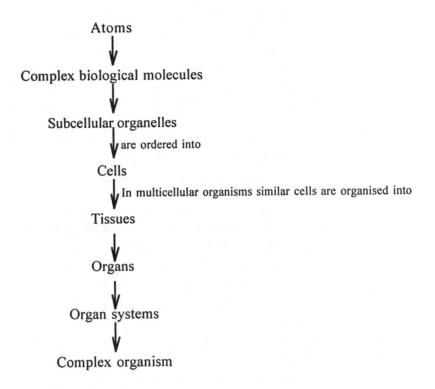

There are levels of organization beyond the individual organism:

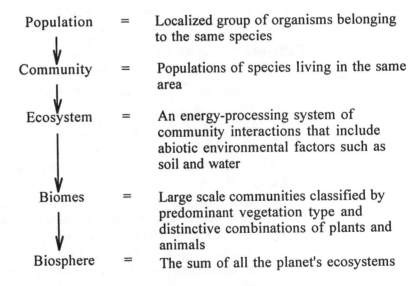

B. Each level of biological organization has emergent properties

Emergent property = Property that emerges as a result of interactions between components.

- With each step upward in the biological hierarchy, new properties emerge that were not present at the simpler organizational levels.
- Life is difficult to define because it is associated with numerous emergent properties that reflect a hierarchy of structural organization.

Some of the emergent properties and processes associated with life are the following:

1. *Order.* Organisms are highly ordered, and other characteristics of life emerge from this complex organization.

2. *Reproduction.* Organisms reproduce; life comes only from life (*biogenesis*).

3. *Growth and Development.* Heritable programs stored in DNA direct the species-specific pattern of growth and development.

4. *Energy Utilization.* Organisms take in and transform energy to do work, including the maintenance of their ordered state.

5. *Response to Environment.* Organisms respond to stimuli from their environment.

6. *Homeostasis.* Organisms regulate their internal environment to maintain a steady-state, even in the face of a fluctuating external environment.

7. *Evolutionary Adaptation.* Life evolves in response to interactions between organisms and their environment.

Because properties of life emerge from complex organization, it is impossible to fully explain a higher level of order by breaking it into its parts.

Holism = The principle that a higher level of order cannot be meaningfully explained by examining component parts in isolation.

- An organism is a living whole greater than the sum of its parts.
- For example, a cell dismantled to its chemical ingredients is no longer a cell.

It is also difficult to analyze a complex process without taking it apart.

Reductionism = The principle that a complex system can be understood by studying its component parts.

- Has been a powerful strategy in biology
- Example: Watson and Crick deduced the role of DNA in inheritance by studying its molecular structure.

The study of biology balances the reductionist strategy with the goal of understanding how the parts of cells, organisms, and populations are functionally integrated.

C. Cells are an organism's basic units of structure and function

The cell is an organism's basic unit of structure and function.

- Lowest level of structure capable of performing all activities of life.
- All organisms are composed of cells.
- May exist singly as unicellular organisms or as subunits of multicellular organisms.

The invention of the microscope led to the discovery of the cell and the formulation of the cell theory.

- Robert Hooke (1665) reported a description of his microscopic examination of cork. Hooke described tiny boxes which he called "cells" (really cell walls). The significance of this discovery was not recognized until 150 years later.
- Antonie van Leeuwenhok (1600's) used the microscope to observe living organisms such as microorganisms in pond water, blood cells, and animal sperm cells.
- Matthias Schleiden and Theodor Schwann (1839) reasoned from their own microscopic studies and those of others, that all living things are made of cells. This formed the basis for the *cell theory*.
- The cell theory has since been modified to include the idea that all cells come from preexisting cells.

Over the past 40 years, use of the electron microscope has revealed the complex ultrastructure of cells.

- Cells are bounded by *plasma membranes* that regulate passage of materials between the cell and its surroundings.
- All cells, at some stage, contain DNA.

Based on structural organization, there are two major kinds of cells: *prokaryotic* and *eukaryotic*.

Prokaryotic cell = Cell lacking membrane-bound organelles and a membrane-enclosed nucleus.

- Found only in the archaebacteria and bacteria
- Generally much smaller than eukaryotic cells
- Contains DNA that is *not* separated from the rest of the cell, as there is no membrane-bound nucleus
- Lacks membrane-bound organelles
- Almost all have tough external walls

Eukaryotic cell = Cell with a membrane-enclosed nucleus and membrane-enclosed organelles.

- Found in protists, plants, fungi, and animals
- Subdivided by internal membranes into different functional compartments called *organelles*
- Contains DNA that is segregated from the rest of the cell. DNA is organized with proteins into *chromosomes* that are located within the *nucleus*, the largest organelle of most cells.
- *Cytoplasm* surrounds the nucleus and contains various organelles of different functions
- Some cells have a tough *cell wall* outside the plasma membrane (e.g., plant cells). Animal cells lack cell walls.

Though structurally different, eukaryotic and prokaryotic cells have many similarities, especially in their chemical processes.

D. The continuity of life is based on heritable information in the form of DNA

Biological instructions for an organism's complex structure and function are encoded in DNA.

- Each DNA molecule is made of four types of chemical building blocks called *nucleotides.*
- The linear sequence of these four nucleotides encode the precise information in a gene, the unit of inheritance from parent to offspring.
- An organism's complex structural organization is specified by an enormous amount of coded information.

Inheritance is based on:

- A complex mechanism for copying DNA.
- Passing the information encoded in DNA from parent to offspring.

All forms of life use essentially the same genetic code.

- A particular nucleotide sequence provides the same information to one organism as it does to another.
- Differences among organisms reflect differences in nucleotide sequence.

E. Structure and function are correlated at all levels of biological organization

There is a relationship between an organism's structure and how it works. Form fits function.

- Biological structure gives clues about what it does and how it works.
- Knowing a structure's function gives insights about its construction.

- **This correlation is apparent at many levels of biological organization.**

F. Organisms are open systems that interact continuously with their environments

Organisms interact with their environment, which includes other organisms as well as abiotic factors.

- Both organism and environment are affected by the interaction between them.
- Ecosystem dynamics include two major processes:
  1. Nutrient cycling
  2. Energy flow (see Campbell, Figure 1.7)

G. Regulatory mechanisms ensure a dynamic balance in living systems

Regulation of biological processes is critical for maintaining the ordered state of life.

Many biological processes are self-regulating; that is, the product of a process regulates that process (= feedback regulation; see Campbell, Figure 1.8).

- Positive feedback speeds a process up
- Negative feedback slows a process down

Organisms and cells also use chemical mediators to help regulate processes.

- The hormone insulin, for example, signals cells in vertebrate organisms to take up glucose. As a result, blood glucose levels go down.
- In certain forms of diabetes mellitus, insulin is deficient and cells do not take up glucose as they should, and as a result, blood glucose levels remain high.

II.   Evolution, Unity, and Diversity

A. Diversity and unity are the dual faces of life on Earth

Biological diversity is enormous.

- Estimates of total diversity range from five million to over 30 million species.
- About 1.5 million species have been identified and named, including approximately 260,000 plants, 50,000 vertebrates, and 750,000 insects.

To make this diversity more comprehensible, biologists classify species into categories.

Taxonomy = Branch of biology concerned with naming and classifying organisms.

- Taxonomic groups are ranked into a hierarchy from the most to least inclusive category: *domain, kingdom, phylum, class, order, family, genus, species.*
- A six-kingdom system recognizes two prokaryotic groups and divides the Monera into the Archaebacteria and Eubacteria.
- The kingdoms of life recognized in the traditional five-kingdom system are Monera, Protista, Plantae, Fungi, and Animalia (see Campbell, Figure 1.10).

There is unity in the diversity of life forms at the lower levels of organization. Unity of life forms is evident in:

- A universal genetic code.
- Similar metabolic pathways (e.g., glycolysis).
- Similarities of cell structure (e.g., flagella of protozoans and mammalian sperm cells).

B. Evolution is the core theme of biology

Evolution is the one unifying biological theme.

- Life evolves. Species change over time and their history can be described as a branching tree of life.
- Species that are very similar share a common ancestor at a recent branch point on the phylogenetic tree.
- Less closely related organisms share a more ancient common ancestor.

- All life is connected and can be traced back to primeval prokaryotes that existed more than three billion years ago.

In 1859, Charles Darwin published *On the Origin of Species* in which he made two major points:

1. Species change, and contemporary species arose from a succession of ancestors through a process of "descent with modification."
2. A mechanism of evolutionary change is *natural selection*.

Darwin synthesized the concept of natural selection based upon the following observations:

- Individuals in a population of any species vary in many inheritable traits.
- Populations have the potential to produce more offspring than will survive or than the environment can support.
- Individuals with traits best suited to the environment leave a larger number of offspring, which increases the proportion of inheritable variations in the next generation. This differential reproductive success is what Darwin called *natural selection*.

Organisms' adaptations to their environments are the products of natural selection.

- Natural selection *does not create* adaptations; it merely increases the frequency of inherited variants that arise by chance.
- Adaptations are the result of the editing process of natural selection. When exposed to specific environmental pressures, certain inheritable variations favor the reproductive success of some individuals over others.

Darwin proposed that cumulative changes in a population over long time spans could produce a new species from an ancestral one.

Descent with modification accounts for both the unity and diversity of life.

- Similarities between two species may be a reflection of their descent from a common ancestor.
- Differences between species may be the result of natural selection modifying the ancestral equipment in different environmental contexts.

III. Science as a Process

A. Testable hypotheses are the hallmarks of the scientific process

As the science of life, biology has the characteristics associated with science in general.

Science is a way of knowing. It is a human endeavor that emerges from our curiosity about ourselves, the world, and the universe. Good scientists are people who:

- Ask questions about nature and believe those questions are answerable.
- Are curious, observant, and passionate in their quest for discovery.
- Are creative, imaginative, and intuitive.
- Are generally skeptics.

Scientific method = Process which outlines a series of steps used to answer questions.

- Is not a rigid procedure.
- Based on the conviction that natural phenomena have natural causes.
- Requires evidence to logically solve problems.

The key ingredient of the scientific process is the *hypothetico-deductive method*, which is an approach to problem-solving that involves:

1. Asking a question and formulating a tentative answer or hypothesis by inductive reasoning.
2. Using deductive reasoning to make predictions from the hypothesis and then testing the validity of those predictions.

Hypothesis = Educated guess proposed as a tentative answer to a specific question or problem.

Inductive reasoning = Making an inference from a set of specific observations to reach a general conclusion.

Deductive reasoning = Making an inference from general premises to specific consequences, which logically follow if the premises are true.

- Usually takes the form of *If...then* logic.
- In science, deductive reasoning usually involves predicting experimental results that are expected *if* the hypothesis is true.

> Some students cannot make the distinction between inductive and deductive reasoning. An effective teaching strategy is to let them actually experience both processes. To illustrate inductive reasoning, provide an every day scenario with enough pieces of information for student to hypothesize a *plausible* explanation for some event. Demonstrate deductive reasoning by asking students to solve a simple problem, based upon given assumptions.

Useful hypotheses have the following characteristics:

- *Hypotheses are possible causes.* Generalizations formed by induction are not necessarily hypotheses. Hypotheses should also be tentative explanations for observations or solutions to problems.
- *Hypotheses reflect past experience with similar questions.* Hypotheses are not just blind propositions, but are *educated* guesses based upon available evidence.
- *Multiple hypotheses should be proposed whenever possible.* The disadvantage of operating under only one hypothesis is that it might restrict the search for evidence in support of this hypothesis; scientists might bias their search, as well as neglect to consider other possible solutions.
- *Hypotheses must be testable via the hypothetico-deductive method.* Predictions made from hypotheses must be testable by making observations or performing experiments. This limits the scope of questions that science can answer.
- *Hypotheses can be eliminated, but not confirmed with absolute certainty.* If repeated experiments consistently disprove the predictions, then we can assume that the hypothesis is false. However, if repeated experimentation supports the deductions, we can only assume that the hypothesis *may* be true; accurate predictions can be made from false hypotheses. The more deductions that are tested and supported, the more confident we can be that the hypothesis is true.

Another feature of the scientific process is the controlled experiment which includes control and experimental groups.

Control group = In a controlled experiment, the group in which all variables are held constant.

- Controls are a necessary basis for comparison with the experimental group, which has been exposed to a single treatment variable.
- Allows conclusions to be made about the effect of experimental manipulation.
- Setting up the best controls is a key element of good experimental design.

Variable = Condition of an experiment that is subject to change and that may influence an experiment's outcome.

Experimental group = In a controlled experiment, the group in which one factor or treatment is varied.

Science is an ongoing process that is a self-correcting way of knowing. Scientists:

- Build on prior scientific knowledge.
- Try to replicate the observations and experiments of others to check on their conclusions.

- Share information through publications, seminars, meetings, and personal communication.

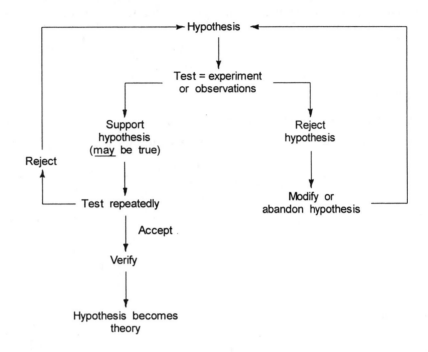

What really advances science is not just an accumulation of facts, but a new concept that collectively explains observations that previously seemed to be unrelated.

- Newton, Darwin, and Einstein stand out in the history of science because they synthesized ideas with great explanatory power.
- *Scientific theories* are comprehensive conceptual frameworks which are well supported by evidence and are widely accepted by the scientific community.

B. Science and technology are functions of society

Science and technology are interdependent.

- Technology extends our ability to observe and measure, which enables scientists to work on new questions that were previously unapproachable.
- Science, in turn, generates new information that makes technological inventions possible.
- Example: Watson and Crick's scientific discovery of DNA structure led to further investigation that enhanced our understanding of DNA, the genetic code, and how to transplant foreign genes into microorganisms. The biotechnology industry has capitalized on this knowledge to produce valuable pharmaceutical products such as human insulin.

We have a love-hate relationship with technology.

- Technology has improved our standard of living.
- The consequence of using technology also includes the creation of new problems such as increased population growth, acid rain, deforestation, global warming, nuclear accidents, ozone holes, toxic wastes, and endangered species.
- Solutions to these problems have as much to do with politics, economics, culture and values as with science and technology.

A better understanding of nature must remain the goal of science. Scientists should:

- Try to influence how technology is used.
- Help educate the public about the benefits and hazards of specific technologies.

C. Biology is a multidisciplinary adventure

Biology is a multidisciplinary science that integrates concepts from chemistry, physics and mathematics. Biology also embraces aspects of humanities and the social sciences.

REFERENCES

Campbell, N. *Biology*. 5th ed. Menlo Park, California: Benjamin/Cummings, 1998.

Moore, J.A. "Science as a Way of Knowing–Evolutionary Biology." *American Zoologist*, 24(2): 470-475, 1980.

CHAPTER 2
THE CHEMICAL
CONTEXT OF LIFE

OUTLINE
I. Chemical Elements and Compounds
   A. Matter consists of chemical elements in pure form and in combinations called compounds
   B. Life requires about 25 chemical elements
II. Atoms and Molecules
   A. Atomic structure determines the behavior of an element
   B. Atoms combine by chemical bonding to form molecules
   C. Weak chemical bonds play important roles in the chemistry of life
   D. A molecule's biological function is related to its shape
   E. Chemical reactions make and break chemical bonds

OBJECTIVES
After reading this chapter and attending lecture, the student should be able to:
1. Define element and compound.
2. State four elements essential to life that make up 96% of living matter.
3. Describe the structure of an atom.
4. Define and distinguish among atomic number, mass number, atomic weight, and valence.
5. Given the atomic number and mass number of an atom, determine the number of neutrons.
6. Explain why radioisotopes are important to biologists.
7. Explain how electron configuration influences the chemical behavior of an atom.
8. Explain the octet rule and predict how many bonds an atom might form.
9. Explain why the noble gases are so unreactive.
10. Define electronegativity and explain how it influences the formation of chemical bonds.
11. Distinguish among nonpolar covalent, polar covalent and ionic bonds.
12. Describe the formation of a hydrogen bond and explain how it differs from a covalent or ionic bond.
13. Explain why weak bonds are important to living organisms.
14. Describe how the relative concentrations of reactants and products affect a chemical reaction.

KEY TERMS

matter
element
trace element
atom
neutron
proton
electron
atomic nucleus
dalton
atomic number
mass number

atomic weight
isotope
radioactive isotope
energy
potential energy
energy level
energy
potential energy
energy level
electron shell
orbital

valence electron
valence shell
chemical bond
covalent bond
molecule
structural formula
molecular formula
double covalent bond
valence
electronegativity
nonpolar covalent bond

polar covalent bond
ion
cation
anion
ionic bond
hydrogen bond
chemical reactions
reactants
products
chemical equilibrium

LECTURE NOTES

I.   **Chemical Elements and Compounds**

   A. **Matter consists of chemical elements in pure form and in combinations called compounds**

   Chemistry is fundamental to an understanding of life, because living organisms are made of matter.

   Matter = Anything that takes up space and has mass.

   Mass = A measure of the amount of matter an object contains.

   > You might want to distinguish between mass and weight for your students. *Mass* is the measure of the amount of matter an object contains, and it stays the same regardless of changes in the object's position. *Weight* is the measure of how strongly an object is pulled by earth's gravity, and it varies with distance from the earth's center. The key point is that the mass of a body does not vary with its position, whereas weight does. So, for all practical purposes—as long as we are earthbound—weight can be used as a measure of mass.

   B. **Life requires about 25 chemical elements**

   Element = A substance that cannot be broken down into other substances by chemical reactions.

   •   All matter is made of elements.
   •   There are 92 naturally occurring elements.
   •   They are designated by a symbol of one or two letters.

   About 25 of the 92 naturally occurring elements are essential to life. Biologically important elements include:

   C = carbon
   O = oxygen   ⎤
   H = hydrogen   ⎥ make up 96% of living matter
   N = nitrogen   ⎦

Ca = calcium ⌉
P = phosphorus |
K = potassium |
S = sulfur | make up remaining 4% of an organism's weight
Na = sodium |
Cl = chlorine |
Mg = magnesium |
Trace elements ⌋

Trace element = Element required by an organism in extremely minute quantities.

- Though required by organisms in small quantity, they are indispensable for life
- Examples: B, Cr, Co, Cu, F, I, Fe, Mn, Mo, Se, Si, Sn, V and Zn

Elements can exist in combinations called compounds.

- *Compound* = A pure substance composed of two or more elements combined in a fixed ratio.
- Example: NaCl (sodium chloride)
- Has unique emergent properties beyond those of its combined elements (Na and Cl have very different properties from NaCl). See Campbell, Figure 2.2.

> Since a compound is the next structural level above element or atom, this is an excellent place to emphasize the concept of emergent properties, an integral theme found throughout the text and course.

II.   Atoms and Molecules

A.   Atomic structure determines the behavior of an element

Atom = Smallest possible unit of matter that retains the physical and chemical properties of its element.

- Atoms of the same element share similar chemical properties.
- Atoms are made up of *subatomic particles*.

1.   Subatomic particles

The three *most stable* subatomic particles are:

1. *Neutrons* [no charge (neutral)].
2. *Protons* [+1 electrostatic charge].
3. *Electrons* [-1 electrostatic charge].

NEUTRON	PROTON	ELECTRON
No charge	+1 charge	-1 charge
Found together in a dense core called the *nucleus* (positively charged because of protons)		Orbits around nucleus (held by electrostatic attraction to positively charged nucleus)
1.009 dalton	1.007 dalton	1/2000 dalton
Masses of both are about the same (about 1 dalton)		Mass is so small, usually not used to calculate atomic mass

NOTE: The *dalton* is a unit used to express mass at the atomic level. One dalton (d) is equal to 1.67×10^{-24} g.

If an atom is electrically neutral, the number of protons equals the number of electrons, which yields an electrostatically balanced charge.

2. **Atomic number and atomic weight**

Atomic number = Number of protons in an atom of a particular element.

- All atoms of an element have the same atomic number.
- Written as a subscript to the left of the element's symbol (e.g., $_{11}Na$)
- In a neutral atom, # protons = # electrons.

Mass number = Number of protons and neutrons in an atom.

- Written as a superscript to left of an element's symbol (e.g., ^{23}Na)
- Is approximate mass of the whole atom, since the mass of a proton and the mass of a neutron are both about 1 dalton
- Can deduce the number of neutrons by subtracting *atomic number* from *mass number*
- Number of neutrons can vary in an element, but number of protons is constant
- Is not the same as an element's *atomic weight*, which is the weighted mean of the masses of an element's constituent isotopes

---

In a large classroom with up to 300 students, it can be difficult to interact. Try putting examples on an overhead transparency and soliciting student input to complete the information. It is a quick way to check for understanding and to actively involve students.

Examples:

$^{(Mass\ \#)\ 23}_{(Atomic\ \#)\ 11}Na$ # of electrons _____

# of protons _____

# of neutrons _____

$^{12}_{6}C$ # of electrons _____

# of protons _____

# of neutrons _____

3. **Isotopes**

Isotopes = Atoms of an element that have the same atomic number but different mass number.

- They have the same number of protons, but a different number of neutrons.
- Under natural conditions, elements occur as mixtures of isotopes.
- Different isotopes of the same element react chemically in same way.
- Some isotopes are radioactive.

Radioactive isotope = Unstable isotope in which the nucleus spontaneously decays, emitting subatomic particles and/or energy as radioactivity.

- Loss of nuclear particles may transform one element to another (e.g., $^{14}_{6}C \rightarrow ^{14}_{7}N$).
- Has a fixed half life.
  - *Half life* = Time for 50% of radioactive atoms in a sample to decay.

Biological applications of radioactive isotopes include:

a. Dating geological strata and fossils

- Has a fixed half life.
  - *Half life* = Time for 50% of radioactive atoms in a sample to decay.

Biological applications of radioactive isotopes include:

a. Dating geological strata and fossils
  - Radioactive decay is at a fixed rate.
  - By comparing the ratio of radioactive and stable isotopes in a fossil with the ratio of isotopes in living organisms, one can estimate the age of a fossil.
  - The ratio of ^{14}C to ^{12}C is frequently used to date fossils less than 50,000 years old.

b. Radioactive tracers
  - Chemicals labelled with radioactive isotopes are used to trace the steps of a biochemical reaction or to determine the location of a particular substance within an organism (see Campbell, p. XX, Methods: The Use of Radioactive Tracers in Biology).
  - Radioactive isotopes are useful as biochemical tracers because they chemically react like the stable isotopes and are easily detected at low concentrations.
  - Isotopes of P, N, and H were used to determine DNA structure.
  - Used to diagnose disease (e.g., PET scanner)
  - Because radioactivity can damage cell molecules, radioactive isotopes can also be hazardous

c. Treatment of cancer
  - e.g., radioactive cobalt

4. The energy levels of electrons

Electrons = Light negatively charged particles that orbit around nucleus.

- Equal in mass and charge
- Are the only stable subatomic particles directly involved in chemical reactions
- Have *potential energy* because of their position relative to the positively charged nucleus

Energy = Ability to do work

Potential energy = Energy that matter stores because of its position or location.

- There is a natural tendency for matter to move to the lowest state of potential energy.
- Potential energy of electrons is not infinitely divisible, but exists only in discrete amounts called *quanta*.
- Different fixed potential energy states for electrons are called *energy levels* or *electron shells* (see Campbell, Figure 2.7).
- Electrons with lowest potential energy are in energy levels closest to the nucleus.
- Electrons with greater energy are in energy levels further from nucleus.

Electrons may move from one energy level to another. In the process, they gain or lose energy equal to the difference in potential energy between the old and new energy level.

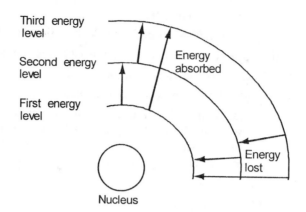

5. Electron orbitals

Orbital = Three-dimensional space where an electron will most likely be found 90% of the time (see Campbell, Figure 2.8).

- Viewed as a three-dimensional probability cloud (a statistical concept)
- No more than two electrons can occupy same orbital.

First energy level:

- Has one spherical *s* orbital (1*s* orbital)
- Holds a maximum of two electrons

Second energy level

- Holds a maximum of eight electrons
- One spherical *s* orbital (2*s* orbital)
- Three dumbbell-shaped *p* orbitals each oriented at right angles to the other two ($2p_x$, $2p_y$, $2p_z$ orbitals)

Higher energy levels:

- Contain *s* and *p* orbitals
- Contain additional orbitals with more complex shapes

6. Electron configuration and chemical properties

An atom's electron configuration determines its chemical behavior.

- *Electron configuration* = Distribution of electrons in an atom's electron shells

The first 18 elements of a periodic chart are arranged sequentially by atomic number into three rows (periods). In reference to these *representative* elements, note the following:

- Outermost shell of these atoms never have more than four orbitals (one *s* and three *p*) or eight electrons.
- Electrons must first occupy lower electron shells before the higher shells can be occupied. (This is a reflection of the natural tendency for matter to move to the lowest possible state of potential energy—the most stable state.)
- Electrons are added to each of the *p* orbitals singly, before they can be paired.
- If an atom does not have enough electrons to fill all shells, the outer shell will be the only one partially filled. Example: O_2 with a total of eight electrons:

OXYGEN

$_8O$

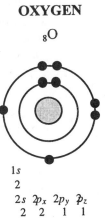

Two electrons have the 1*s* orbital of the first electron shell.

First two electrons in the second shell are both in the 2*s* orbital.

Next three electrons each have a *p* orbital ($2p_x$, $2p_y$, $2p_z$).

Eighth electron is paired in the $2p_x$ orbital.

1s			
2			
2s	$2p_x$	$2p_y$	p_z
2	2	1	1

Chemical properties of an atom depend upon the number of valence electrons.

- *Valence electrons* = Electrons in the outermost energy shell (valence shell).

Octet rule = Rule that a valence shell is complete when it contains eight electrons (except H and He).

- An atom with a complete valence shell is unreactive or *inert*.
- Noble elements (e.g., helium, argon, and neon) have filled outer shells in their elemental state and are thus inert.
- An atom with an incomplete valence shell is chemically reactive (tends to form chemical bonds until it has eight electrons to fill the valence shell).
- Atoms with the same number of valence electrons show similar chemical behavior.

NOTE: The consequence of this unifying chemical principle is that the valence electrons are responsible for the atom's bonding capacity. This rule applies to most of the representative elements, but *not all*.

B. Atoms combine by chemical bonding to form molecules

Atoms with incomplete valence shells tend to fill those shells by interacting with other atoms. These interactions of electrons among atoms may allow atoms to form chemical bonds.

- *Chemical bonds* = Attractions that hold molecules together

Molecules = Two or more atoms held together by chemical bonds.

1. Covalent bonds

Covalent bond = Chemical bond between atoms formed by *sharing* a pair of valence electrons.

- Strong chemical bond
- Example: molecular hydrogen (H_2); when two hydrogen atoms come close

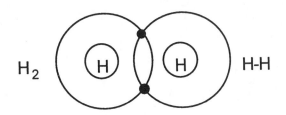

Structural formula = Formula which represents the atoms and bonding within a molecule (e.g., H-H). The line represents a shared pair of electrons.

Molecular formula = Formula which indicates the number and type of atoms (e.g., H_2).

Single covalent bond = Bond between atoms formed by sharing a single pair of valence electrons.

- Atoms may freely rotate around the axis of the bond.

Double covalent bond = Bond formed when atoms share *two* pairs of valence electrons (e.g., O_2).

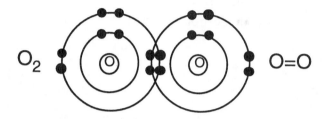

Molecules = Two or more atoms held together by chemical bonds.

Triple covalent bond = Bond formed when atoms share three pairs of valence electrons (e.g., N_2 or N°N).

NOTE: Double and triple covalent bonds are rigid and do not allow rotation.

Valence = Bonding capacity of an atom which is the number of covalent bonds that must be formed to complete the outer electron shell.

- Valences of some common elements: hydrogen = 1, oxygen = 2, nitrogen = 3, carbon = 4, phosphorus = 3 (sometimes 5 as in biologically important compounds, e.g., ATP), sulfur = 2.

Compound = A pure substance composed of two or more elements combined in a fixed ratio.

- Example: water (H_2O), methane (CH_4)
- Note that two hydrogens are necessary to complete the valence shell of oxygen in water, and four hydrogens are necessary for carbon to complete the valence shell in methane.

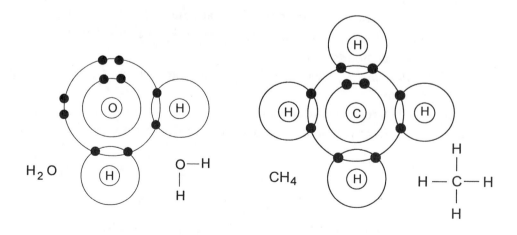

2. **Nonpolar and polar covalent bonds**

Electronegativity = Atom's ability to attract and hold electrons.

- The more electronegative an atom, the more strongly it attracts shared electrons.
- Scale determined by Linus Pauling:

  O = 3.5
  N = 3.0
  S and C = 2.5
  P and H = 2.1

Nonpolar covalent bond = Covalent bond formed by an equal sharing of electrons between atoms.

- Occurs when electronegativity of both atoms is about the same (e.g., CH_4)
- Molecules made of one element usually have nonpolar covalent bonds (e.g., H_2, O_2, Cl_2, N_2).

Polar covalent bond = Covalent bond formed by an unequal sharing of electrons between atoms.

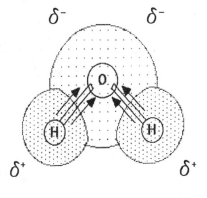

- Occurs when the atoms involved have different electronegativities.
- Shared electrons spend more time around the more electronegative atom.
- In H_2O, for example, the oxygen is strongly electronegative, so negatively charged electrons spend more time around the oxygen than the hydrogens. This causes the oxygen atom to have a slight negative charge and the hydrogens to have a slight positive charge (see also Campbell, Figure 2.11).

3. **Ionic bonds**

Ion = Charged atom or molecule.

Anion = An atom that has gained one or more electrons from another atom and has become negatively charged; a negatively charged ion.

Cation = An atom that has lost one or more electrons and has become positively charged; a positively charged ion.

Ionic bond = Bond formed by the electrostatic attraction after the complete transfer of an electron from a donor atom to an acceptor.

- The acceptor atom attracts the electrons because it is much more electronegative than the donor atom.
- Are strong bonds in crystals, but are fragile bonds in water; salt crystals will readily dissolve in water and dissociate into ions.
- Ionic compounds are called salts (e.g., NaCl or table salt) (see Campbell, Figure 2.13).

NOTE: The *difference* in electronegativity between interacting atoms determines if electrons are shared equally (nonpolar covalent), shared unequally (polar covalent), gained or lost (ionic bond). Nonpolar covalent bonds and ionic bonds are two extremes of a continuum from interacting atoms with similar electronegativities to interacting atoms with very different electronegativities.

C. Weak chemical bonds play important roles in the chemistry of life

Biologically important weak bonds include the following:

- Hydrogen bonds, ionic bonds in aqueous solutions, and other weak forces such as Van der Waals and hydrophobic interactions
- Make chemical signaling possible in living organisms because they are only temporary associations. Signal molecules can briefly and reversibly bind to receptor molecules on a cell, causing a short-lived response.
- Can form between molecules or between different parts of a single large molecule.
- Help stabilize the three-dimensional shape of large molecules (e.g., DNA and proteins).

1. Hydrogen bonds

Hydrogen bond = Bond formed by the charge attraction when a hydrogen atom covalently bonded to one electronegative atom is attracted to another electronegative atom.

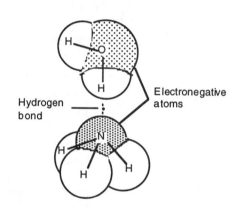

- Weak attractive force that is about 20 times easier to break than a covalent bond
- Is a charge attraction between oppositely charged portions of polar molecules
- Can occur between a hydrogen that has a slight positive charge when covalently bonded to an atom with high electronegativity (usually O and N)
- Example: NH_3 in H_2O (see Campbell, Figure 2.14)

2. Van der Waals interactions

Weak interactions that occur between atoms and molecules that are very close together and result from charge asymmetry in electron clouds.

D. A molecule's biological function is related to its shape

A molecule has a charasteric size and shape.

The function of many molecules depends upon their shape

> Insulin causes glucose uptake into liver and muscle cells of veterbrates because the shape of the insulin molecule is recognized by specific receptors on the target cell.

- Molecules with only two atoms are linear.
- Molecules with more than two atoms have more complex shapes.

When an atom forms covalent bonds, orbitals in the valence shell rearrange into the most stable configuration. To illustrate, consider atoms with valence electrons in the *s* and three *p* orbitals:

- The *s* and three *p* orbitals *hybridize* into four new orbitals.
- The new orbitals are teardrop shaped, extend from the nucleus and spread out as far apart as possible.
- Example: If outer tips of orbitals in methane (CH_4) are connected by imaginary lines, the new molecule has a tetrahedral shape with C at the center (see Campbell, Figure 2.15).

E. Chemical reactions make and break chemical bonds

Chemical reactions = process of making and breaking chemical bonds leading to changes in the composition of matter.

- Process where *reactants* undergo changes into *products*.
- Matter is conserved, so all reactant atoms are only rearranged to form products.
- Some reactions go to completion (all reactants converted to products), but most reactions are *reversible*. For example:

$$3H_2 + N_2 \rightleftharpoons 2NH_3$$

- The relative concentration of reactants and products affects reaction rate (the higher the concentration, the greater probability of reaction).

Chemical equilibrium = Equilibrium established when the rate of forward reaction equals the rate of the reverse reaction.

- Is a *dynamic* equilibrium with reactions continuing in both directions
- Relative concentrations of reactants and products stop changing.

> Point out to students that chemical equilibrium does NOT mean that the concentrations of reactants and products are equal.

REFERENCES

Atkins, P.W. *Atoms, Electrons and Change*. W.H. Freeman and Company, 1991.

Campbell, N., et al. *Biology*. 5th ed. Menlo Park, California: Benjamin/Cummings, 1998.

Weinberg, S. *The Discovery of Subatomic Particles*. New York, San Francisco: W.H. Freeman and Company, 1983.

Brown, T.L., H. E. Le May, Jr., and B. Bursten. *Chemistry: The Central Science*. 7th ed. Upper Saddle River, New Jersey: Prentice Hall, 1997.

CHAPTER 3

WATER AND THE FITNESS OF THE ENVIRONMENT

OUTLINE

I. Water's Polarity and Its Effects
   A. The polarity of water molecules results in hydrogen bonding
   B. Organisms depend on the cohesion of water molecules
   C. Water moderates temperatures on Earth
   D. Oceans and lakes don't freeze solid because ice floats
   E. Water is the solvent of life

II. The Dissociation of Water
   A. Organisms are sensitive to changes in pH

III. Acid Precipitation Threatens the Fitness of the Environment

OBJECTIVES

After reading this chapter and attending lecture, the student should be able to:

1. Describe how water contributes to the fitness of the environment to support life.

2. Describe the structure and geometry of a water molecule, and explain what properties emerge as a result of this structure.

3. Explain the relationship between the polar nature of water and its ability to form hydrogen bonds.

4. List five characteristics of water that are emergent properties resulting from hydrogen bonding.

5. Describe the biological significance of the cohesiveness of water.

6. Distinguish between heat and temperature.

7. Explain how water's high specific heat, high heat of vaporization and expansion upon freezing affect both aquatic and terrestrial ecosystems.

8. Explain how the polarity of the water molecule makes it a versatile solvent.

9. Define molarity and list some advantages of measuring substances in moles.

10. Write the equation for the dissociation of water, and explain what is actually transferred from one molecule to another.

11. Explain the basis for the pH scale.

12. Explain how acids and bases directly or indirectly affect the hydrogen ion concentration of a solution.

13. Using the bicarbonate buffer system as an example, explain how buffers work.

14. Describe the causes of acid precipitation, and explain how it adversely affects the fitness of the environment.

KEY TERMS

polar molecule	Celsius scale	solute	hydrogen ion
cohesion	calorie	solvent	molarity
adhesion	kilocalorie	aqueous solution	hydroxide ion
surface tension	joule	hydrophilic	acid
kinetic energy	specific heat	hydrophobic	base
heat	evaporative cooling	mole	pH scale
temperature	solution	molecular weight	buffer
acid precipitation			

LECTURE NOTES

Water contributes to the fitness of the environment to support life.

- Life on earth probably evolved in water.
- Living cells are 70%-95% H_2O.
- Water covers about 3/4 of the earth.
- In nature, water naturally exists in all three physical states of matter—solid, liquid and gas.

Water's extraordinary properties are emergent properties resulting from water's structure and molecular interactions.

I.   Water's Polarity and Its Effects

A. The polarity of water molecules results in hydrogen bonding

Water is a *polar* molecule. Its polar bonds and asymmetrical shape give water molecules opposite charges on opposite sides.

- Four valence orbitals of O point to corners of a tetrahedron.
- Two corners are orbitals with unshared pairs of electrons and weak negative charge.
- Two corners are occupied by H atoms which are in polar covalent bonds with O. Oxygen is so electronegative, that shared electrons spend more time around the O causing a weak positive charge near H's.

Unbonded electron pairs

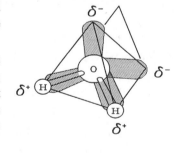

Hydrogen bonding orders water into a higher level of structural organization.

- The polar molecules of water are held together by hydrogen bonds.
- Positively charged H of one molecule is attracted to the negatively charged O of another water molecule.
- Each water molecule can form a maximum of four hydrogen bonds with neighboring water molecules.

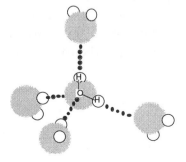

Water has extraordinary properties that emerge as a consequence of its polarity and hydrogen-bonding. Some of these properties are that water:

- has cohesive behavior
- resists changes in temperature
- has a high heat of vaporization and cools surfaces as it evaporates
- expands when it freezes
- is a versatile solvent

B. Organisms depend on the cohesion of water molecules.

Cohesion = Phenomenon of a substance being held together by hydrogen bonds.

- Though hydrogen bonds are transient, enough water molecules are hydrogen bonded at any given time to give water more structure than other liquids.
- Contributes to upward water transport in plants by holding the water column together. *Adhesion* of water to vessel walls counteracts the downward pull of gravity.

Surface tension = Measure of how difficult it is to stretch or break the surface of a liquid.

- Water has a greater surface tension than most liquids; function of the fact that at the air/H_2O interface, surface water molecules are hydrogen bonded to each other and to the water molecules below.
- Causes H_2O to bead (shape with smallest area to volume ratio and allows maximum hydrogen bonding).

C. Water moderates temperatures on Earth

1. Heat and temperature

Kinetic energy = The energy of motion.

Heat = Total kinetic energy due to molecular motion in a body of matter.

Temperature = Measure of heat intensity due to the *average* kinetic energy of molecules in a body of matter.

Calorie (cal) = Amount of heat it takes to raise the temperature of one gram of water by one degree Celsius. Conversely, one calorie is the amount of heat that one gram of water releases when it cools down by one degree Celsius. NOTE: The "calories" on food packages are actually kilocalories (kcal).

Kilocalorie (kcal or Cal) = Amount of heat required to raise the temperature of one kilogram of water by one degree Celsius (1000 cal).

Celsius Scale at Sea Level		Scale Conversion		
100°C (212°F)	= water boils	°C	=	$\dfrac{5(°F- 32)}{9}$
37°C (98.6°F)	= human body temperature	°F	=	$\dfrac{9°C}{5}+ 32$
23°C (72°F)	= room temperature	°K	=	°C + 273
0°C (32°F)	= water freezes			

2. Water's high specific heat

Water has a high *specific heat*, which means that it resists temperature changes when it absorbs or releases heat.

Specific heat = Amount of heat that must be absorbed or lost for one gram of a substance to change its temperature by one degree Celsius.

Specific heat of water = One calorie per gram per degree Celsius (1 cal/g/°C).

- As a result of hydrogen bonding among water molecules, it takes a relatively large heat loss or gain for each 1°C change in temperature.
- Hydrogen bonds must absorb heat to break, and they release heat when they form.
- Much absorbed heat energy is used to disrupt hydrogen bonds before water molecules can move faster (increase temperature).

A large body of water can act as a heat sink, absorbing heat from sunlight during the day and summer (while warming only a few degrees) and releasing heat during the night and winter as the water gradually cools. As a result:

- Water, which covers three-fourths of the planet, keeps temperature fluctuations within a range suitable for life.
- Coastal areas have milder climates than inland.
- The marine environment has a relatively stable temperature.

3. Evaporative cooling

Vaporization (evaporation) = transformation from liquid to a gas.

- Molecules with enough kinetic energy to overcome the mutual attraction of molecules in a liquid, can escape into the air.

Heat of vaporization = Quantity of heat a liquid must absorb for 1 g to be converted to the gaseous state.

- For water molecules to evaporate, hydrogen bonds must be broken which requires heat energy.
- Water has a relatively high heat of vaporization at the boiling point (540 cal/g or 2260 J/g; Joule = 0.239 cal).

Evaporative cooling = Cooling of a liquid's surface when a liquid evaporates (see Campbell, Figure 3.4).

- The surface molecules with the highest kinetic energy are most likely to escape into gaseous form; the average kinetic energy of the remaining surface molecules is thus lower.

Water's high heat of vaporization:

- Moderates the Earth's climate.
    - Solar heat absorbed by tropical seas dissipates when surface water evaporates (evaporative cooling).
    - As moist tropical air moves poleward, water vapor releases heat as it condenses into rain.
- Stabilizes temperature in aquatic ecosystems (evaporative cooling).
- Helps organisms from overheating by *evaporative cooling*.

D. Oceans and lakes don't freeze solid because ice floats

Because of hydrogen bonding, water is less dense as a solid than it is as a liquid. Consequently, ice floats.

- Water is densest at 4°C.
- Water contracts as it cools to 4°C.
- As water cools from 4°C to freezing (0°C), it expands and becomes *less dense* than liquid water (ice floats).
- When water begins to freeze, the molecules do not have enough kinetic energy to break hydrogen bonds.
- As the crystalline lattice forms, each water molecule forms a maximum of four hydrogen bonds, which keeps water molecules further apart than they would be in the liquid state; see Campbell, Figure 3.5.

Expansion of water contributes to the fitness of the environment for life:
- Prevents deep bodies of water from freezing solid from the bottom up.
- Since ice is less dense, it forms on the surface first. As water freezes it releases heat to the water below and insulates it.
- Makes the transitions between seasons less abrupt. As water freezes, hydrogen bonds form releasing heat. As ice melts, hydrogen bonds break absorbing heat.

E. Water is the solvent of life

Solution = A liquid that is a completely homogenous mixture of two or more substances.

Solvent = Dissolving agent of a solution.

Solute = Substance dissolved in a solution.

Aqueous solution = Solution in which water is the solvent.

Water is a versatile solvent owing to the *polarity* of the water molecule.

Hydrophilic

Ionic compounds dissolve in water (see Campbell, Figure 3.8).
- Charged regions of polar water molecules have an electrical attraction to charged ions.
- Water surrounds individual ions, separating and shielding them from one another.

Polar compounds in general, are water-soluble.
- Charged regions of polar water molecules have an affinity for oppositely charged regions of other polar molecules.

Hydrophobic

Nonpolar compounds (which have symmetric distribution in charge) are NOT water-soluble.

1. Hydrophilic and hydrophobic substances

Ionic and polar substances are *hydrophilic*, but nonpolar compounds are hydrophobic.

Hydrophilic = (Hydro = water; philo = loving); property of having an affinity for water.
- Some large hydrophilic molecules can absorb water without dissolving.

Hydrophobic = (Hydro = water; phobos = fearing); property of not having an affinity for water, and thus, not being water-soluble.

2. Solute concentration in aqueous solutions

Most biochemical reactions involve solutes dissolved in water. There are two important quantitative properties of aqueous solutions: solute concentration and pH.

Molecular weight = Sum of the weight of all atoms in a molecule (expressed in daltons).

Mole = Amount of a substance that has a mass in grams numerically equivalent to its molecular weight in daltons.

For example, to determine a mole of sucrose ($C_{12}H_{22}O_{11}$):
- Calculate molecular weight:

C = 12 dal	12 dal × 12 =	144 dal
H = 1 dal	1 dal × 22 =	22 dal
O = 16 dal	16 dal × 11 =	176 dal
		342 dal

- Express it in grams (342 g).

Molarity = Number of moles of solute per liter of solution
- To make a 1M sucrose solution, weigh out 342 g of sucrose and add water up to 1L.

Advantage of measuring in moles:
- Rescales weighing of single molecules in daltons to grams, which is more practical for laboratory use.
- A mole of one substance has the *same* number of molecules as a mole of any other substance (6.02×10^{23}; Avogadro's number).
- Allows one to combine substances in fixed ratios of molecules.

II.   The Dissociation of Water

Occasionally, the hydrogen atom that is shared in a hydrogen bond between two water molecules, shifts from the oxygen atom to which it is covalently bonded to the unshared orbitals of the oxygen atom to which it is hydrogen bonded.
- Only a *hydrogen ion* (proton with a +1 charge) is actually transferred.
- Transferred proton binds to an unshared orbital of the second water molecule creating a *hydronium ion* (H_3O^+).
- Water molecule that lost a proton has a net negative charge and is called a *hydroxide ion* (OH^-).

$$H_2O + H_2O \rightleftharpoons H_3O^+ + OH^-$$

- By convention, ionization of H_2O is expressed as the *dissociation* into H^+ and OH^-.

$$H_2O \rightleftharpoons H^+ + OH^-$$

- Reaction is reversible.
- At equilibrium, most of the H_2O is *not* ionized.

A.   Organisms are sensitive to changes in pH
1.   Acids and bases
At equilibrium in pure water at 25°C:
- Number of H+ ions = number of OH- ions.
- $[H^+] = [OH^-] = \dfrac{1}{10,000,000}$ M = 10^{-7} M
- Note that brackets indicate molar concentration.

> This is a good place to point out how *few* water molecules are actually dissociated (only 1 out of 554,000,000 molecules).

ACID	BASE
Substance that *increases* the relative $[H^+]$ of a solution.	Substance that *reduces* the relative $[H^+]$ of a solution.
Also removes OH^- because it tends to combine with H^+ to form H_2O.	May alternately increase $[OH^-]$.
For example: (in water)	For example:
$$HCl \rightarrow H^+ + Cl^-$$	A base may reduce $[H^+]$ directly: $$NH_3 + H^+ \rightleftharpoons NH_4^+$$ A base may reduce $[H^+]$ indirectly: $$NaOH \rightarrow Na^+ + OH^-$$ $$OH^- + H^+ \rightarrow H_2O$$

A solution in which:

- $[H^+] = [OH^-]$ is a neutral solution.
- $[H^+] > [OH^-]$ is an acidic solution.
- $[H^+] < [OH^-]$ is a basic solution.

Strong acids and bases dissociate completely in water.

- Example: HCl and NaOH
- Single arrows indicate complete dissociation.

$$NaOH \rightarrow Na^+ + OH^-$$

Weak acids and bases dissociate only partially and reversibly.

- Examples: NH_3 (ammonia) and H_2CO_3 (carbonic acid)
- Double arrows indicate a reversible reaction; at equilibrium there will be a fixed ratio of reactants and products.

$$\begin{array}{ccccc} H_2CO_3 & & HCO_3^- & + & H^+ \\ \text{Carbonic} & \rightleftharpoons & \text{Bicarbonate} & & \text{Hydrogen} \\ \text{acid} & & \text{ion} & & \text{ion} \end{array}$$

2. **The pH scale**

In any aqueous solution:

$$[H^+][OH^-] = 1.0 \times 10^{-14}$$

For example:

- In a neutral solution, $[H^+] = 10^{-7}$ M and $[OH^-] = 10^{-7}$ M.
- In an acidic solution where the $[H^+] = 10^{-5}$ M, the $[OH^-] = 10^{-9}$ M.
- In a basic solution where the $[H^+] = 10^{-9}$ M, the $[OH^-] = 10^{-5}$ M.

pH scale = Scale used to measure degree of acidity. It ranges from 0 to 14.

pH = Negative $\log_{10}$ of the $[H^+]$ expressed in moles per liter.

- pH of 7 is a neutral solution.
- pH < 7 is an acidic solution.
- pH > 7 is a basic solution.

- Most biological fluids are within the pH range of 6 to 8. There are some exceptions such as stomach acid with pH = 1.5. (See Campbell, Figure 3.9)
- Each pH unit represents a *tenfold* difference (scale is logarithmic), so a slight change in pH represents a large change in actual $[H^+]$.

> To illustrate this point, project the following questions on a transparency and cover the answer. The students will frequently give the wrong response (3×), and they are surprised when you unveil the solution.
>
> How much greater is the $[H^+]$ in a solution with pH 2 than in a solution with pH 6?
>
> ANS: pH 2 = $[H^+]$ of 1.0×10^{-2} = $\dfrac{1}{100}$ M
>
> pH 6 = $[H^+]$ of 1.0×10^{-6} = $\dfrac{1}{1,000,000}$ M
>
> <u>10,000</u> times greater.

3. Buffers

By minimizing wide fluctuations in pH, buffers help organisms maintain the pH of body fluids within the narrow range necessary for life (usually pH 6-8).

Buffer = Substance that minimizes large sudden changes in pH.

- Are combinations of H^+-donor and H^+-acceptor forms in a solution of weak acids or bases
- Work by accepting H^+ ions from solution when they are in excess and by donating H^+ ions to the solution when they have been depleted

Example: Bicarbonate buffer

$$\underset{\substack{H^+ \text{ donor} \\ \text{(weak acid)}}}{H_2CO_3} \xrightleftharpoons[\substack{\text{response to a} \\ \text{drop in pH}}]{\substack{\text{response to a} \\ \text{rise in pH}}} \underset{\substack{H^+ \text{ acceptor} \\ \text{(weak base)}}}{HCO_3^-} + \underset{\substack{\text{Hydrogen} \\ \text{ion}}}{H^+}$$

$$\underset{\substack{\text{strong} \\ \text{acid}}}{HCl} + NaHCO_3 \longrightarrow \underset{\substack{\text{weak} \\ \text{acid}}}{H_2CO_3} + NaCl$$

$$\underset{\substack{\text{strong} \\ \text{base}}}{NaOH} + H_2CO_3 \longrightarrow \underset{\substack{\text{weak} \\ \text{base}}}{NaHCO_3} + H_2O$$

III. Acid Precipitation Threatens the Fitness of the Environment

Acid precipitation = Rain, snow, or fog more strongly acidic than pH 5.6.

- Has been recorded as low as pH 1.5 in West Virginia
- Occurs when sulfur oxides and nitrogen oxides in the atmosphere react with water in the air to form acids which fall to Earth in precipitation
- Major oxide source is the combustion of fossil fuels by industry and cars
- Acid rain affects the fitness of the environment to support life.
  - Lowers soil pH which affects mineral solubility. May leach out necessary mineral nutrients and increase the concentration of minerals that are potentially toxic to vegetation in higher concentration (e.g., aluminum). This is contributing to the decline of some European and North American forests.

- Lowers the pH of lakes and ponds, and runoff carries leached out soil minerals into aquatic ecosystems. This adversely affects aquatic life. Example: In the Western Adirondack Mountains, there are lakes with a pH < 5 that have no fish.

What can be done to reduce the problem?

- Add industrial pollution controls.
- Develop and use antipollution devices.
- Increase involvement of voters, consumers, politicians, and business leaders.

> The political issues surrounding acid rain can be used to enhance student awareness and make this entire topic more relevant and interesting to the students.

REFERENCES

Campbell, N., et al. *Biology*. 5th ed. Menlo Park, California: Benjamin/Cummings, 1998.

Gould, R. *Going Sour: Science and Politics of Acid Rain*. Boston: Birkhauser, 1985.

Henderson, L. J. *The Fitness of the Environment*. Boston: Beacon Press, 1958.

Mohnen, V.A. "The Challenge of Acid Rain." *Scientific American*, August 1988.

CHAPTER 4
CARBON AND
MOLECULAR DIVERSITY

OUTLINE

OBJECTIVES

After reading this chapter and attending lecture, the student should be able to:

1. Summarize the philosophies of *vitalism* and *mechanism*, and explain how they influenced the development of organic chemistry, as well as mainstream biological thought.

2. Explain how carbon's electron configuration determines the kinds and number of bonds carbon will form.

3. Describe how carbon skeletons may vary, and explain how this variation contributes to the diversity and complexity of organic molecules.

4. Distinguish among the three types of isomers: structural, geometric and enantiomers.

5. Recognize the major functional groups, and describe the chemical properties of organic molecules in which they occur.

KEY TERMS

organic chemistry	enantiomer	aldehyde	amine
hydrocarbon	functional group	ketone	sulfhydryl group
isomer	hydroxyl group	carboxyl group	thiol
structural isomer	alcohol	carboxylic acid	phosphate group
geometric isomer	carbonyl group	amino group	

LECTURE NOTES

Aside from water, most biologically important molecules are carbon-based (*organic*).

The structural and functional diversity of organic molecules emerges from the ability of carbon to form large, complex and diverse molecules by bonding to itself and to other elements such as H, O, N, S, and P.

I.   The Importance of Carbon

A.   Organic chemistry is the study of carbon compounds

Organic chemistry = The branch of chemistry that specializes in the study of carbon compounds.

Organic molecules = Molecules that contain carbon

Vitalism = Belief in a life force outside the jurisdiction of chemical/physical laws.

- Early 19th century organic chemistry was built on a foundation of vitalism because organic chemists could not artificially synthesize organic compounds. It was believed that only living organisms could produce organic compounds.

Mechanism = Belief that all natural phenomena are governed by physical and chemical laws.

- Pioneers of organic chemistry began to synthesize organic compounds from inorganic molecules. This helped shift mainstream biological thought from vitalism to mechanism.
- For example, Friedrich Wohler synthesized urea in 1828; Hermann Kolbe synthesized acetic acid.
- Stanley Miller (1953) demonstrated the possibility that organic compounds could have been produced under the chemical conditions of primordial Earth.

B.   Carbon atoms are the most versatile building blocks of molecules

The carbon atom:

- Usually has an atomic number of 6; therefore, it has 4 valence electrons.
- Usually completes its outer energy shell by sharing valence electrons in four covalent bonds. (Not likely to form ionic bonds.)

Emergent properties, such as the kinds and number of bonds carbon will form, are determined by their *tetravalent* electron configuration.

- It makes large, complex molecules possible. The carbon atom is a central point from which the molecule branches off into four directions.
- It gives carbon covalent compatibility with many different elements. The four major atomic components of organic molecules are as follows:

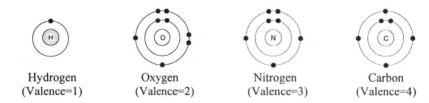

| Hydrogen | Oxygen | Nitrogen | Carbon |
| (Valence=1) | (Valence=2) | (Valence=3) | (Valence=4) |

- It determines an organic molecule's three-dimensional shape, which may affect molecular function. For example, when carbon forms four single covalent bonds, the four valence orbitals hybridize into teardrop-shaped orbitals that angle from the carbon atoms toward the corners of an imaginary tetrahedron.

> Students have problems visualizing shapes of organic molecules in three dimensions. Specific examples can be enhanced by an overhead transparency of ball-and-stick or space-filling models. A large three-dimensional molecular model that can be held up in front of class works best (see Campbell, Figure 4.2)

C. Variation in carbon skeletons contributes to the diversity of organic molecules

Covalent bonds link carbon atoms together in long chains that form the skeletal framework for organic molecules. These carbon skeletons may vary in:

- Length
- Shape (straight chain, branched, ring)
- Number and location of double bonds
- Other elements covalently bonded to available sites

This variation in carbon skeletons contributes to the complexity and diversity of organic molecules (see Campbell, Figure 4.4).

Hydrocarbons = Molecules containing only carbon and hydrogen

- Are major components of fossil fuels produced from the organic remains of organisms living millions of years ago, though they are not prevalent in living organisms.
- Have a diversity of carbon skeletons which produce molecules of various lengths and shapes.
- As in hydrocarbons, a carbon skeleton is the framework for the large diverse organic molecules found in living organisms. Also, some biologically important molecules may have regions consisting of hydrocarbon chains (e.g. fats).
- Hydrocarbon chains are hydrophobic because the C–C and C–H bonds are nonpolar.

1. Isomers

Isomers = Compounds with the same molecular formula but with different structures and hence different properties. Isomers are a source of variation among organic molecules.

There are three types of isomers (see Campbell, Figure 4.6):

Structural isomers = Isomers that differ in the covalent arrangement of their atoms.

- Number of possible isomers increases as the carbon skeleton size increases.
- May also differ in the location of double bonds.

Geometric isomers = Isomers which share the same covalent partnerships, but differ in their spatial arrangements.

- Result from the fact that double bonds will not allow the atoms they join to rotate freely about the axis of the bonds.
- Subtle differences between isomers affects their biological activity.

Enantiomers = Isomers that are mirror images of each other.

- Can occur when four different atoms or groups of atoms are bonded to the same carbon (*asymmetric carbon*).
- There are two different spatial arrangements of the four groups around the asymmetric carbon. These arrangements are mirror images.
- Usually one form is biologically active and its mirror image is not.

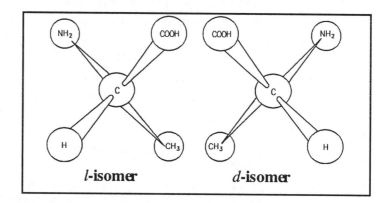

l-isomer   *d*-isomer

It is often helpful to point at the pharmacological significance of enantiomers, e.g., Campbell, Figure 4.7.

II.   Functional Groups

A.   Functional groups also contribute to the molecular diversity of life

Small characteristic groups of atoms (functional groups) are frequently bonded to the carbon skeleton of organic molecules. These functional groups:

- Have specific chemical and physical properties.
- Are the regions of organic molecules which are commonly chemically reactive.
- Behave consistently from one organic molecule to another.
- Depending upon their number and arrangement, determine unique chemical properties of organic molecules in which they occur.

As with hydrocarbons, diverse organic molecules found in living organisms have carbon skeletons. In fact, these molecules can be viewed as hydrocarbon derivatives with functional groups in place of H, bonded to carbon at various sites along the molecule.

1.   The hydroxyl group

Hydroxyl group = A functional group that consists of a hydrogen atom bonded to an oxygen atom, which in turn is bonded to carbon (–OH).

- Is a *polar* group; the bond between the oxygen and hydrogen is a polar covalent bond.
- Makes the molecule to which it is attached *water soluble*. Polar water molecules are attracted to the polar hydroxyl group which can form hydrogen bonds.
- Organic compounds with hydroxyl groups are called *alcohols*.

2.   The carbonyl group

Carbonyl group = Functional group that consists of a carbon atom double-bonded to oxygen (–CO).

- Is a *polar* group. The oxygen can be involved in hydrogen bonding, and molecules with his functional group are *water soluble*.
- Is a functional group found in sugars.

- If the carbonyl is at the end off the carbon skeleton, the compound is an *aldehyde*.

$$
\begin{array}{ccc}
\text{OH} & \text{OH} & \text{O} \\
| & | & \| \\
\text{H–C} & \text{—C} & \text{—C} \\
| & | & | \\
\text{H} & \text{H} & \text{H}
\end{array}
$$

Glyceraldehyde

- If the carbonyl is at the end of the carbon skeleton, the compound is a *ketone*.

$$
\begin{array}{ccc}
\text{H} & \text{O} & \text{H} \\
| & \| & | \\
\text{H–C} & \text{—C} & \text{—C–H} \\
| & & | \\
\text{H} & & \text{H}
\end{array}
$$

Acetone

3. The carboxyl group

Carboxyl group = Functional group that consists of a carbon atom which is both double-bonded to an oxygen and single-bonded to the oxygen of a hydroxyl group (–COOH).

- Is a polar group and water soluble. The covalent bond between oxygen and hydrogen is so polar, that the hydrogen reversibly dissociates as H^+. This polarity results from the combined effect of the two electronegative oxygen atoms bonded to the same carbon.

$$
\begin{array}{ccc}
\text{H} & \text{O} & \\
| & \| & \\
\text{H–C–C} & & \\
| & \backslash & \\
\text{H} & \text{OH} &
\end{array}
\rightleftharpoons
\begin{array}{cc}
\text{H} & \text{O} \\
| & \| \\
\text{H–C–C} & \\
| & \backslash \\
\text{H} & \text{O}^-
\end{array}
\quad + \quad H^+
$$

Acetic acid   Acetate ion   Hydrogen ion

- Since it donates protons, this group has acidic properties. Compounds with this functional group are called *carboxylic acids*.

4. The amino group

Amino group = Functional group that consists of a nitrogen atom bonded to two hydrogens and to the carbon skeleton (–NH$_2$).

- Is a polar group and soluble in water.
- Acts as a weak base. The unshared pair of electrons on the nitrogen can accept a proton, giving the amino group a +1 charge.

$$
\begin{array}{c}
\text{H} \\
/ \\
\text{R–N} \\
\backslash \\
\text{H}
\end{array}
\quad + \quad H^+ \quad \rightleftharpoons \quad
\begin{array}{c}
\text{H} \\
/ \\
\text{R–}^+\text{N–H} \\
\backslash \\
\text{H}
\end{array}
$$

Amine   Ammonium ion

- Organic compounds with this function group are called *amines*.

5. The Sulfhydryl group

Sulfhydryl group = Functional group which consists of an atom of sulfur bonded to an atom of hydrogen (–SH).

- Help stabilize the structure of proteins. (Disulfide bridges will be discussed with tertiary structure of proteins in Chapter 5, Structure and Function of Macromolecules.)
- Organic compounds with this functional group are called *thiols*.

6. The phosphate group

Phosphate group = Functional group which is the dissociated form of phosphoric acid (H_3PO_4).

- Loss of two protons by dissociation leaves the phosphate group with a negative charge.

$$R-O-\overset{\overset{\displaystyle O}{\|}}{\underset{\underset{\displaystyle OH}{|}}{P}}-OH \rightleftharpoons R-O-\overset{\overset{\displaystyle O}{\|}}{\underset{\underset{\displaystyle O^-}{|}}{P}}-O^- \; + \; 2H^+$$

- Has acid properties since it loses protons.
- Polar group and soluble in water.
- Organic phosphates are important in cellular energy storage and transfer. (ATP is discussed with energy for cellular work in Chapter 6: Introduction to Metabolism.)

In lecture, you may also choose to include the methyl group ($-CH_3$) as an example of a nonpolar hydrophobic functional group. This is helpful later in the course in explaining how nonpolar amino acids contribute to the tertiary structure of proteins including integral membrane proteins.

To impress upon students how important functional groups are in determining chemical behavior of organic molecules, use the following demonstration: show a comparison of estradiol and testosterone and ask students to find the differences in functional groups. Ask one male and female student to stand up or show pictures of sexual dimorphism in other vertebrates. Point out that differences between males and females are due to slight variation in functional groups attached to sex hormones.

REFERENCES

Campbell, N. et al. *Biology*. 5th ed. Menlo Park, California: Benjamin/Cummings, 1998.

Lehninger, A.L., D.L. Nelson and M.M. Cox. *Principles of Biochemistry*. 2nd ed. New York: Worth, 1993.

Whitten, K.W. and K.D. Gailey. *General Chemistry*. 4th ed. New York: Saunders, 1992.

CHAPTER 5

THE STRUCTURE AND FUNCTION OF MACROMOLECULES

OUTLINE

I. Polymer Principles
   A. Most macromolecules are polymers
   B. A limitless variety of polymers can be built from a small set of monomers
II. Carbohydrates: Fuel and Building Material
   A. Sugars, the smallest carbohydrates, serve as fuel and carbon sources
   B. Polysaccharides, the polymers of sugars, have storage and structural roles
III. Lipids: Diverse Hydrophobic Molecules
   A. Fats store large amounts of energy
   B. Phospholipids are major components of cell membranes
   C. Steroids include cholesterol and certain hormones
IV. Proteins: The Molecular Tools of the Cell
   A. A polypeptide is a polymer of amino acids connected in a specific sequence
   B. A protein's function depends on its specific conformation
V. Nucleic Acids: Informational Polymers
   A. Nucleic acids store and transmit hereditary information
   B. A nucleic acid strand is a polymer of nucleotides
   C. Inheritance is based on replication of the DNA double helix
   D. We can use DNA and proteins as tape measures of evolution

OBJECTIVES

After reading this chapter and attending lecture, the student should be able to:

1. List the four major classes of biomolecules.
2. Explain how organic polymers contribute to biological diversity.
3. Describe how covalent linkages are formed and broken in organic polymers.
4. Describe the distinguishing characteristics of carbohydrates, and explain how they are classified.
5. List four characteristics of a sugar.
6. Identify a glycosidic linkage and describe how it is formed.
7. Describe the important biological functions of polysaccharides.
8. Distinguish between the glycosidic linkages found in starch and cellulose, and explain why the difference is biologically important.
9. Explain what distinguishes lipids from other major classes of macromolecules.
10. Describe the unique properties, building block molecules and biological importance of the three important groups of lipids: fats, phospholipids and steroids.

11. Identify an ester linkage and describe how it is formed.

12. Distinguish between a saturated and unsaturated fat, and list some unique emergent properties that are a consequence of these structural differences.

13. Describe the characteristics that distinguish proteins from the other major classes of macromolecules, and explain the biologically important functions of this group.

14. List and recognize four major components of an amino acid, and explain how amino acids may be grouped according to the physical and chemical properties of the side chains.

15. Identify a peptide bond and explain how it is formed.

16. Explain what determines protein conformation and why it is important.

17. Define primary structure and describe how it may be deduced in the laboratory.

18. Describe the two types of secondary protein structure, and explain the role of hydrogen bonds in maintaining the structure.

19. Explain how weak interactions and disulfide bridges contribute to tertiary protein structure.

20. Using collagen and hemoglobin as examples, describe quaternary protein structure.

21. Define denaturation and explain how proteins may be denatured.

22. Describe the characteristics that distinguish nucleic acids from the other major groups of macromolecules.

23. Summarize the functions of nucleic acids.

24. List the major components of a nucleotide, and describe how these monomers are linked together to form a nucleic acid.

25. Distinguish between a pyrimidine and a purine.

26. List the functions of nucleotides.

27. Briefly describe the three-dimensional structure of DNA.

KEY TERMS

polymer	cellulose	polypeptide	quaternary structure
monomer	chitin	amino acid	denaturation
condensation reaction	lipid	protein	chaperone proteins
dehydration reaction	fat	conformation	gene
hydrolysis	fatty acid	peptide bond	nucleic acid
carbohydrate	triacylglycerol	primary structure	deoxyribonucleic acid
monosaccharide	saturated fatty acid	secondary structure	ribonucleic acid
disaccharide	unsaturated fatty acid	alpha (α) helix	nucleotide
glycosidic linkage	steroid	pleated sheet	pyrimidine
polysaccharide	cholesterol	tertiary structure	purine
starch	protein	hydrophobic interaction	ribose
glycogen	conformation	disulfide bridges	polynucleotide
double helix			

LECTURE NOTES

The topic of macromolecules lends itself well to illustrate three integral themes that permeate the text and course:

1. There is a natural hierarchy of structural level in biological organization.

2. As one moves up the hierarchy, new properties emerge because of interactions among subunits at the lower levels.

3. Form fits function.

I.   **Polymer Principles**

A. **Most macromolecules are polymers**

Polymer = (Poly = many; mer = part); large molecule consisting of many identical or similar subunits connected together.

Monomer = Subunit or building block molecule of a polymer

Macromolecule = (Macro = large); large organic polymer

- Formation of macromolecules from smaller building block molecules represents another level in the hierarchy of biological organization.
- There are four classes of macromolecules in living organisms:
  1. Carbohydrates
  2. Lipids
  3. Proteins
  4. Nucleic acids

Most polymerization reactions in living organisms are condensation reactions.

- *Polymerization reactions* = Chemical reactions that link two or more small molecules to form larger molecules with repeating structural units.
- *Condensation reactions* = Polymerization reactions during which monomers are covalently linked, producing net removal of a water molecule for each covalent linkage.
  - One monomer loses a hydroxyl (–OH), and the other monomer loses a hydrogen (–H).
  - Removal of water is actually indirect, involving the formation of "activated" monomers (discussed in Chapter 6, Introduction to Metabolism).
  - Process requires energy.
  - Process requires biological catalysts or enzymes.

Hydrolysis = (Hydro = water; lysis = break); a reaction process that breaks covalent bonds between monomers by the addition of water molecules.

- A hydrogen from the water bonds to one monomer, and the hydroxyl bonds to the adjacent monomer.
- Example: Digestive enzymes catalyze hydrolytic reactions which break apart large food molecules into monomers that can be absorbed into the bloodstream.

B. **An immense variety of polymers can be built from a small set of monomers**

Structural variation of macromolecules is the basis for the enormous diversity of life.

- There is *unity* in life as there are only about 40 to 50 common monomers used to construct macromolecules.
- There is *diversity* in life as new properties emerge when these universal monomers are arranged in different ways.

II.   **Carbohydrates: Fuel and Building Material**

A. **Sugars, the smallest carbohydrates, serve as fuel and carbon sources**

Carbohydrates = Organic molecules made of sugars and their polymers

- Monomers or building block molecules are simple sugars called *monosaccharides*.
- Polymers are formed by condensation reactions.
- Carbohydrates are classified by the number of simple sugars.

1. Monosaccharides

Monosaccharides = (Mono = single; sacchar = sugar); simple sugar in which C, H, and O occur in the ratio of (CH_2O).

- Are major nutrients for cells; glucose is the most common
- Can be produced (glucose) by photosynthetic organisms from CO_2, H_2O, and sunlight
- Store energy in their chemical bonds which is harvested by cellular respiration
- Their carbon skeletons are raw material for other organic molecules.
- Can be incorporated as monomers into disaccharides and polysaccharides

Characteristics of a sugar:

   a. An –OH group is attached to each carbon except one, which is double bonded to an oxygen (*carbonyl*).

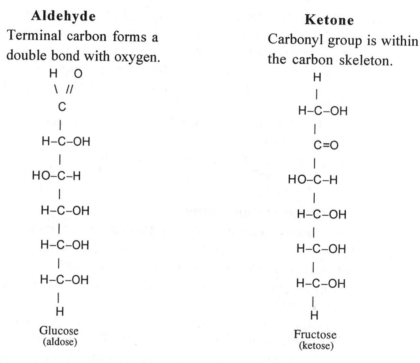

   b. Size of the carbon skeleton varies from three to seven carbons. The *most common* monosaccharides are:

Classification	Number of Carbons	Example
Triose	3	Glyceraldehyde
Pentose	5	Ribose
Hexose	6	Glucose

c. Spatial arrangement around asymmetric carbons may vary. For example, glucose and galactose are enantiomers.

Glucose   Galactose

The small difference between isomers affects molecular shape which gives these molecules distinctive biochemical properties.

d. In aqueous solutions, many monosaccharides form rings. Chemical equilibrium favors the ring structure.

Linear Form   Ring Form of Glucose
of Glucose

2. Disaccharides

Disaccharide = (Di = two; sacchar = sugar); a double sugar that consists of two monosaccharides joined by a *glycosidic linkage*.

Glycosidic linkage = Covalent bond formed by a condensation reaction between two sugar monomers; for example, maltose:

MALTOSE

Examples of disaccharides include:

Disaccharide	Monomers	General Comments
Maltose	Glucose + Glucose	Important in brewing beer
Lactose	Glucose + Galactose	Present in milk
Sucrose	Glucose + Fructose	Table sugar; most prevalent disaccharide; transport form in plants

B. Polysaccharides, the polymers of sugars, have storage and structural roles

Polysaccharides = Macromolecules that are polymers of a few hundred or thousand monosaccharides.

- Are formed by linking monomers in enzyme-mediated condensation reactions
- Have two important biological functions:
    1. Energy storage (starch and glycogen)
    2. Structural support (cellulose and chitin)

1. Storage polysaccharides

Cells hydrolyze storage polysaccharides into sugars as needed. Two most common storage polysaccharides are *starch* and *glycogen*.

Starch = Glucose polymer that is a storage polysaccharide in plants.

- Helical glucose polymer with α 1-4 linkages (see Campbell, Figure 5.6)
- Stored as granules within plant organelles called *plastids*
- *Amylose*, the simplest form, is an unbranched polymer.
- *Amylopectin* is branched polymer.
- Most animals have digestive enzymes to hydrolyze starch.
- Major sources in the human diet are potato tubers and grains (e.g., wheat, corn, rice, and fruits of other grasses).

Glycogen = Glucose polymer that is a storage polysaccharide in animals.

- Large glucose polymer that is more highly branched (α 1-4 and 4-6 linkages) than amylopectin
- Stored in the muscle and liver of humans and other vertebrates

2. Structural polysaccharides

Structural polysaccharides include *cellulose* and *chitin*.

Cellulose = Linear unbranched polymer of *D*-glucose in (α 1-4, β 4-6) linkages.

- A major structural component of plant cell walls
- Differs from starch (also a glucose polymer) in its glycosidic linkages (see Campbell, Figure 5.7)

STARCH	CELLULOSE
Glucose monomers are in α *configuration* (–OH group on carbon one is *below* the ring's plane).	Glucose monomers are in β *configuration* (–OH group on carbon one is *above* the ring's plane).
Monomers are connected with α 1-4 linkage.	Monomers are connected with β 1-4 linkage.

- Cellulose and starch have different three-dimensional shapes and properties as a result of differences in glycosidic linkages.
- Cellulose reinforces plant cell walls. Hydrogen bonds hold together parallel cellulose molecules in bundles of *microfibrils* (see Campbell, Figure 5.8)
- Cellulose cannot be digested by most organisms, including humans, because they lack an enzyme that can hydrolyze the β 1-4 linkage. (Exceptions are some symbiotic bacteria, other microorganisms and some fungi.)

Chitin = A structural polysaccharide that is a polymer of an amino sugar (see Campbell, Figure 5.9).

- Forms exoskeletons of arthropods
- Found as a building material in the cell walls of some fungi
- Monomer is an *amino sugar*, which is similar to *beta*-glucose with a nitrogen-containing group replacing the hydroxyl on carbon 2.

III.   Lipids: Diverse Hydrophobic Molecules

Lipids = Diverse group of organic compounds that are insoluble in water, but will dissolve in nonpolar solvents (e.g., ether, chloroform, benzene). Important groups are *fats*, *phospholipids*, and *steroids*.

A.   Fats store large amounts of energy

Fats = Macromolecules are constructed from (see Campbell, Figure 5.10):

1. Glycerol, a three-carbon alcohol
2. Fatty acid (carboxylic acid)
   - Composed of a carboxyl group at one end and an attached *hydrocarbon chain* ("tail")
   - Carboxyl functional group ("head") has properties of an acid.
   - Hydrocarbon chain has a long carbon skeleton usually with an even number of carbon atoms (most have 16 – 18 carbons).
   - Nonpolar C–H bonds make the chain hydrophobic and not water soluble.

During the formation of a fat, enzyme-catalyzed condensation reactions link glycerol to fatty acids by an ester linkage.

Ester linkage = Bond formed between a hydroxyl group and a carboxyl group.

Each of glycerol's three hydroxyl groups can bond to a fatty acid by an ester linkage producing a fat.

Triacylglycerol = A fat composed of three fatty acids bonded to one glycerol by ester linkages (triglyceride).

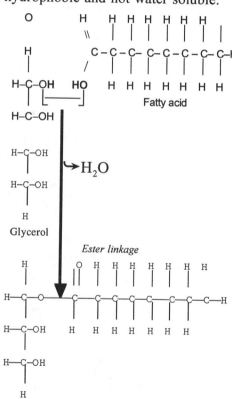

Some characteristics of fat include:

- Fats are insoluble in water. The long fatty acid chains are hydrophobic because of the many nonpolar C–H bonds.
- The source of variation among fat molecules is the fatty acid composition.
- Fatty acids in a fat may all be the same, or some (or all) may differ.
- Fatty acids may vary in length.
- Fatty acids may vary in the number and location of carbon-to-carbon double bonds.

SATURATED FAT	UNSATURATED FAT
No double bonds between carbons in fatty acid tail	One or more double bonds between carbons in fatty acid tail
Carbon skeleton of fatty acid is bonded to maximum number of hydrogens (*saturated* with hydrogens)	Tail kinks at each C=C, so molecules do not pack closely enough to solidify at room temperature
Usually a solid at room temperature	Usually a liquid at room temperature
Most animal fats	Most plant fats
e.g., bacon grease, lard and butter (see Campbell, Figure 5.11)	e.g., corn, peanut and olive oil

- In many commercially prepared food products, unsaturated fats are artificially hydrogenated to prevent them from separating out as oil (e.g., peanut butter and margarine).

Fat serves many useful functions:

- Energy storage. One gram of fat stores twice as much energy as a gram of polysaccharide. (Fat has a higher proportion of energy rich C–H bonds.)
- More compact fuel reservoir than carbohydrate. Animals store more energy with less weight than plants which use starch, a bulky form of energy storage.
- Cushions vital organs in mammals (e.g., kidney).
- Insulates against heat loss (e.g., in mammals such as whales and seals).

B. Phospholipids

Phospholipids = Compounds with molecular building blocks of glycerol, two fatty acids, a phosphate group, and usually, an additional small chemical group attached to the phosphate (see Campbell , Figure 5.12)

- Differ from fat in that the third carbon of glycerol is joined to a *negatively charged* phosphate group
- Can have small variable molecules (usually charged or polar) attached to phosphate
- Are diverse depending upon differences in fatty acids and in phosphate attachments
- Show ambivalent behavior toward water. Hydrocarbon tails are hydrophobic and the polar head (phosphate group with attachments) is hydrophilic.
- Cluster in water as their hydrophobic portions turn away from water. One such cluster, a *micelle*, assembles so the hydrophobic tails turn toward the water-free interior and the hydrophilic phosphate heads arrange facing outward in contact with water (see Campbell, Figure 5.13).
- Are major constituents of cell membranes. At the cell surface, phospholipids form a bilayer held together by hydrophobic interactions among the hydrocarbon tails. Phospholipids in water will spontaneously form such a bilayer.

C. Steroids

Steroids = Lipids which have four fused carbon rings with various functional groups attached.

Cholesterol is an important steroid.

- Is the precursor to many other steroids including vertebrate sex hormones and bile acids.
- Is a common component of animal cell membranes.
- Can contribute to atherosclerosis.

IV.   Proteins: The Molecular Tools of the Cell

Polypeptide chains = Polymers of amino acids that are arranged in a specific linear sequence and are linked by peptide bonds.

Protein = A macromolecule that consists of one or more *polypeptide chains* folded and coiled into specific conformations.

- Are abundant, making up 50% or more of cellular dry weight
- Have important and varied functions in the cell:
  1. Structural support
  2. Storage (of amino acids)
  3. Transport (e.g., hemoglobin)
  4. Signaling (chemical messengers)
  5. Cellular response to chemical stimuli (receptor proteins)
  6. Movement (contractile proteins)
  7. Defense against foreign substances and disease-causing organisms (antibodies)
  8. Catalysis of biochemical reactions (enzymes)
- Vary extensively in structure; each type has a unique three-dimensional shape (conformation)
- Though they vary in structure and function, they are commonly made of only 20 amino acid monomers.

A. A polypeptide is a polymer of amino acids connected in a specific sequence

Amino acid = Building block molecule of a protein; most consist of an asymmetric carbon, termed the *alpha carbon*, which is covalently bonded to a(n):

1. Hydrogen atom.
2. Carboxyl group.
3. Amino group.
4. Variable R group (side chain) specific to each amino acid. Physical and chemical properties of the side chain determine the uniqueness of each amino acid.

(At pH's normally found in the cell, both the carboxyl and amino groups are ionized.)

Amino acids contain both carboxyl and amino functional groups. Since one group acts as a weak acid and the other group acts as a weak base, an amino acid can exist in three ionic states. The pH of the solution determines which ionic state predominates.

Cation — as pH increases / as pH decreases — Zwitterion (dipolar ion) + H$^+$ — as pH increases / as pH decreases — Anion + H$^+$

The twenty common amino acids can be grouped by properties of side chains (see Campbell, Figure 5.15):

1. Nonpolar side groups (hydrophobic). Amino acids with nonpolar groups are less soluble in water.

2. Polar side groups (hydrophilic). Amino acids with polar side groups are soluble in water. Polar amino acids can be grouped further into:

   a. Uncharged polar

   b. Charged polar

      • Acidic side groups. Dissociated carboxyl group gives these side groups a negative charge.

      • Basic side groups. An amino group with an extra proton gives these side groups a net positive charge.

Polypeptide chains are polymers that are formed when amino acid monomers are linked by peptide bonds (see Campbell, Figure 5.16).

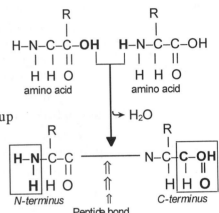

Peptide bond = Covalent bond formed by a condensation reaction that links the carboxyl group of one amino acid to the amino group of another.

• Has polarity with an amino group on one end (*N-terminus*) and a carboxyl group on the other (*C-terminus*).

• Has a backbone of the repeating sequence –N–C–C–N–C–C–.

Polypeptide chains:

   • Range in length from a few monomers to more than a thousand.

   • Have unique linear sequences of amino acids.

B. A protein's function depends on its specific conformation

A protein's function depends upon its unique *conformation.*

Protein conformation = Three-dimensional shape of a protein.

Native conformation = Functional conformation of a protein found under normal biological conditions.

   • Enables a protein to recognize and bind specifically to another molecule (e.g., hormone/receptor, enzyme/substrate, and antibody/antigen)

   • Is a consequence of the specific linear sequence of amino acids in the polypeptide

   • Is produced when a newly formed polypeptide chain coils and folds spontaneously, mostly in response to hydrophobic interactions

   • Is stabilized by chemical bonds and weak interactions between neighboring regions of the folded protein

1. Four levels of protein structure

The correlation between form and function in proteins is an emergent property resulting from superimposed levels of protein structure (see Campbell, Figure 5.24):

   • Primary structure

   • Secondary structure

- Tertiary structure
- When a protein has two or more polypeptide chains, it also has quaternary structure.

a. Primary structure

Primary structure = Unique sequence of amino acids in a protein.

- Determined by genes
- Slight change can affect a protein's conformation and function (e.g., sickle-cell hemoglobin; see Campbell, Figure 5.19).
- Can be sequenced in the laboratory. A pioneer in this work was Frederick Sanger who determined the amino acid sequence in insulin (late 1940s and early 1950s). This laborious process involved:

  1) Determination of amino acid composition by complete acid hydrolysis of peptide bonds and separation of resulting amino acids by chromatography. Using these techniques, Sanger identified the amino acids and determined the relative proportions of each.

  2) Determination of amino acid sequence by partial hydrolysis with enzymes and other catalysts to break only specific peptide bonds. Sanger deductively reconstructed the primary structure from fragments with overlapping segments.

- Most of the sequencing process is now automated.

b. Secondary structure

Secondary structure = Regular, repeated coiling and folding of a protein's polypeptide backbone (see Campbell, Figure 5.20).

- Contributes to a protein's overall conformation.
- Stabilized by hydrogen bonds between peptide linkages in the protein's backbone (carbonyl and amino groups).
- The major types of secondary structure are alpha (α) helix and beta (β) pleated sheet.

  1) Alpha (α) helix

  Alpha (α) helix = Secondary structure of a polypeptide that is a helical coil stabilized by hydrogen bonding between every fourth peptide bond (3.6 amino acids per turn).

    - Described by Linus Pauling and Robert Corey in 1951.
    - Found in fibrous proteins (e.g., α-keratin and collagen) for most of their length and in some portions of globular proteins.

  2) Beta (β) pleated sheet

  Beta (β) pleated sheet = Secondary protein structure which is a sheet of antiparallel chains folded into accordion pleats.

    - Parallel regions are held together by either intrachain or interchain hydrogen bonds (between adjacent polypeptides).
    - Make up the dense core of many globular proteins (e.g., lysozyme) and the major portion of some fibrous proteins (e.g., fibroin, the structural protein of silk).

c. Tertiary structure

Tertiary structure = The three-dimensional shape of a protein. The irregular contortions of a protein are due to bonding between and among side chains (R groups) and to interaction between R groups and the aqueous environment (see Campbell, Figure 5.22).

Types of bonds contributing to tertiary structure are weak interactions and covalent linkage (both may occur in the same protein).

1) Weak interactions

Protein shape is stabilized by the cumulative effect of weak interactions. These weak interactions include:

- Hydrogen bonding between polar side chains.
- Ionic bonds between charged side chains.
- *Hydrophobic interactions* between nonpolar side chains in protein's interior.

Hydrophobic interactions = (Hydro = water; phobos = fear); the clustering of hydrophobic molecules as a result of their mutual exclusion from water.

2) Covalent linkage

Disulfide bridges form between two cysteine monomers brought together by folding of the protein. This is a strong bond that reinforces conformation.

$$H_3N^+-\overset{\overset{\textstyle H}{|}}{\underset{\underset{\textstyle CH_2}{|}}{C}}-\overset{\overset{\textstyle O}{\sslash}}{\underset{\underset{\textstyle O^-}{\diagdown}}{C}}$$

$$\underset{\underset{\textstyle SH}{|}}{}$$

Cysteine

$$-\overset{|}{C}-CH_2-S \longrightarrow S-CH_2-\overset{|}{C}-$$

Disulfide Bridge
(S of one cysteine sulfhydryl, bonds
to the S of a second cysteine.)

d. Quaternary structure

Quaternary structure = Structure that results from the interactions between and among several polypeptides chains (subunits) (see Campbell, Figure 5.23).

- Example: Collagen, a fibrous protein with three helical polypeptides supercoiled into a triple helix; found in animal connective tissue, collagen's supercoiled quaternary structure gives it strength.
- Some globular proteins have subunits that fit tightly together. Example: Hemoglobin, a globular protein that has four subunits (two α chains and two β chains)

2. What determines protein conformation?

A protein's three-dimensional shape is a consequence of the interactions responsible for secondary and tertiary structure.

- This conformation is influenced by physical and chemical environmental conditions.
- If a protein's environment is altered, it may become *denatured* and lose its native conformation.

Denaturation = A process that alters a protein's native conformation and biological activity. Proteins can be denatured by:

- Transfer to an organic solvent. Hydrophobic side chains, normally inside the protein's core, move towards the outside. Hydrophilic side chains turn away from the solvent towards the molecule's interior.
- Chemical agents that disrupt hydrogen bonds, ionic bonds and disulfide bridges.
- Excessive heat. Increased thermal agitation disrupts weak interactions (see Campbell, Figure 5.25).

The fact that some denatured proteins return to their native conformation when environmental conditions return to normal is evidence that a protein's amino acid sequence (primary structure) determines conformation. It influences where and which interactions will occur as the molecule arranges into secondary and tertiary structure.

3. The protein-folding problem

Even though primary structure ultimately determines a protein's conformation, three-dimensional shape is difficult to predict solely on the basis of amino acid sequence. It is difficult to find the rules of protein folding because:

- Most proteins pass through several intermediate stages in the folding process; knowledge of the final conformation does not reveal the folding process required to create it.
- A protein's native conformation may be dynamic, alternating between several shapes.

Using recently developed techniques, researchers hope to gain new insights into protein folding:

- Biochemists can now track a protein as it passes through its intermediate stages during the folding process.
- *Chaperone proteins* have just been discovered that temporarily brace a folding protein.

Rules of protein folding are important to molecular biologists and the biotechnology industry. This knowledge should allow the design of proteins for specific purposes.

V.   Nucleic Acids: Informational Polymers

A.   Nucleic acids store and transmit hereditary information

Protein conformation is determined by primary structure. Primary structure, in turn, is determined by *genes*; hereditary units that consist of DNA, a type of *nucleic acid*.

There are two types of nucleic acids.

1.   Deoxyribonucleic acid (DNA)

- Contains coded information that programs all cell activity.
- Contains directions for its own replication.
- Is copied and passed from one generation of cells to another.
- In eukaryotic cells, is found primarily in the nucleus.
- Makes up *genes* that contain instructions for protein synthesis. Genes do not directly make proteins, but direct the synthesis of mRNA.

2.   Ribonucleic acid (RNA)

- Functions in the actual synthesis of proteins coded for by DNA.
- Sites of protein synthesis are on *ribosomes* in the cytoplasm.
- Messenger RNA (mRNA) carries encoded genetic message from the nucleus to the cytoplasm.
- The flow of genetic information goes from DNA → RNA → protein (see Campbell, Figure 5.26).

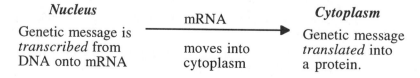

Nucleus	mRNA	*Cytoplasm*
Genetic message is *transcribed* from DNA onto mRNA	moves into cytoplasm	Genetic message *translated* into a protein.

B. A nucleic acid strand is a polymer of nucleotides

Nucleic acid = Polymer of *nucleotides* linked together by condensation reactions.

Nucleotide = Building block molecule of a nucleic acid; made of (1) a five-carbon sugar covalently bonded to (2) a phosphate group and (3) a nitrogenous base.

1. Pentose (5-carbon sugar)

There are two pentoses found in nucleic acids: ribose and deoxyribose.

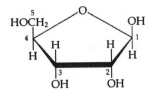

Ribose is the pentose in RNA.

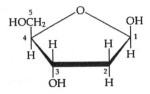

Deoxyribose is the pentose in DNA. (It lacks the –OH group at the number two carbon.)

2. Phosphate

The phosphate group is attached to the number 5 carbon of the sugar.

3. Nitrogenous base

There are two families of *nitrogenous bases*:

Pyrimidine = Nitrogenous base characterized by a six-membered ring made up of carbon and nitrogen atoms. For example:

- Cytosine (C)
- Thymine (T); found only in DNA
- Uracil (U); found only in RNA

Purine = Nitrogenous base characterized by a five-membered ring fused to a six-membered ring. For example:

- Adenine (A)
- Guanine (G)

Nucleotides have various functions:

- Are monomers for nucleic acids.
- Transfer chemical energy from one molecule to another (e.g., ATP).
- Are electron acceptors in enzyme-controlled redox reactions of the cell (e.g., NAD).

A nucleic-acid polymer or polynucleotise, results from joining nucleotides together by covalent bonds called *phosphodiester linkages*. The bond is formed between the phosphate of one nucleotide and the sugar of the next.

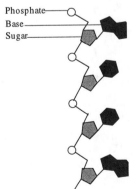

- Results in a backbone with a repeating pattern of sugar-phosphate-sugar-phosphate.
- Variable nitrogenous bases are attached to the sugar-phosphate backbone.
- Each gene contains a unique linear sequence of nitrogenous bases which codes for a unique linear sequence of amino acids in a protein.

C. Inheritance is based on precise replication of the DNA double helix

In 1953, James Watson and Francis Crick proposed the *double helix* as the three dimensional structure of DNA.

- Consists of two nucleotide chains wound in a double helix.

- Sugar-phosphate backbones are on the outside of the helix.

- The two polynucleotide strands of DNA are held together by hydrogen bonds between the paired nitrogenous bases and by van der Waals attraction between the stacked bases (see Campbell, Figure 5.28).

- Base-pairing rules are that adenine (A) always pairs with thymine (T); guanine (G) always pairs with cytosine (C).

- Two strands of DNA are complimentary and thus can serve as templates to make new complementary strands. It is this mechanism of precise copying that makes inheritance possible.

- Most DNA molecules are long, containing thousands or millions of base pairs.

D. We can use DNA and proteins as tape measures of evolution

Closely related species have more similar sequences of DNA and amino acids, than more distantly related species. Using this type of molecular evidence, biologists can deduce evolutionary relationships among species.

> Chapters 16 and 17 are devoted to DNA and protein synthesis. Since any discussion of DNA function must include the details of DNA structure, it may be more practical and less time-consuming to cover nucleic acids later in the course.

REFERENCES

Alberts, B., et al. *Essential Cell Biology: An Introduction to the Molecular Biology of the Cell.* New York: Garland Publishing, Inc., 1998.

Campbell, N., et al. *Biology.* 5th ed. Menlo Park, California: Benjamin/Cummings, 1998.

Lehninger, A.L., D.L. Nelson and M.M. Cox. *Principles of Biochemistry.* 2nd ed. New York: Worth, 1993.

Brown, T.L., H. E. Le May, Jr., and B. Bursten. *Chemistry: The Central Science.* 7th Ed. Upper Saddle River, New Jersey: Prentice Hall, 1997.

CHAPTER 6
AN INTRODUCTION TO METABOLISM

OUTLINE

I. Metabolism, Energy and Life
   A. The chemistry of life is organized into metabolic pathways
   B. Organisms transform energy
   C. The energy transformations of life are subject to two laws of thermodynamics
   D. Organisms live at the expense of free energy
   E. ATP powers cellular work by coupling exergonic to endergonic reactions

II. Enzymes
   A. Enzymes speed up metabolic reactions by lowering energy barriers
   B. Enzymes are substrate-specific
   C. The active site is an enzyme's catalytic center
   D. A cell's physical and chemical environment affects enzyme activity

III. The Control of Metabolism
   A. Metabolic control often demands on allosteric regulation
   B. The location of enzymes within a cell helps order metabolism

OBJECTIVES

After reading this chapter and attending lecture, the student should be able to:

1. Explain the role of catabolic and anabolic pathways in the energy exchanges of cellular metabolism.
2. Distinguish between kinetic and potential energy.
3. Distinguish between open and closed systems.
4. Explain, in their own words, the First and Second Laws of Thermodynamics.
5. Explain why highly ordered living organisms do not violate the Second Law of Thermodynamics.
6. Distinguish between entropy and enthalpy.
7. Write the Gibbs equation for free energy change.
8. Explain how changes in enthalpy, entropy and temperature influence the maximum amount of usable energy that can be harvested from a reaction.
9. Explain the usefulness of free energy.
10. List two major factors capable of driving spontaneous processes.
11. Distinguish between exergonic and endergonic reactions.
12. Describe the relationship between equilibrium and free energy change for a reaction.

13. Describe the function of ATP in the cell.

14. List the three components of ATP and identify the major class of macromolecules to which it belongs.

15. Explain how ATP performs cellular work.

16. Explain why chemical disequilibrium is essential for life.

17. Describe the energy profile of a chemical reaction including activation energy (E_A), free energy change (ΔG) and transition state.

18. Describe the function of enzymes in biological systems.

19. Explain the relationship between enzyme structure and enzyme specificity.

20. Explain the *induced fit* model of enzyme function and describe the catalytic cycle of an enzyme.

21. Describe several mechanisms by which enzymes lower activation energy.

22. Explain how substrate concentration affects the rate of an enzyme-controlled reaction.

23. Explain how enzyme activity can be regulated or controlled by environmental conditions, cofactors, enzyme inhibitors and allosteric regulators.

24. Distinguish between allosteric activation and cooperativity.

25. Explain how metabolic pathways are regulated.

KEY TERMS

metabolism	first law of thermodynamics	catalyst	noncompetitive inhibitors
catabolic pathways	second law of thermodynamics	activation energy	allosteric site
anabolic pathways	free energy	substrate	feedback inhibition
bioenergetics	exergonic reaction	active site	cooperativity
energy	endergonic reaction	induced fit	entropy
kinetic energy	energy coupling	cofactors	spontaneous reaction
potential energy	ATP	coenzymes	
thermodynamics	phosphorylated intermediate	competitive inhibitors	

LECTURE NOTES

I.   **Metabolism, Energy and Life**

A. **The chemistry of life is organized into metabolic pathways**

Metabolism = Totality of an organism's chemical processes (see Campbell, Figure 6.1).

- Property emerging from specific molecular interactions within the cell.
- Concerned with managing cellular resources: material and energy.

Metabolic reactions are organized into pathways that are orderly series of enzymatically controlled reactions. *Metabolic pathways* are generally of two types:

Catabolic pathways = Metabolic pathways that *release energy* by breaking down complex molecules to simpler compounds (e.g., cellular respiration which degrades glucose to carbon dioxide and water; provides energy for cellular work).

Anabolic pathways = Metabolic pathways that *consume energy* to build complicated molecules from simpler ones (e.g., photosynthesis which synthesizes glucose from CO_2 and H_2O; any synthesis of a macromolecule from its monomers).

Metabolic reactions may be coupled, so that energy released from a catabolic reaction can be used to drive an anabolic one.

> It may be useful at this point to illustrate energy exchanges in metabolic reactions.. When respiration is introduced in Chapters 9 and 10, you can use this concept again as a transition.

B. Organisms transform energy

Energy = Capacity to do work

Kinetic energy = Energy in the process of doing work (energy of motion). For example:

- Heat (thermal energy) is kinetic energy expressed in random movement of molecules.
- Light energy from the sun is kinetic energy which powers photosynthesis.

Potential energy = Energy that matter possesses because of its location or arrangement (energy of position). For example:

- In the earth's gravitational field, an object on a hill or water behind a dam have potential energy.
- Chemical energy is potential energy stored in molecules because of the arrangement of nuclei and electrons in its atoms.

Energy can be transformed from one form to another. For example:

- Kinetic energy of sunlight can be transformed into the potential energy of chemical bonds during photosynthesis.
- Potential energy in the chemical bonds of gasoline can be transformed into kinetic mechanical energy which pushes the pistons of an engine.

C. The energy transformations of life are subject to two laws of thermodynamics

Thermodynamics = Study of energy transformations

First Law of Thermodynamics = Energy can be transferred and transformed, but it cannot be created or destroyed (energy of the universe is constant).

Second Law of Thermodynamics = Every energy transfer or transformation makes the universe more disordered (every process increases the *entropy* of the universe).

Entropy = Quantitative measure of disorder that is proportional to randomness (designated by the letter S).

Closed system = Collection of matter under study which is isolated from its surroundings.

Open system = System in which energy can be transferred between the system and its surroundings.

> It is important to distinguish between open and closed systems and to spend lecture time on the second law of thermodynamics. Students often ask: "How is the evolution of complex life forms possible if it violates the second law of thermodynamics?" Thoughtful preparation of an answer beforehand will be well worth the effort.

The entropy of a system may decrease, but the entropy of the system *plus its surroundings* must always increase. Highly ordered living organisms do not violate the second law because they are open systems. For example, animals:

- Maintain highly ordered structure at the expense of increased entropy of their surroundings.
- Take in complex high energy molecules as food and extract chemical energy to create and maintain order.
- Return to the surroundings simpler low energy molecules (CO_2 and water) and heat.

Energy can be transformed, but part of it is dissipated as heat which is largely unavailable to do work. Heat energy *can* perform work if there is a heat gradient resulting in heat flow from warmer to cooler.

Combining the first and second laws; the *quantity* of energy in the universe is constant, but its *quality* is not.

D. Organisms live at the expense of free energy

1. Free energy: a criterion for spontaneous change

Not all of a system's energy is available to do work. The amount of energy that is available to do work is described by the concept of *free energy*. Free energy (G) is related to the system's total energy (H) and its entropy (S) in the following way:

$$G = H - TS$$

where:

G = Gibbs free energy (energy available to do work)

H = enthalpy or total energy

T = temperature in °K

S = entropy

Free energy (G) = Portion of a system's energy available to do work; is the difference between the total energy (*enthalpy*) and the energy *not* available for doing work (TS).

The maximum amount of usable energy that can be harvested from a particular reaction is the system's free energy change from the initial to the final state. This change in free energy (ΔG) is given by the Gibbs-Helmholtz equation at constant temperature and pressure:

$$\Delta G = \Delta H - T\Delta S$$

where:

ΔG = change in free energy

ΔH = change in total energy (enthalpy)

ΔS = change in entropy

T = absolute temperature in °K (which is °C + 273)

> To put these thermodynamic concepts in the context of chemical reactions, you also may briefly discuss the other component of the Gibbs-Helmholtz equation – ΔH or change in enthalpy measured as the *heat of reaction*. Students should understand that during a chemical reaction, reactant molecules must absorb energy for their bonds to break, and that energy is released when bonds form between the rearranged atoms of the products. Consequently, the net energy consumed or released when reactants are converted to products is the *net difference* between the energy consumed to break chemical bonds of reactants and the energy released from the formation of the products.

Significance of free energy:

  a. Indicates the maximum amount of a system's energy which is available to do work.

  b. Indicates whether a reaction will occur spontaneously or not.

   • A *spontaneous reaction* is one that will occur without additional energy.

   • In a spontaneous process, ΔG or free energy of a system *decreases* (ΔG<0).

- A decrease in enthalpy ($-\Delta H$) and an increase in entropy ($+\Delta S$) reduce the free energy of a system and contribute to the spontaneity of a process.
- A higher temperature enhances the effect of an entropy change. Greater kinetic energy of molecules tends to disrupt order as the chances for random collisions increase.
- When enthalpy and entropy changes in a system have an opposite effect on free energy, temperature may determine whether the reaction will be spontaneous or not (e.g., protein denaturation by increased temperature).
- High energy systems, including high energy chemical systems, are unstable and tend to change to a more stable state with a lower free energy.

2. Free energy and equilibrium

There is a relationship between chemical equilibrium and the free energy change (ΔG) of a reaction:

- As a reaction approaches equilibrium, the free energy of the system decreases (spontaneous and exergonic reaction).
- When a reaction is pushed away from equilibrium, the free energy of system increases (non-spontaneous and endergonic reaction).
- When a reaction reaches equilibrium, $\Delta G = 0$, because there is no net change in the system.

3. Free energy and metabolism

a. Reactions can be classified based upon their free energy changes:

Exergonic reaction = A reaction that proceeds with a net loss of free energy.

Endergonic reaction = An energy-requiring reaction that proceeds with a net gain of free energy; a reaction that absorbs free energy from its surroundings.

Exergonic Reaction	Endergonic Reaction
Chemical products have *less* free energy than the reactant molecules.	Products store *more* free energy than reactants.
Reaction is energetically downhill.	Reaction is energetically uphill.
Spontaneous reaction.	Non-spontaneous reaction (requires energy input).
ΔG is negative.	ΔG is positive.
$-\Delta G$ is the maximum amount of work the reaction can perform.	$+\Delta G$ is the minimum amount of work required to drive the reaction.

If a chemical process is exergonic, the reverse process must be endergonic. For example:

- For each mole of glucose oxidized in the exergonic process of cellular respiration, 2870 kJ are released ($\Delta G = -2870$ kJ/mol or -686 kcal/mol).
- To produce a mole of glucose, the endergonic process of photosynthesis requires an energy input of 2870 kJ ($\Delta G = +2870$ kJ/mol or $+686$ kcal/mol).

From this point on, the text uses joules and kilojoules as energy units and puts the caloric equivalent in parentheses. The *joule* (J) is the metric unit of energy; some handy conversions follow:

joule (*J*)	=	0.239 cal
Kilojoule (*kJ*)	=	1000 J or 0.239 kcal
calorie (cal)	=	4.184 J

In cellular metabolism, endergonic reactions are driven by coupling them to reactions with a greater negative free energy (exergonic). ATP plays a critical role in this energy coupling.

b. Metabolic disequilibrium

Since many metabolic reactions are reversible, they have the potential to reach equilibrium.

- At equilibrium, $\Delta G = 0$, so the system can do no work.
- Metabolic disequilibrium is a necessity of life; a cell at equilibrium is dead.
- In the cell, these potentially reversible reactions are pulled forward away from equilibrium, because the products of some reactions become reactants for the next reaction in the metabolic pathway.
- For example, during cellular respiration a steady supply of high energy reactants such as glucose and removal of low energy products such as CO_2 and H_2O, maintain the disequilibrium necessary for respiration to proceed.

E. ATP powers cellular work by coupling exergonic to endergonic reactions

ATP is the immediate source of energy that drives most cellular work, which includes:

- *Mechanical work* such as beating of cilia, muscle contraction, cytoplasmic flow, and chromosome movement during mitosis and meiosis.
- *Transport work* such as pumping substances across membranes.
- *Chemical work* such as the endergonic process of polymerization.

1. The structure and hydrolysis of ATP

ATP (adenosine triphosphate) = Nucleotide with unstable phosphate bonds that the cell hydrolyzes for energy to drive endergonic reactions. ATP consists of:

- Adenine, a nitrogenous base.
- Ribose, a five-carbon sugar.
- Chain of three phosphate groups.

Unstable bonds between the phosphate groups can be hydrolyzed in an exergonic reaction that releases energy.

- When the terminal phosphate bond is hydrolyzed, a phosphate group is removed producing ADP (adenosine diphosphate).

$$ATP + H_2O \longrightarrow ADP + Ⓟ$$

- Under standard conditions in the laboratory, this reaction releases −31 kJ/mol (−7.3 kcal/mol).
- In a living cell, this reaction releases −55 kJ/mol (−13 kcal/mol)—about 77% more than under standard conditions.

The terminal phosphate bonds of ATP are unstable, so:

- The products of the hydrolysis reaction are more stable than the reactant.
- Hydrolysis of the phosphate bonds is thus exergonic as the system shifts to a more stable state.

2   How ATP performs work

Exergonic hydrolysis of ATP is coupled with endergonic processes by transferring a phosphate group to another molecule.

- Phosphate transfer is enzymatically controlled.
- The molecule acquiring the phosphate (*phosphorylated* or *activated intermediate*) becomes more reactive.

For example, conversion of glutamic acid to glutamine (see Campbell, Figure 6.7):

$$\underset{\text{glutamic acid}}{Glu} + \underset{\text{ammonia}}{NH_3} \longrightarrow \underset{\text{glutamine}}{Gln} \quad \underset{\text{(endergonic)}}{\Delta G = +14.2 \text{ kJ/mol } (+3.4 \text{ kcal/mol})}$$

Two step process of energy coupling with ATP hydrolysis:

1. Hydrolysis of ATP and phosphorylation of glutamic acid.

$$Glu + ATP \longrightarrow \underset{\substack{\text{unstable} \\ \text{phosphorylated} \\ \text{intermediate}}}{Glu–Ⓟ} + ADP$$

2. Replacement of the phosphate with the reactant ammonia.

$$Glu–Ⓟ + NH_3 \longrightarrow Gln + Ⓟ$$

Overall ΔG:

$$Glu + NH_3 \longrightarrow Gln \qquad \Delta G = +14.2 \text{ kJ/mol}$$
$$ATP \longrightarrow ADP + Ⓟ \qquad \Delta G = \underline{-31.0 \text{ kJ/mol}}$$
$$\text{Net } \Delta G = - \quad 16.8 \text{ kJ/mol}$$
$$\text{(Overall process is exergonic)}$$

3.   The regeneration of ATP

ATP is continually regenerated by the cell.

- Process is rapid (10^7 molecules used and regenerated/sec/cell).
- Reaction is endergonic.

$$ADP + Ⓟ \longrightarrow ATP \quad \Delta G = +31 \text{ kJ/mol } (+7.3 \text{ kcal/mol})$$

- Energy to drive the endergonic regeneration of ATP comes from the exergonic process of cellular respiration.

II.   Enzymes

A.   Enzymes speed up metabolic reactions by lowering energy barriers

Free energy change indicates whether a reaction will occur spontaneously, but does not give information about the speed of reaction.

- A chemical reaction will occur spontaneously if it releases free energy ($-\Delta G$), but it may occur too slowly to be effective in living cells.
- Biochemical reactions require *enzymes* to speed up and control reaction rates.

Catalyst = Chemical agent that accelerates a reaction without being permanently changed in the process, so it can be used over and over.

Enzymes = Biological catalysts made of protein.

Before a reaction can occur, the reactants must absorb energy to break chemical bonds. This initial energy investment is the *activation energy*.

Free energy of activation (*activation energy*) = Amount of energy that reactant molecules must absorb to start a reaction (E_A).

Transition state = Unstable condition of reactant molecules that have absorbed sufficient free energy to react.

Energy profile of an exergonic reaction:

1. Reactants must absorb enough energy (E_A) to reach the transition state (uphill portion of the curve). Usually the absorption of thermal energy from the surroundings is enough to break chemical bonds.

2. Reaction occurs and energy is released as new bonds form (downhill portion of the curve).

3. ΔG for the overall reaction is the difference in free energy between products and reactants. In an exergonic reaction the free energy of the products is less than reactants.

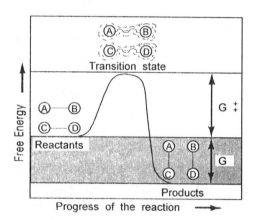

Even though a reaction is energetically favorable, there must be an initial investment of activation energy (E_A).

The breakdown of biological macromolecules is exergonic. However, these molecules react *very slowly* at cellular temperatures because they cannot absorb enough thermal energy to reach transition state.

In order to make these molecules reactive when necessary, cells use biological catalysts called *enzymes*, which:

- Are proteins.
- Lower E_A, so the transition state can be reached at cellular temperatures.
- Do *not change* the nature of a reaction (ΔG), but only speed up a reaction that would have occurred anyway.
- Are very selective for which reaction they will catalyze.

B. Enzymes are substrate-specific

Enzymes are specific for a particular *substrate*, and that specificity depends upon the enzyme's three-dimensional shape.

Substrate = The substance an enzyme acts on and makes more reactive.

- An enzyme binds to its substrate and catalyzes its conversion to product. The enzyme is released in original form.

  Substrate + enzyme ⟶ enzyme-substrate complex ⟶ product + enzyme

- The substrate binds to the enzyme's *active site*.

Active site = Restricted region of an enzyme molecule which binds to the substrate.

- Is usually a pocket or groove on the protein's surface.
- Formed with only a few of the enzyme's amino acids.
- Determines enzyme specificity which is based upon a compatible fit between the shape of an enzyme's active site and the shape of the substrate.
- Changes its shape in response to the substrate.
    - As substrate binds to the active site, it *induces* the enzyme to change its shape.
    - This brings its chemical groups into positions that enhance their ability to interact with the substrate and catalyze the reaction.

Induced fit = Change in the shape of an enzyme's active site, which is induced by the substrate (see Campbell, Figure 6.11).

C. The active site is an enzyme's catalytic center

The entire enzymatic cycle is quite rapid (see Campbell, Figure 6.12).

Steps in the catalytic cycle of enzymes:

1. Substrate binds to the active site forming an *enzyme-substrate complex*. Substrate is held in the active site by weak interactions (e.g., hydrogen bonds and ionic bonds).
2. *Induced fit* of the active site around the substrate. Side chains of a few amino acids in the active site catalyze the conversion of substrate to product.
3. Product departs active site and the enzyme emerges in its original form. Since enzymes are used over and over, they can be effective in very small amounts.

Enzymes lower activation energy and speed up reactions by several mechanisms:

- Active site can hold two or more reactants in the proper position so they may react.
- Induced fit of the enzyme's active site may distort the substrate's chemical bonds, so less thermal energy (lower ΔG) is needed to break them during the reaction.
- Active site might provide a micro-environment conducive to a particular type of reaction (e.g., localized regions of low pH caused by acidic side chains on amino acids at the active site).
- Side chains of amino acids in the active site may participate directly in the reaction.

The initial substrate concentration partly determines the rate of an enzyme controlled reaction.

- The higher the substrate concentration, the faster the reaction - up to a limit.
- If substrate concentration is high enough, the enzyme becomes *saturated* with substrate. (The active sites of all enzymes molecules are engaged.)
- When an enzyme is saturated, the reaction rate depends upon how fast the active sites can convert substrate to product.
- When enzyme is saturated, reaction rate may be increased by adding more enzyme.

D. A cell's physical and chemical environment affects enzyme activity

Each enzyme has optimal environmental conditions that favor the most active enzyme conformation.

1. Effects of temperature and pH

Optimal temperature allows the greatest number of molecular collisions without denaturing the enzyme.

- Enzyme reaction rate increases with increasing temperature. Kinetic energy of reactant molecules increases with rising temperature, which increases substrate collisions with active sites.
- Beyond the optimal temperature, reaction rate slows. The enzyme denatures when increased thermal agitation of molecules disrupts weak bonds that stabilize the active conformation.
- Optimal temperature range of most human enzymes is 35°– 40°C.

Optimal pH range for most enzymes is pH 6 – 8.

- Some enzymes operate best at more extremes of pH.
- For example, the digestive enzyme, pepsin, found in the acid environment of the stomach has an optimal pH of 2.

2. Cofactors

Cofactors = Small nonprotein molecules that are required for proper enzyme catalysis.

- May bind tightly to active site.
- May bind loosely to both active site and substrate.
- Some are inorganic (e.g., metal atoms of zinc, iron or copper).
- Some are organic and are called *coenzymes* (e.g., most vitamins).

3. Enzyme inhibitors

Certain chemicals can selectively inhibit enzyme activity (see Campbell, Figure 6.14).

- Inhibition may be *irreversible* if the inhibitor attaches by covalent bonds.
- Inhibition may be *reversible* if the inhibitor attaches by weak bonds.

Competitive inhibitors = Chemicals that resemble an enzyme's normal substrate and compete with it for the active site.

- Block active site from the substrate.
- If reversible, the effect of these inhibitors can be overcome by increased substrate concentration.

Noncompetitive inhibitors = Enzyme inhibitors that do not enter the enzyme's active site, but bind to another part of the enzyme molecule.

- Causes enzyme to change its shape so the active site cannot bind substrate.
- May act as metabolic poisons (e.g., DDT, many antibiotics).
- Selective enzyme inhibition is an essential mechanism in the cell for regulating metabolic reactions.

III.   The Control of Metabolism

A. Metabolic pathways are regulated by controlling enzyme activity.

Metabolic control often depends on allosteric regulation

1. Allosteric regulation

Allosteric site = Specific receptor site on some part of the enzyme molecule other than the active site.

- Most enzymes with allosteric sites have two or more polypeptide chains, each with its own active site. Allosteric sites are often located where the subunits join.

- Allosteric enzymes have two conformations, one catalytically active and the other inactive (see Campbell, Figure 6.15) .

- Binding of an *activator* to an allosteric site stabilizes the active conformation.

- Binding of an *inhibitor* (noncompetitive inhibitor) to an allosteric site stabilizes the inactive conformation.

- Enzyme activity changes continually in response to changes in the relative proportions of activators and inhibitors (e.g., ATP/ADP).

- Subunits may interact so that a single activator or inhibitor at one allosteric site will affect the active sites of the other subunits.

2. Feedback inhibition

Feedback inhibition = Regulation of a metabolic pathway by its end product, which inhibits an enzyme within the pathway. For example:

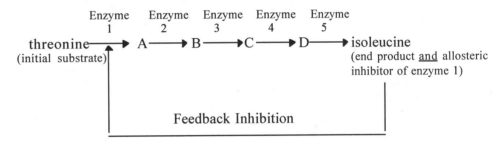

Prevents the cell from wasting chemical resources by synthesizing more product than is necessary (see also Campbell, Figure 6.16).

3. Cooperativity

Substrate molecules themselves may enhance enzyme activity.

Cooperativity = The phenomenon where substrate binding to the active site of one subunit induces a conformational change that enhances substrate binding at the active sites of the other subunits (see Campbell, Figure 6.17).

B. The localization of enzymes within the cell helps order metabolism

Cellular structure orders and compartmentalizes metabolic pathways (see Campbell, Figure 6.18).

- Some enzymes and enzyme complexes have fixed locations in the cell because they are incorporated into a membrane.

- Other enzymes and their substrates may be localized within membrane-enclosed eukaryotic organelles (e.g., chloroplasts and mitochondria).

REFERENCES

Atkins, P.W. *The Second Law*. New York, Oxford: W.H. Freeman and Company, 1984. A beautifully written, understandable description of the Second Law of Thermodynamics; addresses the role of the Second Law in life processes.

Campbell, N., et al. *Biology*. 5th ed. Menlo Park, California: Benjamin/Cummings, 1998.

Lehninger, A.L., D.L. Nelson and M.M. Cox. *Principles of Biochemistry*. 2nd ed. New York: Worth, 1993.

CHAPTER 7
A TOUR OF THE CELL

OUTLINE

OBJECTIVES

After reading this chapter and attending lecture, the student should be able to:

1. Describe techniques used to study cell structure and function.
2. Distinguish between magnification and resolving power.
3. Describe the principles, advantages and limitations of the light microscope, transmission electron microscope and the scanning electron microscope.
4. Describe the major steps of cell fractionation and explain why it is a useful technique.

5.   Distinguish between prokaryotic and eukaryotic cells.

6.   Explain why there are both upper and lower limits to cell size.

7.   Explain why compartmentalization is important in eukaryotic cells.

8.   Describe the structure and function of the nucleus, and briefly explain how the nucleus controls protein synthesis in the cytoplasm.

9.   Describe the structure and function of a eukaryotic ribosome.

10. List the components of the *endomembrane system*, describe their structures and functions and summarize the relationships among them.

11. Explain how impaired lysosomal function causes the symptoms of storage diseases.

12. Describe the types of vacuoles and explain how their functions differ.

13. Explain the role of *peroxisomes* in eukaryotic cells.

14. Describe the structure of a *mitochondrion* and explain the importance of compartmentalization in mitochondrial function.

15. Distinguish among *amyloplast*, *chromoplast* and *chloroplast*.

16. Identify the three functional compartments of a chloroplast, and explain the importance of compartmentalization in chloroplast function.

17. Describe probable functions of the cytoskeleton.

18. Describe the structure, monomers and functions of microtubules, microfilaments and intermediate filaments.

19. Explain how the ultrastructure of cilia and flagella relates to their function.

20. Describe the development of plant cell walls.

21. Describe the structure and list some functions of the extracellular matrix in animal cells.

22. Describe the structure of intercellular junctions found in plant and animal cells, and relate their structure to function.

KEY TERMS

light microscope	nucleolus	thylakoid	middle lamella
resolving power	ribosome	granlakoids	secondary cell wall
organelle	endomembrane system	stroma	extracellular matrix
electron microscope	endoplasmic reticulum (ER)	cytoskeleton	collagen
TEM	smooth ER	microtubules	proteoglycan
SEM	rough ER	microfilaments	fibronectin
cell fractionation	glycoprotein	integrin	intermediate filaments
ultracentrifuges	transport vesicles	centrosome	plasmodesmata
cytoplasm	Golgi apparatus	centriole	tight junctions
prokaryotic cell	phagocytosis	flagella	desmosomes
nucleoid	food vacuole	cilia	gap junctions
eukaryotic cell	contractile vacuole	basal body	
cytoplasm	central vacuole	dynein	
cytosol	peroxisome	actin	
plasma membrane	mitochondria	myosin	
nucleus	chloroplast	pseudopodia	
nuclear lamina	cristae	cytoplasmic streaming	
chromatin	mitochondrial matrix	cell wall	
chromosome	plastid	primary cell wall	

LECTURE NOTES

All organisms are made of cells, the organism's basic unit of structure and function.

The cell as a microcosm can be used to illustrate four themes integral to the text and course:

1. Theme of emergent properties. Life at the cellular level arises from interactions among cellular components.

2. Correlation of structure and function. Ordered cellular processes (e.g., protein synthesis, respiration, photosynthesis, cell-cell recognition, cellular movement, membrane production and secretion) are based upon ordered structures.

3. Interaction of organisms within their environment. Cells are excitable responding to environmental stimuli. In addition, cells are open systems that exchange materials and energy with their environment.

4. Unifying theme of evolution. Evolutionary adaptations are the basis for the correlation between structure and function.

> Students often find this material boring. A good set of micrographs and line drawings in the form of slides or transparencies will help. If the class size is small enough, a tour of an electron microscopy facility will help stimulate interest.

I.   How We Study Cells

A.   Microscopes provide windows to the world of the cell

The microscope's invention and improvement in the seventeenth century led to the discovery and study of cells.

In 1665, Robert Hooke described cells using a *light microscope*. Modern light microscopy is based upon the same principles as microscopy first used by Renaissance scientists.

- Visible light is focused on a specimen with a *condenser lens*.
- Light passing through the specimen is refracted with an *objective lens* and an *ocular lens*. The specimen's image is thus magnified and inverted for the observer.

Two important concepts in microscopy are *magnification* and *resolving power*.

- *Magnification* = How much larger an object is made to appear compared to its real size.
- *Resolving power* = Minimum distance between two points that can still be distinguished as two separate points.
- Resolution of a light microscope is limited by the wavelength of visible light. Maximum possible resolution of a light microscope is 0.2 μm.
- Highest magnification in a light microscope with maximum resolution is about 1000 times.
- By the early 1900s, optics in light microscopes were good enough to achieve the best resolution, so improvements since then have focused on improving contrast.

In the 1950s, researchers began to use the *electron microscope* which far surpassed the resolving power of the light microscope.

- Resolving power is inversely related to wavelength. Instead of light, electron microscopes use electron beams which have much shorter wavelengths than visible light.
- Modern electron microscopes have a practical resolving power of about 2 nm.
- Enhanced resolution and magnification allowed researchers to clearly identify subcellular *organelles* and to study cell *ultrastructure*.

- Two types of electron microscopes are the *transmission electron microscope* (TEM) and the *scanning electron microscope.*

The *transmission electron microscope* (TEM) aims an electron beam at a thin section of specimen which may be stained with metals to absorb electrons and enhance contrast.

- Electrons *transmitted* through the specimen are focused and the image magnified by using electromagnetic lenses (rather than glass lenses) to bend the trajectories of the charged electrons.
- Image is focused onto a viewing screen or film.
- Used to study internal cellular ultrastructure.

The *scanning electron microscope* (SEM) is useful for studying the surface of a specimen.

- Electron beam *scans* the surface of the specimen usually coated with a thin film of gold.
- Scanning beam excites secondary electrons on the sample's surface.
- Secondary electrons are collected and focused onto a viewing screen.
- SEM has a great depth of field and produces a three-dimensional image.

Disadvantages of an electron microscope:

- Can usually only view dead cells because of the elaborate preparation required.
- May introduce structural artifacts.

In laboratory, it would be useful to give students electron micrographs of organelles to identify and label. Many are disappointed when they view wet mounts of cells or prepared slides with their light microscopes and cannot find the detail seen in the micrographs. Clearly, some students have no conception of the resolution and magnifying power of an electron microscope. It would be helpful to indicate a size scale on micrographs you might use in lecture.

B. Cell biologists can isolate organelles to study their function

Modern cell biology integrates the study of cell structure (*cytology*) with the study of cell function. Cell fractionation is a technique that enables researchers to isolate organelles without destroying their function (see Campbell, Figure 7.3).

Cell fractionation = Technique which involves centrifuging disrupted cells at various speeds and durations to isolate components of different sizes, densities, and shapes.

- Development of the *ultracentrifuge* made this technique possible.
- Ultracentrifuges can spin as fast as 80,000 rpm, applying a force of 500,000 g.

The process of cell fractionation involves the following:

- Homogenization of tissue and its cells using pistons, blenders, or ultrasound devices.
- Centrifugation of the resulting homogenate at a slow speed. Nuclei and other larger particles settle at the bottom of the tube, forming a *pellet.*
- The unpelleted fluid or *supernatant* is decanted into another tube and centrifuged at a faster speed, separating out smaller organelles.
- The previous step is repeated, increasing the centrifugation speed each time to collect smaller and smaller cellular components from successive pellets.
- Once the cellular components are separated and identified, their particular metabolic functions can be determined.

II.   A Panoramic View of the Cell

A.   Prokaryotic and eukaryotic cells differ in size and complexity

Living organisms are made of either prokaryotic or eukaryotic cells—two major kinds of cells, which can be distinguished by structural organization.

Prokaryotic (pro = before; karyon = kernel)	Eukaryotic (Eu = true; karyon = kernel)
Found only in bacteria and archaebacteria	Found in the Kingdoms Protista, Fungi, Plantae, and Animalia
No true nucleus; lacks nuclear envelope	True nucleus; bounded by nuclear envelope
Genetic material in *nucleoid* region	Genetic material within nucleus
No membrane-bound organelles (see Campbell, Figure 7.4)	Contains cytoplasm with *cytosol* and membrane-bound *organelles*

Cytoplasm = Entire region between the nucleus and cell membrane

Cytosol = Semi-fluid medium found in the cytoplasm

1.   Cell size

Size ranges of cells:

Cell Type	Diameter
Mycoplasmas	0.1 - 1.0 μm
Most bacteria	1.0 - 10.0 μm
Most eukaryotic cells	10.0 - 100.0 μm

Range of cell size is limited by metabolic requirements. The lower limits are probably determined by the smallest size with enough:

- DNA to program metabolism.
- ribosomes, enzymes and cellular components to sustain life and reproduce.

The upper limits of size are imposed by the surface area to volume ratio. As a cell increases in size, its volume grows proportionately more than its surface area (see Campbell, Figure 7.5).

- The surface area of the plasma membrane must be large enough for the cell volume, in order to provide an adequate exchange surface for oxygen, nutrients and wastes.

B.   Internal membranes compartmentalize the functions of a eukaryotic cell

The average eukaryotic cell has a thousand times the volume of the average prokaryotic cell, but only a hundred times the surface area. Eukaryotic cells compensate for the small surface area to volume ratio by having internal membranes which:

- Partition the cell into compartments.
- Have unique lipid and protein compositions depending upon their specific functions.
- May participate in metabolic reactions since many enzymes are incorporated directly into the membrane.
- Provide localized environmental conditions necessary for specific metabolic processes.

- Sequester reactions, so they may occur without interference from incompatible metabolic processes elsewhere in the cell (see Campbell, Figure 7.6).

III.   The Nucleus and Ribosomes

A. The nucleus contains a eukaryotic cell's genetic library

Nucleus = A generally conspicuous membrane-bound cellular organelle in a eukaryotic cell; contains most of the genes that control the entire cell (see Campbell, Figure 7.9).

- Averages about 5 μm diameter.
- Enclosed by a *nuclear envelope*.

Nuclear envelope = A double membrane which encloses the nucleus in a eukaryotic cell.

- Is two lipid bilayer membranes separated by a space of about 20 to 40 nm. Each lipid bilayer has its own specific proteins.
- Attached to proteins on the envelope's nuclear side is a network of protein filaments, the *nuclear lamina*, which stabilizes nuclear shape.
- Is perforated by pores (100 nm diameter), which are ordered by an octagonal array of protein granules.
    - The envelope's inner and outer membranes are fused at the lip of each pore.
    - Pore complex regulates molecular traffic into and out of the nucleus.
- There is new evidence of an intranuclear framework of fibers, the *nuclear matrix*.

The nucleus contains most of the cell's DNA which is organized with proteins into a complex called *chromatin*.

Chromatin = Complex of DNA and histone proteins, which makes up *chromosomes* in eukaryotic cells; appears as a mass of stained material in nondividing cells.

Chromosomes = Long threadlike association of genes, composed of *chromatin* and found in the nucleus of eukaryotic cells.

- Each species has a characteristic chromosome number.
- Human cells have 46 chromosomes, except egg and sperm cells, which have half or 23.

The most visible structure within the nondividing nucleus is the *nucleolus*.

Nucleolus = Roughly spherical region in the nucleus of nondividing cells, which consists of *nucleolar organizers* and ribosomes in various stages of production.

- May be two or more per cell.
- Packages ribosomal subunits from:
    - rRNA transcribed in the nucleolus.
    - RNA produced elsewhere in the nucleus.
    - Ribosomal proteins produced and imported from the cytoplasm.
- Ribosomal subunits pass through nuclear pores to the cytoplasm, where their assembly is completed.

Nucleolar organizers = Specialized regions of some chromosomes, with multiple copies of genes for rRNA (ribosomal RNA) synthesis.

The nucleus controls protein synthesis in the cytoplasm:

Messenger RNA (mRNA) *transcribed* in the nucleus
from DNA instructions.

Passes through nuclear pores into cytoplasm.

Attaches to ribosomes where the genetic message
is *translated* into primary protein structure.

B. Ribosomes build a cell's proteins

Ribosome = A cytoplasmic organelle that is the site for protein synthesis (see Campbell, Figure 7.10).

- Are complexes of RNA and protein
- Constructed in the nucleolus in eukaryotic cells
- Cells with high rates of protein synthesis have prominent nucleoli and many ribosomes (e.g., human liver cell has a few million).

> Since most organelles are membrane-bound, students frequently ask if the ribosome has a membrane. They can deductively answer the question themselves if they are reminded that prokaryotes have ribosomes as well.

Ribosomes function either free in the cytosol or bound to endoplasmic reticulum. Bound and free ribosomes are structurally identical and interchangeable.

Free ribosomes = Ribosomes suspended in the cytosol.

- Most proteins made by free ribosomes will function in the cytosol.

Bound ribosomes = Ribosomes attached to the outside of the endoplasmic reticulum.

- Generally make proteins that are destined for membrane inclusion or export.
- Cells specializing in protein secretion often have many bound ribosomes (e.g., pancreatic cells).

IV. The Endomembrane System

Biologists consider many membranes of the eukaryotic cell to be part of an *endomembrane system*.

- Membranes may be interrelated *directly* through physical contact.
- Membranes may be related *indirectly* through *vesicles*.

Vesicles = Membrane-enclosed sacs that are pinched off portions of membranes moving from the site of one membrane to another.

Membranes of the endomembrane system vary in structure and function, and the membranes themselves are dynamic structures changing in composition, thickness and behavior.

The endomembrane system includes:

- Nuclear envelope
- Endoplasmic reticulum
- Golgi apparatus
- Lysosomes
- Vacuoles

- Plasma membrane (not actually an *endomembrane*, but related to endomembrane system)

A. The endoplasmic reticulum manufactures membranes and performs many other biosynthetic functions

Endoplasmic reticulum (ER) = (Endoplasmic = within the cytoplasm; reticulum = network); extensive membranous network of tubules and sacs (*cisternae*) which sequesters its internal lumen (*cisternal space*) from the cytosol.

- Most extensive portion of endomembrane system.
- Continuous with the outer membrane of the nuclear envelope; therefore, the space between the membranes of the nuclear envelope is continuous with cisternal space.

There are two distinct regions of ER that differ in structure and function: smooth ER and rough ER (see Campbell, Figure 7.11).

1. Functions of smooth ER

Appears smooth in the electron microscope because its cytoplasmic surface lacks ribosomes. Smooth ER functions in diverse metabolic processes:

a. Participates in the synthesis of lipids, phospholipids and steroids

- For example, vertebrate, particularly mammalian sex hormones and steroids secreted by the adrenal gland.
- Cells that produce and secrete these products are rich in smooth ER (e.g., testes, ovaries, skin oil glands).

b. Participates in carbohydrate metabolism

- Smooth ER in liver contains an embedded enzyme that catalyzes the final step in the conversion of glycogen to glucose (removes the phosphate from glucose-phosphate).

c. Detoxifies drugs and poisons

- Smooth ER, especially in the liver, contains enzymes which detoxify drugs and poisons.
- Enzymes catalyze the addition of hydroxyl groups to drugs and poisons. This makes them soluble in the cytosol, so they may be excreted from the body.
- Smooth ER in liver cells proliferates in response to barbiturates, alcohol and other drugs. This, in turn, may increase drug tolerance.

d. Stores calcium ions necessary for muscle contraction

- In a muscle cell, the ER membrane pumps Ca^{++} from the cytosol into the cisternal space.
- In response to a nerve impulse, Ca^{++} leaks from the ER back into the cytosol, which triggers muscle cell contraction.

2. Rough ER and protein synthesis

Rough ER:

- Appears rough under an electron microscope because the cytoplasmic side is studded with ribosomes.
- Is continuous with outer membrane of the nuclear envelope (which may also be studded with ribosomes on the cytoplasmic side).
- Manufactures secretory proteins and membrane.

Proteins destined for secretion are synthesized by ribosomes attached to rough ER:

Ribosomes attached to rough ER synthesize secretory proteins.

↓

Growing polypeptide is threaded through ER membrane into the lumen or *cisternal space.*

↓

Protein folds into its native conformation.

↓

If destined to be a *glycoprotein,* enzymes localized in the ER membrane catalyze the covalent bonding of an *oligosaccharide* to the secretory protein.

↓

Protein departs in a *transport vesicle* pinched off from *transitional ER* adjacent to the rough ER site of production.

Glycoprotein = Protein covalently bonded to carbohydrate.

Oligosaccharide = Small polymer of sugar units.

Transport vesicle = Membrane vesicle in transit from one part of the cell to another.

It may be useful to point out the protein that will be packaged into vesicles (e.g., hydrolytic enzymes within lysosomes) to be inserted into membranes (e.g., membrane-bound enzymes, receptors) is also synthesized by ribosomes attached to the ER.

---

3. **Rough ER and membrane production**

Membranes of rough ER grow *in place* as newly formed proteins and phospholipids are assembled:

- Membrane proteins are produced by ribosomes. As a polypeptide grows, it is inserted directly into the rough ER membrane where it is anchored by hydrophobic regions of the proteins.
- Enzymes within the ER membrane synthesize phospholipids from raw materials in the cytosol.
- Newly expanded ER membrane can be transported as a vesicle to other parts of the cell.

B. The Golgi apparatus finishes, sorts, and ships cell products

Many transport vesicles leave the ER and travel to the *Golgi apparatus*.

Golgi apparatus = Organelle made of stacked, flattened membranous sacs (*cisternae*), that modifies, stores and routes products of the endoplasmic reticulum (see Campbell, Figure 7.12).

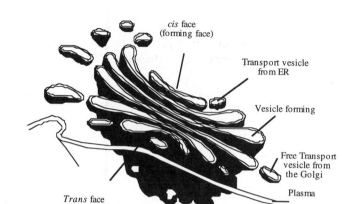

- Membranes of the cisternae sequester cisternal space from the cytosol.
- Vesicles may transport macromolecules between the Golgi and other cellular structures.
- Has a distinct polarity. Membranes of cisternae at opposite ends differ in thickness and composition.
- Two poles are called the *cis face* (forming face) and the *trans face* (maturing face).
- *Cis face*, which is closely associated with transitional ER, receives products by accepting transport vesicles from the ER. A vesicle fuses its membrane to the cis face of the Golgi and empties its soluble contents into the Golgi's cisternal space.
- *Trans face* pinches off vesicles from the Golgi and transports molecules to other sites.

Enzymes in the Golgi modify products of the ER in stages as they move through the Golgi stack from the *cis* to the *trans* face:

- Each cisternae between the *cis* and *trans* face contains unique combinations of enzymes.
- Golgi products in transit from one cisternae to the next, are carried in transport vesicles.

During this process, the Golgi:

- Alters some membrane phospholipids.
- Modifies the oligosaccharide portion of glycoproteins.
- Manufactures certain macromolecules itself (e.g., hyaluronic acid).
- Targets products for various parts of the cell.
  - Phosphate groups or oligosaccharides may be added to Golgi products as molecular identification tags.
  - Membranous vesicles budded from the Golgi may have external molecules that recognize docking sites on the surface of certain other organelles.
- Sorts products for secretion. Products destined for secretion leave the *trans* face in vesicles which eventually fuse with the plasma membrane.

C. Lysosomes are digestive compartments

Lysosome = An organelle which is a membrane-enclosed bag of hydrolytic enzymes that digest all major classes of macromolecules (see Campbell, Figure 7.13).

- Enzymes include lipases, carbohydrases, proteases, and nucleases.
- Optimal pH for lysosomal enzymes is about pH 5.
- Lysosomal membrane performs two important functions:
    - Sequesters potentially destructive hydrolytic enzymes from the cytosol.
    - Maintains the optimal acidic environment for enzyme activity by pumping H^+s inward from the cytosol to the lumen.
- Hydrolytic enzymes and lysosomal membrane are synthesized in the rough ER and processed further in the Golgi apparatus.
- Lysosomes probably pinch off from the *trans* face of the Golgi apparatus (see Campbell, Figure 7.14).

1. Functions of lysosomes

a. Intracellular digestion

Phagocytosis = (Phago = to eat; cyte = cell); cellular process of ingestion, in which the plasma membrane engulfs particulate substances and pinches off to form a particle-containing *vacuole*.

- Lysosomes may fuse with food-filled vacuoles, and their hydrolytic enzymes digest the food.
- Examples are *Amoeba* and other protists which eat smaller organisms or food particles.
- Human cells called *macrophages* phagocytize bacteria and other invaders.

b. Recycle cell's own organic material

- Lysosomes may engulf other cellular organelles or part of the cytosol and digest them with hydrolytic enzymes (autophagy).
- Resulting monomers are released into the cytosol where they can be recycled into new macromolecules.

c. Programmed cell destruction

Destruction of cells by their own lysosomes is important during metamorphosis and development.

2. Lysosomes and human disease

Symptoms of inherited *storage diseases* result from impaired lysosomal function. Lack of a specific lysosomal enzyme causes substrate accumulation which interferes with lysosomal metabolism and other cellular functions.

- In Pompe's disease, the missing enzyme is a carbohydrase that breaks down glycogen. The resulting glycogen accumulation damages the liver.
- Lysosomal lipase is missing or inactive in Tay-Sachs disease, which causes lipid accumulation in the brain.

D. Vacuoles have diverse functions in cell maintenance

Vacuole = Organelle which is a membrane-enclosed sac that is larger than a vesicle (transport vesicle, lysosome, or microbody).

Vacuole types and functions:

Food vacuole = Vacuole formed by phagocytosis which is the site of intracellular digestion in some protists and macrophages (see Campbell, Figure 7.14).

Contractile vacuole = Vacuole that pumps excess water from the cell; found in some freshwater protozoa.

Central vacuole = Large vacuole found in most mature plant cells (see Campbell, Figure 7.15)

- Is enclosed by a membrane called the *tonoplast* which is part of the endomembrane system
- Develops by the coalescence of smaller vacuoles derived from the ER and Golgi apparatus
- Is a versatile compartment with many functions:
    - Stores organic compounds (e.g., protein storage in seeds)
    - Stores inorganic ions (e.g., K^+ and Cl^-)
    - Sequesters dangerous metabolic by-products from the cytoplasm
    - Contains soluble pigments in some cells (e.g., red and blue pigments in flowers)
    - May protect the plant from predators by containing poisonous or unpalatable compounds
    - Plays a role in plant growth by absorbing water and elongating the cell
    - Contributes to the large ratio of membrane surface area to cytoplasmic volume. (There is only a thin layer of cytoplasm between the tonoplast and plasma membrane.)

E. A summary of relationships among endomembranes

Components of the endomembrane system are related through direct contact or through vesicles (see Campbell, Figure 7.16).

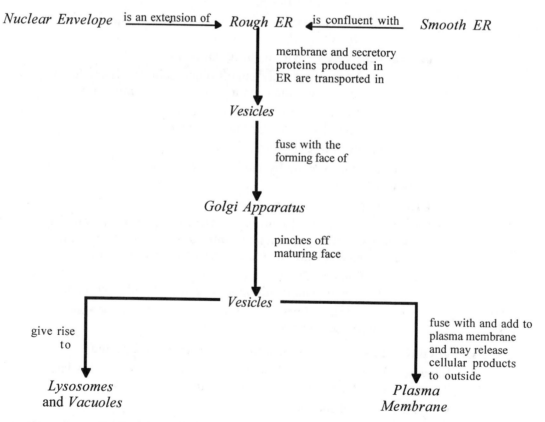

V.   Other Membranous Organelles

A. Peroxisomes consume oxygen in various metabolic functions

Peroxisomes = Membrane-bound organelles that contain specialized teams of enzymes for specific metabolic pathways; all contain peroxide-producing oxidases.

- Bound by a single membrane
- Found in nearly all eukaryotic cells
- Often have a granular or crystalline core which is a dense collection of enzymes (see Campbell, Figure 7.17)
- Contain peroxide-producing oxidases that transfer hydrogen from various substrates to oxygen, producing hydrogen peroxide

$$RH_2 \;+\; O_2 \xrightarrow{\text{oxidase}} R \;+\; H_2O_2$$

- Contain catalase, an enzyme that converts toxic hydrogen peroxide to water

$$2H_2O_2 \xrightarrow{\text{catalase}} 2H_2O \;+\; O_2$$

- Peroxisomal reactions have many functions, some of which are:
  - Breakdown of fatty acids into smaller molecules (acetyl CoA). The products are carried to the mitochondria as fuel for cellular respiration.
  - Detoxification of alcohol and other harmful compounds. In the liver, peroxisomes enzymatically transfer H from poisons to O_2.
- Specialized peroxisomes (*glyoxysomes*) are found in heterotrophic fat-storing tissue of germinating seeds.
  - Contain enzymes that convert lipid to carbohydrate.
  - These biochemical pathways make energy stored in seed oils available for the germinating seedling.
- Current thought is that peroxisome biogenesis occurs by pinching off from preexisting peroxisomes. Necessary lipids and enzymes are imported from the cytosol.

B. Mitochondria and chloroplasts are the main energy transformers of cells

Mitochondria and chloroplasts are organelles that transduce energy acquired from the surroundings into forms useable for cellular work.

- Enclosed by *double* membranes (see Campbell, Figure 7.18).
- Membranes are not part of endomembrane system. Rather than being made in the ER, their membrane proteins are synthesized by free ribosomes in the cytosol and by ribosomes located within these organelles themselves.
- Contain ribosomes and some DNA that programs a small portion of their own protein synthesis, though most of their proteins are synthesized in the cytosol programmed by nuclear DNA.
- Are semiautonomous organelles that grow and reproduce within the cell.

> You may want to just *briefly* mention mitochondria and chloroplasts at this point in the course. Because structure is so closely tied to function, the organelle structure must be covered again in detail with cellular respiration and photosynthesis. In deference to time, it may be more practical to discuss it just once with the metabolism lectures.

1. Mitochondria

Mitochondria = Organelles which are the sites of cellular respiration, a catabolic oxygen-requiring process that uses energy extracted from organic macromolecules to produce ATP.

- Found in nearly all eukaryotic cells
- Number of mitochondria per cell varies and directly correlates with the cell's metabolic activity
- Are about 1 mm in diameter and 1-10 mm in length
- Are dynamic structures that move, change their shape and divide

Structure of the mitochondrion:

- Enclosed by two membranes that have their own unique combination of proteins embedded in phospholipid bilayers (see Campbell, Figure 7.18)

- Smooth *outer membrane* is highly permeable to small solutes, but it blocks passage of proteins and other macromolecules

- Convoluted *inner membrane* contains embedded enzymes that are involved in cellular respiration. The membrane's many infoldings or *cristae* increase the surface area available for these reactions to occur.

- The inner and outer membranes divide the mitochondrion into two internal compartments:

a. Intermembrane space

- Narrow region between the inner and outer mitochondrial membranes.

- Reflects the solute composition of the cytosol, because the outer membrane is permeable to small solute molecules.

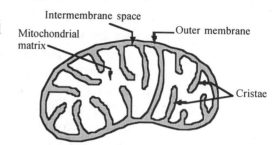

b. Mitochondrial matrix

- Compartment enclosed by the inner mitochondrial membrane

- Contains enzymes that catalyze many metabolic steps of cellular respiration

- Some enzymes of respiration and ATP production are actually embedded in the inner membrane.

2. Chloroplasts

Plastids = A group of plant and algal membrane-bound organelles that include *amyloplasts*, *chromoplasts* and *chloroplasts*.

Amyloplasts = (Amylo = starch); colorless plastids that store starch; found in roots and tubers.

Chromoplasts = (Chromo = color); plastids containing pigments other than chlorophyll; responsible for the color of fruits, flowers and autumn leaves.

Chloroplasts = (Chloro = green); chlorophyll-containing plastids which are the sites of photosynthesis.

- Found in eukaryotic algae, leaves and other green plant organs.

- Are lens-shaped and measure about 2 mm by 5 mm.

- Are dynamic structures that change shape, move and divide.

Structure of the chloroplast:

Chloroplasts are divided into three functional compartments by a system of membranes (see also Campbell, Figure 7.19):

a. Intermembrane space

The chloroplast is bound by a double membrane which partitions its contents from the cytosol. A narrow *intermembrane space* separates the two membranes.

b. Thylakoid space

Thylakoids form another membranous system within the chloroplast. The thylakoid membrane segregates the interior of the chloroplast into two compartments: *thylakoid space* and *stroma*.

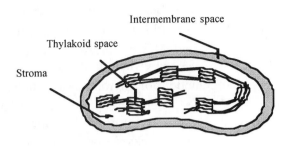

- *Thylakoid space* = Space inside the thylakoid
- *Thylakoids* = Flattened membranous sacs inside the chloroplast
- Chlorophyll is found in the thylakoid membranes.
- Thylakoids function in the steps of photosynthesis that initially convert light energy to chemical energy.
- Some thylakoids are stacked into *grana*.

Grana = (Singular, *granum*); stacks of thylakoids in a chloroplast.

c. Stroma

Photosynthetic reactions that use chemical energy to convert carbon dioxide to sugar occur in the stroma.

Stroma = Viscous fluid outside the thylakoids

VI. The Cytoskeleton

A. Provides structural support to the cells for cell motility and regulation

It was originally thought that organelles were suspended in a formless cytosol. Technological advances in both light and electron microscopy (e.g., high voltage E.M.) revealed a three-dimensional view of the cell, which showed a network of fibers throughout the cytoplasm—the *cytoskeleton*. The cytoskeleton plays a major role in organizing the structures and activities of the cell.

Cytoskeleton = A network of fibers throughout the cytoplasm that forms a dynamic framework for support and movement and regulation (see Campbell, Figure 7.20).

- Gives mechanical support to the cell and helps maintain its shape
- Enables a cell to change shape in an adaptive manner
- Associated with motility by interacting with specialized proteins called *motor molecules* (e.g., organelle movement, muscle contraction, and locomotor organelles)
- Play a regulatory role by mechanically transmitting signals from cell's surface to its interior
- Constructed from at least three types of fibers: *microtubules* (thickest), *microfilaments* (thinnest), and *intermediate filaments* (intermediate in diameter) (see Campbell, Table 7.2)

1. Microtubules

Found in cytoplasm of all eukaryotic cells, *microtubules*:

- Are straight *hollow* fibers about 25 nm in diameter and 200 nm – 25 μm in length
- Are constructed from globular proteins called *tubulin* that consists of one α-tubulin and one β-tubulin molecule

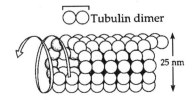

- Begin as two-dimensional sheets of tubulin units, which roll into tubes
- Elongate by adding tubulin units to its ends
- May be disassembled and the tubulin units recycled to build microtubules elsewhere in the cell

Functions of microtubules include:

- Cellular support; these microtubule function as compression-resistant girders to reinforce cell shape
- Tracks for organelle movement (see Campbell, Figure 7.21). Protein *motor molecules* (e.g., kinesin) interact with microtubules to translocate organelles (e.g., vesicles from the Golgi to the plasma membrane).
- Separation of chromosomes during cell division

a. Centrosomes and centrioles

Centriole = Pair of cylindrical structures located in the centrosome of in animal cells, composed of nine sets of triplet microtubules arranged in a ring (see Campbell, Figure 7.22).

- Are about 150 nm in diameter and are arranged at right angles to each other.
- Pair of centrioles located within the centrosome, replicate during cell division.
- May organize microtubule assembly during cell division, but must not be mandatory for this function since plants lack centrioles.

b. Cilia and flagella

Cilia and *flagella* = Locomotor organelles found in eukaryotes that are formed from a specialized arrangement of microtubules.

- Many unicellular eukaryotic organisms are propelled through the water by cilia or flagella and motile sperm cells (animals, algae, some plants) are flagellated.
- May function to draw fluid across the surface of stationary cells (e.g., ciliated cells lining trachea).

Cilia (singular, cilium)	Flagella (singular, flagellum)
Occur in large numbers on cell surface.	One or a few per cell.
Shorter; 2-20 mm in length.	Longer; 10-200 mm in length.
Work like oars, alternating power with recovery strokes. Creates force in a direction perpendicular to the axis of the cilium.	Undulating motion that creates force in the same direction as the axis of the flagellum.

Ultrastructure of cilia and flagella:

- Are extensions of plasma membrane with a core of microtubules (see Campbell, Figure 7.24)

- Microtubular core is made of nine doublets of microtubules arranged in a ring with two single microtubules in the center (*9 + 2 pattern*).

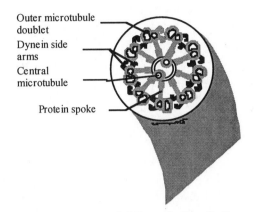

9+2 Pattern in Cross Section

- Each doublet is a pair of attached microtubules. One of the pair shares a portion of the other's wall.

- Each doublet is connected to the center of the ring by *radial spokes* that end near the central microtubules.

- Each doublet is attached to the neighboring doublet by a pair of *side arms*. Many pairs of side arms are evenly spaced along the doublet's length.

- Structurally identical to centrioles, *basal bodies* anchor the microtubular assemblies.

Basal body = A cellular structure, identical to a centriole, that anchors the microtubular assembly of cilia and flagella.

- Can convert into a centriole and vice versa

- May be a template for ordering tubulin into the microtubules of *newly* forming cilia or flagella. As cilia and flagella continue to grow, new tubulin subunits are added to the tips, rather than to the bases.

The unique ultrastructure of cilia and flagella is necessary for them to function:

- Sidearms are made of *dynein*, a large protein motor molecule that changes its conformation in the presence of ATP as an energy source.

- A complex cycle of movements caused by dynein's conformational changes, makes the cilium or flagellum bend (see Campbell, Figure 7.25):

- In cilia and flagella, linear displacement of dynein sidearms is translated into a *bending* by the resistance of the *radial spokes*. Working against this resistance, the "dynein-walking" distorts the microtubules, causing them to bend.

Sidearms of one doublet attach to the
adjacent doublet.

↓

Sidearms swing and the two doublets slide
past one another.

↓

Sidearms release.

↓

Sidearms reattach to the adjacent doublet
farther along its length.

↓

Cycle is repeated.

2. Microfilaments (actin filaments)

Structure of microfilaments (see Campbell, Figure 7.26):

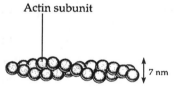

- Solid rods about 7 nm in diameter
- Built from globular protein monomers, *G-actin*, which are linked into long chains
- Two actin chains are wound into a helix

Function of microfilaments:

a. Provide cellular support

- Bear tension (pulling forces)
- In combination with other proteins, they form a three-dimensional network just inside plasma membrane that helps support cell shape.
- In animal cells specialized for transport, bundles of microfilaments make up the core of microvilli (e.g., intestinal epithelial wall).

b. Participate in muscle contraction

- Along the length of a muscle cell, parallel actin microfilaments are interdigitated with thicker filaments made of the protein *myosin*, a motor molecule (see Campbell, Figure 7.27a).
- With ATP as the energy source, a muscle cell shortens as the thin actin filaments slide across the myosin filaments. Sliding results from the swinging of myosin cross-bridges intermittently attached to actin.

c. Responsible for localized contraction of cells

Small actin-myosin aggregates exist in some parts of the cell and cause localized contractions. Examples include:

- Contracting ring of microfilaments pinches an animal cell in two during cell division

- Elongation and contraction of *pseudopodia* during *amoeboid movement*
- Involved in cytoplasmic streaming (cyclosis) found in plant cells

Cytoplasmic streaming (cyclosis) = Flowing of the entire cytoplasm around the space between the vacuole and plasma membrane in a plant cell (see Campbell, Figure 7.27c).

3. Intermediate filaments

Structure of intermediate filaments:

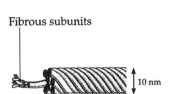

Fibrous subunits

10 nm

- Filaments that are intermediate in diameter (8-12 nm) between microtubules and microfilaments (see Campbell, Figure 7.26)
- Diverse class of cytoskeletal elements that differ in diameter and composition depending upon cell type
- Constructed from *keratin* subunits
- More permanent than microfilaments and microtubules

Function of intermediate filaments:

1. Specialized for bearing tension; may function as the framework for the cytoskeleton
2. Reinforce cell shape (e.g., nerve axons)
3. Probably fix organelle position (e.g., nucleus)
4. Compose the nuclear lamina, lining the nuclear envelope's interior

VII. Cell Surfaces and Junctions

A. Plant cells are encased by cell walls

Most cells produce coats that are external to the plasma membrane.

1. Cell walls

Plant cells can be distinguished from animal cells by the presence of a *cell wall*:

- Thicker than the plasma membrane (0.1–2 μm)
- Chemical composition varies from cell to cell and species to species.
- Basic design includes strong *cellulose* fibers embedded in a matrix of other polysaccharides and proteins.
- Functions to protect plant cells, maintain their shape, and prevent excess water uptake
- Has membrane-lined channels, *plasmodesmata*, that connect the cytoplasm of neighboring cells

Plant cells develop as follows:

- Young plant cell secretes a thin, flexible *primary cell wall*. Between primary cell walls of adjacent cells is a *middle lamella* made of *pectins*, a sticky polysaccharide that cements cells together.
- Cell stops growing and strengthens its wall. Some cells:
  1. secrete hardening substances into primary wall.
  2. add a *secondary cell wall* between plasma membrane and primary wall.

Secondary cell wall is often deposited in layers with a durable matrix that supports and protects the cell (see Campbell, Figure 7.28).

B. The extracellular matrix (ECM) of animal cells functions in support, adhesion, movement, and development

Animal cells lack walls, but they do have an elaborate *extracellular matrix (ECM)*.

Extracellular matrix (ECM) = Meshwork of macromolecules outside the plasma membrane of animal cells. This ECM is:

- locally secreted by cells.
- composed mostly of glycoproteins, the most abundant of which is *collagen* that:
    - accounts for about half of the total protein in the vertebrate body.
    - forms strong extracellular fibers embedded in a meshwork of carbohydrate-rich glycoproteins called *proteoglycans*.

Some cells are attached:

- directly to the collagen and proteoglycan of their extracellular matrix.
- or to the ECM by another class of glycoproteins–*fibronectins*.

Fibronectins bind to transmembrane receptor proteins called *integrins* that:

- bond on their cytoplasmic side to microfilaments of the cytoskeleton.
- integrate cytoskeletal responses to ECM changes and vice versa.

The extracellular matrix:

- provides support and anchorage for cells.
- functions in a cell's dynamic behavior. For example, some embryonic cells migrate along specific pathways by orienting their intracellular microfilaments to the pattern of extracellular fibers in the ECM (see Campbell, Figure 7.29).
- helps control gene activity in the cell's nucleus. Perhaps the transcription of specific genes is a response to chemical signals triggered by communication of mechanical stimuli across the plasma membrane from the ECM through integrins to the cytoskeleton.

C. Intercellular junctions help integrate cells into higher levels of structure and function

Neighboring cells often adhere and interact through special patches of direct physical contact.

Intercellular junctions in plants:

Plasmodesmata (singular, plasmodesma) = Channels that perforate plant cell walls, through which cytoplasmic strands communicate between adjacent cells.

- Lined by plasma membrane. Plasma membranes of adjacent cells are continuous through a plasmodesma.
- Allows free passage of water and small solutes. This transport is enhanced by *cytoplasmic streaming*.

Intercellular junctions in animals (see Campbell, Figure 7.30):

Tight junctions = Intercellular junctions that hold cells together tightly enough to block transport of substances through the intercellular space.

- Specialized membrane proteins in adjacent cells bond directly to each other allowing no space between membranes.
- Usually occur as belts all the way around each cell, that block intercellular transport.
- Frequently found in epithelial layers that separate two kinds of solutions.

Desmosomes = Intercellular junctions that rivet cells together into strong sheets, but still permit substances to pass freely through intracellular spaces. The desmosome is made of:

- Intercellular glycoprotein filaments that penetrate and attach the plasma membrane of both cells.
- A dense disk inside the plasma membrane that is reinforced by *intermediate filaments* made of *keratin* (a strong structural protein).

Gap junctions = Intercellular junctions specialized for material transport between the cytoplasm of adjacent cells.

- Formed by two connecting protein rings (*connexon*), each embedded in the plasma membrane of adjacent cells. The proteins protrude from the membranes enough to leave an intercellular gap of 2–4 nm.
- Have pores with diameters (1.5 nm) large enough to allow cells to share smaller molecules (e.g., inorganic ions, sugars, amino acids, vitamins), but not macromolecules such as proteins.
- Common in animal embryos and cardiac muscle where chemical communication between cells is essential.

REFERENCES

Alberts, B., D. Bray, J. Lewis, M. Raff, K. Roberts and J.D. Watson. *Molecular Biology of the Cell*. 2nd ed. New York: Garland, 1994.

Becker, W.M. and D.W. Deamer. *The World of the Cell*. 3rd ed. Redwood City, California: Benjamin/Cummings, 1996.

Campbell, N., et al. *Biology*. 5th ed. Menlo Park, California: Benjamin/Cummings, 1998.

deDuve, C. *A Guided Tour of the Living Cell*. Volumes I and II. New York: Scientific American Books, 1984. Literally, a guided tour of the cell with the reader as "cytonaut." This is an excellent resource for lecture material and enjoyable reading.

CHAPTER 8

MEMBRANE STRUCTURE AND FUNCTION

OUTLINE
I.   Membrane Structure
    A. Membrane models have evolved to fit new data: *science as a process*
    B. A membrane is a fluid mosaic of lipids, proteins, and carbohydrates
II. Traffic Across Membranes
    A. A membrane's molecular organization results in selective permeability
    B. Passive transport is diffusion across a membrane
    C. Osmosis is the passive transport of water
    D. Cell survival depends on balancing water uptake and loss
    E. Specific proteins facilitate the passive transport of selected solutes
    F. Active transport is the pumping of solutes against their gradients
    G. Some ion pumps generate voltage across membranes
    H. In cotransport, a membrane protein couples the transport of one solute to another
    I. Exocytosis and endocytosis transport large molecules

OBJECTIVES
After reading this chapter and attending lecture, the student should be able to:
1.   Describe the function of the plasma membrane.
2.   Explain how scientists used early experimental evidence to make deductions about membrane structure and function.
3.   Describe the Davson-Danielli membrane model and explain how it contributed to our current understanding of membrane structure.
4.   Describe the contribution J.D. Robertson, S.J. Singer, and G.L. Nicolson made to clarify membrane structure.
5.   Describe the fluid properties of the cell membrane and explain how membrane fluidity is influenced by membrane composition.
6.   Explain how hydrophobic interactions determine membrane structure and function.
7.   Describe how proteins are spatially arranged in the cell membrane and how they contribute to membrane function.
8.   Describe factors that affect selective permeability of membranes.
9.   Define diffusion; explain what causes it and why it is a spontaneous process.
10. Explain what regulates the rate of passive transport.
11. Explain why a concentration gradient across a membrane represents potential energy.
12. Define osmosis and predict the direction of water movement based upon differences in solute concentration.
13. Explain how bound water affects the osmotic behavior of dilute biological fluids.
14. Describe how living cells with and without walls regulate water balance.

15. Explain how transport proteins are similar to enzymes.
16. Describe one model for facilitated diffusion.
17. Explain how active transport differs from diffusion.
18. Explain what mechanisms can generate a membrane potential or electrochemical gradient.
19. Explain how potential energy generated by transmembrane solute gradients can be harvested by the cell and used to transport substances across the membrane.
20. Explain how large molecules are transported across the cell membrane.
21. Give an example of receptor-mediated endocytosis.
22. Explain how membrane proteins interface with and respond to changes in the extracellular environment.

KEY TERMS

selective permeability	hypotonic	membrane potential
amphipathic	isotonic	electrochemical gradient
fluid mosaic model	osmosis	electrogenic pump
integral proteins	osmoregulation	proton pump
peripheral proteins	turgid	cotransport
transport proteins	plasmolysis	exocytosis
diffusion	facilitated diffusion	phagocytosis
concentration gradient	gated channels	pinocytosis
passive transport	active transport	receptor-mediated endocytosis
hypertonic	sodium-potassium pump	ligands

LECTURE NOTES

I.   **Membrane Structure**

The *plasma membrane* is the boundary that separates the living cell from its nonliving surroundings. It makes life possible by its ability to discriminate in its chemical exchanges with the environment. This membrane:

- Is about 8 nm thick
- Surrounds the cell and controls chemical traffic into and out of the cell
- Is *selectively permeable*; it allows some substances to cross more easily than others
- Has a unique structure which determines its function and solubility characteristics

> This is an opportune place to illustrate how form fits function. It is remarkable how much early models contributed to the understanding of membrane structure, since biologists proposed these models without the benefit of "seeing" a membrane with an electron microscope.

A.   **Membrane models have evolved to fit new data:** *science as a process*

Membrane function is determined by its structure. Early models of the plasma membrane were deduced from *indirect* evidence:

1.   Evidence: Lipid and lipid soluble materials enter cells more rapidly than substances that are insoluble in lipids (C. Overton, 1895).

Deduction: Membranes are made of lipids.

Deduction: Fat-soluble substance move through the membrane by dissolving in it ("like dissolves like").

2. Evidence: Amphipathic phospholipids will form an artificial membrane on the surface of water with only the hydrophilic heads immersed in water (Langmuir, 1917).

   Amphipathic = Condition where a molecule has both a hydrophilic region and a hydrophobic region.

   Deduction: Because of their molecular structure, phospholipids can form membranes (see also Campbell, Figure 8.1a).

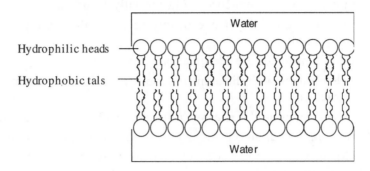

3. Evidence: Phospholipid content of membranes isolated from red blood cells is just enough to cover the cells with two layers (Gorter and Grendel, 1925).

   Deduction: Cell membranes are actually phospholipid bilayers, two molecules thick (see Campbell, Figure 8.1b).

4. Evidence: Membranes isolated from red blood cells contain proteins as well as lipids.

   Deduction: There is protein in biological membranes.

5. Evidence: Wettability of the surface of an actual biological membrane is greater than the surface of an artificial membrane consisting only of a phospholipid bilayer.

   Deduction: Membranes are coated on both sides with proteins, which generally absorb water.

Incorporating results from these and other solubility studies, J.F. Danielli and H. Davson (1935) proposed a model of cell membrane structure (see Campbell, Figure 8.2a):

- Cell membrane is made of a phospholipid bilayer sandwiched between two layers of globular protein.

- The polar (hydrophilic) heads of phospholipids are oriented towards the protein layers forming a hydrophilic zone.

- The nonpolar (hydrophobic) tails of phospholipids are oriented in between polar heads forming a hydrophobic zone.

- The membrane is approximately 8 nm thick.

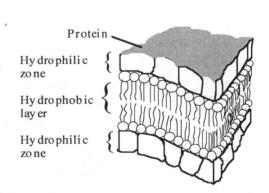

In the 1950s, electron microscopy allowed biologists to visualize the plasma membrane for the first time and provided support for the Davson-Danielli model. Evidence from electron micrographs:

1. Confirmed the plasma membrane was 7 to 8 nm thick (close to the predicted size if the Davson-Danielli model was modified by replacing globular proteins with protein layers in pleated-sheets).

2. Showed the plasma membrane was trilaminar, made of two electron-dense bands separated by an unstained layer. It was assumed that the heavy metal atoms of the stain adhered to the hydrophilic proteins and heads of phospholipids and not to the hydrophobic core.

3. Showed internal cellular membranes that looked similar to the plasma membrane. This led biologists (J.D. Robertson) to propose that all cellular membranes were symmetrical and virtually identical.

Though the phospholipid bilayer is probably accurate, there are problems with the Davson-Danielli model:

1. Not all membranes are identical or symmetrical.
   - Membranes with different functions also differ in chemical composition and structure.
   - Membranes are bifacial with distinct inside and outside faces.

2. A membrane with an outside layer of proteins would be an unstable structure.
   - Membrane proteins are not soluble in water, and, like phospholipid, they are *amphipathic*.
   - Protein layer not likely because its hydrophobic regions would be in an aqueous environment, and it would also separate the hydrophilic phospholipid heads from water.

In 1972, S.J. Singer and G.L. Nicolson proposed the *fluid mosaic model* which accounted for the amphipathic character of proteins (see Campbell, Figure 8.2b). They proposed:

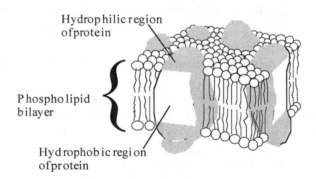

- Proteins are individually embedded in the phospholipid bilayer, rather than forming a solid coat spread upon the surface.
- Hydrophilic portions of both proteins and phospholipids are maximally exposed to water resulting in a stable membrane structure.
- Hydrophobic portions of proteins and phospholipids are in the nonaqueous environment inside the bilayer.
- Membrane is a mosaic of proteins bobbing in a fluid bilayer of phospholipids.
- Evidence from freeze fracture techniques have confirmed that proteins are embedded in the membrane. Using these techniques, biologists can delaminate membranes along the middle of the bilayer. When viewed with an electron microscope, proteins appear to penetrate into the hydrophobic interior of the membrane (see Campbell, Methods Box).

B. A membrane is a fluid mosaic of lipids, proteins and carbohydrates

1. The fluid quality of membranes

Membranes are held together by hydrophobic interactions, which are weak attractions (see Campbell, Figure 8.3).

- Most membrane lipids and some proteins can drift laterally within the membrane.
- Molecules rarely flip transversely across the membrane because hydrophilic parts would have to cross the membrane's hydrophobic core.
- Phospholipids move quickly along the membrane's plane averaging 2 μm per second.
- Membrane proteins drift more slowly than lipids (see Campbell, Figure 8.4). The fact that proteins drift laterally was established experimentally by fusing a human and mouse cell (Frye and Edidin, 1970):

Membrane proteins of a human and mouse
cell were labeled with different green and
red fluorescent dyes.

↓

Cells were fused to form a hybrid cell with
a continuous membrane.

↓

Hybrid cell membrane had initially distinct
regions of green and red dye.

↓

In less than an hour, the two colors were
intermixed.

- Some membrane proteins are tethered to the cytoskeleton and cannot move far.

Membranes must be fluid to work properly. Solidification may result in permeability changes and enzyme deactivation.

- Unsaturated hydrocarbon tails enhance membrane fluidity, because kinks at the carbon-to-carbon double bonds hinder close packing of phospholipids.
- Membranes solidify if the temperature decreases to a critical point. Critical temperature is lower in membranes with a greater concentration of unsaturated phospholipids.
- Cholesterol, found in plasma membranes of eukaryotes, modulates membrane fluidity by making the membrane:
  - Less fluid at warmer temperatures (e.g., 37°C body temperature) by restraining phospholipid movement.
  - More fluid at lower temperatures by preventing close packing of phospholipids.
- Cells may alter membrane lipid concentration in response to changes in temperature. Many cold tolerant plants (e.g., winter wheat) increase the unsaturated phospholipid concentration in autumn, which prevents the plasma membranes from solidifying in winter.

2. **Membranes as mosaics of structure and function**

A membrane is a *mosaic* of different proteins embedded and dispersed in the phospholipid bilayer (see Campbell, Figure 8.5). These proteins vary in both structure and function, and they occur in two spatial arrangements:

    a. *Integral proteins* are generally transmembrane protein with hydrophobic regions that completely span the hydrophobic interior of the membrane (see Campbell, Figure 8.6).

    b. *Peripheral proteins*, which are not embedded but attached to the membrane's surface.

- May be attached to integral proteins or held by fibers of the ECM
- On cytoplasmic side, may be held by filaments of cytoskeleton

Membranes are bifacial. The membrane's synthesis and modification by the ER and Golgi determines this asymmetric distribution of lipids, proteins and carbohydrates:

- Two lipid layers may differ in lipid composition.
- Membrane proteins have distinct directional orientation.
- When present, carbohydrates are restricted to the membrane's exterior.
- Side of the membrane facing the lumen of the ER, Golgi and vesicles is topologically the same as the plasma membrane's outside face (see Campbell, Figure 8.7).
- Side of the membrane facing the cytoplasm has always faced the cytoplasm, from the time of its formation by the endomembrane system to its addition to the plasma membrane by the fusion of a vesicle.
- Campbell, Figure 8.8, provides an overview of the six major kinds of function exhibited by proteins of the plasma membrane.

3. **Membrane carbohydrates and cell-cell recognition**

Cell-cell recognition = The ability of a cell to determine if other cells it encounters are alike or different from itself.

Cell-cell recognition is crucial in the functioning of an organism. It is the basis for:

- Sorting of an animal embryo's cells into tissues and organs
- Rejection of foreign cells by the immune system

The way cells recognize other cells is probably by keying on cell markers found on the external surface of the plasma membrane. Because of their diversity and location, likely candidates for such cell markers are membrane carbohydrates:

- Usually branched *oligosaccharides* (<15 monomers)
- Some covalently bonded to lipids (*glycolipids*)
- Most covalently bonded to proteins (*glycoproteins*)
- Vary from species to species, between individuals of the same species and among cells in the same individual

II. **Traffic Across Membranes**

A. **A membrane's molecular organization results in selective permeability**

The *selectively permeable* plasma membrane regulates the type and rate of molecular traffic into and out of the cell.

Selective permeability = Property of biological membranes which allows some substances to cross more easily than others. The selective permeability of a membrane depends upon:

- Membrane solubility characteristics of the phospholipid bilayer
- Presence of specific integral transport proteins

1. **Permeability of the lipid bilayer**

    The ability of substances to cross the hydrophobic core of the plasma membrane can be measured as the rate of transport through an artificial phospholipid bilayer:

    a. **Nonpolar (hydrophobic) molecules**

    - Dissolve in the membrane and cross it with ease (e.g., hydrocarbons, O, CO_2)
    - If two molecules are equally lipid soluble, the smaller of the two will cross the membrane faster.

    b. **Polar (hydrophilic) molecules**

    - Small, polar uncharged molecules (e.g., H_2O, ethanol) that are small enough to pass between membrane lipids, will easily pass through synthetic membranes.
    - Larger, polar uncharged molecules (e.g., glucose) will *not* easily pass through synthetic membranes.
    - All ions, even small ones (e.g., Na^+, H^+) have difficulty penetrating the hydrophobic layer.

2. **Transport proteins**

    Small polar molecules and nonpolar molecules rapidly pass through the plasma membrane as they do an artificial membrane.

    Unlike artificial membranes, however, biological membranes *are* permeable to specific ions and certain polar molecules of moderate size. These hydrophilic substances avoid the hydrophobic core of the bilayer by passing through *transport proteins*.

    Transport proteins = Integral membrane proteins that transport specific molecules or ions across biological membranes (see Campbell, Figure 8.8a)

    - May provide a hydrophilic tunnel through the membrane.
    - May bind to a substance and physically move it across the membrane.
    - Are specific for the substance they translocate.

B. **Passive transport is diffusion across a membrane**

> Students have particular trouble with the concepts of *gradient* and *net* movement, yet their understanding of diffusion depends upon having a working knowledge of these terms.

Concentration gradient = Regular, graded concentration change over a distance in a particular direction.

Net directional movement = Overall movement away from the center of concentration, which results from random molecular movement in *all* directions.

Diffusion = The *net* movement of a substance down a *concentration gradient* (see Campbell, Figure 8.9).

- Results from the intrinsic kinetic energy of molecules (also called thermal motion, or heat)
- Results from random molecular motion, even though the *net* movement may be directional
- Diffusion continues until a dynamic equilibrium is reached—the molecules continue to move, but there is no net directional movement.

In the absence of other forces (e.g., pressure) a substance will diffuse from where it is more concentrated to where it is less concentrated.

- A substance diffuses down its concentration gradient.
- Because it decreases free energy, diffusion is a spontaneous process ($-\Delta G$). It increases entropy of a system by producing a more random mixture of molecules.
- A substance diffuses down its own concentration gradient and is not affected by the gradients of other substances.

Much of the traffic across cell membranes occurs by diffusion and is thus a form of *passive transport*.

Passive transport = Diffusion of a substance across a biological membrane.

- Spontaneous process which is a function of a concentration gradient when a substance is more concentrated on one side of the membrane.
- Passive process which does not require the cell to expend energy. It is the potential energy stored in a concentration gradient that drives diffusion.
- Rate of diffusion is regulated by the permeability of the membrane, so some molecules diffuse more freely than others.
- Water diffuses freely across most cell membranes.

C. Osmosis is the passive transport of water

Hypertonic solution = A solution with a greater solute concentration than that inside a cell.

Hypotonic solution = A solution with a lower solute concentration compared to that inside a cell.

Isotonic solution = A solution with an equal solute concentration compared to that inside a cell.

> These terms are a source of confusion for students. It helps to point out that these are only relative terms used to compare the osmotic concentration of a solution to the osmotic concentration of a cell.

Osmosis = Diffusion of water across a selectively permeable membrane (see Campbell, Figure 8.10).

- Water diffuses down its concentration gradient.
- Example: If two solutions of different concentrations are separated by a selectively permeable membrane that is permeable to water but not to solute, water will diffuse from the hypoosmotic solution (solution with the lower osmotic concentration) to the hyperosmotic solution (solution with the higher osmotic concentration).
- Some solute molecules can reduce the proportion of water molecules that can freely diffuse. Water molecules form a hydration shell around hydrophilic solute molecules and this bound water cannot freely diffuse across a membrane.
- In dilute solutions including most biological fluids, it is the different in the proportion of the unbound water that causes osmosis, rather than the actual difference in water concentration.
- Direction of osmosis is determined by the difference in total solute concentration, regardless of the type or diversity of solutes in the solutions.
- If two isotonic solutions are separated by a selectively permeable membrane, water molecules diffuse across the membrane in both directions at an equal rate. There is no net movement of water.

> Clarification of this point is often necessary. Students may need to be reminded that even though there is no *net* movement of water across the membrane (or osmosis), the water molecules do not stop moving. At equilibrium, the water molecules move in both directions at the same rate.

Osmotic concentration = Total solute concentration of a solution

Osmotic pressure = Measure of the tendency for a solution to take up water when separated from pure water by a selectively permeable membrane.

- Osmotic pressure of pure water is zero.
- Osmotic pressure of a solution is proportional to its osmotic concentration. (The greater the solute concentration, the greater the osmotic pressure.)

Osmotic pressure can be measured by an *osmometer*:

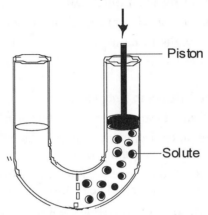

- In one type of osmometer, pure water is separated from a solution by a selectively permeable membrane that is permeable to water but not solute.
- The tendency for water to move into the solution by osmosis is counteracted by applying enough pressure with a piston so the solution's volume will stay the same.
- The amount of pressure required to prevent net movement of water into the solution is the *osmotic pressure*.

D. Cell survival depends on balancing water uptake and loss

1. Water balance of cells without walls

Since animal cells lack cell walls, they are not tolerant of excessive osmotic uptake or loss of water (see Campbell, Figure 8.11).

- In an isotonic environment, the volume of an animal cell will remain stable with no net movement of water across the plasma membrane.
- In a hypertonic environment, an animal cell will lose water by osmosis and *crenate* (shrivel).
- In a hypotonic environment, an animal cell will gain water by osmosis, swell and perhaps *lyse* (cell destruction).

Organisms without cell walls prevent excessive loss or uptake of water by:

- Living in an isotonic environment (e.g., many marine invertebrates are isosmotic with sea water).
- Osmoregulating in a hypo- or hypertonic environment. Organisms can regulate water balance (osmoregulation) by removing water in a hypotonic environment (e.g., *Paramecium* with contractile vacuoles in fresh water) or conserving water and pumping out salts in a hypertonic environment (e.g., bony fish in seawater) (see Campbell, Figure 8.12).

2. Water balance of cells with walls

Cells of prokaryotes, some protists, fungi and plants have cell walls outside the plasma membrane.

- In a hypotonic environment, water moves by osmosis into the plant cell, causing it to swell until internal pressure against the cell wall equals the osmotic pressure of the cytoplasm. A dynamic equilibrium is established (water enters and leaves the cell at the same rate and the cell becomes turgid).
- *Turgid* = Firmness or tension such as found in walled cells that are in a hypoosmotic environment where water enters the cell by osmosis.
  - Ideal state for most plant cells.
  - Turgid cells provide mechanical support for plants.
  - Requires cells to be hyperosmotic to their environment.
- In an isotonic environment, there is no net movement of water into or out of the cell.
  - Plant cells become *flaccid* or limp.
  - Loss of structural support from turgor pressure causes plants to wilt.
- In a hypertonic environment, walled cells will lose water by osmosis and will *plasmolyze*, which is usually lethal.

Plasmolysis = Phenomenon where a walled cell shrivels and the plasma membrane pulls away from the cell wall as the cell loses water to a hypertonic environment.

E. Specific proteins facilitate the passive transport of selected solutes

Facilitated diffusion = Diffusion of solutes across a membrane, with the help of transport proteins.

- Is passive transport because solute is transported down its concentration gradient.
- Helps the diffusion of many polar molecules and ions that are impeded by the membrane's phospholipid bilayer.

Transport proteins share some properties of enzymes:

- Transport proteins are *specific* for the solutes they transport. There is probably a specific binding site analogous to an enzyme's active site.
- Transport proteins can be *saturated* with solute, so the maximum transport rate occurs when all binding sites are occupied with solute.
- Transport proteins can be inhibited by molecules that resemble the solute normally carried by the protein (similar to competitive inhibition in enzymes).

Transport proteins differ from enzymes in they do not usually catalyze chemical reactions.

One model for facilitated diffusion (see Campbell, Figure 8.13):

- Transport protein most likely remains in place in the membrane and translocates solute by alternating between two conformations.
- In one conformation, transport protein binds solute; as it changes to another conformation, transport protein deposits solute on the other side of the membrane.
- The solute's binding and release may trigger the transport protein's conformational change.

Other transport proteins are selective channels across the membrane.

- The membrane is thus permeable to specific solutes that can pass through these channels.
- Some selective channels (gated channels) only open in response to electrical or chemical stimuli. For example, binding of neurotransmitter to nerve cells opens gated channels so that sodium ions can diffuse into the cell.

In some inherited disorders, transport proteins are missing or are defective (e.g., cystinuria, a kidney disease caused by missing carriers for cystine and other amino acids which are normally reabsorbed from the urine).

F. Active transport is the pumping of solutes against their gradients

Active transport = Energy-requiring process during which a transport protein pumps a molecule across a membrane, *against* its concentration gradient.

- Is energetically uphill ($+\Delta G$) and requires the cell to expend energy.
- Helps cells maintain steep ionic gradients across the cell membrane (e.g., Na^+, K^+, Mg^{++}, Ca^{++} and Cl^-).
- Transport proteins involved in active transport harness energy from ATP to pump molecules against their concentration gradients.

An example of an active transport system that translocates ions against steep concentration gradients is the *sodium-potassium pump*. Major features of the pump are:

1. The transport protein oscillates between two conformations:
    a. High affinity for Na^+ with binding sites oriented towards the cytoplasm.
    b. High affinity for K^+ with binding sites oriented towards the cell's exterior.
2. ATP phosphorylates the transport protein and powers the conformational change from Na^+ receptive to K^+ receptive.
3. As the transport protein changes conformation, it translocates bound solutes across the membrane.
4. Na^+K^+-pump translocates three Na^+ ions out of the cell for every two K^+ ions pumped into the cell. (Refer to Campbell, Figure 8.14 for the specific sequence of events.)

G. Some ion pumps generate voltage across membranes

Because anions and cations are unequally distributed across the plasma membrane, all cells have voltages across their plasma membranes.

Membrane potential = Voltage across membranes

- Ranges from -50 to -200 mv. As indicated by the negative sign, the cell's inside is negatively charged with respect to the outside.
- Affects traffic of charged substances across the membrane
- Favors diffusion of cations into cell and anions out of the cell (because of electrostatic attractions)

Two forces drive passive transport of ions across membranes:

1. Concentration gradient of the ion
2. Effect of membrane potential on the ion

Campbell, Figure 8.15, reviews the distinction between active and passive transport.

Electrochemical gradient = Diffusion gradient resulting from the combined effects of membrane potential and concentration gradient.

- Ions may not always diffuse down their concentration gradients, but they *always* diffuse down their electrochemical gradients.
- At equilibrium, the distribution of ions on either side of the membrane may be different from the expected distribution when charge is not a factor.
- Uncharged solutes diffuse down concentration gradients because they are unaffected by membrane potential.

Factors which contribute to a cell's membrane potential (net negative charge on the inside):

1. Negatively charged proteins in the cell's interior.

2. Plasma membrane's selective permeability to various ions. For example, there is a net loss of positive charges as K^+ leaks out of the cell faster than Na^+ diffuses in.

3. The sodium-potassium pump. This electrogenic pump translocates 3 Na^+ out for every 2 K^+ in - a net loss of one positive charge per cycle.

Electrogenic pump = A transport protein that generates voltage across a membrane (see Campbell, Figure 8.16).

- Na^+/K^+ ATPase is the major electrogenic pump in animal cells.

- A *proton pump* is the major electrogenic pump in plants, bacteria, and fungi. Also, mitochondria and chloroplasts use a proton pump to drive ATP synthesis.

- Voltages created by electrogenic pumps are sources of potential energy available to do cellular work.

> This is a good place to emphasize that electrochemical gradients represent potential energy. Spending lecture time on cotransport and the proton pump will help prepare your students for the upcoming topic of chemiosmosis.

H. In cotransport, a membrane protein couples the transport of one solute to another

Cotransport = Process where a single ATP-powered pump actively transports one solute and indirectly drives the transport of other solutes against their concentration gradients.

One mechanism of cotransport involves two transport proteins:

1. ATP-powered pump actively transports one solute and creates potential energy in the gradient it creates.

2. Another transport protein couples the solute's downhill diffusion as it leaks back across the membrane with a second solute's uphill transport against its concentration gradient.

For example, plants use a proton pump coupled with sucrose-H^+ symport to load sucrose into specialized cells of vascular tissue. Both solutes, H^+ and sucrose, must bind to the transport protein for cotransport to take place (see Campbell, Figure 8.17).

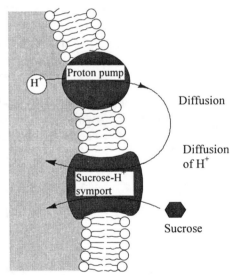

I. Exocytosis and endocytosis transport large molecules

Water and small molecules cross membranes by:

1. Passing through the phospholipid bilayer.

2. Being translocated by a transport protein.

Large molecules (e.g., proteins and polysaccharides) cross membranes by the processes of *exocytosis* and *endocytosis*.

Exocytosis	Endocytosis
Process of exporting macromolecules from a cell by fusion of vesicles with the plasma membrane.	Process of importing macromolecules into a cell by forming vesicles derived from the plasma membrane.
Vesicle usually budded from the ER or Golgi and migrates to plasma membrane.	Vesicle forms from a localized region of plasma membrane that sinks inward; pinches off into the cytoplasm.
Used by secretory cells to export products (e.g., insulin in pancreas, or neuro-transmitter from neuron).	Used by cells to incorporate extracellular substances.

There are three types of endocytosis: (1) *phagocytosis*, (2) *pinocytosis* and (3) *receptor-mediated endocytosis* (see Campbell, Figure 8.18).

Phagocytosis = (cell eating); endocytosis of solid particles
- Cell engulfs particle with *pseudopodia* and pinches off a food vacuole.
- Vacuole fuses with a lysosome containing hydrolytic enzymes that will digest the particle.

Pinocytosis = (cell drinking); endocytosis of fluid droplets
- Droplets of extracellular fluid are taken into small vesicles.
- The process is not discriminating. The cell takes in all solutes dissolved in the droplet.

Receptor-mediated endocytosis = The process of importing specific macromolecules into the cell by the inward budding of vesicles formed from *coated pits*; occurs in response to the binding of specific *ligands* to receptors on the cell's surface.
- More discriminating process than pinocytosis.
- A molecule that binds to a specific receptor site of another molecule is called a *ligand*.
- Membrane-embedded proteins with specific receptor sites exposed to the cell's exterior, cluster in regions called *coated pits*.
- A layer of *clathrin*, a fibrous protein, lines and reinforces the *coated pit* on the cytoplasmic side and probably helps deepen the pit to form a vesicle.

Progressive stages of receptor-mediated endocytosis:

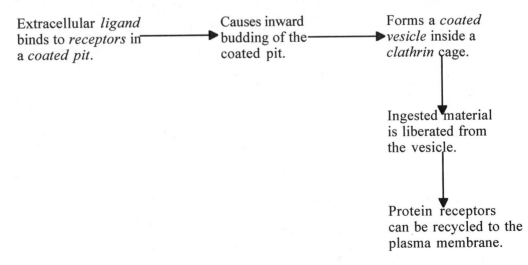

Receptor-mediated endocytosis enables cells to acquire bulk quantities of specific substances, even if they are in low concentration in extracellular fluid. For example, cholesterol enters cells by receptor-mediated endocytosis.

- In the blood, cholesterol is bound to lipid and protein complexes called *low-density lipoproteins* (LDLs).

- These LDLs bind to LDL receptors on cell membranes, initiating endocytosis.

- An inherited disease call familial hypercholesterolemia is characterized by high cholesterol levels in the blood. The LDL receptors are defective, so cholesterol cannot enter the cells by endocytosis and thus accumulates in the blood, contributing to the development of atherosclerosis.

In a nongrowing cell, the amount of plasma membrane remains relatively constant.

- Vesicle fusion with the plasma membrane offsets membrane loss through endocytosis.

- Vesicles provide a mechanism to rejuvenate or remodel the plasma membrane.

REFERENCES

Alberts, B., D. Bray, J. Lewis, M. Raff, K. Roberts and J.D. Watson. *Molecular Biology of the Cell*. 3rd ed. New York: Garland, 1994.

Becker, W.M. and D.W. Deamer. *The World of the Cell*. 3rd ed. Redwood City, California: Benjamin/Cummings, 1996.

Campbell, N. et al. *Biology*. 5th ed. Menlo Park, California: Benjamin/Cummings, 1998.

deDuve, C. *A Guided Tour of the Living Cell*. Volumes I and II. New York: Scientific American Books, 1984. Literally a guided tour of the cell with the reader as "cytonaut." This is an excellent resource for lecture material and enjoyable reading.

Kleinsmith, L.J. and V.M. Kish. *Principles of Cell Biology*. New York: Harper and Row, Publ., 1988.

CHAPTER 9
CELLULAR RESPIRATION:
HARVESTING CHEMICAL ENERGY

OUTLINE

   I.   Principles of Energy Conservation
- A. Cellular respiration and fermentation are catabolic (energy-yielding) pathways
- B. Cells must recycle the ATP they use for work
- C. Redox reactions release energy when electrons move closer to electronegative atoms
- D. Electrons "fall" from organic molecules to oxygen during cellular respiration
- E. The "fall" of electrons during respiration is stepwise, via NAD^+ and an electron transport chain

   II.   The Process of Cellular Respiration
- A. Respiration involves glycolysis, the Krebs cycle, and electron transport: *an overview*
- B. Glycolysis harvests chemical energy by oxidizing glucose to pyruvate: *a closer look*
- C. The Krebs cycle completes the energy-yielding oxidation of organic molecules: *a closer look*
- D. The inner mitochondrial membrane couples electron transport to ATP synthesis: *a closer look*
- E. Cellular respiration generates many ATP molecules for each sugar molecule it oxidizes: *a review*

   III.   Related Metabolic Processes
- A. Fermentation enables some cells to produce ATP without the help of oxygen
- B. Glycolysis and the Krebs cycle connect to many other metabolic pathways
- C. Feedback mechanisms control cellular respiration

OBJECTIVES

After reading this chapter and attending lecture, the student should be able to:

1. Diagram energy flow through the biosphere.
2. Describe the overall summary equation for cellular respiration.
3. Distinguish between substrate-level phosphorylation and oxidative phosphorylation.
4. Explain how exergonic oxidation of glucose is coupled to endergonic synthesis of ATP.
5. Define oxidation and reduction.
6. Explain how redox reactions are involved in energy exchanges.
7. Define coenzyme and list those involved in respiration.
8. Describe the structure of coenzymes and explain how they function in redox reactions.

9. Describe the role of ATP in coupled reactions.
10. Explain why ATP is required for the preparatory steps of glycolysis.
11. Describe how the carbon skeleton of glucose changes as it proceeds through glycolysis.
12. Identify where in glycolysis that sugar oxidation, substrate-level phosphorylation and reduction of coenzymes occur.
13. Write a summary equation for glycolysis and describe where it occurs in the cell.
14. Describe where pyruvate is oxidized to acetyl CoA, what molecules are produced and how it links glycolysis to the Krebs cycle.
15. Describe the location, molecules in and molecules out for the Krebs cycle.
16. Explain at what point during cellular respiration glucose is completely oxidized.
17. Explain how the exergonic "slide" of electrons down the electron transport chain is coupled to the endergonic production of ATP by chemiosmosis.
18. Describe the process of chemiosmosis.
19. Explain how membrane structure is related to membrane function in chemiosmosis.
20. Summarize the net ATP yield from the oxidation of a glucose molecule by constructing an ATP ledger which includes coenzyme production during the different stages of glycolysis and cellular respiration.
21. Describe the fate of pyruvate in the absence of oxygen.
22. Explain why fermentation is necessary.
23. Distinguish between aerobic and anaerobic metabolism.
24. Describe how food molecules other than glucose can be oxidized to make ATP.
25. Describe evidence that the first prokaryotes produced ATP by glycolysis.
26. Explain how ATP production is controlled by the cell and what role the allosteric enzyme, phosphofructokinase, plays in this process.

KEY TERMS

fermentation	Krebs cycle	anaerobic
cellular respiration	oxidative phosphorylation	alcohol fermentation
redox reactions	substrate-level phosphorylation	lactic acid fermentation
oxidation	acetyl CoA	facultative anaerobe
reduction	cytochrome (cyt)	beta oxidation
reducing agent	ATP synthase	
oxidizing agent	chemiosmosis	
NAD+	proton-motive force	
glycolysis	aerobic	

LECTURE NOTES

I. **Principles of Energy Conservation**

As open systems, cells require outside energy sources to perform cellular work (e.g., chemical, transport, and mechanical).

Energy flows into most ecosystems as sunlight.

Photosynthetic organisms trap a portion of the light energy and transform it into chemical bond energy of organic molecules. O_2 is released as a byproduct.

Cells use some of the chemical bond energy in organic molecules to make ATP–the energy source for cellular work.

Energy leaves living organisms as it dissipates as heat.

The products of respiration (CO_2 and H_2O) are the raw materials for photosynthesis. Photosynthesis produces glucose and oxygen, the raw materials for respiration.

Chemical elements essential for life are recycled, but energy is not.

How do cells harvest chemical energy?

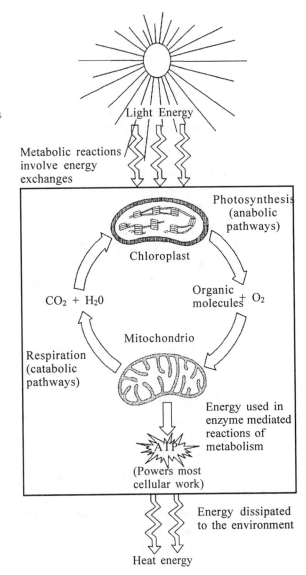

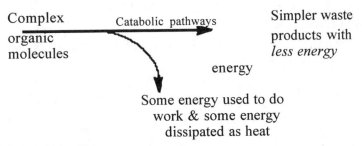

A. Cellular respiration and fermentation are catabolic (energy-yielding) pathways

Fermentation = An ATP-producing catabolic pathway in which both electron donors and acceptors are organic compounds.

- Can be an anaerobic process
- Results in a partial degradation of sugars

Cellular respiration = An ATP-producing catabolic process in which the ultimate electron acceptor is an inorganic molecule, such as oxygen.

- Most prevalent and efficient catabolic pathway
- Is an exergonic process ($\Delta G = -2870$ kJ/mol or -686 kcal/mol)
- Can be summarized as:

Organic compounds + Oxygen $\longrightarrow$ Carbon dioxide + Water + Energy
(food)

- Carbohydrates, proteins, and fats can all be metabolized as fuel, but cellular respiration is most often described as the oxidation of glucose:

$$C_6H_{12}O_6 + 6\,O_2 \longrightarrow 6\,CO_2 + 6\,H_2O + \text{Energy (ATP + Heat)}$$

B. Cells recycle the ATP they use for work

The catabolic process of cellular respiration transfers the energy stored in food molecules to *ATP*.

ATP (adenosine triphosphate) = Nucleotide with high energy phosphate bonds that the cell hydrolyzes for energy to drive endergonic reactions.

- The cell taps energy stored in ATP by enzymatically transferring terminal phosphate groups from ATP to other compounds. (Recall that direct hydrolysis of ATP would release energy as heat, a form unavailable for cellular work. See Chapter 6.)
- The compound receiving the phosphate group from ATP is said to be *phosphorylated* and becomes more reactive in the process.
- The phosphorylated compound loses its phosphate group as cellular work is performed; inorganic phosphate and ADP are formed in the process (see Campbell, Figure 9.2).
- Cells must replenish the ATP supply to continue cellular work. Cellular respiration provides the energy to regenerate ATP from ADP and inorganic phosphate.

C. Redox reactions release energy when electrons move closer to electronegative atoms

1. An introduction to redox reactions

Oxidation-reduction reactions = Chemical reactions which involve a partial or complete transfer of electrons from one reactant to another; called *redox reactions* for short.

Oxidation = Partial or complete loss of electrons

Reduction = Partial or complete gain of electrons

Generalized redox reaction:

Electron transfer requires both a donor and acceptor, so when one reactant is oxidized the other is reduced.

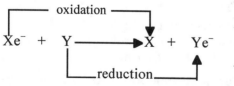

Where:

X = Substance being oxidized; acts as a *reducing agent* because it reduces Y.

Y = Substance being reduced; as an *oxidizing agent* because it oxidizes X.

Not all redox reactions involve a complete transfer of electrons, but, instead, may just change the degree of sharing in covalent bonds (see Campbell, Figure 9.3).

- Example: Covalent electrons of methane are equally shared, because carbon and hydrogen have similar electronegativities.

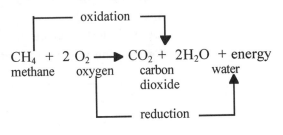

- As methane reacts with oxygen to form carbon dioxide, electrons shift away from carbon and hydrogen to the more electronegative oxygen.

- Since electrons lose potential energy when they shift toward more electronegative atoms, redox reactions that move electrons closer to oxygen release energy.

- Oxygen is a powerful oxidizing agent because it is so electronegative.

D. Electrons "fall" from organic molecules to oxygen during cellular respiration

Cellular respiration is a redox process that transfers hydrogen, including electrons with high potential energy, from sugar to oxygen.

$$\text{oxidation}$$
$$C_6H_{12}O_6 + 6\ O_2 \longrightarrow 6\ CO_2 + 6\ H_2O + \text{energy (used to make ATP)}$$
$$\text{reduction}$$

- Valence electrons of carbon and hydrogen lose potential energy as they shift toward electronegative oxygen.
- Released energy is used by cells to produce ATP.
- Carbohydrates and fats are excellent energy stores because they are rich in C to H bonds.

Without the activation barrier, glucose would combine spontaneously with oxygen.
- Igniting glucose provides the activation energy for the reaction to proceed; a mole of glucose yields 686 kcal (2870 kJ) of heat when burned in air.
- Cellular respiration does not oxidize glucose in one explosive step, as the energy could not be efficiently harnessed in a form available to perform cellular work.
- Enzymes lower the activation energy in cells, so glucose can be slowly oxidized in a stepwise fashion during glycolysis and Krebs cycle.

E. The "fall" of electrons during respiration is stepwise, via NAD$^+$ and an electron transport chain

Hydrogens stripped from glucose are not transferred directly to oxygen, but are first passed to a special electron acceptor—*NAD$^+$*.

Nicotinamide adenine dinucleotide (NAD$^+$) = A *dinucleotide* that functions as a *coenzyme* in the redox reactions of metabolism (see Campbell, Figure 9.4).

- Found in all cells
- Assists enzymes in electron transfer during redox reactions of metabolism

Coenzyme = Small nonprotein organic molecule that is required for certain enzymes to function.

Dinucleotide = A molecule consisting of two nucleotides.

During the oxidation of glucose, NAD^+ functions as an oxidizing agent by trapping energy-rich electrons from glucose or food. These reactions are catalyzed by enzymes called *dehydrogenases*, which:

- Remove a pair of hydrogen atoms (two electrons and two protons) from substrate
- Deliver the two electrons and *one* proton to NAD^+
- Release the remaining proton into the surrounding solution

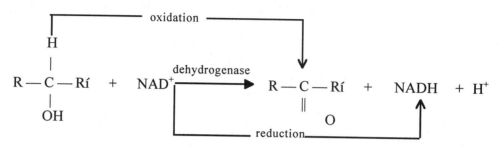

Where:

$$R \overset{\overset{\displaystyle H}{|}}{\underset{\underset{\displaystyle OH}{|}}{C}} Rí \quad = \quad \text{Substrate that is oxidized by enzymatic transfer of electrons to } NAD^+$$

NAD^+ = Oxidized coenzyme (net positive charge)

NADH = Reduced coenzyme (electrically neutral)

The high energy electrons transferred from substrate to NAD^+ are then passed down the *electron transport chain* to oxygen, powering ATP synthesis (*oxidative phosphorylation*).

> Some instructors find it difficult to drive this point home. Surprisingly, some students can recall the intermediate steps of glycolysis or the Krebs cycle, but cannot explain in general terms how energy from food is transferred to ATP. Campbell, Figure 9.16 can be used to give students an overview when respiration is introduced; it is useful to refer to it here so students can place the process you are describing in context. It can be used again later as a summary to bring closure to the topic.

Electron transport chains convert some of the chemical energy extracted from food to a form that can be used to make ATP (see Campbell, Figure 9.5). These transport chains:

- Are composed of electron-carrier molecules built into the inner mitochondrial membrane. Structure of this membrane correlates with its functional role (form fits function).
- Accept energy-rich electrons from reduced coenzymes (NADH and $FADH_2$); and during a series of redox reactions, pass these electrons down the chain to oxygen, the final electron acceptor. The electronegative oxygen accepts these electrons, along with hydrogen nuclei, to form water.
- Release energy from energy-rich electrons in a controlled stepwise fashion; a form that can be harnessed by the cell to power ATP production. If the reaction between hydrogen and oxygen during respiration occurred in a single explosive step, much of the energy released would be lost as heat, a form unavailable to do cellular work.

Electron transfer from NADH to oxygen is exergonic, having a free energy change of −222 kJ/mole (−53 kcal/mol).

- Since electrons lose potential energy when they shift toward a more electronegative atom, this series of redox reactions releases energy.
- Each successive carrier in the chain has a higher electronegativity than the carrier before it, so the electrons are pulled downhill towards oxygen, the final electron acceptor and the molecule with the highest electronegativity.

II.   The Process of Cellular Respiration

A.   Respiration involves glycolysis, the Krebs cycle, and electron transport: *an overview*

There are three metabolic stages of cellular respiration (see Campbell, Figure 9.6):

1. Glycolysis
2. Krebs cycle
3. Electron transport chain (ETC) and oxidative phosphorylation

Glycolysis is a catabolic pathway that:

- Occurs in the cytosol
- Partially oxidizes glucose (6C) into two *pyruvate* (3C) molecules

The *Krebs cycle* is a catabolic pathway that:

- Occurs in the mitochondrial matrix
- Completes glucose oxidation by breaking down a *pyruvate* derivative (acetyl CoA) into carbon dioxide

Glycolysis and the Krebs cycle produce:

- A small amount of ATP by substrate-level phosphorylation
- NADH by transferring electrons from substrate to NAD^+ (Krebs cycle also produces $FADH_2$ by transferring electrons to FAD)

The *electron transport chain*:

- Is located at the inner membrane of the mitochondrion
- Accepts energized electrons from reduced coenzymes (NADH and $FADH_2$) that are harvested during glycolysis and Krebs cycle. Oxygen pulls these electrons down the electron transport chain to a lower energy state.
- Couples this exergonic slide of electrons to ATP synthesis or oxidative phosphorylation. This process produces *most* (90%) of the ATP.

Oxidative phosphorylation = ATP production that is coupled to the exergonic transfer of electrons from food to oxygen.

A small amount of ATP is produced directly by the reactions of glycolysis and Krebs cycle. This mechanism of producing ATP is called substrate-level phosphorylation.

Substrate-level phosphorylation = ATP production by direct enzymatic transfer of phosphate from an intermediate substrate in catabolism to ADP.

B.   Glycolysis harvests chemical energy by oxidizing glucose to pyruvate: *a closer look*

Glycolysis = (Glyco = sweet, sugar; lysis = to split); catabolic pathway during which six-carbon glucose is split into two three-carbon sugars, which are then oxidized and rearranged by a step-wise process that produces two pyruvate molecules.

- Each reaction is catalyzed by specific enzymes dissolved in the cytosol.
- No CO_2 is released as glucose is oxidized to pyruvate; all carbon in glucose can be accounted for in the two molecules of pyruvate.
- Occurs whether or not oxygen is present.

The reactions of glycolysis occur in two phases:

Energy-investment phase. The cell uses ATP to phosphorylate the intermediates of glycolysis.

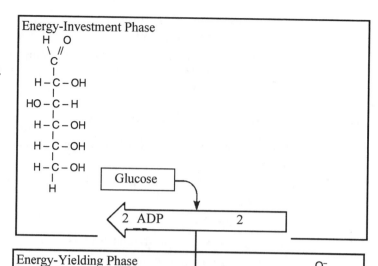

Energy-Investment Phase

Glucose

2 ADP   2

Energy-yielding phase. Two three-carbon intermediates are oxidized. For each glucose molecule entering glycolysis:

1. A net gain of two ATPs is produced by substrate-level phosphorylation.

2. Two molecules of NAD⁺ are reduced to NADH. Energy conserved in the high-energy electrons of NADH can be used to make ATP by oxidative phosphorylation.

> You may not want students to memorize the structures or steps of glycolysis, but you should expect them to understand the process, where it occurs, and the major molecules required and produced. It may be helpful to summarize a lecture with an overhead transparency.

Energy-investment phase:

The *energy investment phase* includes five preparatory steps that split glucose in two. This process actually *consumes* ATP.

Step 1: Glucose enters the cell, and carbon six is phosphorylated. This ATP-coupled reaction:

- Is catalyzed by *hexokinase* (*kinase* is an enzyme involved in phosphate transfer)
- Requires an initial <u>investment</u> of ATP
- Makes glucose more chemically reactive
- Produces glucose-6-phosphate; since the plasma membrane is relatively impermeable to ions, addition of an electrically charged phosphate group traps the sugar in the cell.

Step 2: An *isomerase* catalyzes the rearrangement of glucose-6-phosphate to its isomer, fructose-6-phosphate.

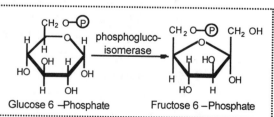

Step 3: Carbon one of fructose-6-phosphate is phosphorylated. This reaction:

- Requires an investment of another ATP.
- Is catalyzed by

phosphofructokinase, an allosteric enzyme that controls the rate of glycolysis. This step commits the carbon skeleton to glycolysis, a catabolic process, as opposed to being used to synthesize glycogen, an anabolic process.

Step 4: *Aldolase* cleaves the six-carbon sugar into two isomeric three-carbon sugars.

- This is the reaction for which glycolysis is named.
- For each glucose molecule that begins glycolysis, there are *two* product molecules for this and each succeeding step.

Step 5: An isomerase catalyzes the reversible conversion between the two three-carbon sugars. This reaction:

- Never reaches equilibrium because only one isomer, *glyceraldehyde phosphate*, is used in the next step of glycolysis.
- Is thus pulled towards the direction of glyceraldehyde phosphate, which is removed as fast as it forms.
- Results in the net effect that, for each glucose molecule, *two* molecules of glyceraldehyde phosphate progress through glycolysis.

Energy-yielding phase:

The *energy-yielding phase* occurs after glucose is split into two three-carbon sugars. During these reactions, sugar is oxidized, and ATP and NADH are produced.

Step 6: An enzyme catalyzes two sequential reactions:

1. Glyceraldehyde phosphate is oxidized and NAD$^+$ is reduced to NADH + H$^+$.

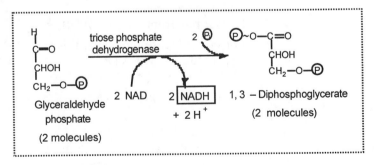

- This reaction is very exergonic and is coupled to the endergonic phosphorylation phase ($\Delta G = -10.3$ kcal/ mol).
- For every glucose molecule, 2 NADH are produced.

2. Glyceraldehyde phosphate is phosphorylated on carbon one.
   - The phosphate source is inorganic phosphate, which is always present in the cytosol.
   - The new phosphate bond is a high energy bond with even more potential to transfer a phosphate group than ATP.

Step 7: ATP is produced by substrate-level phosphorylation.

- In a very exergonic reaction, the phosphate group with the high energy bond is transferred from 1,3-diphosphoglycerate to ADP.

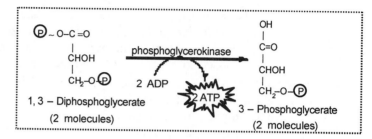

- For each glucose molecule, two ATP molecules are produced. The ATP ledger now stands at zero as the initial debt of two ATP from steps one and three is repaid.

Step 8: In preparation for the next reaction, a phosphate group on carbon three is enzymatically transferred to carbon two.

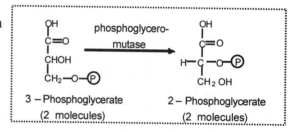

Step 9: Enzymatic removal of a water molecule:

- Creates a double bond between carbons one and two of the substrate.
- Rearranges the substrate's electrons, which transforms the remaining phosphate bond into an unstable bond.

Step 10: ATP is produced by substrate-level phosphorylation.

- In a highly exergonic reaction, a phosphate group is transferred from PEP to ADP.
- For each glucose molecule, this step produces two ATP.

Summary equation for glycolysis:

$$
\begin{aligned}
&C_6H_{12}O_6 && 2\ C_3H_4O_3\ \text{(Pyruvate)} \\
&+\ 2\ NAD^+ && +\ 2\ NADH + 2\ H^+ \\
&+\ 2\ ADP\ +\ 2\,\textcircled{P} && +\ 2\ ATP \\
& && +\ 2\ H_2O
\end{aligned}
$$

- Glucose has been oxidized into two pyruvate molecules.

- The process is exergonic ($\Delta G = -140$ kcal/mol or -586 kJ/mol); most of the energy harnessed is conserved in the high-energy electrons of NADH and in the phosphate bonds of ATP.

C. The Krebs cycle completes the energy-yielding oxidation of organic molecules: *a closer look*

Most of the chemical energy originally stored in glucose still resides in the two pyruvate molecules produced by glycolysis. The fate of pyruvate depends upon the presence or absence of oxygen. If oxygen is present, pyruvate *enters the mitochondrion* where it is *completely oxidized* by a series of enzyme-controlled reactions.

1. Formation of acetyl CoA

The junction between glycolysis and the Krebs cycle is the oxidation of pyruvate to acetyl CoA (see Campbell, Figure 9.10):

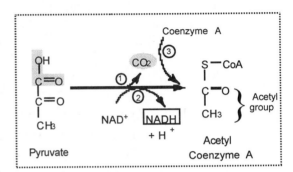

- Pyruvate molecules are translocated from the cytosol into the mitochondrion by a carrier protein in the mitochondrial membrane.

- This step is catalyzed by a *multienzyme complex* which:
    1. Removes CO_2 from the carboxyl group of pyruvate, changing it from a three-carbon to a two-carbon compound. This is the first step where CO_2 is released.
    2. Oxidizes the two-carbon fragment to acetate, while reducing NAD^+ to NADH. Since glycolysis produces two pyruvate molecules per glucose, there are *two* NADH molecules produced.
    3. Attaches coenzyme A to the acetyl group, forming acetyl CoA. This bond is unstable, making the acetyl group very reactive.

2. Krebs cycle

The Krebs cycle reactions oxidize the remaining acetyl fragments of acetyl CoA to CO_2. Energy released from this exergonic process is used to reduce coenzyme (NAD^+ and FAD) and to phosphorylate ATP (substrate-level phosphorylation).

NOTE: The FAD dinucleotide upon reduction accepts two electrons and two protons)

- A German-British scientist, Hans Krebs, elucidated this catabolic pathway in the 1930s.
- The Krebs cycle, which is also known as the *citric acid cycle* or *TCA cycle*, has eight enzyme-controlled steps that occur in the *mitochondrial matrix* (see Campbell, Figure 9.11).

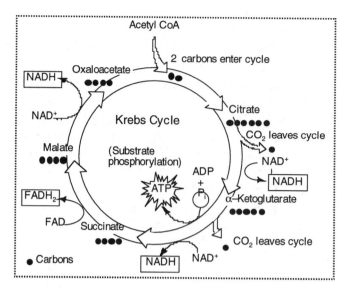

For every turn of Krebs cycle:

- Two carbons enter in the acetyl fragment of acetyl CoA.
- Two different carbons are oxidized and leave as CO_2.
- Coenzymes are reduced; three NADH and one $FADH_2$ are produced.
- One ATP molecule is produced by substrate-level phosphorylation.
- Oxaloacetate is regenerated.

For every glucose molecule split during glycolysis:

- Two acetyl fragments are produced.
- It takes two turns of Krebs cycle to complete the oxidation of glucose.

Steps of the Krebs cycle (see Campbell, Figure 9.12):

Step 1: The unstable bond of acetyl CoA breaks, and the *two-carbon* acetyl group bonds to the *four-carbon* oxaloacetate to form *six-carbon* citrate.

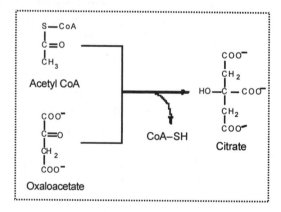

Step 2: Citrate is isomerized to isocitrate.

Step 3: Two major events occur during this step:

- Isocitrate loses CO_2 leaving a *five-carbon* molecule.
- The five-carbon compound is oxidized and NAD^+ is reduced.

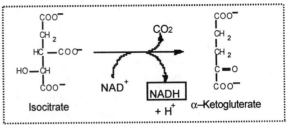

Step 4: A multienzyme complex catalyzes:

- Removal of CO_2
- Oxidation of the remaining *four-carbon* compound and reduction of NAD^+
- Attachment of CoA with a high energy bond to form succinyl CoA

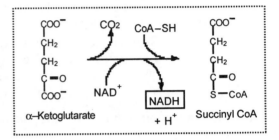

Step 5: Substrate-level phosphorylation occurs in a series of enzyme catalyzed reactions:

- The high energy bond of succinyl CoA breaks, and some energy is conserved as CoA is displaced by a phosphate group.
- The phosphate group is transferred to GDP to form GTP and succinate.
- GTP donates a phosphate group to ADP to form ATP.

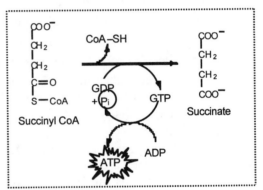

Step 6: Succinate is oxidized to fumarate and FAD is reduced.

- Two hydrogens are transferred to FAD to form $FADH_2$.
- The dehydrogenase that catalyzes this reaction is bound to the inner mitochondrial membrane.

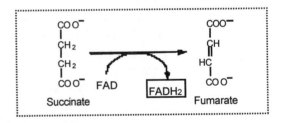

Step 7: Water is added to fumarate which rearranges its chemical bonds to form malate.

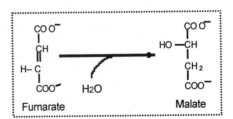

Step 8: Malate is oxidized and NAD^+ is reduced.

- A molecule of NADH is produced.
- Oxaloacetate is regenerated to begin the cycle again.

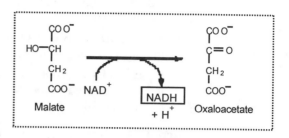

Two turns of the Krebs cycle produce two ATPs by substrate-level phosphorylation. However, *most* ATP output of respiration results from *oxidative phosphorylation*.

- Reduced coenzymes produced by the Krebs cycle (six NADH and two $FADH_2$ per glucose) carry high energy electrons to the electron transport chain.
- The ETC couples electron flow down the chain to ATP synthesis.

D. The inner mitochondrial membrane couples electron transport to ATP synthesis: *a closer look*

Only a few molecules of ATP are produced by substrate-level phosphorylation:

- Two net ATPs per glucose from glycolysis
- Two ATPs per glucose from the Krebs cycle

Most molecules of ATP are produced by oxidative phosphorylation.

- At the end of the Krebs cycle, most of the energy extracted from glucose is in molecules of NADH and $FADH_2$.
- These reduced coenzymes link glycolysis and the Krebs cycle to oxidative phosphorylation by passing their electrons down the electron transport chain to oxygen. (Though the Krebs cycle occurs only under aerobic conditions, it does not use oxygen directly. The ETC and oxidative phosphorylation require oxygen as the final electron acceptor.)
- This exergonic transfer of electrons down the ETC to oxygen is coupled to ATP synthesis.

1. The pathway of electron transport

The *electron transport chain* is made of electron carrier molecules embedded in the inner mitochondrial membrane.

- Each successive carrier in the chain has a higher electronegativity than the carrier before it, so the electrons are pulled downhill towards oxygen, the final electron acceptor and the molecule with the highest electronegativity.
- Except for ubiquinone (Q), most of the carrier molecules are proteins and are tightly bound to *prosthetic groups* (nonprotein cofactors).
- Prosthetic groups alternate between reduced and oxidized states as they accept and donate electrons.

Protein Electron Carriers	Prosthetic Group
flavoproteins	flavin mononucleotide (FMN)
iron-sulfur proteins	iron and sulfur
cytochromes	heme group

Heme group = Prosthetic group composed of four organic rings surrounding a single iron atom

Cytochrome = Type of protein molecule that contains a heme prosthetic group and that functions as an electron carrier in the electron transport chains of mitochondria and chloroplasts

- There are several cytochromes, each a slightly different protein with a heme group.
- It is the iron of cytochromes that transfers electrons.

Sequence of electron transfers along the electron transport chain (see also, Campbell, Figure 9.13):

NADH is oxidized and *flavoprotein* is reduced as high energy electrons from NADH are transferred to FMN.

↓

Flavoprotein is oxidized as it passes electrons to an *iron-sulfur protein*, Fe•S.

↓

Iron-sulfur protein is oxidized as it passes electrons to *ubiquinone* (Q).

↓

Ubiquinone passes electrons on to a succession of electron carriers, most of which are cytochromes.

↓

Cyt a_3, the last cytochrome passes electrons to molecular oxygen, O_2.

↓

As molecular oxygen is reduced it also picks up two protons from the medium to form water. For every two NADHs, one O_2 is reduced to two H_2O molecules.

- $FADH_2$ also donates electrons to the electron transport chain, but those electrons are added at a lower energy level than NADH.
- The electron transport chain does not make ATP directly. It *generates a proton gradient across the inner mitochondrial membrane*, which stores potential energy that can be used to phosphorylate ADP.

2. Chemiosmosis: the energy-coupling mechanism

The mechanism for coupling exergonic electron flow from the oxidation of food to the endergonic process of oxidative phosphorylation is *chemiosmosis*.

Chemiosmosis = The coupling of exergonic electron flow down an electron transport chain to endergonic ATP production by the creation of a proton gradient across a membrane. The proton gradient drives ATP synthesis as protons diffuse back across the membrane.

- Proposed by British biochemist, Peter Mitchell (1961)

- The term *chemiosmosis* emphasizes a coupling between (1) chemical reactions (phosphorylation) and (2) transport processes (proton transport).
- Process involved in oxidative phosphorylation and photophosphorylation.

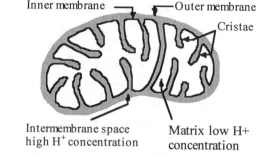

Inner membrane — Outer membrane — Cristae

Intermembrane space high H^+ concentration

Matrix low H+ concentration

The site of oxidative phosphorylation is the inner mitochondrial membrane, which has many copies of a protein complex, *ATP synthase*. This complex:

- Is an enzyme that makes ATP
- Uses an existing *proton gradient* across the inner mitochondrial membrane to power ATP synthesis

Cristae, or infoldings of the inner mitochondrial membrane, increase the surface area available for chemiosmosis to occur.

Membrane structure correlates with the prominent functional role membranes play in chemiosmosis:

- Using energy from exergonic electron flow, the *electron transport chain* creates the proton gradient by pumping H^+s from the *mitochondrial matrix*, across the *inner membrane* to the *intermembrane space*.
- This proton gradient is maintained, because the membrane's phospholipid bilayer is impermeable to H^+s and prevents them from leaking back across the membrane by diffusion.
- *ATP synthases* use the potential energy stored in a proton gradient to make ATP by allowing H^+ to diffuse down the gradient, back across the membrane. Protons diffuse through the ATP synthase complex, which causes the phosphorylation of ADP (see Figure 9.15).

How does the electron transport chain pump hydrogen ions from the matrix to the intermembrane space? The process is based on spatial organization of the electron transport chain in the membrane. Note that:

- Some electron carriers of the transport chain transport only electrons.
- Some electron carriers accept and release protons along with electrons. These carriers are spatially arranged so that protons are picked up from the matrix and are released into the intermembrane space.

Most of the electron carriers are organized into three complexes: 1) NADH dehydrogenase complex; 2) cytochrome b-c_1 complex; and 3) cytochrome oxidase complex (see Campbell, Figure 9.14).

- Each complex is an asymmetric particle that has a specific orientation in the membrane.
- As complexes transport electrons, they also harness energy from this exergonic process to pump protons across the inner mitochondrial membrane.

Mobile carriers transfer electrons between complexes. These mobile carriers are:

1. Ubiquinone (Q). Near the matrix, Q accepts electrons from the NADH dehydrogenase complex, diffuses across the lipid bilayer, and passes electrons to the cytochrome b-c_1 complex.
2. Cytochrome c (Cyt c). Cyt c accepts electrons from the cytochrome b-c_1 complex and conveys them to the cytochrome oxidase complex.

When the transport chain is operating:

- The pH in the intermembrane space is one or two pH units lower than in the matrix.
- The pH in the intermembrane space is the same as the pH of the cytosol because the outer mitochondrial membrane is permeable to protons.

The H^+ gradient that results is called a *proton-motive force* to emphasize that the gradient represents potential energy.

Proton motive force = Potential energy stored in the proton gradient created across biological membranes that are involved in chemiosmosis

- This force is an *electrochemical gradient* with two components:
  1. Concentration gradient of protons (chemical gradient)
  2. Voltage across the membrane because of a higher concentration of positively charged protons on one side (electrical gradient)
- It tends to drive protons across the membrane back into the matrix.

Chemiosmosis couples exergonic chemical reactions to endergonic H^+ transport, which creates the proton-motive force used to drive cellular work, such as:

- ATP synthesis in mitochondria (*oxidative phosphorylation*). The energy to create the proton gradient comes from the oxidation of glucose and the ETC.
- ATP synthesis in chloroplasts (*photophosphorylation*). The energy to create the proton gradient comes from light trapped during the energy-capturing reactions of photosynthesis.
- ATP synthesis, transport processes, and rotation of flagella in bacteria. The proton gradient is created across the plasma membrane. Peter Mitchell first postulated chemiosmosis as an energy-coupling mechanism based on experiments with bacteria.

3. **Biological themes and oxidative phosphorylation**

   The working model of how mitochondria harvest the energy of food illustrates many of the text's integrative themes in the study of life:

   - Energy conversion and utilization
   - Emergent properties - Oxidative phosphorylation is an emergent property of the intact mitochondrion that uses a precise interaction of molecules.
   - Correlation of structure and function - The chemiosmotic model is based upon the spatial arrangement of membrane proteins.
   - Evolution - In an effort to reconstruct the origin of oxidative phosphorylation and the evolution of cells, biologists compare similarities in the chemiosmotic machinery of mitochondria to that of chloroplasts and bacteria.

E. **Cellular respiration generates many ATP molecules for each sugar molecule it oxidizes:** *a review*

During cellular respiration, *most* energy flows in this sequence:

Glucose $\Rightarrow$ NADH $\Rightarrow$ electron transport chain $\Rightarrow$ proton motive force $\Rightarrow$ ATP

The *net* ATP yield from the oxidation of one glucose molecule to six carbon dioxide molecules can be estimated by adding:

1. ATP produced directly by substrate-level phosphorylation during glycolysis and the Krebs cycle.

   - A net of two ATPs is produced during glycolysis. The debit of two ATPs used during the investment phase is subtracted from the four ATPs produced during the energy-yielding phase.

- Two ATPs are produced during the Krebs cycle.

2. ATP produced when chemiosmosis couples electron transport to oxidative phosphorylation.

- The electron transport chain creates enough proton-motive force to produce a maximum of *three ATPs* for each electron pair that travels from NADH to oxygen. The average yield is actually between two and three ATPs per NADH (2.7).

- $FADH_2$ produced during the Krebs cycle is worth a maximum of only *two ATPs*, since it donates electrons at a lower energy level to the electron transport chain.

- In most eukaryotic cells, the ATP yield is lower due to a NADH produced during glycolysis. The mitochondrial membrane is impermeable to NADH, so its electrons must be carried across the membrane in by one of several "shuttle" reactions. Depending on which shuttle is operating, electrons can be transferred to either NAD^+ or FAD^+. A pair of electrons passed to FAD^+ yields about two ATP, whereas a pair of electrons passed to NAD^+ yields about 13 ATP.

- Maximum ATP yield for each glucose oxidized during cellular respiration:

Process	ATP Produced Directly by Substrate-level Phosphorylation	Reduced Coenzyme	ATP Produced by Oxidative Phosphorylation	Total
Glycolysis	*Net* 2 ATP	2 NADH	4 to 6 ATP	6-8
Oxidation of Pyruvate	_____	2 NADH	6 ATP	6
Krebs cycle	2 ATP	6 NADH 2 $FADH_2$	18 ATP 4 ATP	24
			Total	36-38

- This tally only *estimates* the ATP yield from respiration (see Campbell, Figure 9.15). Some variables that affect ATP yield include:

  - The proton-motive force may be used to drive other kinds of cellular work such as active transport.

  - The total ATP yield is inflated (~10%) by rounding off the number of ATPs produced per NADH to three.

Cellular respiration is remarkably efficient in the transfer of chemical energy from glucose to ATP.

- Estimated efficiency in eukaryotic cells is about 40%.

- Energy lost in the process is released as heat.

Calculated by $\dfrac{7.3 \text{ kcal/mol ATP} \times 38 \text{ mol ATP/mol glucose}}{686 \text{ kcal/mol glucose}} \times 100$

III.   Related Metabolic Processes

A. Fermentation enables some cells to produce ATP without the help of oxygen

Food can be oxidized under *anaerobic* conditions.

Aerobic = (Aer = air; bios = life); existing in the presence of oxygen

Anaerobic = (An = without; aer = air); existing in the absence of free oxygen

Fermentation = Anaerobic catabolism of organic nutrients

Glycolysis oxidizes glucose to two pyruvate molecules, and the oxidizing agent for this process is NAD⁺, *not* oxygen.

- Some energy released from the exergonic process of glycolysis drives the production of two net ATPs by substrate-level phosphorylation.
- Glycolysis produces a net of two ATPs whether conditions are aerobic or anaerobic.
    - *Aerobic conditions*: Pyruvate is *oxidized* further by substrate-level phosphorylation and by oxidative phosphorylation and more ATP is made as NADH passes electrons to the electron transport chain. NAD⁺ is regenerated in the process.
    - *Anaerobic conditions*: Pyruvate is *reduced*, and NAD⁺ is regenerated. This prevents the cell from depleting the pool of NAD⁺, which is the oxidizing agent necessary for glycolysis to continue. No additional ATP is produced.

Fermentation recycles NAD⁺ from NADH. This process consists of anaerobic glycolysis plus subsequent reactions that regenerate NAD⁺ by reducing pyruvate. Two of the most common types of fermentation are (1) alcohol fermentation and (2) lactic acid fermentation (see Campbell, Figure 9.16).

Alcohol fermentation:

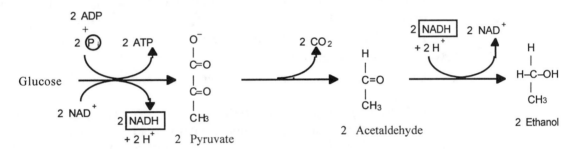

Pyruvate is converted to ethanol in two steps:

- a. Pyruvate loses carbon dioxide and is converted to the two-carbon compound acetaldehyde.
- b. NADH is oxidized to NAD⁺ and acetaldehyde is reduced to ethanol.

Many bacteria and yeast carry out alcohol fermentation under anaerobic conditions.

Lactic acid fermentation:

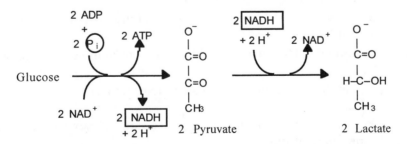

NADH is oxidized to NAD⁺ and pyruvate is reduced to lactate.

- Commercially important products of lactic acid fermentation include cheese and yogurt.
- When oxygen is scarce, human muscle cells switch from aerobic respiration to lactic acid fermentation. Lactate accumulates, but it is gradually carried to the liver where it is converted back to pyruvate when oxygen becomes available.

1. Fermentation and respiration compared

The anaerobic process of fermentation and aerobic process of cellular respiration are similar in that both metabolic pathways:

- Use glycolysis to oxidize glucose and other substrates to pyruvate, producing a net of two ATPs by substrate phosphorylation
- Use NAD^+ as the oxidizing agent that accepts electrons from food during glycolysis

Fermentation and cellular respiration differ in:

- How NADH is oxidized back to NAD^+. Recall that the oxidized form, NAD^+, is necessary for glycolysis to continue.
    - During fermentation, NADH passes electrons to pyruvate or some derivative. As pyruvate is reduced, NADH is oxidized to NAD^+. Electrons transferred from NADH to pyruvate or other substrates are not used to power ATP production.
    - During cellular respiration, the stepwise electron transport from NADH to oxygen not only drives oxidative phosphorylation, but regenerates NAD^+ in the process.
- Final electron acceptor
    - In fermentation, the final electron acceptor is pyruvate (lactic acid fermentation), acetaldehyde (alcohol fermentation), or some other organic molecule.
    - In cellular respiration, the final electron acceptor is oxygen.
- Amount of energy harvested
    - During fermentation, energy stored in pyruvate is unavailable to the cell.
    - Cellular respiration yields 18 times more ATP per glucose molecule than does fermentation. The higher energy yield is a consequence of the Krebs cycle which completes the oxidation of glucose and thus taps the chemical bond energy still stored in pyruvate at the end of glycolysis.
- Requirement for oxygen
    - Fermentation does not require oxygen.
    - Cellular respiration occurs only in the presence of oxygen.

Organisms can be classified based upon the effect oxygen has on growth and metabolism.

Strict (obligate) aerobes = Organisms that require oxygen for growth and as the final electron acceptor for aerobic respiration.

Strict (obligate) anaerobes = Microorganisms that only grow in the absence of oxygen and are, in fact, poisoned by it.

Facultative anaerobes = Organisms capable of growth in either aerobic or anaerobic environments.

- Yeasts, many bacteria, and mammalian muscle cells are facultative anaerobes.
- Can make ATP by fermentation in the absence of oxygen or by respiration in the presence of oxygen.
- Glycolysis is common to both fermentation and respiration, so pyruvate is a key juncture in catabolism (see Campbell, Figure 9.18).

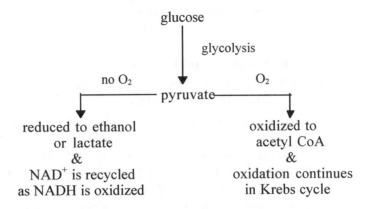

3. The evolutionary significance of glycolysis

The first prokaryotes probably produced ATP by glycolysis. Evidence includes the following:

- Glycolysis does not require oxygen, and the oldest known bacterial fossils date back to 3.5 billion years ago when oxygen was not present in the atmosphere.
- Glycolysis is the most widespread metabolic pathway, so it probably evolved early.
- Glycolysis occurs in the cytosol and does not require membrane-bound organelles. Eukaryotic cells with organelles probably evolved about two billion years after prokaryotic cells.

B. Glycolysis and the Krebs cycle connect to many other metabolic pathways

1. The versatility of catabolism

Respiration can oxidize organic molecules other than glucose to make ATP. Organisms obtain most calories from fats, proteins, disaccharides and polysaccharides. These complex molecules must be enzymatically hydrolyzed into simpler molecules or monomers that can enter an intermediate reaction of glycolysis or the Krebs cycle (see Campbell, Figure 9.19).

Glycolysis can accept a wide range of carbohydrates for catabolism.

- Starch is hydrolyzed to glucose in the digestive tract of animals.
- In between meals, the liver hydrolyzes glycogen to glucose.
- Enzymes in the small intestine break down disaccharides to glucose or other monosaccharides.

Proteins are hydrolyzed to amino acids.

- Organisms synthesize new proteins from some of these amino acids.
- Excess amino acids are enzymatically converted to intermediates of glycolysis and the Krebs cycle. Common intermediates are pyruvate, acetyl CoA, and α-ketoglutarate.
- This conversion process deaminates amino acids, and the resulting nitrogenous wastes are excreted and the carbon skeleton can be oxidized.

Fats are excellent fuels because they are rich in hydrogens with high energy electrons. Oxidation of one gram of fat produces twice as much ATP as a gram of carbohydrate.

- Fat sources may be from the diet or from storage cells in the body.
- Fats are digested into glycerol and fatty acids.
- Glycerol can be converted to glyceraldehyde phosphate, an intermediate of glycolysis.

- Most energy in fats is in fatty acids, which are converted into acetyl CoA by *beta oxidation*. The resulting two-carbon fragments can enter the Krebs cycle.

2. Biosynthesis (anabolic pathways)

Some organic molecules of food provide the carbon skeletons or raw materials for the synthesis of new macromolecules.

- Some organic monomers from digestion can be used *directly* in anabolic pathways.
- Some precursors for biosynthesis do not come directly from digested food, but instead come from glycolysis or Krebs cycle intermediates which are diverted into anabolic pathways.
- These anabolic pathways *require energy (*ATP) produced by catabolic pathways of glycolysis and respiration.
- Glycolysis and the Krebs cycle are metabolic interchanges that can convert one type of macromolecule to another in response to the cell's metabolic demands.

C. Feedback mechanisms control cellular respiration

Cells respond to changing metabolic needs by controlling reaction rates.

- Anabolic pathways are switched off when their products are in ample supply. The most common mechanism of control is *feedback inhibition* (see Campbell, Chapter 6).
- Catabolic pathways, such as glycolysis and Krebs cycle, are controlled by regulating enzyme activity at strategic points.

A key control point of catabolism is the third step of glycolysis, which is catalyzed by an allosteric enzyme, *phosphofructokinase* (see Campbell, Figure 9.20).

- The ratio of ATP to ADP and AMP reflects the energy status of the cell, and phosphofructokinase is sensitive to changes in this ratio.
- Citrate (produced in Krebs cycle) and ATP are *allosteric inhibitors* of phosphofructokinase, so when their concentrations rise, the enzyme slows glycolysis. As the rate of glycolysis slows, Krebs cycle also slows since the supply of acetyl CoA is reduced. This synchronizes the rates of glycolysis and Krebs cycle.
- ADP and AMP are *allosteric activators* for phosphofructokinase, so when their concentrations relative to ATP rise, the enzyme speeds up glycolysis which speeds up the Krebs cycle.
- There are other allosteric enzymes that also control the rates of glycolysis and the Krebs cycle.

REFERENCES

Campbell, N., et al. *Biology*. 5th ed. Menlo Park, California: Benjamin/Cummings, 1998.

Lehninger, A.L., D.L. Nelson and M.M. Cox. *Principles of Biochemistry*. 2nd ed. New York: Worth, 1993.

Matthews, C.K. and K.E. van Holde. *Biochemistry*. 2nd ed. Redwood City, California: Benjamin/Cummings, 1996.

CHAPTER 10
PHOTOSYNTHESIS

OUTLINE

OBJECTIVES

After reading this chapter and attending lecture, the student should be able to:

1. Distinguish between autotrophic and heterotrophic nutrition.
2. Distinguish between photosynthetic autotrophs and chemosynthetic autotrophs.
3. Describe the location and structure of the chloroplast.
4. Explain how chloroplast structure relates to its function.
5. Write a summary equation for photosynthesis.
6. Explain van Niel's hypothesis and describe how it contributed to our current understanding of photosynthesis.
7. Explain the role of REDOX reactions in photosynthesis.
8. Describe the wavelike and particlelike behaviors of light.
9. Describe the relationship between an action spectrum and an absorption spectrum.
10. Explain why the absorption spectrum for chlorophyll differs from the action spectrum for photosynthesis.
11. List the wavelengths of light that are most effective for photosynthesis.
12. Explain what happens when chlorophyll or accessory pigments absorb photons.
13. List the components of a photosystem and explain their function.
14. Trace electron flow through photosystems II and I.
15. Compare cyclic and noncyclic electron flow and explain the relationship between these components of the light reactions.

16. Summarize the light reactions with an equation and describe where they occur.
17. Describe important differences in chemiosmosis between oxidative phosphorylation in mitochondria and photophosphorylation in chloroplasts.
18. Summarize the carbon-fixing reactions of the Calvin cycle and describe changes that occur in the carbon skeleton of the intermediates.
19. Describe the role of ATP and NADPH in the Calvin cycle.
20. Describe what happens to rubisco when the O_2 concentration is much higher than CO_2.
21. Describe the major consequences of photorespiration.
22. Describe two important photosynthetic adaptations that minimize photorespiration.
23. Describe the fate of photosynthetic products.

KEY TERMS

photosynthesis	visible light	noncyclic photophosphorylation
autotrophs	photons	cyclic electron flow
heterotrophs	spectrophotometer	cyclic photophosphorylation
chlorophyll	absorption spectrum	glyceraldehyde 3-phophate (G3P)
mesophyll	chlorophyll *a*	rubisco
stomata	action spectrum	C_3 plants
stroma	chlorophyll *b*	photorespiration
light reactions	carotenoids	C_4 plants
Calvin cycle	photo systems	bundle-sheath cells
NADP+	reaction center	mesophyll cells
photophosphorylation	primary electron acceptor	PEP carboxylase
carbon fixation	photosystem I	crassulacean acid metabolism
wavelength	photosystem II	CAM plants
electromagnetic spectrum	noncyclic electron flow	

LECTURE NOTES

I. Photosynthesis in Nature

Photosynthesis transforms solar light energy trapped by chloroplasts into chemical bond energy stored in sugar and other organic molecules. This process:

- Synthesizes energy-rich organic molecules from the energy-poor molecules, CO_2 and H_2O
- Uses CO_2 as a carbon source and light energy as the energy source
- Directly or indirectly supplies energy to most living organisms

A. Plants and other autotrophs are the producers of the biosphere

Organisms acquire organic molecules used for energy and carbon skeletons by one of two nutritional modes: 1) autotrophic nutrition or 2) heterotrophic nutrition.

Autotrophic nutrition = (Auto = self; trophos = feed); nutritional mode of synthesizing organic molecules from inorganic raw materials

- Examples of autotrophic organisms are plants, which require only CO_2, H_2O and minerals as nutrients.
- Because autotrophic organisms produce organic molecules that enter an ecosystem's food store, autotrophs are also known as *producers*.

- Autotrophic organisms require an energy source to synthesize organic molecules. That energy source may be from light (*photoautotrophic*) or from the oxidation of inorganic substances (*chemoautotrophic*).
  - *Photoautotrophs* = Autotrophic organisms that use light as an energy source to synthesize organic molecules. Examples are photosynthetic organisms such as plants, algae, and some prokaryotes.
  - *Chemoautotrophs* = Autotrophic organisms that use the oxidation of inorganic substances, such as sulfur or ammonia, as an energy source to synthesize organic molecules. Unique to some bacteria, this is a rarer form of autotrophic nutrition.

Heterotrophic nutrition = (Heteros = other; trophos = feed); nutritional mode of acquiring organic molecules from compounds produced by other organisms. Heterotrophs are unable to synthesize organic molecules from inorganic raw materials.

- Heterotrophs are also known as *consumers*.
- Examples are animals that eat plants or other animals.
- Examples also include *decomposers*, heterotrophs that decompose and feed on organic litter. Most fungi and many bacteria are decomposers.
- Most heterotrophs depend on photoautotrophs for food and oxygen (a by-product of photosynthesis).

B. Chloroplasts are the sites of photosynthesis in plants

Although all green plant parts have chloroplasts, leaves are the major sites of photosynthesis in most plants (see Campbell, Figure 10.2).

- *Chlorophyll* is the green pigment in chloroplasts that gives a leaf its color and that absorbs the light energy used to drive photosynthesis.

Leaf cross-section:

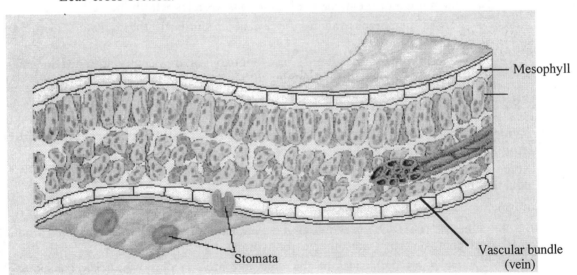

Mesophyll

Stomata

Vascular bundle (vein)

- Chloroplasts are primarily in cells of *mesophyll*, green tissue in the leaf's interior.
- CO_2 enters and O_2 exits the leaf through microscopic pores called *stomata*.
- Water absorbed by the roots is transported to leaves through veins or *vascular bundles* which also export sugar from leaves to nonphotosynthetic parts of the plant.

Chloroplasts are lens-shaped organelles measuring about $2 - 4$ μm by $4 - 7$ μm. These organelles are divided into three functional compartments by a system of membranes:

1. **Intermembrane space**

   The chloroplast is bound by a double membrane which partitions its contents from the cytosol. A narrow *intermembrane space* separates the two membranes.

2. **Thylakoid space**

   Thylakoids form another membranous system within the chloroplast. The thylakoid membrane segregates the interior of the chloroplast into two compartments: *thylakoid space* and *stroma*.

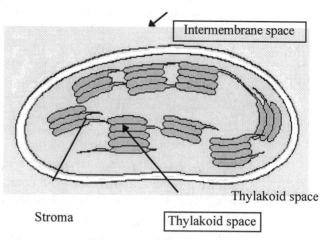

Thylakoids = Flattened membranous sacs inside the chloroplast

   • Chlorophyll is found in the thylakoid membranes.
   • Thylakoids function in the steps of photosynthesis that initially convert light energy to chemical energy.

Thylakoid space = Space inside the thylakoid

Grana = (Singular, granum); stacks of thylakoids in a chloroplast

3. **Stroma**

   Reactions that use chemical energy to convert carbon dioxide to sugar occur in the *stroma*, viscous fluid outside the thylakoids.

   Photosynthetic prokaryotes lack chloroplasts, but have chlorophyll built into the plasma membrane or membranes of numerous vesicles within the cell.

   • These membranes function in a manner similar to the thylakoid membranes of chloroplasts.
   • Photosynthetic membranes of cyanobacteria are usually arranged in parallel stacks of flattened sacs similar to the thylakoids of chloroplasts.

II. **The Pathways of Photosynthesis**

A. **Evidence that chloroplasts split water molecules enabled researchers to track atoms through photosynthesis:** *science as a process*

   Some steps in photosynthesis are not yet understood, but the following summary equation has been known since the early 1800s:

   $$6\,CO_2 + 12\,H_2O + \text{light energy} \longrightarrow C_6H_{12}O_6 + 6\,O_2 + 6\,H_2O$$

   • Glucose ($C_6H_{12}O_6$) is shown in the summary equation, though the main products of photosynthesis are other carbohydrates.
   • Water is on both sides of the equation because photosynthesis consumes 12 molecules and forms 6.

   Indicating the net consumption of water simplifies the equation:

   $$6\,CO_2 + 6\,H_2O + \text{light energy} \longrightarrow C_6H_{12}O_6 + 6\,O_2$$

   • In this form, the summary equation for photosynthesis is the reverse of that for cellular respiration.
   • Photosynthesis and cellular respiration both occur in plant cells, but plants do not simply reverse the steps of respiration to make food.

The simplest form of the equation is: $CO_2 + H_2O \longrightarrow CH_2O + O_2$

- CH_2O symbolizes the general formula for a carbohydrate.
- In this form, the summary equation emphasizes the production of a sugar molecule, one carbon at a time. Six repetitions produces a glucose molecule.

1. The splitting of water

The discovery that O_2 released by plants is derived from H_2O and not from CO_2, was one of the earliest clues to the mechanism of photosynthesis.

- In the 1930s, C.B. van Niel from Stanford University challenged an early model that predicted that:
  a. O_2 released during photosynthesis came from CO_2.

     $CO_2 \longrightarrow C + O_2$

  b. CO_2 was split and water was added to carbon.

     $C + H_2O \longrightarrow CH_2O$

- Van Niel studied bacteria that use hydrogen sulfide (H_2S) rather than H_2O for photosynthesis and produce yellow sulfur globules as a by-product.

     $CO_2 + 2\,H_2S \longrightarrow CH_2O + H_2O + 2\,S$

- Van Niel deduced that these bacteria split H_2S and used H to make sugar. He generalized that all photosynthetic organisms required hydrogen, but that the source varied:

  general: $\quad CO_2 + 2\,H_2X \longrightarrow CH_2O + H_2O + 2\,X$

  sulfur bacteria: $\quad CO_2 + 2\,H_2S \longrightarrow CH_2O + H_2O + 2\,S$

  plants: $\quad CO_2 + 2\,H_2O \longrightarrow CH_2O + H_2O + O_2$

- Van Niel thus hypothesized that plants split water as a source of hydrogen and release oxygen as a by-product.

Scientists later confirmed van Niel's hypothesis by using a heavy isotope of oxygen (^{18}O) as a tracer to follow oxygen's fate during photosynthesis.

- If water was labeled with tracer, released oxygen was ^{18}O:

  Experiment 1: $\quad CO_2 + 2\,H_2O* \longrightarrow CH_2O + H_2O + O_2$

- If the ^{18}O was introduced to the plant as CO_2, the tracer did not appear in the released oxygen:

  Experiment 2: $\quad CO_2* + 2H_2O \longrightarrow CH_2O* + H_2O* + O_2$

An important result of photosynthesis is the extraction of hydrogen from water and its incorporation into sugar.

- Electrons associated with hydrogen have more potential energy in organic molecules than they do in water, where the electrons are closer to electronegative oxygen.
- Energy is stored in sugar and other food molecules in the form of these high-energy electrons.

2. Photosynthesis as a redox process

Respiration is an exergonic redox process; energy is *released* from the oxidation of sugar.

- Electrons associated with sugar's hydrogens lose potential energy as carriers transport them to oxygen, forming water.
- Electronegative oxygen pulls electrons down the electron transport chain, and the potential energy released is used by the mitochondrion to produce ATP.

Photosynthesis is an endergonic redox process; energy is *required* to reduce carbon dioxide.

- Light is the energy source that boosts potential energy of electrons as they are moved from water to sugar.
- When water is split, electrons are transferred from the water to carbon dioxide, reducing it to sugar.

B. The light reactions and the Calvin cycle cooperate in transforming light to the chemical energy of food: *an overview*

Photosynthesis occurs in two stages: the *light reactions* and the *Calvin cycle*.

Light reactions = In photosynthesis, the reactions that convert light energy to chemical bond energy in ATP and NADPH. These reactions:

- Occur in the thylakoid membranes of chloroplasts
- Reduce $NADP^+$ to NADPH
  - Light absorbed by chlorophyll provides the energy to reduce $NADP^+$ to NADPH, which temporarily stores the energized electrons transferred from water.
  - $NADP^+$ (nicotinamide adenine dinucleotide phosphate), a coenzyme similar to NAD^+ in respiration, is reduced by adding a pair of electrons along with a hydrogen nucleus, or H^+.
- Give off O_2 as a by-product from the splitting of water
- Generate ATP. The light reactions power the addition of a phosphate group to ADP in a process called *photophosphorylation*.

Calvin cycle = In photosynthesis, the carbon-fixation reactions that assimilate atmospheric CO_2 and then reduce it to a carbohydrate; named for Melvin Calvin. These reactions:

- Occur in the stroma of the chloroplast
- First incorporate atmospheric CO_2 into existing organic molecules by a process called *carbon fixation*, and then reduce fixed carbon to carbohydrate

Carbon fixation = The process of incorporating CO_2 into organic molecules.

The Calvin cycle reactions do not require light directly, but reduction of CO_2 to sugar requires the *products* of the light reactions:

- NADPH provides the reducing power.
- ATP provides the chemical energy.

Chloroplasts thus use light energy to make sugar by coordinating the two stages of photosynthesis (see Campbell, Figure 10.4).

- Light reactions occur in the thylakoids of chloroplasts.
- Calvin cycle reactions occur in the stroma.
- As $NADP^+$ and ADP contact thylakoid membranes, they pick up electrons and phosphate respectively, and then transfer their high-energy cargo to the Calvin cycle.

C. The light reactions transform solar energy to the chemical energy of ATP and NADPH: *a closer look*

To understand how the thylakoids of chloroplasts transform light energy into the chemical energy of ATP and NADPH, it is necessary to know some important properties of light.

1. The nature of sunlight

Sunlight is *electromagnetic energy*. The quantum mechanical model of electromagnetic radiation describes light as having a behavior that is both wavelike and particlelike.

a. Wavelike properties of light

- *Electromagnetic energy* is a form of energy that travels in rhythmic waves which are disturbances of electric and magnetic fields.
- A *wavelength* is the distance between the crests of electromagnetic waves.
- The electromagnetic spectrum ranges from wavelengths that are less than a nanometer (gamma rays) to those that are more than a kilometer (radio waves) (see Campbell, Figure 10.5).
- *Visible light*, which is detectable by the human eye, is only a small portion of the electromagnetic spectrum and ranges from about 380 to 750 nm. The wavelengths most important for photosynthesis are within this range of visible light.

b. Particlelike properties of light

- Light also behaves as if it consists of discrete particles or quanta called *photons*.
- Each photon has a fixed quantity of energy which is *inversely* proportional to the wavelength of light. For example, a photon of violet light has nearly twice as much energy as a photon of red light.

The sun radiates the full spectrum of electromagnetic energy.

- The atmosphere acts as a selective window that allows visible light to pass through while screening out a substantial fraction of other radiation.
- The visible range of light is the radiation that drives photosynthesis.
- Blue and red, the two wavelengths most effectively absorbed by chlorophyll, are the colors most useful as energy for the light reactions.

2. Photosynthetic pigments: the light receptors

Light may be reflected, transmitted, or absorbed when it contacts matter (see Campbell, Figure 10.6).

Pigments = Substances which absorb visible light

- Different pigments absorb different wavelengths of light.
- Wavelengths that are absorbed disappear, so a pigment that absorbs all wavelengths appears black.
- When white light, which contains all the wavelengths of visible light, illuminates a pigment, the color you see is the color most reflected or transmitted by the pigment. For example, a leaf appears green because chlorophyll absorbs red and blue light but transmits and reflects green light.

Each pigment has a characteristic *absorption spectrum* or pattern of wavelengths that it absorbs. It is expressed as a graph of absorption versus wavelength.

- The absorption spectrum for a pigment in solution can be determined by using a *spectrophotometer*, an instrument used to measure what proportion of a specific wavelength of light is absorbed or transmitted by the pigment (see Campbell Methods Box).

- Since chlorophyll *a* is the light-absorbing pigment that participates directly in the light reactions, the absorption spectrum of chlorophyll *a* provides clues as to which wavelengths of visible light are most effective for photosynthesis (see Campbell, Figure 10.7a).

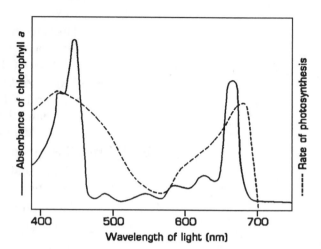

A graph of wavelength versus rate of photosynthesis is called an *action spectrum* and profiles the relative effectiveness of different wavelengths of visible light for driving photosynthesis (see Campbell, Figure 10.7b).

- The action spectrum of photosynthesis can be determined by illuminating chloroplasts with different wavelengths of light and measuring some indicator of photosynthetic rate, such as oxygen release or carbon dioxide consumption (see Campbell, Figure 10.7c).

- It is apparent from the action spectrum of photosynthesis that blue and red light are the most effective wavelengths for photosynthesis and green light is the least effective.

The *action spectrum* for photosynthesis does not exactly match the *absorption spectrum* for chlorophyll *a*.

- Since chlorophyll *a* is not the only pigment in chloroplasts that absorb light, the absorption spectrum for chlorophyll *a* underestimates the effectiveness of some wavelengths.

- Even though only special chlorophyll *a* molecules can participate directly in the light reactions, other pigments, called *accessory pigments*, can absorb light and transfer the energy to chlorophyll *a*.

The *accessory pigments* expand the range of wavelengths available for photosynthesis. These pigments include:

- *Chlorophyll b*, a yellow-green pigment with a structure similar to chlorophyll *a*. This minor structural difference gives the pigments slightly different absorption spectra (see Campbell, Figure 10.8).

- *Carotenoids*, yellow and orange hydrocarbons that are built into the thylakoid membrane with the two types of chlorophyll (see Campbell, Figure 10.7a).

3. **Photoexcitation of chlorophyll**

What happens when chlorophyll or accessory pigments absorb photons (see Campbell, Figure 10.9)?

- Colors of absorbed wavelengths disappear from the spectrum of transmitted and reflected light.

- The absorbed photon boosts one of the pigment molecule's electrons in its lowest-energy state (*ground state*) to an orbital of higher potential energy (*excited state*).

The only photons absorbed by a molecule are those with an energy state equal to the difference in energy between the ground state and excited state.

- This energy difference varies from one molecule to another. Pigments have unique absorption spectra because pigments only absorb photons corresponding to specific wavelengths.
- The photon energy absorbed is converted to potential energy of an electron elevated to the excited state.

The excited state is unstable, so excited electrons quickly fall back to the ground state orbital, releasing excess energy in the process. This released energy may be:

- Dissipated as heat
- Reradiated as a photon of lower energy and longer wavelength than the original light that excited the pigment. This afterglow is called *fluorescence*.

Pigment molecules do not fluoresce when in the thylakoid membranes, because nearby *primary electron acceptor* molecules trap excited state electrons that have absorbed photons.

- In this redox reaction, chlorophyll is photo-oxidized by the absorption of light energy and the electron acceptor is reduced.
- Because no primary electron acceptor is present, *isolated* chlorophyll fluoresces in the red part of the spectrum and dissipates heat.

4. **Photosystems: light-harvesting complexes of the thylakoid membrane**

Chlorophyll *a*, chlorophyll *b* and the carotenoids are assembled into *photosystems* located within the thylakoid membrane. Each photosystem is composed of:

a. **Antenna complex**

- Several hundred chlorophyll *a*, chlorophyll *b* and carotenoid molecules are light-gathering antennae that absorb photons and pass the energy from molecule to molecule (see Campbell, Figure 10.10). This process of resonance energy transfer is called *inductive resonance*.
- Different pigments within the antennal complex have slightly different absorption spectra, so collectively they can absorb photons from a wider range of the light spectrum than would be possible with only one type of pigment molecule.

b. **Reaction-center chlorophyll**

Only one of the many chlorophyll *a* molecules in each complex can actually *transfer* an excited electron to initiate the light reactions. This specialized chlorophyll *a* is located in the *reaction center*.

c. **Primary electron acceptor**

- Located near the reaction center, a primary electron acceptor molecule traps excited state electrons released from the reaction center chlorophyll.
- The transfer of excited state electrons from chlorophyll to primary electron acceptor molecules is the first step of the light reactions. The energy stored in the trapped electrons powers the synthesis of ATP and NADPH in subsequent steps.

Two types of photosystems are located in the thylakoid membranes, *photosystem I* and *photosystem II*.

- The reaction center of photosystem I has a specialized chlorophyll *a* molecule known as *P700*, which absorbs best at 700 nm (the far red portion of the spectrum).
- The reaction center of photosystem II has a specialized chlorophyll *a* molecule known as *P680*, which absorbs best at a wavelength of 680 nm.

- P700 and P680 are identical chlorophyll *a* molecules, but each is associated with a different protein. This affects their electron distribution and results in slightly different absorption spectra.

5. Noncyclic electron flow

There are two possible routes for electron flow during the light reactions: *noncyclic flow* and *cyclic flow*.

Both photosystem I and photosystem II function and cooperate in noncyclic electron flow, which transforms light energy to chemical energy stored in the bonds of NADPH and ATP (see Campbell, Figure 10.11). This process:

- Occurs in the thylakoid membrane
- Passes electrons continuously from water to NADP$^+$
- Produces ATP by *noncyclic photophosphorylation*
- Produces NADPH.
- Produces O$_2$

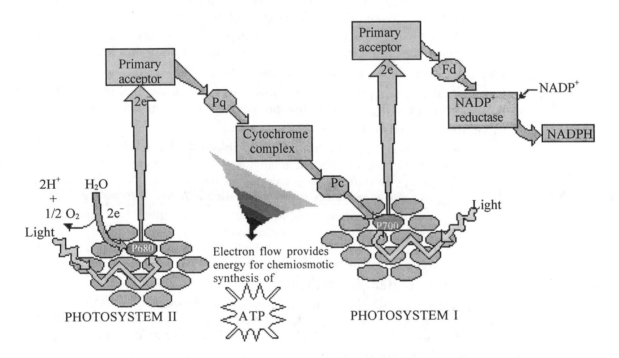

Light excites electrons from P700, the reaction center chlorophyll in photosystem I. These excited state electrons do not return to the reaction center chlorophyll, but are ultimately stored in NADPH, which will later be the electron donor in the Calvin cycle.

- Initially, the excited state electrons are transferred from P700 to the primary electron acceptor for photosystem I.
- The primary electron acceptor passes these excited state electrons to *ferredoxin* (Fd), an iron-containing protein.
- *NADP$^+$ reductase* catalyzes the redox reaction that transfers these electrons from ferredoxin to NADP$^+$, producing reduced coenzyme – NADPH.
- The oxidized P700 chlorophyll becomes an oxidizing agent as its electron "holes" must be filled; photosystem II supplies the electrons to fill these holes.

When the antenna assembly of photosystem II absorbs light, the energy is transferred to the P680 reaction center .

- Electrons ejected from P680 are trapped by the photosystem II primary electron acceptor.
- The electrons are then transferred from this primary electron acceptor to an electron transport chain embedded in the thylakoid membrane. The first carrier in the chain, *plastoquinone* (Pq) receives the electrons from the primary electron acceptor. In a cascade of redox reactions, the electrons travel from Pq to a complex of two cytochromes to plastocyanin (Pc) to P700 of photosystem I.
- As these electrons pass down the electron transport chain, they lose potential energy until they reach the ground state of P700.
- These electrons then fill the electron vacancies left in photosystem I when $NADP^+$ was reduced.

Electrons from P680 flow to P700 during noncyclic electron flow, restoring the missing electrons in P700. This, however, leaves the P680 reaction center of photosystem II with missing electrons; the oxidized P680 chlorophyll thus becomes a strong oxidizing agent.

- A water-splitting enzyme extracts electrons from water and passes them to oxidized P680, which has a high affinity for electrons.
- As water is oxidized, the removal of electrons splits water into two hydrogen ions and an oxygen atom.
- The oxygen atom immediately combines with a second oxygen atom to form O_2. It is this water-splitting step of photosynthesis that releases O_2.

As excited electrons give up energy along the transport chain to P700, the thylakoid membrane couples the exergonic flow of electrons to the endergonic reactions that phosphorylate ADP to ATP.

- This coupling mechanism is *chemiosmosis*.
- Some electron carriers can only transport electrons in the company of protons.
- The protons are picked up on one side of the thylakoid membrane and deposited on the opposite side as the electrons move to the next member of the transport chain.
- The electron flow thus stores energy in the form of a proton gradient across the thylakoid membrane – a *proton-motive force*.
- An ATP synthase enzyme in the thylakoid membrane uses the proton-motive force to make ATP. This process is called *photophosphorylation* because the energy required is light.
- This form of ATP production is called *noncyclic photophosphorylation*.

6. Cyclic electron flow

Cyclic electron flow is the simplest pathway, but involves only photosystem I and generates ATP without producing NADPH or evolving oxygen.

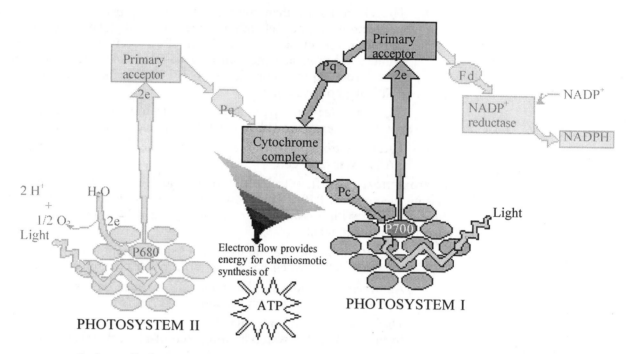

- It is cyclic because excited electrons that leave from chlorophyll *a* at the reaction center return to the reaction center.

- As photons are absorbed by Photosystem I, the P700 reaction center chlorophyll releases excited-state electrons to the primary electron acceptor; which, in turn, passes them to ferredoxin. From there the electrons take an alternate path that sends them tumbling down the electron transport chain to P700. This is the same electron transport chain used in noncyclic electron flow.

- With each redox reaction along the electron transport chain, electrons lose potential energy until they return to their ground-state orbital in the P700 reaction center.

- The exergonic flow of electrons is coupled to ATP production by the process of chemiosmosis. This process of ATP production is called *cyclic photophosphorylation*.

- Absorption of another two photons of light by the pigments send a second pair of electrons through the cyclic pathway.

The function of the cyclic pathway is to produce additional ATP.

- It does so *without* the production of NADPH or O_2.

- Cyclic photophosphorylation supplements the ATP supply required for the Calvin cycle and other metabolic pathways. The noncyclic pathway produces approximately equal amounts of ATP and NADPH, which is not enough ATP to meet demand.

- NADPH concentration might influence whether electrons flow through cyclic or noncyclic pathways.

7. A comparison of chemiosmosis in chloroplasts and mitochondria

Chemiosmosis = The coupling of exergonic electron flow down an electron transport chain to endergonic ATP production by the creation of an electrochemical proton gradient across a membrane. The proton gradient drives ATP synthesis as protons diffuse back across the membrane.

Chemiosmosis in chloroplasts and chemiosmosis in mitochondria are similar in several ways:

- An electron transport chain assembled in a membrane translocates protons across the membrane as electrons pass through a series of carriers that are progressively more electronegative.

- An ATP synthase complex built into the same membrane, couples the diffusion of hydrogen ions down their gradient to the phosphorylation of ADP.

- The ATP synthase complexes and some electron carriers (including quinones and cytochromes) are very similar in both chloroplasts and mitochondria.

Oxidative phosphorylation in mitochondria and photophosphorylation in chloroplasts differ in the following ways:

a. Electron transport chain

- Mitochondria transfer chemical energy from food molecules to ATP. The high-energy electrons that pass down the transport chain are extracted by the oxidation of food molecules.

- Chloroplasts transform light energy into chemical energy. Photosystems capture light energy and use it to drive electrons to the top of the transport chain.

b. Spatial organization

- The inner mitochondrial membrane pumps protons from the matrix out to the intermembrane space, which is a reservoir of protons that power ATP synthase.

- The chloroplast's thylakoid membrane pumps protons from the stroma into the thylakoid compartment, which functions as a proton reservoir. ATP is produced as protons diffuse from the thylakoid compartment back to the stroma through ATP synthase complexes that have catalytic heads on the membrane's stroma side. Thus, ATP forms in the stroma where it drives sugar synthesis during the Calvin cycle (see Campbell, Figure 10.14).

There is a large proton or pH gradient across the thylakoid membrane.

- When chloroplasts are illuminated, there is a thousand-fold difference in H^+ concentration. The pH in the thylakoid compartment is reduced to about 5 while the pH in the stroma increases to about 8.

- When chloroplasts are in the dark, the pH gradient disappears, but can be reestablished if chloroplasts are illuminated.

- Andre Jagendorf (1960s) produced compelling evidence for chemiosmosis when he induced chloroplasts to produce ATP in the dark by using artificial means to create a pH gradient. His experiments demonstrated that during photophosphorylation, the function of the photosystems and the electron transport chain is to create a proton-motive force that drives ATP synthesis.

A tentative model for the organization of the thylakoid membrane includes the following:

- Proton pumping by the thylakoid membrane depends on an asymmetric placement of electron carriers that accept and release protons (H^+).
- There are three steps in the light reactions that contribute to the proton gradient across the thylakoid membrane:
  1. Water is split by Photosystem II on the thylakoid side, releasing protons in the process.
  2. As plastoquinone (Pq), a mobile carrier, transfers electrons to the cytochrome complex, it translocates protons from the stroma to the thylakoid space.
  3. Protons in the stroma are removed from solution as $NADP^+$ is reduced to NADPH.
- NADPH and ATP are produced on the side of the membrane facing the stroma where sugar is synthesized by the Calvin cycle.

> Students must be able to visualize the spatial arrangement of electron carriers in the membrane, since this arrangement is a crucial component of the chemiosmosis model. Figure 10.15 nicely illustrates this spatial arrangement.

8. **Summary of light reactions**

During *noncyclic electron flow*, the photosystems of the thylakoid membrane transform light energy to the chemical energy stored in NADPH and ATP. This process:

- Pushes low energy-state electrons from water to NADPH, where they are stored at a higher state of potential energy. NADPH, in turn, is the electron donor used to reduce carbon dioxide to sugar (Calvin cycle).
- Produces ATP from this light driven electron current
- Produces oxygen as a by-product

During *cyclic electron flow*, electrons ejected from P700 reach ferredoxin and flow *back* to P700. This process:

- Produces ATP
- Unlike noncyclic electron flow, does *not* produce NADPH or O_2

D. **The Calvin cycle uses ATP and NADPH to convert CO_2 to sugar:** *a closer look*

ATP and NADPH produced by the light reactions are used in the Calvin cycle to reduce carbon dioxide to sugar.

- The Calvin cycle is similar to the Krebs cycle in that the starting material is regenerated by the end of the cycle.
- Carbon enters the Calvin cycle as CO_2 and leaves as sugar.
- ATP is the energy source, while NADPH is the reducing agent that adds high-energy electrons to form sugar.
- The Calvin cycle actually produces a three-carbon sugar *glyceraldehyde 3-phosphate* (G3P).

> Students can easily follow the Calvin cycle if you use a diagram for reference, such as Figure 10.16. This figure is especially helpful because you can go through the cycle twice; once to count carbons and once to follow the reactions pointing out where ATP and NADPH are used, where glyceraldehyde phosphate is produced and how RuBP is regenerated.

PHASE I: CARBON FIXATION

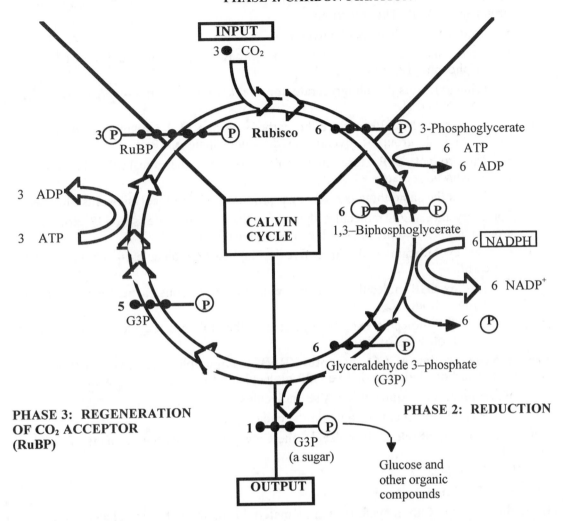

PHASE 3: REGENERATION OF CO₂ ACCEPTOR (RuBP)

PHASE 2: REDUCTION

For the Calvin cycle to synthesize one molecule of sugar (G3P), three molecules of CO_2 must enter the cycle. The cycle may be divided into three phases:

Phase 1: Carbon Fixation. The Calvin cycle begins when each molecule of CO_2 is attached to a five-carbon sugar, *ribulose biphosphate* (*RuBP*).

- This reaction is catalyzed by the enzyme *RuBP carboxylase* (*rubisco*) – one of the most abundant proteins on Earth..

- The product of this reaction is an unstable six-carbon intermediate that immediately splits into two molecules of 3-phosphoglycerate.

- For every three CO_2 molecules that enter the Calvin cycle via rubisco, three RuBP molecules are carboxylated forming six molecules of 3-phosphoglycerate.

Phase 2: Reduction. This endergonic reduction phase is a two-step process that couples ATP hydrolysis with the reduction of 3-phosphoglycerate to glyceraldehyde phosphate.

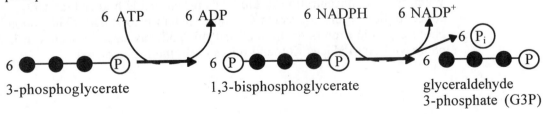

- An enzyme phosphorylates 3-phosphoglycerate by transferring a phosphate group from ATP. This reaction:
  - Produces 1, 3-bisphosphoglycerate
  - Uses six ATP molecules to produce six molecules of 1,3-bisphosphoglycerate.
  - Primes 1,3-bisphosphoglycerate for the addition of high-energy electrons from NADPH.
- Electrons from NADPH reduce the carboxyl group of 1,3-bisphosphoglycerate to the aldehyde group of glyceraldehyde 3-phosphate (G3P).
  - The product, G3P, stores more potential energy than the initial reactant, 3-phosphoglycerate.
  - G3P is the same three-carbon sugar produced when glycolysis splits glucose.
- For every three CO_2 molecules that enter the Calvin cycle, six G3P molecules are produced, only one of which can be counted as net gain.
  - The cycle begins with three five-carbon RuBP molecules – a total of 15 carbons.
  - The six G3P molecules produced contain 18 carbons, a net gain of three carbons from CO_2.
  - One G3P molecule exits the cycle; the other five are recycled to regenerate three molecules of RuBP.

Phase 3: Regeneration of CO_2 acceptor (RuBP). A complex series of reactions rearranges the carbon skeletons of five G3P molecules into three RuBP molecules.

- These reactions require three ATP molecules.
- RuBP is thus regenerated to begin the cycle again.

For the net synthesis of one G3P molecule, the Calvin cycle uses the products of the light reactions:

- 9 ATP molecules
- 6 NADPH molecules

G3P produced by the Calvin cycle is the raw material used to synthesize glucose and other carbohydrates.

- The Calvin cycle uses 18 ATP and 12 NADPH molecules to produce one glucose molecule.

E. Alternative mechanisms of carbon fixation have evolved in hot, arid climates

1. Photorespiration: an evolutionary relic?

A metabolic pathway called *photorespiration* reduces the yield of photosynthesis.

Photorespiration = In plants, a metabolic pathway that consumes oxygen, evolves carbon dioxide, produces no ATP and decreases photosynthetic output.

- Occurs because the active site of rubisco can accept O_2 as well as CO_2
- Produces no ATP molecules
- Decreases photosynthetic output by reducing organic molecules used in the Calvin cycle

When the O_2 concentration in the leaf's air spaces is higher than CO_2 concentration, rubisco accepts O_2 and transfers it to RuBP. (The "photo" in photorespiration refers to the fact that this pathway usually occurs in light when photosynthesis reduces CO_2 and raises O_2 in the leaf spaces.)

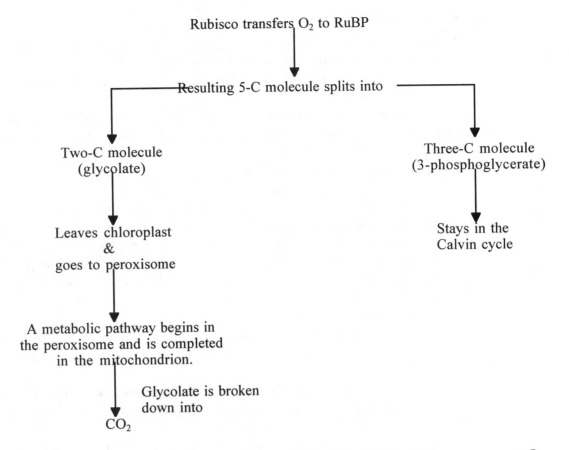

Rubisco transfers O_2 to RuBP

Resulting 5-C molecule splits into

Two-C molecule
(glycolate)

Three-C molecule
(3-phosphoglycerate)

Leaves chloroplast
&
goes to peroxisome

Stays in the
Calvin cycle

A metabolic pathway begins in
the peroxisome and is completed
in the mitochondrion.

Glycolate is broken
down into

CO_2

(The "respiration" in photorespiration refers to the fact that this process uses O_2 and releases CO_2.)

Some scientists believe that photorespiration is a metabolic relic from earlier times when the atmosphere contained less oxygen and more carbon dioxide than is present today.

- Under these conditions, when rubisco evolved, the inability of the enzyme's active site to distinguish carbon dioxide from oxygen would have made little difference.
- This affinity for oxygen has been retained by rubisco and some photorespiration is bound to occur.

Whether photorespiration is beneficial to plants is not known.

- It is known that some crop plants (e.g., soybeans) lose as much as 50% of the carbon fixed by the Calvin cycle to photorespiration.
- If photorespiration could be reduced in some agricultural plants, crop yields and food supplies would increase.

Photorespiration is fostered by hot, dry, bright days.

- Under these conditions, plants close their stomata to prevent dehydration by reducing water loss from the leaf.
- Photosynthesis then depletes available carbon dioxide and increases oxygen within the leaf air spaces. This condition favors photorespiration.

Certain species of plants, which live in hot arid climates, have evolved alternate modes of carbon fixation that minimize photorespiration. C_4 and CAM are the two most important of these photosynthetic adaptations.

2. C$_4$ plants

The Calvin cycle occurs in most plants and produces 3-phosphoglycerate, a three-carbon compound, as the first stable intermediate.

- These plants are called *C$_3$ plants* because the first stable intermediate has three carbons.
- Agriculturally important C$_3$ plants include rice, wheat, and soybeans.

Many plant species preface the Calvin cycle with reactions that incorporate carbon dioxide into four-carbon compounds.

- These plants are called *C$_4$ plants*.
- The C$_4$ pathway is used by several thousand species in at least 19 families including corn and sugarcane, important agricultural grasses.
- This pathway is adaptive, because it enhances carbon fixation under conditions that favor photorespiration, such as hot, arid environments.

Leaf anatomy of C$_4$ plants spatially segregates the Calvin cycle from the initial incorporation of CO$_2$ into organic compounds. There are two distinct types of photosynthetic cells:

1. Bundle-sheath cells
   - Arranged into tightly packed sheaths around the veins of the leaf
   - Thylakoids in the chloroplasts of bundle-sheath cells are not stacked into grana.
   - The Calvin cycle is confined to the chloroplasts of the bundle sheath.
2. Mesophyll cells are more loosely arranged in the area between the bundle sheath and the leaf surface.

The Calvin cycle of C4 plants is preceded by incorporation of CO$_2$ into organic compounds in the mesophyll (see Campbell, Figure 10.18)

Step 1: CO$_2$ is added to phosphoenolpyruvate (PEP) to form oxaloacetate, a four-carbon product.

- *PEP carboxylase* is the enzyme that adds CO$_2$ to PEP. Compared to rubisco, it has a much greater affinity for CO$_2$ and has *no* affinity for O$_2$.
- Thus, PEP carboxylase can fix CO$_2$ efficiently when rubisco cannot — under hot, dry conditions that cause stomata to close, CO$_2$ concentrations to drop and O$_2$ concentrations to rise.

Step 2: After CO$_2$ has been fixed by mesophyll cells, they convert oxaloacetate to another four-carbon compound (usually malate).

Step 3: Mesophyll cells then export the four-carbon products (e.g., malate) through plasmodesmata to bundle-sheath cells.

- In the bundle-sheath cells, the four carbon compounds release CO$_2$, which is then fixed by rubisco in the Calvin cycle.
- Mesophyll cells thus pump CO$_2$ into bundle-sheath cells, minimizing photorespiration and enhancing sugar production by maintaining a CO$_2$ concentration sufficient for rubisco to accept CO$_2$ rather than oxygen.

3. CAM plants

A second photosynthetic adaptation exists in succulent plants adapted to very arid conditions. These plants open their stomata primarily at night and close them during the day (opposite of most plants).

- This conserves water during the day, but prevents CO$_2$ from entering the leaves.

- When stomata are open at night, CO_2 is taken up and incorporated into a variety of organic acids. This mode of carbon fixation is called *crassulacean acid metabolism (CAM)*.
- The organic acids made at night are stored in vacuoles of mesophyll cells until morning, when the stomata close.
- During daytime, light reactions supply ATP and NADPH for the Calvin cycle. At this time, CO_2 is released from the organic acids made the previous night and is incorporated into sugar in the chloroplasts.

The CAM and C_4 pathways:

- Are similar in that CO_2 is first incorporated into organic intermediates before it enters the Calvin cycle.
- Differ in that the initial steps of carbon fixation in C_4 plants are structurally separate from the Calvin cycle; in CAM plants, the two steps occur at separate times.

Regardless of whether the plant uses a C_3, C_4 or CAM pathway, all plants use the Calvin cycle to produce sugar from CO_2.

F. Photosynthesis is the biosphere's metabolic foundation: *a review*

On a global scale, photosynthesis makes about 160 billion metric tons of carbohydrate per year. No other chemical process on Earth is more productive or is as important to life.

- Light reactions capture solar energy and use it to:
  - Produce ATP
  - Transfer electrons from water to $NADP^+$ to form NADPH
- The Calvin cycle uses ATP and NADPH to fix CO_2 and produce sugar.

Photosynthesis transforms light energy to chemical bond energy in sugar molecules.

- Sugars made in chloroplasts supply the entire plant with chemical energy and carbon skeletons to synthesize organic molecules.
- Nonphotosynthetic parts of a plant depend on organic molecules exported from leaves in veins.
  - The disaccharide *sucrose* is the transport form of carbohydrate in most plants.
  - Sucrose is the raw material for cellular respiration and many anabolic pathways in nonphotosynthetic cells.
- Much of the sugar is *glucose* – the monomer linked to form *cellulose*, the main constituent of plant cell walls.

Most plants make more organic material than needed for respiratory fuel and for precursors of biosynthesis.

- Plants consume about 50% of the photosynthate as fuel for cellular respiration.
- Extra sugars are synthesized into starch and stored in storage cells of roots, tubers, seeds, and fruits.
- Heterotrophs also consume parts of plants as food.

Photorespiration can reduce photosynthetic yield in hot dry climates. Alternate methods of carbon fixation minimize photorespiration.

- C4 plants spatially separate carbon fixation from the Calvin cycle.
- CAM plants temporally separate carbon fixation from the Calvin cycle.

REFERENCES

Atkins, P.W. *Atoms, Electrons, and Change*. New York, Oxford: W.H. Freeman and Company, 1991. Chapter 9, "Light and Life" is a witty, imaginative description of photosynthesis. Though written for a lay audience, it is probably best appreciated by someone already familiar with photosynthesis.

Campbell, N., et al. *Biology*. 5th ed. Menlo Park, California: Benjamin/Cummings, 1998.

Lehninger, A.L., D.L. Nelson and M.M. Cox. *Principles of Biochemistry*. 2nd ed. New York: Worth, 1993.

Matthews, C.K. and K.E. van Holde. *Biochemistry*. 2nd ed. Redwood City, California: Benjamin/Cummings, 1996.

OUTLINE

I. An Overview of Cell Signaling
   A. Cell signaling evolved early in the history of life
   B. Communicating cells may be close together or far apart
   C. The three stages of cell signaling are reception, transduction, and response

II. Signal Reception and the Initiation of Transduction
   A. A chemical signal binds to a receptor protein, causing the protein to change shape
   B. Most signal receptors are plasma-membrane proteins

III. Signal Transduction Pathways
   A. Pathways relay signals from receptors to cellular responses
   B. Protein phosphorylation, a common mode of regulation in cells, is a major mechanism of signal transduction
   C. Certain small molecules and ions are key components of signaling pathways (second messengers)

IV. Cellular Responses to Signals
   A. In response to a signal, a cell may regulate activities in the cytoplasm or transcription in the nucleus
   B. Elaborate pathways amplify and specify the cell's responses to signals

OBJECTIVES

After reading the chapter and attending lecture, the student should be able to :

1. Categorize chemical signals in terms of the proximity of the communicating cells.

2. Overview the basic elements of a signaling system of a target cell.

3. Describe the nature of a ligand-receptor interaction and state how such interactions initiate a signal transduction system.

4. Compare and contrast G-protein-linked receptors, tyrosine-kinase receptors, and ligand-gated ion channels.

5. Describe how phosphorylation propagates signal information.

6. Describe how cAMP is formed and how it propagates signal information.

7. Describe how the cytoplasmic concentration of Ca^{2+} can be altered and how this increased pool of Ca^{2+} is involved with signal transduction.

8. Describe how signal information is transduced into cellular responses in the cytoplasm and in the nucleus.

9. Describe how signal amplification is accomplished in target cells.

10. Describe how target cells discriminate among signals and how the same signal can elicit multiple cellular responses.

KEY WORDS

signal transduction pathway	tyrosine kinase	cyclic AMP (cAMP)
local regulator	tyrosine-kinase receptor	adenylyl cyclase
hormone	ligand-gated ion channels	diacylglycerol (DAG)
ligand	protein kinase	inositol triphosphate (IP_3)
G-protein-linked receptor	protein phosphatase	calmodulin
G-protein	second messenger	

LECTURE NOTES

Regulation is an essential feature of life. It unifies the various levels of biological organization by embracing the fields of molecular and cell biology, organismal biology, and population biology and ecology. It provides the necessary coordination for all aspects of life, including metabolism, growth, development, and reproduction.

Chemical substances are the principal agents of biological regulation and they exert their effects on cells through signaling systems. This chapter describes the fundamental components of cell signaling systems.

I. **An Overview of Cell Signaling**

A. **Cell signaling evolved early in the history of life**

Yeast mating behavior is coordinated by chemical signaling.

- Yeast (unicellular eukaryotes) have two mating types: *a* and *α*.

- Type *a* cells secrete an *a*-factor chemical signal; type *α* cells secrete *α*-factor.

- The binding of *a*-factor to type *α* cells and the binding of *α*-factor to type *a* cells induces type *a* and type *α* to move toward one another and fuse.

The steps by which yeast mating signals are converted into yeast cell responses are similar to how chemical signals in prokaryotes (bacteria), plants, and animals are converted to specific cell responses.

In general, the steps by which a chemical signal is converted to a specific cell response is called a *signal transduction pathway*.

B. **Communicating cells may be close together or far apart**

A chemical signal that communicates between two nearby cells is called a *local regulator*. Two types of local signaling have been described in animals: paracrine signaling and synaptic signaling.

- In paracrine signaling, one cell secretes the signal into extracellular fluid and the signal acts on a nearby target cell. Examples of signals which act in a paracrine fashion are growth factors, a group of factors which stimulate cells to divide and grow.

- In synaptic signaling, a nerve cell releases a signal (e.g., neurotransmitter) into a synapse, the narrow space between the transmitting cell and a target cell, such as another nerve cell or muscle cell.

A chemical signal which communicates between cells some distance apart is called a *hormone*.

Hormones have been described in both plants (e.g., ethylene, a gas which promotes growth and fruit ripening) and animals (e.g., insulin, a protein which controls various aspects of metabolism, including the regulation of blood glucose levels).

The distinction between local regulators and hormones is for convenience. A particular chemical signal may act both as a local regulator and as a hormone.

> Insulin, for example, may act in a paracrine fashion on adjacent cells (e.g., other insulin cells in the pancreas, acting to inhibit the further release of insulin in a negative feedback manner) and in a hormonal fashion on distant cells (e.g., liver cells, which store carbohydrate as glycogen).

Cells also may communicate by direct contact. Some plant and animal cell possess junctions though which signals can travel between adjacent cells.

C. The three stages of cell signaling are reception, transduction, and response

In order for a chemical signal to elicit a specific response, the target cell must possess a signaling system for the signal. Cells which do not possess the appropriate signaling system do not respond to the signal.

The signaling system of a target cell consists of the following elements:

- *Signal reception.* The signal binds to a specific cellular protein called a receptor, which is often located on the surface of the cell.
- *Signal transduction.* The binding of the signal changes the receptor in some way, usually a change in conformation or shape. The change in receptor initiates a process of converting the signal into a specific cellular response; this process is called signal transduction. The transduction system may have one or many steps.
- *Cellular response.* The transduction system triggers a specific cellular response. The response can be almost any cellular activity, such as activation of an enzyme or altered gene expression.

The critical features of the target cell signaling system were elucidated by Earl Sutherland (awarded the Nobel Prize in 1971 for his contributions to the understanding of signal transduction) and colleagues who were working on how the hormone, epinephrine, affects carbohydrate metabolism (e.g., glycogen breakdown to glucose-1-phosphate) in liver cells.

- Epinephrine stimulates glycogen breakdown by stimulating the cytosolic enzyme, glycogen phosphorylase (cellular response).
- Epinephrine could only stimulate glycogen phosphorylase activity when presented to intact cells, suggesting that:
  - The plasma membrane is critical for transmitting the signal (reception)
  - Activation of glycogen phosphorylase required the presence of an intermediate step or steps in side the cell (signal transduction)

The mechanisms of the cell signaling process help ensure that important processes occur in the right cells, at the right time, and in proper coordination with other cells of the organism.

II.   Signal Reception and the Initiation of Transduction

A. A chemical signal binds to a receptor protein, causing the protein to change shape

Chemical signals bind to specific receptors.

- The signal molecule is complementary to a specific region of the receptor protein; this interaction is similar to that between a substrate and an enzyme.
- The signal behaves as a *ligand,* a term for a small molecule that binds to another, larger molecule.

Binding of the ligand to the receptor can lead to the following events:

- Alteration in receptor conformation or shape; such alterations may lead to the activation of the receptor which enables it to interact with other cellular molecules
- Aggregation of receptor complexes

B. Most signal receptors are plasma-membrane proteins

Many signal molecules cannot pass freely through the plasma membrane. The receptors for such signal molecules are located on the plasma membrane. Three families of plasma-membrane receptors—*G-protein-linked receptors, tyrosine kinase receptors,* and ion channel receptors— will be described.

1. G-protein-linked receptors

The structure of a ***G-protein-linked receptor*** is characterized by a single polypeptide chain that is threaded back and forth through the plasma membrane in such a way as to possess seven transmembrane domains. An example of a G-protein-linked receptor is the epinephrine receptor.

The receptor propagates the signal by interacting with a variety of proteins on the cytoplasmic side of the membrane called G-proteins, so named because they bind guanine nucleotides, GTP and GDP.

- The function of the G-protein is influenced by the nucleotide to which it is bound:
    - G-proteins bound to GDP are inactive.
    - G-proteins bound to GTP are active.
- When a ligand binds to a G-protein-linked receptor, the receptor changes its conformation and interacts with a G-protein. This interaction causes the GDP bound to the inactive G-protein to be displaced by GTP, thereby activating the G-protein.
- The activated G-protein binds to another protein, usually an enzyme, resulting in the activation of a subsequent target protein.
- The activation state of the G-protein is only temporary, because the active G-protein possesses endogenous GTPase activity, which hydrolyzes the bound GTP to GDP.

- G-protein-linked receptors and G-proteins mediate a host of critical metabolic and developmental processes (e.g., blood vessel growth and development). Defects in the G-protein signaling system form the bases of many human disease states (e.g., cholera).

2. Tyrosine-kinase receptors

The structure of a tyrosine-kinase receptor is characterized by an extracellular ligand-binding domain and a cytosolic domain possessing tyrosine kinase enzyme activity. Examples of tyrosine-kinase receptors are the receptors for numerous growth factors, such as PDGFs, the family of factors which serve as external modulators of the cell-cycle control system.

Propagation of the signal involves several steps as follows:

- Ligand binding causes aggregation of two receptor units, forming receptor dimers.
- Aggregation activates the endogenous tyrosine kinase activity on the cytoplasmic domains.
- The endogenous tyrosine kinase catalyzes the transfer of phosphate groups from ATP to the amino acid tyrosine contained in a particular protein. In this case, the tyrosines which are phosphoryled are in the cytoplasmic domain of the tyrosine-kinase receptor itself (thus, this step is an autophosphorylation).

- The phosphorylated domain of the receptor interacts with other cellular proteins, resulting in the activation of a second, or relay, protein. The relay proteins may or may not be phosphorylated by the tyrosine kinase of the receptor. Many different relay proteins may be activated, each leading to the initiation of many, possibly different, transduction systems.

- One of the activated relay proteins may be *protein phosphatase,* an enzyme which hydrolyzes phosphate groups off of proteins. The dephosphorylation of the tyrosines on the tyrosine kinase domain of the receptor results in inactivating the receptor and the termination of the signal process.

3. Ion-channel receptors

Some chemical signals bind to *ligand-gated ion channels.* These are protein pores in the membrane that open or close in response to ligand binding, allowing or blocking the flow of specific ions (e.g., Na^+, Ca^{2+}). An example of an ion-gated channel would be the binding of a neurotransmitter to a neuron, allowing the inward flow of Na^{2+} that leads to the depolarization of the neuron and the propagation of a nervous impulse to adjacent cells.

Not all signal receptors are located on the plasma membrane. Some are proteins located in the cytoplasm or nucleus of target cells.

- In order for a chemical signal to bind to these intracellular receptors, the signal molecule must be able to pass through plasma membrane. Examples of signals which bind to intracellular receptors include the following:
  - Nitric oxide (NO)
  - Steroid (e.g., estradiol, progesterone, testosterone) and thyroid hormones of animals

III. Signal Transduction Pathways

A. Pathways relay signals from receptors to cellular responses

Ligand binding to a receptor triggers the first step in the chain of reactions—the signal transduction pathway— that leads to the conversion of the signal to a specific cellular response.

- The transduction system does not physically pass along the signal molecule, rather the *information* is passed along. At each step of the process, the nature of the information is converted, or transduced, into a different form.

B. Protein phosphorylation, a common mode of regulation in cells, is a major mechanism of signal transduction

The process of phosphorylation, or the transferring a phosphate group from ATP to a protein substrate, which is catalyzed by enzymes called *protein kinases,* is a common cellular mechanism for regulating the functional activity of proteins.

Protein phosphorylation is commonly used in signal pathways in the cytoplasm of cells. Unlike the case with tyrosine-kinase receptors, protein kinases in the cytoplasm do not act on themselves, but rather on other proteins (sometimes enzymes) and attach the phosphate group to serine or threonine residues.

- Some phosphorylations result in activation of the target protein (increased catalytic activity in the case of an enzyme target). An example of a stimulatory phosphorylation cascade is the pathway involved in the breakdown of glycogen as elucidated by Sutherland, et al. (see Campbell, Figures 11.10; 11.15).

- Some phosphorylations result in inactivation (decreased catalytic activity in the case of an enzyme target).

Cells turn off the signal transduction pathway when the initial signal is no longer present. The effects of protein kinases are reversed by another class of enzymes known as ***protein phosphatases.***

C. Certain small molecules and ions are key components of signaling pathways (second messengers)

Not all of the components of a signal transduction pathways are proteins. Some signaling systems rely on small, nonprotein, water soluble molecules or ions. Such signaling components are called *second messengers*. Two second messenger systems are the *cyclic AMP (cAMP)* system and the Ca^{2+}- *inositol triphosphate (IP₃)* system.

1. Cyclic AMP

Sutherland's group ultimately found that the substance mediating the action of epinephrine on liver glycogen breakdown was cAMP (second messenger). Our present understanding of the transduction steps associated with cAMP is as follows:

- Ligand (first message) binds to a receptor.
- Receptor conformation changes; G-protein complex is activated.
- The active G-protein in turn activates the enzyme, adenylyl cyclase, which is associated with the cytoplasmic side of the plasma membrane.
- Adenylyl cyclase converts ATP to cAMP.
- cAMP binds to and activates a cytoplasmic enzyme, protein kinase A.
- Protein kinase A, as was the case for protein kinases mentioned previously, propagates the message by phosphorylating various other proteins that lead to the cellular response (e.g., glycogen breakdown; see Campbell Figure 11.15).

The pool of cAMP in the cytoplasm is transient because of the breakdown of cAMP by another enzyme to an inactive form (AMP). This conversion provides a shut-off mechanism to the cell to ensure that the target responses ceases in the absence of ligand.

A number of hormones in addition to epinephrine (e.g., glucagon) use cAMP as a second messenger.

2. Calcium ions and inositol triphosphate

Many signaling molecules induce their specific responses in target cells by increasing the cytoplasm's concentration of Ca^{2+}. The Ca^{2+} pool can be affected in two ways:

- Ligand binding to a Ca^{2+}-gated ion channel (discussed above)
- Activation of the *inositol triphosphate (IP₃)* signaling pathway

Activation of the IP₃ pathway involves the following steps:

- Ligand binding results in a conformation change in the receptor.
- The altered receptor activates an enzyme associated with the cytoplasmic side of the plasma membrane (phospholipase). The activated enzyme hydrolyzes membrane phospholipids, giving rise to two important second messengers: IP₃ and *diacylglycerol*.
- Diacylglycerol is linked to a signaling pathway that involves another protein kinase.
- IP₃ is linked to a Ca^{2+} signaling pathway. IP₃ binds to Ca^{2+}-gated channels. A large number of such channels are located on the ER, in the lumen of which high amounts of Ca^{2+} are sequestered. IP₃ binding to these receptors and increases the cytoplasmic concentration of Ca^{2+} (in this case Ca^{2+} could be considered a tertiary messenger; however, by convention, all post-receptor small molecules in the transduction system are referred to as second messengers).

Ca^{2+} acts to affect signal transduction in two ways:

- Directly by affecting the activity or function of target proteins
- Indirectly by first binding to a relay protein, *calmodulin.* Calmodulin, in turn, principally affects transduction systems by modulating the activities of protein kinases and protein phosphatases.

IV.   Cellular Responses to Signals

A.   In response to a signal, a cell may regulate activities in the cytoplasm or transcription in the nucleus

The signal transduction system ultimately brings about the specific cellular response by regulating specific processes in the cytoplasm or in the nucleus.

In the cytoplasm, the signaling can affect the *function or activity* of proteins which carry out various processes, including:

- Rearrangement of the cytoskeleton
- Opening or closing of an ion channel
- Serve at key points in metabolic pathways (e.g., glycogen phosphorylase in the glycogen breakdown scheme; see Campbell Figure 11.15)

In the nucleus, the signaling system affects the *synthesis* of new proteins and enzymes by modulating the expression (turn on or turn off) specific genes. Gene expression involves transcription of DNA into mRNA as well as the translation of mRNA into protein.

- Signal transduction systems can modulate virtually every aspect of gene expression. One example is the regulation of the activity of transcription factors, proteins required for appropriate transcription.
- Dysfunction of signaling pathways that affect gene regulation (e.g., pathways that transduce growth factor action) can have serious consequences and may even lead to cancer.

B.   Elaborate pathways amplify and specify the cell's responses to signals

The elaborate nature of cellular signal transduction systems functions to: amplify signal (and, thus, response) and contribute to the specificity of the response.

1.   Signal amplification

The production of second messengers such cAMP provides a built in means of signal amplification in that the binding of one ligand (first message) can lead to the production of many second messages. The degree of amplification is heightened when the second messenger system is linked to a phosphorylation cascade as in the case of the process of glycogen breakdown. As a result of this inherent amplification, the binding of very few epinephrine molecules to the surface of a liver cell can result in the release of millions of glucose molecules resulting from glycogen breakdown (see Campbell Figure 11.15).

2.   Signal specificity

Only target cells with the appropriate receptor bind to a particular signaling molecule to initiate the transduction of a signal into a specific cellular response.

A particular signal can bind to different cell types and result in different responses in each of the cell types. This is possible because each of the different cell types can express a unique collection of proteins. As a result, the receptor on (or in) each of the different cell types can be linked to variant signal transduction pathways, each leading to a different response. An example is epinephrine action on vertebrate liver and cardiac muscle cells. In liver cells, the principal response is glycogen breakdown (see Campbell Figure 11.15); whereas, in cardiac muscle cells, epinephrine stimulates contraction.

A single cell type may possess divergent and/or convergent ("cross-talk") signal transduction pathways. Such schemes facilitate coordination of cellular responses and economize on the number of required transduction elements (see Campbell Figure 11.17). The diverse symptoms of the human inherited disorder Wiscott-Aldrich syndrome stem from a single defect in a relay protein of a transduction system.

An important feature of cell signaling systems is that there exists mechanisms to both turn-on and turn-off the system. The turn-off mechanisms ensure that cells respond appropriately to changing conditions.

REFERENCES

Alberts, B., D. Bray, J. Lewis, M. Raff, K. Roberts, and J. Watson. *Molecular Biology of the Cell*, 3rd ed. New York: Garland, 1994.

Bolander, F. *Molecular Endocrinology*, 2nd ed. New York: Academic Press, 1994.

Hadley, M. *Endocrinology*, 3rd ed. Englewood Cliffs: Prentice-Hall, 1992.

Norman, A. and G. Litwack. *Hormones*, 2nd ed. New York: Academic Press, 1997

Norris, D. *Vertebrate Endocrinology*, 3rd ed. New York: Academic Press, 1997.

CHAPTER 12
THE CELL CYCLE

OUTLINE

I. The Key Roles of Cell Division
   A. Cell division functions in reproduction, growth, and repair
   B. Cell division distributes identical sets of chromosomes to daughter cells

II. The Mitotic Cell Cycle
   A. The mitotic phase alternates with interphase in the cell cycle: *an overview*
   B. The mitotic spindle distributes chromosomes to daughter cells: *a closer look*
   C. Cytokinesis divides the cytoplasm: *a closer look*
   D. Mitosis in eukaryotes may have evolved from binary fission in bacteria

III. Regulation of the Cell Cycle
   A. A molecular control system drives the cell cycle
   B. Internal and external cues help regulate the cell cycle
   C. Cancer cells have escaped from cell-cycle controls

OBJECTIVES

After reading this chapter and attending lecture, the student should be able to:

1. Describe the structural organization of the genome.

2. Overview the major events of cell division that enable the genome of one cell to be passed on to two daughter cells.

3. Describe how chromosome number changes throughout the human life cycle.

4. List the phases of the cell cycle and describe the sequence of events that occurs during each phase.

5. List the phases of mitosis and describe the events characteristic of each phase.

6. Recognize the phases of mitosis from diagrams or micrographs.

7. Draw or describe the spindle apparatus including centrosomes, nonkinetochore microtubules, kinetochore microtubules, asters, and centrioles (in animal cells).

8. Describe what characteristic changes occur in the spindle apparatus during each phase of mitosis.

9. Explain the current models for poleward chromosomal movement and elongation of the cell's polar axis.

10. Compare cytokinesis in animals and plants.

11. Describe the process of binary fission in bacteria and how this process may have evolved to mitosis in eukaryotes.

12. Describe the roles of checkpoints, cyclin, Cdk, and MPF, in the cell-cycle control system.

13. Describe the internal and external factors which influence the cell-cycle control system.
14. Explain how abnormal cell division of cancerous cells differs from normal cell division.

KEY TERMS

cell cycle	chromosomes	kinetochore	growth factor
cell division	interphase	metaphase plate	density-dependent inhibition
genome	G_1 phase	cleavage furrow	anchorage dependence
somatic cell	S phase	cell plate	transformation
gametes	G_2 phase	binary fission	tumor
chromatin	prophase	cell-cycle control system	benign tumor
sister chromatids	prometaphase	checkpoint	malignant tumor
centromere	metaphase	G_0 phase	metastasis
mitosis	anaphase	cyclin	
cytokinesis	telophase	cyclin-dependent kinase	
mitotic (M) phase	mitotic spindle	MPF	

LECTURE NOTES

The ability to reproduce distinguishes living organisms from nonliving objects; this ability has a cellular basis.

All cells arise from preexisting cells. This fundamental principle, known as the cell doctrine, was originally postulated by Rudolf Virchow in 1858, and it provides the basis for the continuity of life.

A cell reproduces by undergoing a coordinated sequence of events in which it duplicates its contents and then divides in two. This cycle of duplication and division, known as the *cell cycle*, is the means by which all living things reproduce.

I.   **The Key Roles of Cell Division**

A. **Cell division functions in reproduction, growth, and repair**

Cells reproduce for many reasons.

- In unicellular organisms, the division of one cell to form two reproduces an entire organism (e.g., bacteria, yeast, *Amoeba*) (see Campbell, Figure 12.1a).
- In multicellular organisms, cell division allows:
  - Growth and development from the fertilized egg (see Campbell, Figure 12.1b)
  - Replacement of damaged or dead cells

Cell division is a finely controlled process that results in the distribution of identical hereditary material—DNA—to two daughter cells. A dividing cell:

- Precisely replicates its DNA
- Allocates the two copies of DNA to opposite ends of the cell
- Separates into two daughter cells containing identical hereditary information

B. **Cell division distributes identical sets of chromosomes to daughter cells**

The total hereditary endowment of a cell of a particular species is called its genome.

The genomes of some species are quite small (e.g., prokaryotes), while the genomes of other species are quite large (e.g., eukaryotes).

The replication, division, and distribution of the large genomes of eukaryotes is possible because the genomes are organized into multiple functional units called *chromosomes* (see Campbell, Figure 12.2).

Eukaryotic chromosomes have the following characteristics:
- They are supercoils of a DNA-protein complex called *chromatin*. Each chromosome consists of the following:
  - A single, long, double-stranded molecule of DNA, segments of which are called *genes*
  - Various proteins which serve to maintain the structure of the chromosome or are involved with the expression of genes, DNA replication, and DNA repair
- They exist in a characteristic number in different species (e.g., human somatic cells have 46); gamete cells (sperm or ova) possess half the number of chromosomes of somatic cells (e.g., human gametes have 23)
- The exist in different states at different stages of the cell cycle.
  - During interphase, the chromosomes are loosely folded; cannot be seen with a light microscope
  - During the mitotic phase, chromosomes are highly folded and condensed; can be seen with a light microscope

In preparation for eukaryotic cell division, the complete genome is duplicated. As a result of this duplication, each chromosome consists of two *sister chromatids*. The two chromatids possess identical copies of the chromosome's DNA and are initially attached to each other at a specialized region called the *centromere* (see Campbell, Figure 12.3).

Cell division usually proceeds in two sequential steps: nuclear division (*mitosis*) and division of the cytoplasm (*cytokinesis*). Not all cells undergo cytokinesis following mitosis.

In mitosis, the sister chromatids are pulled apart, and this results in the segregation of two sets of chromosomes, one set at each end of the cell.

In cytokinesis, the cytoplasm is divided and two separate daughter cells are formed, each containing a single nucleus with one set of chromosomes.

HUMAN LIFE CYCLE

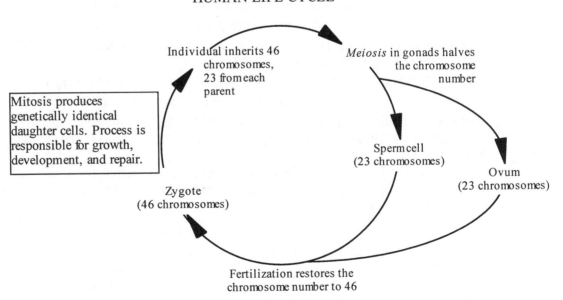

In plant cells, cytokinesis occurs by *cell plate* formation across the parent cell's midline (old metaphase plate).
- Golgi-derived vesicles move along microtubules to the cell's center, where they fuse into a disc-like cell plate.

- Additional vesicles fuse around the edge of the plate, expanding it laterally until its membranes touch and fuse with the existing parent cell's plasma membrane.
- A new cell wall forms as cellulose is deposited between the two membranes of the cell plate.

II.   The Mitotic Cell Cycle

A.   The mitotic phase alternates with interphase in the cell cycle: *an overview*

Cell division is just a portion of the life, or *cell cycle,* of a cell (see Campbell, Figure 12.4).

The cell cycle is a well-ordered sequence of events in which a cell duplicates its contents and then divides in two.

- Some cells go through repeated cell cycles.
- Other cells never or rarely divide once they are formed (e.g., vertebrate nerve and muscle cells).

The cell cycle alternates between the *mitotic (M) phase*, or dividing phase, and *interphase*, the nondividing phase:

- M phase, the shortest part of the cell cycle and the phase during which the cell divides, includes:
    1. *Mitosis* - Division of the nucleus
    2. *Cytokinesis* - Division of the cytoplasm
- *Interphase*, the nondividing phase, includes most of a cell's growth and metabolic activities.
    - Is about 90% of the cell cycle
    - Is a period of intense biochemical activity during which the cell grows and copies its chromosomes in preparation for cell division
    - Consists of three periods:
        1. *G_1 phase* - First growth phase (G stands for "gap")
        2. *S phase* - Synthesis phase occurs when DNA is synthesized as chromosomes are duplicated (S stands for "synthesis")
        3. *G_2 phase* - Second growth phase

Mitosis is unique to eukaryotes and may be an evolutionary adaptation for distributing a large amount of genetic material.

- Details may vary, but overall process is similar in most eukaryotes.
- It is a reliable process with only one error per 100,000 cell divisions.

Mitosis is a continuous process, but for ease of description, mitosis is usually divided into five stages: *prophase, prometaphase, metaphase, anaphase,* and *telophase* (see Campbell, Figures 12.5 and 12.9):

When cytokinesis occurs, it usually is concomitant with telophase of mitosis. The details of mitosis and cytokinesis follow (as exemplified by the pattern of cell division displayed by animal cells):

G₂ of interphase

A G₂ cell is characterized by:

- A well-defined nucleus bounded by a nuclear envelope

- One or more nucleoli

- Two centrosomes adjacent to the nucleus (formed earlier by replication of a single centrosome)

- In animals, a pair of centrioles in each centrosome

- In animals, a radial microtubular array (*aster*) around each pair of centrioles

- Duplicated chromosomes that cannot be distinguished individually due to loosely packed chromatin fibers. (Chromosomes were duplicated earlier in S phase.)

- See also Campbell, Figure 12.5

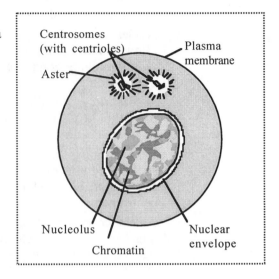

Prophase

In the nucleus:

- Nucleoli disappear

- Chromatin fibers condense into discrete, observable chromosomes, composed of two identical sister chromatids joined at the centromere.

In the cytoplasm:

- Mitotic spindle forms. It is composed of microtubules between the two *centrosomes* or microtubule-organizing centers.

- Centrosomes move apart, apparently propelled along the nuclear surface by lengthening of the microtubule bundles between them.

- See also Campbell, Figure 12.5

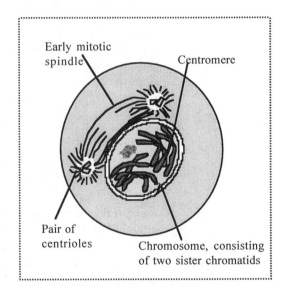

Prometaphase

During prometaphase:

- Nuclear envelope fragments, which allows microtubules to interact with the highly condensed chromosomes.
- *Spindle fibers* (bundles of microtubules) extend from each pole toward the cell's equator.
- Each chromatid now has a specialized structure, the *kinetochore*, located at the centromere region.
- *Kinetochore microtubules* become attached to the kinetochores and put the chromosomes into agitated motion.
- *Nonkinetochore microtubules* radiate from each centrosome toward the metaphase plate without attaching to chromosomes. Nonkinetochore microtubules radiating from one pole overlap with those from the opposite pole.
- See also Campbell, Figure 12.5

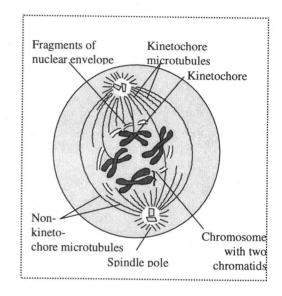

Metaphase

During metaphase:

- Centrosomes are positioned at opposite poles of the cell.
- Chromosomes move to the *metaphase plate*, the plane equidistant between the spindle poles.
- Centromeres of all chromosomes are aligned on the metaphase plate.
- The long axis of each chromosome is roughly at a right angle to the spindle axis.
- Kinetochores of sister chromatids face opposite poles, so identical chromatids are attached to kinetochore fibers radiating from opposite ends of the parent cell.
- Entire structure formed by nonkinetochore microtubules plus kinetochore microtubules is called the *spindle*.
- See also Campbell, Figure 12.6

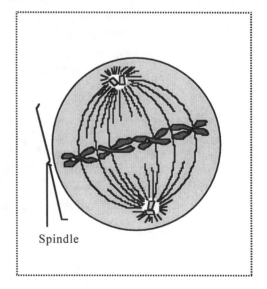

Anaphase

Anaphase is characterized by movement. It begins when paired centromeres of each chromosome move apart.

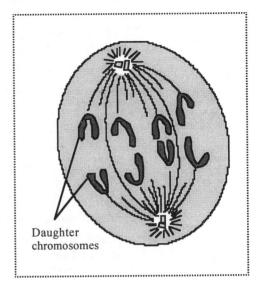

Daughter chromosomes

- Sister chromatids split apart into separate chromosomes and move towards opposite poles of the cell.
- Because kinetochore fibers are attached to the centromeres, the chromosomes move centromere first in a "V" shape.
- Kinetochore microtubules shorten at the kinetochore end as chromosomes approach the poles (see Campbell, Figure 12.7).
- Simultaneously, the poles of the cell move farther apart, elongating the cell.

At the end of anaphase, the two poles have identical collections of chromosomes.

Telophase and Cytokinesis

During telophase:

- Nonkinetochore microtubules further elongate the cell.
- Daughter nuclei begin to form at the two poles.
- Nuclear envelopes form around the chromosomes from fragments of the parent cell's nuclear envelope and portions of the endomembrane system.
- Nucleoli reappear.
- Chromatin fiber of each chromosome uncoils and the chromosomes become less distinct.

By the end of telophase:

- Mitosis, the equal division of one nucleus into two genetically identical nuclei, is complete.
- Cytokinesis has begun and the appearance of two separate daughter cells occurs shortly after mitosis is completed.

> A lecture on mitosis may not last the entire period if it is limited to just a description of mitotic stages. Though it may be tempting to continue with meiosis during the same class period, it is not recommended. Students easily confuse the two processes because they are somewhat similar, so it helps to allow some time for students to assimilate the mitosis material, before discussing meiosis. It is effective to summarize with a comparison of the two processes after the topic of meiosis has been discussed.

B. The mitotic spindle distributes chromosomes to daughter cells: *a closer look*

Many of the events of mitosis depend on the formation of a *mitotic spindle*. The mitotic spindle forms in the cytoplasm from *microtubules* and associated proteins.

- Microtubules of the cytoskeleton are partially disassembled during spindle formation.
  - Spindle microtubules are aggregates of two proteins, α- and β-tubulin.
  - Spindle microtubules elongate by the adding tubulin subunits at one end.
- The assembly of spindle microtubules begins in the *centrosome* or microtubule organizing center.

- In animal cells, a pair of centrioles is in the center of the centrosome, but there is evidence that centrioles are not essential for cell division:
  - If the centrioles of animal cells are destroyed with a laser microbeam, spindles still form and function during mitosis.
  - Plant centrosomes generally lack centrioles.

The chronology of mitotic spindle formation is as follows:

Interphase. The centrosome replicates to form two centrosomes located just outside the nucleus.

Prophase. The two centrosomes move farther apart.

- Spindle microtubules radiate from the centrosomes, elongating at the end away from their centrosome.

Prometaphase. By the end of prometaphase, the two centrosomes are at opposite poles and the chromosomes have moved to the cell's midline.

- Each chromatid of a replicated chromosome develops its own *kinetochore*, a structure of proteins and chromosomal DNA on the centromere. The chromosome's two distinct kinetochores face opposite directions.
- Some spindle microtubules attach to the kinetochores and are called kinetochore microtubules.
- Some spindle microtubules extend from the centrosomes and overlap with those radiating from the cell's opposite pole. These are called nonkinetochore microtubules.

Kinetochore microtubules interact to: (1) arrange the chromosomes so kinetochores face the poles and (2) align the chromosomes at the cell's midline.

The most stable arrangement occurs when sister kinetochores are attached by microtubules to opposite spindle poles.

- Initially, kinetochore microtubules from one pole may attach to a kinetochore, moving the chromosome toward that pole. This movement is checked when microtubules from the opposite pole attach to the chromosome's other kinetochore.
- The chromosome oscillates back and forth until it stabilizes and aligns at the cell's midline.
- Microtubules can remain attached to a kinetochore only if there is opposing tension from the other side. It is this opposing tension that stabilizes the microtubule-kinetochore connection and allows the proper alignment and movement of chromosomes at the cell's midline.

Metaphase. All the duplicated chromosomes align on the cell's midline, or *metaphase plate.*

Anaphase. The chromosome's centromeres split and the sister chromatids move as separate chromosomes toward opposite ends of the cell. The kinetochore and nonkinetochore microtubules direct the segregation of the chromosomes (see Campbell, Figure 12.7).

The kinetochore microtubules function in the poleward movement of chromosomes. Based on experimental evidence, the current model is that:

- Kinetochore microtubules shorten during anaphase by depolymerizing at their kinetochore ends; pulling the chromosomes poleward.
- The mechanism of this interaction between kinetochores and microtubules may involve microtubule-walking proteins similar to dynein that "walk" a chromosome along the shortening microtubules.

The function of the nonkinetochore microtubules:

- Nonkinetochore tubules elongate the whole cell along the polar axis during anaphase.

- These tubules overlap at the middle of the cell and slide past each other away from the cell's equator, reducing the degree of overlap.

- It is hypothesized that dynein cross-bridges may form between overlapping tubules to slide them past one another. Alternatively, motor molecules may link the microtubules to other cytoskeletal elements to drive the sliding.

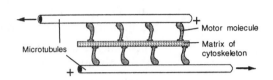

- ATP provides the energy for this endergonic process.

Telophase. At the end of anaphase, the duplicate sets of chromosomes are clustered at opposite ends of the elongated parent cell.

- Nuclei reform during telophase.

- Cytokinesis usually divides the cell's cytoplasm and is coincident with telophase of mitosis. In some exceptional cases, mitosis is not followed by cytokinesis (e.g., certain slime molds form multinucleated masses called *plasmodia*).

C. Cytokinesis divides the cytoplasm: *a closer look*

Cytokinesis, the process of cytoplasmic division, begins during telophase of mitosis. The process by which cytokinesis is accomplished differs in animal and plant cells. In animal cells, cytokinesis occurs by a process called *cleavage*:

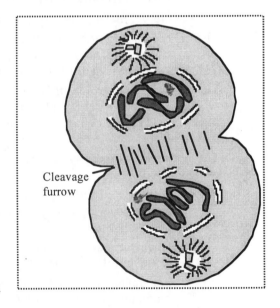

- First, a *cleavage furrow* forms as a shallow groove in the cell surface near the old metaphase plate (see Campbell, Figure 12.8).

- A contractile ring of actin microfilaments forms on the cytoplasmic side of the furrow; this ring contracts until it pinches the parent cell in two.

- Finally, the remaining mitotic spindle breaks, and the two cells become completely separate.

Campbell, Figure 12.9 shows mitosis in a plant cell.

D. Mitosis in eukaryotes may have evolved from binary fission in bacteria

Because prokaryotes (bacteria) are smaller and simpler than eukaryotes and because they preceded eukaryotes on Earth by billions of years, it is reasonable to suggest that the carefully orchestrated process of mitosis had its origins in prokaryotes. Prokaryotes contain:

- Most genes in a single circular chromosome composed of a double-stranded DNA molecule and associated proteins.

- Only about 1/1000 the DNA of eukaryotes, but prokaryotic chromosomes still contain a large amount of DNA relative to the small prokaryotic cell. Consequently, bacterial chromosomes are highly folded and packed within the cell.

Prokaryotes reproduce by *binary fission,* a process during which bacteria replicate their chromosomes and equally distribute copies between the two daughter cells (see Campbell, Figure 12.10).

- The chromosome is replicated; each copy remains attached to the plasma membrane at adjacent sites.
- Between the attachment sites the membrane grows and separates the two copies of the chromosome.
- The bacterium grows to about twice its initial size, and the plasma membrane pinches inward.
- A cell wall forms across the bacterium between the two chromosomes, dividing the original cell into two daughter cells.

Certain modern algae display unusual patterns of nuclear division which may represent intermediate stages between bacterial binary fission and eukaryotic mitosis (see Campbell, Figure 12.11).

III. Regulation of the Cell Cycle

A. A molecular control system drives the cell cycle

Normal growth, development and maintenance depend on the timing and rate of mitosis. Various cell types differ in their pattern of cell division; for example:

- Human skin cells divide frequently.
- Liver cells only divide in appropriate situations, such as wound repair.
- Nerve, muscle and other specialized cells do not divide in mature humans.

The cell cycle is coordinated by the *cell-cycle control system,* a molecular signaling system which cyclically switches on the appropriate parts of the cell-cycle machinery and then switches them off (see Campbell, Figure 12.13).

The cell-cycle control system consists of a cell-cycle molecular clock and a set of *checkpoints,* or switches, that ensure that appropriate conditions have been met before the cycle advances. When the control system malfunctions, as will be seen later, cancer may result.

The cell-cycle control system has checkpoints in the G_1, G_2, and M phases of the cell cycle.

- Signals registered at the checkpoints report the status of various cellular conditions (e.g., Is the environment favorable? Is the cell big enough? Are all DNA replicated?
- Checkpoints integrate a variety of internal (intracellular) and external (extracellular) information.
- For many cells, the G_1 checkpoint (known as the "restriction point" in mammalian cells) is the most important.
  - A go-ahead signal usually indicates that the cell will complete the cycle and divide.
  - In the absence of a go-ahead signal, the cell may exit the cell cycle, switching to the nondividing state called *G_0 phase.*
  - Many cells of the human body are in the G_0 phase. Muscle and nerve cells will remain in G_0 until they die. Liver cells may be recruited back to the cell cycle under certain cues, such as growth factors.

The ordered sequence of cell cycle events is synchronized by rhythmic changes in the activity of certain *protein kinases.*

- Protein kinases are enzymes that catalyze the transfer of a phosphate group from ATP to a target protein.

- Phosphorylation, in turn, induces a conformational change that either activates or inactivates a target protein.
- Changes in target proteins affect the progression through the cell cycle.

Cyclical changes in kinase activity are controlled by another class of regulatory proteins called *cyclins*.

- These regulatory proteins are named cyclins, because their concentrations change cyclically during the cell cycle.
- Protein kinases that regulate cell cycles are *cyclin-dependent kinases* (*Cdks*); they are active only when attached to a particular cyclin.
- Even though Cdk concentration stays the same throughout the cell cycle, its activity changes in response to the changes in cyclin concentration (see Campbell, Figure 12.14a).

An example of a cyclin-Cdk complex is *MPF (maturation promoting factor)*, which controls the cell's progress through the G_2 checkpoint to mitosis (see Campbell, Figure 12.14b).

Cyclin's rhythmic changes in concentration regulate MPF activity, and thus acts as a mitotic clock that regulates the sequential changes in a dividing cell.

- Cyclin is produced at a uniform rate throughout the cell cycle, and it accumulates during interphase.
- Cyclin combines with Cdk to form active MPF, so as cyclin concentration rises and falls, the amount of active MPF changes in a similar way.
- MPF phosphorylates proteins that participate in mitosis and initiates the following process:
  - Chromosome condensation during prophase
  - Nuclear envelope dispersion during prometaphase
- In the latter half of mitosis, MPF activates proteolytic enzymes.
  - The proteolytic enzymes destroy cyclin which leads to the reduction of MPF activity (the Cdk portion of MPF is not degraded).
  - The proteolytic enzymes also are involved in driving the cell cycle past the M-phase checkpoint, which controls the onset of anaphase.
- Continuing cyclin synthesis raises the concentration again during interphase. This newly synthesized cyclin binds to Cdk to form MPF, and mitosis begins again.

Rhythmic changes in different cyclin-Cdk complexes regulate other cell cycle stages.

B. Internal and external cues help regulate the cell cycle

The cell-cycle control system integrates a variety of internal (intracellular) and external (extracellular) information. Knowledge of the chemical signaling pathways that transduce this information into modulation of the cell-cycle machinery is just emerging.

The kinetochores provide internal cues that signal the M-phase checkpoint about the status of chromosome-spindle interactions. All chromosomes must be attached to spindle microtubules before the M-phase checkpoint allows the cycles to proceed to anaphase. This ensures that daughter cells do not end up with missing or extra chromosomes.

- Kinetochores not attached to spindles trigger a signaling pathway that keeps the anaphase promoting complex (APC) in an inactive state.
- Once all kinetochores are attached, the wait signal stops, and the APC complex becomes active. The APC complex contains proteolytic enzymes which break down cyclin.

Using tissue culture, researchers have identified several external factors, both chemical and physical, that can influence cell division:

1. **Chemical factors**

   - If essential nutrients are left out of the culture medium, cells will not divide.

   - Specific regulatory substances called *growth factors* are necessary for most cultured mammalian cells to divide, even if all other conditions are favorable. For example:

     - Binding of platelet-derived growth factor (PDGF) to cell membrane receptors, stimulates cell division in fibroblasts. This regulation probably occurs not only in cell culture, but in the animal's body as well—a response that helps heal wounds.

     - Other cell types may have membrane receptors for different growth factors or for different combinations of several growth factors.

2. **Physical factors**

   - Crowding inhibits cell division in a phenomenon called *density-dependent inhibition*. Cultured cells stop dividing when they form a single layer on the container's inner surface. If some cells are removed, those bordering the open space divide again until the vacancy is filled (see Campbell, Figure 12.15a).

     - Density-dependent inhibition is apparently a consequence of the fact that quantities of nutrients and growth regulators may be insufficient to support cell division, if cell density is too high.

     - Most animal cells also exhibit anchorage dependence. To divide, normal cells must adhere to a substratum, such as the surface of a culture dish or the extracellular matrix of a tissue. anchorage is signaled to the cell-cycle control system via pathways involving membrane proteins and elements of the cytoskeleton that are linked to them.

   - Density-dependent and anchorage-dependent inhibition probably occur in the body's tissues as well as in cell culture. Cancer cells are abnormal and do not exhibit density-dependent or anchorage-dependent inhibition.

C. Cancer cells have escaped from cell-cycle controls

Cancer cells do not respond normally to the body's control mechanisms. They divide excessively, invade other tissues and, if unchecked, can kill the whole organism.

   - Cancer cells in culture do not stop growing in response to cell density (density-dependent inhibition); they do not stop dividing when growth factors are depleted (see Campbell, Figure 12.15b).

   - Cancer cells may make growth factors themselves.

   - Cancer cells may have an abnormal growth factor signaling system.

   - Cancer cells in culture are immortal in that they continue to divide indefinitely, as long as nutrients are available. Normal mammalian cells in culture divide only about 20 to 50 times before they stop.

   - Cancer cells that stop dividing do so at random points in the cycle instead of at checkpoints.

Abnormal cells which have escaped normal cell-cycle controls are the products of mutate or *transformed* normal cells.

The immune system normally recognizes and destroys transformed cells that have converted from normal to cancer cells.

   - If abnormal cells evade destruction, they may proliferate to form a *tumor*, an unregulated growing mass of cells within otherwise normal tissue.

- If the cells remain at this original site, the mass is called a *benign tumor* and can be completely removed by surgery.
- A *malignant tumor* is invasive enough to impair normal function of one or more organs of the body. Only an individual with a malignant tumor is said to have cancer (see Campbell, Figure 12.16).

Properties of malignant (cancerous) tumors include:

- Anomalous cell cycle; excessive proliferation
- May have unusual numbers of chromosomes
- May have aberrant metabolism
- Lost attachments to neighboring cells and extracellular matrix—usually a consequence of abnormal cell surface changes.

Cancer cells also may separate from the original tumor and spread into other tissues, possibly entering the blood and lymph vessels of the circulatory system.

- Migrating cancer cells can invade other parts of the body and proliferate to form more tumors.
- This spread of cancer cells beyond their original sites is called *metastasis*.
- If a tumor metastasizes, it is usually treated with radiation and chemotherapy, which is especially harmful to actively dividing cells.

Researchers are beginning to understand how a normal cell is transformed into a cancerous one. Although the causes of cancer may be diverse, cellular transformation always involves the alteration of genes that somehow influence the cell-cycle control system.

REFERENCES

Alberts, B., D. Bray, A. Johnson, J. Lewis, M. Raff, K. Roberts and P. Walter. *Essential Cell Biology: An Introduction to the Molecular Biology of the Cell.* New York: Garland, 1997.

Alberts, B., D. Bray, J. Lewis, M. Raff, K. Roberts and J.D. Watson. *Molecular Biology of the Cell.* 2nd ed. New York: Garland, 1989.

Becker, W.M., J.B. Reece and M.F. Puente. *The World of the Cell.* 3rd ed. Redwood City, California: Benjamin/Cummings, 1996.

Campbell, N., et al. *Biology.* 5th ed. Menlo Park, California: Benjamin/Cummings, 1998.

Kleinsmith, L.J. and V.M. Kish. *Principles of Cell Biology.* New York: Harper and Row, Publ., 1988.

Varmus, H. and R.A. Weinberg. *Genes and the Biology of Cancer.* New York: Scientific American Books, 1993.

CHAPTER 13
MEIOSIS AND SEXUAL LIFE CYCLES

OUTLINE

I.   An Introduction to Heredity
    A.   Offspring acquire genes from parents by inheriting chromosomes
    B.   Like begets like, more or less: a comparison of asexual versus sexual reproduction
II. The Role of Meiosis in Sexual Life Cycles
    A.   Fertilization and meiosis alternate in sexual life cycles
    B.   Meiosis reduces chromosome number from diploid to haploid: *a closer look*
III. Origins of Genetic Variation
    A.   Sexual life cycles produce genetic variation among offspring
    B.   Evolutionary adaptation depends on a population's genetic variation

OBJECTIVES

After reading this chapter and attending lecture, the student should be able to:

1.   Explain why organisms only reproduce their own kind, and why offspring more closely resemble their parents than unrelated individuals of the same species.
2.   Explain what makes heredity possible.
3.   Distinguish between asexual and sexual reproduction.
4.   Diagram the human life cycle and indicate where in the human body that mitosis and meiosis occur; which cells are the result of meiosis and mitosis; and which cells are haploid.
5.   Distinguish among the life cycle patterns of animals, fungi, and plants.
6.   List the phases of meiosis I and meiosis II and describe the events characteristic of each phase.
7.   Recognize the phases of meiosis from diagrams or micrographs.
9.   Describe the process of synapsis during prophase I, and explain how genetic recombination occurs.
10. Describe key differences between mitosis and meiosis; explain how the end result of meiosis differs from that of mitosis.
11. Explain how independent assortment, crossing over, and random fertilization contribute to genetic variation in sexually reproducing organisms.
12. Explain why inheritable variation was crucial to Darwin's theory of evolution.
13. List the sources of genetic variation.

KEY TERMS

heredity
variation
genetics
gene
asexual reproduction
clone
sexual reproduction
life cycle
somatic cell

karyotype
homologous
   chromosomes
sex chromosomes
autosome
gamete
haploid cell
fertilization
syngamy

zygote
diploid cells
meiosis
alternation of
   generations
sporophyte
spores
gametophyte
meiosis I

meiosis II
synapsis
tetrad
chiasmata
chiasma
crossing over

LECTURE NOTES

Reproduction is an emergent property associated with life. The fact that organisms reproduce their own kind is a consequence of heredity.

Heredity = Continuity of biological traits from one generation to the next
- Results from the transmission of hereditary units, or *genes*, from parents to offspring.
- Because they share similar genes, offspring more closely resemble their parents or close relatives than unrelated individuals of the same species.

Variation = Inherited differences among individuals of the same species
- Though offspring resemble their parents and siblings, they also diverge somewhat as a consequence of inherited differences among them.
- The development of *genetics* in this century has increased our understanding about the mechanisms of variation and heredity.

Genetics = The scientific study of heredity and hereditary variation.

> Beginning students often compartmentalize their knowledge, which makes it difficult to transfer and apply information learned in one context to a new situation. Be forewarned that unless you point it out, some students will never make the connection that meiosis, sexual reproduction, and heredity are all aspects of the same process.

I. **An Introduction to Heredity**

   A. **Offspring acquire genes from parents by inheriting chromosomes**

     DNA = Type of nucleic acid that is a polymer of four different kinds of nucleotides.

     Genes = Units of hereditary information that are made of *DNA* and are located on *chromosomes*.
- Have specific sequences of nucleotides, the monomers of DNA
- Most genes program cells to synthesize specific proteins; the action of these proteins produce an organism's inherited traits.

     Inheritance is possible because:
- DNA is precisely replicated producing copies of genes that can be passed along from parents to offspring.
- Sperm and ova carrying each parent's genes are combined in the nucleus of the fertilized egg.

     The actual transmission of genes from parents to offspring depends on the behavior of *chromosomes*.

Chromosomes = Organizational unit of heredity material in the nucleus of eukaryotic organisms

- Consist of a single long DNA molecule (double helix) that is highly folded and coiled along with proteins
- Contain genetic information arranged in a linear sequence
- Contain hundreds or thousands of genes, each of which is a specific region of the DNA molecule, or *locus*

Locus = Specific location on a chromosome that contains a gene

- Each species has a characteristic number of chromosomes; humans have 46 (except for their reproductive cells).

B. Like begets like, more or less: a comparison of asexual versus sexual reproduction

Asexual Reproduction	Sexual Reproduction
Single individual is the sole parent.	Two parents give rise to offspring.
Single parent passes on *all* its genes to its offspring.	Each parent passes on *half* its genes, to its offspring.
Offspring are genetically identical to the parent.	Offspring have a unique combination of genes inherited from both parents.
Results in a *clone*, or genetically identical individual. Rarely, genetic differences occur as a result of *mutation*, a change in DNA (see Campbell, Figure 13.1).	Results in greater genetic variation; offspring vary genetically from their siblings and parents (see Campbell, Figure 13.2).

What generates this genetic variation during sexual reproduction? The answer lies in the process of meiosis.

III. The Role of Meiosis in Sexual Life Cycles

A. Fertilization and meiosis alternate in sexual life cycles

1. The human life cycle

Follows the same basic pattern found in all sexually reproducing organisms; *meiosis* and *fertilization* result in alternation between the haploid and diploid condition (see Campbell, Figure 13.3).

Life cycle = Sequence of stages in an organism's reproductive history, from conception to production of its own offspring

Somatic cell = Any cell other than a sperm or egg cell

- Human somatic cells contain 46 chromosomes distinguishable by differences in size, position of the centromere, and staining or banding pattern.
- Using these criteria, chromosomes from a photomicrograph can be matched into *homologous* pairs and arranged in a standard sequence to produce a karyotype.

Karyotype = A display or photomicrograph of an individual's somatic-cell metaphase chromosomes that are arranged in a standard sequence. (See Campbell, Methods Box: *Preparation of a Karyotype*)

- Human karyotypes are often made with lymphocytes.
- Can be used to screen for chromosomal abnormalities

Homologous chromosomes (homologues) = A pair of chromosomes that have the same size, centromere position, and staining pattern.

- With one exception, homologues carry the same genetic loci.

- Homologous *autosomes* carry the same genetic loci; however, human *sex chromosomes* carry different loci even though they pair during prophase of meiosis I.

Autosome = A chromosome that is not a sex chromosome.

Sex chromosome = Dissimilar chromosomes that determine an individual's sex

- Females have a homologous pair of X chromosomes.
- Males have one X and one Y chromosome.
- Thus, humans have 22 pairs of autosomes and one pair of sex chromosomes.

Chromosomal pairs in the human karyotype are a result of our sexual origins.

- One homologue is inherited from each parent.
- Thus, the 46 somatic cell chromosomes are actually two sets of 23 chromosomes; one a maternal set and the other a paternal set.
- Somatic cells in humans and most other animals are *diploid*.

Diploid = Condition in which cells contain two sets of chromosomes; abbreviated as 2*n*

Haploid = Condition in which cells contain one set of chromosomes; it is the chromosome number of *gametes* and is abbreviated as *n*

Gamete = A haploid reproductive cell

- *Sperm cells* and *ova* are gametes, and they differ from somatic cells in their chromosome number. Gametes only have one set of chromosomes.
- Human gametes contain a single set of 22 autosomes and one sex chromosome (either an X or a Y).
- Thus, the haploid number of humans is 23.

The diploid number is restored when two haploid gametes unite in the process of *fertilization*. Sexual intercourse allows a haploid sperm cell from the father to reach and fuse with an ovum from the mother.

Fertilization = The union of two gametes to form a *zygote*

Zygote = A diploid cell that results from the union of two haploid gametes

- Contains the maternal and parental haploid sets of chromosomes from the gametes and is diploid (2*n*)
- As humans develop from a zygote to sexually mature adults, the zygote's genetic information is passed with precision to all somatic cells by mitosis.

Gametes are the only cells in the body that are *not* produced by mitosis.

- Gametes are produced in the ovaries or testes by the process of *meiosis*.
- *Meiosis* is a special type of cell division that produces haploid cells and compensates for the doubling of chromosome number that occurs at fertilization.
- Meiosis in humans produces sperm cells and ova which contain 23 chromosomes.
- When fertilization occurs, the diploid condition (2*n*=46) is restored in the zygote.

2. **The variety of sexual life cycles**

Alternation of meiosis and fertilization is common to all sexually reproducing organisms; however, the timing of these two events in the life cycle varies among species. There are three basic patterns of sexual life cycles (see Campbell, Figure 13.4):

a. **Animal**

In animals, including humans,

gametes are the only haploid cells.

- Meiosis occurs during gamete production. The resulting gametes undergo no further cell division before fertilization.
- Fertilization produces a diploid zygote that divides by mitosis to produce a diploid multicellular animal.

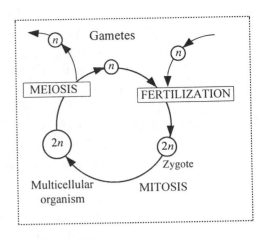

b. Fungi and some protists

In many fungi and some protists, the only diploid stage is the zygote.

- Meiosis occurs immediately after the zygote forms.
- Resulting haploid cells divide by *mitosis* to produce a haploid multicellular organism.
- Gametes are produced by *mitosis* from the already haploid organism.

c. Plants and some algae

Plants and some species of algae alternate between multicellular haploid and diploid generations.

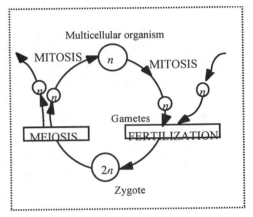

- This type of life cycle is called an *alternation of generations*.
- The multicellular diploid stage is called a *sporophyte*, or spore-producing plant. Meiosis in this stage produces haploid cells called *spores*.
- Haploid spores divide mitotically to generate a multicellular haploid stage called a *gametophyte*, or gamete-producing plant.

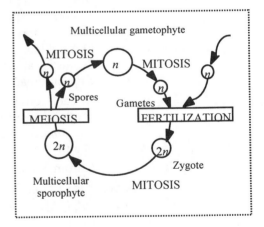

- Haploid gametophytes produce gametes by mitosis.
- Fertilization produces a diploid zygote which develops into the next sporophyte generation.

B. Meiosis reduces chromosome number from diploid to haploid: *a closer look*

Meiosis and sexual reproduction significantly contribute to genetic variation among offspring.

Meiosis includes steps that closely resemble corresponding steps in mitosis (see Campbell, Figure 13.5).

- Like mitosis, meiosis is preceded by replication of the chromosomes.
- Meiosis differs from mitosis in that this single replication is followed by two consecutive cell divisions: *meiosis I* and *meiosis II*.

- These cell divisions produce *four* daughter cells instead of two as in mitosis.
- The resulting daughter cells have *half* the number of chromosomes as the original cell; whereas, daughter cells of mitosis have the same number of chromosomes as the parent cell.
- Campbell, Figure 13.6 shows mitosis and meiosis in animals.

The stages of meiotic cell division:

Interphase I. Interphase I precedes meiosis.

- Chromosomes replicate as in mitosis.
- Each duplicated chromosome consists of two identical sister chromatids attached at their centromeres.
- Centriole pairs in animal cells also replicate into two pairs.

Meiosis I. This cell division segregates the two chromosomes of each homologous pair and reduces the chromosome number by one-half. It includes the following four phases:

Prophase I. This is a longer and more complex process than prophase of mitosis.

- Chromosomes condense.
- *Synapsis* occurs. During this process, homologous chromosomes come together as pairs.
- Chromosomes condense further until they are distinct structures that can be seen with a microscope. Since each chromosome has two chromatids, each homologous pair in synapsis appears as a complex of four chromatids or a *tetrad*.
- In each tetrad, sister chromatids of the same chromosome are attached at their centromeres. Nonsister chromatids are linked by X-shaped *chiasmata*, sites where homologous strand exchange or *crossing-over* occurs.
- Chromosomes thicken further and detach from the nuclear envelope.

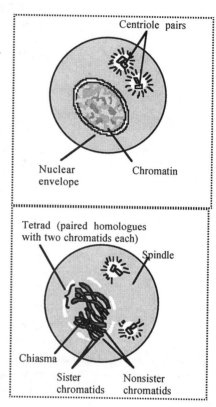

As prophase I continues, the cell prepares for nuclear division.

- Centriole pairs move apart and spindle microtubules form between them.
- Nuclear envelope and nucleoli disperse.
- Chromosomes begin moving to the metaphase plate, midway between the two poles of the spindle apparatus.
- Prophase I typically occupies more than 90% of the time required for meiosis.

Metaphase I. Tetrads are aligned on the metaphase plate.

- Each synaptic pair is aligned so that centromeres of homologues point toward opposite poles.
- Each homologue is thus attached to kinetochore microtubules emerging from the pole it faces, so that the two homologues are destined to separate in anaphase and move toward opposite poles.

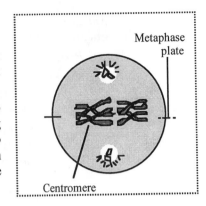

Anaphase I. Homologues separate and are moved toward the poles by the spindle apparatus.

- Sister chromatids remain attached at their centromeres and move as a unit toward the same pole, while the homologue moves toward the opposite pole.
- This differs from mitosis during which chromosomes line up individually on the metaphase plate (rather than in pairs) and sister chromatids are moved apart toward opposite poles of the cell.

Telophase I and Cytokinesis. The spindle apparatus continues to separate homologous chromosome pairs until the chromosomes reach the poles.

- Each pole now has a haploid set of chromosomes that are each still composed of two sister chromatids attached at the centromere.
- Usually, cytokinesis occurs simultaneously with telophase I, forming two haploid daughter cells. *Cleavage furrows* form in animal cells, and cell plates form in plant cells.
- In some species, nuclear membranes and nucleoli reappear, and the cell enters a period of *interkinesis* before meiosis II. In other species, the daughter cells immediately prepare for meiosis II.
- Regardless of whether a cell enters interkinesis, *no DNA replication occurs before meiosis II.*

Meiosis II. This second meiotic division separates sister chromatids of each chromosome.

Prophase II

- If the cell entered interkinesis, the nuclear envelope and nucleoli disperse.
- Spindle apparatus forms and chromosomes move toward the metaphase II plate.

Metaphase II

- Chromosomes align singly on the metaphase plate.
- Kinetochores of sister chromatids *point toward opposite poles.*

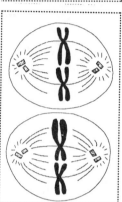

Anaphase II

- Centromeres of sister chromatids separate.
- Sister chromatids of each pair (now individual chromosomes) move toward opposite poles of the cell.

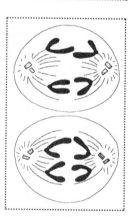

Telophase II and Cytokinesis

- Nuclei form at opposite poles of the cell.
- Cytokinesis occurs producing four haploid daughter cells.

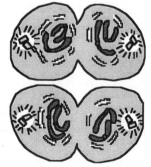

Haploid daughter cells

1. Mitosis and meiosis compared

> Spending class time on a comparison of mitosis and meiosis is really worth the effort. It not only brings closure to the topic, but also provides an opportunity to check for understanding. One check is to ask students to identify unlabeled diagrams of various stages in mitosis and meiosis. The ability to distinguish metaphase of mitosis from metaphase of meiosis I, is particularly diagnostic of student understanding.
>
> If you are fortunate enough to have video capability in your classroom, you can show moving sequences of mitosis and meiosis. The fact that these are dynamic processes involving chromosomal movement is not a trivial point, but is often lost in the course of a lecture where the only visuals are drawings or micrographs.

Though the processes of mitosis and meiosis are similar in some ways, there are some key differences (see Campbell, Figure 13.7):

- *Meiosis is a reduction division.* Cells produced by mitosis have the same number of chromosomes as the original cell, whereas cells produced by meiosis have *half* the number of chromosomes as the parent cell.
- *Meiosis creates genetic variation.* Mitosis produces *two* daughter cells genetically identical to the parent cell and to each other. Meiosis produces *four* daughter cells genetically different from the parent cell and from each other.
- *Meiosis is two successive nuclear divisions.* Mitosis, on the other hand, is characterized by just one nuclear division.

Comparison of Meiosis I and Mitosis

	Meiosis I	Mitosis
Prophase	*Synapsis* occurs to form tetrads. *Chiasmata* appear as evidence that crossing over has occurred.	Neither synapsis nor crossing over occurs.
Metaphase	Homologous pairs (tetrads) align on the metaphase plate.	Individual chromosomes align on the metaphase plate.
Anaphase	Meiosis I separates pairs of chromosomes. Centromeres do *not* divide and sister chromatids stay together. Sister chromatids of each chromosome move to the *same* pole of the cell; only the homologues separate.	Mitosis separates sister chromatids of individual chromosomes. Centromeres divide and sister chromatids move to *opposite* poles of the cell.

Meiosis II is virtually identical in mechanism to mitosis, separating sister chromatids.

III. Origins of Genetic Variation

A. Sexual life cycles produce genetic variation among offspring

Meiosis and fertilization are the primary sources of genetic variation in sexually reproducing organisms. Sexual reproduction contributes to genetic variation by:

- Independent assortment
- Crossing over during prophase I of meiosis

- Random fusion of gametes during fertilization

1. Independent assortment of chromosomes

At metaphase I, each homologous pair of chromosomes aligns on the metaphase plate. Each pair consists of one maternal and one paternal chromosome.

- The orientation of the homologous pair to the poles is random, so there is a 50-50 chance that a particular daughter cell produced by meiosis I will receive the maternal chromosome of a homologous pair, and a 50-50 chance that it will receive the paternal chromosome.
- Each homologous pair of chromosomes orients independently of the other pairs at metaphase I; thus, the first meiotic division results in *independent assortment* of maternal and paternal chromosomes (see Campbell, Figure 13.8)
- A gamete produced by meiosis contains just one of all the possible combinations of maternal and paternal chromosomes.

Independent assortment = The random distribution of maternal and paternal homologues to the gametes. (In a more specific sense, assortment refers to the random distribution of genes located on different chromosomes.)

- Since each homologous pair assorts independently from all the others, the process produces 2^n possible combinations of maternal and paternal chromosomes in gametes, where n is the haploid number.
- In humans, the possible combinations would be 2^{23}, or about eight million.
- Thus, each human gamete contains one of eight million possible assortments of chromosomes inherited from that person's mother and father.
- Genetic variation results from this reshuffling of chromosomes, because the maternal and paternal homologues will carry different genetic information at many of their corresponding loci.

2. Crossing over

Another mechanism that increases genetic variation is the process of *crossing over*, during which homologous chromosomes exchange genes.

Crossing over = The exchange of genetic material between homologues; occurs during prophase of meiosis I. This process:

- Occurs when homologous portions of two nonsister chromatids trade places. During prophase I, X-shaped *chiasmata* become visible at places where this homologous strand exchange occurs.
- Produces chromosomes that contain genes from *both* parents.
- In humans, there is an average of two or three crossovers per chromosome pair.
- Synapsis during prophase I is precise, so that homologues align gene by gene. The exact mechanism of synapsis is still unknown, but involves a protein apparatus, the *synaptonemal complex*, that joins the chromosomes closely together.
- Campbell, Figure 13.9 shows the results of crossing over during meiosis.

3. Random fertilization

Random fertilization is another source of genetic variation in offspring.

- In humans, when individual ovum representative of one of eight million possible chromosome combinations is fertilized by a sperm cell, which also represents one of eight million possibilities, the resulting zygote can have one of 64 trillion possible diploid combinations (without considering variations from crossing over!).

B. Evolutionary adaptation depends on a population's genetic variation

Heritable variation is the basis for Charles Darwin's theory that natural selection is the mechanism for evolutionary change. Natural selection:

- Increases the frequency of heritable variations that favor the reproductive success of some individuals over others
- Results in *adaptation*, the accumulation of heritable variations that are favored by the environment
- In the face of environmental change, genetic variation increases the likelihood that some individuals in a population will have heritable variations that help them cope with the new conditions.

There are two sources of genetic variation:

1. Sexual reproduction. Results from independent assortment in meiosis I, crossing over in prophase of meiosis I, and random fusion of gametes during fertilization.
2. Mutation, which are random and relatively rare structural changes made during DNA replication in a gene could result from mistakes.

REFERENCES

Alberts, B., D. Bray, J. Lewis, M. Raff, K. Roberts and J.D. Watson. *Molecular Biology of the Cell*. 3rd ed. New York: Garland, 1994.

Becker, W.M., J.B. Reece, and M.F. Puente. The World of the Cell. 3rd. ed. Redwood City, California: Benjamin/Cummings, 1996.

Campbell, N, et al. *Biology*. 5th ed. Menlo Park, California: Benjamin/Cummings, 1998.

Kleinsmith, L.J. and V.M. Kish. *Principles of Cell Biology*. New York: Harper and Row, Publ., 1988.

CHAPTER 14
MENDEL AND THE GENE IDEA

OUTLINE

I. Gregor Mendel's Discoveries
    A. Mendel brought an experimental and quantitative approach to genetics: *science as a process*
    B. By the law of segregation, the two alleles for a character are packaged into separate gametes
    C. By the law of independent assortment, each pair of alleles segregates into gametes independently
    D. Mendelian inheritance reflects rules of probability
    E. Mendel discovered the particulate behavior of genes: *a review*
II. Extending Mendelian Genetics
    A. The relationship between genotype and phenotype is rarely simple
III. Mendelian Inheritance in Humans
    A. Pedigree analysis reveals Mendelian patterns in human inheritance
    B. Many human disorders follow Mendelian patterns of inheritance
    C. Technology is providing new tools for genetic testing and counseling

OBJECTIVES

After reading this chapter and attending lecture, the student should be able to:

1. Describe the favored model of heredity in the 19th century prior to Mendel, and explain how this model was inconsistent with observations.
2. Explain how Mendel's hypothesis of inheritance differed from the blending theory of inheritance.
3. List several features of Mendel's methods that contributed to his success.
4. List four components of Mendel's hypothesis that led him to deduce the law of segregation.
5. State, in their own words, Mendel's law of segregation.
6. Use a Punnett square to predict the results of a monohybrid cross and state the phenotypic and genotypic ratios of the F_2 generation.
7. Distinguish between genotype and phenotype; heterozygous and homozygous; dominant and recessive.
8. Explain how a testcross can be used to determine if a dominant phenotype is homozygous or heterozygous.
9. Define random event, and explain why it is significant that allele segregation during meiosis and fusion of gametes at fertilization are random events.
10. Use the rule of multiplication to calculate the probability that a particular F_2 individual will be homozygous recessive or dominant.

11.   Given a Mendelian cross, use the rule of addition to calculate the probability that a particular F_2 individual will be heterozygous.

12.   Describe two alternate hypotheses that Mendel considered for how two characters might segregate during gamete formation, and explain how he tested those hypotheses.

13.   State, in their own words, Mendel's law of independent assortment.

14.   Use a Punnett square to predict the results of a dihybrid cross and state the phenotypic and genotypic ratios of the F_2 generation.

15.   Using the laws of probability, predict from a trihybrid cross between two individuals that are heterozygous for all three traits, what expected proportion of the offspring would be:

   a.   Homozygous for the three dominant traits
   b.   Heterozygous for all three traits
   c.   Homozygous recessive for two specific traits and heterozygous for the third

16.   Give an example of incomplete dominance and explain why it is not evidence for the blending theory of inheritance.

17.   Explain how the phenotypic expression of the heterozygote is affected by complete dominance, incomplete dominance and codominance.

18.   Describe the inheritance of the ABO blood system and explain why the I^A and I^B alleles are said to be *codominant*.

19.   Define and give examples of pleiotropy.

20.   Explain, in their own words, what is meant by "one gene is epistatic to another."

21.   Explain how epistasis affects the phenotypic ratio for a dihybrid cross.

22.   Describe a simple model for polygenic inheritance, and explain why most polygenic characters are described in quantitative terms.

23.   Describe how environmental conditions can influence the phenotypic expression of a character.

24.   Given a simple family pedigree, deduce the genotypes for some of the family members.

25.   Describe the inheritance and expression of cystic fibrosis, Tay-Sachs disease, and sickle-cell disease.

26.   Explain how a lethal recessive gene can be maintained in a population.

27.   Explain why consanguinity increases the probability of homozygosity in offspring.

28.   Explain why lethal dominant genes are much more rare than lethal recessive genes.

29.   Give an example of a late-acting lethal dominant in humans and explain how it can escape elimination.

30.   Explain how carrier recognition, fetal testing and newborn screening can be used in genetic screening and counseling.

KEY TERMS

character	dominant allele	law of independent	polygenic inheritance
trait	recessive allele	assortment	norm of reaction
true-breeding	law of segregation	incomplete dominance	multifactorial
hybridization	homozygous	complete dominance	carriers
monohybrid cross	heterozygous	codominance	cystic fibrosis
P generation	phenotype	multiple alleles	Tay-Sachs disease
F_1 generation	genotype	pleiotropy	sickle-cell disease
F_2 generation	testcross	epistasis	Huntington's disease
alleles	dihybrid cross	quantitative character	

LECTURE NOTES

Listed below are a few suggestions for teaching Mendelian genetics:

1. There is a certain baseline working vocabulary that students need in order to follow your lecture, understand the text and solve problems. It is more economical to recognize this fact and just begin with some "definitions you should know." Once that is done, you can use the terms in context during the lecture and focus attention on the major points rather than on defining terms.

2. Demonstrating how to work a Punnett square and how to solve genetics problems is obviously necessary. But your students will learn best if they actively participate in the process. You can structure opportunities for students to solve problems during lecture and then let them participate in the explanation. If time does not allow this, it is highly recommended that there be an additional recitation or problem-solving session outside of class.

3. After Mendel's laws of segregation and independent assortment have been introduced, it is extremely useful to put up a transparency of meiosis and ask students to identify where in meiosis that segregation and assortment occur. Many students will not make this connection on their own.

I.   Gregor Mendel's Discoveries

Based upon their observations from ornamental plant breeding, biologists in the 19th century realized that both parents contribute to the characteristics of offspring. Before Mendel, the favored explanation of heredity was the *blending theory*.

Blending theory of heredity = Pre-Mendelian theory of heredity proposing that hereditary material from each parent mixes in the offspring; once blended like two liquids in solution, the hereditary material is inseparable and the offspring's traits are some intermediate between the parental types. According to this theory:

- Individuals of a population should reach a uniform appearance after many generations.
- Once hereditary traits are blended, they can no longer be separated out to appear again in later generations.

This blending theory of heredity was inconsistent with the observations that:

- Individuals in a population do not reach a uniform appearance; inheritable variation among individuals is generally preserved.
- Some inheritable traits skip one generation only to reappear in the next.

Modern genetics began in the 1860s when Gregor Mendel, an Augustinian monk, discovered the fundamental principles of heredity. Mendel's great contribution to modern genetics was to replace the blending theory of heredity with the *particulate theory of heredity*.

Particulate theory of heredity = Gregor Mendel's theory that parents transmit to their offspring discrete inheritable factors (now called genes) that remain as separate factors from one generation to the next.

A. Mendel brought an experimental and quantitative approach to genetics: *science as a process*

While attending the University of Vienna from 1851-1853, Mendel was influenced by two professors:

- Doppler, a physicist, trained Mendel to apply a *quantitative experimental* approach to the study of natural phenomena.
- Unger, a botanist, interested Mendel in the causes of inheritable variation in plants.

These experiences inspired Mendel to use key elements of the scientific process in the study of heredity. Unlike most nineteenth century biologists, he used a quantitative approach to his experimentation.

In 1857, Mendel was living in an Augustinian monastery, where he bred garden peas in the abbey garden. He probably chose garden peas as his experimental organisms because:

- They were available in many easily distinguishable varieties.
- Strict control over mating was possible to ensure the parentage of new seeds. Petals of the pea flower enclose the pistil and stamens, which prevents cross-pollination. Immature stamens can be removed to prevent self-pollination. Mendel hybridized pea plants by transferring pollen from one flower to another with an artist's brush (see Campbell, Figure 14.1).

Character = Detectable inheritable feature of an organism

Trait = Variant of an inheritable character

Mendel chose characters in pea plants that differed in a relatively clear-cut manner. He chose seven characters, each of which occurred in two alternative forms:

1. Flower color (purple or white)
2. Flower position (axial or terminal)
3. Seed color (yellow or green)
4. Seed shape (round or wrinkled)
5. Pod shape (inflated or constricted)
6. Pod color (green or yellow)
7. Stem length (tall or dwarf)

True breeding = Always producing offspring with the same *traits* as the parents when the parents are self-fertilized

Mendel started his experiments with true-breeding plant varieties, which he hybridized (cross-pollinated) in experimental crosses.

- The true-breeding parental plants of such a *cross* are called the *P generation* (parental).
- The hybrid offspring of the P generation are the F_1 *generation* (first filial).
- Allowing F_1 generation plants to self-pollinate, produces the next generation, the F_2 *generation* (second filial).

Mendel observed the transmission of selected traits for at least three generations and arrived at two principles of heredity that are now known as the *law of segregation* and the *law of independent assortment*.

B. By the law of segregation, the two alleles for a character are packaged into separate gametes

When Mendel crossed true-breeding plants with different character traits, he found that the traits did not blend.

- Using the scientific process, Mendel designed experiments in which he used large sample sizes and kept accurate quantitative records of the results.
- For example, a cross between true-breeding varieties, one with purple flowers and one with white flowers, produced F_1 *progeny* (offspring) with only purple flowers.

Hypothesis: Mendel hypothesized that if the inheritable factor for white flowers had been lost, then a cross between F_1 plants should produce only purple-flowered plants.

Experiment: Mendel allowed the F_1 plants to self-pollinate.

Results: There were 705 purple-flowered and 224 white-flowered plants in the F_2 generation—a ratio of 3:1. The inheritable factor for white flowers was not lost, so the hypothesis was rejected (see Campbell, Figure 14.2).

Conclusions: From these types of experiments and observations, Mendel concluded that since the inheritable factor for white flowers was not lost in the F_1 generation, it must have been masked by the presence of the purple-flower factor. Mendel's factors are now called *genes*; and in Mendel's terms, purple flower is the *dominant trait* and white flower is the *recessive trait*.

Mendel repeated these experiments with the other six characters and found similar 3:1 ratios in the F_2 generations (see Campbell, Table 14.1). From these observations he developed a hypothesis that can be divided into four parts:

1. Alternative forms of genes are responsible for variations in inherited characters.

    • For example, the gene for flower color in pea plants exists in two alternative forms; one for purple color and one for white color.

    • Alternative forms for a gene are now called *alleles* (see Campbell, Figure 14.3).

2. For each character, an organism inherits two alleles, one from each parent.

    • Mendel deduced that each parent contributes one "factor," even though he did not know about chromosomes or meiosis.

    • We now know that Mendel's factors are genes. Each genetic locus is represented twice in diploid organisms, which have homologous pairs of chromosomes, one set for each parent. Homologous loci may have identical alleles as in Mendel's true-breeding organisms, or the two alleles may differ, as in the F_1 hybrids.\

3. If the two alleles differ, one is fully expressed (dominant allele); the other is completely masked (recessive allele).

    • Dominant alleles are designated by a capital letter: P = purple flower color.

    • Recessive alleles are designated by a lowercase letter: p = white flower color.

4. The two alleles for each character segregate during gamete production.

    • Without any knowledge of meiosis, Mendel deduced that a sperm cell or ovum carries only one allele for each inherited characteristic, because allele pairs separate (segregate) from each other during gamete production.

    • Gametes of true-breeding plants will all carry the same allele. If different alleles are present in the parent, there is a 50% chance that a gamete will receive the dominant allele, and a 50% chance that it will receive the recessive allele.

    • This sorting of alleles into separate gametes is known as Mendel's law of segregation.

Mendel's law of segregation = Allele pairs segregate during gamete formation (meiosis), and the paired condition is restored by the random fusion of gametes at fertilization (see Campbell, Figure 14.4).

This law predicts the 3:1 ratio observed in the F_2 generation of a monohybrid cross.

    • F_1 hybrids (Pp) produce two classes of gametes when allele pairs segregate during gamete formation. Half receive a purple-flower allele (P) and the other half the white-flower allele (p).

    • During self-pollination, these two classes of gametes unite randomly. Ova containing purple-flower alleles have equal chances of being fertilized by sperm carrying purple-flower alleles or sperm carrying white-flower alleles.

    • Since the same is true for ova containing white-flower alleles, there are four equally likely combinations of sperm and ova.

The combinations resulting from a genetic cross may be predicted by using a *Punnett square*.

The F$_2$ progeny would include:

- One-fourth of the plants with two alleles for purple flowers.
- One-half of the plants with one allele for purple flowers and one allele for white flowers. Since the purple-flower allele is dominant, these plants have purple flowers.
- One-fourth of the plants with two alleles for white flower color, which will have white flowers since no dominant allele is present.

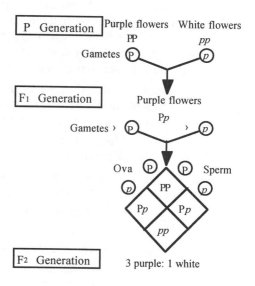

The pattern of inheritance for all seven of the characteristics studied by Mendel was the same: one parental trait disappeared in the F$_1$ generation and reappeared in one-fourth of the F$_2$ generation.

1. Some useful genetic vocabulary

Homozygous = Having two identical alleles for a given trait (e.g., PP or pp).

- All gametes carry that allele.
- Homozygotes are *true-breeding*.

Heterozygous = Having two different alleles for a trait (e.g., Pp).

- Half of the gametes carries one allele (P) and the remaining half carries the other (p).
- Heterozygotes are not true-breeding.

Phenotype = An organism's expressed traits (e.g., purple or white flowers).

- In Mendel's experiment above, the F$_2$ generation had a 3:1 *phenotypic ratio* of plants with purple flowers to plants with white flowers.

Genotype = An organism's genetic makeup (e.g., PP, Pp, or pp).

- The *genotypic ratio* of the F$_2$ generation was 1:2:1 (1 PP:2 Pp:1 pp).
- Campbell, Figure 14.5 compares genotype to phenotype.

2. The testcross

Because some alleles are dominant over others, the genotype of an organism may not be apparent. For example:

- A pea plant with purple flowers may be either homozygous dominant (PP) or heterozygous (Pp).

To determine whether an organism with a dominant phenotype (e.g., purple flower color) is homozygous dominant or heterozygous, you use a *testcross*.

Testcross = The breeding of an organism of unknown genotype with a homozygous recessive (see also Campbell, Figure 14.6).

- For example, if a cross between a purple-flowered plant of unknown genotype (P___) produced only purple-flowered plants, the parent was probably homozygous dominant since a PP × pp cross produces all purple-flowered progeny that are heterozygous (Pp).

- If the progeny of the testcross contains both purple and white phenotypes, then the purple-flowered parent was heterozygous since a Pp × pp cross produces Pp and pp progeny in a 1:1 ratio.

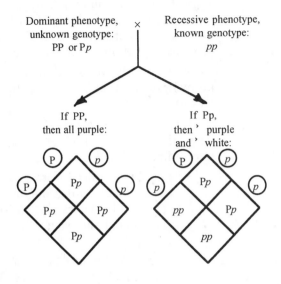

C. By the law of independent assortment, each pair of alleles segregates into gametes independently

Mendel deduced the law of segregation from experiments with *monohybrid crosses*, breeding experiments that used parental varieties differing in a single trait. He then performed crosses between parental varieties that differed in two characters or *dihybrid crosses*.

Dihybrid cross = A mating between parents that are heterozygous for two characters (dihybrids).

- Mendel began his experiments by crossing true-breeding parent plants that differed in two characters such as seed color (yellow or green) and seed shape (round or wrinkled). From previous monohybrid crosses, Mendel knew that yellow seed (Y) was dominant to green (y), and that round (R) was dominant to wrinkled (r).

- Plants homozygous for round yellow seeds (RRYY) were crossed with plants homozygous for wrinkled green seeds (rryy).

- The resulting F_1 dihybrid progeny were heterozygous for both traits (RrYy) and had round yellow seeds, the dominant phenotypes.

- From the F_1 generation, Mendel could not tell if the two characters were inherited independently or not, so he allowed the F_1 progeny to self-pollinate. In the following experiment, Mendel considered two alternate hypotheses (see also Campbell, Figure 14.7):

Hypothesis 1: If the two characters segregate *together*, the F_1 hybrids can only produce the same two classes of gametes (RY and ry) that they received from the parents, and the F_2 progeny will show a 3:1 phenotypic ratio.

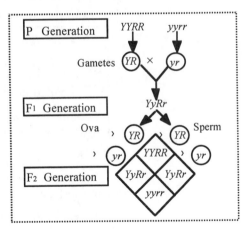

Hypothesis 2: If the two characters segregate *independently*, the F_1 hybrids will produce four classes of gametes (RY, Ry, rY, ry), and the F_2 progeny will show a 9:3:3:1 ratio.

Experiment: Mendel performed a dihybrid cross by allowing self-pollination of the F_1 plants (RrYy × RrYy).

Results: Mendel categorized the F_2 progeny and determined a ratio of 315:108:101:32, which approximates 9:3:3:1.

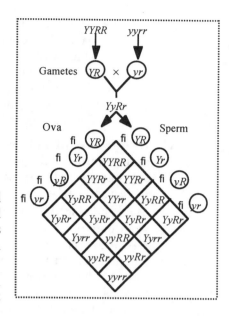

- These results were repeatable. Mendel performed similar dihybrid crosses with all seven characters in various combinations and found the same 9:3:3:1 ratio in each case.

- He also noted that the ratio for each individual gene pair was 3:1, the same as that for a monohybrid cross.

Conclusions: The experimental results supported the hypothesis that *each allele pair segregates independently during gamete formation.*

This behavior of genes during gamete formation is referred to as *Mendel's law of independent assortment.*

Mendel's law of independent assortment = Each allele pair segregates independently of other gene pairs during gamete formation.

D. Mendelian inheritance reflects rules of probability

Segregation and independent assortment of alleles during gamete formation and fusion of gametes at fertilization are random events. Thus, if we know the genotypes of the parents, we can predict the most likely genotypes of their offspring by using the simple *laws of probability:*

- The probability scale ranges from 0 to 1; an event that is certain to occur has a probability of 1, and an event that is certain *not* to occur has a probability of 0.

- The probabilities of all possible outcomes for an event must add up to 1.

- For example, when tossing a coin or rolling a six-sided die:

Event		Probability
Tossing heads with a two-headed coin	1	
Tossing tails with a two-headed coin	0	1 + 0 = 1
Tossing heads with a normal coin	1/2	
Tossing tails with a normal coin	1/2	1/2 + 1/2 = 1
Rolling 3 on a six-sided die	1/6	
Rolling a number other than 3	5/6	1/6 + 5/6 = 1

Random events are *independent* of one another.

- The outcome of a random event is unaffected by the outcome of previous such events.
- For example, it is possible that five successive tosses of a normal coin will produce five heads; however, the probability of heads on the sixth toss is still 1/2.

Two basic rules of probability are helpful in solving genetics problems: the *rule of multiplication* and the *rule of addition*.

1. Rule of multiplication

Rule of multiplication = The probability that independent events will occur simultaneously is the product of their individual probabilities (see Campbell, Figure 14.8). For example:

Question: In a Mendelian cross between pea plants that are heterozygous for flower color (Pp), what is the probability that the offspring will be homozygous recessive?

Answer:

Probability that an egg from the F1 (Pp) will receive a p allele = 1/2.

Probability that a sperm from the F1 will receive a p allele = 1/2.

The overall probability that two recessive alleles will unite at fertilization: $1/2 \times 1/2 = 1/4$.

This rule also applies to dihybrid crosses. For example:

Question: For a dihybrid cross, YyRr × YyRr, what is the probability of an F_2 plant having the genotype YYRR?

Answer:

Probability that an egg from a YyRr parent will receive the Y and R alleles = $1/2 \times 1/2 = 1/4$.

Probability that a sperm from a YyRr parent will receive a the Y and R alleles = $1/2 \times 1/2 = 1/4$.

The overall probability of an F_2 plant having the genotype YYRR: $1/4 \times 1/4 = \underline{1/16}$.

2. Rule of addition

Rule of addition = The probability of an event that can occur in two or more independent ways is the sum of the separate probabilities of the different ways. For example:

Question: In a Mendelian cross between pea plants that are heterozygous for flower color (Pp), what is the probability of the offspring being a heterozygote?

Answer: There are two ways in which a heterozygote may be produced: the dominant allele (P) may be in the egg and the recessive allele (p) in the sperm, or the dominant allele may be in the sperm and the recessive in the egg. Consequently, the probability that the offspring will be heterozygous is the sum of the probabilities of those two possible ways:

Probability that the dominant allele will be in the egg with the recessive in the sperm is $1/2 \times 1/2 = 1/4$.

Probability that the dominant allele will be in the sperm and the recessive in the egg is $1/2 \times 1/2 = 1/4$.

Therefore, the probability that a heterozygous offspring will be produced is 1/4 + 1/4 = $\underline{1/2}$.

3. **Using rules of probability to solve genetics problems**

The rules of probability can be used to solve complex genetics problems. For example, Mendel crossed pea varieties that differed in three characters (*trihybrid crosses*).

Question: What is the probability that a trihybrid cross between two organisms with the genotypes AaBbCc and AaBbCc will produce an offspring with the genotype aabbcc?

Answer: Because segregation of each allele pair is an independent event, we can treat this as three separate monohybrid crosses:

Aa × Aa: probability for aa offspring   =   1/4
Bb × Bb: probability for bb offspring   =   1/4
Cc × Cc: probability for cc offspring   =   1/4

The probability that these independent events will occur simultaneously is the product of their independent probabilities (rule of multiplication). So the probability that the offspring will be aabbcc is:

1/4 aa × 1/4 bb × 1/4 cc   =   <u>1/64</u>

For another example, consider a trihybrid cross of garden peas, where:

Character	Trait & Genotype
Flower Color	Purple: PP, Pp
	White: pp
Seed Color	Yellow: YY, Yy
	Green: yy
Seed Shape	Round: RR, Rr
	Wrinkled: rr

Question: phenotypes for *at least* two of the three traits?

PpYyRr × Ppyyrr

Answer: First list those genotypes that are homozygous recessive for at least two traits, (note that this includes the homozygous recessive for all three traits). Use the *rule of multiplication* to calculate the probability that offspring would be one of these genotypes. Then use the *rule of addition* to calculate the probability that two of the three traits would be homozygous recessive.

Genotypes with at least two homozygous recessives			Probability of genotype
ppyyRr	1/4 × 1/2 × 1/2	=	1/16
ppYyrr	1/4 × 1/2 × 1/2	=	1/16
Ppyyrr	1/2 × 1/2 × 1/2	=	2/16
PPyyrr	1/4 × 1/2 × 1/2	=	1/16
ppyyrr	1/4 × 1/2 × 1/2	=	<u>1/16</u>
		=	6/16 or
			<u>3/8</u> chance of two recessive traits

D. **Mendel discovered the particulate behavior of genes: *a review***

If a seed is planted from the F_2 generation of a monohybrid cross, we cannot predict with absolute certainty that the plant will grow to produce white flowers (pp). We *can* say that there is a 1/4 chance that the plant will have white flowers.

- Stated in statistical terms: among a large sample of F_2 plants, 25% will have white flowers.
- The larger the sample size, the closer the results will conform to predictions.

Mendel's quantitative methods reflect his understanding of this statistical feature of inheritance. Mendel's laws of segregation and independent assortment are based on the premise that:

- Inheritance is a consequence of discrete factors (genes) that are passed on from generation to generation.
- Segregation and assortment are random events and thus obey the simple laws of probability.

II. Extending Mendelian Genetics

A. The relationship between genotype and phenotype is rarely simple

As Mendel described it, characters are determined by one gene with two alleles; one allele completely dominant over the other. There are other patterns of inheritance not described by Mendel, but his laws of segregation and independent assortment can be extended to these more complex cases.

1. Incomplete dominance

In cases of *incomplete dominance*, one allele is not completely dominant over the other, so the heterozygote has a phenotype that is intermediate between the phenotypes of the two homozygotes (see Campbell, Figure 14.9).

Incomplete dominance = Pattern of inheritance in which the dominant phenotype is not fully expressed in the heterozygote, resulting in a phenotype intermediate between the homozygous dominant and homozygous recessive.

- For example, when red snapdragons (RR) are crossed with white snapdragons (rr), all F_1 hybrids (Rr) have pink flowers. (The heterozygote produces half as much red pigment as the homozygous red-flowered plant.)
- Since the heterozygotes can be distinguished from homozygotes by their phenotypes, the phenotypic and genotypic ratios from a monohybrid cross are the same— 1:2:1.
- Incomplete dominance is *not* support for the blending theory of inheritance, because alleles maintain their integrity in the heterozygote and segregate during gamete formation. Red and white phenotypes reappear in the F_2 generation.

2. What is a dominant allele?

Dominance/recessiveness relationships among alleles vary in a continuum from *complete dominance* on one end of the spectrum to *codominance* on the other, with various degrees of incomplete dominance in between these extremes.

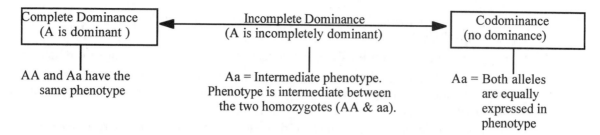

Complete dominance = Inheritance characterized by an allele that is fully expressed in the phenotype of a heterozygote and that masks the phenotypic expression of the recessive allele; state in which the phenotypes of the heterozygote and dominant homozygote are indistinguishable.

Codominance = Inheritance characterized by full expression of both alleles in the heterozygote.

- For example, the MN blood-group locus codes for the production of surface glycoproteins on the red blood cell. In this system, there are three blood types: M, N and MN.

Blood Type	Genotype
M	MM
N	NN
MN	MN

- The MN blood type is the result of full phenotypic expression of *both* alleles in the heterozygote; both molecules, M and N, are produced on the red blood cell.

Apparent dominance/recessiveness relationships among alleles reflect the level at which the phenotype is studied. For example:

- *Tay-Sachs disease* is a recessively inherited disease in humans; only children who are homozygous recessive for the Tay-Sachs allele have the disease.

- Brain cells of Tay-Sachs babies lack a crucial lipid-metabolizing enzyme. Thus, lipids accumulate in the brain, causing the disease symptoms and ultimately leading to death.

- At the *organismal level*, since heterozygotes are symptom free, it appears that the normal allele is completely dominant and the Tay-Sachs allele is recessive.

- At the *biochemical level*, inheritance of Tay-Sachs seems to be incomplete dominance of the normal allele, since there is an intermediate phenotype. Heterozygotes have an enzyme activity level that is intermediate between individuals homozygous for the normal allele and individuals with Tay-Sachs disease.

- At the *molecular level*, the normal allele and the Tay-Sachs allele are actually codominant. Heterozygotes produce equal numbers of normal and dysfunctional enzymes. They lack disease symptoms, because half the normal amount of functional enzyme is sufficient to prevent lipid accumulation in the brain.

Dominance/recessiveness relationships among alleles:

- Are a consequence of the mechanism that determines phenotypic expression, not the ability of one allele to subdue another at the level of the DNA

- Do not determine the relative abundance of alleles in a population
  - In other words, dominant alleles are not necessarily more common and recessive alleles more rare.

- For example, the allele for polydactyly is quite rare in the U.S. (1 in 400 births), yet it is caused by a dominant allele. (Polydactyly is the condition of having extra fingers or toes.)

3. Multiple alleles

Some genes may have *multiple alleles*; that is, more than just two alternative forms of a gene. The inheritance of the ABO blood group is an example of a locus with three alleles (see Campbell, Figure 14.10).

Paired combinations of three alleles produce four possible phenotypes:

- Blood type A, B, AB, or O.
- A and B refer to two genetically determined polysaccharides (A and B antigens) which are found on the surface of red blood cells different from the (different from the MN characters).

There are three alleles for this gene: I^A, I^B, and i.

- The I^A allele codes for the production of A antigen, the I^B allele codes for the production of B antigen, and the i allele codes for *no* antigen production on the red blood cell (neither A or B).
- Alleles I^A and I^B are *codominant* since both are expressed in heterozygotes.
- Alleles I^A and I^B are dominant to allele i, which is recessive.
- Even though there are three possible alleles, every person carries only two alleles which specify their ABO blood type; one allele is inherited from each parent.

Since there are three alleles, there are six possible genotypes:

Blood Type	Possible Genotypes	Antigens on the Red Blood Cell	Antibodies in the Serum
A	$I^A I^A$ $I^A i$	A	anti-B
B	$I^B I^B$ $I^B i$	B	anti-A
AB	$I^A I^B$	A, B	----
O	ii	----	anti-A anti-B

Foreign antigens usually cause the immune system to respond by producing *antibodies*, globular proteins that bind to the foreign molecules causing a reaction that destroys or inactivates it. In the ABO blood system:

- The antigens are located on the red blood cell and the antibodies are in the serum.
- A person produces antibodies against foreign blood antigens (those not possessed by the individual). These antibodies react with the foreign antigens causing the blood cells to clump or *agglutinate*, which may be lethal.
- For a blood transfusion to be successful, the red blood cell *antigens of the donor* must be compatible with the *antibodies of the recipient*.

4. Pleiotropy

Pleiotropy = The ability of a single gene to have multiple phenotypic effects.

- There are many hereditary diseases in which a single defective gene causes complex sets of symptoms (e.g., sickle-cell anemia).

- One gene can also influence a combination of seemingly unrelated characteristics. For example, in tigers and Siamese cats, the gene that controls fur pigmentation also influences the connections between a cat's eyes and the brain. A defective gene causes both abnormal pigmentation and cross-eye condition.

5. Epistasis

Different genes can interact to control the phenotypic expression of a single trait. In some cases, a gene at one locus alters the phenotypic expression of a second gene, a condition known as *epistasis* (see Campbell, Figure 14.12).

Epistasis = (Epi=upon; stasis=standing) Interaction between two nonallelic genes in which one modifies the phenotypic expression of the other.

- If one gene suppresses the phenotypic expression of another, the first gene is said to be *epistatic* to the second.

- If epistasis occurs between two nonallelic genes, the phenotypic ratio resulting from a dihybrid cross will deviate from the 9:3:3:1 Mendelian ratio.

- For example, in mice and other rodents, the gene for pigment deposition (C) is epistatic to the gene for pigment (melanin) production. In other words, whether the pigment can be deposited in the fur determines whether the coat color can be expressed. Homozygous recessive for pigment deposition (cc) will result in an albino mouse regardless of the genotype at the black/brown locus (BB, Bb or bb):

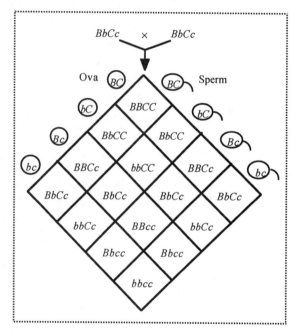

CC, Cc = Melanin deposition

cc = Albino

BB, Bb = Black coat color

bb = Brown coat color

- Even though both genes affect the same character (coat color), they are inherited separately and will assort independently during gamete formation. A cross between black mice that are heterozygous for the two genes results in a 9:3:4 phenotypic ratio:

9 Black (B_C_)

3 Brown (bbC_)

4 Albino (__cc)

6. Polygenic inheritance

Mendel's characters could be classified on an either-or basis, such as purple versus white flower. Many characters, however, are *quantitative characters* that vary in a continuum within a population.

Quantitative characters = Characters that vary by degree in a continuous distribution rather than by discrete (either-or) qualitative differences.

- Usually, continuous variation is determined not by one, but by many segregating loci or *polygenic inheritance*.

Polygenic inheritance = Mode of inheritance in which the additive effect of two or more genes determines a single phenotypic character.

For example, skin pigmentation in humans appears to be controlled by at least three separately inherited genes. The following is a simplified model for the polygenic inheritance of skin color:

- Three genes with the dark-skin allele (*A*, *B*, *C*) contribute one "unit" of darkness to the phenotype. These alleles are incompletely dominant over the other alleles (*a*, *b*, *c*).
- An *AABBCC* person would be very dark and an aabbcc person would be very light.
- An *AaBbCc* person would have skin of an intermediate shade.
- Because the alleles have a cumulative effect, genotypes *AaBbCc* and *AABbcc* make the same genetic contribution (three "units") to skin darkness. (See Campbell, Figure 14.12)
- Environmental factors, such as sun exposure, could also affect the phenotype.

7. **Nature versus nurture: the environmental impact on phenotype**

Environmental conditions can influence the phenotypic expression of a gene, so that a single genotype may produce a range of phenotypes. This environmentally-induced phenotypic range is the *norm of reaction* for the genotype.

Norm of reaction = Range of phenotypic variability produced by a single genotype under various environmental conditions (see Campbell, Figure 14.13). Norms of reaction for a genotype:

- May be quite limited, so that a genotype only produces a specific phenotype, such as the blood group locus that determines ABO blood type.
- May also include a wide range of possibilities. For example, an individual's blood cell count varies with environmental factors such as altitude, activity level or infection.
- Are generally broadest for polygenic characters, including behavioral traits.

The expression of most polygenic traits, such as skin color, is *multifactorial*; that is, it depends upon many factors - a variety of possible genotypes, as well as a variety of environmental influences.

8. **Integrating a Mendelian view of heredity and variation**

These patterns of inheritance that are departures from Mendel's original description, can be integrated into a comprehensive theory of Mendelian genetics.

- Taking a holistic view, an organism's entire phenotype reflects its overall genotype and unique environmental history.
- Mendelism has broad applications beyond its original scope; extending the principles of segregation and independent assortment helps explain more complex hereditary patterns such as epistasis and quantitative characters.

III. Mendalian Inheritance in Humans

A. Pedigree analysis reveals Mendelian patterns in human inheritance

Mendelian inheritance in humans is difficult to study because:

- The human generation time is about 20 years.
- Humans produce relatively few offspring compared to most other species.
- Well-planned breeding experiments are impossible.

Our understanding of Mendelian inheritance in humans is based on the analysis of family pedigrees or the results of matings that have already occurred.

Pedigree = A family tree that diagrams the relationships among parents and children across generations and that shows the inheritance pattern of a particular phenotypic character (see Campbell, Figure 14.14). By convention:

- Squares symbolize males and circles represent females.
- A horizontal line connecting a male and female indicates a mating; offspring are listed below in birth order, from left to right.
- Shaded symbols indicate individuals showing the trait being traced.

Following a dominant trait. For example, family members' genotypes can be deduced from a pedigree that traces the occurrence of widow's peak, the expression of a dominant allele.

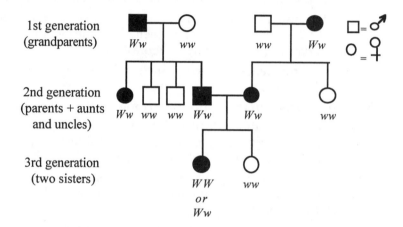

- If a widow's peak results from a dominant allele, *W*, then all individuals that do not have a widow's peak hairline must be homozygous recessive (*ww*). The genotypes of all recessives can be written on the pedigree.
- If widow's peak results from a dominant allele, W, then individuals that have a widow's peak hairline must be either homozygous dominant (*WW*) or heterozygous (*Ww*).
- If only some of the second generation offspring have a widow's peak, then the grandparents that show the trait must be heterozygous (*Ww*). (Note: if the grandparents with widow's peak were homozygous dominant, then all their respective offspring would show the trait.)
- Second generation offspring with widow's peaks must be heterozygous, because they are the result of *Ww* × *ww* matings.
- The third generation sister with widow's peak may be either homozygous dominant (*WW*) or heterozygous (*Ww*), because her parents are both heterozygous.

Following a recessive trait. For example, the same family can be used to trace a recessive trait such as attached ear lobes.

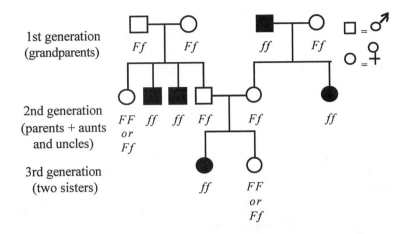

1st generation (grandparents)

2nd generation (parents + aunts and uncles)

3rd generation (two sisters)

- If attached earlobes is due to a recessive allele (*f*), then all individuals with attached earlobes must be homozygous recessive (*ff*).
- Since attached earlobes appears in second generation offspring, the grandparents with free earlobes are heterozygous (*Ff*) since they must be capable of passing on a recessive allele (*f*).
- Since one of the third generation sisters has attached earlobes (ff), her parents are heterozygous; they have free earlobes (dominant trait) and yet must be able to contribute a recessive allele to their daughter. The other sister shows the dominant trait, so her genotype is unknown; it is possible that she may be either homozygous dominant or heterozygous.

Pedigree analysis can also be used to:

- Deduce whether a trait is determined by a recessive or dominant allele. Using the example above:
  - The first-born third generation daughter has attached earlobes. Since both parents *lack* the trait, it must not be determined by a dominant allele.
- Predict the occurrence of a trait in future generations. For example, if the second generation couple decide to have another child,
  - What is the probability the child will have a widow's peak? From a mating of *Ww* × *Ww*:

Probability of a child being *WW*	=	1/4
Probability of a child being *Ww*	=	2/4
Probability of widow's peak	=	3/4

  - What is the probability the child will have attached earlobes? From a mating of *Ff* × *Ff*:

    probability of a child being *ff* =   1/4

  - What is the probability the child will have a widow's peak and attached earlobes? From a cross of *WwFf* × *WwFf*, use the rule of multiplication:

    3/4 (probability of widow's peak) × 1/4 (probability of attached earlobes) = 3/16

This type of analysis is important to geneticists and physicians, especially when the trait being analyzed can lead to a disabling or lethal disorder.

B. Many human disorders follow Mendelian patterns of inheritance

1. Recessively inherited disorders

Recessive alleles that cause human disorders are usually defective versions of normal alleles.

- Defective alleles code for either a malfunctional protein or no protein at all.
- Heterozygotes can be phenotypically normal, if one copy of the normal allele is all that is needed to produce sufficient quantities of the specific protein.

Recessively inherited disorders range in severity from nonlethal traits (e.g., albinism) to lethal diseases (e.g., cystic fibrosis). Since these disorders are caused by recessive alleles:

- The phenotypes are expressed only in homozygotes (aa) who inherit one recessive allele from each parent.
- Heterozygotes (Aa) can be phenotypically normal and act as *carriers*, possibly transmitting the recessive allele to their offspring.

Most people with recessive disorders are born to normal parents, both of whom are carriers.

- The probability is 1/4 that a mating of two carriers (Aa × Aa) will produce a homozygous recessive zygote.
- The probability is 2/3 that a normal child from such a mating will be a heterozygote, or a carrier.

Human genetic disorders are not usually evenly distributed among all racial and cultural groups due to the different genetic histories of the world's people. Three examples of such recessively inherited disorders are *cystic fibrosis*, *Tay-Sachs disease* and *sickle-cell disease*.

Cystic fibrosis, the most common lethal genetic disease in the United States, strikes 1 in every 2,500 Caucasians (it is much rarer in other races).

- Four percent of the Caucasian population are carriers.
- The dominant allele codes for a membrane protein that controls chloride traffic across the cell membrane. Chloride channels are defective or absent in individuals that are homozygous recessive for the cystic fibrosis allele.
- Disease symptoms result from the accumulation of thickened mucus in the pancreas, intestinal tract and lungs, a condition that favors bacterial infections.

Tay-Sachs disease occurs in 1 out of 3,600 births. The incidence is about 100 times higher among Ashkenazic (central European) Jews than among Sephardic (Mediterranean) Jews and non-Jews.

- Brain cells of babies with this disease are unable to metabolize gangliosides (a type of lipid), because a crucial enzyme does not function properly.
- As lipids accumulate in the brain, the infant begins to suffer seizures, blindness and degeneration of motor and mental performance. The child usually dies after a few years.

Sickle-cell disease is the most common inherited disease among African Americans. It affects 1 in 400 African Americans born in the United States (see Campbell, Figure 14.15).

- The disease is caused by a single amino acid substitution in hemoglobin.
- The abnormal hemoglobin molecules tend to link together and crystallize, especially when blood oxygen content is lower than normal. This causes red blood cells to deform from the normal disk-shape to a sickle-shape.

- The sickled cells clog tiny blood vessels, causing the pain and fever characteristic of a sickle-cell crisis.

About 1 in 10 African Americans are heterozygous for the sickle-cell allele and are said to have the *sickle-cell trait*.

- These carriers are usually healthy, although some suffer symptoms after an extended period of low blood oxygen levels.
- Carriers can function normally because the two alleles are codominant (heterozygotes produce not only the abnormal hemoglobin but also normal hemoglobin).
- The high incidence of heterozygotes is related to the fact that in tropical Africa where malaria is endemic, heterozygotes have enhanced resistance to malaria compared to normal homozygotes. Thus, heterozygotes have an advantage over both homozygotesÑthose who have sickle cell disease and those who have normal hemoglobin.

The probability of inheriting the same rare harmful allele from both parents, is greater if the parents are closely related.

Consanguinity = A genetic relationship that results from shared ancestry

- The probability is higher that consanguinous matings will result in homozygotes for harmful recessives, since parents with recently shared ancestry are more likely to inherit the same recessive alleles than unrelated persons.
- It is difficult to accurately assess the extent to which human consanguinity increases the incidence of inherited diseases, because embryos homozygous for deleterious mutations are affected so severely that most are spontaneously aborted before birth.
- Most cultures forbid marriage between closely related adults. This may be the result of observations that stillbirths and birth defects are more common when parents are closely related.

2. Dominantly inherited disorders

Some human disorders are dominantly inherited.

- For example, *achondroplasia* (a type of dwarfism) affects 1 in 10,000 people who are heterozygous for this gene.
- Homozygous dominant condition results in spontaneous abortion of the fetus, and homozygous recessives are of normal phenotype (99.9% of the population).

Lethal dominant alleles are much rarer than lethal recessives, because they:

- Are always expressed, so their effects are not masked in heterozygotes.
- Usually result from new genetic mutations that occur in gametes and later kill the developing embryo.

Late-acting lethal dominants can escape elimination if the disorder does not appear until an advanced age after afflicted individuals may have transmitted the lethal gene to their children. For example,

- *Huntington's disease*, a degenerative disease of the nervous system, is caused by a late-acting lethal dominant allele. The phenotypic effects do not appear until 35 to 40 years of age. It is irreversible and lethal once the deterioration of the nervous system begins.
- Molecular geneticists have recently located the gene for Huntington's near the tip of chromosome #4.
- Children of an afflicted parent have a 50% chance of inheriting the lethal dominant allele. A newly developed test can detect the Huntington's allele before disease symptoms appear.

3. Multifactorial disorders

Not all hereditary diseases are simple Mendelian disorders; that is, diseases caused by the inheritance of certain alleles at a single locus. More commonly, people are afflicted by *multifactorial* disorders, diseases that have both genetic and environmental influences.

- Examples include heart disease, diabetes, cancer, alcoholism and some forms of mental illness.
- The hereditary component is often polygenic and poorly understood.
- The best public-health strategy is to educate people about the role of environmental and behavioral factors that influence the development of these diseases.

C. Technology is providing new tools for genetic testing and counseling

Genetic counselors in many hospitals can provide information to prospective parents concerned about a family history for a genetic disorder.

- This preventative approach involves assessing the risk that a particular genetic disorder will occur.
- Risk assessment includes studying the family history for the disease using Mendel's law of segregation to deduce the risk.

For example, a couple is planning to have a child, and both the man and woman had siblings who died from the same recessively inherited disorder. A genetic counselor could deduce the risk of their first child inheriting the disease by using the laws of probability:

Question: What is the probability that the husband and wife are each carriers?

Answer: The genotypic ratio from an Aa × Aa cross is 1 AA:2 Aa:1 aa. Since the parents are normal, they have a 2/3 of being carriers.

Question: What is the chance of two carriers having a child with the disease?

Answer: 1/2 (mother's chance of passing on the gene) × 1/2 (father's chance of passing on the gene) = 1/4

Question: What is the probability that their firstborn will have the disorder?

Answer: (Chance that the father is a carrier) × (chance that mother is a carrier) × (chance of two carriers having a child with the disease).

2/3 × 2/3 × 1/4 = 1/9

If the first child is born with the disease, what is the probability that the second child will inherit the disease?

- If the first child is born with the disease, then it is certain that both the man and the woman are carriers. Thus, the probability that other children produced by this couple will have the disease is 1/4.
- The conception of each child is an independent event, because the genotype of one child does not influence the genotype of the other children. So there is a 1/4 chance that any additional child will inherit the disease.

1. Carrier recognition

Several tests are available to determine if prospective parents are carriers of genetic disorders.

- Tests are currently available that can determine heterozygous carriers for the Tay-Sachs allele, cystic fibrosis, and sickle-cell disease.
- Tests such as these enable people to make informed decisions about having children, but they could also be abused. Ethical dilemmas about how this information should be used points to the immense social implications of such technological advances.

2. Fetal testing

A couple that learns they are both carriers for a genetic disease and decide to have a child can determine if the fetus has the disease. Between the fourteenth and sixteenth weeks of pregnancy, *amniocentesis.* can be done to remove amniotic fluid for testing (see Campbell, Figure 14.17).

- During amniocentesis, a physician inserts a needle into the uterus and extracts about ten milliliters of amniotic fluid.
- The presence of certain chemicals in amniotic fluid indicate some genetic disorders.
- Some tests (including one for Tay-Sachs) are performed on cells grown in culture from fetal cells sloughed off in the amniotic fluid. These cells can also be karyotyped to identify chromosomal defects.

Chorionic villus sampling (CVS) is a newer technique during which a physician suctions off a small amount of fetal tissue from the chorionic villi of the placenta.

- These rapidly dividing embryonic cells can be karyotyped immediately, usually providing results in 24 hoursÑa major advantage over amniocentesis which may take several weeks. (Amniocentesis requires that the cells must first be cultured before karyotyping can be done.)
- Another advantage of CVS is that it can be performed at only eight to ten weeks of pregnancy.

Other techniques such as *ultrasound* and *fetoscopy* allow physicians to examine a fetus for major abnormalities.

- Ultrasound is a non-invasive procedure which uses sound waves to create an image of the fetus.
- Fetoscopy involves inserting a thin fiber-optic scope into the uterus.

Amniocentesis and fetoscopy have a 1% risk of complication such as maternal bleeding or fetal death. Thus, they are used only when risk of genetic disorder or birth defect is relatively high.

3. Newborn screening

In most U.S. hospitals, simple tests are routinely performed at birth, to detect genetic disorders such as *phenylketonuria* (PKU).

- PKU is recessively inherited and occurs in about 1 in 15,000 births in the United States.
- Children with this disease cannot properly break down the amino acid phenylalanine.
- Phenylalanine and its by-product (phenylpyruvic acid) can accumulate in the blood to toxic levels, causing mental retardation.
- Fetal screening for PKU can detect the deficiency in a newborn and retardation can be prevented with a special diet (low in phenylalanine) that allows normal development.

REFERENCES

Campbell, N, et al. *Biology.* 5th ed. Menlo Park, California: Benjamin/Cummings, 1998.

Griffith, A.J.E., J.H. Miller, D.T. Suzuki, R.C. Lewontin, and W.M. Gelbart. *An Introduction to Genetic Analysis.* 5th ed. New York: W.H. Freeman, 1993.

Kowles, R.V. *Genetics, Society and Decisions.* 1st ed. Columbus, Ohio: Charles E. Merrill, 1985. Though the target audience for this book is non-majors, it can be a useful lecture supplement for contrasting controversial social issues.

Russell, P.J. *Genetics.* 2nd ed. Glenview, Illinois: Scott, Foresman, 1990.

CHAPTER 15
THE CHROMOSOMAL BASIS OF INHERITANCE

OUTLINE

I. Relating Mendelism to Chromosomes
   A. Mendelian inheritance has its physical basis in the behavior of chromosomes during sexual life cycles
   B. Morgan traced a gene to a specific chromosome: *science as a process*
   C. Linked genes tend to be inherited together because they are located on the same chromosome
   D. Independent assortment of chromosomes and crossing over produce genetic recombinants
   E. Geneticists can use recombination data to map a chromosome's genetic loci

II. Sex Chromosomes
   A. The chromosomal basis of sex varies with the organism
   B. Sex-linked genes have unique patterns of inheritance

III. Errors and Exceptions to Chromosomal Inheritance
   A. Alterations of chromosome number or structure cause some genetic disorders
   B. The phenotypic effects of some genes depend on whether they were inherited from the mother or father
   C. Extranuclear genes exhibit a non-Mendelian pattern of inheritance

OBJECTIVES

After reading this chapter and attending lecture, the student should be able to:

1. Explain how the observations of cytologists and geneticists provided the basis for the chromosome theory of inheritance.
2. Describe the contributions that Thomas Hunt Morgan, Walter Sutton, and A.H. Sturtevant made to current understanding of chromosomal inheritance.
3. Explain why *Drosophila melanogaster* is a good experimental organism.
4. Define linkage and explain why linkage interferes with independent assortment.
5. Distinguish between parental and recombinant phenotypes.
6. Explain how crossing over can unlink genes.
7. Map a linear sequence of genes on a chromosome using given recombination frequencies from experimental crosses.
8. Explain what additional information cytological maps provide over crossover maps.
9. Distinguish between a heterogametic sex and a homogametic sex.
10. Describe sex determination in humans.
11. Describe the inheritance of a sex-linked gene such as color-blindness.
12. Explain why a recessive sex-linked gene is always expressed in human males.

13. Explain how an organism compensates for the fact that some individuals have a double dosage of sex-linked genes while others have only one.

14. Distinguish among nondisjunction, aneuploidy, and polyploidy; explain how these major chromosomal changes occur and describe the consequences.

15. Distinguish between trisomy and triploidy.

16. Distinguish among deletions, duplications, translocations, and inversions.

17. Describe the effects of alterations in chromosome structure, and explain the role of position effects in altering the phenotype.

18. Describe the type of chromosomal alterations implicated in the following human disorders: Down syndrome, Klinefelter syndrome, extra Y, triple-X syndrome, Turner syndrome, *cri du chat* syndrome, and chronic myelogenous leukemia.

19. Define genomic imprinting and provide evidence to support this model.

20. Explain how the complex expression of a human genetic disorder, such as fragile-X syndrome, can be influenced by triplet repeats and genomic imprinting.

21. Give some exceptions to the chromosome theory of inheritance, and explain why cytoplasmic genes are not inherited in a Mendelian fashion.

KEY TERMS

chromosome theory of inheritance
wild type
mutant phenotype
sex-linked genes
linked genes
genetic recombination
parental type
recombinants

linkage map
cytological map
Duchenne muscular dystrophy
hemophilia
Barr body
nondisjunction
aneuploidy
trisomic
monosomic

polyploidy
deletion
duplication
inversion
translocation
Down syndrome
fragile X syndrome

LECTURE NOTES

I.   **Relating Mendelism to Chromosomes**

A.   **Mendelian inheritance has its physical basis in the behavior of chromosomes during sexual life cycles**

Genetics	Cytology
1860s: Mendel proposed that discrete inherited factors segregate and assort independently during gamete formation	
	1875: Cytologists worked out process of mitosis
	1890: Cytologists worked out process of meiosis
1900: Three botanists (Correns, de Vries, and von Seysenegg) independently rediscovered Mendel's principles of segregation and independent assortment	
1902: Cytology and genetics converged as Walter Sutton, Theodor Boveri, and others noticed parallels between the behavior of Mendel's factors and the behavior of chromosomes. For example: • Chromosomes and genes are both paired in diploid cells. • Homologous chromosomes separate and allele pairs segregate during meiosis. • Fertilization restores the paired condition for both chromosomes and genes.	

Based upon these observations, biologists developed the *chromosome theory of inheritance* (see Campbell, Figure 15.1). According to this theory:
  • Mendelian factors or genes are located on chromosomes.
  • It is the chromosomes that segregate and independently assort.

B.   **Morgan traced a gene to a specific chromosome:** *science as a process*

Thomas Hunt Morgan from Columbia University performed experiments in the early 1900s which provided convincing evidence that Mendel's inheritable factors are located on chromosomes.

I.   **Morgan's choice of an experimental organism**

Morgan selected the fruit fly, *Drosophila melanogaster*, as the experimental organism because these flies:
  • Are easily cultured in the laboratory
  • Are prolific breeders
  • Have a short generation time
  • Have only four pairs of chromosomes which are easily seen with a microscope

There are three pairs of autosomes (II, III, and IV) and one pair of sex chromosomes. Females have two X chromosomes, and males have one X and one Y chromosome.

Morgan and his colleagues used genetic symbols that are now convention. For a particular character:

- A gene's symbol is based on the first *mutant*, non-wild type discovered.
- If the mutant is recessive, the first letter is lowercase. (e.g., w = white eye allele in *Drosophila*.)
- If the mutant is dominant, the first letter is capitalized. (e.g., Cy = "curly" allele in *Drosophila* that causes abnormal, curled wings.)
- Wild-type trait is designated by a superscript +. (e.g., Cy^+ = allele for normal, straight wings.)

Wild type = Normal or most frequently observed phenotype (see Campbell, Figure 15.2)

Mutant phenotypes = Phenotypes which are alternatives to the wild type due to mutations in the wild-type gene

2. Discovery of a sex linkage

After a year of breeding *Drosophila* to find variant phenotypes, Morgan discovered a single male fly with white eyes instead of the wild-type red. Morgan mated this mutant white-eyed male with a red-eyed female. The cross is outlined below (see also Campbell, Figure 15.3).

w = white-eye allele

w^+ = red-eye or wild-type allele

Drosophila geneticists symbolize a recessive mutant allele with one or more lower case letters. The corresponding wild-type allele has a superscript plus sign.

P generation:

w^+w^+ $\times$ w
red-eyed ♀   white-eyed ♂

F_1 generation:

w^+w $\times$ w^+
red-eyed ♀   red-eyed ♂

The fact that all the F_1 progeny had red eyes, suggested that the wild-type allele was dominant over the mutant allele.

F_2 generation:

w^+w^+ ww^+
red-eyed ♀   red-eyed ♀

w^+ w
red-eyed ♂   white-eyed ♂

White-eyed trait was expressed only in the male, and all the F_2 females had red eyes.

Morgan deduced that eye color is linked to sex and that the gene for eye color is located only on the X chromosome. Premises for his conclusions were:

- If eye color is located only on the X chromosome, then females (XX) carry two copies of the gene, while males (XY) have only one.
- Since the mutant allele is recessive, a white-eyed female must have that allele on both X chromosomes which was impossible for F_2 females in Morgan's experiment.
- A white-eyed male has no wild-type allele to mask the recessive mutant allele, so a single copy of the mutant allele confers white eyes.

Sex-linked genes = Genes located on sex chromosomes. The term is commonly applied only to genes on the X chromosome.

C. Linked genes tend to be inherited together because they are located on the same chromosome

Genes located on the same chromosome tend to be linked in inheritance and do not assort independently.

Linked genes = Genes that are located on the same chromosome and that tend to be inherited together.

- Linked genes do not assort independently, because they are on the same chromosome and move together through meiosis and fertilization.
- Since independent assortment does not occur, a dihybrid cross following two linked genes will not produce an F_2 phenotypic ratio of 9:3:3:1.

T.H. Morgan and his students performed a dihybrid testcross between flies with autosomal recessive mutant alleles for black bodies and vestigial wings and wild-type flies heterozygous for both traits (see Campbell, Figure 15.4).

b = black body vg = vestigial wings
b^+ = gray body vg^+ = wild-type wings

b^+bvg^+vg × $bbvgvg$
gray, normal wings   black, vestigial wings

- Resulting phenotypes of the progeny did not occur in the expected 1:1:1:1 ratio for a dihybrid testcross.
- A disproportionately large number of flies had the phenotypes of the parents: gray with normal wings and black with vestigial wings.
- Morgan proposed that these unusual ratios were due to linkage. The genes for body color and wing size are on the same chromosome and are usually thus inherited together.

D. Independent assortment of chromosomes and crossing over produce genetic recombinants

Genetic recombination = The production of offspring with new combinations of traits different from those combinations found in the parents; results from the events of meiosis and random fertilization.

1. The recombination of unlinked genes: independent assortment of chromosomes

Mendel discovered that some offspring from dihybrid crosses have phenotypes unlike either parent. An example is the following test cross between pea plants:

YY, Yy = yellow seeds   RR, Rr = round seeds
yy = green seeds   rr = wrinkled seeds

P generation:

$$\begin{array}{ccc}
\text{YyRr} & \times & \text{yyrr} \\
\text{yellow round} & & \text{green wrinkled}
\end{array}$$

Testcross progeny:

_ YyRr yellow, round	_ yyrr green, wrinkled	Parental types (50%)
_ yyRr green, round	_ Yyrr yellow, wrinkled	Recombinant types (50%)

Parental types = Progeny that have the same phenotype as one or the other of the parents.

Recombinants = Progeny whose phenotypes differ from either parent.

In this cross, seed shape and seed color are unlinked.

- One-fourth of the progeny have round yellow seeds, and one-fourth have wrinkled green seeds. Therefore, one-half of the progeny are *parental types*.

- The remaining half of the progeny are *recombinants*. One-fourth are round green and one-fourth are wrinkled yellow – phenotypes not found in either parent.

- When half the progeny are recombinants, there is a 50% frequency of recombination.

- A 50% frequency of recombination usually indicates that the two genes are on different chromosomes, because it is the expected result if the two genes assort randomly.

- The genes for seed shape and seed color assort independently of one another because they are located on different chromosomes which randomly align during metaphase of meiosis I.

2. The recombination of linked genes: crossing over

If genes are totally linked, some possible phenotypic combinations should not appear. Sometimes, however, the unexpected recombinant phenotypes do appear.

As described earlier, T.H. Morgan and his students performed the following dihybrid testcross between flies with autosomal recessive mutant alleles for black bodies and vestigial wings and wild-type flies heterozygous for both traits.

$$\begin{array}{llll}
b & = & \text{black body} & \quad vg & = & \text{vestigial wings} \\
b^+ & = & \text{gray body} & \quad vg^+ & = & \text{wild-type wings}
\end{array}$$

$$\begin{array}{ccc}
b^+bvg^+vg & \times & bbvgvg \\
\text{gray, normal wings} & & \text{black, vestigial wings}
\end{array}$$

Phenotypes	Genotypes	Expected Results If Genes Are Unlinked	Expected Results If Genes Are Totally Linked	Actual Results
Black body, normal wings	$\dfrac{b\ vg^+}{b\ vg}$	575		206
Gray body, normal wings	$\dfrac{b^+vg^+}{b\ vg}$	575	1150	965
Black body, vestigial wings	$\dfrac{b\ vg}{b\ vg}$	575	1150	944
Gray body, vestigial wings	$\dfrac{b^+vg}{b\ vg}$	575		185

$$\text{Recombination Frequency} = \frac{391 \text{ recombinants}}{2300 \text{ total offspring}} \times 100 = 17\%$$

Morgan's results from this dihybrid testcross showed that the two genes were neither unlinked or totally linked.

- If wing type and body color genes were unlinked, they would assort independently, and the progeny would show a 1:1:1:1 ratio of all possible phenotypic combinations.
- If the genes were completely linked, expected results from the testcross would be a 1:1 phenotypic ratio of *parental types only*.
- Morgan's testcross did not produce results consistent with unlinkage or total linkage. The high proportion of parental phenotypes suggested linkage between the two genes.
- Since 17% of the progeny were recombinants, the linkage must be incomplete. Morgan proposed that there must be some mechanism that occasionally breaks the linkage between the two genes (see Campbell, Figure 15.5).
- It is now known that *crossing over* during meiosis accounts for the recombination of linked genes. The exchange of parts between homologous chromosomes breaks linkages in parental chromosomes and forms recombinants with new allelic combinations.

E. Geneticists can use recombination data to map a chromosome's genetic loci

Scientists used recombination frequencies between genes to *map* the sequence of linked genes on particular chromosomes.

Morgan's *Drosophila* studies showed that some genes are linked more tightly than others.

- For example, the recombination frequency between the *b* and *vg* loci is about 17%.
- The recombination frequency is only 9% between *b* and *cn*, a third locus on the same chromosome. (The cinnabar gene, *cn*, for eye color has a recessive allele causing "cinnabar eyes.")

A.H. Sturtevant, one of Morgan's students, assumed that if crossing over occurs randomly, the probability of crossing over between two genes is directly proportional to the distance between them.

- Sturtevant used recombination frequencies between genes to assign them a linear position on a chromosome *map* (see Campbell, Figure 15.6).

- He defined one *map unit* as 1% recombination frequency. (Map units are now called *centimorgans*, in honor of Morgan.)

Using crossover data, a map may be constructed as follows:

Loci	Recombination Frequency	Approximate Map Units
b vg	17.0%	18.5*
cn b	9.0%	9.0
cn vg	9.5%	9.5

1. Establish the relative distance between those genes farthest apart or with the highest recombination frequency.

    b —————— *vg*
    ← 17 →

2. Determine the recombination frequency between the third gene (*cn*) and the first (*b*).

    ← 9 →
    cn ———— *b*

3. Consider the two possible placements of the third gene:

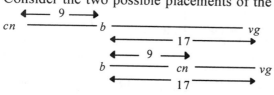

4. Determine the recombination frequency between the third gene (*cn*) and the second (*vg)* to eliminate the incorrect sequence.

    ← 9 → ← 9.5 →
    b ———— *cn* ———— *vg*
    ← 17 →

    So, the correct sequence is *b–cn–vg*.

> Note that there are actually 18.5 map units between *b* and *vg*. This is higher than that predicted from the recombination frequency of 17.0%. Because *b* and *vg* are relatively far apart, double crossovers occur between these loci and cancel each other out, leading us to underestimate the actual map distance.

If linked genes are so far apart on a chromosome that the recombination frequency is 50%, they are indistinguishable from unlinked genes that assort independently.

- Linked genes that are far apart can be mapped, if additional recombination frequencies can be determined between intermediate genes and each of the distant genes.

Sturtevant and his coworkers extended this method to map other *Drosophila* genes in linear arrays (see Campbell, Figure 15.7)

- The crossover data allowed them to cluster the known mutations into four major linkage groups.
- Since *Drosophila* has four sets of chromosomes, this clustering of genes into four linkage groups was further evidence that genes are on chromosomes.

Maps based on crossover data only give information about the relative position of linked genes on a chromosome. Another technique, *cytological mapping*, locates genes with respect to chromosomal features, such as stained bands that can be viewed with a microscope.

- The ultimate genetic maps are constructed by sequences, or DNA; in this case, distances between gene loci can be measured in nucleotides.

II.   Sex Chromosomes

A.   The chromosomal basis of sex varies with the organism

In most species, sex is determined by the presence or absence of special chromosomes. As a result of meiotic segregation, each gamete has one sex chromosome to contribute at fertilization.

Heterogametic sex = The sex that produces two kinds of gametes and determines the sex of the offspring.

Homogametic sex = The sex that produces one kind of gamete.

Campbell, Figure 15.8 shows four chromosomal systems of sex determination.

1.   The chromosomal basis of sex in humans

Mammals, including humans, have an *X-Y* mechanism that determines sex at fertilization.

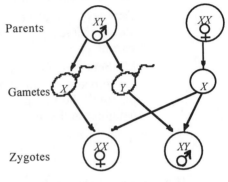

- There are two chromosomes, *X* and *Y*. Each gamete has one sex chromosome, so when sperm cell and ovum unite at fertilization, the zygote receives one of two possible combinations: *XX* or *XY*.
- Males are the heterogametic sex (*XY*). Half the sperm cells contain an *X* chromosome, while the other half contain a *Y* chromosome.
- Females are the homogametic sex (*XX*); all ova carry an *X* chromosome.

Whether an embryo develops into a male or female depends upon the presence of a *Y* chromosome.

- A British research team has identified a gene, *SRY* (sex-determining region of Y), on the *Y* chromosome that is responsible for triggering the complex series of events that lead to normal testicular development. In the absence of SRY, the gonads develop into ovaries.
- *SRY* probably codes for a protein that regulates other genes.

B.   Sex-linked genes have unique patterns of inheritance

Some genes on sex chromosomes play a role in sex determination, but these chromosomes also contain genes for other traits.

1.   Sex-linked disorders in humans

In humans, the term *sex-linked traits* usually refers to *X*-linked traits.

- The human *X*-chromosome is much larger than the *Y*. Thus, there are more *X*-linked than *Y*-linked traits.
- Most *X*-linked genes have no homologous loci on the *Y* chromosome.

- Most genes on the *Y* chromosome not only have no *X* counterparts, but they encode traits found only in males (e.g., testis-determining factor).
- Examples of sex-linked traits in humans are color blindness, *Duchenne muscular dystrophy* and hemophilia.

Fathers pass *X*-linked alleles to all their daughters only.

- Males receive their *X* chromosome only from their mothers.
- Fathers cannot pass sex-linked traits to their sons.

Mothers can pass sex-linked alleles to *both* sons and daughters.

- Females receive two *X* chromosomes, one from each parent.
- Mothers pass on one *X* chromosome (either maternal or paternal homologue) to every daughter and son.

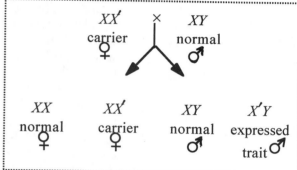

If a sex-linked trait is due to a recessive allele, a female will express the trait only if she is homozygous.

- Females have two *X* chromosomes, therefore they can be either homozygous or heterozygous for sex-linked alleles.
- There are fewer females with sex-linked disorders than males, because even if they have one recessive allele, the other dominant allele is the one that is expressed. A female that is heterozygous for the trait can be a *carrier*, but not show the recessive trait herself.
- A carrier that mates with a normal male will pass the mutation to half her sons and half her daughters.
- If a carrier mates with a male who has the trait, there is a 50% chance that each child born to them will have the trait, regardless of sex.

Campbell, Figure 15.9 depicts the transmission of sex-linked recessive traits.

Because males have only one *X*-linked locus, any male receiving a mutant allele from his mother will express the trait.

- Far more males than females have sex-linked disorders.
- Males are said to be hemizygous.

Hemizygous = A condition where only one copy of a gene is present in a diploid organism.

2. X-inactivation in female mammals

How does an organism compensate for the fact that some individuals have a double dosage of sex-linked genes while others have only one?

In female mammals, most diploid cells have only one fully functional X chromosome.

- The explanation for this process is known as the *Lyon hypothesis*, proposed by the British geneticist Mary F. Lyon.
- In females, each of the embryonic cells inactivates one of the two X chromosomes.
- The inactive X chromosome contracts into a dense object called a *Barr body*.

Barr body = Located inside the nuclear envelope, it is a densely staining object that is an inactivated X chromosome in female mammalian cells.

- Most Barr body genes are not expressed.
- They are reactivated in gonadal cells that undergo meiosis to form gametes.

Female mammals are a *mosaic* of two types of cellsÑthose with an active maternal X and those with an active paternal X.

- Which of the two Xs will be inactivated is determined randomly in embryonic cells.
- After an X is inactivated, all mitotic descendants will have the same inactive X.
- As a consequence, if a female is heterozygous for a sex-linked trait, about half of her cells will express one allele and the other cells well express the alternate allele.

> Examples of this type of mosaicism are coloration in calico cats and normal sweat gland development in humans (see Campbell, Figure 15.10).

X chromosome inactivation is associated with DNA methylation.

- Methyl groups ($-CH_3$) attach to cytosine, one of DNA's nitrogenous bases.
- Barr bodies are highly methylated compared to actively transcribed DNA.

What determines which of the two X chromosomes will be methylated?

- A recently discovered gene, *XIST* is active *only* on the Barr body.
- The product of the *XIST* gene, *X-inactive specific transcript*, is an RNA; multiple copies of *XIST* attach to the X chromosome inactivating it.

Many questions have yet to be answered.

- How does *XIST* initiate X-inactivation?
- What determines which X chromosome in each of a female's cells will have an active *XIST* gene and become a Barr body?

III. Errors and Exceptions in Chromosomal Inheritance

A. Alterations of chromosome number or structure cause some genetic disorders

Meiotic errors and mutagens can cause major chromosomal changes such as altered chromosome numbers or altered chromosomal structure.

1. Alterations of chromosome number: aneuploidy and polyploidy

Nondisjunction = Meiotic or mitotic error during which certain homologous chromosomes or sister chromatids fail to separate.

- Meiotic nondisjunction:
  - May occur during meiosis I so that a homologous pair does not separate (see Campbell, Figure 15.11a)
  - May occur during meiosis II when sister chromatids do not separate (see Campbell, Figure 15.11b)

- Results in one gamete receiving two of the same type of chromosome and another gamete receiving no copy. The remaining chromosomes may be distributed normally.
- Mitotic nondisjunction:
  - Also results in abnormal number of certain chromosomes
  - If it occurs in embryonic cells, mitotic division passes this abnormal chromosome number to a large number of cells, and thus, can have a large effect.

Aneuploidy = Condition of having an abnormal number of certain chromosomes

- Aneuploid offspring may result if a normal gamete unites with an aberrant one produced as a result of *nondisjunction*.
- An aneuploid cell that has a chromosome in triplicate is said to be *trisomic* for that chromosome.
- An aneuploid with a missing chromosome is said to be *monosomic* for that chromosome.
- When an aneuploid zygote divides by mitosis, it transmits the chromosomal anomaly to all subsequent embryonic cells.
- Abnormal gene dosage in aneuploids causes characteristic symptoms in survivors. An example is Down's syndrome which results from trisomy of chromosome 21.

Polyploidy = A chromosome number that is more than two complete chromosome sets.

- *Triploidy* is a polyploid chromosome number with three haploid chromosome sets (3N).
- *Tetraploidy* is polyploidy with four haploid chromosome sets (4N).
- Triploids may be produced by fertilization of an abnormal diploid egg produced by nondisjunction of all chromosomes.
- Tetraploidy may result if a diploid zygote undergoes mitosis without cytokinesis. Subsequent normal mitosis would produce a 4N embryo.
- Polyploidy is common in plants and important in plant evolution.
- Polyploids occur rarely among animals, and they are more normal in appearance than aneuploids. Mosaic polyploids, with only patches of polyploid cells, are more common than complete polyploid animals.

2. **Alterations of chromosome structure**

Chromosome breakage can alter chromosome structure in four ways (see Campbell, Figure 15.12):

- Chromosomes which lose a fragment lacking a centromere will have a deficiency or *deletion*.
- Fragments without centromeres are usually lost when the cell divides, or they may:
  - Join to a homologous chromosome producing a *duplication*.
  - Join to a nonhomologous chromosome (*translocation*).
  - Reattach to the original chromosome in reverse order (*inversion*).

Crossing-over error is another source of deletions and duplications.

- Crossovers are normally reciprocal, but sometimes one sister chromatid gives up more genes than it receives in an unequal crossover.
- A nonreciprocal crossover results in one chromosome with a deletion and one chromosome with a duplication.

Alterations of chromosome structure, can have various effects:

- Homozygous deletions, including a single X in a male, are usually lethal.
- Duplications and translocations tend to have deleterious effects.
- Even if all genes are present in normal dosages, reciprocal translocations between nonhomologous chromosomes and inversions can alter the phenotype because of subtle *position effects*.

Position effect = Influence on a gene's expression because of its location among neighboring genes.

3. **Human disorders due to chromosomal alterations**

Chromosomal alterations are associated with some serious human disorders.

Aneuploidy, resulting from meiotic nondisjunction during gamete formation, usually prevents normal embryonic development and often results in spontaneous abortion.

- Some types of aneuploidy cause less severe problems, and aneuploid individuals may survive to birth and beyond with a set of characteristic symptoms or *syndrome*.
- Aneuploid conditions can be diagnosed before birth by *fetal testing*.

Down syndrome, an aneuploid condition, affects 1 out of 700 U.S. children (see Campbell, Figure 15.13).

- Is usually the result of trisomy 21
- Includes characteristic facial features, short stature, heart defects, mental retardation, susceptibility to respiratory infections, and a proneness to developing leukemia and Alzheimer's disease
- Though most are sexually underdeveloped and sterile, a few women with Down syndrome have had children.
- The incidence of Down syndrome offspring correlates with maternal age.
  - May be related to the long time lag between the first meiotic division during the mother's fetal life and the completion of meiosis at ovulation.
  - May be that older women have less chance of miscarrying a trisomic embryo.

Other rarer disorders caused by autosomal aneuploidy are:

- *Patau syndrome* (trisomy 13)
- *Edwards syndrome* (trisomy 18)

Sex chromosome aneuploidies result in less severe conditions than those from autosomal aneuploidies. This may be because:

- The *Y* chromosome carries few genes.
- Copies of the *X* chromosome become inactivated as Barr bodies.

The basis of sex determination in humans is illustrated by sex chromosome aneuploidies.

- A single *Y* chromosome is sufficient to produce maleness.
- The absence of *Y* is required for femaleness.

Examples of sex chromosome aneuploidy in males are:

Klinefelter Syndrome

Genotype: Usually *XXY*, but may be associated with *XXYY, XXXY, XXXXY, XXXXXY*.

Phenotype: Male sex organs with abnormally small testes; sterile; feminine body contours and perhaps breast enlargement; usually of normal intelligence.

Extra Y

Genotype: *XYY.*

Phenotype: Normal male; usually taller than average; normal intelligence and fertility.

Abnormalities of sex chromosome number in females include:

Triple-X Syndrome

Genotype: *XXX.*

Phenotype: Usually fertile; can show a normal phenotype.

Turner Syndrome

Genotype: *XO* (only known viable human monosomy).

Phenotype: Short stature; at puberty, secondary sexual characteristics fail to develop; internal sex organs do not mature; sterile.

Structural chromosomal alterations such as deletions and translocations can also cause human disorders.

- Deletions in human chromosomes cause severe defects even in the heterozygous state. For example,
  - *Cri du chat* syndrome is caused by a deletion on chromosome 5. Symptoms are mental retardation, a small head with unusual facial features and a cry that sounds like a mewing cat.
- Translocations associated with human disorders include:
  - Certain cancers such as *chronic myelogenous leukemia* (CML). A portion of chromosome 22 switches places with a small fragment from chromosome 9.
- Some cases of Down syndrome. A third chromosome 21 translocates to chromosome 15, resulting in two normal chromosomes 21 plus the translocation.

B. The phenotypic effects of some genes depend on whether they were inherited from the mother or the father

The expression of some traits may depend upon which parent contributes the alleles for those traits.

- Example: Two genetic disorders, *Prader-Willi syndrome* and *Angelman syndrome*, are caused by the same deletion on chromosome 15. The symptoms differ depending upon whether the gene was inherited from the mother or from the father.
- Prader-Willi syndrome is caused by a deletion from the *paternal* version of chromosome 15. The syndrome is characterized by mental retardation, obesity, short stature, and unusually small hands and feet.
- Angelman syndrome is caused by a deletion from the *maternal* version of chromosome 15. This syndrome is characterized by uncontrollable spontaneous laughter, jerky movements, and other motor and mental symptoms.
- The Prader-Willi/Angelman syndromes imply that the deleted genes normally behave differently in offspring, depending on whether they belong to the maternal or the paternal homologue.

- In other words, homologous chromosomes inherited from males and females are somehow differently *imprinted*, which causes them to be functionally different in the offspring.

Genomic imprinting = Process that induces intrinsic changes in chromosomes inherited from males and females; causes certain genes to be differently expressed in the offspring depending upon whether the alleles were inherited from the ovum or from the sperm cell (see Campbell, Figure 15.14).

- According to this hypothesis, certain genes are imprinted in some way each generation, and the imprint is different depending on whether the genes reside in females or in males.
- The same alleles may have different effects on offspring depending on whether they are inherited from the mother or the father.
- In the new generation, both maternal and paternal imprints can be reversed in gamete-producing cells, and all the chromosomes are re-coded according to the sex of the individual in which they now reside.
- *DNA methylation* may be one mechanism for genomic imprinting

Affecting about one in every 1500 males and one in every 2500 females, *fragile X syndrome* is the most common genetic cause of mental retardation.

- The "fragile *X*" is an abnormal *X* chromosome, the tip of which hangs on the rest of the chromosome by a thin DNA thread.

Fragile X syndrome's complex expression may be a consequence of maternal genomic imprinting.

- The syndrome is more likely to appear if the abnormal *X* chromosome is inherited from the mother rather than the father; this is consistent with the disorder being more common in males.
- Fragile x is unusual in that maternal imprinting (methylation) does not silence the abnormal allele but rather, somehow causes the syndrome.

C. Extranuclear genes exhibit a non-Mendelian pattern of inheritance

There are some exceptions to the chromosome theory of inheritance.

- Extranuclear genes are found in cytoplasmic organelles such as plastids and mitochondria.
- These cytoplasmic genes are not inherited in Mendelian fashion, because they are not distributed by segregating chromosomes during meiosis.

In plants, a zygote receives its plastids from the ovum, not from pollen. Consequently, offspring receive only maternal cytoplasmic genes.

- Cytoplasmic genes in plants were first descaribed by Karl Corens (1909) when he noticed that plant coloration of an ornamental species was determined by the seed bearing plants and not by the pollen producing plants (see Campbell, Figure 15.15).
- It is now known that maternal plastid genes control variegation of leaves.

In mammals, inheritance of mitochondrial DNA is also exclusively maternal.

- Since the ovum contributes most of the cytoplasm to the zygote, the mitochondria are all maternal in origin.

REFERENCES

Campbell, N., et al. *Biology*. 5th ed. Menlo Park, California: Benjamin/Cummings, 1998.

Griffith, A.J.E., J.H. Miller, D.T. Suzuki, R.C. Lewontin, and W.M. Gelbart. *An Introduction to Genetic Analysis*. 5th ed. New York: W.H. Freeman, 1993.

Kowles, R.V. *Genetics, Society and Decisions*. 1st ed. Columbus, Ohio: Charles E. Merrill, 1985.

CHAPTER 16
THE MOLECULAR BASIS
OF INHERITANCE

OUTLINE

I. DNA as the Genetic Material
    A. The search for the genetic material led to DNA: *science as a process*
    B. Watson and Crick discovered the double helix by building models to conform to X-ray data: *science as a process*
II. DNA Replication and Repair
    A. During DNA replication, base-pairing enables existing DNA strands to serve as templates for new complementary strands
    B. A large team of enzymes and other proteins carries out DNA replication
    C. Enzymes proofread DNA during its replication and repair damage to existing DNA
    D. The ends of DNA molecules pose a special function.

OBJECTIVES

After reading this chapter and attending lecture, the student should be able to:

1. Explain why researchers originally thought protein was the genetic material.
2. Summarize experiments performed by the following scientists, which provided evidence that DNA is the genetic material:
    a. Frederick Griffith
    b. Alfred Hershey and Martha Chase
    c. Erwin Chargaff
3. List the three components of a nucleotide.
4. Distinguish between deoxyribose and ribose.
5. List the nitrogen bases found in DNA, and distinguish between pyrimidine and purine.
6. Explain how Watson and Crick deduced the structure of DNA, and describe what evidence they used.
7. Explain the "base-pairing rule" and describe its significance.
8. Describe the structure of DNA, and explain what kind of chemical bond connects the nucleotides of each strand and what type of bond holds the two strands together.
9. Explain, in their own words, semiconservative replication, and describe the Meselson-Stahl experiment.
10. Describe the process of DNA replication, and explain the role of helicase, single strand binding protein, DNA polymerase, ligase, and primase.
11. Explain what energy source drives endergonic synthesis of DNA.
12. Define antiparallel, and explain why continuous synthesis of both DNA strands is not possible.

13. Distinguish between the leading strand and the lagging strand.

14. Explain how the lagging strand is synthesized when DNA polymerase can add nucleotides only to the 3′ end.

15. Explain the role of DNA polymerase, ligase, and repair enzymes in DNA proofreading and repair.

KEY TERMS

phages	DNA polymerase	primase	nuclease
double helix	leading strand	helicase	excision repair
semiconservative model	lagging strand	single-strand binding	telomerase
origins of replication	DNA ligase	protein	
replication fork	primer	mismatch repair	

LECTURE NOTES

Deoxyribonucleic acid, or DNA, is genetic material. DNA is the substance of Mendel's heritable factors and of Morgan's genes on chromosomes. Inheritance has its molecular basis in the precise replication and transmission of DNA from parent to offspring.

I. DNA as the Genetic Material

A. The search for the genetic material led to DNA: *science as a process*

By the 1940s, scientists knew that chromosomes carried hereditary material and consisted of DNA and protein. Most researchers thought protein was the genetic material because:

- Proteins are macromolecules with great heterogeneity and functional specificity.
- Little was known about nucleic acids.
- The physical and chemical properties of DNA seemed too uniform to account for the multitude of inherited traits.

1. Evidence that DNA can transform bacteria

In 1928, Frederick Griffith performed experiments that provided evidence that genetic material is a specific molecule.

Griffith was trying to find a vaccine against *Streptococcus pneumoniae*, a bacterium that causes pneumonia in mammals. He knew that:

- There were two distinguishable strains of the pneumococcus: one produced smooth colonies (S) and the other rough colonies (R).
- Cells of the smooth strain were encapsulated with a polysaccharide coat and cells of the rough strain were not.
- These alternative phenotypes (S and R) were inherited.

Griffith performed four sets of experiments:

Experiment: Griffith injected live S strain of *Streptococcus pneumoniae* into mice.

> *Results*: Mice died of pneumonia.

> *Conclusions*: Encapsulated strain was pathogenic.

Experiment: Mice were injected with live R strain.

> *Results*: Mice survived and were healthy.

> *Conclusions*: The bacterial strain lacking the polysaccharide coat was non-pathogenic.

Experiment: Mice were injected with heat-killed S strain of pneumococcus.

Results: Mice survived and were healthy.

Conclusions: Polysaccharide coat did not cause pneumonia because it was still present in heat-killed bacteria which proved to be non-pathogenic.

Experiment: Heat-killed S cells mixed with live R cells were injected into mice.

Results: Mice developed pneumonia and died. Blood samples from dead mice contained live S cells.

Conclusions: R cells had acquired from the dead S cells the ability to make polysaccharide coats. Griffith cultured S cells from the dead mice. Since the dividing bacteria produced encapsulated daughter cells, he concluded that this newly acquired trait was inheritable. This phenomenon is now called transformation.

Transformation = Change in phenotype due to the assimilation of external genetic material by a cell

What was the chemical nature of the transforming agent?

- Griffith was unable to answer this question, but other scientists continued the search.

- Griffith's experiments hinted that protein is not the genetic material. Heat denatures protein, yet it did not destroy the transforming ability of the genetic material in the heat-killed S cells.

- In 1944, after a decade of research, Oswald Avery, Maclyn McCarty, and Colin MacLeod discovered that the transforming agent had to be DNA.

- The discovery by Avery and his coworkers was met with skepticism by other scientists, because they still believed protein was a better candidate for the genetic material and so little was known about DNA.

2. Evidence that viral DNA can program cells

More evidence that DNA is the genetic material came from studies of bacteriophages.

Bacteriophage (phage) = Virus that infects bacteria

In 1952, Alfred Hershey and Martha Chase discovered that DNA was the genetic material of a phage known as T2. They knew that T2:

- Was one of many phages to infect the enteric bacterium *Escherichia coli* (*E. coli*).

- Like many other viruses, was little more than DNA enclosed by a protein coat.

- Could quickly reprogram an *E. coli* cell to produce T2 phages and release the viruses when the cell lysed.

What Hershey and Chase did not know was which viral component—DNA or protein—was responsible for reprogramming the host bacterial cell. They answered this question by performing the following experiment (see Campbell, Figure 16.1):

Experiment:

Step 1: Viral protein and DNA were tagged with different radioactive isotopes.

- *Protein tagging*: T2 and *E. coli* were grown in media with radioactive sulfur (^{35}S) which was incorporated only into the phage protein.

- *DNA tagging*: T2 and *E. coli* were grown in media containing radioactive phosphorus (^{32}P) which was incorporated only into the phage DNA.

Step 2: Protein-labeled and DNA-labeled T2 phages were allowed to infect separate samples of nonradioactive *E. coli* cells.

Step 3: Cultures were agitated to shake loose phages that remained outside the bacterial cells.

Step 4: Mixtures were centrifuged forcing the heavier bacterial cells into a pellet on the bottom of the tubes. The lighter viruses remained in the supernatant.

Step 5: Radioactivity in the pellet and supernatant was measured and compared.

Results:

1. In tubes with E. coli infected with protein-labeled T2, most of the radioactivity was in the supernatant with viruses.
2. In tubes with E. coli infected with DNA-labeled T2, most of the radioactivity was in the pellet with the bacterial cells.
3. When the bacteria containing DNA-labeled phages were returned to culture medium, the bacteria released phage progeny which contained 32P in their DNA.

Conclusions:

1. Viral proteins remain outside the host cell.
2. Viral DNA is injected into the host cell.
3. Injected DNA molecules cause cells to produce additional viruses with more viral DNA and proteins.
4. These data provided evidence that nucleic acids rather than proteins were the hereditary material.

3. Additional evidence that DNA is the genetic material of cells

Hershey and Chase's experiments provided evidence that DNA is the hereditary material in viruses. Additional evidence pointed to DNA as the genetic material in eukaryotes as well.

Some circumstantial evidence was:

- A eukaryotic cell doubles its DNA content prior to mitosis.
- During mitosis, the doubled DNA is equally divided between two daughter cells.
- An organism's diploid cells have twice the DNA as its haploid gametes.

Experimental evidence for DNA as the hereditary material in eukaryotes came from the laboratory of Erwin Chargaff. In 1947, he analyzed the DNA composition of different organisms. Using paper chromatography to separate nitrogenous bases, Chargaff reported the following:

- DNA composition is species-specific.
- The amount and ratios of nitrogenous bases vary from one species to another.
- This source of molecular diversity made it more credible that DNA is the genetic material.
- In every species he studied, there was a regularity in base ratios.
  - The number of adenine (A) residues approximately equaled the number of thymines (T), and the number of guanines (G) equaled the number of cytosines (C).
  - The A=T and G≡C equalities became known later as *Chargaff's rules*. Watson and Crick's structural model for DNA explained these rules.

B. Watson and Crick discovered the double helix by building models to conform to x-ray data: *science as a process*

> Before this discussion, it may be helpful to review material from Chapter 7, such as: components of nucleotides; nitrogenous bases in DNA; difference between purine and pyrimidine; the kinds of chemical bonds that connect nucleotides of each DNA strand; and the type of bond that holds the two strands together. Because so much time elapses between the lectures on organic molecules and molecular genetics, students may forget crucial information necessary to understand this material. For this reason, you may find it best to defer discussion of nucleic acids (from Chapter 7) until this point in the course.

By the 1950s, DNA was accepted as the genetic material, and the covalent arrangement in a nucleic acid polymer was well established (see Campbell, Figure 16.2). The three-dimensional structure of DNA, however, was yet to be discovered.

Among scientists working on the problem were the following:

    Linus Pauling, California Institute of Technology

    Maurice Wilkins and Rosalind Franklin, King's College in London

    James D. Watson (American) and Francis Crick, Cambridge University

James Watson went to Cambridge to work with Francis Crick who was studying protein structure with *X-ray crystallography*.

Watson saw an X-ray photo of DNA produced by Rosalind Franklin at King's College, London (see Campbell, Figure 16.3). Watson and Crick deduced from Franklin's X-ray data that:

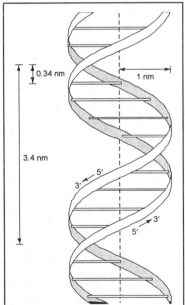

    a. DNA is a helix with a uniform width of 2 nm. This width suggested that it had two strands.

    b. Purine and pyrimidine bases are stacked .34 nm apart.

    c. The helix makes one full turn every 3.4 nm along its length.

    d. There are ten layers of nitrogenous base pairs in each turn of the helix.

Watson and Crick built scale models of a double helix that would conform to the X-ray data and the known chemistry of DNA.

- One of their unsuccessful attempts placed the sugar-phosphate chains inside the molecule.
- Watson next put the sugar-phosphate chains on the outside which allowed the more hydrophobic nitrogenous bases to swivel to the interior away from the aqueous medium (see Campbell, Figure 16.4).
- Their proposed structure was a ladder-like molecule twisted into a spiral, with sugar-phosphate backbones as uprights and pairs of nitrogenous bases as rungs.

- The two sugar-phosphate backbones of the helix were *antiparallel*; that is, they ran in opposite directions.

Watson and Crick finally solved the problem of DNA structure by proposing that there is a specific pairing between nitrogenous bases. After considering several arrangements, they concluded:

- To be consistent with a 2 nm width, a purine on one strand must pair (by hydrogen bonding) with a pyrimidine on the other.
- Base structure dictates which pairs of bases can hydrogen bonds. The base pairing rule is that adenine can only pair with thymine, and guanine with cytosine (see Campbell, Figure 16.5).

Purines	Pyrimidines	Possible Base Pairs	Number of Hydrogen Bonds
Adenine (A)	Thymine (T)	A – T	2
Guanine (G)	Cytosine (C)	G – C	3

ADENINE (A)   THYMINE (T)   GUANINE (G)   CYTOSINE (C)

The base-pairing rule is significant because:

- It explains Chargaff's rules. Since A must pair with T, their amounts in a given DNA molecule will be about the same. Similarly, the amount of G equals the amount of C.
- It suggests the general mechanisms for DNA replication. If bases form specific pairs, the information on one strand complements that along the other.
- It dictates the combination of complementary base pairs, but places no restriction on the linear sequence of nucleotides along the length of a DNA strand. The sequence of bases can be highly variable which makes it suitable for coding genetic information.
- Though hydrogen bonds between paired bases are weak bonds, collectively they stabilize the DNA molecule. Van der Waals forces between stacked bases also help stabilize DNA.

I.   **DNA Replication and Repair**

A. **During DNA replication, base pairing enables existing DNA strands to serve as templates for new complementary strands**

In April 1953, Watson and Crick's new model for DNA structure, the double helix, was published in the British journal *Nature*. This model of DNA structure suggested a *template* mechanism for DNA replication.

- Watson and Crick proposed that genes on the original DNA strand are copied by a specific pairing of complementary bases, which creates a complementary DNA strand.

- The complementary strand can then function as a template to produce a copy of the original strand.

In a second paper, Watson and Crick proposed that during DNA replication (see also Campbell, Figure 16.6):

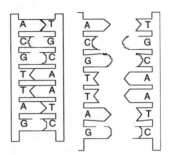

1. The two DNA strands separate.

2. Each strand is a template for assembling a complementary strand.

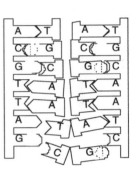

3. Nucleotides line up singly along the template strand in accordance with the base-pairing rules (A-T and G-C).

4. Enzymes link the nucleotides together at their sugar-phosphate groups.

Watson and Crick's model is a *semiconservative model* for DNA replication.

- They predicted that when a double helix replicates, each of the two daughter molecules will have one old or *conserved* strand from the parent molecule and one newly created strand.

- In the late 1950s, Matthew Meselson and Franklin Stahl provided the experimental evidence to support the semiconservative model of DNA replication. A brief description of the experimental steps follows (see Campbell, Figure 16.8.).

Hypotheses:

There were three alternate hypotheses for the pattern of DNA replication (see Campbell, Figure 16.7).:

a. If DNA replication is *conservative*, then the parental double helix should remain intact and the second DNA molecule should be constructed as entirely new DNA.

b. If DNA replication is *semiconservative*, then each of the two resulting DNA molecules should be composed of one original or conserved strand (template) and one newly created strand.

c. If DNA replication is *dispersive*, then both strands of the two newly produced DNA molecules should contain a mixture of old and new DNA.

Experiment:

Step 1: Labeling DNA strands with ^{15}N.

E. coli was grown for several generations in a medium with heavy nitrogen (^{15}N). As *E. coli* cells reproduced, they incorporated the ^{15}N into their nitrogenous bases.

Step 2: Transfer of *E. coli* to a medium with ^{14}N.

E. coli cells grown in heavy medium were transferred to a medium with light nitrogen, ^{14}N. There were three experimental classes of DNA based upon times after the shift from heavy to light medium:

a. Parental DNA from *E. coli* cells grown in heavy medium

b. First-generation DNA extracted from *E. coli* after one generation of growth in the light medium

c. Second-generation DNA extracted from *E. coli* after two replications in the light medium

Step 3: Separation of DNA classes based upon density differences.

Meselson and Stahl used a new technique to separate DNA based on density differences between ^{14}N and ^{15}N.

Isolated DNA was mixed with a CsCl
solution and placed in an ultracentrifuge.

↓

DNA-CsCl solution was centrifuged at high
speed for several days.

↓

Centrifugal force created a CsCl gradient
with increased concentration toward the
bottom of the centrifuge tube.

↓

DNA molecules moved to a position in the
tube where their density equaled that of the
CsCl solution. Heavier DNA molecules were
closer to the bottom and lighter DNA
molecules were closer to the top where the
CsCl solution was less dense.

Results:

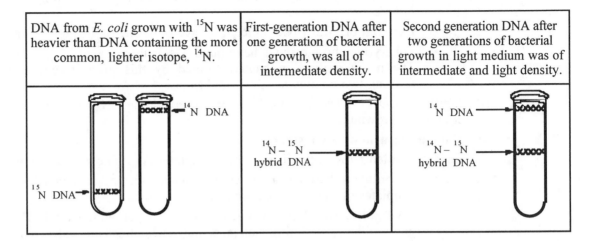

| DNA from *E. coli* grown with ^{15}N was heavier than DNA containing the more common, lighter isotope, ^{14}N. | First-generation DNA after one generation of bacterial growth, was all of intermediate density. | Second generation DNA after two generations of bacterial growth in light medium was of intermediate and light density. |

Conclusions:

Their results supported the hypothesis of semiconservative replication for DNA.

- The first generation DNA was all hybrid DNA containing one heavy parental strand and one newly synthesized light strand.

- These results were predicted by the semiconservative model. If DNA replication was conservative, two classes of DNA would be produced. The heavy parental DNA would be conserved and newly synthesized DNA would be light. There would be no intermediate density hybrid DNA.

- The results for first-generation DNA eliminated the possibility of conservative replication, but did not eliminate the possibility of dispersive replication.

- Consequently, Meselson and Stahl examined second-generation DNA. The fact that there were two bands, one light and one hybrid band supported the semiconservative model and allowed the investigators to rule out dispersive replication. If DNA replication was dispersive, there would have been only one intermediate density band between light and hybrid DNA.

B. A large team of enzymes and other proteins carries out DNA replication

The general mechanism of DNA replication is conceptually simple, but the actual process is:

- *Complex.* The helical molecule must untwist while it copies its two antiparallel strands simultaneously. This requires the cooperation of over a dozen enzymes and other proteins.

- *Extremely rapid.* In prokaryotes, up to 500 nucleotides are added per second. It takes only a few hours to copy the 6 billion bases of a single human cell.

- *Accurate.* Only about one in a billion nucleotides is incorrectly paired.

Generally similar in prokaryotes and eukaryotes.

1. Getting started: origins of replication

DNA replication begins at special sites called *origins of replication* that have a specific sequence of nucleotides (see Campbell, Figure 16.9).

- Specific proteins required to initiate replication bind to each origin.

- The DNA double helix opens at the origin and *replication forks* spread in both directions away from the central initiation point creating a *replication bubble*.
- Bacterial or viral DNA molecules have only one replication origin.
- Eukaryotic chromosomes have hundreds or thousands of replication origins. The many replication bubbles formed by this process, eventually merge forming two continuous DNA molecules.

Replication forks = The Y-shaped regions of replicating DNA molecules where new strands are growing.

2. Elongating a new DNA strand

Enzymes called *DNA polymerases* catalyze synthesis of a new DNA strand.

- According to base-pairing rules, new nucleotides align themselves along the templates of the old DNA strands.
- *DNA polymerase* links the nucleotides to the growing strand. These strands grow in the 5' → 3' direction since new nucleotides are added only to the 3' end of the growing strand.

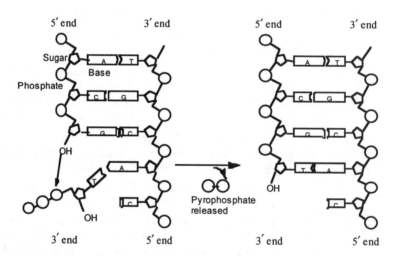

Hydrolysis of *nucleoside triphosphates* provides the energy necessary to synthesize the new DNA strands.

Nucleoside triphosphate = Nucleotides with a triphosphate (three phosphates) covalently linked to the 5' carbon of the pentose.

- Recall that the pentose in DNA is deoxyribose, and the pentose in RNA is ribose.
- Nucleoside triphosphates that are the building blocks for DNA lose two phosphates (pyrophosphate group) when they form covalent linkages to the growing chain (see Campbell, Figure 16.10).
- Exergonic hydrolysis of this phosphate bond drives the endergonic synthesis of DNA; it provides the required energy to form the new covalent linkages between nucleotides.

3. The problem of antiparallel DNA strands

Continuous synthesis of both DNA strands at a replication fork is not possible, because:

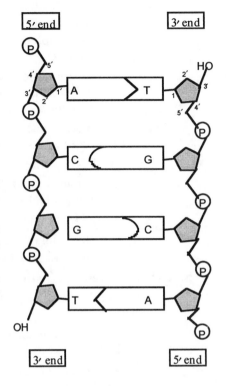

- The sugar phosphate backbones of the two complementary DNA strands run in opposite directions; that is, they are *antiparallel* (see Campbell, Figure 16.11).

- Recall that each DNA strand has a distinct polarity. At one end (3′ end), a hydroxyl group is attached to the 3′ carbon of the terminal deoxyribose; at the other end (5′ end), a phosphate group is attached to the 5′ carbon of the terminal deoxyribose.

- DNA polymerase can only elongate strands in the 5′ to 3′ direction.

The problem of antiparallel DNA strands is solved by the continuous synthesis of one strand (*leading strand*) and discontinuous synthesis of the complementary strand (*lagging strand*).

Leading strand = The DNA strand which is synthesized as a single polymer in the 5′ → 3′ direction towards the replication fork.

Lagging strand = The DNA strand that is discontinuously synthesized against the overall direction of replication.

- Lagging strand is produced as a series of short segments called *Okazaki fragments* which are each synthesized in the 5′ → 3′ direction.

- Okazaki fragments are 1000 to 2000 nucleotides long in bacteria and 100 to 200 nucleotides long in eukaryotes.

- The many fragments are ligated by *DNA ligase*, a linking enzyme that catalyzes the formation of a covalent bond between the 3′ end of each new Okazaki fragment to the 5′ end of the growing chain.

4. Priming DNA synthesis

Before new DNA strands can form, there must be small preexisting primers to start the addition of new nucleotides.

Primer = Short RNA segment that is complementary to a DNA segment and that is necessary to begin DNA replication.

- Primers are short segments of RNA polymerized by an enzyme called *primase* (see Campbell, Figure 16.13).
- A portion of the parental DNA serves as a template for making the primer with a complementary base sequence that is about ten nucleotides long in eukaryotes.
- Primer formation must precede DNA replication, because DNA polymerase can only add nucleotides to a polynucleotide that is already correctly base-paired with a complementary strand.

Only one primer is necessary for replication of the leading strand, but many primers are required to replicate the lagging strand.

- An RNA primer must initiate the synthesis of each Okazaki fragment.
- The many Okazaki fragments are ligated in two steps to produce a continuous DNA strand:
  - DNA polymerase removes the RNA primer and replaces it with DNA.
  - DNA ligase catalyzes the linkage between the 3' end of each new Okazaki fragment to the 5' end of the growing chain.

5. Other proteins assisting DNA replication

There are two types of proteins involved in the separation of parental DNA strands:

a. *Helicases* are enzymes which catalyze unwinding of the parental double helix to expose the template.

b. *Single-strand binding proteins* are proteins which keep the separated strands apart and stabilize the unwound DNA until new complementary strands can be synthesize.

Campbell, Figure 16.14 summarizes the functions of the main proteins that cooperate in DNA replication.

Campbell, Figure 16.15 is a visual summary of DNA replication.

C. Enzymes proofread DNA during its replication and repair damage in existing DNA

DNA replication is highly accurate, but this accuracy is not solely the result of base-pairing specificity.

- Initial pairing errors occur at a frequency of about one in ten thousand, while errors in a complete DNA molecule are only about one in one billion.
- DNA can be repaired as it is being synthesized (e.g., *mismatch repair*) or after accidental changes in existing DNA (e.g., *excision repair*).

Mismatch repair, corrects mistakes when DNA is synthesized.

- In bacteria, DNA polymerase proofreads each newly added nucleotide against its template. If polymerase detects an incorrectly paired nucleotide, the enzyme removes and replaces it before continuing with synthesis.
- In eukaryotes, additional proteins as well as polymerase participate in mismatch repair. A hereditary defect in one of these proteins has been associated with a form of colon cancer. Apparently, DNA errors accumulate in the absence of adequate proofreading.

Excision repair, corrects accidental changes that occur in existing DNA (see Campbell, Figure 16.17).

- Accidental changes in DNA can result from exposure to reactive chemicals, radioactivity, X-rays and ultraviolet light.
- There are more than fifty different types of DNA repair enzymes that repair damage. For example, in *excision repair*:

- The damaged segment is excised by one *repair enzyme* and the remaining gap is filled in by base-pairing nucleotides with the undamaged strand.
- *DNA polymerase* and *DNA ligase* are enzymes that catalyze the filling-in process.

D. The ends of DNA molecules pose a special problem

Because DNA polymerases can only add nucleotides to the 3' end of a preexisting polynucleotide, repeated replication of linear DNA, such as that possessed by all eukaryotes, would result in successively shorter molecules, potentially deleting genes (see Campbell, Figure 16.17).

- Prokaryotes don't have this problem because they possess circular DNA.
- Eukaryotic DNA is flanked by *telomeres,* repeats of short noncoding nucleotide sequences (see Campbell, Figure 16.18).

REFERENCES

Alberts, B., D. Bray, J. Lewis, M. Raff, K. Roberts and J.D. Watson. *Molecular Biology of the Cell.* 3rd ed. New York: Garland, 1994.

Becker, W.M. and D.W. Deamer. *The World of the Cell.* 3rd ed. Redwood City, California: Benjamin/Cummings, 1996.

Campbell, N., et al. *Biology.* 5th ed. Menlo Park, California: Benjamin/Cummings, 1998.

Culotta, E. and D.E. Koshland, Jr. "DNA Repair Works its Way to the Top." *Science.* December 23, 1994.

CHAPTER 17
FROM GENE TO PROTEIN

OUTLINE

I. The Connection between Genes and Proteins
   A. The study of metabolic defects provided evidence that genes specify proteins: *science as a process*
   B. Transcription and translation are the two main processes linking gene to protein: *an overview*
   C. In the genetic code, nucleotide triplets specify amino acids
   D. The genetic code must have evolved very early in the history of life

II. The Synthesis and Processing of RNA
   A. Transcription is the DNA-directed synthesis of RNA: *a closer look*
   B. Eukaryotic cells modify RNA after transcription

III. The Synthesis of Protein
   A. Translation is the RNA-directed synthesis of a polypeptide: *a closer look*
   B. Signal peptides target some eukaryotic polypeptides to specific locations in the cell
   C. RNA plays multiple roles in the cell: a review
   D. Comparing protein synthesis in prokaryotes and eukaryotes: *a review*
   E. Point mutations can affect protein structure and function
   F. What is a gene? *revisiting the question*

OBJECTIVES

After reading this chapter and attending lecture, the student should be able to:

1. Give early experimental evidence that implicated proteins as the links between genotype and phenotype.

2. Describe Beadle and Tatum's experiments with *Neurospora*, and explain the contribution they made to our understanding of how genes control metabolism.

3. Distinguish between "one gene—one enzyme" hypothesis and "one gene—one polypeptide," and explain why the original hypothesis was changed.

4. Explain how RNA differs from DNA.

5. In their own words, briefly explain how information flows from gene to protein.

6. Distinguish between transcription and translation.

7. Describe where transcription and translation occur in prokaryotes and in eukaryotes; explain why it is significant that in eukaryotes, transcription and translation are separated in space and time.

8. Define codon, and explain what relationship exists between the linear sequence of codons on mRNA and the linear sequence of amino acids in a polypeptide.

9. List the three stop codons and the one start codon.

10. Explain in what way the genetic code is redundant and unambiguous.

11. Explain the evolutionary significance of a nearly universal genetic code.
12. Explain the process of transcription including the three major steps of initiation, elongation, and termination.
13. Describe the general role of RNA polymerase in transcription.
14. Explain how RNA polymerase recognizes where transcription should begin.
15. Specifically, describe the primary functions of RNA polymerase II.
16. Distinguish among mRNA, tRNA, and rRNA.
17. Describe the structure of tRNA and explain how the structure is related to function.
18. Given a sequence of bases in DNA, predict the corresponding codons transcribed on mRNA and the corresponding anticodons of tRNA.
19. Describe the wobble effect.
20. Explain how an aminoacyl-tRNA synthetase matches a specific amino acid to its appropriate tRNA; describe the energy source that drives this endergonic process.
21. Describe the structure of a ribosome, and explain how this structure relates to function.
22. Describe the process of translation including initiation, elongation, and termination and explain what enzymes, protein factors, and energy sources are needed for each stage.
23. Explain what determines the primary structure of a protein and describe how a polypeptide must be modified before it becomes fully functional.
24. Describe what determines whether a ribosome will be free in the cytosol or attached to rough ER.
25. Explain how proteins can be targeted for specific sites within the cell.
26. Describe the difference between prokaryotic and eukaryotic mRNA.
27. Explain how eukaryotic mRNA is processed before it leaves the nucleus.
28. Describe some biological functions of introns and gene splicing.
29. Explain why base-pair insertions or deletions usually have a greater effect than base-pair substitutions.
30. Describe how mutagenesis can occur.

KEY TERMS

auxotroph	transcription unit	anticodon	point mutation
one gene–one polypeptide	transcription factors	wobble	base-pair substitution
transcription	transcription initiation complex	aminoacyl-tRNA synthetases	missense mutation
messenger RNA (mRNA)	TATA box	ribosomal RNA (rRNA)	nonsense mutation
translation	terminator	P site	insertion
RNA processing	5' cap	A site	deletion
primary transcript	poly (A) tail	E site	frameshift mutation
triplet code	RNA splicing	polyribosome	mutagens
template strand	intron	signal peptide	Ames test
codon	exon	signal-recognition particle (SRP)	
reading frame	spliceosome	mutation	
RNA polymerase	domain	point mutation	
	transfer RNA (tRNA)		

LECTURE NOTES

Inherited instructions in DNA direct protein synthesis. Thus, proteins are the links between genotype and phenotype, since proteins are directly involved in the expression of specific phenotypic traits.

I.   The Connection between Genes and Proteins

A.   The study of metabolic defects provided evidence that genes specify proteins: *science as a process*

Archibald Garrod was the first to propose the relationship between genes and proteins (1909).

- He suggested that genes dictate phenotypes through enzymes that catalyze reactions.
- As a physician, Garrod was familiar with inherited diseases which he called "inborn errors in metabolism." He hypothesized that such diseases reflect the patient's inability to make particular enzymes.
- One example he studied was *alkaptonuria*, which causes the afflicted person's urine to turn black.
    - People with alkaptonuria accumulate alkapton in their urine, causing it to darken on contact with air.
    - Garrod reasoned that alkaptonurics, unlike normal individuals, lack the enzyme that breaks down alkapton.

1.   How genes control metabolism

Garrod's hypothesis was confirmed several decades later by research which determined that specific genes direct production of specific enzymes.

- Biochemists found that cells synthesize and degrade organic compounds via metabolic pathways, with each sequential step catalyzed by a specific enzyme.
- Geneticists George Beadle and Boris Ephrussi (1930s) studied eye color in *Drosophila*. They speculated that mutations affecting eye color block pigment synthesis by preventing enzyme production at certain steps in the pigment synthesis pathway.

George Beadle and Edward Tatum were later able to demonstrate the relationship between genes and enzymes by studying mutants of a bread mold, *Neurospora crassa* (see Campbell, Figure 17.1).

- Wild-type *Neurospora* in laboratory colonies can survive on *minimal medium* All other molecules needed by the mold are produced by its own metabolic pathways from this minimal nutrient source.
- Beadle and Tatum searched for mutants or *auxotrophs* that could not survive on minimal medium because they lacked the ability to synthesize essential molecules.
- Mutants were identified by transferring fragments of growing fungi (in complete medium) to vials containing minimal medium. Fragments that didn't grow were identified as auxotrophic mutants.

Auxotroph = (Auxo = to augment; troph = nourishment); nutritional mutants that can only be grown on *minimal medium* augmented with nutrients not required by the wild type

Minimal medium = Support medium that is mixed only with molecules required for the growth of wild-type organisms

- Minimal medium for *Neurospora* contains inorganic salts, sucrose, and the vitamin biotin.
- Nutritional mutants cannot survive only on minimal medium.

Complete growth medium= Minimal medium supplemented with all 20 amino acids and some other nutrients

- Nutritional mutants can grow on complete growth medium, since all essential nutrients are provided.

Beadle and Tatum then identified specific metabolic defects (from mutations) by transferring fragments of auxotrophic mutants growing on complete growth medium to vials containing minimal medium each supplemented with only *one* additional nutrient.

- Vials where growth occurred indicated the metabolic defect, since the single supplement provided the necessary component.
- For example, if a mutant grew on minimal medium supplemented with only arginine, it could be concluded that the mutant was defective in the arginine synthesis pathway.

Experiment:

Beadle and Tatum experimented further to more specifically describe the defect in the multistep pathway that synthesizes the amino acid arginine.

- Arginine synthesis requires three steps each catalyzed by a specific enzyme:

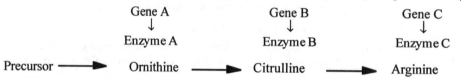

- They distinguished between three classes of arginine auxotrophs by adding either arginine, citrulline, or ornithine to the medium and seeing if growth occurred.

Results:

Some mutants required arginine, some either arginine or citrulline, and others could grow when any of the three were added.

	Minimal Medium (MM)	MM plus Ornithine	Mm plus Citrulline	MM plus Arginine
Wild Type	+	+	+	+
Class I Mutants	–	+	+	+
Class II Mutants	–	–	+	+
Class III Mutants	–	–	–	+

+ = growth, – = no growth

Conclusions:

Beadle and Tatum deduced from their data that the three classes of mutants each lacked a different enzyme and were thus blocked at different steps in the arginine synthesis pathway.

- Class I mutants lacked enzyme A; Class II mutants lacked enzyme B; and Class III mutants lacked enzyme C.

- Assuming that each mutant was defective in a single gene, they formulated the *one gene-one enzyme* hypothesis, which states that the function of a gene is to dictate the production of a specific enzyme.

2. **One gene—one polypeptide**

Beadle and Tatum's one gene-one enzyme hypothesis has been slightly modified:

- While most enzymes are proteins, many proteins are not enzymes. Proteins that are not enzymes are still, nevertheless, gene products.
- Also, many proteins are comprised of two or more polypeptide chains, each chain specified by a different gene (e.g., globulin chains of hemoglobin).

As a result of this new information, Beadle and Tatum's hypothesis has been restated as *one gene-one polypeptide*.

> As we will see later, even this notion is no longer tenable given that a) differential processing of a single RNA transcript can lead to the synthesis of numerous different proteins, and b) not all RNA is translated into protein.

B. **Transcription and translation are the two main processes linking gene to protein:** *an overview*

Ribonucleic acid (RNA) links DNA's genetic instructions for making proteins to the process of protein synthesis. It copies or transcribes the message from DNA and then translates that message into a protein.

- RNA, like DNA, is a nucleic acid or polymer of nucleotides.
- RNA structure differs from DNA in the following ways:
    - The five-carbon sugar in RNA nucleotides is *ribose* rather than deoxyribose.
    - The nitrogenous base *uracil* is found in place of thymine.

The linear sequence of nucleotides in DNA ultimately determines the linear sequence of amino acids in a protein.

- Nucleic acids are made of four types of nucleotides which differ in their nitrogenous bases. Hundreds or thousands of nucleotides long, each gene has a specific linear sequence of the four possible bases.
- Proteins are made of 20 types of amino acids linked in a particular linear sequence (the protein's primary structure).
- Information flows from gene to protein through two major processes, *transcription* and *translation* (see Campbell, Figure 17.2).

Transcription = The synthesis of RNA using DNA as a template

- A gene's unique nucleotide sequence is transcribed from DNA to a complementary nucleotide sequence in messenger RNA (mRNA).
- The resulting mRNA carries this transcript of protein-building instructions to the cell's protein-synthesizing machinery.

Translation = Synthesis of a polypeptide, which occurs under the direction of messenger RNA (mRNA)

- During this process, the linear sequence of bases in mRNA is translated into the linear sequence of amino acids in a polypeptide.
- Translation occurs on *ribosomes*, complex particles composed of ribosomal RNA (rRNA) and protein that facilitate the orderly linking of amino acids into polypeptide chains.

Prokaryotes and eukaryotes differ in how protein synthesis is organized within their cells.

- Prokaryotes lack nuclei, so DNA is not segregated from ribosomes or the protein-synthesizing machinery. Thus, transcription and translation occur in rapid succession.

- Eukaryotes have nuclear envelopes that segregate transcription in the nucleus from translation in the cytoplasm; mRNA, the intermediary, is modified before it moves from the nucleus to the cytoplasm where translation occurs. This *RNA processing* occurs only in eukaryotes.

C. In the genetic code, nucleotide triplets specify amino acids

There is not a one-to-one correspondence between the nitrogenous bases and the amino acids they specify, since there are only 4 nucleotides and 20 amino acids.

- A two-to-one correspondence of bases to amino acids would only specify 16 (4^2) of the 20 amino acids.
- A three-to-one correspondence of bases to amino acids would specify 64 (4^3) amino acids.

Researchers have verified that the flow of information from a gene to a protein is based on a triplet code (see Campbell, Figure 17.3).

- Triplets of nucleotides are the smallest units of uniform length to allow translation into all 20 amino acids with plenty to spare.
- These three-nucleotide "words" are called *codons*.

Codon = A three-nucleotide sequence in mRNA that specifies which amino acid will be added to a growing polypeptide or that signals termination; the basic unit of the genetic code

Genes are not directly translated into amino acids, but are first transcribed as codons into mRNA.

- For each gene, only one of the two DNA strands (the *template strand*) is transcribed.
- The complementary nontemplate strand is the parental strand for making a new template when DNA replicates.
- The same DNA strand can be the template strand for some genes and the nontemplate strand for others.

An mRNA is complementary to the DNA template from which it is transcribed.

- For example, if the triplet nucleotide sequence on the template DNA strand is CCG; GGC, the codon for glycine, will be the complementary mRNA transcript.
- Recall that according to the base-pairing rules, uracil (U) in RNA is used in place of thymine (T); uracil thus base-pairs with adenine (A).

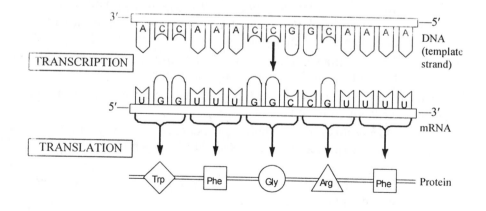

During translation, the linear sequence of codons along mRNA is translated into the linear sequence of amino acids in a polypeptide.

- Each mRNA codon specifies which one of 20 amino acids will be incorporated into the corresponding position in a polypeptide.
- Because codons are base triplets, the number of nucleotides making up a genetic message is three times the number of amino acids making up the polypeptide product.

1. Cracking the genetic code

The first codon was deciphered in 1961 by Marshall Nirenberg of the National Institutes of Health.

- He synthesized an mRNA by linking only uracil-bearing RNA nucleotides, resulting in UUU codons.
- Nirenberg added this "poly U" to a test-tube mixture containing the components necessary for protein synthesis. The artificial mRNA (poly U) was translated into a polypeptide containing a string of only one amino acid, phenylalanine.
- Nirenberg concluded that the mRNA codon UUU specifies the amino acid phenylalanine.
- These same techniques were used to determine amino acids specified by the codons AAA, GGG, and CCC.

More elaborate techniques allowed investigators to determine all 64 codons by the mid-1960s (see Campbell, Figure 17.4).

- 61 of the 64 triplets code for amino acids.
- The triplet AUG has a *dual* function—it is the start signal for translation and codes for methionine.
- Three codons do not code for amino acids, but signal termination (UAA, UAG, and UGA).

There is redundancy in the genetic code, but no ambiguity.

- *Redundancy* exists since two or more codons differing only in their third base can code for the same amino acid (UUU and UUC both code for phenylalanine).
- *Ambiguity* is absent, since codons code for only one amino acid.

The correct ordering and grouping of nucleotides is important in the molecular language of cells. This ordering is called the *reading frame.*

Reading frame = The correct grouping of adjacent nucleotide triplets into codons that are in the correct sequence on mRNA.

- For example, the sequence of amino acids Trp–Phe–Gly–Arg–Phe can be assembled in the correct order only if the mRNA codons UGGUUUGGCCGUUUU are read in the correct sequence and groups.
- The cell reads the message in the correct frame as a series of *nonoverlapping* three-letter words: UGG–UUU–GGC–CGU–UUU.

D. The genetic code must have evolved very early in the history of life

The genetic code is shared nearly universally among living organisms.

- For example, the RNA codon CCG is translated into proline in all organisms whose genetic codes have been examined.
- The technology exists to transfer genes from one species to another. For example, the human gene for insulin can be inserted into bacteria where it is successfully expressed. Campbell, Figure 17.5 shows the incorporation of a firefly gene into a tobacco plant.

There are some exceptions to this universality:

- Several ciliates (e.g., *Paramecium* and *Tetrahymena*) depart from standard code; codons UAA and UAG are not stop signals, but code for glutamine.
- Mitochondria and chloroplasts have their own DNA that codes for some proteins.
- Mitochondrial genetic codes vary even among organisms; for example, CUA codes for threonine in yeast mitochondria and leucine in mammalian mitochondria.

The fact that the genetic code is shared nearly universally by all organisms indicates that this code was established very early in life's history.

II.   The Synthesis and Processing of RNA

A.   Transcription is the DNA-directed synthesis of RNA: *a closer look*

Transcription of messenger RNA (mRNA) from template DNA is catalyzed by *RNA polymerases*, which:

- Separate the two DNA strands and link RNA nucleotides as they base-pair along the DNA template.
- Add nucleotides only to the 3′ end; thus, mRNA molecules grow in the 5′ to 3′ direction.

There are several types of RNA polymerase.

- Prokaryotes have only one type of RNA polymerase that synthesizes all types of RNA—mRNA, rRNA, and tRNA.
- Eukaryotes have three RNA polymerases that transcribe genes. *RNA polymerase II* is the polymerase that catalyzes mRNA synthesis; it transcribes genes that will be translated into proteins.

Specific DNA nucleotide sequences mark where transcription of a gene begins (*initiation*) and ends (*termination*). Initiation and termination sequences plus the nucleotides in between are called a transcription unit.

Transcription unit = Nucleotide sequence on the template strand of DNA that is transcribed into a single RNA molecule by RNA polymerase; it includes the initiation and termination sequences, as well as the nucleotides in between

- In eukaryotes, a transcription unit contains a single gene, so the resulting mRNA codes for synthesis of only one polypeptide.
- In prokaryotes, a transcription unit can contain several genes, so the resulting mRNA may code for different, but functionally related, proteins.

Transcription occurs in three stages: a) polymerase binding and initiation; b) elongation; and c) termination (see Campbell, Figure 17.6).

1.   RNA polymerase binding and initiation of transcription

RNA polymerases bind to DNA at regions called *promoters*.

Promoter = Region of DNA that includes the site where RNA polymerase binds and where transcription begins (*initiation site*). In eukaryotes, the promoter is about 100 nucleotides long and consists of:

a.   The initiation site, where transcription begins (including which DNA strand serves as template).

b.   A few nucleotide sequences recognized by specific DNA-binding proteins (transcription factors) that help initiate transcription.

In eukaryotes, RNA polymerases cannot recognize the promoter without the help of *transcription factors* (see Campbell, Figure 17.7).

Transcription factors = DNA-binding proteins which bind to specific DNA nucleotide sequences at the promoter and help RNA polymerase recognize and bind to the promoter region, so transcription can begin.

- RNA polymerase II, the enzyme that synthesizes mRNA in eukaryotes, usually cannot recognize a promoter unless a specific transcription factor binds to a region on the promoter called a TATA box.

TATA box = A short nucleotide sequence at the promoter which is rich in thymine (T) and adenine (A) and located about 25 nucleotides upstream from the initiation site.

- RNA polymerase II recognizes the complex between the bound TATA transcription factor and the DNA binding site.
- Once RNA polymerase recognizes and attaches to the promoter region, it probably associates with other transcription factors before RNA synthesis begins.

When active RNA polymerase binds to a promoter, the enzyme separates the two DNA strands at the initiation site, and transcription begins.

2. Elongation of the RNA strand

Once transcription begins, RNA polymerase II moves along DNA and performs two primary functions:

a. It untwists and opens a short segment of DNA exposing about ten nucleotide bases; one of the exposed DNA strands is the template for base-pairing with RNA nucleotides.

b. It links incoming RNA nucleotides to the 3′ end of the elongating strand; thus, RNA grows one nucleotide at a time in the 5′ to 3′ direction.

During transcription, mRNA grows about 30 to 60 nucleotides per second. As the mRNA strand elongates:

- It peels away from its DNA template.
- The nontemplate strand of DNA re-forms a DNA-DNA double helix by pairing with the template strand.

Following in series, several molecules of RNA polymerase II can simultaneously transcribe the same gene.

- Cells can thus produce particular proteins in large amounts.
- The growing RNA strands hang free from each polymerase. The length of each strand varies and reflects how far the enzyme has traveled from the initiation site on template DNA.

3. Termination of transcription

Transcription proceeds until RNA polymerase transcribes a DNA sequence called a terminator. The transcribed terminator functions as the actual termination signal.

- Additional proteins may cooperate with RNA polymerase in termination.
- In eukaryotes, the most common terminator sequence is AAUAAA.

Prokaryotic mRNA is ready for translation as soon as it leaves the DNA template. Eukaryotic mRNA, however, must be processed before it leaves the nucleus and becomes functional.

B. Eukaryotic cells modify RNA after transcription

RNA transcripts in eukaryotes are modified, or processed, before leaving the nucleus to yield functional mRNA. Eukaryotic RNA transcripts can be processed in two ways: a) covalent alteration of both the 3' and 5' ends and b) removal of intervening sequences.

Primary transcript = General term for initial RNA transcribed from DNA

Pre-mRNA = Primary transcript that will be processed to functional mRNA

1. Alteration of pre-mRNA ends

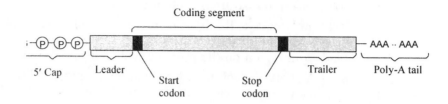

During pre-mRNA processing, both the 5′ and 3′ ends are covalently modified.

5′ cap = Modified guanine nucleotide (guanosine triphosphate) that is added to the 5′ end of mRNA shortly after transcription begins; has two important functions:

- Protects the growing mRNA from degradation by hydrolytic enzymes.
- Helps small ribosomal subunits recognize the attachment site on mRNA's 5′ end. A *leader* segment of mRNA may also be part of the ribosome recognition signal.

Leader sequence = Noncoding (untranslated) sequence of mRNA from the 5′ end to the start codon.

The 3′ end, which is transcribed last, is modified by enzymatic addition of a *poly-A tail*, before the mRNA exits the nucleus.

Poly (A) tail = Sequence of about 30 to 200 adenine nucleotides added to the 3′ end of mRNA before it exits the nucleus.

- May inhibit degradation of mRNA in the cytoplasm.
- May facilitate attachment to small ribosomal subunit
- May regulate protein synthesis by facilitating mRNA's export from the nucleus to the cytoplasm.
- Is not attached directly to the stop codon, but to an untranslated *trailer* segment of mRNA.

Trailer sequence = Noncoding (untranslated) sequence of mRNA from the stop codon to the poly (A) tail.

2. Split genes and RNA splicing

Genes that code for proteins in eukaryotes may not be continuous sequences.

- Coding sequences of a gene are interrupted by noncoding segments of DNA called intervening sequences, or introns.

Introns = Noncoding sequences in DNA that intervene between coding sequences (exons); are initially transcribed, but not translated, because they are excised from the transcript before mature RNA leaves the nucleus. Not all genes possess introns.

- Coding sequences of a gene are called *exons*, because they are eventually expressed (translated into protein).

Exons = Coding sequences of a gene that are transcribed and expressed

- In 1977, Richard Roberts and Philip Sharp independently found evidence for "split genes"; they received a Nobel Prize in 1993 for their discovery.

Introns and exons are both transcribed to form pre-mRNA, but the introns are subsequently removed and the remaining exons linked together during the process of RNA splicing.

RNA splicing = RNA processing that removes introns and joins exons from eukaryotic pre-mRNA; produces mature mRNA that will move into the cytoplasm from the nucleus (see Campbell, Figure 17.9).

- Enzymes excise introns and splice together exons to form an mRNA with a continuous coding sequence.

- RNA splicing also occurs during post-transcriptional processing of tRNA and rRNA.

Though there is much left to be discovered, some details of RNA splicing are now known.

- Each end of an intron has short boundary sequences that accurately signal the RNA splicing sites.
- Small nuclear ribonucleoproteins (snRNPs), play a key role in RNA splicing.

Small nuclear ribonucleoproteins (snRNPs) = Complexes of proteins and small nuclear RNAs that are found only in the nucleus; some participate in RNA splicing; snRNPs is pronounced "snurps".

- These small nuclear particles are composed of:
  1. *Small nuclear RNA (snRNA)*. This small RNA molecule has less than 300 nucleotides—much shorter than mRNA.
  2. *Protein*. Each snRNP possesses several different proteins.
- involved in RNA splicing are part of a larger, more complex assembly called There are various types of snRNPs with different functions; those a spliceosome.

Spliceosome = A large molecular complex that catalyzes RNA splicing reactions; composed of small nuclear ribonucleoproteins (snRNPs) and other proteins (see Campbell, Figure 17.10)

- As the spliceosome is assembled, one type of snRNP base pairs with a complementary sequence at the 5′ end of the intron.
- The spliceosome precisely cuts the RNA transcript at specific splice sites at either end of the intron, which is excised as a lariat-shaped loop.
- The intron is released and the adjacent exons are immediately spliced together by the spliceosome.

3. **Ribozymes**

Other kinds of RNA primary transcripts, such as those giving rise to tRNA and rRNA, are spliced by mechanisms that do not involve spliceosomes; however, as with mRNA splicing, RNA is often involved in catalyzing the reactions.

Ribozymes = RNA molecules that can catalyze reactions by breaking and forming covalent bonds; called ribozymes to emphasize their catalytic activity.

- Ribozymes were first discovered in *Tetrahymena*, a ciliated protozoan that has self-splicing rRNA. That is, intron rRNA itself catalyzes splicing, which occurs completely without proteins or extra RNA molecules.
- Since RNA is acting as a catalyst, it can no longer be said that "All biological catalysts are proteins."
- It has since been discovered that rRNA also functions as a catalyst during translation.

4. **The functional and evolutionary importance of introns**

Introns may play a regulatory role in the cell.

- Intron DNA sequences may control gene activity.
- The splicing process itself may help regulate the export of mRNA to the cytoplasm.

Introns may allow a single gene to direct the synthesis of different proteins.

- This can occur if the same RNA transcript is processed differently among various cell types in the same organism.

- For example, all introns may be removed from a particular transcript in one case; but in another, one or more of the introns may be left in place to be translated. Thus, the resulting proteins in each case would be different.

Introns play an important role in the evolution of protein diversity; they increase the probability that recombination of exons will occur between alleles.

- In split genes, coding sequences can be separated by long distances, so they have higher recombination frequencies than continuously coded genes without introns.
- Exons of a "split gene" may code for different *domains* of a protein that have specific functions, such as, an enzyme's active site or a protein's binding site.

Protein domains = Continuous polypeptide sequences that are structural and functional units in proteins with a modular architecture

- Genetic recombination can occur in just one exon resulting in the synthesis of a novel protein with only one altered domain.

Introns also may increase the likelihood of genetic exchange between and among non-allelic genes.

III.   The Synthesis of Protein

A.   Translation is the RNA-directed synthesis of a polypeptide: *a closer look*

During translation, proteins are synthesized according to a genetic message of sequential codons along mRNA (see Campbell, Figure 17.11).

- *Transfer RNA* (*tRNA*) is the interpreter between the two forms of information—base sequence in mRNA and amino acid sequence in polypeptides.
- tRNA aligns the appropriate amino acids to form a new polypeptide. To perform this function, tRNA must:
  - Transfer amino acids from the cytoplasm's amino acid pool to a ribosome.
  - Recognize the correct codons in mRNA.

Molecules of tRNA are specific for only one particular amino acid. Each type of tRNA associates a distinct mRNA codon with one of the 20 amino acids used to make proteins.

- One end of a tRNA molecule attaches to a specific amino acid.
- The other end attaches to an mRNA codon by base-pairing with its anticodon.

Anticodon = A nucleotide triplet in tRNA that base pairs with a complementary nucleotide triplet (codon) in mRNA.

tRNAs decode the genetic message, codon by codon. For example:

- The mRNA codon UUU is translated as the amino acid phenylalanine (see Campbell, Figure 17.4)
- The tRNA that transfers phenylalanine to the ribosome has an anticodon of AAA.
- When the codon UUU is presented for translation, phenylalanine will be added to the growing polypeptide.
- As tRNAs deposit amino acids in the correct order, ribosomal enzymes link them into a chain.

1.   The structure and function of transfer RNA

All types of RNA, including tRNA, are transcribed from template DNA.

- In eukaryotes, tRNA, like mRNA, must travel from the nucleus to the cytoplasm, where translation occurs.
- In prokaryotes and eukaryotes, each tRNA molecule can be used repeatedly.

The ability of tRNA to carry specific amino acids and to recognize the correct codons depends upon its structure; its form fits function (see also Campbell, Figure 17.12).

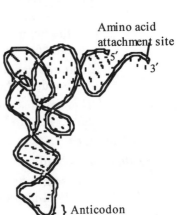

- tRNA is a single-stranded RNA only about 80 nucleotides long.
- The strand is folded, forming several double-stranded regions where short base sequences of hydrogen bond with complementary base sequences.
- A single-plane view reveals a clover leaf shape.

The three-dimensional structure is roughly L-shaped.

- A loop protrudes at one end of the L and has a specialized sequence of three bases called the *anticodon*.
- At the other end of the L protrudes the 3′ end of the tRNA molecule —the attachment site for an amino acid.

There are only about 45 distinct types of tRNA. However, this is enough to translate the 64 codons, since some tRNAs recognize two or three mRNA codons specifying the same amino acid.

- This is possible because the base-pairing rules are relaxed between the third base of an mRNA codon and the corresponding base of a tRNA anticodon.
- This exception to the base-pairing rule is called *wobble*.

Wobble = The ability of one tRNA to recognize two or three different mRNA codons; occurs when the third base (5' end) of the tRNA anticodon has some play or wobble, so that it can hydrogen bond with more than one kind of base in the third position (3' end) of the codon.

- For example, the base U in the wobble position of a tRNA anticodon can pair with either A or G in the third position of an mRNA codon.
- Some tRNAs contain a modified base called inosine (I), which is in the anticodon's wobble position and can base pair with U, C, or A in the third position of an mRNA codon.
- Thus, a single tRNA with the anticodon CCI will recognize three mRNA codons: GGU, GGC, or GGA, all of which code for glycine.

2. Aminoacyl-tRNA synthetases

The correct linkage between tRNA and its designated amino acid must occur before the anticodon pairs with its complementary mRNA codon. This process of correctly pairing a tRNA with its appropriate amino acid is catalyzed by an aminoacyl-tRNA synthetase.

Aminoacyl-tRNA synthetase = A type of enzyme that catalyzes the attachment of an amino acid to its tRNA

- Each of the 20 amino acids has a specific aminoacyl-tRNA synthetase.
- In an endergonic reaction driven by the hydrolysis of ATP, the proper synthetase attaches an amino acid to its tRNA in two steps (see Campbell, Figure 17.13):
    1. *Activation of the amino acid with AMP.* The synthetase's active site binds the amino acid and ATP; the ATP loses two phosphate groups and attaches to the amino acid as AMP (adenosine monophosphate).
    2. *Attachment of the amino acid to tRNA.* The appropriate tRNA covalently bonds to the amino acid, displacing AMP from the enzyme's active site.
- The aminoacyl-tRNA complex releases from the enzyme and transfers its amino acid to a growing polypeptide on the ribosome.

3. Ribosomes

Ribosomes coordinate the pairing of tRNA anticodons to mRNA codons.

- Ribosomes have two subunits (small and large) which are separated when not involved in protein synthesis (see Campbell, Figure 17.14a).
- Ribosomes are composed of about 60% *ribosomal RNA* (*rRNA*) and 40% protein.

The large and small subunits of eukaryotic ribosomes are:
- Constructed in the nucleolus
- Dispatched through nuclear pores to the cytoplasm
- Once in the cytoplasm, are assembled into functional ribosomes only when attached to an mRNA

Compared to eukaryotic ribosomes, prokaryotic ribosomes are smaller and have a different molecular composition.

- Selection of effective drug therapies against bacterial pathogens capitalizes on this difference.

- For example, the antibiotics tetracycline and streptomycin can be used to combat bacterial infections, because they inhibit bacterial protein synthesis without affecting the ribosomes of the eukaryotic host.

In addition to an mRNA binding site, each ribosome has three tRNA binding sites (P, A, and E) (see Campbell, Figure 17.14b).

- The *P site* holds the tRNA carrying the growing polypeptide chain.
- The *A site* holds the tRNA carrying the next amino acid to be added.
- Discharged tRNAs exit the ribosome from the *E site*.

As the ribosome holds the tRNA and mRNA molecules together, enzymes transfer the new amino acid from its tRNA to the carboxyl end of the growing polypeptide (see Campbell, Figure 17.14c).

4. **Building a polypeptide**

The building of a polypeptide, or translation, occurs in three stages: 1) initiation, 2) elongation, and 3) termination.

- All three stages require enzymes and other protein factors.
- Initiation and elongation also require energy provided by GTP (a molecule closely related to ATP).

a. **Initiation**

Initiation brings together mRNA, a tRNA attached to the first amino acid (initiator tRNA; the first amino acid is always methionine), and the two ribosomal subunits.

The first step of initiation involves the binding of the small ribosomal subunit to mRNA and initiator tRNA (see Campbell, Figure 17.15a).

- In prokaryotes, rRNA in the small subunit base-pairs with specific nucleotides in the leader sequence of the mRNA.
- In eukaryotes, the 5' cap of the mRNA aids in binding of the leader sequence to the small ribosomal subunit.
- With help from the small ribosomal subunit, methionine-bound initiator tRNA finds and base-pairs with the initiation or start codon on mRNA. This start codon, AUG, marks the place where translation will begin and is located just downstream from the leader sequence.
- Assembly of the initiation complex—small ribosomal subunit, initiator tRNA and mRNA—requires:
  - Protein *initiation factors* that are bound to the small ribosomal subunit
  - One GTP molecule that probably stabilizes the binding of initiation factors, and upon hydrolysis, drives the attachment of the large ribosomal subunit.

In the second step, a large ribosomal subunit binds to the small one to form a functional translation complex (see Campbell, Figure 17.15b).

- Initiation factors attached to the small ribosomal subunit are released, allowing the large subunit to bind with the small subunit.
- The initiator tRNA fits into the P site on the ribosome.
- The vacant A site is ready for the next aminoacyl-tRNA.

b. **Elongation**

Several proteins called *elongation factors* take part in this three-step cycle which adds amino acids one by one to the initial amino acid (see Campbell, Figure 17.16).

1. *Codon recognition.* The mRNA codon in the A site of the ribosome forms hydrogen bonds with the anticodon of an entering tRNA carrying the next amino acid in the chain.
   - An elongation factor directs tRNA into the A site.
   - Hydrolysis of GTP provides energy for this step.
2. *Peptide bond formation.* A peptide bond is formed between the polypeptide in the P site and the new amino acid in the A site by a *peptidyl transferase.*
   - The peptidyl transferase activity appears to be one of the rRNAs in the large ribosomal subunit (ribozyme).
   - The polypeptide separates from its tRNA and is transferred to the new amino acid carried by the tRNA in the A site.
3. *Translocation.* The tRNA in the A site, which is now attached to the growing peptide, is translocated to the P site. Simultaneously, the tRNA that was in the P site is translocated to the E site and from there it exits the ribosome.
   - During this process, the codon and anticodon remain bonded, so the mRNA and the tRNA move as a unit, bringing the next codon to be translated into the A site.
   - The mRNA is moved through the ribosome only in the 5′ to 3′ direction.
   - GTP hydrolysis provides energy for each translocation step.

> Some students have trouble visualizing translocation, especially how the tRNA and mRNA move as a unit, exposing a new codon in the A site. Showing the class an animated sequence would no doubt solve the problem. However, if you do not have a monitor or video-projection capability, a paper simulation is just as effective. Paper models can be tacked to a large bulletin board or small cutouts can be moved on an overhead projector. Students may also actively participate by practicing their own simulations.

c. Termination

Each iteration of the elongation cycle takes less than a tenth of a second and is repeated until synthesis is complete and a termination codon reaches the ribosome's A site (see Campbell, Figure 17.7).

Termination codon (stop codon) = Base triplet (codon) on mRNA that signals the end of translation

- Stop codons are UAA, UAG, and UGA.
- Stop codons do not code for amino acids.

> Students often confuse terminator sequence (on DNA), which signals the end of transcription, with termination or stop codons (on mRNA), which signal the end of translation.

When a stop codon reaches the ribosome's A site, a protein *release factor* binds to the codon and initiates the following sequence of events:

- Release factor hydrolyzes the bond between the polypeptide and the tRNA in the P site.
- The polypeptide and tRNA are released from the ribosome.
- The remainder of the translation complex dissociates, including separation of the small and a large ribosomal subunits.

5. Polyribosomes

Single ribosomes can make average-sized polypeptides in less than a minute; usually, however, clusters of ribosomes simultaneously translate an mRNA.

Polyribosome = A cluster of ribosomes simultaneously translating an mRNA molecule (see Campbell, Figure 17.8)

- Once a ribosome passes the initiation codon, a second ribosome can attach to the leader sequence of the mRNA.
- Several ribosomes may translate an mRNA at once, making many copies of a polypeptide.
- Polyribosomes are found in both prokaryotes and eukaryotes.

6. From polypeptide to functional protein

The biological activity of proteins depends upon a precise folding of the polypeptide chain into a native three-dimensional conformation.

- Genes determine *primary structure*, the linear sequence of amino acids.
- Primary structure determines how a polypeptide chain will spontaneously coil and fold to form a three-dimensional molecule with *secondary* and *tertiary structure*; chaperone proteins facilitate polypeptide folding

Some proteins must undergo *post-translational modification* before they become fully functional in the cell. Post-translational modification affects function by affecting protein structure.

- Chemical modification
  - Sugars, lipids, phosphate groups, or other additives may be attached to some amino acids.
- Chain-length modification
  - One or more amino acids may be enzymatically cleaved from the leading (amino) end of the polypeptide chain.
  - Single polypeptide chains may be divided into two or more pieces. The translated product of the insulin gene is a large protein precursor (preproinsulin). The precursor is modified by removal of N-terminal fragments and by internal enzymatic cleavage to yield to separate chains held together by disulfide bonds.
  - Two or more polypeptides may join as subunits of a protein that has quaternary structure (e.g., hemoglobin).

B. Signal peptides target some eukaryotic polypeptides to specific destinations in the cell

Eukaryotic ribosomes function either free in the cytosol or bound to endomembranes.

- Bound and free ribosomes are structurally identical and interchangeable.
- Most proteins made by free ribosomes will function in the cytosol.
- Attached to the outside of the endoplasmic reticulum, bound ribosomes generally make proteins that are destined for:
  - Membrane inclusion in membrane component of the endomembrane system (e.g., membrane-bound enzymes of the nuclear envelope, ER, Golgi, lysosomes, vacuoles, and plasma membrane)
  - Partitioning into the lumenal component of the endomembrane system (e.g., lysozymes)
  - Secretion from the cell (e.g., hormones such as insulin)

There is only one type of ribosome, and synthesis of all proteins begins in the cytosol. What determines whether a ribosome will be free in the cytosol or attached to rough ER?

- Messenger RNA for secretory proteins code for an initial *signal sequence* of 16 to 20 hydrophobic amino acids at the amino end of the newly forming polypeptide (see Campbell, Figure 17.19).
- When a ribosome begins to synthesize a protein with a signal sequence, it moves to the ER membrane by a mechanism that involves two other components.
  - *Signal recognition particle (SRP)*. SRPs are a complex of protein and RNA (SRP RNA). They serve as an adaptor between the translation complex and the ER. SRPs first attach to the signal sequence of a growing polypeptide and link the translation complex to a receptor protein on the ER membrane (SRP receptor).
  - *SRP receptor*. This receptor protein is built into the ER membrane. The signal recognition particle docks with the receptor, and the ribosome thus becomes bound to the ER membrane.
- The ribosome continues protein synthesis and the leading end of the new polypeptide (N-terminus) threads into the cisternal space.
- The signal sequence is removed by an enzyme.
- Newly formed polypeptide is released from the ribosome and folds into its native conformation.
- If an mRNA does not code for a signal sequence, the ribosome remains free and synthesizes its protein in the cytosol.

Different signal sequences may also dispatch proteins to specific sites other than the ER. For example, newly formed proteins may be targeted for mitochondria or chloroplasts. In these cases, however, translation is completed in the cytoplasm.

C. RNA plays multiple roles in the cell: a review

Three-dimensional conformations vary among the types of RNA. These differences in shape give RNA its ability to perform a variety of functions, such as:

1. *Information carrier*. Messenger RNA (mRNA) carries genetic information from DNA to ribosomes; this genetic message specifies a protein's primary structure.

2. *Adaptor molecule*. Transfer RNA (tRNA) acts as an adaptor in protein synthesis by translating information from one form (mRNA nucleotide sequence) into another (protein amino acid sequence). SRP RNA helps direct translation complexes to the ER.

3. *Catalyst and structural molecule*. During translation, ribosomal RNA (rRNA) plays structural and probably enzymatic roles in ribosomes. Small nuclear RNA (snRNA) in snRNP particles also plays structural and enzymatic roles within spliceosomes that catalyze RNA splicing reactions.

4. *Viral genomes*. Some viruses use RNA as their genetic material.

D. Comparing protein synthesis in prokaryotes and eukaryotes: *a review*

While transcription and translation are similar in prokaryotes and eukaryotes, there are some notable differences in the cellular machinery and in some of the details of the processes. The following differ in prokaryotes and eukaryotes:

- RNA polymerases; those of eukaryotes depend on transcription factors
- Termination of transcription
- Ribosomes
- Location (see also Campbell, Figure 17.20)

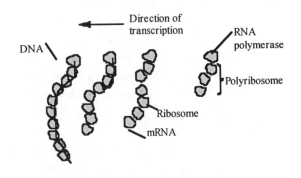

  - Prokaryotes lack nuclei, so transcription is not segregated from translation; consequently, translation may begin as soon as the 5′ end of mRNA peels away from template DNA, even before transcription is complete.
  - The significance of a eukaryotic cell's compartmental organization is that transcription and translation are segregated by the nuclear envelope. This allows mRNA to be modified before it moves from the nucleus to the cytoplasm. Such *RNA processing* occurs only in eukaryotes.

E. Point mutations can affect protein structure and function

Knowing how genes are translated into proteins, scientists can give a molecular description of heritable changes that occur in organisms.

Mutation = A change in the genetic material of a cell (or virus)

Point mutation = A mutation limited to about one or a few base pairs in a single gene

1. Types of point mutations

There are two categories of point mutations: 1) base-pair substitutions and 2) base-pair insertions or deletions (see Campbell, Figure 17.22).

a. Substitutions

Base-pair substitution = The replacement of one base pair with another; occurs when a nucleotide and its partner in the complementary DNA strand are replaced with another pair of nucleotides according to base-pairing rules.

Depending on how base-pair substitutions are translated, they can result in little or no change in the protein encoded by the mutated gene.

- Redundancy in the genetic code is why some substitution mutations have no effect. A base pair change may simply transform one codon into another that codes for the same amino acid (*silent substitution*).
- Even if the substitution alters an amino acid, the new amino acid may have similar properties to the one it replaces, or it may be in a part of the protein where the exact amino acid sequence is not essential to its activity (*conservative substitution*).

Some base-pair substitutions result in readily detectable changes in proteins.

- Alteration of a single amino acid in a crucial area of a protein will significantly alter protein activity.
- On rare occasions, such a mutation will produce a protein that is improved or has capabilities that enhance success of the mutant organism and its descendants.

- More often, such mutations produce a less active or inactive protein that impairs cell function.

Base-pair substitutions are usually missense mutations or nonsense mutations.

Missense mutation = Base-pair substitution that alters an amino acid codon (sense codon) to a new codon that codes for a different amino acid

- Altered codons make sense (are translated), but not necessarily the right sense.
- Base-pair substitutions are usually missense mutations.

Nonsense mutation = Base-pair substitution that changes an amino acid codon (sense codon) to a chain termination codon, or vice versa

- Nonsense mutations can result in premature termination of translation and the production of a shorter than normal polypeptide.
- Nearly all nonsense mutations lead to nonfunctional proteins.

b. Insertions or deletions

Base-pair insertions or deletions usually have a greater negative effect on proteins than substitutions.

Base-pair insertion = The insertion of one or more nucleotide pairs into a gene

Base-pair deletion = The deletion of one or more nucleotide pairs from a gene

Because mRNA is read as a series of triplets during translation, insertion or deletion of nucleotides may alter the reading frame (triplet grouping) of the genetic message. This type of *frameshift mutation* will occur whenever the number of nucleotides inserted or deleted is not 3 or a multiple of 3.

Frameshift mutation = A base-pair insertion or deletion that causes a shift in the reading frame, so that codons beyond the mutation will be the wrong grouping of triplets and will specify the wrong amino acids

- A frameshift mutation causes the nucleotides following the insertion or deletion to be improperly grouped into codons.
- This results in extensive missense, which will sooner or later end in nonsense (premature termination).
- Frameshift will produce a nonfunctional protein unless the insertion or deletion is very near the end of the gene.

2. Mutagens

Mutagenesis = The creation of mutations

- Mutations can occur as errors in DNA replication, repair, or recombinations that result in base-pair substitutions, insertions, or deletions.
- Mutagenesis may be a naturally occurring event causing *spontaneous mutations* or mutations may be caused by exposure to mutagens.

Mutagen = Physical or chemical agents that interact with genetic material to cause mutations

- Radiation is the most common physical mutagen in nature and has been used in the laboratory to induce mutations.
- Several categories of chemical mutagens are known including *base analogues*, which are chemicals that mimic normal DNA bases, but base-pair incorrectly.
- The Ames test, developed by Bruce Ames, is one of the most widely used tests for measuring the mutagenic strength of various chemicals. Since most mutagens are carcinogenic, this test is also used to screen for chemical carcinogens.

F. What is a gene? *revisiting the question*

The concept of the gene has emerged as the history of genetics has unfolded.

- The Mendelian concept of a gene was that it served as a discrete unit of inheritance that affected a phenotypic character.

- Morgan and colleagues assigned such units of inheritance (or genes) to specific loci on chromosomes.

- In molecular terms, a gene is a specific sequence of nucleotides at a given location in the genome of an organism. Depending on the gene, the final gene product may be RNA or a specific polypeptide.

REFERENCES

Alberts, B., D. Bray, J. Lewis, M. Raff, K. Roberts and J.D. Watson. *Molecular Biology of the Cell*. 3rd ed. New York: Garland, 1994.

Campbell, N., et al. *Biology*. 5th ed. Menlo Park, California: Benjamin/Cummings, 1998.

Tijan, R. "Molecular Machines that Control Genes." *Scientific American*, February, 1995.

CHAPTER **18**

MICROBIAL MODELS: THE GENETICS OF VIRUSES AND BACTERIA

OUTLINE

I. The Genetics of Viruses
   - A. Researchers discovered viruses by studying a plant disease: *science as a process*
   - B. A virus is a genome enclosed in a protective coat.
   - C. Viruses can reproduce only within a host cell: *an overview*
   - D. Phages reproduce using lytic or lysogenic cycles
   - E. Animal viruses are diverse in their modes of infection and of replication
   - F. Plant viruses are serious agricultural pests
   - G. Viroids and prions are infectious agents even simpler than viruses
   - H. Viruses may have evolved from other mobile genetic elements

II. The Genetics of Bacteria
   - A. The short generation span of bacteria facilitates their evolutionary adaptation to changing environments
   - B. Genetic recombination produces new bacterial strains
   - C. The control of gene expression enables individual bacteria to adjust their metabolism to environmental change

OBJECTIVES

After reading this chapter and attending lecture, the student should be able to:

1. Recount the history leading up to the discovery of viruses and discuss the contributions of A. Mayer, D. Ivanowsky, Martinus Beijerinck, and Wendell Stanley.
2. List and describe structural components of viruses.
3. Explain why viruses are obligate parasites.
4. Describe three patterns of viral genome replication.
5. Explain the role of reverse transcriptase in retroviruses.
6. Describe how viruses recognize host cells.
7. Distinguish between lytic and lysogenic reproductive cycles using phage T_4 and phage λ as examples.
8. Outline the procedure for measuring phage concentration in a liquid medium.
9. Describe several defenses bacteria have against phage infection.
10. Using viruses with envelopes and RNA viruses as examples, describe variations in replication cycles of animal viruses.
11. Explain how viruses may cause disease symptoms, and describe some medical weapons used to fight viral infections.
12. List some viruses that have been implicated in human cancers, and explain how tumor viruses transform cells.
13. Distinguish between horizontal and vertical routes of viral transmission in plants.

14. List some characteristics that viruses share with living organisms, and explain why viruses do not fit our usual definition of life.
15. Provide evidence that viruses probably evolved from fragments of cellular nucleic acid.
16. Describe the structure of a bacterial chromosome.
17. Describe the process of binary fission in bacteria, and explain why replication of the bacterial chromosome is considered to be semiconservative.
18. List and describe the three natural processes of genetic recombination in bacteria.
19. Distinguish between general transduction and specialized transduction.
20. Explain how the F plasmid controls conjugation in bacteria.
21. Explain how bacterial conjugation differs from sexual reproduction in eukaryotic organisms.
22. For donor and recipient bacterial cells, predict the consequences of conjugation between the following: 1) F^+ and F^- cell, 2) Hfr and F^- cell.
23. Define transposon, and describe two essential types of nucleotide sequences found in transposon DNA.
24. Distinguish between an insertion sequence and a complex transposon.
25. Describe the role of transposase and DNA polymerase in the process of transposition.
26. Explain how transposons can generate genetic diversity.
27. Briefly describe two main strategies cells use to control metabolism.
28. Explain why grouping genes into an operon can be advantageous.
29. Using the *trp* operon as an example, explain the concept of an operon and the function of the operator, repressor, and corepressor.
30. Distinguish between structural and regulatory genes.
31. Describe how the *lac* operon functions and explain the role of the inducer allolactose.
32. Explain how repressible and inducible enzymes differ and how these differences reflect differences in the pathways they control.
33. Distinguish between positive and negative control, and give examples of each from the *lac* operon.
34. Explain how CAP is affected by glucose concentration.
35. Describe how *E. coli* uses the negative and positive controls of the *lac* operon to economize on RNA and protein synthesis.

KEY TERMS

capsid	provirus	transformation	operator
viral envelope	retrovirus	transduction	operon
bacteriophage (phage)	reverse transcriptase	conjugation	repressor
host range	HIV	F factor	regulatory gene
lytic cycle	AIDS	episome	corepressor
virulent virus	vaccine	F plasmid	inducer
lysogenic cycle	virion	R plasmid	cyclic amp (cAMP)
temperate virus	prion	transposon	cAMP receptor protein
prophage	nucleoid	insertion sequence	(CRP)

LECTURE NOTES

Scientists discovered the role of DNA in heredity by studying the simplest of biological systems—viruses and bacteria. Most of the molecular principles discovered through microbe research applies to higher organisms, but viruses and bacteria also have unique genetic features.

- Knowledge of these unique genetic features has helped scientists understand how viruses and bacteria cause disease.
- Techniques for gene manipulation emerged from studying genetic peculiarities of microorganisms.

I.   The Genetics of Viruses

A. Researchers discovered viruses by studying a plant disease: *science as a process*

The discovery of viruses resulted from the search for the infectious agent causing tobacco mosaic disease. This disease stunts the growth of tobacco plants and gives their leaves a mosaic coloration (see Campbell, Figure 18.8a).

1883: A. Mayer, a German scientist demonstrated that the disease was contagious and proposed that the infectious agent was an unusually small bacterium that could not be seen with a microscope.

- He successfully transmitted the disease by spraying sap from infected plants onto the healthy ones.
- Using a microscope, he examined the sap and was unable to identify a microbe.

1890s: D. Ivanowsky, a Russian scientist proposed that tobacco mosaic disease was caused by a bacterium that was either too small to be trapped by a filter or that produced a filterable toxin.

- To remove bacteria, he filtered sap from infected leaves.
- Filtered sap still transmitted disease to healthy plants.

1897: Martinus Beijerinck, a Dutch microbiologist proposed that the disease was caused by a reproducing particle much smaller and simpler than a bacterium.

- He ruled out the theory that a filterable toxin caused the disease by demonstrating that the infectious agent in filtered sap could reproduce.

Plants were sprayed with filtered sap from
diseased plant.

↓

Sprayed plants developed tobacco mosaic
disease.

↓

Sap from newly infected plants was used to
infect others.

- This experiment was repeated for several generations. He concluded that the pathogen must be reproducing because its ability to infect was undiluted by transfers from plant to plant.
- He also noted that unlike bacteria, the pathogen:
  - Reproduced only within the host it infected
  - Could not be cultured on media
  - Could not be killed by alcohol

1935: Wendell M. Stanley, an American biologist, crystallized the infectious particle now known as *tobacco mosaic virus* (*TMV*).

B. A virus is a genome enclosed in a protective coat

In the 1950s, TMV and other viruses were finally observed with electron microscopes. Viral structure appeared to be unique from the simplest of cells.

- The smallest viruses are only 20 nm in diameter.
- The virus particle, consists of nucleic acid enclosed by a protein coat and sometimes a membranous envelope.

1. Viral genomes

Depending upon the virus, viral genomes:

- May be double-stranded DNA, single-stranded DNA, double-stranded RNA, or single-stranded RNA
- Are organized as single nucleic acid molecules that are linear or circular
- May have as few as four genes or as many as several hundred

2. Capsids and envelopes

Capsid = Protein coat that encloses the viral genome

- Its structure may be rod-shaped, polyhedral, or complex
- Composed of many *capsomeres*, protein subunits made from only one or a few types of protein.

Envelope = Membrane that cloaks some viral capsids

- Helps viruses infect their host
- Derived from host cell membrane which is usually virus-modified and contains proteins and glycoproteins of viral origin

The most complex capsids are found among *bacteriophages* or bacterial viruses.

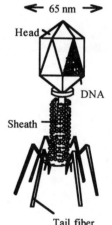

Structure of the
T-Even Phage

- Of the first phages studied, seven infected *E. coli*. These were named types 1 – 7 (T1, T2, T3, ... T7).
- The T-even phages – T2, T4, and T6—are structurally very similar.
    - The icosohedral head encloses the genetic material.
    - The protein tailpiece with tail fibers attaches the phage to its bacterial host and injects its DNA into the bacterium.

Campbell, Figure 18.2 shows the structure of various viruses.

C. Viruses can reproduce only within a host cell: *an overview*

Viral reproduction differs markedly from cellular reproduction, because viruses are *obligate intracellular parasites* which can express their genes and reproduce only within a living cell. Each virus has a specific *host range*.

Host range = Limited number or range of host cells that a parasite can infect

- Viruses recognize host cells by a complementary fit between external viral proteins and specific cell surface *receptor sites*.
- Some viruses have broad host ranges which may include several species (e.g., swine flu and rabies).
- Some viruses have host ranges so narrow that they can:
    - Infect only one species (e.g., phages of *E. coli*)
    - Infect only a single tissue type of one species (e.g., human cold virus infects only cells of the upper respiratory tract; AIDS virus binds only to specific receptors on certain white blood cells)

There are many patterns of viral life cycles, but they all generally involve:
- Infecting the host cell with viral genome
- Co-opting host cell's resources to:
  - Replicate the viral genome
  - Manufacture capsid protein
- Assembling newly produced viral nucleic acid and capsomeres into the next generation of viruses (see Campbell, Figure 18.3)

There are several mechanisms used to infect host cells with viral DNA.
- For example, T-even phages use an elaborate tailpiece to inject DNA into the host cell.
- Once the viral genome is inside its host cell, it commandeers the host's resources and reprograms the cell to copy the viral genes and manufacture capsid protein.

There are three possible patterns of viral genome replication:
1. *DNA → DNA.* If viral DNA is double-stranded, DNA replication resembles that of cellular DNA, and the virus uses DNA polymerase produced by the host.
2. *RNA → RNA.* Since host cells lack the enzyme to copy RNA, most RNA viruses contain a gene that codes for *RNA replicase*, an enzyme that uses viral RNA as a template to produce complementary RNA.
3. *RNA → DNA → RNA.* Some RNA viruses encode *reverse transcriptase*, an enzyme that transcribes DNA from an RNA template.

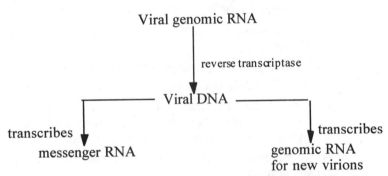

Regardless of how viral genomes replicate, all viruses divert host cell resources for viral production.
- Viral genes use the host cell's enzymes, ribosomes, tRNAs, amino acids, ATP, and other resources to make copies of the viral genome and produce viral capsid proteins.
- These viral components—nucleic acid and capsids—are assembled into hundreds or thousands of virions, which leave to parasitize new hosts.

Viral nucleic acid and capsid proteins assemble spontaneously into new virus particles, a process called *self-assembly.*
- Since most viral components are held together by weak bonds (e.g., hydrogen bonds and Van der Waals forces), enzymes are not usually necessary for assembly.
- For example, TMV can be disassembled in the laboratory. When mixed together, the RNA and capsids spontaneously reassemble to form complete TMV virions.

D. Phages reproduce using lytic or lysogenic cycles

Bacteriophages are the best understood of all viruses, and many of the important discoveries in molecular biology have come from bacteriophage studies.

- In the 1940s, scientists determined how the T phages reproduce within a bacterium; this research:
- Demonstrated that DNA is the genetic material
- Established the phage-bacterium system as an important experimental tool
- Studies on lambda (λ) phage of *E. coli* showed that some double-stranded DNA viruses reproduce by two alternative mechanisms: the *lytic cycle* and the *lysogenic cycle*.

1. The lytic cycle

Virulent bacteriophages reproduce only by a *lytic* replication cycle.

Virulent phages = Phages that lyse their host cells

Lytic cycle = A viral replication cycle that results in the death or lysis of the host cell

The lytic cycle of phage T4 illustrates this type of replication cycle (see Campbell, Figure 18.4):

1. Phage attaches to cell surface.

   - T4 recognizes a host cell by a complementary fit between proteins on the virion's tail fibers and specific receptor sites on the outer surface of an *E. coli* cell.

2. Phage contracts sheath and injects DNA.

   - ATP stored in the phage tailpiece is the energy source for the phage to: a) pierce the *E. coli* wall and membrane, b) contract its tail sheath, and c) inject its DNA.
   - The genome separates from the capsid leaving a capsid "ghost" outside the cell.

3. Hydrolytic enzymes destroy host cell's DNA.

   - The *E. coli* host cell begins to transcribe and translate the viral genome.
   - One of the first viral proteins produced is an enzyme that degrades host DNA. The phage's own DNA is protected, because it contains modified cytosine not recognized by the enzyme.

4. Phage genome directs the host cell to produce phage components: DNA and capsid proteins.

   - Using nucleotides from its own degraded DNA, the host cell makes many copies of the phage genome.
   - The host cell also produces three sets of capsid proteins and assembles them into phage tails, tail fibers, and polyhedral heads.
   - Phage components spontaneously assemble into virions.

5. Cell lyses and releases phage particles.

   - Lysozymes specified by the viral genome digest the bacterial cell wall.
   - Osmotic swelling lyses the cell which releases hundreds of phages from their host cell.
   - Released virions can infect nearby cells.
   - Lytic cycle takes only 20 to 30 minutes at 37°C. In that period, a T4 population can increase a hundredfold.

Bacteria have several defenses against destruction by phage infection.

- Bacterial mutations can change receptor sites used by phages for recognition, and thus avoid infection.

- Bacterial *restriction nucleases* recognize and cut up foreign DNA, including certain phage DNA. Bacterial DNA is chemically altered, so it is not destroyed by the cell's own restriction enzymes.

Restriction enzymes = Naturally occurring bacterial enzymes that protect bacteria against intruding DNA from other organisms. The enzymes also catalyze restriction, the process of cutting foreign DNA into small segments.

Bacterial hosts and their viral parasites are continually *coevolving*.

- Most successful bacteria have effective mechanisms for preventing phage entry or reproduction.
- Most successful phages have evolved ways around bacterial defenses.
- Many phages check their own destructive tendencies and may coexist with their hosts.

2. The lysogenic cycle

Some viruses can coexist with their hosts by incorporating their genome into the host's genome.

Temperate viruses = Viruses that can integrate their genome into a host chromosome and remain latent until they initiate a lytic cycle

- They have two possible modes of reproduction, the lytic cycle and the lysogenic cycle.
- An example is phage λ, discovered by E. Lederberg in 1951 (see Campbell, Figure 18.5)

Lysogenic cycle = A viral replication cycle that involves the incorporation of the viral genome into the host cell genome

Details of the lysogenic cycle were discovered through studies of phage λ life cycle:

1. Phage λ binds to the surface of an *E. coli* cell.
2. Phage λ injects its DNA into the bacterial host cell.
3. λ DNA forms a circle and either begins a lytic or lysogenic cycle.
4. During a lysogenic cycle, λ DNA inserts by genetic recombination (crossing over) into a specific site on the bacterial chromosome and becomes a prophage.

Prophage = A phage genome that is incorporated into a specific site on the bacterial chromosome

- Most prophage genes are inactive.
- One active prophage gene codes for the production of *repressor protein* which switches off most other prophage genes.
- Prophage genes are copied along with cellular DNA when the host cell reproduces. As the cell divides, both prophage and cellular DNA are passed on to daughter cells.
- A prophage may be carried in the host cell's chromosomes for many generations.

Occasionally, a prophage may leave the bacterial chromosome.

- This may be spontaneous or caused by environmental factors (e.g., radiation).
- The excision process may begin the phage's lytic reproductive cycle.
- Virions produced during the lytic cycle may begin either a lytic or lysogenic cycle in their new host cells.

Lysogenic cell = Host cell carrying a prophage in its chromosome

- It is called lysogenic because it has the potential to lyse.

- Some prophage genes in a lysogenic cell may be expressed and change the cell's phenotype in a process called *lysogenic conversion*.
- Lysogenic conversion occurs in bacteria that cause diphtheria, botulism, and scarlet fever. Pathogenicity results from toxins coded for by prophage genes.

E. Animal viruses are diverse in their modes of infection and replication

1. Reproductive cycles of animal viruses

Replication cycles of animal viruses may show some interesting variations from those of other viruses. Two examples are the replication cycles of: 1) viruses with envelopes, and 2) viruses with RNA genomes that serve as the genetic material. (See Campbell, Table 18.1 for families of animal viruses grouped by type of nucleic acid.)

a. Viral envelopes

Some animal viruses are surrounded by a membranous envelope, which is unique to several groups of animal viruses. This envelope is:

- Outside the capsid and helps the virus enter host cells.
- A lipid bilayer with glycoprotein spikes protruding from the outer surface.

Enveloped viruses have replication cycles characterized by (see Campbell, Figure 18.6):

1. *Attachment.* Glycoprotein spikes protruding from the viral envelope attach to receptor sites on the host's plasma membrane.

2. *Entry.* As the envelope fuses with the plasma membrane, the entire virus (capsid and genome) is transported into the cytoplasm by receptor-mediated endocytosis.

3. *Uncoating.* Cellular enzymes uncoat the genome by removing the protein capsid from viral RNA.

4. *Viral RNA and protein synthesis.* Viral enzymes are required to replicate the RNA genome and to transcribe mRNA.

   - Some viral RNA polymerase is packaged in the virion.
   - Viral RNA polymerase (transcriptase) replicates the viral genome and transcribes viral mRNA. Note that the viral genome is a strand complementary to mRNA.
   - Viral mRNA is translated into viral proteins including:
     - Capsid proteins synthesized in the cytoplasm by free ribosomes
     - Viral-envelope glycoproteins synthesized by ribosomes bound to rough ER. Glycoproteins produced in the host's ER are sent to the Golgi apparatus for further processing. Golgi vesicles transport the glycoproteins to the plasma membrane, where they cluster at exit sites for the virus.

5. *Assembly and release.* New capsids surround viral genomes. Once assembled, the virions envelop with host plasma membrane as they bud off from the cell's surface. The viral envelope is derived from:

   - Host cell's plasma membrane lipid
   - Virus-specific glycoprotein

Some viral envelopes are not derived from host plasma membrane.

For example, herpesviruses are double-stranded DNA viruses which:

- Contain envelopes derived from the host cell's nuclear envelope rather than from the plasma membrane
- Reproduce within the host cell's nucleus
- Use both viral and cellular enzymes to replicate and transcribe their genomic DNA
- May integrate their DNA into the cell's genome as a *provirus*. Evidence comes from the nature of herpes infections, which tend to recur. After a period of latency, physical or emotional stress may cause the proviruses to begin a productive cycle again.

Provirus = Viral DNA that inserts into a host cell chromosome

b. RNA as viral genetic material

All possible types of viral genomes are represented among animal viruses. Since mRNA is common to all types, DNA and RNA viruses are classified according to the relationship of their mRNA to the genome. In this classification:

- mRNA or the strand that corresponds to mRNA is the *plus (+) strand*; it has the nucleotide sequence that codes for proteins.
- The *minus (–) strand* is a template for synthesis of a plus strand; it is complementary to the sense strand or mRNA.

Animal RNA viruses are classified as following:

- *Class III RNA viruses.* Double-stranded RNA genome; the minus strand is the template for mRNA. (Reoviruses)
- *Class IV RNA viruses.* Single plus strand genome; the plus strand can function directly as mRNA, but also is a template for synthesis of minus RNA. (Minus RNA is a template for synthesis of additional plus strands.) Viral enzymes are required for RNA synthesis from RNA templates. (Picornavirus, Togavirus)
- *Class V RNA viruses.* Single minus strand genome; mRNA is transcribed directly from this genomic RNA. (Rhabdovirus, Paramyxovirus, Orthomyxovirus)
- *Class VI RNA viruses.* Single plus strand genome; the plus strand is a template for complementary DNA synthesis. *Reverse transcriptase* catalyzes this reverse transcription from RNA to DNA. mRNA is then transcribed from a DNA template. (*Retroviruses*)

Retrovirus = (Retro = backward) RNA virus that uses *reverse transcriptase* to transcribe DNA from the viral RNA genome.

- *Reverse transcriptase* is a type of DNA polymerase that transcribes DNA from an RNA template.
- *HIV* (*human immunodeficiency virus*), the virus that causes *AIDS* (*acquired immunodeficiency syndrome*) is a retrovirus.

RNA viruses with the most complicated reproductive cycles are the retroviruses, because retroviruses must first carry out *reverse transcription*: (See Campbell, Figure 18.7)

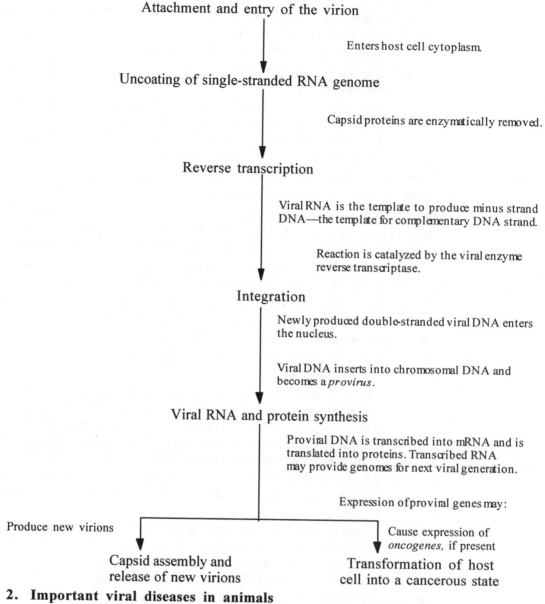

Attachment and entry of the virion

Enters host cell cytoplasm.

Uncoating of single-stranded RNA genome

Capsid proteins are enzymatically removed.

Reverse transcription

Viral RNA is the template to produce minus strand DNA—the template for complementary DNA strand.

Reaction is catalyzed by the viral enzyme reverse transcriptase.

Integration

Newly produced double-stranded viral DNA enters the nucleus.

Viral DNA inserts into chromosomal DNA and becomes a *provirus*.

Viral RNA and protein synthesis

Proviral DNA is transcribed into mRNA and is translated into proteins. Transcribed RNA may provide genomes for next viral generation.

Expression of proviral genes may:

Produce new virions

Capsid assembly and release of new virions

Cause expression of *oncogenes*, if present

Transformation of host cell into a cancerous state

2. Important viral diseases in animals

It is often unclear how certain viruses cause disease symptoms. Viruses may:

- Damage or kill cells. In response to a viral infection, lysosomes may release hydrolytic enzymes.
- Be toxic themselves or cause infected cells to produce toxins.
- Cause varying degrees of cell damage depending upon regenerative ability of the infected cell. We recover from colds because infected cells of the upper respiratory tract can regenerate by cell division. Poliovirus, however, causes permanent cell damage because the virus attacks nerve cells which cannot divide.
- Be indirectly responsible for disease symptoms. Fever, aches and inflammation may result from activities of the immune system.

Medical weapons used to fight viral infections include *vaccines* and *antiviral drugs*.

Vaccines = Harmless variants or derivatives of pathogenic microbes that mobilize a host's immune mechanism against the pathogen

- Edward Jenner developed the first vaccine (against smallpox) in 1796. According to the WHO, a vaccine has almost completely eradicated smallpox.
- Effective vaccines now exist for polio, rubella, measles, mumps, and many other viral diseases.

While vaccines can prevent some viral illnesses, little can be done to cure a viral disease once it occurs. Some *antiviral drugs* have recently been developed.

- Several are analogs of purine nucleosides that interfere with viral nucleic acid synthesis (e.g., adenine arabinoside and acyclovir).

3. Emerging viruses

Emerging viruses are viruses that make an apparent sudden appearance. In reality, they are not likely to be new viruses, but rather existing ones that have expanded their host territory.

Emerging viral diseases can arise if an existing virus:

1. Evolves and thus causes disease in individuals who have immunity only to the ancestral virus (e.g., influenza virus)
2. Spreads from one host species to another
   - For example, the 1993 hantavirus outbreak in New Mexico was the result of a population explosion in deer mice that are the viral reservoirs. Humans became infected by inhaling airborne hantavirus that came from the excreta of deer mice.
3. Disseminates from a small population to become more widespread
   - AIDS, once a rare disease, has become a global epidemic. Technological and social factors influenced the spread of AIDS virus.

Environmental disturbances can increase the viral traffic responsible for emerging diseases. For example:

- Traffic on newly cut roads through remote areas can spread viruses among previously isolated human populations.
- Deforestation activities brings humans into contact with animals that may host viruses capable of infecting humans.

4. Viruses and cancer

Some tumor viruses cause cancer in animals.

- When animal cells grown in tissue culture are infected with tumor viruses, they *transform* to a cancerous state.
- Examples are members of the retrovirus, papovavirus, adenovirus and herpesvirus groups.
- Certain viruses are implicated in human cancers:

Viral Group	Examples/Diseases	Cancer Type
Retrovirus	HTLV-1/adult leukemia	Leukemia
Herpesvirus	Epstein-Barr/infectious mononucleosis	Burkitt's lymphoma
Papovavirus	Papilloma/human warts	Cervical cancer
Hepatitis B virus	Chronic hepatitis	Liver cancer

Tumor viruses transform cells by inserting viral nucleic aids into host cell DNA.

- This insertion is permanent as the provirus never excises.
- Insertion for DNA tumor viruses is straightforward.

Several viral genes have been identified as oncogenes.

Oncogenes = Genes found in viruses or as part of the normal eukaryotic genome, that trigger transformation of a cell to a cancerous state.

- Code for cellular growth factors or for proteins involved in the function of growth factors.
- Are not unique to tumor viruses, but are found in the normal cells of many species. In fact, some tumor viruses transform cells by activating cellular oncogenes.

More than one oncogene must usually be activated to completely transform a cell.

- Indications are that tumor viruses are effective only in combination with other events such as exposure to carcinogens.
- Carcinogens probably also act by turning on cellular oncogenes.

F. Plant viruses are serious agricultural pests

As serious agricultural pests, many of the plant viruses:

- Stunt plant growth and diminish crop yields (see Campbell, Figure 18.8a)
- Are RNA viruses
- Have rod-shaped capsids with capsomeres arranged in a spiral

Capsomere = Complex capsid subunit consisting of several identical or different protein molecules

Plant viruses spread from plant to plant by two major routes: horizontal transmission and vertical transmission.

Horizontal transmission = Route of viral transmission in which an organism receives the virus from an external source

- Plants are more susceptible to viral infection if their protective epidermal layer is damaged.
- Insects may be *vectors* that transmit viruses from plant to plant and can inject the virus directly into the cytoplasm.
- By using contaminated tools, gardeners and farmers may transmit plant viruses.

Vertical transmission = Route of viral transmission in which an organism inherits a viral infection from its parent

- Can occur in asexual propagation of infected plants (e.g., by taking cuttings)
- Can occur in sexual reproduction via infected seeds

Once a plant is infected, viruses reproduce and spread from cell to cell by passing through plasmodesmata (see Campbell, Figure 18.8b).

Most plant viral diseases have no cure, so current efforts focus on reducing viral propagation and breeding resistant plant varieties.

G. Viroids and prions are infectious agents even simpler than viruses

Another class of plant pathogens called *viroids* are smaller and simpler than viruses.

- They are small, naked, circular RNA molecules that do not encode protein, but can replicate in host plant cells.
- It is likely that viroids disrupt normal plant metabolism, development, and growth by causing errors in regulatory systems that control gene expression.
- Viroid diseases affect many commercially important plants such as coconut palms, chrysanthemums, potatoes, and tomatoes.

Some scientists believe that viroids originated as escaped introns.

- Nucleotide sequences of viroid RNA are similar to self-splicing introns found within some normal eukaryotic genes, including rRNA genes.

- An alternative hypothesis is that viroids and self-splicing introns share a common ancestral molecule.

As nucleic acids, viroids self-direct their replication and thus are not diluted during transmission from host to host. Molecules other than nucleic acids can be infectious agents even though they cannot self-replicate.

- *Prions* are pathogens that are proteins, and they appear to cause a number of degenerative brain diseases, such as:
  - Scrapie in sheep
  - "Mad cow" disease
  - Creutzfeldt-Jakob disease in humans
- How can a protein which cannot replicate itself be an infectious pathogen? According to one hypothesis:
  - Prions are defective versions (misfolded) of normally occurring cellular proteins.
  - When prions infect normal cells, they somehow convert the normal protein to the prion version (see Campbell, Figure 18.9).
  - Prions could thus trigger chain reactions that increase their numbers and allow them to spread through a host population without dilution.

H. Viruses may have evolved from other mobile genetic elements

Viruses do not fit our usual definitions of living organisms. They cannot reproduce independently, yet they:

- Have a genome with the same genetic code as living organisms
- Can mutate and evolve

Viruses probably evolved after the first cells, from fragments of cellular nucleic acid that were mobile genetic elements. Evidence to support this includes:

- Genetic material of different viral families is more similar to host genomes than to that of other viral families.
- Some viral genes are identical to cellular genes (e.g., oncogenes in retroviruses).
- Viruses of eukaryotes are more similar in genomic structure to their cellular hosts than to bacterial viruses.
- Viral genomes are similar to certain cellular genetic elements such as plasmids and transposons; they are all *mobile* genetic elements.

II.   The Genetics of Bacteria

A. The short generation span of bacteria facilitates their evolutionary adaptation to changing environments

The average bacterial genome is larger than a viral genome, but much smaller than a typical eukaryotic genome.

The major component of the bacterial genome is the *bacterial chromosome*. This structure is:

- Composed of one double-stranded, circular molecule of DNA
- Structurally simpler and has fewer associated proteins than a eukaryotic chromosome
- Found in the *nucleoid* region; since this region is not separated from the rest of the cell (by a membrane), transcription and translation can occur simultaneously.

Many bacteria also contain extrachromosomal DNA in plasmids.

Plasmid = A small double-stranded ring of DNA that carries extrachromosomal genes in some bacteria

Most bacteria can rapidly reproduce by *binary fission*, which is preceded by DNA replication.

- Semi-conservative replication of the bacterial chromosome begins at a single origin of replication.
- The two replication forks move bidirectionally until they meet and replication is complete (see Campbell, Figure 18.10).
- Under optimal conditions, some bacteria can divide in twenty minutes. Because of this rapid reproductive rate, bacteria are useful for genetic studies.

Binary fission is asexual reproduction that produces clones, or daughter cells that are genetically identical to the parent.

- Though mutations are rare events, they can impact genetic diversity in bacteria because of their rapid reproductive rate.
- Though mutation can be a major source of genetic variation in bacteria, it is *not* a major source in more slowly reproducing organisms (e.g., humans). In most higher organisms, genetic recombination from sexual reproduction is responsible for most of the genetic diversity within populations.

B. Genetic recombination produces new bacterial strains

There are three natural processes of genetic recombination in bacteria: *transformation, transduction,* and *conjugation*. These mechanisms of gene transfer occur separately from bacterial reproduction, and in addition to mutation, are another major source of genetic variation in bacterial populations.

1. Transformation

Transformation = Process of gene transfer during which a bacterial cell assimilates foreign DNA from the surroundings

- Some bacteria can take up naked DNA from the surroundings. (Refer to Avery's experiments with *Streptococcus pneumoniae* in Chapter 16.)
- Assimilated foreign DNA may be integrated into the bacterial chromosome by recombination (crossing over).
- Progeny of the recipient bacterium will carry a new combination of genes.

Many bacteria have surface proteins that recognize and import naked DNA from closely related bacterial species.

- Though lacking such proteins, *E. coli* can be artificially induced to take up foreign DNA by incubating the bacteria in a culture medium that has a high concentration of calcium ions.
- This technique of artificially inducing transformation is used by the biotechnology industry to introduce foreign genes into bacterial genomes, so that bacterial cells can produce proteins characteristic of other species (e.g., human insulin and human growth hormone).

2. Transduction

Transduction = Gene transfer from one bacterium to another by a bacteriophage (see Campbell, Figure 18.12)

Generalized transduction = Transduction that occurs when random pieces of host cell DNA are packaged within a phage capsid during the lytic cycle of a phage

- This process can transfer almost any host gene and little or no phage genes.
- When the phage particle infects a new host cell, the donor cell DNA can recombine with the recipient cell DNA.

Specialized transduction = Transduction that occurs when a prophage excises from the bacterial chromosome and carries with it only certain host genes adjacent to the excision site. Also known as *restricted transduction*.

- Carried out only by temperate phages
- Differs from general transduction in that:
  - Specific host genes and most phage genes are packed into the same virion.
  - Transduced bacterial genes are restricted to specific genes adjacent to the prophage insertion site. In general transduction, host genes are randomly selected and almost any host gene can be transferred.

3. Conjugation and plasmids

Conjugation = The direct transfer of genes between two cells that are temporarily joined.

- Discovered by Joshua Lederberg and Edward Tatum
- Conjugation in *E. coli* is one of the best-studied examples:

A DNA-donating *E. coli* cell extends
external appendages called *sex pili*
(see Campbell, Figure 18.13)

↓

Sex pili attach to a DNA-receiving cell

↓

A cytoplasmic bridge forms through which
DNA transfer occurs

The ability to form sex pili and to transfer DNA is conferred by genes in a plasmid called the *F plasmid*.

a. General characteristics of plasmids

Plasmid = A small, circular, double-stranded, self-replicating molecule ring of DNA that carries extrachromosomal genes in some bacteria.

- Plasmids have only a few genes, and they are not required for survival and reproduction.
- Plasmid genes can be beneficial in stressful environments. Examples include the F plasmid, which confers ability to conjugate; and the R plasmid, which confers antibiotic resistance.

These small circular DNA molecules replicate independently:

- Some plasmids replicate in synchrony with the bacterial chromosome, so only a few are present in the cell.
- Some plasmids under more relaxed control can replicate on their own schedule, so the number of plasmids in the cell at any one time can vary from only a few to as many as 100.

Some plasmids are episomes that can reversibly incorporate in the cell's chromosome.

Episomes = Genetic elements that can replicate either independently as free molecules in the cytoplasm or as integrated parts of the main bacterial chromosome.

- Examples include some plasmids and temperate viruses such as lambda phage.
- Temperate phage genomes replicate separately in the cytoplasm during a lytic cycle and as an integral part of the host's chromosome during a lysogenic cycle.

While plasmids and viruses can both be episomes, they differ in that:

- Plasmids, unlike viruses, lack an extracellular stage.

- Plasmids are generally beneficial to the cell, while viruses are parasites that usually harm their hosts.

b. **The F plasmid and conjugation**

The F plasmid (F for fertility) has about 25 genes, most of which are involved in the production of sex pili.

- Bacterial cells that contain the F factor and can donate DNA ("male") are called *F⁺ cells*.
- The F factor replicates in synchrony with chromosomal DNA, so the F⁺ factor is heritable; that is, division of an F⁺ cell results in two F⁺ daughter cells.
- Cells without the F factor are designated *F⁻* ("female").

During conjugation between an F⁺ and an F⁻ bacterium:

- The F factor replicates by *rolling circle replication*. The 5′ end of the copy peels off the circular plasmid and is transferred in linear form.
- The F⁺ cell transfers a copy of its F factor to the F⁻ partner, and the F⁻ cell *becomes* F⁺ (see Campbell, Figure 18.14)
- The donor cell remains F⁺, with its original DNA intact.

The F factor is an episome and occasionally inserts into the bacterial chromosome.

- Integrated F factor genes are still expressed.
- Cells with integrated F factors are called *Hfr cells* (high frequency of recombination).

Conjugation can occur between an Hfr and an F⁻ bacterium.

- As the integrated F factor of the Hfr cell transfers to the F⁻ cell, it pulls the bacterial chromosome behind its leading end.
- The F factor always opens up at the same point for a particular Hfr strain. As rolling circle replication proceeds, the sequence of chromosomal genes behind the leading 5′ end is always the same.
- The conjugation bridge usually breaks before the entire chromosome and tail end of the F factor can be transferred. As a result:
  - Only some bacterial genes are donated.
  - The recipient F⁻ cell does not become an F⁺ cell, because only part of the F factor is transferred.
  - The recipient cell becomes a partial diploid.
  - Recombination occurs between the Hfr chromosomal fragment and the F⁻ cell. Homologous strand exchange results in a *recombinant F⁻ cell*.
  - Asexual reproduction of the recombinant F⁻ cell produces a bacterial colony that is genetically different from both original parental cells.

c. **R plasmids and antibiotic resistance**

One class of nonepisomal plasmids, the *R plasmids* (for resistance), carry genes that confer resistance to certain antibiotics.

- Some carry up to ten genes for resistance to antibiotics.
- During conjugation, some mobilize their own transfer to nonresistant cells.
- Increased antibiotic use has selected for antibiotic resistant bacterial strains carrying the R plasmid.

- Additionally, R plasmids can transfer resistance genes to bacteria of different species including pathogenic strains. As a consequence, resistant strains of pathogens are becoming more common.

4. Transposons

Pieces of DNA called transposons, or transposable genetic elements, can actually move from one location to another in a cell's genome.

Transposons = DNA sequences that can move from one chromosomal site to another.

- Occur as natural agents of genetic change in both prokaryotic and eukaryotic organisms.
- Were first proposed in the 1940s by Barbara McClintock, who deduced their existence in maize. Decades later, the importance of her discovery was recognized; in 1983, at the age of 81, she received the Nobel Prize for her work.

There are two patterns of transposition: a) conservative transposition and b) replicative transposition.

Conservative transposition = Movement of preexisting genes from one genomic location to another; the transposon's genes are not replicated before the move, so the number of gene copies is conserved.

Replicative transposition = Movement of gene copies from their original site of replication to another location in the genome, so the transposon's genes are inserted at some new site without being lost from the original site.

Transposition is fundamentally different from all other mechanisms of genetic recombination, because transposons may scatter certain genes throughout the genome with no apparent single, specific target.

- All other mechanisms of genetic recombination depend upon homologous strand exchange: meiotic crossing over in eukaryotes; and transformation, transduction, and conjugation in prokaryotes.
- Insertion of episomic plasmids into chromosomes is also site specific, even though it does not require an extensive stretch of DNA homologous to the plasmid.

a. Insertion sequences

The simplest transposons are insertion sequences (see Campbell, Figure 18.15).

Insertion sequences = The simplest transposons, which contain only the genes necessary for the process of transposition. Insertion sequence DNA includes two essential types of nucleotide sequences:

   a. Nucleotide sequence coding for *transposase*

   b. Inverted repeats

Transposase = Enzyme that catalyzes insertion of transposons into new chromosomal sites.

- The transposase gene in an insertion sequence is flanked by inverted repeats.

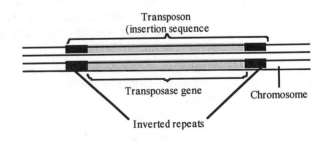

Inverted repeats (IR) = Short noncoding nucleotide sequences of DNA that are repeated in reverse order on opposite ends of a transposon. For example:

DNA strand #1 ... A T C C G G T ...   ... A C C G G A T ...

DNA strand #2 ___ A T C C G G T ___   ___ A C C G G A T ___

Note that each base sequence (IR) is repeated in reverse, on the DNA strand *opposite* the inverted repeat at the other end. Inverted repeats:

- Contain only 20 to 40 nucleotide pairs
- Are recognition sites for transposase

Transposase catalyzes the recombination by:

- Binding to the inverted repeats and holding them close together
- Cutting and resealing DNA required for insertion of the transposon at a new site

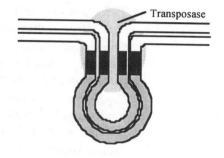

Insertion of transposons also requires other enzymes, such as DNA polymerase. For example,

- At the target site, transposase makes staggered cuts in the two DNA strands, leaving short segments of unpaired DNA at each end of the cut.
- Transposase inserts the transposon into the open target site.
- DNA polymerase helps form direct repeats, which flank transposons in their target site. Gaps in the two DNA strands fill in when nucleotides base pair with the exposed single-stranded regions.

Direct repeats = Two or more identical DNA sequences in the same molecule.

- The transposition process creates direct repeats that flank transposons in their target site (see Campbell, Figure 18.16).

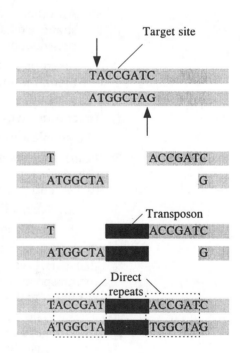

Transposed insertion sequences are likely to somehow alter the cell's phenotype; they may:

- Cause mutations by interrupting coding sequences for proteins.
- Increase or decrease a protein's production by inserting within regulatory regions that control transcription rates.

Transposition of insertion sequences probably plays a significant role in bacterial evolution as a source of genetic variation.

- Though insertion sequences only rarely cause mutations (about one in every 10^6 generations), the mutation rate from transpositions is about the same as the mutation rate from extrinsic causes, such as radiation and chemical mutagens.

b. Composite transposons

Composite (complex) transposons = Transposons which include additional genetic material besides that required for transposition; consist of one or more genes flanked by insertion sequences (see Campbell, Figure 18.17).

- The additional DNA may have any nucleotide sequence.
- Can insert into almost any stretch of DNA since their insertion is not dependent upon DNA sequence homology.
- Generate genetic diversity in bacteria by moving genes from one chromosome, or even one species, to another. This diversity may help bacteria adapt to new environmental conditions.

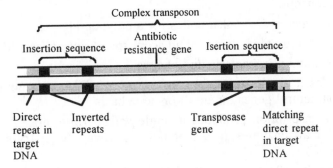

An example is a transposon that carries a bacterial gene for antibiotic resistance.

Examples of genetic elements that contain one or more complex transposons include:

- F factor.
- DNA version of the retrovirus genome.

C. The control of gene expression enables individual bacteria to adjust their metabolism to environmental change

> This material is difficult to teach. The problem is that there are so many components to track, and the students are not familiar enough with the vocabulary. Figures 18.17 and 18.18 are good teaching aids for the *trp* and *lac* operons. It is also helpful to construct flow charts where there are decision points (e.g., glucose present—glucose absent), so that students can visually follow alternate paths and their respective consequences.

Genes switch on and off as conditions in the intracellular environment change. Bacterial cells have two main ways of controlling metabolism:

1. *Regulation of enzyme activity.* The catalytic activity of many enzymes increases or decreases in response to chemical cues.
   - For example, the end product of an anabolic pathway may turn off its own production by inhibiting activity of an enzyme at the beginning of the pathway (*feedback inhibition*).
   - Useful for immediate short-term response.
2. *Regulation of gene expression.* Enzyme concentrations may rise and fall in response to cellular metabolic changes that switch genes on or off.

- For example, accumulation of product may trigger a mechanism that inhibits transcription of mRNA production by genes that code for an enzyme at the beginning of the pathway (*gene repression*).
- Slower to take effect than feedback inhibition, but is more economical for the cell. It prevents unneeded protein synthesis for enzymes, as well as, unneeded pathway product.

An example illustrating regulation of a metabolic pathway is the tryptophan pathway in *E. coli*. (See Campbell, Figure 18.18) Mechanisms for gene regulation were first discovered for *E. coli*, and current understanding of such regulatory mechanisms at the molecular level is still limited to bacterial systems.

1. **Operons: the basic concept**

Regulated genes can be switched on or off depending on the cell's metabolic needs. From their research on the control of lactose metabolism in *E. coli*, Francois Jacob and Jacques Monod proposed a mechanism for the control of gene expression, the *operon* concept.

Structural gene = Gene that codes for a polypeptide

Operon = A regulated cluster of adjacent *structural genes* with related functions

- Common in bacteria and phages
- Has a single promoter region, so an RNA polymerase will transcribe all structural genes on an all-or-none basis
- Transcription produces a single *polycistronic* mRNA with coding sequences for all enzymes in a metabolic pathway (e.g., tryptophan pathway in *E. coli*)

Polycistronic mRNA = A large mRNA molecule that is a transcript of several genes

- Is translated into separate polypeptides
- Contains stop and start codons for the translation of each polypeptide

Grouping structural genes into operons is advantageous because:

- Expression of these genes can be coordinated. When a cell needs the product of a metabolic pathway, all the necessary enzymes are synthesized at one time.
- The entire operon can be controlled by a single *operator*.

Operator = A DNA segment located within the promotor or between the promoter and structural genes, which controls access of RNA polymerase to structural genes.

- Sometimes overlaps the transcription starting point for the operon's first structural gene
- Acts as an on/off switch for movement of RNA polymerase and transcription of the operon's structural genes

What determines whether an operator is in the "on" or "off" mode? By itself, the operator is on; it is switched off by a protein repressor.

Repressor = Specific protein that binds to an operator and blocks transcription of the operon

- Blocks attachment of RNA polymerase to the promoter
- Is similar to an enzyme, in that it:
  - Has an active site with a specific conformation, which discriminates among operators. Repressor proteins are specific only for operators of certain operons.
  - Binds *reversibly* to DNA
  - May have an allosteric site in addition to its DNA-binding site
- Repressors are encoded by *regulatory genes*.

Regulatory genes = Genes that code for repressor or regulators of other genes
- Are often located some distance away from the operons they control and has its own promotor
- Are involved in switching on or off the transcription of structural genes by the following process:

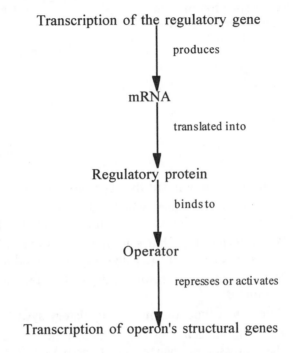

Transcription of the regulatory gene

produces

mRNA

translated into

Regulatory protein

binds to

Operator

represses or activates

Transcription of operon's structural genes

Regulatory genes are continually transcribed, so their activity depends upon how efficient their promoters are in binding RNA polymerase.
- They produce repressor molecules continuously, but slowly.
- Operons are still expressed even though repressor molecules are always present, because repressors are not always capable of blocking transcription; they alternate between inactive and active conformations.

A repressor's activity depends upon the presence of key metabolites in the cell.
- Regulation of the *trp* operon in *E. coli* is an example of how a metabolite cues a repressor (see Campbell, Figure 18.19):
- Repressible enzymes catalyze the anabolic pathway that produces tryptophan, an amino acid.
- Tryptophan accumulation represses synthesis of the enzymes that catalyze its production.

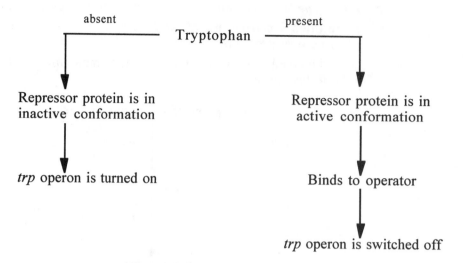

How does tryptophan activate the repressor protein?

- The repressor protein, which normally has a low affinity for the operator, has a DNA binding site plus an allosteric site specific for tryptophan.
- When tryptophan binds to the repressor's allosteric site, it activates the repressor causing it to change its conformation.
- The activated repressor binds to the operator, which switches the *trp* operon off.
- Tryptophan functions in this regulatory system as a *corepressor*.

Corepressor = A molecule, usually a metabolite, that binds to a repressor protein, causing the repressor to change into its active conformation

- Only the *repressor-corepressor complex* can attach to the operator and turn off the operon.
- When tryptophan concentrations drop, it is less likely to be bound to repressor protein. The *trp* operon, once free from repression, begins transcription.
- As concentrations of tryptophan rise, it turns off its own production by activating the repressor.
- Enzymes of the tryptophan pathway are said to be *repressible*.

2. **Repressible versus inducible operons: two types of negative gene regulation**

Repressible operon = Operons which have their transcription inhibited. Usually associated with anabolic processes, (e.g., tryptophan synthesis via *trp* operon).

Inducible operons = Operons which have their transcription stimulated. Usually associated with catabolic processes.

Some operons can be switched on or *induced* by specific metabolites (e.g., *lac* operon in *E. coli*).

- *E. coli* can metabolize the disaccharide lactose. Once lactose is transported into the cell, β-galactosidase cleaves lactose into glucose and galactose:

$$\text{lactose} \xrightarrow{\;\beta\text{-galactosidase}\;} \text{glucose + galactose}$$
$$\text{(disaccharide)} \qquad\qquad \text{(monosaccharides)}$$

- When *E. coli* is in a lactose-free medium, it only contains a few β-galactosidase molecules.

- When lactose is added to the medium, *E. coli* increases the number of mRNA molecules coding for β-galactosidase. These mRNA molecules are quickly translated into thousands of β-galactosidase molecules.

- Lactose metabolism in *E. coli* is programmed by the *lac* operon which has three structural genes:

  1. *lac Z* - Codes for β-galactosidase which hydrolyzes lactose
  2. *lac Y* - Codes for a permease, a membrane protein that transports lactose into the cell
  3. *lac A* - Codes for transacetylase, an enzyme that has no known role in lactose metabolism

- The *lac* operon has a single promoter and operator. The *lac* repressor is innately active, so it attaches to the operon without a corepressor.

- Allolactose, an isomer of lactose, acts as an *inducer* to turn on the *lac* operon (see Campbell, Figure 18.20):

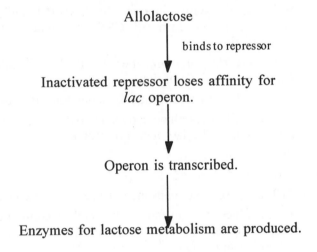

Allolactose

binds to repressor

Inactivated repressor loses affinity for *lac* operon.

Operon is transcribed.

Enzymes for lactose metabolism are produced.

Differences between repressible and inducible operons reflect differences in the pathways they control.

Repressible Operons	Inducible Operons
Their genes are switched on until a specific metabolite activates the repressor.	Their genes are switched off until a specific metabolite inactivates the repressor.
They generally function in anabolic pathways.	They function in catabolic pathways
Pathway end product switches off its own production by repressing enzyme synthesis.	Enzyme synthesis is switched on by the nutrient the pathway uses.

Repressible and inducible operons share similar features of gene regulation. In both cases:

- Specific repressor proteins control gene expression.
- Repressors can assume an active conformation that blocks transcription and an inactive conformation that allows transcription.
- Which form the repressor assumes depends upon cues from a metabolite.

Both systems are thus examples of *negative control*.

- Binding of active repressor to an operator always turns off structural gene expression.
- The *lac* operon is a system with negative control, because allolactose does not interact directly with the genome. The derepression allolactose causes is indirect, by freeing the *lac* operon from the repressor's negative effect.

Positive control of a regulatory system occurs only if an activator molecule interacts directly with the genome to turn on transcription.

3. **An example of positive gene regulation**

The *lac* operon is under dual regulation which includes negative control by repressor protein and positive control by cAMP receptor protein (CRP).

CRP (cAMP receptor protein) = An allosteric protein that binds cAMP and activates transcription binding to an operon's promoter region (enhances the promoter's affinity for RNA polymerase) (see Campbell, Figure 18.21)

- Exists in two states: inactive (no cAMP bound) and active (cAMP bound). Only the active form of CRP can bind to the promoter to stimulate transcription.
- It is a *positive regulator* because it *directly* interacts with the genome to stimulate gene expression.
- CRP binding to a promoter is dependent on glucose concentration.

E. coli preferentially uses glucose over lactose as a substrate for glycolysis. So normal expression of the lac operon requires:

- Presence of lactose
- *Absence* of glucose

How is CRP affected by the absence or presence of glucose?

- When glucose is missing, the cell accumulates *cyclic AMP* (*cAMP*), a nucleotide derived from ATP. cAMP activates CRP so that it can bind to the *lac* promoter.
- When glucose concentration rises, glucose catabolism decreases the intracellular concentration of cAMP. Thus, cAMP releases CRP.

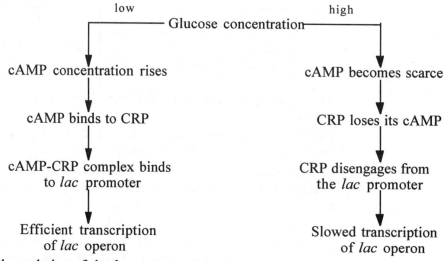

In this dual regulation of the *lac* operon:

- Negative control by the repressor determines whether or not the operon will transcribe the structural genes.
- Positive control by CRP determines the rate of transcription.

E. coli economizes on RNA and protein synthesis with the help of these negative and positive controls.

- CRP is an activator of several different operons that program catabolic pathways.
- Glucose's presence deactivates CRP. This, in turn, slows synthesis of those enzymes a cell needs to use catabolites other than glucose.
- *E. coli* preferentially uses glucose as its primary carbon and energy source, and the enzymes for glucose catabolism are coded for by unregulated genes that are continuously transcribed (constitutive).
- Consequently, when glucose is present, CRP does not work and the cell's systems for using secondary energy sources are inactive.

When glucose is absent, the cell metabolizes alternate energy sources.

- The cAMP level rises, CRP is activated and transcription begins of operons that program the use of alternate energy sources (e.g., lactose).
- Which operon is actually transcribed depends upon which nutrients are available to the cell. For example, if lactose is present, the *lac* operon will be switched on as allolactose inactivates the repressor.

REFERENCES

Alberts, B., D. Bray, J. Lewis, M. Raff, K. Roberts and J.D. Watson. *Molecular Biology of the Cell*. 3rd ed. New York: Garland, 1994.

Campbell, N., et al. *Biology*. 5th ed. Menlo Park, California: Benjamin/Cummings, 1998.

CHAPTER 19

THE ORGANIZATION AND CONTROL OF EUKARYOTIC GENOMES

OUTLINE

I.   The Structure of Chromatin
     A. Chromatin structure is based on successive levels of DNA packing

II. Genome Organization at the DNA Level
     A. Repetitive DNA and other noncoding sequences account for much of a eukaryotic genome
     B. Gene families have evolved by duplication of ancestral genes
     C. Gene amplification, loss, or rearrangement can alter a cell's genome

III. The Control of Gene Expression
     A. Each cell of a multicellular eukaryote expresses only a small fraction of its genome
     B. The control of gene expression can occur at any step in the pathway from gene to functional protein
     C. Chromatin modifications affect the availability of genes for transcription
     D. Transcription initiation is controlled by proteins that interact with DNA and with each other
     E. Posttranscriptional mechanisms play supporting roles in the control of gene expression

IV. The Molecular Biology of Cancer
     A. Cancer results from genetic changes that affect the cell cycle
     B. Oncogene proteins and faulty tumor-suppressor proteins
     C. Multiple mutations underlie the development of cancer

OBJECTIVES

After reading this chapter and attending lecture, the student should be able to:

1.   Compare the organization of prokaryotic and eukaryotic genomes.
2.   Describe the current model for progressive levels of DNA packing.
3.   Explain how histones influence folding in eukaryotic DNA.
4.   Distinguish between heterochromatin and euchromatin.
5.   Using the Barr body as an example, describe the function of heterochromatin in interphase cells.
6.   Describe where satellite DNA is found and what role it may play in the cell.
7.   Describe the role of telomeres in solving the end-replication problem with the lagging DNA strand.
8.   Using the genes for rRNA as an example, explain how multigene families of identical genes can be advantageous for a cell.

9. Using α-globin and β-globin genes as examples, describe how multigene families of nonidentical genes probably evolve, including the role of transposition.

10. Explain the potential role that promoters and enhancers play in transcriptional control.

11. Explain why the nuclear envelope in eukaryotes offers a level of post-transcriptional control beyond that found in prokaryotes.

12. Explain why the ability to rapidly degrade mRNA can be an adaptive advantage for prokaryotes.

13. Describe the importance of mRNA degradation in eukaryotes, describe how it can be prevented.

14. Explain how gene expression may be controlled at the translational and post-translational level.

15. Compare the arrangement of coordinately controlled genes in prokaryotes and eukaryotes.

16. Explain how eukaryotic genes can be coordinately expressed and give some examples of coordinate gene expression in eukaryotes.

17. Provide evidence from studies of polygene chromosomes, that eukaryotic gene expression is controlled at transcription and that gene regulation responds to chemical signals such as steroid hormones.

18. Describe the key steps of steroid hormone action on gene expression in vertebrates.

19. In general terms, explain how genome plasticity can influence gene expression.

20. Describe the effects of gene amplification, selective gene loss and DNA methylation.

21. Explain how rearrangements in the genome can activate or inactivate genes.

22. Explain the genetic basis for antibody diversity.

23. Explain how DNA methylation may be a cellular mechanism for long-term control of gene expression and how it can influence early development.

24. Describe the normal control mechanisms that limit cell growth and division.

25. Briefly describe the four mechanisms that can convert proto-oncogenes to oncogenes.

26. Explain how changes in tumor-suppressor genes can be involved in transforming normal cells into cancerous cells.

27. Explain how oncogenes are involved in virus-induced cancers.

KEY TERMS

histones	multigene family	genomic imprinting	proteasomes
nucleosome	pseudogene	histone acetylation	oncogenes
heterochromatin	gene amplification	control elements	proto-oncogenes
euchromatin	retrotransposons	enhancers	tumor-suppressor genes
repetitive DNA	immunoglobulins	activator	*ras* gene
satellite DNA	differentiation	DNA-binding domain	p53 gene
Alu elements	DNA methylation	alternative splicing	

LECTURE NOTES

Eukaryotic gene regulation is more complex than in prokaryotes, because eukaryotes:

- Have larger, more complex genomes. This requires that eukaryotic DNA be more complexly organized than prokaryotic DNA.
- Require cell specialization or *differentiation*.

I.   The Structure of Chromatin

A. Chromatin Structure is Based on Successive Levels of DNA Packing

Prokaryotic and eukaryotic cells both contain double-stranded DNA, but their genomes are organized differently.

Prokaryotic DNA is:

- Usually circular
- Much smaller than eukaryotic DNA; it makes up a small nucleoid region only visible with an electron microscope
- Associated with only a few protein molecules
- Less elaborately structured and folded than eukaryotic DNA; bacterial chromosomes have some additional structure as the DNA-protein fiber forms loops that are anchored to the plasma membrane

Eukaryotic DNA is:

- Complexed with a large amount of protein to form *chromatin*
- Highly extended and tangled during interphase
- Condensed into short, thick, discrete *chromosomes* during mitosis; when stained, chromosomes are clearly visible with a light microscope

Eukaryotic chromosomes contain an enormous amount of DNA, which requires an elaborate system of DNA packing to fit all of the cell's DNA into the nucleus.

B. Nucleosomes, or "beads on a string"

Histone proteins associated with DNA are responsible for the first level of DNA packing in eukaryotes.

Histones = Small proteins that are rich in basic amino acids and that bind to DNA, forming chromatin.

- Contain a high proportion of positively charged amino acids (arginine and lysine), which bind tightly to the negatively charged DNA
- Are present in approximately equal amounts to DNA in eukaryotic cells
- Are similar from one eukaryote to another, suggesting that histone genes have been highly conserved during evolution. There are five types of histones in eukaryotes.

Partially unfolded *chromatin* (DNA and its associated proteins) resembles beads spaced along the DNA string. Each beadlike structure is a nucleosome (see Campbell, Figure 19.1a).

Nucleosome = The basic unit of DNA packing; it is formed from DNA wound around a protein core that consists of two copies each of four types of histone (H2A, H2B, H3, H4). A fifth histone (H1) attaches near the bead when the chromatin undergoes the next level of packing.

- Nucleosomes may control gene expression by controlling access of transcription proteins to DNA.
- Nucleosome heterogeneity may also help control gene expression; nucleosomes may differ in the extent of amino acid modification and in the type of nonhistone proteins present.

C. Higher levels of DNA packing

The *30-nm chromatin fiber* is the next level of DNA packing (see Campbell, Figure 19.1b).

- This structure consists of a tightly wound coil with six nucleosomes per turn.
- Molecules of histone H1 pull the nucleosomes into a cylinder 30nm in diameter.

In the next level of higher-order packing, the 30-nm chromatin fiber forms *looped domains*, which:

- Are attached to a nonhistone protein scaffold
- Contain 20,000 to 100,000 base pairs
- Coil and fold, further compacting the chromatin into a mitotic chromosome characteristic of metaphase

Interphase chromatin is much less condensed than mitotic chromatin, but it still exhibits higher-order packing.

- Its nucleosome string is usually coiled into a 30-nm fiber, which is folded into looped domains.
- Interphase looped domains attach to a scaffolding inside the nuclear envelope (nuclear lamina); this helps organize areas of active transcription.
- Chromatin fibers of different chromosomes do not become entangled as they occupy restricted areas within the nucleus.

Portions of some chromosomes remain highly condensed throughout the cell cycle, even during interphase. Such heterochromatin is not transcribed.

Heterochromatin = Chromatin that remains highly condensed during interphase and that is not actively transcribed

Euchromatin = Chromatin that is less condensed during interphase and is actively transcribed; euchromatin becomes highly condensed during mitosis

What is the function of heterochromatin in interphase cells?

- Since most heterochromatin is not transcribed, it may be a coarse control of gene expression.
- For example, Barr bodies in mammalian cells are *X* chromosomes that are mostly condensed into heterochromatin. In female somatic cells, one *X* chromosome is a Barr body, so the other *X* chromosome is the only one transcribed.

II. Genome Organization at the DNA Level

An organism's genome is plastic, or changeable, in ways that affect the availability of specific genes for expression.

- Genes may be available for expression in some cells and not others, or at some time in the organism's development and not others.
- Genes may, under some conditions, be amplified or made more available than usual.
- Changes in the physical arrangement of DNA, such as levels of DNA packing, affect gene expression. For example, genes in heterochromatin and mitotic chromosomes are not expressed.

The structural organization of an organism's genome is also somewhat plastic; movement of DNA within the genome and chemical modification of DNA influence gene expression.

A. Repetitive DNA and noncoding sequences account for much of a eukaryotic genome

DNA in eukaryotic genomes is organized differently from that in prokaryotes.

- In prokaryotes, most DNA codes for protein (mRNA), tRNA or rRNA, and coding sequences are uninterrupted. Small amounts of noncoding DNA consist mainly of control sequences, such as promoters.
- In eukaryotes, most DNA does *not* encode protein or RNA, and coding sequences may be interrupted by long stretches of noncoding DNA (introns). Certain DNA sequences may be present in multiple copies.

1. **Tandemly repetitive DNA**

   About 10–25% of total DNA in higher eukaryotes is *satellite DNA* that consists of short (five to 10 nucleotides) sequences that are tandemly repeated thousands of times.

   Satellite DNA = In eukaryotic chromosomes, highly repetitive DNA consisting of short unusual nucleotide sequences that are tandemly repeated thousands of times.

   - Called satellite DNA because its unusual nucleotide ratio gives it a density different from the rest of the cell's DNA. Thus, during ultracentrifugation, satellite DNA separates out in a cesium chloride gradient as a "satellite" band separate from the rest of the DNA.

   - Is not transcribed and its function is not known. Since most satellite DNA in chromosomes is located at the tips and the centromere, scientists speculate that it plays a structural role during chromosome replication and chromatid separation in mitosis and meiosis.

   It is known that short tandem repeats called telomeres—at the ends of eukaryotic chromosomes—are important in maintaining the integrity of the lagging DNA strand during replication.

   Telomere = Series of short tandem repeats at the ends of eukaryotic chromosomes; prevents chromosomes from shortening with each replication cycle

   - Before an Okazaki fragment of the lagging DNA strand can be synthesized, RNA primers must be produced on a DNA template ahead of the sequence to be replicated.

   - Since such a template is not possible for the end of a linear DNA molecule, there must be a mechanism to prevent DNA strands from becoming shorter with each replication cycle.

   - This end-replication problem is solved by the presence of special repeating telomeric sequences on the ends of linear chromosomes.

   - To compensate for the loss of telomeric nucleotides that occurs each replication cycle, the enzyme *telomerase* periodically restores this repetitive sequence to the ends of DNA molecules.

   - Telomeric sequences are similar among many organisms and contain a block of G nucleotides. For example, human chromosomes have 250–1500 repetitions of the base sequence TTAGGG (AATCCC on the complementary strand).

   There are other highly repetitive sequences in eukaryotic genomes. For example,

   - Some are transposons; generally regarded as nonfunctional, they are associated with some diseases (e.g., neurofibromatosis-1 or elephant man's disease and some cancers).

   - Mutations can extend the repetitive sequences normally found within the boundary of genes and cause them to malfunction. (e.g., fragile X syndrome and Huntington's disease.)

2. **Interspersed repetitive DNA**

   Eukaryotes also possess large amounts (25-40% in mammals) of repeated units, hundreds or thousands of base pairs long, dispersed at random intervals throughout the genome.

B. **Gene families have evolved by duplication of ancestral genes**

   Most eukaryotic genes are *unique sequences* present as single copies in the genome, but some genes are part of a multigene family.

Multigene family = A collection of genes that are similar or identical in sequence and presumably of common ancestral origin; such genes may be clustered or dispersed in the genome.

Families of *identical* genes:

- Probably arise from a single ancestral gene that has undergone repeated duplication. Such *tandem gene duplication* results from mistakes made during DNA replication and recombination.

- Are usually clustered and almost exclusively genes for RNA products. (One exception is the gene family coding for histone proteins.)

- Include genes for the major rRNA molecules; huge tandem repeats of these genes enable cells to make millions of ribosomes during active protein synthesis (see Campbell, Figure 19.2).

Families of *nonidentical* genes:

- Arise over time from mutations that accumulate in duplicated genes.

- Can be clustered on the same chromosome or scattered throughout the genome. (Note that for various reasons, gene sequences in tandem arrays on the same chromosome tend to stay very similar to one another. Transposition events that translocate variants of duplicated genes to different chromosomes, help stabilize their differences and thus promote diversity.)

- May include pseudogenes or nonfunctional versions of the duplicated gene.

Pseudogene = Nonfunctional gene that has a DNA sequence similar to a functional gene; but as a consequence of mutation, lacks sites necessary for expression.

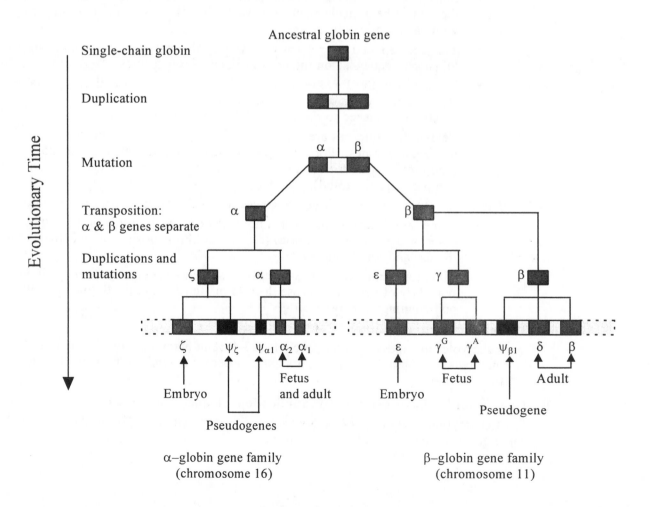

A good example of how multigene families can evolve from a single ancestral gene is the globin gene family—actually two related families of genes that encode globins, the α and β polypeptide subunits of hemoglobin.

Based on amino acid homologies, the evolutionary history has been reconstructed as follows:

- The original α and β genes evolved from duplication of a common ancestral globin gene. Gene duplication was followed by mutation.
- Transposition separated the α globin and β globin families, so they exist on different chromosomes.
- Subsequent episodes of gene duplication and mutation resulted in new genes and pseudogenes in each family.

The consequence is that each globin gene family consists of a group of similar, but not identical genes clustered on a chromosome.

- In each gene family, the genes are arranged in order of their expression. During development, genes are turned on or off in response to the organism's changing environment as it develops from an embryo into a fetus, and then into an adult.
- At all times during development, functional hemoglobin consists of two α-like and two β-like polypeptides.
  - During embryogenesis, the ζ and ε forms predominate.
  - About 10 weeks after fertilization, the products of the α-globin genes replace that of the ζ gene, and the γ-chain gene products—γ^G and the γ^A globins—become more prevalent.
  - Prior to birth, the β-gene product begins to replace the γ^G and γ^A globins, so by six months of age, the adult β-like globins, δ and β, are present.
- In humans, embryonic and fetal hemoglobins have a higher affinity for oxygen than the adult forms, allowing efficient oxygen exchange between mother and developing fetus.

C. Gene amplification, loss, or rearrangement can alter a cell's genome

1. Gene amplification and selective gene loss

Gene amplification may temporarily increase the number of gene copies at certain times in development.

Gene amplification = Selective synthesis of DNA, which results in multiple copies of a single gene.,

- For example, amphibian rRNA genes are selectively amplified in the oocyte, which:
  - Results in a million or more additional copies of the rRNA genes that exist as extrachromosomal circles of DNA in the nucleoli.
  - Permits the oocyte to make huge numbers of ribosomes that will produce the vast amounts of proteins needed when the egg is fertilized.
- Gene amplification occurs in cancer cells exposed to high concentrations of chemotherapeutic drugs.
  - Some cancer cells survive chemotherapy, because they contain amplified genes conferring drug resistance.
  - Increased drug resistance can be created experimentally by exposing a cell population to increasing drug doses and artificially selecting for surviving cells that have amplified drug-resistance genes.

Genes may also be selectively lost in certain tissues by elimination of chromosomes.

Chromosome diminution = Elimination of whole chromosomes or parts of chromosomes from certain cells early in embryonic development.

- For example, chromosome diminution occurs in gall midges during early development; all but two cells lose 32 of their 40 chromosomes during the first mitotic division after the 16-cell stage.

- The two cells that retain the complete genome are germ cells that will produce gametes in the adult. The other 14 cells become somatic cells with only eight chromosomes.

2. Rearrangements in the genome

Substantial stretches of DNA can be re-shuffled within the genome; these rearrangements are more common that gene amplification or gene loss.

a. Transposons

All organisms probably have transposons that move DNA from one location to another within the genome (see Campbell, Chapter 17). Transposons can rearrange the genome by:

- Inserting into the middle of a coding sequence of another gene; it can prevent the interrupted gene from functioning normally (see Campbell, Figure 19.4).

- Inserting within a sequence that regulates transcription; the transposition may increase or decrease a protein's production.

- Inserting its own gene just downstream from an active promoter that activates its transcription.

Retrotranposons = Transposable elements that move within a genome by means of an RNA intermediate (see Campbell, Figure 19.5).

Retrotransposons insert at another site by utilizing reverse transcriptase to convert back to DNA.

b. Immunoglobulin genes

During cellular differentiation in mammals, permanent rearrangements of DNA segments occur in those genes that encode antibodies, or *immunoglobulins*.

Immunoglobulins = A class of proteins (antibodies) produced by B lymphocytes that specifically recognize and help combat viruses, bacteria, and other invaders of the body. Immunoglobulin molecules consist of:

- Four polypeptide chains held together by disulfide bridges
- Each chain has two major parts:
  - A *constant region*, which is the same for all antibodies of a particular class
  - A *variable region*, which gives an antibody the ability to recognize and bind to a specific foreign molecule

B lymphocytes, which produce immunoglobulins, are a type of white blood cell found in the mammalian immune system.

- The human immune system contains millions of subpopulations of B lymphocytes that produce different antibodies.

- B lymphocytes are very specialized; each differentiated cell and its descendants produce only one specific antibody.

Antibody specificity and diversity are properties that emerge from the unique organization of the antibody gene, which is formed by a rearrangement of the genome during B cell development (see Campbell, Figure 19.6).

- As an unspecialized cell differentiates into a B lymphocyte, its antibody gene is pieced together randomly from several DNA segments that are physically separated in the genome.

- In the genome of an embryonic cell, there is an intervening DNA sequence between the sequence coding for an antibody's constant region and the site containing hundreds of coding sequences for the variable regions.
- As a B cell differentiates, the intervening DNA is deleted, and the DNA sequence for a variable region connects with the DNA sequence for a constant region, forming a continuous nucleotide sequence that will be transcribed.
- The primary RNA transcript is processed to form mRNA that is translated into one of the polypeptides of an antibody molecule.
- Antibody variation results from:
  - Different combinations of variable and constant regions in the polypeptides
  - Different combinations of polypeptides

III.   The Control of Gene Expression

A. Each cell of a multicellular eukaryote expresses only a small fraction of its genes

Cellular differentiation = Divergence in structure and function of different cell types, as they become specialized during an organism's development

- Cell differentiation requires that gene expression must be regulated on a long-term basis.
- Highly specialized cells, such as muscle or nerve, express only a small percentage of their genes, so transcription enzymes must locate the right genes at the right time.
- Uncontrolled or incorrect gene action can cause serious imbalances and disease, including cancer. Thus, eukaryotic gene regulation is of interest in medical as well as basic research.

DNA-binding proteins regulate gene activity in all organisms—prokaryotes as well as eukaryotes.

- Usually, it is DNA transcription that is controlled.
- Eukaryotes have more complex chromosomal structure, gene organization and cell structure than prokaryotes, which offer added opportunities for controlling gene expression.

B. The control of gene expression can occur at any step in the pathway from gene to functional protein: *an overview*

Complexities in chromosome structure, gene organization and cell structure provide opportunities for the control of gene expression in eukaryotic cells. The steps of gene expression where gene regulation can occur are outlined below (see also Campbell, Figure 19.7).

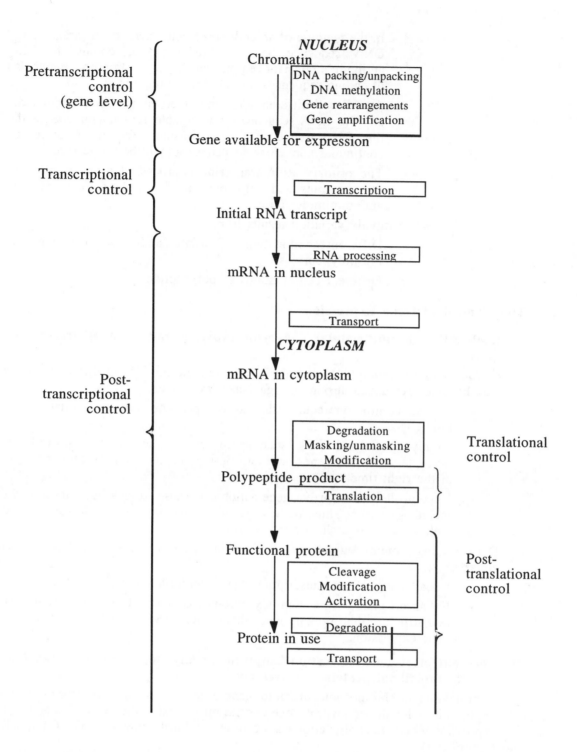

C. Chromatin modifications affect the availability of genes for transcription

Chromatin organization:

- Packages DNA into a compact form that can be contained by the cell's nucleus.
- Controls which DNA regions are available for transcription.
    - Condensed heterochromatin is not expressed.
    - A gene's location relative to nucleosomes and to scaffold attachment sites influences its expression.

Chemical modifications of chromatin play key roles in both chromatin structure and the regulation of transcription.

1. DNA methylation

DNA methylation = The addition of methyl groups ($-CH_3$) to bases of DNA, after DNA synthesis

- Most plant and animal DNA contains methylated bases (usually cytosine); about 5% of the cytosine residues are methylated.
- May be a cellular mechanism for long-term control of gene expression. When researchers examine the same genes from different types of cells, they find:
  - Genes that are not expressed (e.g., Barr bodies) are more heavily methylated than those that are expressed.
  - Drugs that inhibit methylation can induce gene reactivation, even in Barr bodies.
- In vertebrates, DNA methylation reinforces earlier developmental decisions made by other mechanisms.
  - For example, genes must be selectively turned on or off for normal cell differentiation to occur. DNA methylation ensures that once a gene is turned off, it stays off.
  - DNA methylation patterns are inherited and thus perpetuated as cells divide; clones of a cell lineage forming specialized tissues have a chemical record of regulatory events that occurred during early development.

2. Histone acetylation

- Acetylation enzymes attach $-COCH_3$ groups to certain amino acids of histone proteins

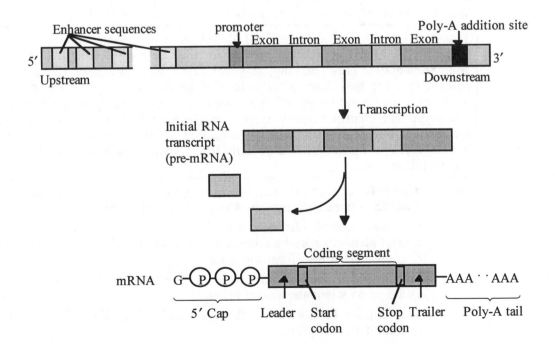

- Acetylated histone proteins have altered conformation and bind to DNA less tightly; as a result, transcription proteins have easier access to genes in the acetylated region.

D. Transcription initiation is controlled by proteins that interact with DNA and with each other

1. Organization of a typical eukaryotic gene

The following is a brief review of a eukaryotic gene and its transcript (see also Campbell, Figure 19.8).

Eukaryotic genes:

- Contain introns, noncoding sequences that intervene within the coding sequence
- Contain a promoter sequence at the 5′ upstream end; a transcription initiation complex, including RNA polymerase, attaches to a promoter sequence and transcribes introns along with the coding sequences, or exons
- May be regulated by control elements, other noncoding control sequences that can be located thousands of nucleotides away from the promoter

Control element = Segments of noncoding DNA that help regulate the transcription of a gene by binding specific proteins (transcription factors).

The primary RNA transcript (pre-mRNA) is processed into mature mRNA by:

- Removal of introns
- Addition of a modified guanosine triphosphate cap at the 5′ end
- Addition of a poly-A tail at the 3′ end

2. The roles of transcription factors

In both prokaryotes and eukaryotes, transcription requires that RNA polymerase recognize and bind to DNA at the promoter. However, transcription in eukaryotes requires the presence of proteins known as *transcription factors*; transcription factors augment transcription by binding:

- Directly to DNA (protein-DNA interactions)
- To each other and/or to RNA polymerase (protein-protein interactions)

Eukaryotic RNA polymerase cannot recognize the promoter without the help of a specific transcription factor that binds to the TATA box of the promoter.

> Transcription factors and the TATA box were discussed with the material in Chapter 17. This is another point in the course where students, if they compartmentalize their learning, may not relate your discussion of transcriptional control to what they have already learned about transcription. Experiment with organizing the presentation of this material. It may be more prudent to cover transcriptional control of gene expression (both prokaryotic and eukaryotic) with the lectures on protein synthesis. If not, you should probably begin your lecture with a review of transcription, especially RNA polymerase binding and initiation of transcription.

Associations between transcription factors and control elements (specific segments of DNA) are important transcriptional controls in eukaryotes.

- Proximal control elements are close to or within the promoter; distal control elements may be thousands of nucleotides away from the promoter or even downstream from the gene.
- Transcription factors known as activators bind to enhancer control elements to stimulate transcription.
- Transcription factors known as repressors bind to silencer control elements to inhibit transcription

How do activators stimulate transcription?

- One hypothesis is that a hairpin loop forms in DNA, bringing the activator bound to an enhancer into contact with other transcription factors and polymerase at the promoter (see Campbell, Figure 19.9).
- Diverse activators may selectively stimulate gene expression at appropriate stages in cell development.

The involvement of transcription factors in eukaryotes offers additional opportunities for transcriptional control. This control depends on selective binding of specific transcription factors to specific DNA sequences and/or other proteins; the highly selective binding depends on molecular structure.

- There must be a complementary fit between the surfaces of a transcription factor and its specific DNA-binding site.
- Hundreds of transcription factors have been discovered; and though each of these proteins is unique, many recognize their DNA-binding sites with only one of a few possible structural motifs or *domains* containing α helices or β sheets (see Campbell, Figure 19.10).

3. Coordinately controlled genes

Coordinately controlled genes are arranged differently in a eukaryotic chromosome than in prokaryotic genomes.

- Prokaryotic genes that are turned on and off together are often clustered into operons; these adjacent genes share regulatory sites located at one end of the cluster. All genes of the operon are transcribed into one mRNA molecule and are translated together.
- Eukaryotic genes coding for enzymes of a metabolic pathway are often scattered over different chromosomes. Even functionally related genes on the same chromosome have their own promoters and are individually transcribed.

Eukaryotic genes can be coordinately expressed, even though they may be scattered throughout the genome.

- Coordinately controlled genes are each associated with specific regulatory DNA sequences or enhancers. These sequences are recognized by a single type of transcription factor that activates or represses a group of genes in synchrony.
- Examples of coordinate gene expression in eukaryotes include:
  - *Heat shock response.* Exposure to high temperature activates genes coding for heat shock proteins, which help stabilize and repair heat-denatured proteins in the cell.
  - *Steroid hormone action.* Steroids activate protein receptors, and the protein-receptor complex, in turn, activates genes. In a secondary response, proteins produced this way can activate another group of genes (see Campbell Chapter 45).
  - *Cellular differentiation.* During cellular differentiation, coordinately controlled genes producing particular sets of proteins are switched on and off.

E. Posttranscriptional mechanisms play supporting roles in the control of gene expression

Transcription produces a primary transcript, but gene expression—the production of protein, tRNA, or rRNA—may be stopped or enhanced at any posttranscriptional step.

Because eukaryotic cells have a nuclear envelope, translation is segregated from transcription. This offers additional opportunities for controlling gene expression.

1. **Regulation of mRNA degradation**

   Protein synthesis is also controlled by mRNA's lifespan in the cytoplasm.

   - Prokaryotic mRNA molecules are degraded by enzymes after only a few minutes. Thus, bacteria can quickly alter patterns of protein synthesis in response to environmental change.
   - Eukaryotic mRNA molecules can exist for several hours or even weeks.
   - The longevity of a mRNA affects how much protein synthesis it directs. Those that are viable longer can produce more of their protein.
   - For example, long-lived mRNAs for hemoglobin are repeatedly translated in developing vertebrate red blood cells.

2. **Control of translation**

   Gene expression can also be regulated by mechanisms that control translation of mRNA into protein. Most of these translational controls repress initiation of protein synthesis; for example.

   - Binding of translation repressor protein to the 5'-end of a particular mRNA can prevent ribosome attachment.
   - Translation of all mRNAs can be blocked by the inactivation of certain initiation factors. Such global translational control occurs during early embryonic development of many animals.
     - Prior to fertilization, the ovum produces and stores inactive mRNA to be used later during the first embryonic cleavage.
     - The inactive mRNA is stored in the ovum's cytosol until fertilization, when the sudden activation of an initiation factor triggers translation.
     - Delayed translation of stockpiled mRNA allows cells to respond quickly with a burst of protein synthesis when it is needed.

3. **Protein processing and degradation**

   Posttranslational control is the last level of control for regulating gene expression.

   - Many eukaryotic polypeptides must be modified or transported before becoming biologically active. Such modifications include:
     - Adding phosphate groups
     - Adding chemical groups, such as sugars
     - Dispatching proteins targeted by signal sequences for specific sites
   - Selective degradation of particular proteins and regulation of enzyme activity are also control mechanisms of gene expression.
     - Cells attach ubiquitin to proteins to mark them for destruction
     - Proteasomes recognize the ubiquitin and degrade the tagged protein (see Campbell, Figure 19.11)
     - Mutated cell-cycle proteins that are impervious to proteasome degradation can lead to cancer

IV. **The Molecular Biology of Cancer**

   A. **Cancer results from genetic changes that affect the cell cycle**

   Cancer is a variety of diseases in which cells escape from the normal controls on growth and division—the cell cycle—and it can result from mutations that alter normal gene expression in somatic cells. These mutations:

   - Can be random and spontaneous
   - Most likely occur as a result of environmental influences, such as:
     - Infection by certain viruses
     - Exposure to carcinogens

Carcinogens = Physical agents such as X-rays and chemical agents that cause cancer by mutating DNA

Whether cancer is caused by physical agents, chemicals or viruses, the mechanism is the same—the activation of *oncogenes* that are either native to the cell or introduced in viral genomes.

Oncogene = Cancer-causing gene

- Discovered during the study of tumors induced by specific viruses
- Harold Varmus and Michael Bishop won a Nobel Prize for their discovery of oncogenes in RNA viruses (retroviruses) that cause uncontrolled growth of infected cells in culture.

Researchers later discovered that some animal genomes, including human, contain genes that closely resemble viral oncogenes. These proto-oncogenes normally regulate growth, division and adhesion in cells.

Proto-oncogenes = Gene that normally codes for regulatory proteins controlling cell growth, division and adhesion, and that can be transformed by mutation into an oncogene.

Three types of mutations can convert proto-oncogenes to oncogenes:

1. *Movement of DNA within the genome.* Malignant cells frequently contain chromosomes that have broken and rejoined, placing pieces of different chromosomes side-by-side and possibly separating the oncogene from its normal control regions. In its new position, an oncogene may be next to active promoters or other control sequences that enhance transcription. Abnormal expression of an oncogene may occur if the oncogene is transposed to a new locus that has a highly active promoter.

2. *Gene amplification.* Sometimes more copies of oncogenes are present in a cell than is normal.

3. *Point mutation.* A slight change in the nucleotide sequence might produce a growth-stimulating protein that is more active or more resistant to degradation than the normal protein.

In addition to mutations affecting growth-stimulating proteins, changes in *tumor-suppressor genes* coding for proteins that normally *inhibit* growth can also promote cancer.

- The protein products of tumor-supressor genes have several functions:
  - Cooperate in DNA repair (helping obviate cancer-causing mutations)
  - Control cell anchorage (cell-cell adhesion; cell interaction with extracellular matrix)
  - Components of cell-signaling pathways that inhibit the cell-cycle

B. Oncogene proteins and faulty tumor-suppressor proteins interfere with normal signaling pathways

Mutation in the *ras* proto-oncogene and the *p*53 tumor suppressor gene are very common in human cancers; the frequency of mutation for *ras* is about 30 % and close to 50% for *p*53.

The *ras* protein is a G protein that relays a growth signal from a growth factor receptor to a cascade of protein kinases. The cellular response is the synthesis of a protein that stimulates the cell cycle (see Campbell Fig 19.13a).

- Under normal conditions, the pathway will not operate unless triggered by the appropriate growth factor.
- A mutated *ras* gene can produce a hyperactive version of the *ras* protein that stimulates the signal transduction cascade on it own, leading to excessive cell division.

The tumor-suppressor protein encoded by the wild-type *p*53 gene is a transcription factor of several genes that promotes the synthesis of growth-inhibiting proteins. Expression of this protective "guardian angel" protein prevents a cell from passing on mutations due to DNA damage in three ways:

- Activates the *p*21 gene, whose product allosterically binds to cyclin-dependent kinases, halting the cell cycle.
- Activates genes directly involved in DNA repair
- When DNA damage is irreparable, activates "suicide" genes, whose products cause cell death (apoptosis)

Mutations in the *p*53 gene can lead to excessive cell growth and cancer.

- For example, mutant tumor-suppressor genes are associated with inherited forms of colorectal cancer, Wilm's tumor, and breast cancer.

C. Multiple mutations underlie the development of cancer

More than one somatic mutation is probably needed to transform normal cells into cancerous cells. One of the best understood examples is colorectal cancer (see Campbell, Figure 19.14).

- Development of metastasizing colorectal cancer is gradual, and the first sign is unusually rapid cell division of apparently normal cells in the colon lining; a benign tumor (polyp) appears, and eventually a malignant tumor may develop.
- During this process, mutations in oncogenes and tumor-suppressor genes gradually accumulate. After a number of genes have changed, the tumor becomes malignant.
- About half a dozen changes must occur at the DNA level for the cell to become fully cancerous: usually, the appearance of at least one active oncogene and the mutation or loss of several tumor-supressor genes.

Viruses play a role in about 15% of human cancer cases worldwide, e.g., some types of leukemia, liver cancer, cervical cancer.

- Viruses might add oncogenes to cells, disrupt tumor-supressor genes DNA, or convert proto-oncogenes to oncogenes.

Breast cancer, the second most common type of cancer in women, is associated with somatic mutations of tumor-suppressor genes.

- Inherited breast cancer accounts for 5–10% of all breast cancer cases.
  - Mutations in either the BRCA1 or BRCA2 (stands for BReast CAncer) gene increase the risk of developing breast cancer. BRCA1 mutations also increase the risk of ovarian cancer.
  - Another locus, accounts for most of the remaining breast cancer cases linked to family history.

The study of genes associated with inherited cancer may lead to early diagnosis and treatment.

REFERENCES

Alberts, B., D. Bray, J. Lewis, M. Raff, K. Roberts and J.D. Watson. *Molecular Biology of the Cell*. 3rd ed. New York: Garland, 1994.

Campbell, N., et al. *Biology*. 5th ed. Menlo Park, California: Benjamin/Cummings, 1998.

Cavenee, W.K., and R.L. White. "The Genetic Basis of Cancer." *Scientific American*, March, 1995.

CHAPTER 20
DNA TECHNOLOGY

OUTLINE

OBJECTIVES

After reading this chapter and attending lecture, the student should be able to:

1. Explain how advances in recombinant DNA technology have helped scientists study the eukaryotic genome.
2. Describe the natural function of restriction enzymes.
3. Describe how restriction enzymes and gel electrophoresis are used to isolate DNA fragments.
4. Explain how the creation of sticky ends by restriction enzymes is useful in producing a recombinant DNA molecule.
5. Outline the procedures for producing plasmid and phage vectors.
6. Explain how vectors are used in recombinant DNA technology.
7. List and describe the two major sources of genes for cloning.
8. Describe the function of reverse transcriptase in retroviruses and explain how they are useful in recombinant DNA technology.
9. Describe how "genes of interest" can be identified with the use of a probe.
10. Explain the importance of DNA synthesis and sequencing to modern studies of eukaryotic genomes.
11. Describe how bacteria can be induced to produce eukaryotic gene products.
12. List some advantages for using yeast in the production of gene products.
13. List and describe four complementary approaches used to map the human genome.

14. Explain how RFLP analysis and PCR can be applied to the Human Genome Project.
15. Describe how recombinant DNA technology can have medical applications such as diagnosis of genetic disease, development of gene therapy, vaccine production, and development of pharmaceutical products.
16. Describe how gene manipulation has practical applications for agriculture.
17. Describe how plant genes can be manipulated using the Ti plasmid carried by *Agrobacterium* as a vector.
18. Explain how foreign DNA may be transferred into monocotyledonous plants.
19. Describe how recombinant DNA studies and the biotechnology industry are regulated with regards to safety and policy matters.

KEY TERMS

genetic engineering
recombinant DNA
biotechnology
nucleic acid probe
gene cloning
restriction enzymes
antisense nucleic
  acid
restriction fragments
  length polymorphisms
  (RFLPs)

cloning vector
nucleic acid
  hybridization
denaturation
expression vector
restriction site
complementary DNA
  (cDNA)
sticky ends
DNA ligase
electroporation

genomic library
cDNA library
polymerase chain reaction
  (PCR)
in vitro mutagenesis
gel electrophoresis
Southern blotting
restriction fragment
artificial chromosomes
in situ hybridization
Ti plasmid

Human Genome
  Project
chromosome
  walking
DNA microarray
  assays
vaccine
DNA fingerprint
simple tandem
  repeats (STRs)

LECTURE NOTES

Recombinant DNA technology refers to the set of techniques for recombining genes from different sources *in vitro* and transferring this recombinant DNA into a cell where it may be expressed.

*   These techniques were first developed around 1975 for basic research in bacterial molecular biology, but this technology has also led to many important discoveries in basic eukaryotic molecular biology.
*   Such discoveries resulted in the appearance of the *biotechnology* industry. Biotechnology refers to the use of living organisms or their components to do practical tasks such as:
    ⇒ The use of microorganisms to make wine and cheese
    ⇒ Selective breeding of livestock and crops
    ⇒ Production of antibiotics from microorganisms
    ⇒ Production of monoclonal antibodies

The use of recombinant DNA techniques allows modern biotechnology to be a more precise and systematic process than earlier research methods.

*   It is also a powerful tool since it allows genes to be moved across the species barrier.
*   Using these techniques, scientists have advanced our understanding of eukaryotic molecular biology.
*   The *Human Genome Project* is an important application of this technology. This project's goal is to transcribe and translate the entire human genome in order to better understand the human organism.
*   A variety of applications are possible for this technology, and the practical goal is the improvement of human health and food production.

- A variety of applications are possible for this technology, and the practical goal is the improvement of human health and food production.

I.   DNA Cloning

A. DNA technology makes it possible to clone genes for basic research and commercial applications: *an overview*

Prior to the discovery of recombinant DNA techniques, procedures for altering the genes of organisms were constrained by the need to find and propagate desirable mutants.

- Geneticists relied on either natural processes, mutagenic radiation, or chemicals to induce mutations.
- In a laborious process, each organism's phenotype was checked to determine the presence of the desired mutation.
- Microbial geneticists developed techniques for screening mutants. For example, bacteria was cultured on media containing an antibiotic to isolate mutants which were antibiotic resistant.

Before 1975, transferring genes between organisms was accomplished by cumbersome and nonspecific breeding procedures. The only exception to this was the use of bacteria and their phages.

- Genes can be transferred from one bacterial strain to another by the natural processes of transformation, conjugation or transduction.
- Geneticists used these processes to carry out detailed molecular studies on the structure and functioning of prokaryotic and phage genes.
- Bacteria and phages are ideal for laboratory experiments because they are relatively small, have simple genomes, and are easily propagated.
- Although the technique was available to grow plant and animal cells in culture, the workings of their genomes could not be examined using existing methods.

Campbell Figure 20.1 provides an overview of how bacterial plasmids are used to clone genes for biotechnology.

Recombinant DNA technology now makes it possible for scientists to examine the structure and function of the eukaryotic genome, because it contains several key components:

- Biochemical tools that allow construction of recombinant DNA
- Methods for purifying DNA molecules and proteins of interest
- Vectors for carrying recombinant DNA into cells and replicating it
- Techniques for determining nucleotide sequences of DNA molecules.

B. Restriction enzymes are used to make recombinant DNA

Restriction enzymes are major tools in recombinant DNA technology.

- First discovered in the late 1960s, these enzymes occur naturally in bacteria where they protect the bacterium against intruding DNA from other organisms.
- This protection involves *restriction*, a process in which the foreign DNA is cut into small segments.
- Most restriction enzymes only recognize short, specific nucleotide sequences called *recognition sequences* or restriction sites. They only cut at specific points within those sequences.

Bacterial cells protect their own DNA from restriction through *modification* or methylation of DNA.

- Methyl groups are added to nucleotides within the recognition sequences.

There are several hundred restriction enzymes and about a hundred different specific recognition sequences.

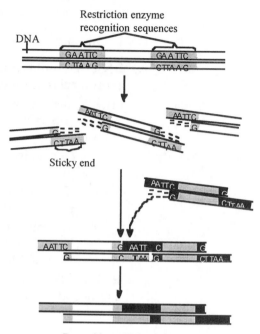

Restriction enzyme recognition sequences

Sticky end

Recombinant DNA molecule

- Recognition sequences are symmetric in that the same sequence of four to eight nucleotides is found on both strands, but run in opposite directions.
- Restriction enzymes usually cut phosphodiester bonds of both strands in a staggered manner, so that the resulting double-stranded DNA fragments have single-stranded ends, called *sticky ends*.
- The single-stranded short extensions form hydrogen-bonded base pairs with complementary single-stranded stretches on other DNA molecules.

Sticky ends of *restriction fragments* are used in the laboratory to join DNA pieces from different sources (cells or even different organisms).

- These unions are temporary since they are only held by a few hydrogen bonds.
- These unions can be made permanent by adding the enzyme *DNA ligase*, which catalyzes formation of covalent phosphodiester bonds.

The outcome of this process is the same as natural genetic recombination, the production of recombinant DNA – a DNA molecule carrying a new combination of genes (see Campbell, Figure 20.2)..

C. Gene can be cloned in recombinant DNA vectors: *a closer look*

Most DNA technology procedures use carriers or vectors for moving DNA from test tubes back into cells.

Cloning vector = A DNA molecule that can carry foreign DNA into a cell and replicate there

- Two most often used types of vectors are bacterial plasmids and viruses.
- Restriction fragments of foreign DNA can be spliced into a bacterial plasmid without interfering with its ability to replicate within the bacterial cell. Isolated recombinant plasmids can be introduced into bacterial cells by transformation.

Bacteriophages, such as lambda phage, can also be used as vectors.

- The middle of the linear genome, which contains nonessential genes, is deleted by using restriction enzymes.
- Restriction fragments of foreign DNA are then inserted to replace the deleted area.
- The recombinant phage DNA is introduced into an *E. coli* cell.
- The phage replicates itself inside the bacterial cell.
- Each new phage particle carries the foreign DNA "passenger."

Sometimes it is necessary to clone DNA in eukaryotic cells rather than in bacteria. Under the right conditions, yeast and animal cells growing in culture can also take up foreign DNA from the medium.

- If the new DNA becomes incorporated into chromosomal DNA or *can* replicate itself, it can be cloned with the cell.
- Since yeast cells have plasmids, scientists can construct recombinant plasmids that combine yeast and bacterial DNA and that can replicate in either cell type.
- Viruses can also be used as vectors with eukaryotic cells. For example, retroviruses used as vectors in animal cells can integrate DNA directly into the chromosome.

1. **Procedure for cloning a eukaryotic gene in a bacterial plasmid**

Recombinant DNA molecules are only useful if they can be made to replicate and produce a large number of copies. A typical gene-cloning procedure includes the following steps (see Campbell, Figure 20.3):

Step 1: Isolation of vector and gene-source DNA

- Bacterial plasmids and foreign DNA containing the gene of interest are isolated.
- In this example, the foreign DNA is human, and the plasmid is from *E. coli* and has two genes:
  - *amp*^{R} which confers antibiotic resistance to ampicillin
  - *lacZ* which codes for β-galactosidase, the enzyme that catalyzes the hydrolysis of lactose
- Note that the recognition sequence for the restriction enzyme used in this example is *within* the *lacZ* gene.

Step 2: Insertion of gene-source DNA into the vector

  a. Digestion
  - The restriction enzyme cuts plasmid DNA at the *restriction site*, disrupting the *lacZ* gene.
  - The foreign DNA is cut into thousands of fragments by the same restriction enzyme; one of the fragments contains the gene of interest.
  - When the restriction enzyme cuts, it produces *sticky ends* on both the foreign DNA fragments and the plasmid.
  b. Mixture of foreign DNA fragments with clipped plasmids
  - Sticky ends of the plasmid base pair with complementary sticky ends of foreign DNA fragments.
  c. Addition of DNA ligase
  - DNA ligase catalyzes the formation of covalent bonds, joining the two DNA molecules and forming a new plasmid with recombinant DNA.

Step 3: Introduction of cloning vector into bacterial cells

- The naked DNA is added to a bacterial culture.
- Some bacteria will take up the plasmid DNA by transformation.

Step 4: Cloning of cells (and foreign DNA)

- Bacteria with the recombinant plasmid are allowed to reproduce, cloning the inserted gene in the process.
- Recombinant plasmids can be identified by the fact that they are ampicillin resistant and will grow in the presence of ampicillin.

Step 5: Identification of cell clones carrying the gene of interest

- X-gal, a modified sugar added to the culture medium, turns blue when hydrolyzed by β-galactosidase. It is used as an indicator that cells have been transformed by plasmids containing the foreign insert.

- Since the foreign DNA insert disrupts the *lacZ* gene, bacterial colonies that have successfully acquired the foreign DNA fragment will be white. Those bacterial colonies lacking the DNA insert will have a complete *lacZ* gene that produces β-galactosidase and will turn blue in the presence of X-gal.

- The methods for detecting the DNA of a gene depend directly on base pairing between the gene of interest and a complementary sequence on another nucleic acid molecule, a process called *nucleic acid hybridization*. The complementary molecule, a short piece of RNA or DNA is called a *nucleic acid probe* (see Campbell, Fig. 20.4).

2. Cloning and expressing eukaryotic genes: problems and solutions

Problem: Getting a cloned eukaryotic gene to function in a prokaryotic setting can be difficult because certain details of gene expression are different in the two kinds of cells.

Solution: Expression vectors allow the synthesis of many eukaryotic proteins in bacterial cells.

- Expression vectors contain a prokaryotic promoter just upstream of a restriction site where the eukaryotic gene can be inserted.

- The bacterial host cell recognizes the promoter and proceeds to express the foreign gene that has been linked to it.

Problem: Eukaryotic genes of interest may be too large to clone easily because they contain long noncoding regions (introns), which prevent correct expression of the gene by bacterial cells, which lack RNA-splicing machinery..

Solution: Scientists can make artificial eukaryotic genes that lack introns (see Campbell, Figure 20.5).

Solution: Artificial chromosomes, which combine the essentials of a eukaryotic chromosome with foreign DNA, can carry much more DNA than plasmid vectors, thereby enabling very long pieces of DNA to be cloned.

Bacteria are commonly used hosts in genetic engineering because:

- DNA can be easily isolated from and reintroduced into bacterial cells

- Bacterial cultures grow quickly, rapidly cloning the inserted foreign genes.

Some disadvantages to using bacterial host cells are that bacterial cells:

- May not be able to use the information in a eukaryotic gene, since eukaryotes and prokaryotes use different enzymes and regulatory mechanisms during transcription and translation.

- Cannot make the posttranslational modifications required to produce some eukaryotic proteins (e.g., addition of lipid or carbohydrate groups)

Using eukaryotic cells as hosts can avoid the eukaryotic-prokaryotic incompatibility issue.

- Yeast cells are as easy to grow as bacteria and contain plasmids.

- Some recombinant plasmids combine yeast and bacterial DNA and can replicate in either.

- Posttranslational modifications required to produce some eukaryotic proteins (e.g., addition of lipid or carbohydrate groups) can occur

There are more aggressive techniques for inserting foreign DNA into eukaryotic cells:

- In *electroporation*, a brief electric pulse applied to a cell solution causes temporary holes in the plasma membrane, through which DNA can enter.

- With thin needles, DNA can be injected directly into a eukaryotic cell.

- DNA attached to microscopic metal particles can be fired into plant cells with a gun (see Campbell, Figure 38.13).

Bacteria and yeast are not suitable for every purpose. For certain applications, plant or animal cell cultures must be used.

- Cells of more complex eukaryotes carry out certain biochemical processes not found in yeast (e.g. only animal cells produce antibodies).

D. Cloned genes are stored in DNA libraries

There are two major sources of DNA which can be inserted into vectors and clones:

1. DNA isolated directly from an organism
2. Complementary DNA made in the laboratory from mRNA templates

DNA isolated directly from an organism contains all genes including the gene of interest.

- Restriction enzymes are used to cut this DNA into thousands of pieces which are slightly larger than a gene.
- All of these pieces are then inserted into plasmids or viral DNA.
- These vectors containing the foreign DNA are introduced into bacteria.
- This produces the *genomic library,* the complete set of thousands of recombinant-plasmid clones, each carrying copies of a particular segment from the initial genome (see Campbell, Figure 20.6).
- Libraries can be saved and used as a source of other genes of interest or for genome mapping.

The cDNA method produces a more limited kind of gene library, a cDNA library. A cDNA library represents only part of the cell's genome because it contains only the genes that were transcribed in the starting cells (recall that cDNA is derived from isolated RNA).

- By using cells from specialized tissues or a cell culture used exclusively for making one gene product, the majority of mRNA produced is for the gene of interest.
- For example, most of the mRNA in precursors of mammalian erythrocytes is for the protein hemoglobin.

E. The polymerase chain reaction (PCR) clones DNA entirely in vitro

PCR is a technique that allows any piece of DNA to be quickly amplified (copied many times) in vitro (see Campbell, Methods Box)

- DNA is incubated under appropriate conditions with special primers and DNA polymerase molecules.
- Billions of copies of the DNA are produced in just a few hours.
- PCR is highly specific; primers determine the sequence to be amplified.
- Only minute amounts of DNA are needed.

PCR is presently being applied in many ways for analysis of DNA from a wide variety of sources:

- Ancient DNA fragments from a woolly mammoth; DNA is a stable molecule and can be amplified by PCR from sources thousands, even millions, of years old.
- DNA from tiny amounts of tissue or semen found at crime scenes
- DNA from single embryonic cells for prenatal diagnosis
- DNA of viral genes from cells infected with difficult to detect viruses such as HIV

Amplification of DNA by PCR is being used in the Human Genome Project to produce linkage maps without the need for large family pedigree analysis.

- DNA from sperm of a single donor can be amplified to analyze the immediate products of meiotic recombination.

- This process eliminates the need to rely on the chance that offspring will be produced with a particular type of recombinant chromosome.

- It makes it possible to study genetic markers that are extremely close together.

II.   Analysis of Cloned DNA

Once a gene is cloned, scientists can then analyze the cloned DNA to address numerous questions, such as:

- Does a gene differ in different organisms?

- Are there certain alleles associated with a hereditary disorder?

- Where in the organism is the gene expressed?

- What control the pattern of expression?

- What is the location of the gene within the genome?

A. Restriction fragment analysis detects DNA differences that affect restriction sites

Gel electrophoresis (see Campbell's Methods box) is used to separate either nucleic acids or proteins based upon molecular size, charge, and other physical properties. Using this technique:

- Viral DNA, plasmid DNA, and segments of chromosomal DNA can be identified by their characteristic banding patterns after being cut with various restriction enzymes.

  - Each band corresponds to a DNA restriction fragment of a certain length.

  - DNA segments carrying different alleles of a gene can result in dissimilar banding patterns, since numbers and locations of restriction sites may not be the same in the different nucleotide sequences (see Campbell, Figure 20.7).

  - Similar differences in banding patterns result when *noncoding* segments of DNA are used as starting material.

- DNA fragments containing genes of interest can be isolated, purified, and then recovered from the gel with full biological activity.

The technique of hybridization is used to determine the presence of a specific nucleotide sequence (see Campbell's Methods Box on *Southern blotting* for an explanation of the entire procedure and a demonstration of how it can be used to compare DNA from three individuals).

- Labeled probes complementary to the gene of interest are allowed to bind to DNA from cells being tested (see Campbell, Fig. 20.4).

- Variations of this technique allows researchers to determine whether a:

  - Gene is present in various organisms

  - Sequence is present, how many sequences there are, and the size of the restriction fragments containing these sequences

  - Gene is made into mRNA, how much of that mRNA is present, and whether the amount of that mRNA changes at different stages of development or in response to certain regulatory signals (*Northern blotting*)

Differences in restriction fragment length that reflect variations in homologous DNA sequences are called *restriction fragment length polymorphisms (RFLPs)*.

- DNA sequence differences on homologous chromosomes that result in RFLPs are scattered abundantly throughout genomes, including the human genome.
- RFLPs are not only abundant, but can easily be detected as to whether they affect the organism's phenotype; they can be located in an exon, intron, or any noncoding part of the genome.
- RFLP are detected and analyzed by Southern blotting.
- Because RFLPs can be readily detected, they are extremely useful as *genetic markers* for making linkage maps.
- A RFLP marker is often found in numerous variants in a population. RFLPs have provided many markers for mapping the human genome since geneticists are no longer limited to genetic variations that lead to phenotypic differences or protein products.

RFLPs are proving useful in several areas.

- Disease genes are being located by examining known RFLPs for linkage to them. RFLP markers inherited at a high frequency with a disease are probably located close to the defective gene on the chromosome.
- An individual's RFLP markers provide a "genetic fingerprint" which can be used in forensics, since there is a very low probability that two people would have the same set of RFLP markers.

B. Entire genomes can be mapped at the DNA level

1. Locating genes by *in situ* hybridization

DNA probes can be used to help map genes on eukaryotic chromosomes.

- *In situ hybridization* uses a radioactive DNA probe that base pairs with complementary sequences in the denatured DNA of intact chromosomes.
- Autoradiography and chromosome staining reveal to which band of which chromosome the probe has attached. Alternatively, the probe is labeled with fluorescent dye.

2. The mapping of entire genomes

The Human Genome Project, begun in 1990, is an international effort to map the entire human genome.

Scientists are also mapping the genomes of species that are particularly useful for genetic research, including *E. coli, Saccharomyces cerevisiae* (yeast), *Caenorhabditis elegans* (nematode), *Drosophila melanogaster* (fruit fly), and *Mus musculus* (mouse).

Several complementary approaches are being used to map the precise locations of all of an organism's genes:

a. Genetic (linkage) mapping

- The first step in mapping a large genome is to construct a linkage map of several thousand genetic markers, which can be genes, RFLPs, or microsatellites.
- Relying primarily on microsatellites, researchers have completed a human genetic map with over 5000 markers.
- This map will enable researchers to locate other markers by testing for genetic linkage to the known markers.

b. Physical mapping: ordering the DNA fragments

- A physical map is made by cutting the DNA of each chromosome into a number of identifiable fragments.

- The key is to make fragments that overlap and to find the overlapping ends
- Campbell, Figure 20.8 shows *chromosome walking*, a method which uses probes to find the overlapping ends.
- Researchers carry out several rounds of DNA cutting, cloning, and mapping in order to prepare supplies of DNA fragments to map large genomes.
- The goal is to find the original order of the fragments in the chromosomal DNA.

c. Sequencing DNA

- The complete nucleotide sequence of a genome is the ultimate map.
- This will be the most time consuming part of the Human Genome Project as each haploid set of human chromosomes contains about 3 *billion* nucleotide pairs.

In addition to chromosome walking, Human Genome Project researchers are using PCR amplification.

- PCR can amplify specific portions of DNA from individual sperm cells—the immediate products of meiotic recombination.
- Researchers can amplify DNA in amounts sufficient for study, and analyze samples as large as thousands of sperm.
- Based on the crossover frequencies between genes, researchers can deduce human linkage maps without having to find large families for pedigree analysis.

These approaches will be used to completely map genomes and provide an understanding of how the human genome compares to those of other organisms. Potential benefits include:

- Identification and mapping of genes responsible for genetic diseases will aid diagnosis, treatment and prevention.
- Detailed knowledge of the genomes of humans and other species will give insight into genome organization, control of gene expression, cellular growth and differentiation, and evolutionary biology.

Achieving the goals of the Human Genome Project in a timely way will come from advances in automation and utilization of the latest electronic technology.

3. Genome analysis

Geneticists are also trying to determine phenotype from genotype, or identify genes within a long DNA sequence and determine their function.

a. Analyzing DNA sequences

Many researchers are studying the structure and organization of genes. This type of genome analysis relies of DNA sequencing (see Campbell's Methods box) and other mapping approaches outlined above.

DNA sequencing techniques have enabled scientists to collect thousands of DNA sequences in computer data banks.

- Using a computer, scientists can scan sequences for protein-coding genes and gene-control sequences.
- A list of nucleotide sequences for putative genes is assembled and compared to sequences of known genes.

- In the sequences compiled to date, many putative genes have been found to be entirely new to science; for example, 38% of the genes of E. Coli, the best studied research organism.

DNA sequences confirm the evolutionary connections between distantly related organisms and the relevance of research on simpler organisms to understanding human biology.

b. Studying gene expression

Other researchers are studying patterns of gene expression and how such patterns act to produce and maintain a functioning organism. This type of genome analysis can be performed without knowledge of the complete DNA sequence of an organism.

- One strategy for evaluating gene expression is to isolate the mRNA made in particular cells, use these molecules as templates for making a cDNA library by reverse transcription, and then compare this cDNA with other collections of DNA by hybridization.
  - This approach reveals which genes are active at different stages of development, in different tissues, or in different physiological conditions (or states of health).

Another approach uses DNA microarray assays to detect and measure the expression of thousands of genes at one time (see Campbell, Figure 20.9).

- This method is being used to compare cancerous and noncancerous tissues.
- Studying the differences in gene expression may lead researchers to new diagnostic techniques and biochemically targeted treatments

c. Determining gene function

Still other researchers are studying the function of genes.

In vitro mutagenesis is a technique that can be used to determine the function of a protein product from cDNA cloning of an mRNA.

- Mutations are induced into the sequence of a cloned gene.
- The mutated gene is returned to the host cell.
- If the mutation alters the function of the protein product, it may be possible to determine the function of the protein by examining what changes occur in cell physiology or developmental pattern.

III. Practical Applications of DNA Technology

A. DNA technology is reshaping medicine and the pharmaceutical industry

Modern biotechnology has resulted in significant advances in many areas of medicine.

1. Diagnosis of diseases

Medical scientists currently use DNA technology to diagnose hundreds of human genetic disorders.

- This allows early disease detection and identification of carriers for potentially harmful recessive mutations – even before the onset of symptoms.
- Genes have been cloned for many genetic disorders including hemophilia, phenylketonuria, cystic fibrosis, and Duchenne muscular dystrophy.

- Gene cloning permits direct detection of gene mutations. A cloned normal gene can be used as a probe to find the corresponding gene in cells being tested; the alleles are compared with normal and mutant standards usually by RFLP analysis.

When the normal gene has not been cloned, a closely linked RFLP marker may indicate the presence of an abnormal allele if the RFLP marker is frequently co-inherited with the disease.

- Blood samples from relatives can be used to determine which RFLP marker is linked to the abnormal allele and which is linked to the normal allele.
- The RFLP markers must be different for the normal and abnormal alleles.
- Under these conditions, the RFLP marker variant found in the person being tested can reveal whether the normal or abnormal allele is likely to be present.
- Alleles for cystic fibrosis and Huntington's disease can be detected in this manner.

2. Human gene therapy

Traceable genetic disorders in individuals may eventually be correctable.

- Theoretically, it should be possible to replace or supplement defective genes with functional normal genes using recombinant DNA techniques.
- Correcting somatic cells of individuals with well-defined, life-threatening genetic defects will be the starting place.

The principle behind human gene therapy is that normal genes are introduced into a patient's own somatic cells.

- For this therapy to be permanent, the cells receiving the normal allele must actively reproduce, so the normal allele will be replicated and continually expressed (see Campbell, Figure 20.11).
- Bone marrow cells are prime candidates.

Attempts at human gene therapy have not yet produced any proven benefits to patients, contrary to claims in the popular media.

- The most promising gene therapy trials under way are ones that involve bone marrow cells but are not necessarily aimed at correcting genetic defects; for example, improving the abilities of immune cells to fight off cancer and resist HIV
- Most experiments to date have been designed to test the safety and feasibility of a procedure rather than attempt a cure.

Many technical questions are posed by gene therapy.

- Can the proper genetic control mechanisms be made to operate on the transferred gene so that cells make appropriate amounts of the gene product at the right time and in the right place?
- How can we be sure that the inserted therapeutic gene does not harm some other necessary cell function?

Gene therapy raises difficult social and ethical questions:

- Is it advisable under any circumstances to alter the genomes of human germ lines (eggs) or embryos in hope of correcting the defect in future generations?
  - Some critics believe tampering with human genes in any way is wrong and may lead to eugenics.
  - Others say that genetic engineering of somatic cells is no different than other conventional medical interventions used to save lives.

Treating germ cells is possible and has been used in mice for some time.

- Mice have been created with sickle cell anemia to further the study of the disease.
- Recipient mice and their descendants contain the active human gene not only in the proper location (erythrocytes) but also during the proper stage of development.

3. **Pharmaceutical products**

DNA technology has been used to create many useful pharmaceutical products, mostly proteins.

- Highly active promoters and other gene control elements are put into vector DNA to create expression vectors that enable the host cell to make large amounts of the product of a gene inserted into the vector.
- Host cells can be engineered to secrete a protein as it is made, thereby simplifying the task of purifying it.

Human insulin and growth hormone are early applications of gene splicing..

- Two million individuals with diabetes in the United States have benefited from genetically engineered human insulin.
- Insulin produced this way is chemically identical to that made by the human pancreas, and it causes fewer adverse reactions than insulin extracted from pig and cattle pancreas.
- Human growth hormone has been a boon to children with hypopituitarism (pituitary dwarfism).
    - The growth hormone molecule is much larger than insulin (almost 200 amino acids long) and more species specific.
    - Thus, growth hormone from other animals is not an effective growth stimulator in humans.
    - Previously, these individuals were treated with growth hormone obtained from human cadavers.

Another important product produced by genetic engineering is tissue plasminogen activator (TPA).

- This protein helps dissolve blood clots and reduces the risk of later heart attacks if administered very shortly after an initial heart attack
- TPA illustrates a drawback to genetically engineered products. Because the development costs were high and the market relatively limited, the product has been very expensive.

Recent developments include novel ways to fight diseases that don't respond to traditional drug treatments.

- Antisense nucleic acid is used to base-pair with mRNA molecules and block their translation.
    - This could prevent the spread of diseases by interfering with viral replication or the transformation of cells into a cancerous state.
- Genetically engineered proteins block or mimic surface receptors on cell membranes.
    - For example, an experimental drug mimics a receptor protein that HIV binds to when it attacks white blood cells. The HIV binds to the drug molecules instead and fails to enter the blood cells.

Prevention by vaccine is the only way to fight many viral diseases for which no treatment exists.

Vaccine = A harmless variant or derivative of a pathogen that stimulates the immune system to fight the pathogen

Traditional vaccines for viral diseases are of two types:

- Particles of virulent virus that have been inactivated by chemical or physical means.
- Active virus particles of an attenuated (nonpathogenic) viral strain.

Since the particles in both cases are similar to active virus, both types of vaccine will trigger an animal's immune system to produce antibodies, which react very specifically against invading pathogens.

Biotechnology is being used in several ways to modify current vaccines and to produce new ones. Recombinant DNA techniques can be used to:

- Produce large amounts of specific protein molecules (*subunits*) from the surface of a pathogen. If these protein subunits cause immune responses to the pathogens, they can be used as vaccines.
- Modify genomes of pathogens to directly attenuate them. Vaccination with live, attenuated organisms is more effective than a subunit vaccine. Small amounts of material trigger greater immune response, and pathogens attenuated by gene-splicing may also be safer than using natural mutants.

B. DNA technology offers forensic, environmental, and agricultural applications

1. Forensic uses of DNA technology

Forensic labs can determine blood or tissue type from blood, small fragments of other tissue, or semen left at the scene of violent crime. These tests, however, have limitations:

- They require fresh tissue in sufficient amounts for testing.
- This approach can exclude a suspect but is not evidence of guilt, because many people have the same blood type or tissue type.

DNA testing can identify an individual with a much higher degree of certainty, since everyone's DNA base sequence is unique (except for identical twins).

- RFLP analysis by Southern blotting is a powerful method for the forensic detection of similarities and differences in DNA samples (see Campbell, Figure 20.12).
    - This method is used to compare DNA samples from the suspect, the victim, and a small amount of semen, blood or other tissue found at the scene of the crime.
    - Restriction fragments from the DNA samples are separated by electrophoresis; radioactive probes mark the bands containing RFLP markers.
    - Usually the forensic scientist tests for five markers.
    - Even a small set of RFLP markers from an individual can provide a *DNA fingerprint* that is of forensic use; the probability that two individuals would have the same RFLP markers is quite low.
- Increasingly, variations in the lengths of satellite DNA are used instead of RFLPs in DNA fingerprinting.
    - The most useful satellite sequences for forensic purposes are microsatellites, which are 10 - 100 base pairs long, have repeating units of only 1 - 4 base pairs, and are highly variable from person to person.
    - Individuals have different numbers of repeats at genome loci (*simple tandem repeats (STRs)*)

- Restriction fragments containing STRs vary in size among individuals because of differences in STR lengths rather than because of different numbers of restriction sites within that region of the genome, as in RFLP analysis.
- The greater the number of markers examined in a sample, the more likely it is that the DNA fingerprint is unique to one individual.
- PCR is often used to selectively simplify particular STRs or other markers before electrophoresis. This is especially useful when the DNA is in poor condition or only available in minute quantities (only 20 cells are needed!).

How reliable is DNA fingerprinting?

- Though each individual's DNA fingerprint is unique, most forensic tests do not analyze the entire genome but focus on tiny regions known to be highly variable from one person to another.
- The probability is minute (between on in 100,000 and one in a billion) that two people will have matching DNA fingerprints.
  - The exact figure depends on the number of markers compared and on the frequency of those markers in the population.
  - The frequency of markers varies by ethnic group, which allows forensic scientists to make extremely accurate statistical calculations.
- Problems can arise from insufficient statistical data, human error, or flawed evidence.
- DNA fingerprints are now accepted as compelling evidence by legal and scientific experts.

As with most new technology, forensic applications of DNA fingerprinting raises important ethical questions such as:

- Once collected, what happens to the DNA data?
- Should DNA fingerprints be filed or destroyed? Some states now save DNA data from convicted criminals.

2. Environmental uses of DNA technology

Scientists are engineering metabolic capabilities of organisms so they can transform chemicals and thus help solve environmental problems. For example:

- Some microorganisms can extract heavy metals (e.g., copper, lead, and nickel) from their environments and incorporate them into recoverable compounds such as copper sulfate or lead sulfate.
- As metal reserves are depleted, genetically engineered microbes may perhaps be used in mining and in cleaning up mining waste.

Metabolic diversity of microbes is used in the recycling of wastes and detoxification of toxic chemicals.

- Sewage treatment plants use microorganisms to degrade organic compounds into non-toxic forms.
- Biologically and chemically active compounds that cannot be easily degraded are often released into the environment.
- The intent is to engineer microorganisms that can degrade these compounds and that can be used in waste water treatment plants.
- Such microbes might be incorporated directly into the manufacturing process, preventing toxic chemicals from ever being released as waste in the first place.
- Bacterial strains have been developed to detoxify specific toxic wastes found in spills and waste dumps.

3. **Agricultural uses of DNA technology**

Recombinant DNA techniques are being used to study plants and animals of agricultural importance to improve their productivity.

a. **Animal husbandry**

Products produced by recombinant DNA methods, such as vaccines, antibodies and growth hormone, are already used in animal husbandry. For example,

- *Bovine growth hormone* (*bGH*). Made by *E. coli*, bGH is injected into milk cows to enhance milk production and into beef cattle to increase weight gain.

- *Cellulase.* Also produced by *E. coli*, this enzyme hydrolyzes cellulose making all plant parts useable for animal feed.

Transgenic animals, those that contain DNA from other species, have been commercially produced by injecting foreign DNA into egg nuclei or early embryos.

- Transgenic beef and dairy cattle, hogs, sheep and several species of commercially raised fish have been produced for potential agricultural use.

b. **Genetic engineering in plants**

Plant cells are easier to engineer than animal cells, because an adult plant can be produced from a single cell growing in tissue culture.

- This is important since many types of genetic manipulation are easier to perform and assess on single cells than on whole organisms.

- Asparagus, cabbage, citrus fruits, sunflowers, carrots, alfalfa, millet, tomatoes, potatoes, and tobacco are all commercial plants that can be grown from single somatic cells.

The best developed DNA vector for plant cells is the *Ti plasmid* (tumor inducing), carried by the normally pathogenic bacterium *Agrobacterium tumefaciens*.

- Ti plasmid usually induces tumor formation in infected plants by integrating a segment of its DNA (called T DNA) into the chromosomes of the host plant cell.

- Researchers have turned this plasmid into a useful vector by eliminating its disease-causing ability without interfering with its potential to move genetic material into infected plants.

With recombinant DNA methods, foreign genes can be inserted into Ti plasmid.

- The recombinant plasmid can either be put back into *Agrobacterium tumefaciens*, which is used to infect plant cells in culture, or it can be introduced directly into plant cells.

- Individual modified plant cells are then used to regenerate whole plants that contain, express and pass on to their progeny the foreign gene (see Campbell, Figure 20.13).

Using Ti plasmid as a vector has one major drawback, only dicotyledons are susceptible to infection by *Agrobacterium*; important commercial plants such as corn and wheat are monocotyledons and cannot be infected.

- Newer methods that allow researchers to overcome this limitation are *electroporation* and use of the *DNA particle gun*.

- Electroporation uses high-voltage jolts of electricity to open temporary pores in the cell membrane; foreign DNA can enter the cells through these pores.

- The DNA gun shoots tiny DNA-coated metal pellets through the cell walls into the cytoplasm, where the foreign DNA becomes integrated into the host cell DNA.

Though cloning plant DNA is straightforward, plant molecular geneticists still face several technical problems:

- Identifying genes of interest may be difficult.
- Many important plant traits (such as crop yield) are polygenic.

Genetic engineering of plants has yielded positive results in cases where useful traits are determined by single genes. For example:

- Of the genetically engineered plants now in field trials, over 40% have received genes for herbicide resistance
- A bacterial gene that makes plants resistant to glyphosate (a powerful herbicide) has been successfully introduced into several crop plants; glyphosate-resistant plants makes it easier to grow crops and destroy weeds simultaneously.
- The first gene-spliced fruits approved by the FDA for human consumption were tomatoes engineered with antisense RNA to suppress ripening and retard spoilage. The general process is outlined below:
  1. Researchers clone the tomato gene coding for enzymes responsible for ripening.
  2. The complementary (antisense) gene is cloned.
  3. The antisense gene is spliced into the tomato plant's DNA, where it transcribes mRNA complementary to the ripening genes' mRNA.
  4. When the ripening gene produces a normal mRNA transcript, the antisense mRNA binds to it, blocking synthesis of the ripening enzyme.

Crop plants are being engineered to resist pathogens and pest insects. For example,

- Tomato and tobacco plants can be engineered to carry certain genes of viruses that normally infect and damage plants; expression of these versions of viral genes confer resistance to viral attack.
- Some plants have been engineered to resist insect attack, reducing the need to apply chemical insecticides to crops. Genes for an insecticidal protein have been successfully transferred from the bacterium, *Bacillus thuringiensis*, to corn, cotton and potatoes; field tests show that these engineered plants are insect resistant.

In the near future, crop plants developed with recombinant DNA techniques are likely to:

- Be made more productive by enlarging agriculturally valuable parts - whether they be roots, leaves, flowers, or stems.
- Have an enhanced food value. For example, corn and wheat might be engineered to produce mixes of amino acids optimal for the human diet.

c. **The nitrogen-fixation challenge**

Nitrogen fixation is the conversion of atmospheric, gaseous nitrogen (N_2) into nitrogen-containing compounds.

- Gaseous nitrogen is useless to plants.
- Nitrogen-fixing bacteria live in the soil or are symbiotic within plant roots.

- The nitrogen-containing compounds produced by nitrogen fixation are taken up from the soil by plants and used to make organic molecules such as amino acids and nucleotides.
- A major limiting factor in plant growth and crop yield is availability of useable nitrogen compounds. This is why nitrogenous fertilizers are used in agriculture.

Recombinant DNA technology can possibly be used to increase biological nitrogen fixation of bacterial species living in the soil or in association with plants.

C. DNA technology raises important safety and ethical questions

When scientists realized the potential power of DNA technology, they also became concerned that recombinant microorganisms could create hazardous new pathogens, which might escape from the laboratory.

- In response to these concerns, scientists developed and agreed upon a self-monitoring approach, which was soon formalized into federal regulatory programs.
- Today, governments and regulatory agencies worldwide are monitoring the biotechnology industry - promoting potential industrial, medical, and agricultural benefits, while ensuring that new products are safe.
- In the U.S., the FDA, National Institutes of Health (NIH) Recombinant DNA Advisory Committee, Department of Agriculture (USDA), and Environmental Protection Agency (EPA) set policies and regulate new developments in genetic engineering.

While genetic engineering holds enormous potential for improving human health and increasing agricultural productivity, new developments in DNA technology raise ethical concerns. For example, mapping the human genome will contribute to significant advances in gene therapy, but:

- Who should have the right to examine someone else's genes?
- Should a person's genome be a factor in their suitability for a job?
- Should insurance companies have the right to examine an applicant's genes?
- How do we weigh the benefits of gene therapy against assurances that the gene vectors are safe?

For environmental problems such as oil spills, genetically engineered organisms may be part of the solution, but what is their potential impact on native species?

For new medical products, what is the potential for harmful side effects, both short-term and long-term?

- New medical products must pass exhaustive tests before the FDA approves it for general marketing.
- Currently awaiting federal approval are hundreds of new genetically engineered diagnostic products, vaccines, and drugs - including some to treat AIDS and cancer.

For agricultural products, what are the potential dangers of introducing new genetically engineered organisms into the environment?

- Some argue that producing transgenic organisms is only an extension of traditional hybridization and should not be treated differently from the production of other hybrid crops or animals. The FDA holds that if products of genetic engineering are not significantly different from products already on the market, testing is not required.

- Others argue that creating transgenic organisms by splicing genes from one species to another is radically different from hybridizing closely related species of plants or animals.
- Some concerns are that genetically altered food products may contain new proteins that are toxic or cause severe allergies; genetically engineered crop plants could become superweeds resistant to herbicides, disease and insect pests; and engineered crop plants may hybridize with native plants and pass their new genes to closely related plants in the wild.

REFERENCES

Alberts, B., D. Bray, J. Lewis, M. Raff, K. Roberts and J.D. Watson. *Molecular Biology of the Cell*. 3rd ed. New York: Garland, 1994.

Campbell, N., et al. *Biology*. 5th ed. Menlo Park, California: Benjamin/Cummings, 1998.

Watson, J.D., N.H. Hopkins, J.W. Roberts, J.A. Steitz and A.M. Weiner. *Molecular Biology of the Gene*. 4th ed. Menlo Park, California: Benjamin/Cummings, 1987.

CHAPTER 21
THE GENETIC BASIS OF DEVELOPMENT

OUTLINE

I. From Single Cell to Multicellular Organism
    A. Embryonic development involves cell division, morphogenesis, and cell differentiation
    B. Researchers study development in model organisms to identify general principles: *science as a process*
II. Differential Gene Expression
    B. Different types of cells in an organism have the same DNA
    C. Different cell types make different proteins, usually as a result of transcriptional regulation
    D. Transcriptional regulation is directed by maternal molecules in the cytoplasm and signals from other cells
III. Genetic and Cellular Mechanisms of Pattern Formation
    A. Genetic analysis of development in *Drosophila* reveals how genes control development: *an overview*
    B. Gradients of maternal molecules in the early embryo control axis formation
    C. A cascade of gene activations sets up the segmentation pattern in *Drosophila*: *a closer look*
    D. Homeotic genes direct the identity of body parts
    E. Homeobox genes have been highly conserved in evolution
    F. Neighboring cells instruct other cells to form particular structures: cell signaling and induction in the nematode
    G. Plant development depends on cell signaling and transcriptional regulation: *science as a process*

OBJECTIVES

After reading the chapter and attending lecture, students should be able to do the following:

1. Distinguish between the patterns of morphogesis in plants and in animals.
2. List the animals used as models for developmental biology research and provide a rationale for their choice.
3. Describe how genomic equivalence was determined for plants and animals.
4. Describe what kinds of changes occur to the genome during differentiation.
5. Describe the general processes by which "Dolly" was cloned.
6. Describe the molecular basis of determination.
7. Describe the two sources of information that instruct a cell to express genes at the appropriate time.
8. Describe how *Drosophila* were used to explain basic aspects of pattern formation (axis formation and segmentation).

9.   Describe how homeotic genes serve to identify parts of the developing organism.

10.   Provide evidence of the conservation of homeobox sequences.

11.   Describe how the study of nematodes contributed to the general understanding of embryonic induction.

12.   Describe how apoptosis functions in normal and abnormal development.

13.   Describe how the study of tomatoes has contributed to the understanding of flower development.

14.   Describe how the study of *Arabidopsis* has contributed to the understanding of organ identity in plants.

KEY TERMS

differentiation	determination	maternal effect genes	segment-polarity genes
morphogenesis	cytoplasmic determinants	egg-polarity genes	homeotic genes
apical meristems	pattern formation	morphogens	homeobox
model organism	induction	segmentation genes	apoptosis
cell lineage	positional information	gap genes	chimeras
totipotent	embryonic lethals	pair-rule genes	organ-identity genes

LECTURE NOTES

The study of how a single cell develops into a multicellular organism and the functional maintenance of the developed structures is one of the most intriguing aspects of biology. The complete instructions to execute the developmental program of an organism are encoded in its genes. This chapter discusses how control of the spatial and temporal expression of genes contributes to the development of a multicellular organism.

I.   **From Single Cell to Multicellular Organism**

A. **Embryonic development involves cell division, morphogenesis, and cell differentiation**

A multicellular organism develops from a fertilized egg through three processes: cell division, cell differentiation, and morphogenesis (see Campbell, Figure 21.1).

*   Cell division increases cell number.

*   During cell *differentiation*, the cells become specialized in structure and function.

*   Through a host of processes, collectively referred to as *morphogenesis*, the overall shape of the organism is established.

During development, these three processes overlap in time.

*   Initial aspects of morphogenesis during early development establish the basic body plan (e.g., which end of an animal will give rise to the head).

*   Cell division and differentiation as well as selective cell death are important components of morphogenesis.

Animals and plants differ in their developmental programs (see Campbell, Figure 21.2).

*   In animals, movement of cells and tissues are involved in the development of physical form.

*   Growth in plants is not limited to embryonic and juvenile periods as it is in animals. The root and shoot tips of plants possess perpetual embryonic tissues, known as *apical meristems*, that are responsible for the continuous growth of new organs.

B. Researchers study development in model organisms to identify general principles: *science as a process*

Researchers use *model organisms* to facilitate to discovery of fundamental developmental processes. Model organisms are chosen because they possess features that make the study easier to conduct; the criteria used to select an organism include the following:

- Large eggs (easy to manipulate and observe)
- Readily observable embryos
- Short generation times
- Small genomes
- Preexisting knowledge of organism's genes and life history

Frogs were used widely as models in early studies of development, but they actually have relatively complex genetics. As a result, most current research is conducted on the following organisms because of their unique characteristics (see Campbell, Figure 21.3):

- The fruit fly, *Drosophila melanogaster:* easily grown in the lab, short generation time, embryos grow outside of the mother's body
- The nematode, *Caenorhabditis elegans:* easily grown in lab, transparent body composed of only a few cell types that always arise in the same way, short generation time, hermaphroditic
  ⇒ Researchers have been able to construct the complete *cell lineage* of *C. elegans*, or the ancestry of every cell in the adult body (see Campbell, Figure 21.4).
- The zebrafish, *Danio rerio*: small and easy to breed in the lab, transparent embryo, rapid embryonic development, smaller genome size than that of mice
- The mouse, *Mus musculus*: long used as a vertebrate model, much is known about its genes; gene manipulations and gene "knock out" technologies are available; however, complex genetics and large genome
- The plant, *Arabidopsis thaliana:* easily grown in culture, small genome, cells take up foreign DNA

II.   Differential Gene Expression

Differences among the cells of a multicellular organism arise from different patterns of gene expression and not from differences in the genomes of the cells.

A. Different types of cells in an organism have the same DNA

Nearly all of the cells of an organism have the same genes (genomic equivalence). What happens to these genes as the cells differentiate?

1. Totipotency in plants

Genomic equivalence among the cells of plants was demonstrated by experiments in which entire individuals developed from differentiated somatic cells (see Campbell, Figure 21.5).

- The observation that somatic cells can dedifferentiate and then give rise to all of the various cells of a new individual demonstrates that differentiation does involve irreversible changes to the genome.
- Cells that retain the ability of the zygote to give rise to all the specialized cells of a mature organism are called *totipotent*.

2. Nuclear transplantation in animals

Because the cells of animals will not often divide in culture, scientists have adopted alternative approaches to examine genomic equivalence in animals (see Campbell, Figure 21.6).

- By transplanting the nuclei of differentiated cells into enucleated egg cells of frogs, Briggs and King determined that the genome within the transplanted nuclei could support development; however, normal development was inversely related to the age of donor embryos.

Such transplantation studies lead to the following conclusions:

- Nuclei do change in some ways during differentiation.
- Changes do not occur to the sequence of DNA, but rather, in chromatin structure.
  - ⇒ The age-related relationship in developmental potential of frog nuclei is related to development-related changes in chromatin structure.

Mammals have been successfully cloned form nuclei and cells of early embryos.

- An adult sheep, "Dolly," was cloned by Ian Wilmut and colleagues in Scotland by transplanting the nucleus of a dedifferentiated mammary cell from one sheep into an unfertilized, enucleated egg of another sheep.

B. Different cell types make different proteins, usually as a result of transcription regulation

As cells differentiate, they become obviously different in structure and function.

- The earliest changes are subtle, manifested only at the molecular level; such changes, known as *determination*, irreversibly commit the cell to its final fate.
- The result of determination is the presence of tissue-specific proteins (e.g., crystallins of the vertebrate lens, muscle-specific forms of actin and myosin) characteristic of a cell's structure and function.

The complement of proteins that a cell makes results from the pattern of gene expression in the differentiating cell; a pattern that is, for the most part, regulated at the level of transcription.

Transcription regulation of gene expression during development is exemplified in muscle cell determination (see Campbell, Figure 21.8).

Researchers tested the hypothesis that certain muscle-specific regulatory genes were active in myoblasts in the following way:

- By using reverse transcriptase, a *cDNA* library of genes was generated from RNA isolated from cultured myoblasts. (These cDNAs were intron-lacking versions of the genes that normally occur in myoblasts.)
- The cDNAs were ligated into bioengineered plasmids that contained a promoter that would turn on any kind of gene.
- The plasmids were then inserted into embryonic precursor cells to determine if differentiation into myoblasts and muscle cells would occur.

Researchers determined that the molecular basis of muscle cell determination is the transcription (and translation) of critical muscle-determination genes (a type of "master regulatory gene"). One of these muscle-determination genes is called *myoD*.

- The protein product of *myoD*, called MyoD, is a transcription factor that binds to control elements of DNA, and in turn, enhances the expression of other muscle-specific transcription factors.
- The secondary transcription factors activate genes encoding muscle-protein.

C. Transcriptional regulation is directed by maternal molecules in the cytoplasm and signals from other cells

Explaining the molecular basis of determination of a single cell type, such as the role of *myoD* in muscle cell differentiation, is only part of how a multicellular organism arises.

Lingering questions about how such master regulatory genes themselves are turned on remain. The answers to such questions rest again on understanding control of differential gene expression during early development.

Two sources of information instruct a cell on which genes to express at a given time:

- Information in the cytoplasm of the unfertilized egg, in the form of RNA and protein, that is of maternal origin (*cytoplasmic determinants*) (see Campbell, Figure 21.9).
- Chemical signals produced by neighboring embryonic cells; such signals, through a process known as *induction*, influence the growth and differentiation of adjacent cells.

III.   Genetic and Cellular Mechanisms of Pattern Formation

Cytoplasmic determinants and inductive signals contribute to morphogenesis by modeling *pattern formation*, the spatial organization of tissues and organs characteristic of a mature organism. In plants, pattern formation occurs continuously; in animals, pattern formation is usually restricted to embryos and juveniles.

A.   Genetic analysis of *Drosophila* reveals how genes control development: *an overview*

By studying *Drosophila*, researchers have identified how specific molecules influence position and direct differentiation.

1.   The life cycle of *Drosophila*

Fruit flies and other arthropods are segmented into three major body parts: head, thorax, and abdomen.

The cytoplasmic determinants provide positional information.

- In unfertilized eggs, the placement of the anterior-posterior and dorsal-ventral axes is determined
- After fertilization, orientation of body segments and development of associated structures is initiated.

The developmental stages of *Drosophila* are shown in Campbell, Figure 21.10. Note that by division 13, the basic body plan, including body axes and segmentation, has been determined.

2.   Genetic analysis of early development in *Drosophila*: *science as a process*

By using mutants, E.B. Lewis in the 1940s demonstrated that genes somehow direct development.

In the 1970s, Nusslein-Volhard and Wieschaus (who were awarded a Nobel prize in 1995), studied pattern formation, specifically, the basis of segmentation at the molecular level.

- Their research was fraught with many challenges (see Campbell Methods box on *Drosophila* development):
  ⇒ Segmentation may be influenced a large number of genes (out of a possible 12,000).
  ⇒ Mutations affecting segmentation would be lethal to embryos (*embryonic lethals*).
  ⇒ Because maternally-derived cytoplasmic determinants affected segmentation, the scope of their analysis would have to include maternal genes as well as embryonic genes.
- Eventually, they identified some 1200 genes that were essential for development, of which 120 played a role in segmentation.
- Various cytoplasmic determinants were found to control the expression of segmentation genes.

B. Gradients of maternal molecules in the early embryo control axis formation

Cytoplasmic determinants are encoded by maternal genes called *maternal effect genes* (or sometimes *egg-polarity genes* because of the effects of their products on orientation/polarity).

- One set of genes helps establish the anterior-posterior axis of the embryo.
- A second set of genes is involved with the dorsal-ventral axis of the embryo.

The means by which maternal effect genes influence pattern formation is exemplified by the *bicoid* gene.

- A mother missing the *bicoid* gene gives rise to an embryo that lacks the front half of its body.
  - ⇒ The phenotype of the offspring suggests that the *bicoid* gene is essential for development of the anterior end of the fly, possibly because the gene product, a cytoplasmic determinant, is required at the anterior end.
  - ⇒ The requirement for the appropriate distribution of cytoplasmic determinants is a special version of the gradient hypothesis developed over 100 years ago. It maintained that gradients of substances, or *morphogens*, were required to establish the axes of the embryo.

Recent research indicates that the *bicoid* product is a morphogen that affects head-end development.

- *Bicoid* mRNA is concentrated at the anterior end of unfertilized eggs produced by wild-type mothers. After fertilization, the mRNA is translated and forms a gradient of *bicoid* protein within the embryo.
- Injection of *bicoid* mRNA into early embryos results in the development of anterior structures at the injection sites.

The factors involved with posterior end development, as well as with the development of anterior and posterior surfaces, also have been identified.

C. A cascade of gene activations sets up the segmentation pattern of *Drosophila*: a closer look

The *bicoid* protein and other morphogens are transcription factors that regulate the transcription of selected genes of the embryo.

- The gradients of the morphogens are responsible for the pattern of regional differences in the expression of *segmentation genes* (the genes that control segmentation following the establishment of the major body axes).

The sequential activation of three sets of segmentation genes are responsible for refinement of the body plan; in order of activation, the gene sets are as follows (se Campbell, Figure 21.12):

- Products of *map genes* influence basic subdivision along the anterior-posterior axis.
- *Pair-rule* genes control the pairing of segments.
- *Segment-polarity genes* serve to direct anterior-posterior orientation within each segment.

The products of the segmentation genes operate in numerous ways:

- Many are transcription factors that enhance the expression of the segmentation gene next in the sequence.
- Others are components of signaling pathways, including signal molecules used in the cell-cell communication and the membrane receptors that recognize them.

D. Homeotic genes direct the identity of body parts

Continued morphogenesis, including the appropriate placement of appendages, requires identification of specific regions of the body. The identity of segments is conveyed through master regulatory genes called *homeotic genes*.

- Homeotic genes encode for transcription factors that influence the genes responsible for specific structures.
  - ⇒ For example, homeotic proteins produced in cells of a particular thoracic segment lead to leg development.
  - ⇒ Homeotic mutations replace structures characteristic of one part of an animal with structures normally found at some other location (see Campbell, Figure 21.13).
- Scientists are in the process of identifying the genes activated by homeotic genes.

E. Homeobox genes have been highly conserved in evolution

The homeotic genes of Drosophila all contain a 180-nucleotide sequence called the *homeobox*. (For this reason, all genes that contain the homeobox are referred to as *Hox* genes.)

Sequences identical or very similar to the homeobox of *Drosophila* have been discovered in other invertebrates and vertebrates, as well as yeast and prokaryotes.

- Such sequence similarity suggests that the homeobox sequence emerged early during the evolution of life.
- Animal genes homologous to the homeotic genes of fruit flies have even kept their chromosomal arrangement (see Campbell, Figure 21.14)

Not all homeobox genes serve as homeotic genes, however, most homeobox genes are associated with some aspect of development. For example, in Drosophila, homeoboxes are present in homeotic genes, the bicoid gene, several of the segmentation genes, and in the master regulatory gene for eye development.

What is the role of the protein segment encoded by the homeobox sequence?

- The homeobox encodes for a 60-amino-acid-long homeodomain. Proteins containing homeodomains serve as transcription factors.
- The homeodomain influences protein-protein interactions critical to transcriptional regulation.

F. Neighboring cells instruct other cells to form particular structures: cell signaling and induction in the nematode

Communication between and among cells of the embryo is critical to the development of the organism. The signaling process helps to coordinate the appropriate spatial and temporal expression of genes.

1. Induction in vulval development

Research on the development of the opening (*vulva*) through which nematodes lay their eggs has provided much insight into cell signaling and induction of development. By studying mutants, scientists have identified a number of genes involved in vulval development (see Campbell, Figure 21.15).

The anchor cell releases an inducer that binds to vulval precursor cells. (This inducer is a growth factor that binds to a tyrosine kinase receptor; see Campbell, Figure 19.13a.)

- Initially, all precursor cells are the same. The cell that gives rise to the inner part of the vulva receives a higher concentration of inducer.
- The high concentration of inducer stimulates:
  - ⇒ Division and differentiation that lead to inner vulva formation
  - ⇒ The production of a second inducer
- The second inducer binds to the other precursor cells, stimulating them to form the outer vulva.

Vulva development illustrates several important developmental concepts not only in nematodes, but in animals generally.

- Sequential inductions control organ formation.
- The effect of an inducer can depend on its concentration.
- Inducers operate through signal systems similar to those in adult organisms.
- Induction results in the selective activation or inactivation of specific genes within the target cell.
- Genetics is useful to our understanding of the mechanisms that underlie development.

2. Programmed cell death (apoptosis)

The study of *C. elegans* also has revealed that normal pattern formation depends on selective, programmed cell death (*apoptosis*).

- Selective cell death occurs 131 times during normal development.
- Chemical signals initiate the activation of a cascade "suicide genes."
  - ⇒ Two key suicide genes are *ced-3* and *ced-4*; the protein products of these genes are continuously present in the cell in an inactive form.
  - ⇒ Control of apoptosis, then, depends not on transcription or translation, but on regulating protein activity (see Campbell, Figure 21.16)
- The cell is killed when enzymes are activated to hydrolyze DNA and protein.

Certain degenerative diseases and cancers may have their basis in faulty apoptotic mechanisms.

G. Plant development depends on cell signaling and transcriptional regulation.

Because the last common ancestor of plants and animals was a single-celled organism living millions of years ago, the developmental processes in these two phyla most likely evolved independently.

Despite the differences between plants and animals, some of the basic molecular, cellular, and genetic mechanisms of development are similar.

Clues to the details associated with plant development come from DNA technology, insights from animal research, and the study of the model plant *Arabidopsis*.

1. Cell signaling in flower development

Environmental cues (e.g., day length) initiate processes that convert ordinary shoot meristems to floral meristems. Such induction is exemplified with the development of tomato flowers (see Campbell, Figure 21.17).

- Tomato plants homozygous for a mutant allele, called fasciated (*f*), produce flowers with an abnormally large number of organs.
- Stems from mutant plants grafted onto wild-type plants resulted in new plants that were *chimeras*, organisms with a mix of genetically different cells.
- Some of the chimeras possessed floral meristems in which the three cell layers did not all arise from the same "parent."
- By tracing the sources of the meristem layers, it was determined that the number of organs per flower depended on genes in the L3 (innermost) cell layer.

2. Organ-identity genes in plants

Organ-identity genes determine the type of structure (e.g., petal) that will grow from a meristem. Most of the information on organ-identity genes comes from studies of *Arabidopsis*.

- Organ-identity genes are analogous to homeotic genes.
- Organ-identity genes are divided into three classes: A, B, and C.

- The simple model in Campbell, Figure 21.16 shows how three kinds of genes direct the formation of four type of organs.
- Organ-identity genes appear to be acting like master regulatory genes that control the transcription of other genes directly involved in plant morphogenesis.
    ⇒ The organ-identity genes of plants do not contain the homeobox sequence
    ⇒ A different sequence of about the same length is present; this sequence is also present in some transcription-factor genes of yeast and animals

REFERENCES

Campbell, N., et al. *Biology*. 5th ed. Menlo Park, California: Benjamin/Cummings, 1998.

Fjose, A. "Spatial Expression of Homeotic Genes in *Drosophila*." *Bioscience*, September 1986.

Gehring, Walter J. "The Molecular Basis of Development." *Scientific American*, October 1985.

Gilbert, S.F. *Developmental Biology*. 5th ed. Sunderland, MA: Sinauer Associates, 1997.

Kalthoff, K. *Analysis of Biological Development*. New York, NY: McGraw-Hill, 1996.

OUTLINE

I.   Historical-Context for Evolutionary Theory
    A. Western culture resisted evolutionary views of life
    B. Theories of geological gradualism helped clear the path for evolutionary biologists
    C. Lamarck placed fossils in an evolutionary context

II. The Darwinian Revolution
    A. Field research helped Darwin frame his view of life: *science as a process*
    B. *The Origin of Species* developed two main points: the occurrence of evolution and natural selection as its mechanism

III. Evidence of Evolution
    A. Evidence of evolution pervades biology
    B. What is theoretical about the Darwinian view of life?

OBJECTIVES

After reading this chapter and attending lecture, the student should be able to:

1.   State the two major points Darwin made in *The Origin of Species* concerning the Earth's biota.

2.   Compare and contrast Plato's philosophy of idealism and Aristotle's *scala naturae*.

3.   Describe Carolus Linnaeus' contribution to Darwin's theory of evolution.

4.   Describe Georges Cuvier's contribution to paleontology.

5.   Explain how Cuvier and his followers used the concept of catastrophism to oppose evolution.

6.   Explain how the principle of gradualism and Charles Lyell's theory of uniformitarianism influenced Darwin's ideas about evolution.

7.   Describe Jean Baptiste Lamarck's model for how adaptations evolve.

8.   Describe how Charles Darwin used his observations from the voyage of the HMS *Beagle* to formulate and support his theory of evolution.

9.   Describe how Alfred Russel Wallace influenced Charles Darwin.

10. Explain what Darwin meant by the principle of common descent and "descent with modification".

11. Explain what evidence convinced Darwin that species change over time.

12. State, in their own words, three inferences Darwin made from his observations, which led him to propose natural selection as mechanism for evolutionary change.

13. Explain why variation was so important to Darwin's theory.

14. Explain how Reverend Thomas Malthus' essay influenced Charles Darwin.

15. Distinguish between artificial selection and natural selection.

16. Explain why the population is the smallest unit that can evolve.

17. Using some contemporary examples, explain how natural selection results in evolutionary change.

18. Explain why the emergence of population genetics was an important turning point for evolutionary theory.

19. Describe the lines of evidence Charles Darwin used to support the principle of common descent.

20. Describe how molecular biology can be used to study the evolutionary relationships among organisms.

21. Explain the problem with the statement that Darwinism is "just a theory".

22. Distinguish between the scientific and colloquial use of the word "theory".

KEY TERMS

evolution	fossils	descent with modification	vestigial organs
natural selection	sedimentary rocks	artificial selection	ontogeny
evolutionary adaptations	paleontology	biogeography	phylogeny
natural theology	gradualism	homology	
taxonomy	uniformitarianism	homologous-structures	

LECTURE NOTES

Evolution, the unifying theme woven throughout the text and course, refers to the processes that have transformed life on earth from its earliest forms to the enormous diversity that characterizes it today.

The first convincing case for evolution was published in a book by Charles Darwin on November 24, 1859. In this book, *On the Origin of Species by Means of Natural Selection*, Darwin:

- Synthesized seemingly unrelated facts into a conceptual framework that accounts for both the unity and diversity of life.

- Discussed important biological issues about organisms, such as why there are so many kinds of organisms, their origins and relationships, similarities and differences, geographic distribution, and adaptations to their environment.

- Made two major points:
    1. Species evolved from ancestral species and were not specially created.
    2. *Natural selection* is a mechanism that could result in this evolutionary change.

I.   **Historical Context for Evolutionary Theory**

   A. **Western culture resisted evolutionary views of life**

   The impact of Darwin's ideas partially depended upon historical and social context (see Campbell, Figure 22.1).

   - Darwin's view of life contrasted sharply with the accepted viewpoint: the Earth was only a few thousand years old and was populated by unchanging life forms made by the Creator during a single week.

   - Thus, *On the Origin of Species by Means of Natural Selection* not only challenged prevailing scientific views, but also challenged the roots of Western culture.

1. **The scale of nature and natural theology**

Many Greek philosophers believed in the gradual evolution of life. However, the two that influenced Western culture most, Plato (427 – 347 B.C.) and his student Aristotle (384 – 322 B.C.), held opinions which were inconsistent with a concept of evolution.

- Plato, whose philosophy is known as *idealism* (*essentialism*), believed that there were two coexisting worlds: an ideal, eternal, real world and an illusionary imperfect world that humans perceive with their senses. To Plato,
  - ⇒ Variations in plant and animal populations were merely imperfect representatives of ideal forms; only the perfect ideal forms were real.
  - ⇒ Evolution would be counterproductive in a world where ideal organisms were already perfectly adapted to their environments.
- Aristotle questioned the Platonic philosophy of dual worlds, but his beliefs also excluded evolution.
  - ⇒ Recognizing that organisms vary from simple to complex, he believed that they could be placed on a scale of increasing complexity (*scala naturae*); on this ladder of life, each form had its allotted rung and each rung was occupied.
  - ⇒ In this view of life, species were fixed and did not evolve.
  - ⇒ The *scala naturae* view of life prevailed for over 2000 years.

The *creationist–essentialist* dogma that species were individually created and fixed became embedded in Western thought as the Old Testament account of creation from the Judeo–Christian culture fortified prejudice against evolution.

- *Natural Theology*, a philosophy that the Creator's plan could be revealed by studying nature, dominated European and American biology even as Darwinism emerged.
- For natural theologians, adaptations of organisms were evidence that the Creator had designed every species for a particular purpose.
- Natural theology's major objective was to classify species revealing God's created steps on the ladder of life.

Carolus Linnaeus (1707 – 1778), a Swedish physician and botanist, sought order in the diversity of life *ad majorem Dei gloriam* (for the greater glory of God).

- Known as the father of *taxonomy*—the naming and classifying of organisms—he developed the system of *binomial nomenclature* still used today.
- He adopted a system for grouping species into categories and ranking the categories into a hierarchy. For example, similar species are grouped into a genus; similar genera are grouped into the same order.

Linnaeus found order in the diversity of life with his hierarchy of taxonomic categories.

- The clustering of species in taxonomic groups did not imply evolutionary relationships to Linnaeus, since he believed that species were permanent creations.
- Linnaeus, a natural theologian, developed his classification scheme only to reveal God's plan and even stated *Deus creavit, Linnaeus disposuit* ("God creates, Linnaeus arranges").

2. Cuvier, fossils, and catastrophism

Fossils = Relics or impressions of organisms from the past preserved in rock

- Most fossils are found in *sedimentary rocks*, which:
  - ⇒ Form when new layers of sand and mud settle to the bottom of seas, lakes, and marshes, covering and compressing older layers into rock (e.g. sandstone and shale)
  - ⇒ May be deposited in many layers (*strata*) in places where shorelines repeatedly advance and retreat. Later erosion can wear away the upper (younger) strata, revealing older strata which had been buried.
- The fossil record thus provides evidence that Earth has had a succession of flora and fauna (see Campbell, Figure 22.2).

The study of fossils, *paleontology*, was founded by the French anatomist Georges Cuvier (1769-1832) who:

- Realized life's history was recorded in fossil-containing strata and documented the succession of fossil species in the Paris Basin
- Noted each stratum was characterized by a unique set of fossil species and that the older (deeper) the stratum, the more dissimilar the flora and fauna from modern life forms
- Understood that extinction had been a common occurrence in the history of life since, from stratum to stratum, new species appeared and others disappeared

Even with paleontological evidence, Cuvier was an effective opponent to the evolutionists of his day.

- He reconciled the fossil evidence with his belief in the fixity of species by speculating that boundaries between fossil strata corresponded in time to catastrophic events, such as floods or droughts.
- This view of Earth's history is known as *catastrophism*.

Catastrophism = Theory that major changes in the Earth's crust are the result of catastrophic events rather than from gradual processes of change

Cuvier explained the appearance of new species in younger rock that were absent from older rock by proposing that:

- Periodic localized catastrophes resulted in mass extinctions.
- After the local flora and fauna had become extinct, the region would be repopulated by foreign species immigrating from other areas.

B. Theories of geological gradualism helped clear the path for evolutionary biologists

In the late 18th century, a new theory of geological *gradualism* gained popularity among geologists that would greatly influence Darwin.

Gradualism = Principle that profound change is the cumulative product of slow, continuous processes

- Competed with Cuvier's theory of catastrophism
- Proposed by James Hutton (1975), a Scottish geologist. He proposed that it was possible to explain the various land forms by looking at mechanisms currently operating in the world.

  Example: Canyons form by erosion from rivers, and fossil-bearing sedimentary rocks form from particles eroded from the land and carried by rivers to the sea.

Charles Lyell, a leading geologist of Darwin's time, expanded Hutton's gradualism into the theory known as uniformitarianism.

Uniformitarianism = Theory that geological processes are uniform and have operated from the origin of the Earth to the present

- It was Lyell's extreme idea that geological processes are so uniform that their rates and effects must balance out through time.
- Example: Processes that build mountains are eventually balanced by the erosion of mountains.

Darwin rejected uniformitarianism, but was greatly influenced by conclusions that followed directly from the observations of Hutton and Lyell:

- The Earth must be ancient. If geological change results from slow, gradual processes rather than sudden events, then the Earth must be much older than the 6000 years indicated by many theologians on the basis of biblical inference.
- Very slow and subtle processes persisting over a great length of time can cause substantial change.

C. Lamarck placed fossils in an evolutionary context

Several 18th century naturalists suggested that life had evolved along with Earth's changes. Only Jean Baptiste Lamarck (1744-1829) developed and published (1809) a comprehensive model which attempted to explain how life evolved.

Lamarck was in charge of the invertebrate collection at the Natural History Museum in Paris, which allowed him to:

- Compare modern species to fossil forms, and in the process, identify several lines of descent composed of a chronological series of older fossils to younger fossils to modern species.
- Envision many ladders of life which organisms could climb (as opposed to Aristotle's single ladder without movement).
    - ⇒ The bottom rungs were occupied by microscopic organisms which were continually generated spontaneously from nonliving material.
    - ⇒ At the tops of the ladders were the most complex plants and animals.

Lamarck believed that evolution was driven by an innate tendency toward increasing complexity, which he equated with perfection.

- As organisms attained perfection, they became better and better adapted to their environments.
- Thus, Lamarck believed that evolution responded to organisms' *sentiments interieurs* ("felt needs").

Lamarck proposed a mechanism by which specific adaptations evolve, which included two related principles:

1. *Use and disuse.* Those body organs used extensively to cope with the environment become larger and stronger while those not used deteriorate.
2. *Inheritance of acquired characteristics.* The modifications an organism acquired during its lifetime could be passed along to its offspring.

Although his mechanism of evolution was in error, Lamarck deserves credit for proposing that:

- Evolution is the best explanation for both the fossil record and the extant diversity of life.
- The Earth is ancient.

- Adaptation to the environment is a primary product of evolution.

II. The Darwinian Revolution

At the beginning of the 19th century, natural theology still dominated the European and American intellectual climate. In 1809, the same year Lamarck published his theory of evolution, Charles Darwin was born in Shrewsbury, England.

- Though interested in nature, Charles (at 16) was sent by his physician father to the University of Edinburgh to study medicine, which he found boring and distasteful.
- He left Edinburgh without a degree and enrolled at Christ College, Cambridge University to prepare for the clergy.
  ⇒ Nearly all naturalists and other scientists were clergymen, and a majority held to the philosophy of natural theology.
  ⇒ Charles studied under the Reverend John Henslow, a botany professor at Cambridge, and received his B.A. degree in 1831.
  ⇒ Professor Henslow recommended him to Captain Robert FitzRoy who was preparing the survey ship HMS *Beagle* for an around the world voyage.

A. Field research helped Darwin frame his view of life: *science as a process*
1. The voyage of the Beagle

The HMS *Beagle*, with Darwin aboard, sailed from England in December 1831 (see Campbell Figure 22.3).

- The voyage's mission was to chart the poorly known South American coastline.
- While the ship's crew surveyed the coast, Darwin spent most of his time ashore collecting specimens of the exotic and diverse flora and fauna.

While the ship worked its way around the continent, Darwin observed the various adaptations of plants and animals that inhabited the diverse environments of South America: Brazilian jungles, grasslands of the Argentine pampas, desolate islands of Tierra del Fuego, and the Andes Mountains. Darwin noted the following:

- The South American flora and fauna from different regions were distinct from the flora and fauna of Europe.
- Temperate species were taxonomically closer to species living in tropical regions of South America than to temperate species of Europe.
- The South American fossils he found (while differing from modern species) were distinctly South American in their resemblance to the living plants and animals of that continent.

Geographical distribution was particularly confusing in the case of the fauna of the Galapagos, recently formed volcanic islands which lie on the equator about 900 km west of South America.

- Most animal species on the Galapagos are unique to those islands, but resemble species living on the South American mainland.
- Darwin collected 13 types of finches from the Galapagos, and although they were similar, they seemed to be different species.
  ⇒ Some were unique to individual islands
  ⇒ Others were found on two or more islands that were close together

By the time the *Beagle* left the Galapagos, Darwin had read Lyell's *Principles of Geology*, and was influenced by Lyell's ideas.

- Darwin had begun to doubt the church's position that the Earth was static and had been created only a few thousand years before.
- When Darwin acknowledged that the Earth was ancient and constantly changing, he had taken an important step toward recognizing that life on Earth had also evolved.

2. Darwin focuses on adaptation

Darwin was not sure whether the 13 types of finches he collected on the Galapagos were different species or varieties of the same species.

- After he returned to England in 1836, an ornithologist indicated that they were actually different species.
- He reassessed observations made during the voyage and in 1837 began the first notebook on the origin of species.

Darwin perceived the origin of new species and adaptation as closely related processes; new species could arise from an ancestral population by gradually accumulating adaptations to a different environment. For example,

- Two populations of a species could be isolated in different environments and diverge as each adapted to local conditions.
- Over many generations, the two populations could become dissimilar enough to be designated separate species.
- This is apparently what happened to the Galapagos finches; their different beaks are adaptations to specific foods available on their home islands. (See Campbell, Figure 22.4)

By the early 1840s, Darwin had formed his theory of natural selection as the mechanism of adaptive evolution, but delayed publishing it.

- Reclusive and in poor health, Darwin was well known as a naturalist from the specimens and letters he had sent to Britain from the voyage on the *Beagle*.
- He frequently corresponded and met with Lyell, Henslow, and other scientists.

In 1844, Darwin wrote a long essay on the origin of species and natural selection.

- He realized the importance and subversive nature of his work, but did not publish the information because he wished to gather more evidence in support of his theory.
- Evolutionary thinking was emerging at this time, and Lyell admonished Darwin to publish on the subject before someone else published it first.

In June 1858, Darwin received a letter from Alfred Wallace, who was working as a specimen collector in the East Indies.

- Accompanying the letter was a manuscript detailing Wallace's own theory of natural selection which was almost identical to Darwin's.
- The letter asked Darwin to evaluate the theory and forward the manuscript to Lyell if it was thought worthy of publication.
- Darwin did so, although he felt that his own originality would be "smashed."
- Lyell and a colleague presented Wallace's paper along with excerpts from Darwin's unpublished 1844 essay to the Linnaean Society of London on July 1, 1858.

Darwin finished *The Origin of Species* and published it the next year.

- Darwin is considered the main author of the idea since he developed and supported natural selection much more extensively than Wallace.
- Darwin's book and its proponents quickly convinced the majority of biologists that biodiversity is a product of evolution.
- Darwin succeeded where previous evolutionists had failed not only because science was moving away from natural theology, but because he convinced his readers with logic and evidence.

B. *The Origin of Species* developed two main points: the occurrence of evolution and natural selection as its mechanism

1. Descent with modification

Darwin used the phrase "descent with modification," not evolution, in the first edition of *The Origin of Species*.

- He perceived a unity in life with all organisms related through descent from some unknown ancestral population that lived in the remote past.
- Diverse modifications (*adaptations*) accumulated over millions of years, as descendants from this common ancestor moved into various habitats.

Darwin's metaphor for the history of life was a branching tree with multiple branching from a common trunk to the tips of living twigs, symbolic of the diversity of contemporary organisms.

- At each fork or branch point is an ancestral population common to all evolutionary lines of descent branching from that fork.
- Species that are very similar share a common ancestor at a recent branch point on the phylogenetic tree.
- Less closely related organisms share a more ancient common ancestor at an earlier branch point.
- Most branches of evolution are dead ends since about 99% of all species that ever lived are extinct.

To Darwin, Linnaeus' taxonomic scheme reflected the branching genealogy of the tree of life.

- It recognized that the diversity of organisms could be ordered into "groups subordinate to groups", with organisms at the different taxonomic levels related through descent from common ancestors.
- Classification alone does not confirm the principle of common descent, but when combined with other lines of evidence, the relationships are clear.
- For example, genetic analysis of species that are thought to be closely related on the basis of anatomical features and other criteria reveals a common hereditary background.

2. Natural selection and adaptation

Darwin's book focused on the role of natural selection in adaptation (see Campbell, Figure 22.5). Ernst Mayr of Harvard University dissected the logic of Darwin's theory into three inferences based on five observations:

- *Observation 1*: All species have such great potential fertility that their population size would increase exponentially if all individuals that are born reproduced successfully.
- *Observation 2*: Populations tend to remain stable in size, except for seasonal fluctuations.

- *Observation 3*:   Environmental resources are limited.
  - *Inference 1*: Production of more individuals than the environment can support leads to a struggle for existence among individuals of a population, with only a fraction of offspring surviving each generation.
- *Observation 4*:   Individuals of a population vary extensively in their characteristics; no two individuals are exactly alike.
- *Observation 5*:   Much of this variation is heritable.
  - *Inference 2*: Survival in the struggle for existence is not random, but depends in part on the hereditary constitution of the surviving individuals. Those individuals whose inherited characteristics fit them best to their environment are likely to leave more offspring than less fit individuals.
  - *Inference 3*: This unequal ability of individuals to survive and reproduce will lead to a gradual change in a population, with favorable characteristics accumulating over the generations.

Summarizing Darwin's ideas:

- Natural selection is this differential success in reproduction, and its product is adaptation of organisms to their environment.
- Natural selection occurs from the interaction between the environment and the inherent variability in a population.
- Variations in a population arise by chance, but natural selection is not a chance phenomenon, since environmental factors set definite criteria for reproductive success.

Darwin was already aware of the struggle for existence caused by overproduction, when he read an essay on human population written by the Reverend Thomas Malthus (1798).

- In this essay, Malthus held that much of human suffering was a consequence of human populations growing faster than the food supply.
- This capacity for overproduction is common to all species, and only a fraction of new individuals complete development and leave offspring of their own; the rest die or are unable to reproduce.

Variation and overproduction in populations make natural selection possible.

- On the average, the most fit individuals pass their genes on to more offspring than less fit individuals.
- This results from environmental editing, which favors some variations over others.

From his experiences with *artificial selection*, Darwin inferred that natural selection could cause substantial change in populations.

- Through the breeding of domesticated plants and animals, humans have modified species over many generations by selecting individuals with desired traits as breeding stock.
- The plants and animals we grow for food show little resemblance to their wild ancestors (see Campbell, Figure 22.6).
- Darwin reasoned that if such change could be achieved by artificial selection in a relatively short period of time, then natural selection should be capable of considerable modifications of species over hundreds of thousands of generations.

- Even if the advantages of some heritable traits over others are slight, they will accumulate in the population after many generations of natural selection eliminating less favorable variations.

Gradualism is fundamental to the Darwinian view of evolution. Darwin reasoned that:

- Life did not evolve suddenly by quantum leaps, but instead by a gradual accumulation of small changes.
- Natural selection operating in differing contexts over vast spans of time could account for the diversity of life.

Summarizing Darwin's view of evolution:

- The diverse forms of life have arisen by descent with modification from ancestral species.
- The mechanism of modification has been natural selection working gradually over long periods of time.

a. Some subtleties of natural selection

Populations are important in evolutionary theory, since a population is the smallest unit that can evolve.

Population = A group of interbreeding individuals belonging to a particular species and sharing a common geographic area

Natural selection is a consequence of interactions between individual organisms and their environment, but individuals do not evolve.

- Evolution can only be measured as change in relative proportions of variations in a population over several generations.
- Natural selection can only amplify or diminish heritable variations.
- Organisms can adapt to changes in their immediate environment and can be otherwise modified by life experiences, but these acquired characteristics cannot be inherited.
- Evolutionists must distinguish between adaptations an organism acquires during its lifetime and those inherited adaptations that evolve in a population over many generations as a result of natural selection.

Specifics of natural selection are situational.

- Environmental factors vary from area to area and from time to time.
- An adaptation under one set of conditions may be useless or detrimental in different circumstances.

b. Examples of natural selection in action

In an effort to test Darwin's hypothesis that the beaks of Galapagos finches are evolutionary adaptations to different food sources, Peter and Rosemary Grant of Princeton University have been conducting a long-term study on medium ground finches (*Geospiza fortis*) on Daphne Major, a tiny Galapagos island. They have discovered that:

- Average beak depth (an inherited trait) oscillates with rainfall (see Campbell, Figure 22.7).
  ⇒ In wet years, birds preferentially feed on small seeds, and average beak depth decreases.
  ⇒ In dry years, small seeds are less plentiful, so survival depends on the finches being able to crack the less preferred larger seeds. Average beak depth increases during dry years.

- It can be inferred that the change in beak depth is an adaptive response to the relative availability of small seeds from year to year.

This study illustrates some important points about adaptive change:

- *Natural selection is situational.* What works in one environmental context may not work in another.

- *Beak evolution on Daphne Major does not result from inheritance of acquired characteristics.* The environment did not *create* beaks specialized for large or small seeds, but only acted on inherited variations already present in the population. The proportion of thicker-beaked finches increased during dry periods because, on average, thicker-beaked birds transmitted their genes to more offspring than did thinner-beaked birds.

Michael Singer and Camille Parmesan of the University of Texas, have documented rapid evolutionary adaptation in a butterfly population (Edith's checkerspot) living in a meadow near Carson City, Nevada.

- In only a decade, this butterfly population apparently adapted to changing vegetation by inherited changes in reproductive behavior.

- Females lay eggs preferentially on certain plants which provide food for the larvae after they hatch. In 1983, checkerspots laid about 80% of their eggs on a native plant, *Collinsia parviflora*.

- By 1993, the butterflies were laying about 70% of their eggs on *Plantago lanceolata*, an invading weed from surrounding cattle ranches.

- The researchers demonstrated that the switch in plant preference is genetic; daughters of butterflies that deposited eggs on *Plantago* inherited the taste for that plant, choosing it over *Collinsia* when they laid their eggs.

There are hundreds of examples of natural selection in laboratory populations of such organisms as *Drosophila*. Other examples of natural selection in action include:

- Antibiotic resistance in bacteria (see Campbell, Chapter 18)

- Body size of guppies exposed to different predators (see Campbell, Chapter 1)

III. Evidence of Evolution

A. Evidence of evolution pervades biology

Darwin used several lines of evidence to support his principle of common descent, an evolutionary change. Recent discoveries, including those from molecular biology, lend support to his evolutionary view of life.

1. Biogeography

It was biogeographical evidence that first suggested common descent to Darwin, because the biogeographical patterns he observed only made sense in the light of evolution.

Biogeography = The geographical distribution of species

Islands have many endemic species which are closely related to species on the nearest mainland or neighboring island. Some logical questions follow:

- Why are two islands with similar environments in different parts of the world not populated by closely related species, but rather by species more closely related to those from the nearest mainland even when that environment is quite different?
- Why are South American tropical animals more closely related to South American desert animals than to African tropical animals?
- Why does Australia have such a diversity of marsupial animals and very few placental animals even though the environment can easily support placentals?

2. The fossil record

Darwin was troubled by the absence of transitional fossils linking modern life to ancestral forms.

- Even though the fossil record is still incomplete, paleontologists continue to find important new fossils, and many key links are no longer missing.
- For example, fossilized whales link these aquatic mammals to their terrestrial predecessors (see Campbell, Figure 22.8).

Although still incomplete, the fossil record provides information that supports other types of evidence about the major branches of the phylogenetic tree. For example:

- Prokaryotes are placed as the ancestors of all life by evidence from cell biology, biochemistry, and molecular biology.
- Fossil evidence shows the chronological appearance of the vertebrates as being sequential with fishes first, followed by amphibians, reptiles and then birds and mammals. This sequence is also supported by many other types of evidence.

3. Comparative anatomy

Anatomical similarities among species grouped in the same taxonomic category are a reflection of their common descent.

- The skeletal components of mammalian forelimbs are a good example (see Campbell, Figure 22.9)
  ⇒ Although the limbs are used for different functions, it is obvious that the same skeletal elements are present.
  ⇒ It is logical that whether the forelimb is a foreleg, wing, flipper, or arm, the basic similarity is the consequence of descent from a common ancestor and that the limbs have been modified for different functions. They are homologous structures.

Homologous structures = Structures that are similar because of common ancestry

- Other evidence from comparative anatomy supports that evolution is a remodeling process in which ancestral structures that functioned in one capacity have become modified as they take on new functions.
- Some homologous structures are *vestigial organs*.

Vestigial organs = Rudimentary structures of marginal or no use to an organism

- Vestigial organs are remnants of structures that had important functions in ancestral forms but are no longer essential.
- Example: The remnants of pelvic and leg bones in snakes show descent from a walking ancestor, but have no function in the snake.

- Because it would be wasteful to continue providing blood, nutrients, and space to structures that no longer have a major function, vestigial organs serve evidence of evolution by natural selection.

4. Comparative embryology

Closely related organisms go through similar stages in their embryonic development.

- Vertebrate embryos (fishes, amphibians, reptiles, birds, mammals) go through an embryonic stage in which they possess gill slits on the sides of their throats (see Campbell, Figure 22.10).
- As development progresses, the gill slits develop into divergent structures characteristic of each vertebrate class.
- In fish, the gill slits form gills; in humans, they form the eustachian tubes that connect the middle ear with the throat.

Comparative embryology often establishes homology among structures, such as gill pouches, that become so altered in later development that their common origin is not apparent by comparing their fully developed forms.

In the late nineteenth century, embryologists developed the view that "ontogeny recapitulates phylogeny."

- This view held that the embryonic development of an individual organism (*ontogeny*) is a replay of the evolutionary history of the species (*phylogeny*).
- This is an extreme view; what does occur is a series of similar *embryonic* stages that exhibit the same characteristics, not a sequence of adult-like stages.
- Ontogeny can provide clues to phylogeny, but all stages of development may become modified over the course of evolution.

5. Molecular biology

An organism's hereditary background is reflected in its genes and their protein products.

- Siblings have greater similarity in their DNA and proteins than do two unrelated organisms of the same species.
- Likewise, two species considered to be closely related by other criteria should have a greater proportion of their DNA and proteins in common than more distantly related species.

Molecular taxonomists use a variety of modern techniques to measure the degree of similarity among DNA nucleotide sequences of different species.

- The closer two species are taxonomically, the higher the percentage of common DNA; this evidence supports common descent.
- Common descent is also supported by the fact that closely related species also have proteins of similar amino acid sequence (resulting from inherited genes).
- If two species have many genes and proteins with sequences of monomers that match closely, the sequences must have been copied from a common ancestor.

Molecular biology has also substantiated Darwin's idea that all forms of life are related to some extent through branching descent from the earliest organisms (see Campbell, Figure 22.11).

- Even taxonomically distant organisms (bacteria and mammals) have some proteins in common.
- For example, cytochrome *c* (a respiratory protein) is found in all aerobic species. Cytochrome *c* molecules of all species are very similar in structure and function, even though mutations have substituted amino acids in some areas of the protein during the course of evolution.
- Additional evidence for the unity of life is the common genetic code. This mechanism has been passed through all branches of evolution since its beginning in an early form of life.

B. What is theoretical about the Darwinian view of life?

Dismissing Darwinism as "just a theory" is flawed because:

- Darwin made *two* claims:
  1. Modern species evolved from ancestral forms.
  2. The mechanism for evolution is natural selection.
- The conclusion that species change or evolve is based on historical fact.

What, then, is theoretical about evolution?

- Theories are conceptual frameworks with great explanatory power used to interpret facts.
- That species can evolve is fact, but the *mechanism* Darwin proposed for that change—natural selection—is a theory. Darwin used this theory of natural selection to explain facts of evolution documented by fossils, biogeography, and other historical evidence.

In science, "theory" is very different from the colloquial use of the word, which comes closer to what scientists mean by a hypothesis, or educated guess.

- Unifying concepts do not become scientific theories, unless their predictions stand up to thorough and continuous testing by experiment and observation.
- Good scientists, however, do not allow theories to become dogma; many evolutionary biologists now question whether natural selection alone can account for evolutionary history observed in the fossil record.

REFERENCES

Campbell, N., et al. *Biology*. 5th ed. Menlo Park, California: Benjamin/Cummings, 1998.

Futuyma, D.J. *Evolutionary Biology*. 2nd ed. Sunderland, Massachusetts: Sinauer, 1986. Because it is broad in scope, this advanced textbook is a good reference for background information.

Mayr, E. *The Growth of Biological Thought: Diversity, Evolution and Inheritance*. Cambridge, Massachusetts: Harvard University Press, 1982. This is an excellent reference for historical perspective and could be a useful companion to the text for supplementing lecture material on many topics.

OUTLINE

I. Population Genetics

    A. The modern evolutionary synthesis integrated Darwinian selection and Mendelian inheritance

    B. The genetic structure of a population is defined by its allele and genotype frequencies

    C. The Hardy-Weinberg theorem describes a nonevolving population

II. Causes of Microevolution

    A. Microevolution is a generation–to–generation change in a population's allele or genotype frequencies

    B. The five causes of microevolution are genetic drift, gene flow, mutation, nonrandom mating, and natural selection

III. Genetic Variation, the Substrate for Natural Selection

    A. Genetic variation occurs within and between populations

    B. Mutation and sexual recombination generate genetic variation

    C. Diploidy and balanced polymorphism preserve variation

IV. Natural Selection as the Mechanisms of Adaptive Evolution

    A. Evolutionary fitness is the relative contribution an individual makes to the gene pool of the next generation

    B. The effect of selection on a varying characteristic can be stabilizing, directional, or diversifying

    C. Sexual selection may lead to pronounced secondary differences between the sexes

    D. Natural selection cannot fashion perfect organisms

OBJECTIVES

After reading this chapter and attending lecture, the student should be able to:

1. Explain what is meant by the "modern synthesis".

2. Explain how microevolutionary change can affect a gene pool.

3. In their own words, state the Hardy-Weinberg theorem.

4. Write the general Hardy-Weinberg equation and use it to calculate allele and genotype frequencies.

5. Explain the consequences of Hardy-Weinberg equilibrium.

6. Demonstrate, with a simple example, that a disequilibrium population requires only one generation of random mating to establish Hardy-Weinberg equilibrium.

7. Describe the usefulness of the Hardy-Weinberg model to population geneticists.

8. List the conditions a population must meet in order to maintain Hardy-Weinberg equilibrium.

9. Explain how genetic drift, gene flow, mutation, nonrandom mating and natural selection can cause microevolution.

10. Explain the role of population size in genetic drift.

11. Distinguish between the bottleneck effect and the founder effect.

12. Explain why mutation has little quantitative effect on a large population.

13. Describe how inbreeding and assortative mating affect a population's allele frequencies and genotype frequencies.

14. Explain, in their own words, what is meant by the statement that natural selection is the only agent of microevolution which is adaptive.

15. Describe the technique of electrophoresis and explain how it has been used to measure genetic variation within and between populations.

16. List some factors that can produce geographical variation among closely related populations.

17. Explain why even though mutation can be a source of genetic variability, it contributes a negligible amount to genetic variation in a population.

18. Give the cause of nearly all genetic variation in a population.

19. Explain how genetic variation may be preserved in a natural population.

20. In their own words, briefly describe the neutral theory of molecular evolution and explain how changes in gene frequency may be nonadaptive.

21. Explain what is meant by "selfish" DNA.

22. Explain the concept of relative fitness and its role in adaptive evolution.

23. Explain why the rate of decline for a deleterious allele depends upon whether the allele is dominant or recessive to the more successful allele.

24. Describe what selection acts on and what factors contribute to the overall fitness of a genotype.

25. Give examples of how an organism's phenotype may be influenced by the environment.

26. Distinguish among stabilizing selection, directional selection and diversifying selection.

27. Define sexual dimorphism and explain how it can influence evolutionary change.

28. Give at least four reasons why natural selection cannot breed perfect organisms.

KEY TERMS

population genetics	microevolution	geographical variation	relative fitness
modern synthesis	bottleneck effect	cline	stabilizing selection
population	founder effect	balanced polymorphism	directional selection
species	gene flow	heterozygote advantage	diversifying selection
gene pool	mutation	hybrid vigor	sexual dimorphism
genetic structure	inbreeding	frequency-dependent	sexual selection
Hardy-Weinberg	assortative mating	selection	
theorem, equilibrium,	natural selection	neutral variation	
and equation	polymorphism	Darwinian fitness	

LECTURE NOTES

Natural selection works on individuals, but it is the *population* that evolves (see Campbell Figure 23.1). Darwin understood this, but was unable to determine its genetic basis.

I.   Population Genetics

A. The modern evolutionary synthesis integrated Darwinian selection and Mendelian inheritance

Shortly after the publication of *The Origin of Species*, most biologists were convinced that species evolved. Darwin was less successful in convincing them that natural selection was the mechanism for evolution, because little was known about inheritance.

- An understanding about inheritance was necessary to explain how:
    ⇒ Chance variations arise in populations
    ⇒ These variations are precisely transmitted from parents to offspring
- Though Gregor Mendel was a contemporary of Darwin's, Mendel's principles of inheritance went unnoticed until the early 1900s.

For Darwin, the raw material for natural selection was variation in quantitative characters that vary along a continuum in a population.

- We now know that continuous variation is usually determined by many segregating loci (polygenic inheritance).
- Mendel and geneticists in the early 1900s recognized only discrete characters inherited on an either-or basis. Thus, for them, there appeared to be no genetic basis for the subtle variations that were central to Darwin's theory.

In the 1930s, the science of *population genetics* emerged, which:

- Emphasized genetic variation within populations and recognized the importance of quantitative characters
- Was an important turning point for evolutionary theory, because it reconciled Mendelian genetics with Darwinian evolution

In the 1940s, the genetic basis of variation and natural selection was worked out, and the *modern synthesis* was formulated. This comprehensive theory:

- Integrated discoveries from different fields (e.g., paleontology, taxonomy, biogeography, and population genetics)
- Was collectively developed by many scientists including:
    ⇒ Theodosius Dobzhansky – geneticist
    ⇒ Ernst Mayr – biogeographer and systematist
    ⇒ George Gaylord Simpson – paleontologist
    ⇒ G. Ledyard Stebbins – botanist
- Emphasized the following:
    ⇒ Importance of populations as units of evolution
    ⇒ The central role of natural selection as the primary mechanism of evolutionary change
    ⇒ Gradualism as the explanation of how large changes can result from an accumulation of small changes occurring over long periods of time

Most of Darwin's ideas persisted in the modern synthesis, although many evolutionary biologists are challenging some generalizations of the modern synthesis.

- This debate focuses on the rate of evolution and on the relative importance of evolutionary mechanisms other than natural selection.
- These debates do not question the fact of evolution, only what mechanisms are most important in the process.
- Such disagreements indicate that the study of evolution is very lively and that it continues to develop as a science.

B. The genetic structure of a population is defined by its allele and genotype frequencies

Population = Localized group of organisms which belong to the same species

Species = A group of populations whose individuals have the potential to interbreed and produce fertile offspring in nature

Most species are not evenly distributed over a geographical range, but are concentrated in several localized population centers.

- Each population center is isolated to some extent from other population centers with only occasional gene flow among these groups.
- Obvious examples are isolated populations found on widely separated islands or in unconnected lakes.
- Some populations are not separated by such sharp boundaries.
  - ⇒ For example, a species with two population centers may be connected by an intermediate sparsely populated range.
  - ⇒ Even though these two populations are not absolutely isolated, individuals are more likely to interbreed with others from their population center (see Campbell, Figure 23.2). Gene flow between the two population centers is thus reduced by the intermediate range.

Gene pool = The total aggregate of genes in a population at any one time

- Consists of all the alleles at all gene loci in all individuals of a population. Alleles from this pool will be combined to produce the next generation.
- In a diploid species, an individual may be homozygous or heterozygous for a locus since each locus is represented twice.
- An allele is said to be *fixed* in the gene pool if all members of the population are homozygous for that allele.
- Normally there will be two or more alleles for a gene, each having a relative frequency in the gene pool.

C. The Hardy-Weinberg theorem describes a nonevolving population

> The Hardy-Weinberg model is much easier to teach if the students calculate gene frequencies along with the instructor. This means that you must pause frequently to allow plenty of time for students to actively process the information and practice the calculations.

In the absence of other factors, the segregation and recombination of alleles during meiosis and fertilization will not alter the overall genetic makeup of a population.

- The frequencies of alleles in the gene pool will remain constant unless acted upon by other agents; this is known as the *Hardy-Weinberg theorem*.
- The Hardy-Weinberg model describes the genetic structure of nonevolving populations. This theorem can be tested with theoretical population models.

To test the Hardy-Weinberg theorem, imagine an isolated population of wildflowers with the following characteristics (see Campbell, Figure 23.3):

- It is a diploid species with both pink and white flowers.
- The population size is 500 plants: 480 plants have pink flowers, 20 plants have white flowers.
- Pink flower color is coded for by the dominant allele "A," white flower color is coded for by the recessive allele "a."
- Of the 480 pink-flowered plants, 320 are homozygous (AA) and 160 are heterozygous (Aa). Since white color is recessive, all white flowered plants are homozygous aa.

- There are 1000 genes for flower color in this population, since each of the 500 individuals has two genes (this is a diploid species).
- A total of 320 genes are present in the 160 heterozygotes (Aa): half are dominant (160 A) and half are recessive (160 a).
- 800 of the 1000 total genes are dominant.
- The frequency of the A allele is 80% or 0.8 (800/1000).

Genotypes	# of plants	# of A alleles per individual	Total # A alleles
AA plants	320	× 2 =	640
Aa plants	160		160
		× 1 =	
			800

- 200 of the 1000 total genes are recessive.
- The frequency of the a allele is 20% or 0.2 (200/1000).

Genotypes	# of plants	# of A alleles per individual	Total # A alleles
aa plants	20	× 2 =	40
Aa plants	160		160
		× 1 =	
			200

Assuming that mating in the population is completely random (all male-female mating combinations have equal chances), the frequencies of A and a will remain the same in the next generation.

- Each gamete will carry one gene for flower color, either A or a.
- Since mating is random, there is an 80% chance that any particular gamete will carry the A allele and a 20% chance that any particular gamete will carry the a allele.

The frequencies of the three possible genotypes of the next generation can be calculated using the rule of multiplication (see Campbell, Chapter 14):

- The probability of two A alleles joining is $0.8 \times 0.8 = 0.64$; thus, 64% of the next generation will be AA.
- The probability of two a alleles joining is $0.2 \times 0.2 = 0.04$; thus, 4% of the next generation will be aa.
- Heterozygotes can be produced in two ways, depending upon whether the sperm or ovum contains the dominant allele (Aa or aA). The probability of a heterozygote being produced is thus $(0.8 \times 0.2) + (0.2 \times 0.8) = 0.16 + 0.16 = 0.32$.

The frequencies of possible genotypes in the next generation are 64% AA, 32% Aa and 4% aa.

- The frequency of the A allele in the new generation is 0.64 + (0.32/2) = 0.8, and the frequency of the a allele is 0.04 + (0.32/2) = 0.2. Note that the alleles are present in the gene pool of the new population at the *same* frequencies they were in the original gene pool.

- Continued sexual reproduction with segregation, recombination and random mating would *not alter* the frequencies of these two alleles: the gene pool of this population would be in a state of equilibrium referred to as *Hardy-Weinberg equilibrium.*

- If our original population had not been in equilibrium, only one generation would have been necessary for equilibrium to become established.

From this theoretical wildflower population, a general formula, called the *Hardy-Weinberg equation*, can be derived to calculate allele and genotype frequencies.

- The Hardy-Weinberg equation can be used to consider loci with three or more alleles.

- By way of example, consider the simplest case with only two alleles with one dominant to the other.

- In our wildflower population, let p represent allele A and q represent allele a, thus p = 0.8 and q = 0.2.

- The sum of frequencies from all alleles must equal 100% of the genes for that locus in the population: p + q = 1.

- Where only two alleles exist, only the frequency of one must be known since the other can be derived:

$$1 - p = q \quad \text{or} \quad 1 - q = p$$

When gametes fuse to form a zygote, the probability of producing the AA genotype is p^2; the probability of producing aa is q^2; and the probability of producing an Aa heterozygote is 2pq (remember heterozygotes may be formed in two ways).

- The sum of these frequencies must equal 100%, thus:

$$\underset{\substack{\text{Frequency}\\\text{of AA}}}{p^2} + \underset{\substack{\text{Frequency}\\\text{of Aa}}}{2pq} + \underset{\substack{\text{Frequency}\\\text{of aa}}}{q^2} = 1$$

The Hardy-Weinberg equation permits the calculation of allelic frequencies in a gene pool, if the genotype frequencies are known. Conversely, the genotype can be calculated from known allelic frequencies.

For example, the Hardy-Weinberg equation can be used to calculate the frequency of inherited diseases in humans (e.g., phenylketonuria):

- 1 of every 10,000 babies in the United States is born with phenylketonuria (PKU), a metabolic disorder that, if left untreated, can result in mental retardation.

- The allele for PKU is recessive, so babies with this disorder are homozygous recessive = q^2.

- Thus q^2 = 0.0001, with q = 0.01 (the square root of 0.0001).

- The frequency of p can be determined since p = 1 − q:

$$p = 1 - 0.01 = 0.99$$

- The frequency of carriers (heterozygotes) in the population is 2pq.

$$2pq = 2(0.99)(0.01) = 0.0198$$

- Thus, about 2% of the U.S. population are carriers for PKU.

II. Causes of Microevolution

A. Microevolution is a generation–to–generation change in a population's allele or genotype frequencies

The Hardy-Weinberg equilibrium is important to the study of evolution since it tells us what will happen in a *nonevolving* population.

- This equilibrium model provides a base line from which evolutionary departures take place.
- It provides a reference point with which to compare the frequencies of alleles and genotypes of natural populations whose gene pools may be changing.

If the frequencies of alleles or genotypes deviate from values expected from the Hardy-Weinberg equilibrium, then the population is evolving

- Therefore, a refined definition of evolution at the population level is a generation-to-generation change in a population's frequencies of alleles or genotypes.
- Because such change in a gene pool is evolving on the smallest scale, it is referred to as microevolution.

B. The five causes of microevolution are genetic drift, gene flow, mutation, nonrandom mating, and natural selection

For Hardy-Weinberg equilibrium to be maintained, five conditions *must* be met:

1. Very large population size
2. Isolation from other populations. There is no migration of individuals into or out of the population.
3. No net mutations
4. Random mating
5. No natural selection. All genotypes are equal in survival and reproductive success. Differential reproductive success can alter gene frequencies.

In real populations, several factors can upset Hardy-Weinberg equilibrium and cause *microevolutionary* change.

Microevolution = Small scale evolutionary change represented by a generational shift in a population's relative allelic frequencies.

- Microevolution can be caused by *genetic drift, gene flow, mutation, nonrandom mating*, and *natural selection*; each of these conditions is a deviation from the criteria for Hardy-Weinberg equilibrium.
- Of these five possible agents for microevolution, only natural selection generally leads to an accumulation of favorable adaptations in a population.
- The other four are nonadaptive and are usually called non-Darwinian changes.

1. Genetic drift

Genetic drift = Changes in the gene pool of a small population due to chance

- If a population is small, its existing gene pool may not be accurately represented in the next generation because of sampling error.
- Chance events may cause the frequencies of alleles to drift randomly from generation to generation, since the existing gene pool may not be accurately represented in the next generation.

For example, assume our theoretical wildflower population contains only 25 plants, and the genotypes for flower color occur in the following numbers: 16 AA, 8 Aa and 1 aa. In this case, a chance event could easily change the frequencies of the two alleles for flower color (see Campbell, Figure 23.4).

- A rock slide or passing herbivore which destroys three AA plants would immediately change the frequencies of the alleles from A = 0.8 and a = 0.2, to A = 0.77 and a = 0.23.
- Although this change does not seem very drastic, the frequencies of the two alleles were changed by a chance event.

The larger the population, the less important is the effect of genetic drift.

- Even though natural populations are not infinitely large (in which case genetic drift could be completely eliminated as a cause of microevolution), most are so large that the effect of genetic drift is negligible.
- However, some populations are small enough that genetic drift can play a major role in microevolution, especially when the population has less than 100 individuals.

Two situations which result in populations small enough for genetic drift to be important are the *bottleneck effect* and the *founder effect*.

a.   The bottleneck effect

The size of a population may be reduced drastically by such natural disasters as volcanic eruptions, earthquakes, fires, floods, etc. which kill organisms nonselectively.

- The small surviving population is unlikely to represent the genetic makeup of the original population.
- Genetic drift which results from drastic reduction in population size is referred to as the *bottleneck effect*.
- By chance some individuals survive. In the small remaining population, some alleles may be overrepresented, some underrepresented, and some alleles may be totally absent (see Campbell, Figure 23.5).
- Genetic drift which has occurred may continue to affect the population for many generations, until it is large enough for random drift to be insignificant.

The bottleneck effect reduces overall genetic variability in a population since some alleles may be entirely absent.

- For example, a population of northern elephant seals was reduced to just 20 individuals by hunters in the 1890's.
  - ⇒ Since this time, these animals have been protected and the population has increased to about 30,000 animals.
  - ⇒ Researchers have found that *no* variation exists in the 24 loci examined from the present population. A single allele has been fixed at each of the 24 loci due to genetic drift by the bottleneck effect.
  - ⇒ This contrasts sharply with the large amount of genetic variation found in southern elephant seal populations which did not undergo the bottleneck effect.
- A lack of genetic variation in South African cheetahs may also have resulted from genetic drift, since the large population was severely reduced during the last ice age and again by hunting to near extinction.

b.   The founder effect

When a few individuals colonize a new habitat, genetic drift is also likely to occur. Genetic drift in a new colony is called the *founder effect*.

- The smaller the founding population, the less likely its gene pool will be representative of the original population's genetic makeup.
- The most extreme example would be when a single seed or pregnant female moves into a new habitat.

- If the new colony survives, random drift will continue to affect allele frequencies until the population reaches a large enough size for its influence to be negligible.
- No doubt, the founder effect was instrumental in the evolutionary divergence of the Galapagos finches.

The founder effect probably resulted in the high frequency of *retinitis pigmentosa* (a progressive form of blindness that affects humans homozygous for this recessive allele) in the human population of Tristan da Cunha, a group of small Atlantic islands.

- This area was colonized by 15 people in 1814, and one must have been a carrier.
- The frequency of this allele is much higher on this island than in the populations from which the colonists came.

Although inherited diseases are obvious examples of the founder effect, this form of genetic drift can alter the frequencies of any alleles in the gene pool.

2. Gene flow

Gene flow = The migration of fertile individuals, or the transfer of gametes, between populations

- Natural populations may gain or lose alleles by gene flow, since they do not have gene pools which are closed systems required for Hardy-Weinberg equilibrium.
- Gene flow tends to reduce between-population differences which have accumulated by natural selection or genetic drift.
- An example of gene flow would be if our theoretical wildflower population was to begin receiving wind blown pollen from an all white-flower population in a neighboring field. This new pollen could greatly increase the frequency of the white flower allele, thus also altering the frequency of the red flower allele.
- Extensive gene flow can eventually group neighboring populations into a single population.

3. Mutations

A new mutation that is transmitted in gametes immediately changes the gene pool of a population by substituting one allele for another.

In our theoretical wildflower population, if a mutation in a white flowered plant caused that plant to begin producing gametes which carried a red flower allele, the frequency of the white flower allele is reduced and the frequency of the red flower allele is increased.

Mutation itself has little quantitative effect on large populations in a single generation, since mutation at any given locus is very rare.

- Mutation rates of one mutation per 10^5 to 10^6 gametes are typical, but vary depending on the species and locus.
- An allele with a 0.50 frequency in the gene pool that mutates to another allele at a rate of 10^{-5} mutations per generation would take 2000 generations to reduce the frequency of the original allele from 0.50 to 0.49.
- The gene pool is effected even less, since most mutations are reversible.
- If a new mutation increases in frequency, it is because individuals carrying this allele are producing a larger percentage of offspring in the population due to genetic drift or natural selection, not because mutation is producing the allele in abundance.

Mutation is important to evolution since it is the original source of genetic variation, which is the raw material for natural selection.

4. **Nonrandom mating**

Nonrandom mating increases the number of homozygous loci in a population, but does not in itself alter frequencies of alleles in a population's gene pool. There are two kinds of nonrandom mating: *inbreeding* and *assortative mating*.

a. **Inbreeding**

Individuals of a population usually mate with close neighbors rather than with more distant members of a population, especially if the members of the population do not disperse widely.

- This violates the Hardy-Weinberg criteria that an individual must choose its mate at random from the population.
- Since neighboring individuals of a large population tend to be closely related, inbreeding is promoted.
- Self-fertilization, which is common in plants, is the most extreme example of inbreeding.

Inbreeding results in relative genotypic frequencies (p^2, 2pq, q^2) that deviate from the frequencies predicted for Hardy-Weinberg equilibrium, but does not in itself alter frequencies of alleles (p and q) in the gene pool.

Self-fertilization in our theoretical wildflower population would increase the frequencies of homozygous individuals and reduce the frequency of heterozygotes.

- Selfing of AA and aa individuals would produce homozygous plants.
- Selfing of Aa plants would produce half homozygotes and half heterozygotes.
- Each new generation would see the proportion of heterozygotes decrease, while the proportions of homozygous dominant and homozygous recessive plants would increase.
- Inbreeding without selfing would also result in a reduction of heterozygotes, although it would take much longer.

One effect of inbreeding is that the frequency of homozygous recessive phenotypes increases.

An interesting thing to note is that even if the phenotypic and genotypic ratios change, the values of p and q do not change in these situations, only the way they are combined. A smaller proportion of recessive alleles are masked by the heterozygous state.

b. **Assortative mating**

Assortative mating is another type of nonrandom mating which results when individuals mate with partners that are like themselves in certain phenotypic characters.

Examples:
- Some toads (*Bufo*) most commonly mate with individuals of the same size (see Campbell, Figure 23.6).
- Snow geese occur in a blue variety and a white variety, with the blue color allele being dominant. Birds prefer to mate with those of their own color, blue with blue and white with white; this results in a lower frequency of heterozygotes than predicted by Hardy-Weinberg.
- Blister beetles (*Lytla magister*) in the Sonoran Desert usually mate with a same-size individual.

5. Natural selection

The Hardy-Weinberg equilibrium condition that all individuals in a population have equal ability to produce viable, fertile offspring is probably never met.

- In any sexually reproducing population, variation among individuals exists and some variants leave more offspring than others.
- *Natural selection* is this differential success in reproduction.

Due to selection, alleles are passed on to the next generation in disproportionate numbers relative to their frequencies in the present generation.

- If in our theoretical wildflower population, white flowers are more visible to herbivores than pink flowers, plants with pink flowers (both AA and Aa) would leave more offspring on the average.
- Genetic equilibrium would be disturbed and the frequency of allele A would increase and the frequency of the a allele would decrease.

Natural selection is the only agent of microevolution which is adaptive, since it accumulates and maintains favorable genotypes.

- Environmental change would result in selection favoring genotypes present in the population which can survive the new conditions.
- Variability in the population makes it possible for natural selection to occur.

III.   Genetic Variation, the Substrate for Natural Selection

Members of a population may vary in subtle or obvious ways. It is the genetic basis of this variation that makes natural selection possible.

A. Genetic variation occurs within and between populations

Darwin considered the slight differences between individuals of a population as raw material for natural selection.

While we are more conscious of the variation among humans, an equal if not greater amount of variation exists among the many plant and animals species.

- Phenotypic variation is a product of inherited genotype and numerous environmental influences.
- Only the genetic or inheritable component of variation can have adaptive impact as a result of natural selection.

1. Variation within populations

Both quantitative and discrete characters contribute to variation within a population.

- Polygenic characters, which vary quantitatively within a population, are responsible for much of the inheritable variation.
  - For example, the height of the individuals in our theoretical wildflower population may vary from very short to very tall with all sorts of intermediate heights.
- Discrete characters, such as pink vs. white flowers, can be classified on an either-or basis, usually because they are determined by only one locus with different alleles that produce distinct phenotypes.

a. Polymorphism

In our wildflower population, the red and white flowers would be referred to as different *morphs* (contrasting forms of a Mendelian character).

A population is referred to as *polymorphic* for a character if two or more morphs are present in noticeable frequencies (see Campbell, Figure 23.7).

Polymorphism is found in human populations not only in physical characters (e.g., presence or absence of freckles) but also in biochemical characters (e.g., ABO blood group).

b. **Measuring genetic variation**

Darwin did not realize the extent of genetic variation in populations, since much of the genetic variation can only be determined with biochemical methods.

- Electrophoresis has been used to determine genetic variation among individuals of a population. This technique allows researchers to identify variations in protein products of specific gene loci.

- Electrophoretic studies show that in *Drosophila* populations the gene pool has two or more alleles for about 30% of the loci examined, and each fly is heterozygous at about 12% of its loci.

- Thus, there are about 700 – 1200 heterozygous loci in each fly. Any two flies in a *Drosophila* population will differ in genotype at about 25% of their loci.

- Electrophoretic studies also show comparable variation in the human population.

- Note that electrophoresis underestimates genetic variation:

  - Proteins produced by different alleles may vary in amino acid composition and still have the same overall charge, which makes them indistinguishable by electrophoresis.

  - Also, DNA variation not expressed as protein is not detected by electrophoresis.

2. **Variation between populations**

Geographical variation in allele frequencies exists among populations of most species.

- Natural selection can contribute to geographical variation, since at least some environmental factors are different between two locales. For example, one population of our wildflowers may have a higher frequency of white flowers because of the prevalence in that area of pollinators that prefer white flowers.

- Genetic drift may cause chance variations among different populations.

- Also, subpopulations may appear within a population due to localized inbreeding resulting from a "patchy" environment.

Cline = One type of geographical variation that is a graded change in some trait along a geographic transect

- Clines may result from a gradation in some environmental variable.

- It may be a graded region of overlap where individuals of neighboring populations interbreed.

- For example, the average body size of many North American mammal species gradually increases with increasing latitude. It is presumed that the reduced surface area to volume ratio associated with larger size helps animals in cold environments conserve body heat.

- Studies of geographical variation confirm that genetic variation affects spatial differences of phenotypes in some clines. For example, yarrow plants are shorter at higher elevations, and some of this phenotypic variation has a genetic basis (see Campbell, Figure 23.8)

B. Mutation and sexual recombination generate genetic variation

Two random processes, mutation and sexual recombination (see Chapter 15), create variation in the genetic composition of a population.

1. Mutation

Mutations produce new alleles. They are rare and random events which usually occur in somatic cells and are thus not inheritable.

- Only mutations that occur in cell lines which will produce gametes can be passed to the next generation.
- Geneticists estimate that only an average of one or two mutations occur in each human gamete-producing cell line.

Point mutation = Mutation affecting a single base in DNA

- Much of the DNA in eukaryotes does not code for proteins, and it is uncertain how a point mutation in these regions affect an organism.
- Point mutations in structural genes may cause little effect, partly due to the redundancy of the genetic code.

Mutations that alter a protein enough to affect the function are more often harmful than beneficial, since organisms are evolved products shaped by selection and a chance change is unlikely to improve the genome.

- Occasionally, a mutant allele is beneficial, which is more probable when environmental conditions are changing.
- The mutation which allowed house flies to be resistant to DDT was present in the population and under normal conditions resulted in reduced growth rate. It became beneficial to the house fly population only after a new environmental factor (DDT) was introduced and tipped the balance in favor of the mutant alleles.

Chromosomal mutations usually affect many gene loci and tend to disrupt an organism's development.

- On rare occasions, chromosomal rearrangement may be beneficial. These instances (usually by translocation) may produce a cluster of genes with cooperative functions when inherited together.

Duplication of chromosome segments is nearly always deleterious.

- If the repeated segment does not severely disrupt genetic balance, it may persist for several generations and provide an expanded genome with extra loci.
- These extra loci may take on new functions by mutation while the original genes continue to function.
- Shuffling of exons within the genome (single locus or between loci) may also produce new genes.

Mutation can produce adequate genetic variation in bacteria and other microorganisms which have short generation times.

- Some bacteria reproduce asexually by dividing every 20 minutes, and a single cell can produce a billion descendants in only 10 hours.
- With this type of reproduction, a beneficial mutation can increase in frequency in a bacterial population very rapidly.
- A bacterial cell with a mutant allele which makes it antibiotic resistant could produce an extremely large population of clones in a short period, while other cells without that allele are eliminated.
- Although bacteria reproduce primarily by asexual means, most increase genetic variation by occasionally exchanging and recombining genes through processes such as conjugation, transduction and transformation.

2. Sexual recombination

The contribution of mutations to genetic variation is negligible.

- Mutations are so infrequent at a single locus that they have little effect on genetic variation in a large gene pool.
- Although mutations produce new alleles, nearly all genetic variation in a population results from new combinations of alleles produced by sexual recombination.

Gametes from each individual vary extensively due to crossing over and random segregation during meiosis.

- Thus, each zygote produced by a mating pair possesses a unique genetic makeup.
- Sexual reproduction produces new combinations of old alleles each generation.

Plants and animals depend almost entirely on sexual recombination for genetic variation which makes adaptation possible.

C. Diploidy and balanced polymorphism preserve variation

Natural selection tends to produce genetic uniformity in a population by eliminating unfavorable genotypes. This tendency is opposed by several mechanisms that preserve or restore variation.

1. Diploidy

Diploidy hides much genetic variation from selection by the presence of recessive alleles in heterozygotes.

- Since recessive alleles are not expressed in heterozygotes, less favorable or harmful alleles may persist in a population.
- This variation is only exposed to selection when two heterozygotes mate and produce offspring homozygous for the recessive allele.
- If a recessive allele has a frequency of 0.01 and its dominant counterpart 0.99, then 99% of the recessive allele copies will be protected in heterozygotes. Only 1% of the recessive alleles will be present in homozygotes and exposed to selection.
- The more rare the recessive allele, the greater its protection by heterozygosity. That is, a greater proportion are hidden in heterozygotes by the dominant allele.
- This type of protection maintains a large pool of alleles which may be beneficial if conditions change.

2. Balanced polymorphism

Selection may also preserve variation at some gene loci.

Balanced polymorphism = The ability of natural selection to maintain diversity in a population

One mechanism by which selection preserves variation is *heterozygote advantage*.

- Natural selection will maintain two or more alleles at a locus if heterozygous individuals have a greater reproductive success than any type of homozygote.
- An example is the recessive allele that causes sickle-cell anemia in homozygotes. The locus involved codes for one chain of hemoglobin.
- Homozygotes for this recessive allele develop sickle-cell anemia which is often fatal.

- Heterozygotes are resistant to malaria. Heterozygotes thus have an advantage in tropical areas where malaria is prevalent, since homozygotes for the dominant allele are susceptible to malaria and homozygous recessive individuals are incapacitated by the sickle-cell condition.
- In some African tribes from areas where malaria is common, 20% of the hemoglobin loci in the gene pool is occupied by the recessive allele.

Other examples of heterozygote advantage are found in crop plants (e.g., corn) where inbred lines become homozygous at more loci and show stunted growth and sensitivity to diseases.

- Crossbreeding different inbred varieties often produces hybrids which are more vigorous than the parent stocks.
- This *hybrid vigor* is probably due to:
  1. Segregation of harmful recessives that were homozygous in the inbred varieties.
  2. Heterozygote advantage at many loci in the hybrids.

Balanced polymorphism can also result from patchy environments where different phenotypes are favored in different subregions of a populations geographic range (see Campbell, Figure 23.9).

Frequency-dependent selection also causes balanced polymorphism (see Campbell, Figure 23.10).

- In this situation the reproductive success of any one morph declines if that phenotype becomes too common in the population.
- For example, in *Papilio dardanus*, an African swallowtail butterfly, males have similar coloration but females occur in several morphs.
- The female morphs resemble other butterfly species which are noxious to bird predators. *Papilio* females are not noxious, but birds avoid them because they look like distasteful species.
- This type of protective coloration (*Batesian mimicry*) would be less effective if all the females looked like the same noxious species, because birds would encounter good-tasting mimics as often as noxious butterflies and would not associate a particular color pattern with bad taste.

3. Neutral variation

Some genetic variations found in populations confer no selective advantage or disadvantage. They have little or no impact on reproductive success. This type of variation is called *neutral variation*.

- Much of the protein variation found by electrophoresis is adaptively neutral.
- For example, 99 known mutations affect 71 of 146 amino acids in the beta hemoglobin chain in humans. Some, like the sickle-cell anemia allele, affect the reproductive potential of an individual, while others have no obvious effect.
- The *neutral theory* of molecular evolution states than many variant alleles at a locus may confer no selective advantage or disadvantage.
- Natural selection would not affect the relative frequencies of neutral variations. Frequency of some neutral alleles will increase in the gene pool and others will decrease *due the chance effects of genetic drift*.

Variation in DNA which does not code for proteins may also be nonadaptive.

- Most eukaryotes contain large amounts of DNA in their genomes which have no known function. Such noncoding DNA can be found in varying amounts in closely related species.

- Some scientists speculate that noncoding DNA has resulted from the inherent capacity of DNA to replicate itself and has expanded to the tolerance limits of the each species. The entire genome could exist as a consequence of being self-replicating rather than by providing an adaptive advantage to the organism.

- Transposons might fit this definition of "selfish DNA," although the degree of influence these sequences have on the evolution of genomes is not known.

Evolutionary biologists continue to debate how much variation, or even whether variation, is neutral.

- It is easy to show that an allele is deleterious to an organism.

- It is not easily shown that an allele provides no benefits, since such benefits may occur in immeasurable ways.

- Also, a variation may appear to be neutral under one set of environmental conditions and not neutral under other conditions.

We cannot know how much genetic variation is neutral, but if even a small portion of a population's genetic variation significantly affects the organisms, there is still a tremendous amount of raw material for natural selection and adaptive evolution.

IV. Natural Selection as the Mechanisms of Adaptive Evolution

Adaptive evolution results from a combination of:

- Chance events that produce new genetic variation (e.g., mutation and sexual recombination)

- Natural selection that favors propagation of some variations over others

A. Evolutionary fitness is the relative contribution an individual makes to the gene pool of the next generation

Darwinian fitness is measured by the relative contribution an individual makes to the gene pool of the next generation.

- It is not a measure of physical and direct confrontation, but of the success of an organism in producing progeny.

- Organisms may produce more progeny because they are more efficient feeders, attract more pollinators (as in our wildflowers), or avoid predators.

Survival does not guarantee reproductive success, since a sterile organism may outlive fertile members of the population.

- A long life span may increase fitness if the organism reproduces over a longer period of time (thus leaving more offspring) than other members of the population.

- Even if all members of a population have the same life span, those that mature early and thus have a longer reproductive time span, have increased their fitness.

- Every aspect of survival and fecundity are components of fitness.

Relative fitness = The contribution of a genotype to the next generation compared to the contributions of alternative genotypes for the same locus

- For example, if pink flower plants (AA and Aa) in our wildflower population produce more offspring than white flower plants (aa), then AA and Aa genotypes have a higher relative fitness.

Statistical estimates of fitness can be produced by the relative measure of selection *against* an inferior genotype. This measure is called the *selection coefficient*.

- For comparison, relative fitness of the most fecund variant (AA or Aa in our wildflower population) is set at 1.0.

- If white flower plants produce 80% as many progeny on average, then the white variant relative fitness is 0.8.
- The selection coefficient is the difference between these two values (1.0 − 0.8 = 0.2).
- The more disadvantageous the allele, the greater the selection coefficient.
- Selection coefficients can range up to 1.0 for a lethal allele.

The rate of decline in relative frequencies of deleterious alleles in a population depends on the magnitude of the selection coefficient working against it and whether the allele is dominant or recessive to the more successful allele.

- Deleterious recessives are normally protected from elimination by heterozygote protection.
- Selection against harmful dominant alleles is faster since they are expressed in heterozygotes.

The rate of increase in relative frequencies of beneficial alleles is also affected by whether it is a dominant or recessive.

- New recessive mutations spread slowly in a population (even if beneficial) because selection can not act in its favor until the mutation is common enough for homozygotes to be produced.
- New dominant mutations that are beneficial increase in frequency faster since even heterozygotes benefit from the allele's presence (for example, the mutant dark color producing allele in peppered moths).

Most new mutations, whether harmful or beneficial, probably disappear from the gene pool early due to genetic drift.

Selection acts on phenotypes, indirectly adapting a population to its environment by increasing or maintaining favorable genotypes in the gene pool.

- Since it is the phenotype (physical traits, metabolism, physiology, and behavior) which is exposed to the environment, selection can only act indirectly on genotypes.

The connection between genotype and phenotype may not be as simple and definite as with our wildflower population where pink was dominant to white.

- *Pleiotropy* (the ability of a gene to have multiple effects) often clouds this connection. The overall fitness of a genotype depends on whether its beneficial effects exceed any harmful effects on the organism's reproductive success.
- Polygenic traits also make it difficult to distinguish the phenotype-genotype connection. Whenever several loci influence the same characteristic, the members of the population will not fit into definite categories, but represent a continuum along a range.

An organism is an integrated composite of many phenotypic features, and the fitness of a genotype at any one locus depends upon the entire genetic context. A number of genes may work cooperatively to produce related phenotypic traits.

B. The effect of selection on a varying characteristic can be stabilizing, directional, or diversifying

The frequency of a heritable characteristic in a population may be affected in one of three different ways by natural selection, depending on which phenotypes are favored (see Campbell, Figure 23.11)

1. Stabilizing selection

Stabilizing selection favors intermediate variants by selecting against extreme phenotypes.

- The trend is toward reduced phenotypic variation and greater prevalence of phenotypes best suited to relatively stable environments.

- For example, human birth weights are in the 3 − 4 kg range. Much smaller and much higher birth weight babies have a greater infant mortality.

2. Directional selection

Directional selection favors variants of one extreme. It shifts the frequency curve for phenotypic variations in one direction toward rare variants which deviate from the average of that trait.

- This is most common when members of a species migrate to a new habitat with different environmental conditions or during periods of environmental change.
- For example, fossils show the average size of European black bears increased after periods of glaciation, only to decrease during warmer interglacial periods.

3. Diversifying selection

In diversifying selection, opposite phenotypic extremes are favored over intermediate phenotypes.

- This occurs when environmental conditions are variable in such a way that extreme phenotypes are favored.
- For example, balanced polymorphism of *Papilio* where butterflies with characteristics between two noxious model species (thus not favoring either) gain no advantage from their mimicry.

C. Sexual selection may lead to pronounced secondary differences between the sexes

Sexual dimorphism = Distinction between the secondary sexual characteristics of males and females

- Often seen as differences in size, plumage, lion manes, deer antlers, or other adornments in males.
- In vertebrates it is usually the male that is the "showier" sex.
- In some species, males use their secondary sexual characteristics in direct competition with other males to obtain female mates (especially where harem building is common). These males may defeat other males in actual combat, but more often they use ritualized displays to discourage male competitors.

Darwin viewed sexual selection as a separate selection process leading to sexual dimorphism.

- These enhanced secondary sexual characteristics usually have no adaptive advantage other than attracting mates.
- However, if these adornments increase a males ability to attract more mates, his reproductive success is increased and he contributes more to the gene pool of the next generation.

The evolutionary outcome is usually a compromise between natural selection and sexual selection.

- In some cases the line between these two types of selection is not distinct, as in male deer.
- A stag may use his antlers to defend himself from predators and also to attract females.

Every time a female chooses a mate based particular phenotypic traits, she perpetuates the alleles that caused her to make that choice and allows the male with a particular phenotype to perpetuate his alleles (see Campbell, Figure 23.12).

D. Natural selection cannot fashion perfect organisms

Natural selection *cannot* breed perfect organisms because:

1. *Organisms are locked into historical constraints.* Each species has a history of descent with modification from ancestral forms.

   - Natural selection modifies existing structures and adapts them to new situations, it does not rebuild organisms.

   - For example, back problems suffered by some humans are in part due to the modification of a skeleton and musculature from the anatomy of four-legged ancestors which are not fully compatible to upright posture.

2. *Adaptations are often compromises.*

   - Each organism must be versatile enough to do many different things.

   - For example, seals spend time in the water and on rocks; they could walk better with legs, but swim much better with flippers.

   - Prehensile hands and flexible limbs allow humans to be very versatile and athletic, but they also make us prone to sprains, torn ligaments, and dislocations. Structural reinforcement would prevent many of these disabling occurrences but would limit agility.

3. *Not all evolution is adaptive.*

   - Genetic drift probably affects the gene pool of populations to a large extent.

   - Alleles which become fixed in small populations formed by the founder effect may not be better suited for the environment than alleles that are eliminated.

   - Similarly, small surviving populations produced by bottleneck effect may be no better adapted to the environment or even less well adapted than the original population.

4. Selection can only edit variations that exist.

   - These variations may not represent ideal characteristics.

   - New genes are not formed by mutation on demand.

These limitations allow natural selection to operate on a "better than" basis and subtle imperfections are the best evidence for evolution.

REFERENCES

Campbell, N., et al. *Biology.* 5th ed. Menlo Park, California: Benjamin/Cummings, 1998.

Doolittle, W.F. and C. Sapienza. "Selfish Genes, the Phenotype Paradigm and Genome Evolution." *Nature,* 284: 601-603, 1980.

Mettler, L.E., T.G. Gregg, and H.E. Schaffer. *Population Genetics and Evolution.* 2nd ed. Englewood Ciiffs, New Jersey: Prentice Hall, 1988.

Milkman, R. *Perspectives on Evolution.* Sinauer, Massachusetts: Sunderland, 1982.

CHAPTER 24
THE ORIGIN OF SPECIES

OUTLINE

I. What Is a Species?
    A. The biological species concept emphasizes reproductive isolation
    B. Prezygotic and postzygotic barriers isolate the gene pools of biological species
    C. The biological species concept does not work in all situations
    D. Other species concepts emphasize features and processes that identify and unite species members

II. Modes of Speciation
    A. Geographical isolation can lead to the origin of species: allopatric speciation
    B. A new species can originate in the geographical midst of the parent species: sympatric speciation
    C. Genetic change in populations can account for speciation
    D. The punctuated equilibrium model has stimulated research on the tempo of speciation

The Origin of Evolutionary Novelty
    A. Most evolutionary novelties are modified versions of older structures
    B. Genes that control development play a major role in evolutionary novelty
    C. An evolutionary trend does not mean that evolution is goal oriented

OBJECTIVES

After reading this chapter and attending lecture, the student should be able to:

1. Distinguish between anagenesis and cladogenesis.
2. Define morphospecies and explain how this concept can be useful to biologists.
3. Define biological species (E. Mayr).
4. Describe some limitations of the biological species concept.
5. Explain how gene flow between closely related species can be prevented.
6. Distinguish between prezygotic and postzygotic isolating mechanisms.
7. Describe five prezygotic isolating mechanisms and give an example of each.
8. Explain why many hybrids are sterile.
9. Explain, in their own words, how hybrid breakdown maintains separate species even if gene flow occurs.
10. Distinguish between allopatric and sympatric speciation.
11. Explain, in their own words, the allopatric speciation model and describe the role of intraspecific variation and geographical isolation.

12. Explain why peripheral isolates are susceptible if geographic barriers arise.
13. Describe the adaptive radiation model and use it to describe how it might be possible to have many sympatric closely related species even if geographic isolation is necessary for them to evolve.
14. Define sympatric speciation and explain how polyploidy can cause reproductive isolation.
15. Distinguish between autopolyploidy and allopolyploidy.
16. List some points of agreement and disagreement between the two schools of thought about the tempo of speciation (gradualism vs. punctuated equilibrium).
17. Describe the origins of evolutionary novelty.

KEY TERMS

macroevolution	postzygotic barriers	adaptive radiation	allometric growth
speciation	morphological species concept	polyploidy	heterochrony
anagenesis	recognition species concept	autopolyploid	homeosis
phyletic evolution	cohesion species concept	allopolyploid	
cladogenesis	ecological species concept	hybrid zone	
branching evolution	evolutionary species concept	punctuated equilibrium	
species	allopatric speciation	exaptation	
prezygotic barriers	sympatric speciation	paedomorphosis	

LECTURE NOTES

Evolutionary theory must explain *macroevolution*, the origin of new taxonomic groups (e.g., new species, new genera, new families). *Speciation*, or the origin of new species, is a central process of macroevolution because any genus, family, or higher taxon originates with a new species novel enough to be the first member of the higher taxon

The fossil record provides evidence for two patterns of speciation: *anagenesis* and *cladogenesis* (see Campbell, Figure 24.1).

- *Anagenesis (phyletic evolution)* = The transformation of an unbranched lineage of organisms, sometimes to a state different enough from the ancestral population to justify renaming it as a new species.

- *Cladogenesis (branching evolution)* = The budding of one or more new species from a parent species that continues to exist; is more important than anagenesis in life's history, because it is more common and can promote biological diversity.

I.   **What Is a Species?**

Species = Latin term meaning "kind" or "appearance

Linnaeus (founder of modern taxonomy) described species in terms of their physical form (morphology). Morphology is still the most common method used for describing species.

Modern taxonomists also consider genetic makeup and functional and behavior features when describing species.

A. **The biological species concept emphasizes reproductive isolation**

In 1942, Ernst Mayr proposed the *biological species concept.*

Biological species = A population or group of populations whose members have the potential to interbreed with one another in nature and to produce viable, fertile offspring, but cannot produce viable, fertile offspring with members of other species (see Campbell, Figure 24.2).

- Is the largest unit of population in which gene flow is possible
- Is defined by reproductive isolation from other species in *natural* environments (hybrids may be possible between two species in the laboratory or in zoos)

B. Prezygotic and postzygotic barriers isolate the gene pools of biological species

Any factor that impedes two species from producing viable, fertile hybrids contributes to reproductive isolation.

- Most species are genetically sequestered from other species by more than one type of reproductive barrier.
- Only intrinsic biological barriers to reproduction will be considered here. Geographic segregation (even though it prevents interbreeding) will not be considered.
- Reproductive barriers prevent interbreeding between closely related species.

The various reproductive barriers which isolate the gene pools of species are classified as either prezygotic or postzygotic, depending on whether they function before or after the formation of zygotes.

- *Prezygotic barriers* impede mating between species or hinder fertilization of the ova should members of different species attempt to mate.
- In the event fertilization does occur, *postzygotic barriers* prevent the hybrid zygote from developing into a viable, fertile adult.

1. Prezygotic barriers

a. Habitat isolation

Two species living in different habitats within the same area may encounter each other rarely, if at all, even though they are not technically geographically isolated.

- For example, two species of garter snakes (*Thamnophis*) occur in the same areas but for intrinsic reasons, one species lives mainly in water and the other is mainly terrestrial.
- Since these two species live primarily in separate habitats, they seldom come into contact as they are ecologically isolated.

c. Behavioral isolation

Species-specific signals and elaborate behavior to attract mates are important reproductive barriers among closely related species.

- Male fireflies of different species signal to females of the same species by blinking their lights in a characteristic pattern; females discriminate among the different signals and respond only to flashes of their own species by flashing back and attracting the males.

Many animals recognize mates by sensing pheromones (distinctive chemical signals).

- Female Gypsy moths attract males by emitting a volatile compound to which the olfactory organs of male Gypsy moths are specifically tuned: when a male detects this pheromone, it follows the scent to the female.
- Males of other moth species do not recognize this chemical as a sexual attractant.

Other factors may also act as behavioral isolating mechanisms:

- Eastern and western meadowlarks are almost identical in shape, coloration, and habitat, and their ranges overlap in the central United States (see Campbell, Figure 24.2a).

- They retain their biological species integrity partly because of the difference in their songs, which enables them to recognize potential mates as members of their own kind.

Another form of behavioral isolation is courtship ritual specific to a species (see Campbell, Figure 24.3).

c. Temporal isolation

Two species that breed at different times of the day, seasons, or years cannot mix their gametes.

- For example, brown trout and rainbow trout cohabit the same streams, but brown trout breed in the fall and rainbow trout breed in the spring.
- Since they breed at different times of the year, their gametes have no opportunity to contact each other and reproductive isolation is maintained.

d. Mechanical isolation

Anatomical incompatibility may prevent sperm transfer when closely related species attempt to mate.

- For example, male dragonflies use a pair of special appendages to clasp females during copulation. When a male tries to mount a female of a different species, he is unsuccessful because his clasping appendages do not fit the female's form well enough to grip securely.
- In plants that are pollinated by insects or other animals, the floral anatomy is often adapted to a specific pollinator that transfers pollen only among plants of the same species.

e. Gametic isolation

Gametes of different species that meet rarely fuse to form a zygote.

- For animals that use internal fertilization, the sperm of one species may not be able to survive the internal environment of the female reproductive tract of a different species.
- Cross-specific fertilization is also uncommon for animals that utilize external fertilization due to a lack of gamete recognition.

Gamete recognition may be based on the presence of specific molecules on the coats around the egg which adhere only to complementary molecules on sperm cells of the same species.

- Similar mechanisms of molecular recognition enables a flower to discriminate between pollen of the same species and pollen of different species.

2. Postzygotic barriers

When prezygotic barriers are crossed and a hybrid zygote forms, one of several postzygotic barriers may prevent development of a viable, fertile hybrid.

a. Reduced hybrid viability

Genetic incompatibility between the two species may abort development of the hybrid at some embryonic stage.

- For example, several species of frogs in the genus *Rana* live in the same regions and habitats.
- They occasionally hybridize but the hybrids generally do not complete development, and those that do are frail and soon die.

b. Reduced hybrid fertility

If two species mate and produce hybrid offspring that are viable, reproductive isolation is intact if the hybrids are sterile because genes cannot flow from one species' gene pool to the other.

- One cause of this barrier is that if chromosomes of the two parent species differ in number or structure, meiosis cannot produce normal gametes in the hybrid.
- The most familiar case is the mule which is produced by crossing a donkey and a horse; very rarely are mules able to backbreed with either parent species (see Campbell, Figure 24.4).

c. Hybrid breakdown

When some species cross-mate, the first generation hybrids are viable and fertile, but when these hybrids mate with one another or with either parent species, offspring of the next generation are feeble or sterile.

- For example, different cotton species can produce fertile hybrids, breakdown occurs in the next generation when progeny of the hybrids die in their seeds or grow into weak defective plants.

Campbell, Figure 24.5 summarizes the reproductive barriers between closely related species.

C. The biological species concept does not work in all situations

The biological species concept cannot be applied to:

- Organisms that are completely asexual in their reproduction. Some protists and fungi, some commercial plants (bananas), and many bacteria are exclusively asexual.
  - ⇒ Asexual reproduction effectively produces a series of clones, which genetically speaking, represent a single organism.
  - ⇒ Asexual organisms can be assigned to species only by grouping clones with the same morphology and biochemical characteristics.
- Extinct organisms represented only by fossils. These must be classified by the morphospecies concept.

In some cases, unambiguous determination of species is not possible, even though the populations are sexual, contemporaneous, and contiguous.

- Four phenotypically distinct populations of the deer mouse (*Peromyscus maniculatus*) found in the Rocky Mountains are geographically isolated and referred to as *subspecies*. (see Campbell, Figure 24.6)
- These populations overlap at certain locations and some interbreeding occurs in these areas of cohabitation, which indicates they are the same species by the biological species criteria.
- Two subspecies (*P. m. artemisiae* and *P. m. nebrascensis*) are an exception, since they do not interbreed in the area of cohabitation. However, their gene pools are not completely isolated since they freely interbreed with other neighboring populations.
- This circuitous route could only produce a very limited gene flow, but the route is open and possible between the populations of *P. m. artemisiae* and *P. m. nebrascensis* through the other populations.

- If this route was closed by extinction or geographic isolation of the neighboring populations, then *P. m. artemisiae* and *P. m. nebrascensis* could be named separate species without reservation.

More examples are being discovered where there is a blurry distinction between populations with limited gene flow and full biological species with segregated gene pools.

- If two populations cannot interbreed when in contact, they are clearly distinct species.
- When there is gene flow (even very limited) between two populations that are in contact, it is difficult to apply the biological species concept.
- This is equivalent to finding two populations at different stages in their evolutionary descent from a common ancestor, which is to be expected if new species arise by gradual divergence of populations.

Other species concepts have been developed in an effort to accommodate the dynamic, quantitative aspects of speciation; however, the species problem may never be completely resolved as it is unlikely that a single definition of species will apply to all cases.

D. Other species concepts emphasize features and processes that identify and unite species members

The *morphological species concept* characterizes species on the basis of measurable physical features.

- Can be useful in the field
- Sometimes difficult to apply (e.g., Do physical differences between a set of organisms represent species differences or phenotypic variation within a species?)

In the *recognition species concept*, a species is defined by a unique set of characteristics that maximize successful mating.

- Characteristics may be molecular, morphological, or behavioral in nature
- Characteristics are subject to natural selection

The *cohesion species concept* relies on the mechanisms that maintain species as discrete phenotypic entities.

- Mechanisms may include reproductive barriers, stabilizing selection, and linkages among sets of genes that make a zygote develop into a an adult organism with species-specific characteristics (e.g., sexual reproduction)
- This concept acknowledges that interbreeding between some species produces fertile hybrids (e.g., corn)

The *ecological species concept* defines a species on the basis of where they live and what they do.

The *evolutionary species concept* defines a species as a sequence of ancestral and descendent populations that are evolving independently of other such groups

- Each evolutionary species has its own unique role in the environment; roles are influenced by natural selection.

Campbell, Table 24.1 reviews the species concepts.

II.   Modes of Speciation

Reproductive barriers form boundaries around species, and the evolution of these barriers is the key biological event in the origin of new species.

- An essential episode in the origin of a species occurs when the gene pool of a population is separated from other populations of the parent species.

- This genetically isolated splinter group can then follow its own evolutionary course, as changes in allele frequencies caused by selection, genetic drift, and mutations occur undiluted by gene flow from other populations.

There are two general modes of speciation: allopatric speciation and sympatric speciation.

A. **Geographical isolation can lead to the origin of species: allopatric speciation**

1. **Geographic barriers**

   Allopatric speciation = Speciation that occurs when the initial block to gene flow is a geographical barrier that physically isolates the population

   - Geological processes can fragment a population into two or more allopatric populations (having separate ranges).
     ⇒ Such occurrences include emergence of mountain ranges, movement of glaciers, formation of land bridges, subsidence of large lakes.
     ⇒ Also small populations may become geographically isolated when individuals from the parent population travel to a new location.
   - The extent of development of a geographical barrier necessary to isolate two populations depends on the ability of the organisms to disperse due to the mobility of animals or the dispersibility of spores, pollen and seeds of plants.
     ⇒ For example, the Grand Canyon is an impassable barrier to small rodents, but is easily crossed by birds. As a result, the same bird species populate both rims of the canyon, but each rim has several unique species of rodents (see Campbell, Figure 24.7).

   An example of how geographic isolation can result in allopatric speciation is the pupfish.

   - About 50,000 years ago, during an ice age, the Death Valley region of California and Nevada had a rainy climate and a system of interconnecting lakes and rivers.
   - A drying trend began about 10,000 years ago, and by 4000 years ago, the region had become a desert.
   - Presently, isolated springs in deep clefts between rocky walls are the only remnants of the lake and river networks. Living in many of these isolated springs are small pupfishes (*Cyprinodon* spp.).
   - Each inhabited spring contains its own species of pupfish which is adapted to that pool and found nowhere else in the world.
   - The endemic pupfish species probably descended from a single ancestral species whose range was fragmented when the region became arid, thus isolating several small populations that diverged in their evolution as they adapted to their spring's environment.

2. **Conditions favoring allopatric speciation**

   When populations become allopatric, speciation can potentially occur as the isolated gene pools accumulate differences by microevolution that may cause the populations to diverge in phenotype.

   - A small isolated population is more likely to change substantially enough to become a new species than is a large isolated population.
   - The geographic isolation (peripheral isolate) of a small population usually occurs at the fringe of the parent population's range.
   - As long as the gene pools are isolated from the parental population, *peripheral isolates* are good candidates for speciation for three reasons:

1. *The gene pool of the peripheral isolate probably differs from that of the parent population from the outset.* Since fringe inhabiters usually represent the extremes of any genotypic and phenotypic clines in an original sympatric population. With a small peripheral isolate, there will be a founder effect with chance resulting in a gene pool that is not representative of the gene pool of the parental population.

2. *Genetic drift will continue to cause chance changes in the gene pool of the small peripheral isolate until a large population is formed.* New mutations or combinations of alleles that are neutral in adaptive value may become fixed in the population by chance alone, causing phenotypic divergence from the parent population.

3. *Evolution caused by selection is likely to take a different direction in the peripheral isolate than in the parental population.* Since the peripheral isolate inhabits a frontier with a somewhat different environment, it will probably be exposed to different selection pressures than those encountered by the parental population.

- Due to the severity of a fringe environment, most peripheral isolates do not survive long enough to speciate.

Although most peripheral isolates become extinct, evolutionary biologists agree that a small population can accumulate enough genetic change to become a new species in only hundreds to thousands of generations.

3. Adaptive radiation on island chains

Allopatric speciation occurs on island chains where new populations, which stray or are passively dispersed from their ancestral populations, evolve in isolation.

Adaptive radiation = The evolution of many diversely adapted species from a common ancestor.

Examples of adaptive radiation are the endemic species of the Galapagos Islands which descended from small populations which floated, flew, or were blown from South America to the islands. Darwin's finches can be used to illustrate a model for such adaptive radiation on island chains (see also, Campbell, Figure 24.8).

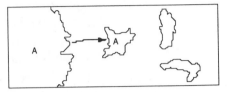
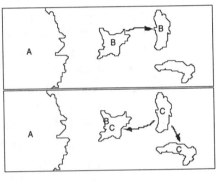
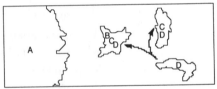
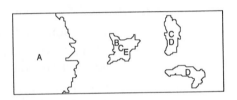

- A single dispersal event may have seeded one island with a peripheral isolate of the ancestral finch which diverged as it underwent allopatric speciation.

- A few individuals of this new species may have reached neighboring islands, forming new peripheral isolates which also speciated (see Campbell, Figure 24.9).

- After diverging on the island it invaded, a new species could re-colonize the island from which its founding population emigrated and coexist with the ancestral species or form still another species.

- Multiple invasions of islands could eventually lead to coexistence of several species on each island since the islands are distant enough from each other to permit geographic isolation, but near enough for occasional dispersal.

Similar evolutionary events have occurred on the Hawaiian Archipelago. These volcanic islands are 3500 km from the nearest continent.

- Hawaii is the youngest (<one million years old), largest island and has active volcanoes.

- The islands grow progressively older in a northwesterly direction away from Hawaii.

- As each island was formed and cooled, flora and fauna carried by ocean and wind currents from other islands and continents became established.

- The physical diversity of each island provided many environmental opportunities for evolutionary divergence by natural selection.

- Multiple invasions and allopatric speciations have permitted such a degree of adaptive radiation that there are thousands of endemic species on the archipelago which are found no where else on Earth.

In contrast to the Hawaiian Archipelago, islands such as the Florida Keys are close enough to a mainland to allow free movement from the island to the mainland.

- Such islands are not characterized by endemic species since there is no long-term isolation of founding populations.

- Intrinsic reproductive barriers that block gene flow do not develop due to a steady influx of immigrants from the mainland parental populations.

B. A new species can originate in the geographical midst of the parent species: sympatric speciation

Sympatric speciation = Formation of new species within the range of parent populations

- Reproductive isolation evolves without geographical isolation.
- This can occur quickly (in one generation) if a genetic change results in a reproductive barrier between the mutants and the parent population.

Many plant species have originated from improper cell division that results in extra sets of chromosomes—a mutant condition called *polyploidy* (see Campbell, Figure 24.10)

Depending on the origin of the extra set of chromosomes, polyploids are classified in two forms: autopolyploids and allopolyploids.

Autopolyploid = An organism that has more than two chromosome sets, all derived from a single species. For example,

- Nondisjunction in the germ cell line (in either mitosis or meiosis) results in diploid gametes.
- Self-fertilization would double the chromosome number to the tetraploid state.
- Tetraploids can self-pollinate or mate with other tetraploids.
- The mutants cannot interbreed with diploids of the parent population because hybrids would be triploid ($3n$) and sterile due to impaired meiosis from unpaired chromosomes.
- An instantaneous special genetic event would thus produce a postzygotic barrier which isolates the gene pool of the mutant in just one generation.
- Sympatric speciation by autopolyploidy was first discovered by Hugo De Vries in the early 20th century while working with *Oenothera*, the evening primrose.

Allopolyploid = A polyploid hybrid resulting from contributions by two different species.

- More common than autopolyploidy.
- Potential evolution of an allopolyploid begins when two different species interbreed and a hybrid is produced (see Campbell, Figure 24.10b).
- Such interspecific hybrids are usually sterile, because the haploid set of chromosomes from one species cannot pair during meiosis with the haploid set of chromosomes from the second species.
- These sterile hybrids may actually be more vigorous than the parent species and propagate asexually.

At least two mechanisms can transform sterile allopolyploid hybrids into fertile polyploids:

1. During the history of the hybrid clone, mitotic nondisjunction in the reproductive tissue may double the chromosome number (see Campbell, Figure 24.10b).
   - The hybrid clone will then be able to produce gametes since each chromosome will have a homologue to synapse with during meiosis.
   - Gametes from this fertile tetraploid could unite and produce a new species of interbreeding individuals, reproductively isolated from both parent species.
2. Meiotic nondisjunction in one species produces an unreduced (diploid) gamete.

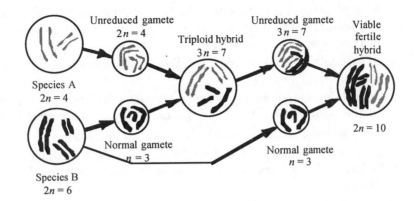

- This abnormal gamete fuses with a normal haploid gamete of a second species and produces a triploid hybrid.
- The triploid hybrid will be sterile, but may propagate asexually.
- During the history of this sterile triploid clone, meiotic nondisjunction again produces an unreduced gamete (triploid).
- Combination of this triploid gamete with a normal haploid gamete from the second parent species would result in a fertile hybrid with homologous pairs of chromosomes.
- his allopolyploid would have a chromosome number equal to the sum of the chromosome numbers of the two ancestral species (as in 1 above).

Speciation of polyploids (especially allopolyploids) has been very important in plant evolution.

- Some allopolyploids are very vigorous because they contain the best qualities of both parent species.
- The accidents required to produce these new plant species (interspecific hybridization coupled with nondisjunction) have occurred often enough that between 25% and 50% of all plant species are polyploids.

Some of these species have originated and spread in relatively recent times and many others are of importance to humans.

- Spartina angelica is a species of salt-marsh grass which evolved as an allopolyploid in the 1870s.
  ⇒ It is derived from a European species (*Spartina maritima*) and an American species (*Spartina alternaflora*).
  ⇒ In addition to being morphologically distinct and reproductively isolated from its parent species, *S. angelica* has a chromosome number ($2n = 122$) indicative of its mechanism of speciation (*S. maritima*, $2n =$, *S. alternaflora*, $2n = 62$).
- *Triticum aestivum*, bread wheat, is a 42-chromosome allopolyploid that is believed to have originated about 8000 years ago as a hybrid of a 28 chromosome cultivated wheat and a 14 chromosome wild grass (see Campbell, Figure 24.11).
- Other important polyploid species include oats, cotton, potatoes, and tobacco.
- Plant geneticists are presently inducing these genetic accidents to produce new polyploids which will combine high yield and disease resistance.

Sympatric speciation may also occur in animal evolution through different mechanisms.

- A group of animals may become isolated within the range of a parent population if genetic factors cause them to become fixed on resources not used by the parent population as a whole. For example,

  ⇒ A particular species of wasp pollinates each species of figs. The wasps mate and lay their eggs in the figs.

  ⇒ A genetic change causing wasps to select a different fig species would segregate mating individuals of the new phenotype from the parental population.

  ⇒ Divergence could then occur after such an isolation.

  ⇒ The great diversity of cichlid fishes in Lake Victoria may have arisen from isolation due to exploitation of different food sources and other resources in the lake.

- Sympatric speciation could also result from a balanced polymorphism combined with assortative mating.

  ⇒ For example, if birds in a population that is dimorphic for beak size began to selectively mate with birds of the same morph, speciation could occur over time.

While both allopatric speciation and sympatric speciation have important roles in plant evolution, allopatric speciation is far more common in animals.

C. Genetic change in populations can account for speciation

Classifying modes of speciation as allopatric or sympatric emphasizes biogeographical factors but does not emphasize the actual genetic mechanisms. An alternative method which takes genetic mechanism into account, groups speciation into two categories: *speciation by adaptive divergence* and *speciation by shifts in adaptive peaks*.

1. Adaptive divergence

Two populations which adapt to different environments accumulate differences in the frequencies of alleles and genotypes.

- During this gradual adaptive divergence of the two gene pools, reproductive barriers may evolve between the two populations.

- Evolution of reproductive barriers would differentiate these populations into two species.

A key point in evolution by divergence is that reproductive barriers can arise without being favored directly by natural selection.

- Divergence of two populations is due to their adaptation to separate environments, with reproductive isolation being a secondary development.

- Postzygotic barriers may be pleiotropic effects of interspecific differences in those genes that control development.

  ⇒ Hybrids may not be viable if both sets of genes for rRNA synthesis are not active (e.g., hybrids between *D. melanogaster* and *D. simulans*).

- Gradual genetic divergence of two populations may also result in the evolution of prezygotic barriers.

  ⇒ For instance, an ecological barrier to inbreeding may secondarily result from the adaptation of an insect population to a new host plant different from the original population's host.

In some isolated populations, reproductive isolation has evolved more directly from sexual selection. For example,

- In *Drosophila heteroneura*, the male's wide head enhances reproductive success with females of the same species while reducing the probability that a male *D. heteroneura* will mate with females of other species.
- Sexual selection, in this case, probably evolved as an adaptation for enhanced reproductive success. A secondary consequence is that it prevents interbreeding with other *Drosophila* species.
- Since reproductive barriers usually evolve when populations are allopatric, they do not function directly to isolate the gene pools of populations.
  ⇒ For this reason, the emphasis on reproductive isolating mechanisms is one criticism of the biological species concept.

2. **Hybrid zones and the cohesion concept of species**

Three possible outcomes are possible when two closely related populations that have been allopatric for some time come back into contact:

- The two populations may interbreed freely.
  ⇒ The gene pools would become incorporated into a single pool indicating that speciation had not occurred during their time of geographical isolation.
- The two populations may not interbreed due to reproductive barriers.
  ⇒ The gene pools would remain separate due to the evolutionary divergence which occurred during the time of geographical isolation. Speciation has taken place.
- A hybrid zone may be established.

Hybrid zone = A region where two related populations that diverged after becoming geographically isolated make secondary contact and interbreed where their geographical ranges overlap

- For example, the red-shafted flicker of western North America and the yellow-shafted flicker of central North America are two phenotypically distinct populations of woodpeckers that interbreed in a hybrid zone stretching from southern Alaska to the Texas panhandle.
- The two populations came into renewed contact a few centuries ago after being separated during the ice ages.
- The hybrid zone is relatively stable and not expanding.
  ⇒ The introgression of alleles between the populations has not penetrated far beyond the hybrid zone, although the two populations have been interbreeding for at least two hundred years.
  ⇒ The genotypic and phenotypic frequencies that distinguish the two populations form steep clines into the hybrid zone.
- Away from the hybrid zone, the two populations remain distinct.

Should two populations which form a hybrid zone be considered subspecies or separate species?

- Some researchers who support species status for such populations, recognize that the presence of stable hybrid zones creates a problem for the biological species concept.
  ⇒ If the taxonomic identity of two species is maintained, even though they hybridize, there must be cohesive forces other than reproductive isolation that maintain the species and prevent their merging into a single species.

- These researchers favor an alternative known as the *cohesion concept of species*.

The cohesion concept of species holds that the cohesion may involve a distinctive, integrated set of adaptations that has been refined during the evolutionary history of a population.

- Phenotypic variation would be restricted by stabilizing selection to a range narrow enough to define the species as separate from other species.
- In the adaptive landscape view:
  ⇒ The red-shafted and yellow-shafted flickers are clustered around different adaptive peaks.
  ⇒ Specific combinations of alleles and specific linkages between gene loci on chromosomes may form a genetic basis for the cohesion of phenotypes.
  ⇒ The clinal change of genetic structure and phenotype noted in the hybrid zone may be correlated with transitions in environmental factors that help shape the two distinct populations.

3. How much genetic change is required for speciation?

No generalizations can be made about genetic distance between closely related species. Reproductive isolation may result from changes in many loci or only in a few.

- Two species of Hawaiian *Drosophila* (*D. silvestris* and *D. heteroneura*) differ at only one locus which determines head shape, an important factor in mate recognition (see Campbell, Figure 24.12).
  ⇒ The phenotypic effect of different alleles at this locus is multiplied by epistasis involving at least ten other loci.
  ⇒ Thus, no more than one mutation was necessary to differentiate the two species.
- Changes in one gene in a coadapted gene complex can substantially impact the development of an organism.

D. The punctuated equilibrium model has stimulated research on the tempo of speciation

Traditional evolutionary trees diagram the descent of species from ancestral forms as branches that gradually diverge with each new species evolving continuously over long spans of time (see Campbell, Figure 24.13a).

- The theory behind such a tree is the extrapolation of microevolutionary processes (allele frequency changes in the gene pool) to the divergence of species.
- Big changes thus occur due to the accumulation of many small changes.

Paleontologists rarely find gradual transitions of fossil forms but often observe species appearing as new forms suddenly in the rock layers.

- These species persist virtually unchanged and then disappear as suddenly as they appeared.
- Even Darwin, who believed species from a common ancestral stock evolve differences gradually, was perplexed by the lack of transitional forms in the fossil record.

Advocates of *punctuated equilibrium* have redrawn the evolutionary tree to represent fossil evidence for evolution occurring in spurts of relatively rapid change instead of gradual divergence (see Campbell, Figure 24.13b).

- This theory was proposed by Niles Eldredge and Stephen Jay Gould in 1972.

- It depicts species undergoing most of their morphological modification as they first separate from the parent species then showing little change as they produce additional species.
- In this theory gradual change is replaced with long periods of stasis punctuated with episodes of speciation.
- The origin of new polyploid plants through genome changes is one mechanism of sudden speciation.
- Allopatric speciation of a splinter population separated from its parent population by geographical barriers may also be rapid.
  ⇒ For a population facing new environmental conditions, genetic drift and natural selection can cause significant change in only a few hundred or thousand generations.

A few thousand generations is considered rapid in reference to the geologic time scale.

- The fossil record indicates that successful species survive for a few million years on average.
- If a species survives for five million years and most of its morphological changes occur in the first 50,000 years, then the speciation episode occurred in just 1% of the species' lifetime.
- With this time scale, a species will appear suddenly in rocks of a certain age, linger relatively unchanged for millions of years, then become extinct.
- While forming, the species may have gradually accumulated modifications, but with reference to its overall history, its formation was sudden.
- An evolutionary spurt preceding a longer period of morphological stasis would explain why paleontologists find so few transitions in the fossils record of a species.

Because "sudden" can refer to thousands of years on the geological time scale, differing opinions of punctuationalists and gradualists about the rate of speciation may be more a function of time perspective than conceptual difference. There is clear disagreement, however, over how much a species changes after its origin.

- In a species adapted to an environment that stays the same, natural selection would counter changes in the gene pool.
  ⇒ Once selection during speciation produces new complexes of coadapted genes, mutations are likely to impose disharmony on the genome and disrupt the development of the organism.
- Stabilizing selection would thus hold a population at one adaptive peak to produce long periods of stasis.

Some gradualists feel that stasis is an illusion since many species may continue to change, after they have diverged from the parent population, in ways undetectable in fossils.

- Changes in internal anatomy, physiology and behavior would go unnoticed by paleontologists as fossils only show external anatomy and skeletons.
- Population geneticists also point out that many microevolutionary changes occur at the molecular level without affecting morphology.

It is obvious that additional extensive studies of fossil morphology where specific lineages are preserved should be carried out to assess the relative importance of gradual and punctuated tempos in the origin of new species.

III. The Origin of Evolutionary Novelty

What processes cause the evolutionary changes that can be traced through the fossil record? How do the novel features that define taxonomic groups above the species level arise? The following concepts address the processes relevant to these questions.

A. Most evolutionary novelties are modified versions of older structures

Higher taxa such as families and classes are defined by evolutionary novelties. For example:

- Birds evolved from dinosaurs, and their wings are homologous to the forelimbs of modern reptiles.
- Birds are adapted to flight, yet their ancestors were earthbound.

How could these new designs evolve?

- One mechanism is the gradual refinement of existing structures for new functions.
- Most biological structures have an evolutionary plasticity that makes alternative functions possible.

Exaptation is a term applied to a structure that evolves in one context and becomes co-opted for another function.

- Natural selection cannot anticipate the future, but it can improve an existing structure in context of its current utility. For example,
  ⇒ The honeycombed bones and feathers of birds did not evolve as adaptations for flight.
  ⇒ They must have been beneficial to the bipedal reptilian ancestors of birds (reduction of weight, gathering food, courtship), and later through modification, became functional for flying.
- Exaptation cannot be proven, but provides an explanation for how novel designs can arise gradually through a series of intermediate stages, each having some function in the organism.
- The evolution of novelties by remodeling of old structures for new functions reflects the Darwinian tradition of large changes being crafted by natural selection through an accumulation of many small changes.

B. Genes that control development play a major role in evolutionary novelty

The evolution of complex structures (e.g., wings) requires such large modifications that changes at many gene loci are probably involved.

- In other cases, relatively few changes in the genome can cause major modifications in morphology (e.g., humans vs. chimpanzees).
- Thus, slight genetic divergence can become magnified into major differences.

In animal development, a system of regulatory genes coordinates activities of structural genes to guide the rate and pattern of development from zygote to adult.

- A slight alteration of development becomes compounded in its effect on adult *allometric growth* (differences in relative rates of growth of various parts of the body) which helps to shape an organism.
- A slight change in these relative rates of growth will result in a substantial change in the adult (see Campbell, Figure 24.14a).
- Thus, altering the parameters of allometric growth is one way relatively small genetic differences can have major morphological impact (see Campbell, Figure 24.14b).

Changes in developmental dynamics, both *temporal* and *spatial*, have played a major role in macroevolution.

Temporal changes in development that create evolutionary novelties are called *heterochrony*.

Heterochrony = Evolutionary changes in the timing or rate of development

- Genetic changes that alter the timing of development can also produce novel organisms.

Paedomorphosis = Retention of ancestral juvenile structures in a sexually mature adult organism

   ⇒ Campbell, Figure 24.15 shows the effect of developmental timing on zebra stripes. Figure 24.16 shows paedomorphosis in salamanders.

   ⇒ A slight change in timing that retards the development of some organs in comparison to others produces a different kind of animal.

- Changes in developmental chronology may have contributed to human evolution.

   ⇒ Humans and chimpanzees are closely related through descent from a common ancestor.

   ⇒ They are much more similar as fetuses than as adults.

   ⇒ Different allometric properties and variations result in the human brain being proportionally larger than that in chimpanzees.

   ⇒ The human brain continues to grow several years longer than the chimpanzee brain.

   ⇒ Thus, the genetic changes responsible for humans are not great, but have profound effects.

Equally important in evolution is the alteration of the spatial pattern of development or homeosis.

Homeosis = Alteration in the placement of different body parts (for example, to the arrangement of different kinds of appendages in animals or the placement of flower parts on a plant)

Since each regulatory gene may influence hundreds of structural genes, there is a potential for evolutionary novelties that define higher taxa to arise much faster than would occur by the accumulation of changes in only structural genes.

C. An evolutionary trend does not mean that evolution is goal oriented

Extracting a single evolutionary progression from the fossil record that is likely to be incomplete is misleading.

- For example, by selecting certain species from available fossils, it is possible to arrange a succession of animals between *Hyracotherium* and modern horses that shows a trend toward increased size, reduced number of toes, and modification of teeth for grazing (see Campbell, Figure 24.17 yellow line). Consideration of all known fossil horses negates this trend, and reveals that the line to modern horses is one of a series of species episodes.

Branching evolution (cladogenesis) can produce a trend even if some new species counter the trend.

- There was an overall trend in reptilian evolution toward large size during the Mesozoic era which eventually produced the dinosaurs.

- This trend was sustained even though some new species were smaller than their parental species.

One view of macroevolution, forwarded by Steven Stanley of Johns Hopkins, holds that species are analogous to individuals.

- In this analogy, speciation is birth and extinction is death.
- An evolutionary trend is produced by *species selection*, which is analogous to the production of a trend within a population by natural selection.
- The species that endure the longest and generate the greatest number of new species determine the direction of major evolutionary trends.
- Differential speciation plays a role in macroevolution similar to the way differential reproduction plays a role in microevolution.

Qualities unrelated to the success of organisms in a specific environment may be equally important in species selection.

- The ability of a species to disperse to new habitats may result in development of new "daughter species" as organisms adapt to new conditions.
- A criticism of species selection is the argument that gradual modification of populations in response to environmental change is the most common stimulus to evolutionary trends.

No intrinsic drive toward a preordained state of being is indicated by the presence of an evolutionary trend.

- Evolution is a response to interactions between organisms and their current environments.
- An evolutionary trend may cease or reverse itself under changing conditions. For example, conditions of the Mesozoic era favored giant reptiles, but by the end of that era, smaller species prevailed.

REFERENCES

Campbell, N., et al. *Biology*. 5th ed. Menlo Park, California: Benjamin/Cummings, 1998.

Futuyma, D.J. *Evolutionary Biology*. 3rd ed. Sunderland, Massachusetts: Sinauer, 1998.

Mayr, E. *Populations, Species and Evolution: An Abridgement of Animal Species and Evolution*. Cambridge, Massachusetts: Harvard University Press, 1970.

CHAPTER 25
TRACING PHYLOGENY

OUTLINE

I. The Fossil Record and Geologic Time
   - A. Sedimentary rocks are the richest source of fossils
   - B. Paleontologists use a variety of methods to date fossils
   - C. The fossil record is a substantial, albeit incomplete, chronicle of evolutionary history
   - D. Phylogeny has a biogeographical basis in continental drift
   - E. The history of life is punctuated by mass extinctions followed by adaptive radiations of the survivors

II. Phylogeny and Systematics
   - A. Taxonomy employs a hierarchical system of classification
   - B. The branching pattern of a phylogenetic tree represents the taxonomic hierarchy
   - C. Determining monophyletic taxa is a key to classifying organisms according to their evolutionary history
   - D. Molecular biology provides powerful new tools for systematics
   - E. The search for fossilized DNA continues despite recent setbacks: *science as a process*

III. The Science of Phylogenetic Systematics
   - A. Phenetics increased the objectivity of systematic analysis
   - B. Cladistic analysis uses novel homologies to define branchpoints on phylogenetic trees
   - C. Phylogenetic systematics relies on both morphology and molecules

OBJECTIVES

After reading this chapter and attending lecture, the student should be able to:
1. Explain the importance of the fossil record to the study of evolution.
2. Describe how fossils form.
3. Distinguish between relative dating and absolute dating.
4. Explain how isotopes can be used in absolute dating.
5. Explain how continental drift may have played a role in history of life.
6. Describe how radiation into new adaptive zones could result in macroevolutionary change.
7. Explain how mass extinctions could occur and affect evolution of surviving forms.
8. List the major taxonomic categories from the most to least inclusive.

9. Explain why it is important when constructing a phylogeny to distinguish between homologous and analogous character traits.
10. Distinguish between homologous and analogous structures.
11. Describe three techniques used in molecular systematics and explain what information each provides.
12. Distinguish between a monophyletic and a polyphyletic group, and explain what is meant by a "natural taxon."
13. Describe the contributions of phenetics and cladistics to phylogenetic systematics.
14. Describe how cladistic analysis uses novel homologies to define branch points on phylogenetic trees.

KEY TERMS

phylogeny	binomial	monophyletic	DNA sequence analysis
systematics	genus (genera, pl.)	polyphyletic	phenetics
fossil record	specific epithet	paraphyletic	cladistic analysis
geological time scale	family	homology	clade
radiometric dating	order	convergent evolution	outgroup
half-life	class	analogy	synapomorphies
Pangaea	phylum (phyla, pl.)	DNA-DNA	parsimony
adaptive zone	kingdom	hybridization	phylogenetic biology
phylogenetic trees	taxon (taxa, pl.)	restriction maps	

LECTURE NOTES

Biologists reconstruct evolutionary history by studying the succession of organisms in the fossil record. Fossils are collected and interpreted by *paleontologists*.

Phylogeny = The evolutionary history of a species or group of related species
- Phylogeny is usually diagrammed with *phylogenetic trees* that trace inferred evolutionary relationships.

Systematics = The study of biological diversity in an evolutionary context; reconstructing phylogeny is part of the scope of systematics
- Biological diversity reflects past episodes of speciation and macroevolution.
- Encompasses the identification and classification of species (*taxonomy*) in its search for evolutionary relationships.

Fossil record = The ordered array in which fossils(any preserved remnant or impression left by an organism that lived in the past) appear within layers of rock that mark the passing of geological time

I. The Fossil Record and Geological Time

A. Sedimentary rocks are the richest sources of fossils

Sedimentary rocks are the richest sources of fossils.
- These rocks form from deposits of sand and silt that have weathered or eroded from the land and are carried by rivers to seas and swamps.
- Aquatic organisms and some terrestrial forms were swept into seas and swamps and became trapped in sediment when they died.
- New deposits pile on and compress older sediments below into rock. Sand is compressed into sandstone and mud into shale.
- A small proportion of the organisms left a fossil record.

Fossils usually form from mineral-rich hard parts of organisms (bones, teeth, shells of invertebrates) since most organic substances usually decay rapidly (see Campbell, Figure 25.1 a and b).

- Paleontologists usually find parts of skulls, bone fragments, or teeth; although nearly complete skeletons of dinosaurs and other forms have been found.
- Many of the parts found have been hardened by *petrification*, which occurs when minerals dissolved in groundwater seep into the tissues of dead organisms and replace organic matter (see Campbell, Figure 25.1c).

Some fossils, found as thin films pressed between layers of sandstone or shale, retain organic material.

- Paleontologists have found leaves millions of years old that are still green with chlorophyll and preserved well enough that their organic composition and ultrastructure could be analyzed (see Campbell, Figure 25.1d).
- One research team was even able to clone a very small sample of DNA from an ancient magnolia leaf.

Other fossils found by paleontologists are replicas formed in molds left when corpses were covered by mud or sand (see Campbell, Figure 25.1e).

- Minerals from the water which filled the mold eventually crystallized in the shape of the organism.

Trace fossils form in footprints, animal burrows, and other impressions left in sediments by animal activity. These can provide a great deal of information.

- Dinosaur tracks can provide information about the animals gait, stride length, and speed.

Rarely has an entire organism been fossilized. This only happens if the organism was buried in a medium that prevented bacteria and fungi from decomposing the body (see Campbell, Figure 25.1g).

B. Paleontologists use a variety of methods to date fossils

Several methods are used to determine the age of fossils, which makes them useful in studies of macroevolution.

1. Relative dating

Sedimentation may occur when the sea-level changes or lakes and swamps dry and refill.

- The rate of sedimentation and the types of particles that sediment vary with time when a region is submerged.
- The different periods of sedimentation resulted in formation of rock layers called *strata*.
- Younger strata are superimposed on top of older ones.
- The succession of fossil species chronicles phylogeny, since fossils in each layer represent organisms present at the time of sedimentation.

Strata from different locations can often be correlated by the presence of similar fossils, known as index fossils.

- The shells of widespread marine organisms are the best index fossils for correlating strata from different areas.
- Gaps in the sequence may appear in an area if it was above sea level (which prevents sedimentation) or if it was subjected to subsequent erosion.

Geologists have formulated a sequence of geological periods by comparing many different sites. This sequence is known as the *geological time scale* (see Campbell, Table 25.1).

- These periods are grouped into four eras with boundaries between the eras marking major transitions in the life forms fossilized in the rocks.
- Periods within each era are subdivided into shorter intervals called *epochs*.
- The divisions are not arbitrary, but are associated with boundaries that correspond to times of change.

This record of the rocks presents a chronicle of the relative ages of fossils, showing the order in which species groups evolved.

2. Absolute dating

Absolute dating is not errorless, but it does give the age in years rather than in relative terms (e.g., before, after).

The most common method for determining the age of rocks and fossils on an absolute time scale is *radiometric dating*.

- Fossils contain isotopes of elements that accumulated in the living organisms.
- Since each radioactive isotope has a fixed half-life, it can be used to date fossils by comparing the ratio of certain isotopes (e.g., ^{14}C and ^{12}C) in a living organism to the ratio of the same isotopes in the fossil.
- *Half-life* = The number of years it takes for 50% of the original sample to decay
- The half-life of an isotope is not affected by temperature, pressure or other environmental variables.

Carbon-14 has a half-life of 5600 years, meaning that one-half of the carbon-14 in a specimen will be gone in 5600 years; half of the remaining carbon-14 would disappear from the specimen in the next 5600 years; this would continue until all of the carbon-14 had disappeared (see Campbell, Figure 25.2).

- Thus a sample beginning with 8g of carbon-14 would have 4g left after 5600 years and 2g after 11,200 years.
- Carbon-14 is useful in dating fossils less than 50,000 years old due to its relatively short half-life.

Paleontologists use other radioactive isotopes with longer half-lives to date older fossils.

- Uranium-238 has a half-life of 4.5 billion years and is reliable for dating rocks (and fossils within those rocks) hundreds of millions of years old.
- This isotope was used to place the oldest fossil-containing rocks in the Cambrian period.
- An error of < 10% is present with radioactive dating.

Other methods may also be used to date some fossils.

- Amino acids can have either left-handed (L-form) or right-handed (D-form) symmetry.
- Living organisms only synthesize L-form amino acids to incorporate into proteins.
- After an organism dies, L-form amino acids are slowly converted to D-form.
- The ratio of L-form to D-form amino acids can be measured in fossils.
- Knowing the rate of chemical conversion (*racemization*) allows this ratio to be used in determining how long the organism has been dead.
- This method is most reliable in environments where the climate has not changed significantly, since the conversion is temperature sensitive.

The dating of rocks and fossils they contain has enabled researchers to determine the geological periods (see Campbell, Table 25.1)

C. The fossil record is a substantial, albeit incomplete, chronicle of evolutionary history

A fossil represents a sequence of improbable events:

- An organism had to die in the right place and at the proper time for burial conditions favoring fossilization.
  - The rock layer containing the fossil had to escape geologic events (erosion, pressure, extreme heat) which would have distorted or destroyed the rock.
  - The fossil had to be exposed and not destroyed.
- Someone who knew what they were doing had to find the fossil.

The fossil record is slanted in favor of species that existed for a long time, were abundant and widespread, and had shells or hard skeletons

Paleontologists thus work with an incomplete record for many reasons.

- A large fraction of species that have lived probably left no fossils.
- Most fossils that were formed have probably been destroyed.
- Only a small number of existing fossils have been discovered.

Even though it is incomplete, the fossil record provides the outline of macroevolution, but the evolutionary relationships between modern organisms must be studied to provide the details.

D. Phylogeny has a biogeographical basis in continental drift

Evolution has dimension in space as well as time.

- Biogeography was a major influence on Darwin and Wallace in developing their views on evolution.
- Drifting of continents is the major geographical factor correlated with the spatial distribution of life (see Campbell, Figure 25.3a).

Continental drift results from the movement of great plates of crust and upper mantle that float on the Earth's molten core.

- The relative positions of two land masses to each other changes unless they are embedded on the same plate.
- North America and Europe are drifting apart at a rate of 2 cm per year.
- Where two plates meet (boundaries), many important geological phenomena occur: mountain building, volcanism, and earthquakes (see Campbell, Figure 25.3b).
  - ⇒ Volcanism, in turn, forms volcanic islands (e.g., Galapagos), which opens new environments for founders and adaptive radiation.

Plate movements continually rearrange geography, however, two occurrences had important impacts on life: the formation of *Pangaea* and the subsequent breakup of Pangaea (see Campbell, Figure 25.4).

At the end of the Paleozoic era (250 million years ago), plate movements brought all land masses together into a super-continent called Pangaea.

- Species evolving in isolation where brought together and competition increased.
- Total shoreline was reduced and the ocean basins became deeper (draining much of the remaining shallow coastal seas).
- Marine species (which inhabit primarily the shallow coastal areas) were greatly affected by reduction of habitat.
- Terrestrial organisms were affected as continental interior habitats (and their harsher environments) increased in size.

- Changes in ocean currents would have affected both terrestrial and marine organisms.
- Overall diversity was thus impacted by extinctions and increased opportunities for surviving species.

During the early Mesozoic era (about 180 million years ago) Pangaea began to breakup due to continuing continental drift.

- This isolated the fauna and flora occupying different plates.
- The biogeographical realms were formed and divergence of organisms in the different realms continued.

Many biogeographical puzzles are explained by the pattern of continental separations. For example,

- Matching fossils recovered from widely separated areas.
    - ⇒ Although Ghana and Brazil are separated by 3000 km of ocean, matching fossils of Triassic reptiles have been recovered from both areas.
- Australia has unique fauna and flora.
    - ⇒ Australian marsupials are very diverse and occupy the same ecological roles as placental mammals on other continents.
    - ⇒ Marsupials probably evolved on the portion of Pangaea that is now North America and migrated into the area that would become Australia.
    - ⇒ The breakup of Pangaea isolated Australia (and its marsupial populations) 50 million years ago, while placental mammals evolved and diversified on the other continents.

E. The history of life is punctuated by mass extinctions followed by adaptive radiations of survivors

The evolution of modern life has included long, relatively quiescent periods punctuated by briefer intervals of more extensive turnover in species composition.

- These intervals of extensive turnover included explosive adaptive radiations of major taxa as well as mass extinctions.

1. Examples of major adaptive radiations

The evolution of some novel characteristics opened the way to new *adaptive zones* allowing many taxa to diversify greatly during their early history. For example,

- Evolution of wings allowed insects to enter an adaptive zone with abundant new food sources and adaptive radiation resulted in hundreds of thousands of variations on the basic insect body plan.
- A large increase in the diversity of sea animals occurred at the boundary between the Precambrian and Paleozoic eras. This was a result, in part, of the origin of shells and skeletons in a few key taxa.
    - ⇒ Precambrian rock contains the oldest animals (700 million years old) which were shell-less invertebrates that differed significantly from their successors found in Paleozoic rock.
    - ⇒ Nearly all the extant animal phyla and many extinct phyla evolved in less than 10 million years during the mid-Cambrian (early Paleozoic era).
    - ⇒ Shells and skeletons opened a new adaptive zone by making many new complex body designs possible and altering the basis of predator-prey relationships.
    - ⇒ It is possible that genes controlling development evolved during this time, resulting in a potential for increased morphological complexity and diversity.

An empty adaptive zone can be exploited only if the appropriate evolutionary novelties arise. For example,

- Flying insects existed for 100 million years before the appearance of flying reptiles and birds that fed on them.

Conversely, an evolutionary novelty cannot enable organisms to exploit adaptive zones that are occupied or that do not exist.

- Mass extinctions have often opened adaptive zones and allowed new adaptive radiations.
- For example, mammals existed 75 million years before their first large adaptive radiation in the early Cenozoic. This may have resulted from the ecological void created with the extinction of the dinosaurs.

2. Examples of mass extinctions

Extinction is inevitable in a changing world. The average rate of extinction has been between 2.0 and 4.6 families (each family may include many species) per million years.

Extinctions may be caused by habitat destruction or by unfavorable environmental changes.

- Many very well adapted marine species would become extinct if the ocean's temperature fell only a few degrees.
- Changes in biological factors may cause extinctions even if physical factors remain stable.
- Since many species coexist in each community, an evolutionary change in one species will probably impact other species. For example,
  ⇒ The evolution of shells by some Cambrian animals may have contributed to the extinction of some shell-less forms.

There have been periods of global environmental change which greatly disrupted life and resulted in mass extinctions.

- During these periods, the rate of extinction escalated to as high 19.3 families per million years.
- These mass extinctions are recognized primarily from the decimation of hard-bodied animals in shallow seas which have the most complete fossil record.
- Two (of about a dozen) mass extinction episodes have been studied extensively by paleontologists.

The Permian extinctions (the boundary between the Paleozoic and Mesozoic eras) eliminated over 90% of the species of marine animals about 250 million years ago (see Campbell, Figure 25.5).

- Terrestrial life was probably also affected greatly. For example, eight of the 27 orders of Permian insects did not survive into the Triassic.
- This mass extinction took place in less than five million years and probably resulted from several factors.
  ⇒ Occurred about the time Pangaea was formed by the merging of continents which disturbed many habitats and altered the climate.
  ⇒ A period of extreme vulcanism and resulting volcanic debris (including carbon dioxide) in the atmosphere may have altered the global temperature.

The Cretaceous extinction (the boundary between the Mesozoic and Cenozoic eras) occurred about 65 million years ago.

- More than 50% of the marine species and many terrestrial plants and animals (including dinosaurs) were eliminated.

- During this time the climate was cooling and many shallow seas receded from continental lowlands.
- Increased volcanic activity during this time may have contributed to the cooling by releasing materials into the atmosphere and blocking the sunlight.

Evidence also indicates that an asteroid or comet struck the Earth (*impact hypothesis*) while the Cretaceous extinctions were in progress.

- Iridium, an element rare on earth but common in meteorites, is found in large quantities in the clay layer separating Mesozoic and Cenozoic sediments.
- Walter and Luis Alvarez (and colleagues), after studying this clay layer, proposed that it is fallout from a huge cloud of dust ejected into the atmosphere when an asteroid collided with the Earth.
- This cloud would have both blocked the sunlight and severely disturbed the climate for several months.
- Although the asteroid hit the earth during this time, some researchers feel it did not cause the mass extinction of this period.

The impact hypothesis consisted of two parts: a large asteroid or comet collided with the Earth and the collision caused the Cretaceous extinctions.

- Many forms of evidence support the idea that a large comet or small asteroid collided with Earth 65 million years ago.
  - ⇒ Many craters have been found and indicate that a large number of objects have fallen to the Earth's surface.
  - ⇒ A large crater ($\approx$ 180 km in diameter) located beneath sediments on the Yucatan coast of Mexico has been located.
- Questions about the impact hypothesis are now concentrated on the second part: the collision caused the Cretaceous extinctions.
- Advocates of the impact hypothesis point to several items in their support (see Campbell, Figure 25.6):
  - ⇒ The large size of the impact would darken the Earth for several years and the reduction in photosynthesis output would be sufficient to cause food chains to collapse.
  - ⇒ Severe acid precipitation would result from the increased mineral content of the atmosphere.
  - ⇒ The content of sediments at the upper Cretaceous boundary indicated global fires were occurring and smoke from these fires would increase the atmospheric effects of the impact.
- Opponents of the impact hypothesis hold that the impact occurred within the period of mass extinction, but that the two occurrences are not a cause and effect event.
  - ⇒ Many paleontologists and geologists believe that the climatic changes which occurred were due to continental drift, increased vulcanism, and other processes.
  - ⇒ They also feel these events were sufficient to cause the mass extinction.

Many paleontologists are now trying to determine how sudden and uniform the Cretaceous extinctions were (on a geological time scale).

- Disappearance of diverse groups (from microscopic marine plankton to dinosaurs) during a short time span would support the impact hypothesis.
- A gradual decline with different groups disappearing at different rates, would support hypotheses emphasizing terrestrial causes.

- It is possible that the impact was a final, sudden event in environmental changes that were affecting the biota of the late Cretaceous.

Mass extinctions, whatever the cause, profoundly affect biological diversity.

- Not only are many species eliminated, but those that survive are able to undergo new adaptive radiations into the vacated adaptive zones and produce new diversity.

II. Phylogeny and Systematics

One of the main goals of systematics is to make biological classification reflect phylogeny.

A. Taxonomy employs a hierarchical system of classification

The taxonomic system used today was developed by Linnaeus in the eighteenth century. This system has two main features: the assignment of a binomial to each species and a filing system for grouping species (see Campbell, Figure 25.7).

The *binomial* (two part Latin name) assigned to each species is unique to that species.

- The first word of the binomial is the *genus* (pl. genera); the second word is the *specific epithet* of the species.
- The scientific name of a species combines the genus and specific epithet.
- Each genus can include many species of related organisms. For example, *Felis silvestris* is the domestic cat; *Felis lynx* is the lynx.
- Use of the scientific name defines the organism referred to and removes ambiguity.

The filing system for grouping species into a hierarchy of increasingly general categories formalizes the grouping of organisms.

- Binomial nomenclature is the first step in grouping: similar species are grouped in the same genus.
- The system then progresses into broader categories:
  ⇒ Similar genera are grouped into the same family.
  ⇒ Families are grouped into orders.
  ⇒ Orders are grouped into classes.
  ⇒ Classes are grouped into phyla.
  ⇒ Phyla are grouped into kingdoms.
- Each taxonomic level is more inclusive than the one below. The more closely related two species are, the more levels they share:

Category	Domestic Cat	Bobcat	Lion	Dog
specific epithet	silvestris	rufus	leo	familiaris
genus	Felis	Felis	Panthera	Canis
family	Felidae	Felidae	Felidae	Canidae
order	Carnivora	Carnivora	Carnivora	Carnivora
class	Mammalia	Mammalia	Mammalia	Mammalia
phylum	Chordata	Chordata	Chordata	Chordata
kingdom	Animalia	Animalia	Animalia	Animalia

The two main objectives of taxonomy are to sort out and identify closely related species and to order species into the broader taxonomic categories.

- In sorting, closely related organisms are assigned to separate species (with the proper binomial) and described using the diagnostic characteristics which distinguish the species from one another.

- In categorizing, the species are grouped into broader categories from genera to kingdoms.

- In some cases, intermediate categories (e.g., subclasses; between orders and classes) are also used.

- The named taxonomic unit at any level is called a *taxon* (pl. taxa).

- Rules of nomenclature have been established: the genus name and specific epithet are italicized, all taxa from the genus level and higher are capitalized.

B. The branching pattern of a phylogenetic tree represents the taxonomic hierarchy

The goal of systematics is to have classification reflect the evolutionary affinities of species. The taxonomic hierarchy is set up to fit evolutionary history (see Campbell, Table 25.2).

- In general, groups subordinate to other groups in the taxonomic hierarchy should represent finer and finer branching of phylogenetic trees (see Campbell, Figure 25.8)

Classification schemes and phylogenetic trees are hypotheses of history based on current data. Like all hypotheses, they may be refined with further study.

C. Determining monophyletic taxa is key to classifying organisms according to their evolutionary history

In order for a classification scheme to reflect the evolutionary history of an organism, the species must be grouped into taxa that are monophyletic.

- A *monophyletic taxon* is one where a single ancestor gave rise to all species in that taxon and to no species placed in any other taxon. For example, Family Ursidae evolved from a common ancestor (Taxon 1) (see also Campbell, Figure 25.9).

By contrast, other kinds of taxa do not accurately reflect evolutionary history

- A *polyphyletic taxon* is one whose members are derived from two or more ancestral forms not common to all members. For example, Kingdom Plantae includes both vascular plants and mosses which evolved from different algal ancestors (Taxon 2).

- A *paraphyletic taxon* is one that excludes species that share a common ancestor that gave rise to the species included in the taxon. For example, Class Reptilia excludes the Class Aves although a reptilian ancestor common to all reptiles is shared (Taxon 3).

1. Sorting homology from analogy

Systematists classify species into higher, preferably monophyletic, taxa based on the extent of similarities in morphology and other characteristics.

Homology = Likeness attributed to shared ancestry

- The forelimbs of mammals are homologous, they share a similarity in the skeletal support that has a genealogical basis.
- Homology must be distinguished from analogy in evolutionary trees.

Analogy = Similarities due to *convergent evolution*, not common ancestry

Convergent evolution = Acquisition of similar characteristics in species from different evolutionary branches due to sharing similar ecological roles with natural selection shaping analogous adaptations (see Campbell, Figure 25.10)

The distinction between homology and analogy is sometimes relative.

> Example: The wings of birds and bats are modifications of the vertebrate forelimb.

- ⇒ The appendages are thus homologous.
- ⇒ As wings they are analogous since they evolved independently from the forelimbs of different flightless ancestors.
- Insect wings and bird wings are analogous.
- ⇒ They evolved independently and are constructed from entirely different structures.
- Convergent evolution has produced analogous similarities between Australian marsupials and placental mammals on other continents.

Homology must be sorted from analogy to reconstruct phylogenetic trees on the basis of homologous similarities.

- Generally, the greater the amount of homology, the more closely related the species and this should be reflected in their classification.
- Adaptation and convergence often obscure homologies, although studies of embryonic development can expose homology that is not apparent in mature structures.
- Additionally, the more complex two similar structures are, the less likely it is that they have evolved independently.

> Example: The skulls of humans and chimpanzees are composed of many bones which are fused together and match almost perfectly. It is unlikely such a complex structure would have evolved independently in separate groups.

D. Molecular biology provides powerful new tools for systematics

Molecular comparisons of proteins and DNA have added another useful method for studying the evolutionary relationships between species.

- Inherited nucleotide sequences in DNA program the corresponding sequences of amino acids in proteins.
- Examination of these macromolecules provide much information about evolutionary relationships.

Molecular comparisons are:

- Objective and quantitative
- Used to assess relationships between species so distantly related that no morphological similarities exist

1. Protein comparisons

The primary structure of proteins is genetically programmed and a similarity in the amino acid sequence of two proteins from different species indicates that the genes for those proteins evolved from a common gene present in a shared ancestor.

Studies of cytochrome *c* (an ancient protein common to all aerobic organisms) have been used to compare many diverse species.

- The amino acid sequence has been determined for species ranging from bacteria to complex plants and animals.
- The sequence in cytochrome *c* is identical in chimpanzees and humans.
- The sequence in humans and chimpanzees differs at only one of the 104 amino acid positions in the rhesus monkeys.
- Chimpanzees, humans, and rhesus monkeys belong to the Order Primates.
    - ⇒ Comparing these sequences with nonprimate species shows greater differences (13 with the dog and 20 with the rattlesnake).
- Phylogenetic trees based on cytochrome *c* are consistent with evidence from comparative anatomy and the fossil record.

One disadvantage to the use of amino acid data for phylogenetic studies is that such sequences provide information only about those genes that code for protein; this can be a small fraction of the genome (e.g., 2% in humans)

2. DNA and RNA comparisons

The most direct measure of common inheritance from shared ancestors is a comparison of the genes or genomes of two species. Comparison can be made by three methods: DNA-DNA hybridization, restriction maps, and DNA sequencing.

DNA-DNA hybridization can compare whole genomes by measuring the degree of hydrogen bonding between single-stranded DNA obtained from two sources.

- DNA is extracted from different species and the complementary strands separated by heating.
- The single-stranded DNA from the two species is mixed and cooled to allow double-stranded DNA reformation which results from hydrogen bonding.
- The hybrid DNA is then reheated to separate the double strands.
- The temperature necessary to separate the hybrid DNA is indicative of the similarity in the DNA from the two species.
    - ⇒ The temperature correlation is based on the degree of bonding between the strands of the two species with more bonding occurring with greater similarity.
    - ⇒ The more extensive the pairing, the more heat is needed to separate the hybrid strand.
- Evolutionary trees constructed through DNA-DNA hybridization usually agree with those based on other methods, however, this technique is very beneficial in settling taxonomic debates that have not been finalized by other methods.

Restriction maps provide precise information about the match-up of specific DNA nucleotide sequences.

- Restriction enzymes are used to cut DNA into fragments which can be separated by electrophoresis and compared to restriction fragments of other species.
- Two samples of DNA with similar maps for the locations of restriction sites will produce similar collections of fragments.
- The greater the divergence of two species from the common ancestor, the greater the differences in restriction sites and less similarity of the restriction fragments.
- This method works best when comparing small fragments of DNA.

- Mitochondrial DNA (mtDNA) is best suited for this type of comparison since it is smaller than nuclear DNA (produces smaller fragments) and mutates about ten times faster than nuclear DNA.
- The faster mutation rate of mtDNA allows it to be used to determine phylogenetic relationships between not only closely related species, but also populations of the same species.
    - ⇒ mtDNA was used to establish the close relationship among the Pima, Mayan, and Yanomami groups of Native Americans.
    - ⇒ The results supported linguistic evidence that these groups descended from the first wave of immigrants to cross the Bering land bridge from Asia during the late Pleistocene.

DNA sequence analysis is the most precise method of comparing DNA as it determines the actual nucleotide sequence of a DNA segment.

- Uses polymerase chain reaction (PCR) technology to clone traces of DNA.
- PCR is coupled with automated sequencing to provide a simpler and faster method of collecting sequence data.
- DNA sequencing and comparisons show exactly how much divergence there has been in the evolution of two genes derived from the same ancestral gene.
- Ribosomal RNA (rRNA) sequencing is a similar technique which can provide information about some of the earliest branching in phylogenetic relationships since DNA coding for rRNA changes very slowly.
- rRNA sequencing has been very useful in examining the relationships among bacteria.

3. Identifying and comparing homologous DNA sequences

Comparing nucleotide sequences between corresponding DNA segments from different species has the potential of telling us how much divergence there has been in the evolution of two genes derived from the same gene.

To measure differences between two species, it is necessary to identify homologous nucleotide sequences Following sequencing, the sequences are aligned and evaluated.

- Common ancestry is clear when two sequences of the same gene from two species that have diverged very recently are the same or differ by only a few bases.
- Mutations tend to accumulate as species diverge; the number of differences is a measure of evolutionary distance.
- Once a match is achieved, usually with the aid of computer programs, the investigator can develop a phylogenetic hypothesis (see Campbell, Figure 25.12).

4. Molecular clocks

Different proteins and nucleic acids evolve at different rates, although each type of molecule evolves at a relatively constant rate over time.

When comparing homologous proteins and nucleotide sequences from taxa that are known to have diverged from common ancestors, the number of amino acid substitutions is proportional to the elapsed time since divergence.

> Example: Homologous proteins of bats and dolphins are more similar than those of sharks and tuna.

- This is consistent with fossil evidence showing that tuna and sharks have been separated much longer than bats and dolphins.

DNA comparisons may be even more reliable than protein comparisons.

- Phylogenetic branching based on nucleotide substitutions in DNA generally approximates dates determined from the fossil record.
- The difference in DNA between two taxa is more closely correlated with the time since divergence than is morphological difference.

Molecular clocks (DNA and protein) are calibrated by graphing the number of nucleotide or amino acid differences against the times for a series of evolutionary branch points known from the fossil record.

- The graph can then be used to determine the time of divergence between taxa for which no substantial fossil record is available.

The assumption that mutation rates for genes (and their protein products) are relatively constant is the basis for using molecular clocks in evolutionary biology.

- This assumption is relatively solid when comparing groups of closely related species.
- Molecular clocks are less reliable when comparing more distantly related groups since differences in generation times and metabolic rates affect mutation rates.

Among closely related species, the constant mutation rate for specific genes implies that the accumulation of selectively neutral mutations changes the genome more than adaptive mutations.

- Many evolutionary biologists doubt the prevalence of neutral variation, so they also question the use of molecular clocks to accurately date the time of divergence.
- There is less skepticism about the value of molecular clocks for determining the relative sequence of branch points in phylogeny.
- Modern systematists use available molecular data along with all other evidence to reconstruct phylogeny.

E. **The search for fossilized DNA continues despite recent setbacks:** *science as a process*

Science is a dynamic process that continuous tests hypotheses; sometime the hypotheses are supported and sometimes they are rejected

The nucleotide sequences in DNA traces recovered from fossils that retain organic material can be analyzed by using PCR.

- Fossilized DNA from 17 million-year-old magnolia leaves was first reported in 1990 .
- Since 1990, DNA fragments have been sequenced from a frozen mammoth (40,000 years old), an insect fossilized in amber (40 million years old), a 65-million-year-old *Tyrannosaurus rex* from Montana, a 30,000-year old fossil arm bone from an extinct member of the human family, and a frozen Stone Age man (5000 years old).

It now appears that most of the DNA first reported to be ancient is actually DNA from contaminating fungus or other organisms.

III. **The Science of Phylogenetic Systematics**

The two significant features of a phylogenetic tree are the location of branch points along the tree and the degree of divergence between branches.

- The locations of branch points along the tree symbolize the relative times of origin for different taxa.
- The degree of divergence between branches represents how different two taxa have become since branching from a common ancestor.

Initially, phylogenies were devised largely on morphology and were considered by many to be too subjective. In the 1960s, new computational technology helped usher in two new, more objective analytical approaches: phenetics and cladistics.

A. Phenetics increased the objectivity of systematic analysis

Phenetics makes no evolutionary assumptions and decides taxonomic affinities entirely on the basis of measurable similarities and differences.

- A comparison is made of as many characters (anatomical characteristics) as possible without attempting to sort homology from analogy.
- Pheneticists feel that the contribution of analogy to overall similarity will be overridden by the degree of homology if enough characters are compared.
- Critics of phenetics argue that overall phenotypic similarity is not a reliable index of phylogenetic proximity.
- While supported by few systematists, the emphasis of phenetics on multiple quantitative comparisons has made important contributions to systematics.
  ⇒ Especially useful for analyzing DNA sequence data and other molecular comparisons between species.

B. Cladistic analysis uses novel homologies to define branchpoints on phylogenetic trees

Cladistic analysis, which has become synonymous with phylogenetic systematics, classifies organisms according to the order in time that branches arise along a phylogenetic tree, without considering the degree of divergence.

- This produces a cladogram, a dichotomous tree that branches repeatedly.
  ⇒ Each branch point is defined by novel homologies unique to the various species on that branch or *clade*.

1. Outgroup comparison

Each taxon has a mixture of primitive characters that existed in the common ancestor and characters that evolved more recently.

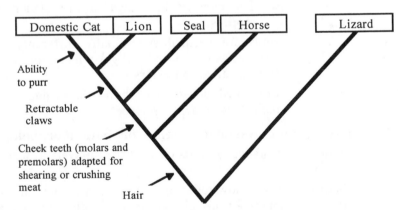

- The sharing of primitive characters indicates nothing about the pattern of evolutionary branching from a common ancestor.

Cladistic analysis uses a concept called outgroup comparison as an objective means of defining the roots of a phylogenetic tree and thereby distinguishing among the shared characters that are more primitive (see Campbell, Figures 25.13 and 25.14).

Outgroup = A species or group of species that is relatively closely related to a group of species being studied, but clearly not as closely related as any study-group member is to any other study-group member

- All members of the study group are compared as a whole to the outgroup

- Characters common to the outgroup and group being studied are likely to have been present in a common ancestor and are considered shared primitive characters

2. Use of synapomorphies (shared derived characters) and parsimony

A major difficulty in cladistic analysis is finding characters that are appropriate for identifying each branch point.

- Branch points of a phylogenetic tree can be identified by finding shared derived characters, or *synapomorphies*.
- These are homologies that evolved in an ancestor common to all species on one branch of a fork of the tree, but not common to the other branch.

A guiding principle of cladistic analysis is parsimony, the quest for the simplest, and probably most likely, explanation for observed phenomena.

- In systematics, parsimony means that a phylogenetic tree using the minimum number of changes to illustrate evolutionary relationships has the greatest likelihood of being correct.

3. Acceptance of only monophyletic taxa

By focusing on phylogenetic branching, cladistic analysis accepts only monophyletic taxa. As a result, some taxonomic surprises are produced by cladistic systematics:

- The branch point between crocodiles and birds is more recent than between crocodiles and other reptiles (a fact also supported by the fossil record).
  - ⇒ Crocodiles and birds have synapomorphies not present in lizards and snakes.
- In a strict cladistic analysis, the Class Aves and the Class Reptilia, as we know them now, would be eliminated.
  - ⇒ The birds would be included in a cladogram of the animals we know as reptiles.
- Birds are deemed superficially different because of the morphological changes associated with flight which have developed since their divergence from reptilian ancestors.
- Continued debate over how life evolved is indicative that evolutionary biology is an active science.

C. Phylogenetic systematics relies on both morphology and molecules

Modern systematics depends on cladistic analysis to formulate hypotheses about the history of life.

Cladistic analysis now relies upon both morphology and on molecular data; the strongest support for any phylogenetic hypothesis is agreement between data derived from both of these sources.

- Scientists have been collecting phenotypic information about living and extinct organisms for centuries. The amount of this morphological information exceeds by far the current molecular data.
- Molecular databases are expanding rapidly. While no means complete, the genome contains more information about an organism's evolutionary history than its anatomical features.

Phylogenetic biology, or the application of cladistic analysis in the study of evolutionary history and its relations to all aspects of life, pervades virtually every field of biology.

- Physiology - evolution of body temperature regulation
- Plant development - evolution of genes regulating flower development

- Behavior - evolutionary history of animal social systems
- Conservation biology - evaluates genetic differences among populations of endangered species

REFERENCES

Campbell, N., et al. *Biology*. 5th ed. Menlo Park, California: Benjamin/Cummings, 1998.

Gould, S.J. "The Wheel of Fortune and the Wedge of Progress." *Natural History*, March 1989.

Stanley, S. *Extinction*. New York: Scientific American Library, 1987.

CHAPTER 26
EARLY EARTH AND
THE ORIGIN OF LIFE

OUTLINE

I. Introduction to the History of Life
   A. Life on Earth originated between 3.5 and 4.0 billion years ago
   B. Major episodes in the history of life: *a preview*

II. Prebiotic Chemical Evolution and the Origin of Life
   A. The first cells may have originated by chemical evolution on a young Earth: *an overview*
   B. Abiotic synthesis of organic monomers is a testable hypothesis: *science as a process*
   C. Laboratory simulations of early Earth conditions have produced organic polymers
   D. Protobionts can form by self-assembly
   E. RNA was probably the first genetic material
   F. The origin of hereditary information made Darwinian evolution possible
   G. Debate about the origin of life abounds

III. The Major Lineages of Life
   A. Arranging the diversity of life into the highest taxa is a work in progress

OBJECTIVES

After reading this chapter and attending lecture, the student should be able to:

1. Provide at least two lines of evidence for the antiquity of life.
2. Describe the contributions that A.I. Oparin, J.B.S. Haldane, Stanley Miller and Harold Urey made towards developing a model for abiotic synthesis of organic molecules.
3. Provide plausible evidence to support the hypothesis that chemical evolution resulting in life's origin occurred in four stages:
   a. Abiotic synthesis of organic monomers
   b. Abiotic synthesis of polymers
   c. Formation of protobionts
   d. Origin of genetic information
4. Describe the basis for Whittaker's five-kingdom system.
5. Describe three alternatives to the five-kingdom system and explain the rationale for each.

KEY TERMS

stromatolites   protobionts   ribozyme

LECTURE NOTES

The history of living organisms and the history of Earth are inextricably linked.

Examples:

* Formation and subsequent breakup of Pangaea affected biotic diversity.
* The first photosynthetic organisms released oxygen into the air and altered Earth's atmosphere.
* *Homo sapiens* has changed the land, water and air on a scale and at a rate unprecedented for a single species.

In order to reconstruct life's history, scientists use evidence from:

* The fossil record, which is less complete the older the strata studied. In fact, there is no fossil record for the seminal episode of the origin of Earth's life.
* Contemporary organisms which, in their molecules and anatomy, carry traces of their evolutionary histories.

I. Introduction to the History of Life

A. Life on Earth originated between 3.5 and 4.0 billion years ago

Life probably appeared relatively early in the Earth's history. Scientists have found isotopes of carbon in 3.8 billion year old rocks in Greenland.

Because of the relatively simple structure of prokaryotes, it is reasonable to assume that the earliest organisms were prokaryotes. The fossil record supports this notion.

* Fossils similar to spherical and filamentous prokaryotes have been recovered from stromatolites 3.5 billion years old in western Australia and southern Africa (see Campbell, Figure 26.2).

Stromatolites = Banded domes of sediment similar to the layered mats constructed by colonies of bacteria and cyanobacteria currently living in salty marshes (see Campbell, Figure 26.1)

* The western Australian fossils appear to be of photosynthetic organisms, which indicates life evolved before these organisms lived—perhaps some 4 billion years ago.
* Other fossils similar to the prokaryotes have been recovered from the Fig Tree Chert rock formation in southern Africa which date to 3.4 billion years.

B. Major episodes in the history of life: *a preview*

Campbell, Figure 26.3, is a diagram of some major episodes in the history of life.

Fossil evidence suggests that prokaryotes appeared at least 2 billion years before the oldest eukaryotes

* Two distinct groups of prokaryotes, Bacteria and Archaea, diverged early, between 2 to 3 billion years ago.
* Photosynthetic bacteria started the production of oxygen about 2.5 billion years ago, setting the stage for aerobic life.

Eukaryotes emerged some 2 billion years ago

* Strong evidence supports the hypothesis that eukaryotic cells evolved from a symbiotic community of prokaryotes

Plants, fungi, and animals arose from distinct groups of unicellular eukaryotes during the Precambrian.

* Plants evolved from green algae.

- Fungi and animals arose from different groups of heterotrophic unicells. Based on molecular evidence, fungi are more closely related to animals than they are to plants.

The oldest fossils of animals are those of soft-bodied invertebrates from about 700 million years ago. The basic body plans of most of the modern animal phyla probably arose in the late Precambrian.

The transition from the aquatic environment to land was a pivotal point in the history of life.

- The first terrestrial colonization was by plants and fungi some 475 million years ago; the move may have depended upon a beneficial association between the two groups.

- The transformation of the landscape by plants created new opportunities for all forms of life.

II.   Prebiotic Chemical Evolution and the Origins of Life

A.   The first cells may have originated by chemical evolution on a young Earth: *an overview*

Life originated between 3.5 and 4.0 billion years ago. During this timespan the Earth's crust began to solidify (4.1 billion) and bacteria advanced enough to build stromatolites (3.5 billion).

The origin of life was possible in Earth's ancient environment, which was different from today:

- There was little atmospheric oxygen.
- Lightning, volcanic activity, meteorite bombardment, and ultraviolet radiation were more intense.

One hypothesis about the first living organisms is that they were the products of a chemical evolution that occurred in four stages:

1. Abiotic synthesis and accumulation of monomers, or small organic molecules, that are the building blocks for more complex molecules
2. Joining of monomers into polymers (e.g., proteins and nucleic acids)
3. Formation of *protobionts*, droplets which formed from aggregates of abiotically produced molecules and which differed chemically from their surroundings
4. Origin of heredity during or before protobiont appearance.

B.   Abiotic synthesis of organic monomers is a testable hypothesis: *science as a process*

In the 1920', A.I. Oparin and J.B.S. Haldane independently postulated that the reducing atmosphere and greater UV radiation on primitive Earth favored reactions that built complex organic molecules from simple monomers as building blocks. This is not possible today because:

- Oxygen in Earth's oxidizing environment attacks chemical bonds, removing electrons. An important characteristic of the early atmosphere must have been the rarity of oxygen.

- The modern atmosphere has a layer of ozone that screens UV radiation, so the energy required to abiotically synthesize organic molecules is not available. On primitive Earth, energy was available from frequent lightning and intense UV. radiation that penetrated the atmosphere.

Stanley Miller and Harold Urey tested the Oparin/Haldane hypothesis (see Campbell, Figure 26.4). They simulated conditions on early Earth by constructing an apparatus containing H_2O, H_2, CH_4 and NH_3.

- Their simulated environment produced some amino acids and other organic molecules.
- Now we know the atmosphere of early Earth probably included CO, CO_2, and N_2, and was less reducing than the Miller-Urey model, and thus, less favorable to formation of organic compounds.
- Additional experiments have produced all 20 amino acids, ATP, some sugars, lipids and purine and pyrimidine bases of RNA and DNA.

C. Laboratory simulations of early Earth conditions have produced organic polymers

The forming of complex organic molecules, or polymers, from simpler building-block molecules may have been inevitable on the primitive Earth.

Polymers = Chains of similar building blocks or monomers

Polymers are synthesized by dehydration (condensation) reactions. For example:

$$\Box - H + OH - \Box \longrightarrow \Box - \Box + H_2O$$

- H and OH groups are removed from the monomers.
- H_2O is produced as a by-product.

Abiotic polymerization reactions in early-Earth conditions must have occurred:

- Without the help of enzymes
- With dilute solutions of monomers (spontaneous dehydration reactions that produce water would be unlikely in already dilute solutions)

Abiotic polymerization does occur with dilute solutions of monomers under certain laboratory conditions:

- Dilute solutions of organic monomers are dripped onto hot sand, clay, or rock. Water vaporizes and concentrates the monomers on the substrate.
- Sidney Fox (University of Miami) used this method to abiotically produce polypeptides called proteinoids.

Clay may have been an important substrate for abiotic synthesis of polymers since:

- Monomers bind to charged sites in clay, concentrating amino acids and other monomers.
- Metal ions (e.g., iron and zinc) could catalyze dehydration reactions.
- The binding sites on clay could have brought many monomers close together and assisted in forming polymers.
- Pyrite (iron and sulfur) may also have been an important substrate. It has a charged surface and electrons freed during its formation could support bonding between molecules.

D. Protobionts can form by self-assembly

Living cells may have been preceded by protobionts.

Protobionts = Aggregates of abiotically produced molecules able to maintain an internal environment different from their surroundings and exhibiting some life properties such as metabolism and excitability

There is experimental evidence for the spontaneous formation of protobionts:

- When mixed with cool water, proteinoids self-assemble into microspheres (see Campbell, Figure 26.5a) surrounded by a selectively permeable protein membrane. These microspheres:
  ⇒ Undergo osmotic swelling and shrinking
  ⇒ Have potential energy in the form of a membrane potential
- Liposomes can form spontaneously when phospholipids form a bilayered membrane similar to those of living cells.

- Coacervates (colloidal drops of polypeptides, nucleic acids, and polysaccharides) self-assemble.

E. RNA was probably the first genetic material

Today's cells transcribe DNA into RNA, which is then translated into proteins. This chain of command must have evolved from a simpler mechanism of heritable control.

- One hypothesis is that before DNA, there existed a primitive mechanism for aligning amino acids along RNA molecules, which were the first genes. Evidence to support this hypothesis includes:
  ⇒ RNA molecules may have been able to self-replicate. Short polymers of ribonucleotides that can base pair (5 – 10 bases without enzyme, up to 40 bases with zinc added as catalyst) have been produced abiotically in test tubes (see Campbell, Figure 26.6).
  ⇒ RNA is autocatalytic, as indicated by *ribozymes* (RNA that acts as a catalyst to remove introns, or catalyze synthesis of mRNA, tRNA or rRNA).
- RNA folds uniquely depending on sequence (unlike DNA), thereby providing raw materials for natural selection—different molecular shapes (phenotypes) varying in stability and catalytic properties. Replication errors (mutations) probably created additional variation within families of closely related sequences.
- In addition to molecular competition, molecular cooperation probably evolved as RNA-directed protein synthesis produced short polypeptides that catalyzed RNA replication.
- Once this simple machinery for replication and translation of genetic information became sequestered into membrane-bound protobionts, molecular cooperation could be refined as natural selection acted on the level of the entire protobiont.

F. The origin of hereditary information made Darwinian evolution possible

Perhaps this hypothetical membrane-bound protobiont:

- Incorporated genetic information
- Selectively accumulated monomers from its surroundings
- Used enzymes programmed by genes to make polymers and carry out other chemical reactions
- Grew and split, distributing copies of its genes to offspring

If these cell precursors could also grow, divide, and distribute genes to offspring, the descendant protobionts would vary because of errors in the copying of RNA (mutations).

- The variation among related protobionts would be subject to natural selection.
- Evolution in the Darwinian sense—differential reproductive success—presumably accumulated refinements to primitive metabolism and inheritance, including the appearance of DNA as the hereditary material.
  ⇒ Initially, RNA could have provided the template to produce DNA.
  ⇒ Because it is more stable, DNA would have replaced RNA as the store of genetic information.
  ⇒ RNA's role would change as it became an intermediate in translation.

G. Debate about the origin of life abounds

No one knows how life actually began on Earth. The chemical evolution described and supporting lab simulations indicate key steps that could have occurred.

Several alternatives have been proposed.

- Panspermia - Some organic compounds may have reached Earth by way of meteorites and comets. Organic compounds (e.g., amino acids) have been recovered from modern meteorites. These extraterrestrial organic compounds may have contributed to the pool of molecules which formed early life.

- Most researchers believe life first appeared in shallow water or moist sediments. Some now feel the first organisms developed on the sea floor due to the harsh conditions on the surface during that time. This position was strengthened in the 1970s by discovery of the deep sea vents. Hot water and minerals emitted from such vents may have provided the energy and chemicals necessary for early protobionts.

- Simpler hereditary systems may have preceded nucleic acid genes. Julius Rebek synthesized a simple organic molecule in 1991. The importance of this molecule was that it served as a template for self-replication (see Campbell, Figure 26.8). This discovery supported the idea held by some biologists that RNA strands are too complicated to be the first self-replicating molecules.

III. The Major Lineages of Life

A. Arranging the diversity of life into the highest taxa is a work in progress

Systematists have traditionally considered the kingdom to be the highest, most inclusive taxonomic category.

- The two kingdom system (animals and plants) long prevailed, but was not suitable as biologists learned more about the structures and life histories of different organisms.

- The five kingdom system was proposed by Robert H. Whittaker (1969) and modified by Lynn Margulis (see also Campbell, Figure 26.9).

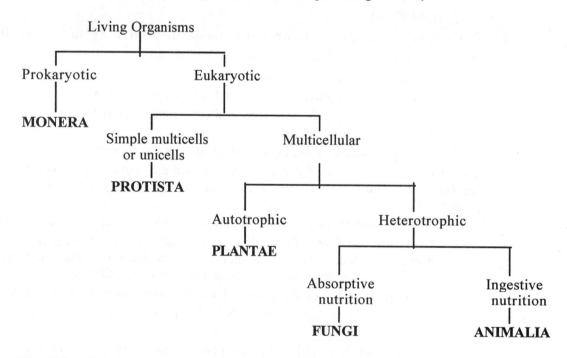

Classifying living systems is a work in progress that reflects our increased understanding of the phylogeny of living organisms.

- Using the tools of molecular systematics, biologists have gathered new data that leads them to challenge the traditional five-kingdom system.

- This new information has reopened issues of biological diversity at the highest taxonomic levels. Three alternative classification systems are outlined below (see also Campbell, Figure 26.10):

Six-kingdom system. The prokaryotes are split into two kingdoms based on molecular evidence for an early evolutionary divergence between eubacteria and archaebacteria.

Eubac-teria	Archae-bacteria	Protista	Plantae	Fungi	Animalia

Three-domain system. This scheme assigns more significance to the ancient evolutionary split between eubacteria and archaebacteria by using a superkingdom taxon called the *domain*. The domain Eukarya includes four kingdoms of eukaryotic organisms.

Bacteria	Archaea	Eukarya (Eukaryotes)

Eight-kingdom system. In addition to two separate prokaryotic kingdoms, this system also splits the protists into three kingdoms.

Bacteria	Archaea	Archae-zoa	Chrom-ista	Pro-tista	Plantae	Fungi	Animalia

REFERENCES

Campbell, N., et al. *Biology*. 5th ed. Menlo Park, California: Benjamin/Cummings, 1998.

Margulis, L., and K.V. Schwartz. *Five Kingdoms: An Illustrated Guide to the Phyla of Life on Earth*. 2nd ed. New York: Freeman, 1987.

Orgel, L.E. "*The Origin of Life on Earth.*" Scientific American, October 1994.

Whittaker, R.H. "New Concepts of Kingdoms of Organisms." *Science* 163: 150-160, 1969.

CHAPTER 27
PROKARYOTES AND THE ORIGINS
OF METABOLIC DIVERSITY

OUTLINE

OBJECTIVES

After reading this chapter and attending lecture, the student should be able to:

1.   List unique characteristics that distinguish archaea from bacteria.

2.   Describe the three-domain system of classification and explain how it differs from previous systems.

3.   Using a diagram or micrograph, distinguish among the three most common shapes of prokaryotes.

4.   Describe the structure and functions of prokaryotic cell walls.

5.   Distinguish between the structure and staining properties of gram-positive and gram-negative bacteria.

6.   Explain why disease-causing gram-negative bacterial species are generally more pathogenic than disease-causing gram-positive bacteria.

7.   Describe three mechanisms motile bacteria use to move.

8.   Explain how prokaryotic flagella work and why they are not considered to be homologous to eukaryotic flagella.

9.   Indicate where photosynthesis and cellular respiration take place in prokaryotic cells.

10.   Explain how organization of the prokaryotic genome differs from that in eukaryotic cells.

11.   Explain what is meant by geometric growth.

12.   List the sources of genetic variation in prokaryotes and indicate which one is the major source.

13.   Distinguish between autotrophs and heterotrophs.

14.   Describe four modes of bacterial nutrition and give examples of each.

15.   Distinguish among obligate aerobes, facultative anaerobes and obligate anaerobes.

16.   Describe, with supporting evidence, plausible scenarios for the evolution of metabolic diversity of prokaryotes

17.   Explain how molecular systematics has been used in developing a classification of prokaryotes.

18.   List the three main groups of archaea, describe distinguishing features among the groups and give examples of each.

19.   List the major groups of bacteria, describe their mode of nutrition, some characteristic features and representative examples.

20.   Explain how endospores are formed and why endospore-forming bacteria are important to the food-canning industry.

21.   Explain how the presence of *E. coli* in public water supplies can be used as an indicator of water quality.

22.   State which organism is responsible for the most common sexually transmitted disease in the United States.

23.   Describe how mycoplasmas are unique from other prokaryotes.

24.   Explain why all life on earth depends upon the metabolic diversity of prokaryotes.

25.   Distinguish among mutualism, commensalism and parasitism.

26.   List Koch's postulates that are used to substantiate a specific pathogen as the cause of a disease.

27.   Distinguish between exotoxins and endotoxins.

28.   Describe how humans exploit the metabolic diversity of prokaryotes for scientific and commercial purposes.

29.   Describe how *Streptomyces* can be used commercially.

KEY TERMS

bacteria	nucleoid region	parasites	decomposers
archaea	binary fission	nitrogen fixation	symbiosis
domains	transformation	obligate aerobes	symbionts
domain Archaea	conjugation	facultative anaerobes	host
domain Bacteria	transduction	obligate anaerobes	mutualism
peptidoglycan	endospores	anaerobic respiration	commensalism
Gram stain	antibiotics	bacteriorhodopsin	parasitism
gram-positive	photoautotrophs	cyanobacteria	parasite
gram-negative	chemoautotrophs	signature sequences	Koch's postulates
capsule	photoheterotrophs	methanogens	exotoxins

| pili (sing., pilus) | chemoheterotrophs | extreme halophiles | endotoxins |
| taxi | saprobes | extreme thermophiles | |

LECTURE NOTES

Appearing about 3.5 billion years ago, prokaryotes were the earliest living organisms and the only forms of life for 2 billion years.

I. The World of Prokaryotes

A. They're (almost) everywhere! *an overview of prokaryotic life*

Prokaryotes dominate the biosphere; they are the most numerous organisms and can be found in all habitats.

- Approximately 4000 species are currently recognized, however, estimates of the actual diversity range from 400,000 – 4 million species
- They are structurally and metabolically diverse.

Prokaryotic cells differ from eukaryotic cells in several ways:

- Prokaryotes are smaller and lack membrane-bound organelles.
- Prokaryotes have cell walls but the composition and structure differ from those found in plants, fungi and protists.
- Prokaryotes have simpler genomes. They also differ in genetic replication, protein synthesis, and recombination.

Prokaryotes, while very small, have a tremendous impact on the Earth.

- A small percentage cause disease.
- Some are decomposers, key organisms in life-sustaining chemical cycles.
- Many form symbiotic relationships with other prokaryotes and eukaryotes. Mitochondria and chloroplasts may have evolved from such symbioses.

B. Bacteria and Archaea are the two main branches of prokaryotic evolution

The traditional five-kingdom system recognizes one kingdom of prokaryotes (Monera) and four kingdoms of eukaryotes (Protista, Plantae, Fungi, and Animalia).

- This system emphasizes the structural differences between prokaryotic and eukaryotic cells.

Recent research in systematics has resulted in questions about the placement of a group as diverse as the prokaryotes in a single kingdom. Two major branches of prokaryotic evolution have been indicated by comparing ribosomal RNA and other genetic products:

- One branch is called the Archaea (formerly archaebacteria).
    - ⇒ Believed to have evolved from the earliest cells
    - ⇒ Inhabit extreme environments which may resemble the Earth's early habitats (hot springs and salt ponds)
- The second branch is called Bacteria (formerly eubacteria).
    - ⇒ Considered the more "modern" prokaryotes, having evolved later in Earth's history
    - ⇒ More numerous than Archaea
    - ⇒ Differ from Archaea in structural, biochemical, and physiological characteristics

This recently acquired molecular data has led to new proposals for the systematic relationships of organisms.

- Initially, researchers including Carl Woese recognized the distinction between Archaea and Bacteria and proposed a six-kingdom system.

- Because the Archaea and Bacteria diverged so early in the history of life, many systematists now favor organizing life into three *domains*, a taxonomic level higher than kingdom (see Campbell, Figure 27.1).

⇒Prokaryotes comprise two of these domains: *domain Archaea* and *domain Bacteria*.

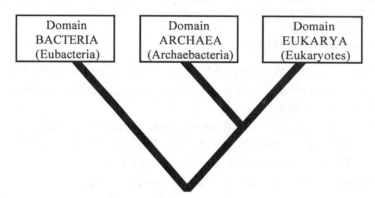

II.   Structure, Function, and Reproduction

A majority of prokaryotes are unicellular, although some aggregate into two-celled to several-celled groups. Others form true, permanent aggregates and some bacterial species have a simple multicellular form with a division of labor between specialized cells.

- Cells have a diversity of shapes, the most common being spheres (cocci), rods (bacilli), and helices (spirilla and spirochetes).
- One rod-shaped species measures 0.5 mm in length, and is the largest prokaryotic cell known (see Campbell, Figure 27.3).
- Most have diameters of 1–5μm (compared to eukaryotic diameters of 10–100μm).

A.   Nearly all prokaryotes have cell walls external to their plasma membranes

A majority of prokaryotes have external cell walls that:

- Maintain the cell shape
- Protect the cell
- Prevent the cell from bursting in a hypotonic environment
- Differ in chemical composition and construction from the cell walls of protists, fungi, and plants

In bacteria, the major cell material is peptidoglycan (cellulose is the main component of plant cell walls); archaea lack peptidoglycan in their cell walls.

- *Peptidoglycan* = Modified sugar polymers cross-linked by short polypeptides
  - Exact composition varies among species.
  - Some antibiotics work by preventing formation of the cross-links in peptidoglycan, thus preventing the formation of a functional cell wall.

Gram stain = A stain used to distinguish two groups of bacteria by virtue of a structural difference in their cell walls

Gram-positive bacteria

- Have simple cell walls with large amounts of peptidoglycan
- Stain blue

Gram-negative bacteria:

- Have more complex cell walls with smaller amounts of peptidoglycan
- Have an outer lipopolysaccharide-containing membrane that covers the cell wall
- Stain pink
- Are more often disease-causing (pathogenic) than gram-positive bacteria
- Lipopolysaccharides:
  - Are often toxic and the outer membrane helps protect these bacteria from host defense systems
  - Impede entry of drugs into the cells, making gram-negative bacteria more resistant to antibiotics

Many prokaryotes also secrete sticky, gelatinous substances that form a layer outside the cell well called a *capsule*. Capsules also aid in adhesion to other cells (to form prokaryotic aggregates or facilitate attachment to host cells).

Some prokaryotes adhere to one another and/or to a substrate by means of a surface appendage called a *pili*. Some pili are specialized for transferring DNA when bacteria conjugate (see Campbell, Figure 27.4).

B. Many prokaryotes are motile

Motile bacteria ($\approx$ 50% of known species) use one of three mechanisms to move:

1. Flagella
    - Prokaryotic flagella differ from eukaryotic flagella in that they are:
      - Unique in structure and function; prokaryotic flagella lack the "9 + 2" microtubular structure and rotate rather than whip back and forth like eukaryotic flagella
      - Not covered by an extension of the plasma membrane
      - One-tenth the width of eukaryotic flagella (see Campbell, Figure 27.5)
    - Filaments, composed of chains of the protein *flagellin*, are attached to another protein hook which is inserted into the basal apparatus.
    - The basal apparatus consist of 35 different proteins arranged in a system of rings which sit in the various cell wall layers.
    - Their rotation is powered by the diffusion of protons into the cell. The proton gradient is maintained by an ATP-driven proton pump.
2. Filaments, which are characteristic of spirochetes, or helical-shaped bacteria
    - Several filaments spiral around the cell inside the cell wall.
    - Similar to prokaryotic flagella in structure, axial filaments are attached to basal motors at either end of the cell. Filaments attached at opposite ends move relative to each other, rotating the cell like a corkscrew.
3. Gliding
    - Some bacteria move by gliding through a layer of slimy chemicals secreted by the organism.
    - The movement may result from flagellar motors that lack flagellar filaments.

Prokaryotic movement is fairly random in homogenous environments but may become directional in a heterogenous environment.

Taxis = Movement to or away from a stimulus

- The stimulus can be light (phototaxis), a chemical (chemotaxis), or a magnetic field (magnetotaxis).

- Movement toward a stimulus is a positive taxis (e.g., positive phototaxis = toward light) while movement away from a stimulus is a negative taxis (e.g., negative phototaxis = away from light).

During taxis (directed movement), bacteria move by running and tumbling movements.

- Enabled by rotation of flagella either counterclockwise or clockwise
- Caused by flagella moving coordinately about each other (for a run), or in separate and randomized movements (for a tumble)

C. The cellular and genomic organization of prokaryotes is fundamentally different from that of eukaryotes

Prokaryotes lack the diverse internal membranes characteristic of eukaryotes. Some prokaryotes, however, do have specialized membranes, formed by invaginations of the plasma membranes (see Campbell, Figure 27.6).

- Infoldings of the plasma membrane function in the cellular respiration of aerobic bacteria.
- Cyanobacteria have thylakoid membranes that contain chlorophyll and that function in photosynthesis.

The prokaryotic genome has only 1/1000 as much DNA as the eukaryotic genome.

Genophore = The bacterial chromosome, usually one double-stranded, circular DNA molecule

- This DNA is concentrated in the *nucleoid region*, and is not surrounded by a membrane; therefore, there is no true nucleus.
- Has very little protein associated with the DNA

Many bacteria also have plasmids.

- *Plasmid* = Smaller rings of DNA having supplemental (usually not essential) genes for functions such as antibiotic resistance or metabolism of unusual nutrients
  - Replicate independently of the genophore
  - Can be transferred between partners during conjugation

While prokaryotic and eukaryotic DNA replication and translation are similar, there are some differences. For example,

- Bacterial ribosomes are smaller and have different protein and RNA content than eukaryotic ribosomes.
  - ⇒ This difference permits some antibiotics (e.g., tetracycline) to block bacterial protein synthesis while not inhibiting the process in eukaryotic cells.

D. Populations of prokaryotes grow and adapt rapidly

Neither mitosis nor meiosis occur in the prokaryotes.

- Reproduction is asexual by *binary fission*.
- DNA synthesis is almost continuous.

Although meiosis and syngamy do not occur in prokaryotes, genetic recombination can take place through three mechanisms that transfer variable amounts of DNA:

- *Transformation* = The process by which external DNA is incorporated by bacterial cells
- *Conjugation* = The direct transfer of genes from one bacterium to another
- *Transduction* = The transfer of genes between bacteria by viruses

Growth in the numbers of cells is geometric in an environment with unlimited resources.

- Generation time is usually one to three hours, although it can be 20 minutes in optimal environments.
- At high concentrations of cells, growth slows due to accumulation of toxic wastes, lack of nourishment, among other things.
- Competition in natural environments is reduced by the release of antibiotic chemicals which inhibit the growth of other species.
- Optimal growth requirements vary depending upon the species.

Some bacteria survive adverse environmental conditions and toxins by producing endospores.

- *Endospore* = Resistant cell formed by some bacteria; contains one chromosome copy surrounded by a thick wall
  - When endospores form, the original cell replicates its chromosome and surrounds one copy with a durable wall. The original surrounding cell disintegrates, releasing the resistant endospore.
  - Since some endospores can survive boiling water for a short time, home canners and food canning industry must take special precautions to kill endospores of dangerous bacteria.
  - May remain dormant for many years until proper environmental conditions return.

Short generation times allow prokaryotic populations to adapt to rapidly changing environmental conditions.

- New mutations and genomes (from recombination) are screened by natural selection very quickly.
- This has resulted in the current diversity and success of prokaryotes as well as the variety of nutritional and metabolic mechanisms found in this group.

III.   Nutritional and Metabolic Diversity

The prokaryotes exhibit some unique modes of nutrition as well as every type of nutrition found in eukaryotes. In addition, metabolic diversity is greater among prokaryotes than eukaryotes.

A. Prokaryotes can be grouped into four categories according to how they obtain energy and carbon

Prokaryotes exhibit a great diversity in how they obtain the necessary resources (energy and carbon) to synthesize organic compounds.

- Some obtain energy from light (phototrophs), while others use chemicals taken from the environment (chemotrophs).
- Many can utilize CO_2 as a carbon source (autotrophs) and others require at least one organic nutrient as a carbon source (heterotrophs).

Depending upon the energy source and the carbon source, prokaryotes have four possible nutritional modes (see Campbell, Table 27.1):

1. *Photoautotrophs* - Use light energy to synthesize organic compounds from CO_2. Include the cyanobacteria. (Actually all photosynthetic eukaryotes fit in this category.)
2. *Chemoautotrophs* - Require only CO_2 as a carbon source and obtain energy by oxidizing inorganic compounds such as H_2S, NH_3 and Fe^{2+}. This mode of nutrition is unique to certain prokaryotes (i.e. archaea of the genus *Sulfobolus*).
3. *Photoheterotrophs* - Use light to generate ATP from an organic carbon source. This mode of nutrition is unique to certain prokaryotes.
4. *Chemoheterotrophs* - Must obtain organic molecules for energy and as a source of carbon. Found in many bacteria as well as most eukaryotes.

1. **Nutritional diversity among chemoheterotrophs**

   Most bacteria are chemoheterotrophs and can be divided into two subgroups: *saprobes* and *parasites.*

   - Saprobes are decomposers that absorb nutrients from dead organic matter.
   - Parasites are bacteria that absorb nutrients from body fluids of living hosts.

   The chemoheterotrophs are a very diverse group, some have very strict requirements, while others are extremely versatile.

   - *Lactobacillus* will grow well only when the medium contains all 20 amino acids, several vitamins, and other organic compounds.
   - *E. coli* will grow on a medium which contains only a single organic ingredient (e.g., glucose or some other substitute).

   Almost any organic molecule can serve as a carbon source for some species.

   - Some bacteria are capable of degrading petroleum and are used to clean oil spills.
   - Those compounds that cannot be used as a carbon source by bacteria are considered non-biodegradable (e.g., some plastics).

2. **Nitrogen metabolism**

   While eukaryotes can only use some forms of nitrogen to produce proteins and nucleic acid, prokaryotes can metabolize most nitrogen compounds.

   Prokaryotes are extremely important to the cycling of nitrogen through ecosystems.

   - Some chemoautotrophic bacteria (*Nitrosomonas*) convert $NH_3 \rightarrow NO_2^-$.
   - Other bacteria, such as *Pseudomonas*, denitrify NO_2^- or NO_3^- to atmospheric N_2.
   - *Nitrogen fixation* ($N_2 \rightarrow NH_3$) is unique to certain prokaryotes (cyanobacteria) and is the only mechanism that makes atmospheric nitrogen available to organisms for incorporation into organic compounds.
   - The nitrogen-fixing cyanobacteria are very self-sufficient, they need only light energy, CO_2, N_2, water, and a few minerals to grow.

3. **Metabolic relationships to oxygen**

   Prokaryotes differ in their growth response to the presence of oxygen.

   - *Obligate aerobes* = Prokaryotes that need O_2 for cellular respiration
   - *Facultative anaerobes* = Prokaryotes that use O_2 when present, but in its absence can grow using fermentation
   - *Obligate anaerobes* = Prokaryotes that are poisoned by oxygen
     - Some species live exclusively by fermentation.
     - Other species use inorganic molecules (other than O_2) as electron acceptors during anaerobic respiration.

B. **The evolution of prokaryotic metabolism was both cause and effect of changing environments on Earth**

   Prokaryotes evolved all forms of nutrition and most metabolic pathways eons before eukaryotes arose.

   - Evolution of these new metabolic capabilities was a response to the changing environment of the early atmosphere.
   - As these new capabilities evolved, they changed the environment for subsequent prokaryotic communities.

Information from molecular systematics, comparisons of energy metabolism, and geological studies about Earth's early atmosphere have resulted in many hypotheses about the evolution of prokaryotes and their metabolic diversity.

1. The origins of metabolism

The first prokaryotes, which evolved 3.5 - 4.0 billion years ago, probably had few enzymes and were very simple. Moreover, living in an environment with virtually no oxygen, they would have been anaerobes.

The universal role of ATP implies that prokaryotes used that molecule for energy very early in their evolution.

- As ATP supplies were depleted, natural selection favored those prokaryotes that could regenerate ATP from ADP, leading to step-by-step evolution of glycolysis and other catabolic pathways.

Glycolysis is the only metabolic pathway common to all modern organisms and does not require O_2 (which was not abundant on early Earth).

- Some extant archaea and other obligate anaerobes that live by fermentation have forms of nutrition believed to be similar to those of the original prokaryotes.

Chemiosmotic ATP synthesis is also an ancient process since it is common to all three domains of life, but it more likely emerged later in prokaryotic evolution .

Many biologists believe that environmental conditions on early Earth would not have generated enough ATP or other organic molecules by abiotic synthesis to support chemoheterotrophs (see Campbell, Figure 27.8).

The most widely accepted view is that the first prokaryotes were chemoautotrophs that obtained their energy from inorganic chemicals and made their own energy currency molecules instead of absorbing ATP.

- Hydrogen sulfide and iron compounds were abundant and early cells could have obtained energy with their use.

2. The origin of photosynthesis

As the supply of free ATP and abiotically produced organic molecules was depleted, natural selection may have favored organisms that could make their own organic molecules from inorganic resources.

Light absorbing pigments in the earliest prokaryotes may have provided protection to the cells by absorbing excess light energy, especially ultraviolet, that could be harmful.

- These energized pigments may have then been coupled with electron transport systems to power ATP synthesis.
- *Bacteriorhodopsin*, the light-energy capturing pigment in the membrane of extreme halophiles (a group of archaea), uses light energy to pump H^+ out of the cell to produce a gradient of hydrogen ions. This gradient provides the power for production of ATP.
- This mechanism is being studied as a model system of solar energy conversion.

Components of electron transport chains that functioned in anaerobic respiration in other prokaryotes also may have been co-opted to provide reducing power. For example, H_2S could be used as a source of electrons and hydrogen for fixing CO_2.

- The nutritional modes of modern purple and green sulfur bacteria are believed the most similar to early prokaryotes.
- The colors of these bacteria are due to bacteriochlorophyll, their main photosynthetic pigment.

3. Cyanobacteria, the oxygen revolution, and the origins of cellular respiration

Eventually, some prokaryotes evolved that could use H_2O as the electron source. Thus evolved *cyanobacteria*, which released oxygen (see Campbell, Figure 27.9).

- Cyanobacteria evolved between 2.5 and 3.4 billion years ago.
- They lived with other bacteria in colonies that resulted in the formation of the stromatolites.

Oxygen released by photosynthesis may have first reacted with dissolved iron ions to precipitate as iron oxide (supported by geological evidence of deposits), preventing accumulation of free O_2.

- Precipitation of iron oxide would have eventually depleted the supply of dissolved iron and O_2 would have accumulated in the seas.
- As seas became saturated with O_2, the gas was released into the atmosphere.
- As O_2 accumulated, many species became extinct while others survived in anaerobic environments (including some archaea) and others evolved with antioxidant mechanisms.
- Aerobic respiration may have originated as a modification of electron transport chains used in photosynthesis. The purple nonsulfur bacteria are photoheterotrophs that still use a hybrid electron transport system between a photosynthetic and respiratory system.
- Other bacterial lineages reverted to chemoheterotrophic nutrition with electron transport chains adapted only to aerobic respiration.

All major forms of nutrition evolved among prokaryotes before the first eukaryotes arose.

IV. Phylogeny of Prokaryotes

A. Molecular systematics is leading to a phylogenetic classification of prokaryotes

The use of molecular systematics (especially ribosomal RNA comparisons) has shown that prokaryotes diverged into the archaea and bacteria lineages very early in prokaryotic evolution.

- Studies of ribosomal RNA indicate the presence of signature sequences.
  - *Signature sequences* = Domain-specific base sequences at comparable locations in ribosomal RNA or other nucleic acids
- Numerous other characteristics differentiate these two domains (see Campbell, Table 27.2).
- A somewhat surprising result of these types of studies has been the realization that archaea have at least as much in common with eukaryotes as they do with bacteria.

1. Domain Archaea

Some unique characteristics of archaea include:

- Cell walls lack peptidoglycan.
- Plasma membranes have a unique lipid composition.
- RNA polymerase and ribosomal protein are more like those of eukaryotes than of bacteria.

The archaea inhabit the most extreme environments of the Earth. Studies of these organisms have identified three main groups:

1. *Methanogens* are named for their unique form of energy metabolism.
   - They use H_2 to reduce CO_2 to CH_4 and are strict anaerobes.

- Some species are important decomposers in marshes and swamps (form marsh gas) and some are used in sewage treatment.
- Other species are important digestive system symbionts in termites and herbivores that subsist on cellulose diets.

2. *Extreme halophiles* inhabit high salinity (15–20%) environments (e.g., Dead Sea).

- Some species simply tolerate extreme salinities while others require such conditions (see Campbell, Figure 27.10).
- They have the pigment bacteriorhodopsin in their plasma membrane which absorbs light to pump H^+ ions out of the cell.
- This pigment is also responsible for the purple-red color of the colonies.

3. *Extreme thermophiles* inhabit hot environments.

- They live in habitats of 60 – 80°C.
- One sulfur-metabolizing thermophile inhabits water of 105°C near deep sea hydrothermal vents.

2. Domain Bacteria

Bacteria comprise a majority of the prokaryotes.

The major groups of bacteria include a very diverse assemblage of organisms. Among the thousands of known species are forms which exhibit every known mode of nutrition and energy metabolism.

Molecular systematics has provided an increased understanding of the once hazy relationships among members of this taxon. At present, most prokaryotic systematists recognize a dozen groups of bacteria (see Campbell, Figures 27.11 and Table 27.3).

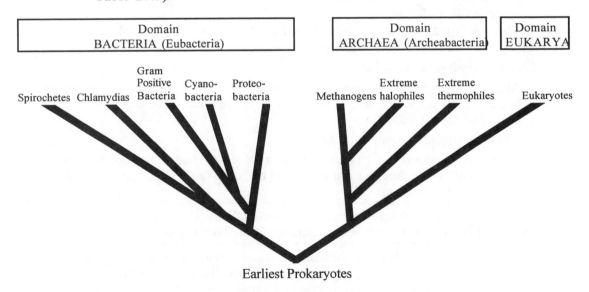

V.   Ecological Impact of Prokaryotes

A. Prokaryotes are indispensable links in the recycling of chemical elements in ecosystems

Prokaryotes are critical links in the recycling of chemical elements between the biological and physical components of ecosystems—a critical element in the continuation of life.

Decomposers = Prokaryotes that decompose dead organisms and waste of live organisms

- Return elements such as carbon and nitrogen to the environment in inorganic forms needed for reassimilation by other organisms, many of which are also prokaryotes.

Autotrophic bacteria = Bacteria that fix CO_2, thus supporting food chains through which organic nutrients pass

Cyanobacteria supplement plants in restoring oxygen to the atmosphere as well as fixing nitrogen into nitrogenous compounds used by other organisms.

Other prokaryotes also support cycling of nitrogen, sulfur, iron and hydrogen.

B. Many prokaryotes are symbiotic

Most prokaryotes form associations with other organisms; usually with other bacterial species possessing complementary metabolisms.

Symbiosis = Ecological relationships between organisms of different species that are in direct contact

- Usually the smaller organism, the *symbiont*, lives within or on the larger *host*.

Three categories of symbiosis:

1. *Mutualism* = Symbiosis in which both symbionts benefit

> Example: Nitrogen-fixing bacteria in root nodules of certain plants fix nitrogen to be used by the plant, which in turn furnishes sugar and other nutrients to the bacteria.

2. *Commensalism* = Symbiosis in which one symbiont benefits while neither helping nor harming the other symbiont

3. *Parasitism* = Symbiosis in which one symbiont (the *parasite*) benefits at the expense of the host

Symbiosis is believed to have played a major role not only in the evolution of prokaryotes, but also in the origin of early eukaryotes.

1.   Prokaryotes and disease

About one-half of all of human disease is caused by bacteria. To cause a disease, the bacteria must invade, evade, or resist the host's internal defenses, long enough to grow and harm the host.

Some pathogens are opportunistic.

- *Opportunistic* = Normal inhabitants of the body that become pathogenic only when defenses are weakened by other factors such as poor nutrition or other infections

> Example: *Streptococcus pneumoniae* lives in the throat of most healthy humans, but can cause pneumonia if the host's defenses are weakened.

Louis Pasteur, Joseph Lister, and others began linking disease to pathogenic microbes in the late 1800s, but Robert Koch was the first to determine a direct connection between specific bacteria and certain diseases.

- Koch identified the bacteria responsible for anthrax and tuberculosis, and his methods established the four criteria to use as guidelines in medical microbiology.

- *Koch's postulates* = Four criteria to substantiate a specific pathogen as the cause for a disease; they are:

  1. Find the same pathogen in each diseased individual.

  2. Isolate the pathogen from a diseased subject and grow it in a pure culture.

3. Use cultured pathogen to induce the disease in experimental animals.

4. Isolate the same pathogen in the diseased experimental animal.

Some pathogens cause disease by growth and invasion of tissues which disrupts the physiology of the host, while others cause disease by production of a toxin. Two major types of toxins have been found:

- *Exotoxins* = Proteins secreted by bacterial cells
  - Can cause disease without the organism itself being present; the toxin is enough
  - Among the most potent poisons known
  - Elicits specific symptoms

> Examples: Botulism toxin from *Clostridium botulinum* and cholera toxin from *Vibrio cholerae*

- *Endotoxins* = Toxic component of outer membranes of some gram-negative bacteria
  - All induce the general symptoms of fever and aches

> Examples: *Salmonella typhi* (typhoid fever) and other species of *Salmonella* that cause food poisoning

Improved sanitation measures and development of antibiotics have greatly reduced mortality due to bacterial diseases.

- Many of the antibiotics now in use are produced naturally by members of the genus *Streptomyces*. In its natural habitat (soil) such materials would reduce competition from other prokaryotes.
- Although beneficial, the excessive and improper use of antibiotics has resulted in the evolution of many antibiotic-resistant bacterial species, which now pose a major health problem.

C. Humans use prokaryotes in research and technology

Humans use the metabolic diversity of bacteria for a multitude of purposes. The range of these purposes has increased through the application of recombinant DNA technology.

- Pharmaceutical companies use cultured bacteria to make vitamins and antibiotics. More than half of the antibiotics used to treat bacterial diseases come from cultures of various species of *Streptomyces* maintained by pharmaceutical companies.
- As simple models of life to learn about metabolism and molecular biology. (*E. coli* is the best understood of all organisms.)
- Methanogens are used to digest organic wastes at sewage treatment plants.
- Some species of pseudomonads are used to decompose pesticides and other synthetic compounds (see Campbell, Figure 27.12).
- Industry uses bacterial cultures to produce products such acetone and butanol.
- The food industry uses bacteria to convert milk into yogurt and cheese.

REFERENCES

Campbell, N., et al. *Biology*. 5th ed. Menlo Park, California: Benjamin/Cummings, 1998.

Tortora, G.J., et al. *Microbiology: An Introduction*. 6th. ed. Redwood City, California: Benjamin/Cummings, 1997.

Woese, Carl R. "Archaeabacteria." *Scientific American*, June 1981.

CHAPTER 28
THE ORIGINS OF EUKARYOTIC DIVERSITY

OUTLINE

I. Introduction to the Protists
   A. Protists are the most diverse of all eukaryotes
   B. Symbiosis was involved in the genesis of eukaryotes from prokaryotes
II. Protist Systematics and Phylogeny
   A. Monophyletic taxa are emerging from modern research in protist systematics
   B. Members of candidate kingdom Archaezoa lack mitochondria and may represent early eukaryotic lineages
   C. Candidate kingdom Euglenozoa includes both autotrophs and heterotrophic flagellates
   D. Surface cavities (alveoli) are diagnostic of candidate kingdom Alveolata
   E. A diverse assemblage of unicellular eukaryotes move by means of pseudopodia
   F. Slime molds have structural adaptations and life cycles that enhance their ecological role as decomposers
   G. Diatoms, golden algae, brown algae, and water molds are members of the candidate kingdom Stramenopila
   H. Structural and biochemical adaptations help seaweeds survive and reproduce at the ocean's margins
   I. Some algae have life cycles with alternating multicellular haploid and diploid generations
   J. Red algae (candidate kingdom Rhodophyta) lack flagella
   K. Green algae and plants probably had a common photoautotrophic ancestor
   L. Multicellularity originated independently many times

OBJECTIVES

After reading this chapter and attending lecture, the student should be able to:
   1. List the characteristics of protists.
   2. Explain why some biologists prefer to use the term *undulipodia* for eukaryotic flagella and cilia.
   3. Briefly summarize and compare the two major models of eukaryotic origins, the autogenous hypothesis and the endosymbiotic hypothesis.
   4. Provide three major lines of evidence for the endosymbiotic hypothesis.
   5. Explain why some critics are skeptical about the bacterial origins for chloroplasts and mitochondria.
   6. Explain why modern biologists recommend expanding the original boundaries of the Kingdom Protista.

7. Explain what is meant by the statement that the Kingdom Protista is a polyphyletic group.

8. List five candidate kingdoms of protists and describe a major feature of each.

9. Describe amoeboid movement.

10. Outline the life cycle of *Plasmodium.*

11. Indicate the organism that causes African sleeping sickness and explain how it is spread and why it is difficult to control.

12. Describe the function of contractile vacuoles in freshwater ciliates.

13. Distinguish between macronuclei and micronuclei.

14. Using diagrams, describe conjugation in *Paramecium caudatum.*

15. Explain how accessory pigments can be used to classify algae and determine phylogenetic relationships among divisions.

16. Distinguish among the following algal groups based upon pigments, cell wall components, storage products, reproduction, number and position of flagella, and habitat:

    a. Dinoflagellata   d. Phaeophyta
    b. Bacillariophyta   e. Rhodophyta
    c. Chrysophyta   f. Chlorophyta

17. Describe three possible evolutionary trends that led to multicellularity in the Chlorophyta.

18. Outline the life cycles of *Chlamydomonas, Ulva,* and *Laminaria* and indicate whether the stages are haploid or diploid.

19. Distinguish between isogamy and oogamy; sporophyte and gametophyte; and isomorphic and heteromorphic generations.

20. Compare the life cycles of plasmodial and cellular slime molds and describe the major differences between them.

21. Provide evidence that the oomycetes are not closely related to true fungi.

22. Give examples of oomycetes and describe their economic importance.

23. Explain the most widely accepted hypothesis for the evolution of multicellularity.

KEY TERMS

acritarchs	apicomplexans	Stramenopila	sporophyte
protozoa	sporozoites	diatoms	gametophyte
algae	ciliates	golden algae	heteromorphic
syngamy	conjugation	water mold	isomorphic
plankton	pseudopodia	white rust	red algae
serial endosymbiosis	detritus	brown algae	green algae
flagellates	amoebas	thallus	lichens
Euglenozoa	heliozoans	holdfast	diatoms
euglenoids	radiolarians	stipe	laminarin
kinetoplastids	forams	blades	isogamy
Alveolata	plasmodial slime molds	alternation of	anisogamy
dinoflagellates	cellular slime molds	generations	oogamy

LECTURE NOTES

Using lenses he developed, Anton von Leeuwenhoek (17th century) was the first to describe the diversity of microscopic protists.

I.   **Introduction to the Protists**

Protists are the earliest eukaryotic descendants of prokaryotes.

Protists arose a billion years before the emergence of other eukaryotes such as plants, fungi, and animals.

Precambrian rock dated to about 2.1 billion years of age contain *acritarchs*, the oldest commonly accepted fossils of protists.

- Remnants of the proper size and structure to be ruptured coats of cysts similar to those of extant protists.
- Adaptive radiation produced a diversity of protists over the next billion years.
- The variations present in these organisms were representative of the structure and function possible in eukaryotic cells.

A.   **Protists are the most diverse of all the eukaryotes**

Because protists vary so much in structure and function, more so than any other group, few other general characteristics besides their being eukaryotes can be cited without exception.

There are about 60,000 extant species of protists.

- Most are unicellular, but colonial forms and even some simple multicellular forms exist.
- Their eukaryotic structure makes even the simplest protist more complex than prokaryotes.
- Primal eukaryotes not only gave rise to current protists, but also to plants, fungi, and animals.

Protists are considered the simplest eukaryotic organisms because most are unicellular.

- At the cellular level, protists are extremely complex.
- The unicellular protist is not analogous to a single plant or animal cell, but is a complete organism.
- The single cell of a protist must perform all the basic functions carried out by the specialized cells of plants and animals.

Protists are metabolically divers, and as a groups, they are the most nutritionally diverse of all eukaryotes.

- Almost all protists are aerobic, using mitochondria for cellular respiration.
- Anaerobic forms lack mitochondria and live in anaerobic environments or have mutalistic respiring bacteria.
- Some may be photoautotrophic, heterotrophic, or mixotrophic (combining photosynthesis and heterotrophy).
- The different modes of nutrition are used to separate protists into three categories: photosynthetic forms are typically referred to as algae, ingestive forms as protozoa, and absorptive protists. (Although the terms protozoa and algae are commonly used, they have no basis in phylogeny and no significance in taxonomy.)

Most protists have flagella or cilia (not homologous to prokaryotic flagella) at some time in life cycle.

- Eukaryotic cilia and flagella are extensions of the cytoplasm. (Prokaryotic flagella are attached to the cell surface.)
- These cilia and flagella have the same basic 9 + 2 microtubular ultrastructure, but cilia are shorter and more numerous.

The life cycles of protists are quite variable.

- Unique mitotic divisions occur in many groups.

- Some can reproduce asexually
- Some can also reproduce sexually or at least use *syngamy* (fusion of gametes) to trade genes between asexual reproductive episodes.
- Some form resistant cysts when stressed by harsh environments.

Protists are found in almost all moist environments: the seas, freshwater systems, and moist terrestrial habitats such as damp soil and leaf litter.

- They are important components of marine and freshwater plankton.
  - *Plankton* = Communities of organisms, mostly microscopic, that drift passively or swim weakly near the surface of oceans, ponds, and lakes
- Many are bottom dwellers in freshwater and marine habitats where they attach to rocks or live in the sand and silt.
- Photosynthetic species form mats at the still-water edges of lakes and ponds where they provide a food source for other protists.

Some protists are free-living, while others are symbiotic species found in the body fluids, tissues, or cells of host organisms.

- The nature of symbiosis ranges from mutualism to parasitism and many are important pathogens.

B. Symbiosis was involved in the genesis of eukaryotes from prokaryotes

There is a greater difference between prokaryotic and eukaryotic cells than between the cells of plants and animals.

During the genesis of eukaryotes, the following cellular structures and process unique to eukaryotes arose:

- A membrane-bound nucleus
- Mitochondria, chloroplasts, and the endomembrane system
- A cytoskeleton
- 9 + 2 flagella
- Multiple chromosomes consisting of linear DNA molecules compactly arranged with proteins
- Life cycles that involve mitosis, meiosis, and sex

The small size and simpler construction of the prokaryotic cell has many advantages but also imposes a number of limitations.

> Examples:
> - The number of metabolic activities that can occur at one time is smaller.
> - The smaller size of the prokaryotic genome limits the number of genes which code for enzymes controlling these activities.

While prokaryotes are extremely successful, natural selection resulted in increasing complexity in some groups, trending toward:

- Multicellular forms, such as the cyanobacteria, which have different cells types with specialized functions
- Evolution of complex prokaryotic communities in which each species benefits from the metabolic activities of other species
- Compartmentalization of different functions within single cells
  ⇒ The first eukaryotes resulted from this solution.

The evolution of the compartmentalized nature of eukaryotic cells may have resulted from two processes.

1. Specialization of plasma membrane invaginations

- Invaginations and subsequent specializations may have given rise to the nuclear envelope, endoplasmic reticulum, Golgi apparatus, and other components of the endomembrane system (see Campbell, Figure 28.2a).
2. Endosymbiotic associations of prokaryotes may have resulted in the appearance of some organelles.
    - Mitochondria, chloroplasts, and some other organelles evolved from prokaryotes living within other prokaryotic cells.

The hypothesis of *serial endosymbiosis* proposes that certain prokaryotic species, called *endosymbionts,* lived within larger prokaryotes. This theory was developed extensively by Lynn Margulis of University of Massachusetts (see Campbell, Figure 28.2b).

- Hypothesis focuses mainly on mitochondria and chloroplasts.
- Chloroplasts are descended from endosymbiotic photosynthesizing prokaryotes, such as cyanobacteria, living in larger cells.
- Mitochondria are postulated to be descendants of prokaryotic aerobic heterotrophs.
    ⇒ May have been parasites or undigested prey of larger prokaryotes.
    ⇒ The association progressed from parasitism or predation to mutualism.
    ⇒ As the host and endosymbiont became more interdependent, they integrated into a single organism.
- Many extant organisms are involved in endosymbiotic relationships.

Evidence for the endosymbiotic origin of mitochondria and chloroplasts includes the similarities between these organelles and prokaryotes.

- Are of appropriate size to be descendants of bacteria
- Have inner membranes containing several enzymes and transport systems similar to those on prokaryotic plasma membranes
- Replicate by splitting processes similar to binary fission present in prokaryotes
- Have DNA that is circular and not associated with histones or other proteins, as in prokaryotes
- Contain their own tRNA, ribosomes, and other components for DNA transcription and translation into proteins
- Chloroplasts have ribosomes more similar to prokaryotic ribosomes (in size, biochemical characteristics, and antibiotic sensitivity) than to eukaryotic ribosomes.
- Mitochondrial ribosomes vary, but are also more similar to prokaryotic ribosomes.

Molecular systematics lends even more evidence to support the endosymbiotic theory.

- The rRNA of chloroplasts is more similar in base sequence to RNA from certain photosynthetic eubacteria than to rRNA in eukaryotic cytoplasm.
    ⇒ Chloroplast rRNA is transcribed from genes in the chloroplast while eukaryotic rRNA is transcribed from nuclear DNA.
- Mitochondrial rRNA also has a base sequence which supports a prokaryotic origin.

A comprehensive theory for the origin of eukaryotic cells must also include the evolution of:

- 9 + 2 flagella and cilia, which are analogous, not homologous, to prokaryotic flagella.
- The origins of mitosis and meiosis which also utilize microtubules.
    ⇒ Mitosis made it possible for large eukaryotic genomes to be reproduced.

⇒ Meiosis is essential to sexual reproduction.

⇒ Protists have the most varied sexual life histories of the eukaryotes.

II.   Protist Systematics and Phylogeny

A. Monophyletic taxa are emerging from modern research in protist systematics

Classification schemes and the phylogeny they reflect are based on available information.

- These presentations are tentative and often change as additional information becomes available.

In 1969, Robert H. Whittaker popularized the five-kingdom taxonomic system and placed only unicellular eukaryotes in the kingdom Protista.

- During the 1970s and 1980s, the Kingdom Protista was expanded to include some multicellular organisms earlier classified as either plants or fungi.
- Studies of cell ultrastructure and life cycle details formed the basis for such taxonomic transfers.
    - ⇒ Seaweeds were found to exhibit characteristics which indicated a closer relationship with certain algae than to plants.
    - ⇒ Slime molds and water molds were found to be more closely related to certain protozoans than to fungi.
- The tendency was to place all eukaryotes that could not comfortably be fitted into the plants, fungi, or animals into this kingdom.

Molecular systematics, especially rRNA comparisons, have stimulated three main trends in eukaryotic systematics and taxonomy over the last decade.

1. Reassessment of the number and membership of protistan phyla
   - The phylum Sarcodina once housed all unicellular organisms that possessed pseudopodia, current classification splits these organisms into several phyla.
2. Arrangement of the phyla into a cladogram based largely on what molecular methods and cell structure comparisons reveal about evolutionary relationships of protists.
3. Reevaluation of the five-kingdom system and debate about the addition of new kingdoms.

At present, most systematists working on the origins of eukaryotes consider the Kingdom Protista and the five-kingdom system obsolete. This is based on the observation that the Kingdom Protista is polyphyletic.

The organization of protists into three groups as in the eight-kingdom system is still polyphyletic.

In light of current research, the organization of protists into five groups, candidate kingdoms (Archaezoa, Euglenozoa, Alveolata, Stramenopila, and Rhodophyta), is indicated. All but one of the five candidate kingdoms, Archaezoa, is monophyletic.

B. Members of candidate kingdom Archaezoa lack mitochondria and may represent early eukaryote lineages

An ancient lineage of eukaryotes branched away from the eukaryotic tree perhaps as early as two billion years ago (see Campbell, Figure 28.2). This group is referred to as the Archaezoa and contains only a few phyla.

- These organisms lack mitochondria and plastids and have relatively simple cytoskeletons.
- Their ribosomes have some characteristics more closely aligned with prokaryotes than with eukaryotes; rRNA sequencing indicates a closer relationship.

Giardia lamblia, a diplomonad, is a modern representative of the archaezoa (see Campbell, Figure 28.4).

- It is a flagellated, unicellular eukaryote that is parasitic in the human intestine.
- It is most commonly transmitted in the cyst form through water contaminated with human feces.

Giardia's importance to evolutionary biologists is related more to its characteristics than to its role as a human parasite.

- Diplomonads have two separate haploid nuclei which produce a "face-like" appearance.
    - Dual nuclei may be a vestige of early eukaryotic evolution.
- Prokaryotes have haploid genomes and some researchers postulate that early eukaryotes had a single haploid nucleus bounded by a nuclear envelope.
- In most modern eukaryotes, the diploid stages in the life cycle result from the fusion of haploid nuclei which form the diploid nucleus.
    - Diplomonads may represent an early mechanism in the evolution of diploidy in eukaryotes.

If the diplomonads diverged from the eukaryotic lineage before the process of nuclear fusion and meiosis evolved, their dual nuclei may be a clue to the past.

- This coupled with the absence of mitochondria in this group and other archaezoans is consistent with an origin occurring before the endosymbiotic relationships that gave rise to mitochondria in aerobic species.

C. Candidate kingdom Euglenozoa includes both autotrophic and heterotrophic flagellates

Protists with flagella are often informally referred to as *flagellates*.

Two groups of flagellates make up the monophyletic candidate kingdom Euglenozoa: euglenoids and kinetoplastids.

Euglenoids (e.g., *Euglena*) have the following characteristics:

- Anterior pocket or chamber from which one or two flagella project
- Production of paramylum, a glucose polymer
- Varying modes of nutrition depending on species
    - Autotrophic
    - Mixotrophic - chiefly autotrophic with some requirement for organic molecules (e.g., vitamins)
    - Heterotrophic

Kinetoplastids have the following characteristics:

- Possess a single large mitochondrion associated with a unique organelle, the kinetoplast, that contains extranuclear DNA
- Symbiotic; some are pathogenic to hosts
    - Species of *Trypanosoma* cause African sleeping sickness and are spread by the bite of the tsetse fly (see Campbell, Figure 28.5).

D. Subsurface cavities (alveoli) are diagnostic of candidate kingdom Alveolata

This candidate kingdom encompasses photosynthetic flagellates (dinoflagellates), a group of parasites (apicomplexans), and a distinctive group that move by means of cilia (ciliates).

All alveolates have small membrane-bound cavities, or alveoli, under their cell surfaces.

The function of alveoli is unknown; however, they may help to:

- Stabilize the cell surface

- Regulate water and ion transport

1. Dinoflagellates

Dinoflagellates are components of phytoplankton that provides the foundation of most marine food chains.

- May cause *red tides* by explosive growth (bloom)
  - ⇒ These dinoflagellates produce a toxin that is concentrated by invertebrates, including shellfish.
  - ⇒ The toxin is dangerous to humans consuming shellfish and causes the condition known as paralytic shellfish poisoning.
- Most are unicellular, some are colonial
- Cell surface is reinforced by cellulose plates with flagella in perpendicular grooves, creating its whirling movement and resulting in a characteristic shape (see Campbell, Figure 28.6)
- Some live as photosynthetic symbionts of the cnidarians that build coral reefs
- Some lack chloroplasts and live as parasites; a few carnivorous species are known
- Have brownish plastids containing chlorophyll *a*, chlorophyll *c* and a mix of carotenoids, including *peridinin* (found only in this phylum)
- Food is stored as starch
- Chromosomes lack histones and are always condensed
- Has no mitotic stages
- Kinetochores are attached to the nuclear envelope and chromosomes distributed to daughter cells by the splitting of the nucleus

2. Apicomplexans

All member of apicomplexans (formerly called sporozoans) are parasites of animals.

- The infectious cells produced in the life cycle are called *sporozoites.*
- The apex of sporozoites has organelles for penetrating host cells and tissues; the phylum is named for these apical organelles.
- Life cycles are intricate having both sexual and asexual reproduction, often requiring two or more different host species.

Several species of *Plasmodium* cause malaria (see Campbell, Figure 28.7).

- *Anopheles* mosquitoes serve as the intermediate host and humans as the final host.
- The incidence of malaria was greatly reduced by the use of insecticides against mosquitoes in the 1960s.
- More recently, incidence of malaria has increased due to insecticide-resistant strains of mosquitoes and drug-resistant strains of *Plasmodium.*
  - ⇒ This is a relatively common (300 million new cases per year) and potentially fatal tropical disease resulting in about two million deaths each year.

There has been little success in developing a vaccine against *Plasmodium.*

- The human immune system has little effect on the parasite.
  - ⇒ *Plasmodium* spends most of its life cycle in blood cells or liver cells.
  - ⇒ *Plasmodium* also has the ability to alter its surface proteins.
- The most promising treatment may lie in inhibiting the function of one or more processes in a *Plasmodium* plastid.

3. Ciliates (Ciliophorans)

Species within this group use cilia to move and feed.

- Most ciliates exist as solitary cells in fresh water.
- Cilia are relatively short and beat in synchrony.
- The cilia are associated with a submembranous system that coordinates the movement of thousands of cilia.
- Cilia may be dispersed over surface, or clustered in fewer rows or tufts.
- Some species move on leg-like *cirri* (many cilia bonded together).
- Other species have rows of tightly packed cilia that function together as locomotor membranelles (e.g., *Stentor*).

Ciliates are among the most complex of all cells (see Campbell, Figure 28.8).

They possess two types of nuclei: one large macronucleus and from one to several small micronuclei.

- Characteristics of the macronucleus:
  - ⇒ It is large and has over 50 copies of the genome.
  - ⇒ Genes are packaged in a large number of small units, each with hundreds of copies of just a few genes.
  - ⇒ It controls everyday functions of the cell by synthesizing RNA.
  - ⇒ It is also necessary for asexual reproduction during binary fission. The macronucleus elongates and splits instead of undergoing mitosis.
- Characteristics of the micronucleus:
  - ⇒ It is small and may number from 1 to 80 micronuclei, depending on the species.
  - ⇒ It does not function in growth, maintenance or asexual reproduction.
  - ⇒ It functions in *conjugation*, a sexual process which produces genetic variation (see Campbell, Figure 28.9).
- Note that in ciliates, meiosis and syngamy are separate from reproduction.

E. A diverse assemblage of unicellular eukaryotes move by means of pseudopodia

Three groups of unicellular eukaryotes move and feed by means of cellular extensions called *pseudopodia*.

The mode of nutrition among member groups varies:

- Most are heterotrophs.
  - Some actively seek and feed on bacteria, other protists, or *detritus* (dead organic matter)
  - Some are symbiotic species, including some parasites that cause human diseases

1. Rhizopods (amoebas)

This group (*Rhizopoda* = rootlike feet) includes the *amoebas* and their relatives.

- Simplest of protists; all are unicellular
- No flagellated stages in life cycle
- *Pseudopodia* form as cellular extensions and function in feeding and movement (see Campbell, Figure 28.10)
  - ⇒ The cytoskeleton of microtubules and microfilaments functions in amoeboid movement.
- All reproduction is by asexual mechanisms: no meiosis or sexual reproduction are known to occur.

- During mitosis, spindle fibers form, but typical stages of mitosis are not apparent in most species.
  ⇒ The nuclear envelope persists during cell division in many genera.
- Rhizopods inhabit freshwater, marine, and soil habitats.
- Most are free-living, although some are parasitic.

2. Actinopods (heliozoans and radiozoans)

Actinopods (= ray feet) possess axopodia, a slender form of pseudopodia.

- *Axopodia* = Projections reinforced by bundles of microtubules thinly covered by cytoplasm
- Axopodia increase the surface area that comes into contact with the surrounding water.
- They help the organisms float and function in feeding.
- Small protists and other microorganisms stick to the axopodia and are phagocytized by the thin layer of cytoplasm. They are carried to the main portion of the cell by cytoplasmic streaming.

The two main groups of actinopods are the heliozoans and the radiolarians.

- Most are planktonic.
- *Heliozoans* live primarily in fresh water.
- *Radiolarians* are primarily marine and have delicate shells, usually made of silica.

3. Foraminiferans (forams)

Forams have porous, multi-chambered shells of organic material hardened by calcium carbonate.

- Forams are almost all marine with most living in the sand or attached to algae and rocks; some are planktonic.
- Cytoplasmic strands extend through the shell's pores and function in swimming, feeding, and shell formation.
- Many have symbiotic algae living beneath the shell that provide nourishment through photosynthesis.
- 90% of the described species are fossils.
- Foram shells are an important component of sediments and sedimentary rocks (see Campbell, Figure 28.12).

F. Slime molds have structural adaptations and life cycles that enhance their ecological role as decomposers

Two groups of protists called slime molds resemble fungi in appearance and lifestyle.

The resemblance of slime molds and water molds to true fungi is a result of convergent evolution of filamentous body structure.

- A filamentous body structure increases exposure to the environment and enhances their roles as decomposers.
- Slime molds differ from true fungi in their cellular organization, reproduction, and life cycles.

1. Plasmodial slime molds (Myxomycota)

The *plasmodial slime molds* are all heterotrophs and many are brightly pigmented.

Plasmodium = Feeding stage of life cycle consisting of an amoeboid, coenocytic (multi-nucleated cytoplasm undivided by membranes) mass (see also Campbell, Figure 28.13).

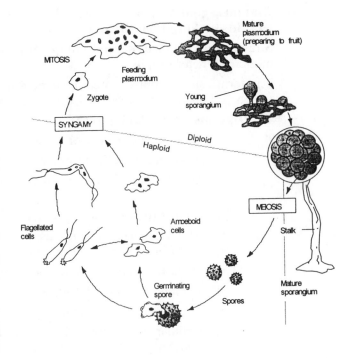

- In most species, the nuclei of plasmodia are diploid and exhibit synchronous mitotic divisions.

- Cytoplasmic streaming within the plasmodium helps distribute nutrients and oxygen.

- Engulfs food by phagocytosis as it grows by extending pseudopodia.

- Live in moist soil, leaf mulch, and rotting logs.

- When stressed by drying or lack of food, the plasmodium ceases growth and forms sexually reproductive structures called fruiting bodies, or sporangia.

2. Cellular slime molds (Acrasiomycota)

This group possesses the following features:

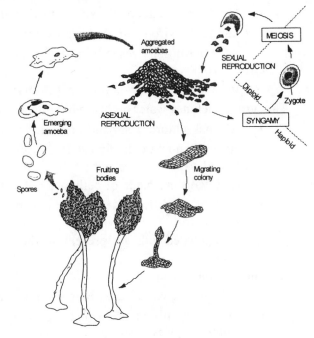

- Feeding stage of life cycle consists of individual, solitary haploid cells.

- When the food supply is depleted, cells aggregate to form a mass similar to those of myxomycota but cells remain separate (not coenocytic) (see also Campbell, Figure 28.14).

- Fruiting bodies function in *asexual* reproduction (unlike plasmodial slime molds).

- Only a few have flagellated stages.

G. Diatoms, golden algae, brown algae, and water molds are members of the candidate kingdom Stramenopila

Stramenopila includes several groups of photosynthetic autotrophs (algae) and numerous heterotrophs.

The term "stramenopila" refers to numerous, fine, hairlike projections on the flagella, which is a characteristic feature of this group

1. Diatoms (Bacillariophyta)

- Diatoms are yellow or brown in color due to the presence of brown plastids.
- Many have a gliding movement produced by chemical secretions
- Usually reproduce asexually; sexual stages (egg and sperm production) are rare
- Some produce resistant cysts
- Mostly unicellular organisms with overlapping glasslike walls of hydrated silica in an organic matrix (see Campbell, Figure 28.15)
- Have the same photosynthetic pigments as in Chrysophyta
- Are components of freshwater and marine plankton
- Store food in laminarin (a glucose polymer) and in the form of oil
- Cellular regulation of ions counteracts weight of walls and maintains buoyancy

2. Golden algae (Chrysophyta)

- Algae named for their color, which results from accessory pigments (yellow and brown carotenoids and xanthophyll)
- Live among freshwater and marine plankton
- Most are unicellular, but some are colonial (see Campbell, Figure 28.16)
- Most are biflagellated, with both flagella attached near one end of the cell
- Survive environmental stress by forming resistant cysts
  - Microfossils resembling these cysts have been found in Precambrian rocks.

3. Water molds and their relatives (Oomycota)

This group includes water molds, white rusts, and downy mildews.

- All lack chloroplasts and are heterotrophic
- Have coenocytic hyphae (fine, branching filaments) that are analogous to fungal hyphae.
- Cell walls are made of cellulose rather than the chitin found in true fungi
- Diploid condition in the life cycle prevails in most species (see Campbell, Figure 28.17)
- Biflagellated cells are present in the life cycles, while fungi lack flagellated cells

In water molds:

- A large egg is fertilized by a smaller sperm cell to form a resistant zygote.
- These organisms are usually decomposers that grow on dead algae and animals in fresh water.
- Some are parasitic and grow on injured tissue, but they may also grow on the skin and gills of fish.

White rusts and downy mildews:

- Are usually parasitic on terrestrial plants

- Disperse by windblown spores, but also form flagellated zoospores at some point in their life cycle
- Some of the most important plant pathogens are members of this phylum.

4. Brown algae (Phaeophyta)

This is the largest and most complex of the algae. Color is due to accessory pigments.

- All are multicellular and most are marine inhabitants
- Have chlorophyll *a*, chlorophyll *c*, and the carotenoid pigment *fucoxanthin*
- Store carbohydrate food reserves in the form of laminarin
- Cell walls made of cellulose and *algin*

Many eukaryotes commonly called seaweeds are brown algae; however, red algae and green algae also are components of seaweeds.

H. Structural and biochemical adaptations help seaweeds survive and reproduce at the ocean's margins

Seaweeds are large, multicellular marine algae, which are found in the intertidal and subtidal zones of coastal waters.

- They are a diverse group and include members of the Phaeophyta (brown algae), Rhodophyta (red algae), and Chlorophyta (green algae).
- **The following emphasizes adaptations found in the red algae, however, many of these adaptations also apply to the brown algae and green algae seaweeds.**

The habitat of seaweeds, particularly the intertidal zone, poses several challenges to the survival of these organisms.

- Movement of the water due to wave action and winds produces a physically active habitat.
- Tidal rhythms result in the seaweeds being alternately covered by seawater and exposed to direct sunlight and the drying conditions of the air.

Seaweeds have evolved several unique structural and biochemical adaptations to survive the conditions of their habitats.

Structural adaptations found in seaweeds are a result of their complex multicellular anatomy. Some forms have differentiated tissues and organs analogous to those of plants.

- The body of a seaweed is called a *thallus*. It is plantlike in appearance but has no true roots, stems, or leaves.
- A thallus consists of a rootlike *holdfast* (maintains position), a stemlike *stipe* (supports the blades), and leaflike *blades* (large surfaces for photosynthesis) (see Campbell, Figure 28.18).
- Floats, which help suspend blades near the water surface, are present in some brown algae.
- Brown algae known as giant kelp occur beyond the intertidal zone where less harsh conditions exist and may have stipes which reach a length of up to 60 m (see Campbell, Figure 28.19).

Biochemical adaptations in some seaweeds reinforce the anatomical adaptations and enhance survival.

- Cellulose cell walls contain gel-forming polysaccharides (algin in brown algae; carageenan in red algae), which cushion the thalli against wave action and prevent desiccation during low tide.

- Some red algae retard grazing by marine invertebrates by incorporating large amounts of calcium carbonate into their cell walls, rendering them unpalatable.

Seaweeds are used by humans in a variety of ways:

- Brown and red alga are used as food in many parts of Asia.
- Algin, agar, and carageenan are extracted and used as thickeners for processed foods and lubricants in oil drilling.
- Agar is also used as a microbiological culture media.

I. Some algae have life cycles with alternating multicellular haploid and diploid generations

A variety of life cycles in the brown algae, the most complex involve alternation of generations. Also found in certain groups of red algae and green algae.

Alternation of generations = Alternation between multicellular haploid forms and multicellular diploid forms in a life history

The life cycle of *Lamanaria* is an example of a complex life cycle with an alteration of generations (see Campbell, Figure 28.20).

- The diploid individual is called a *sporophyte* because it produces reproductive cells called spores.
- The haploid individual is called the *gametophyte* because it produces gametes.
- The sporophyte and gametophyte generations of the life cycle take turns producing one another.
  - Spores released from the sporophyte develop into gametophytes.
  - Gametophytes produce gametes which fuse (fertilization) to form a diploid zygote that develops into a sporophyte.
- In *Laminaria,* the sporophyte and gametophyte generations are said to be *heteromorphic,* because they are morphologically different.
- In *Ulva*, a green algae exhibiting alteration of generations, the generations are referred to as *isomorphic* because they look alike.

J. Red algae (candidate kingdom Rhodophyta) lack flagella

The defining characteristic of red algae is that they do not have flagella in any of their life cycle stages, unlike other eukaryotic algae

- Current data suggests that red algae aren't ancient, but that flagella were lost during their evolution.
- Red algae probably arose about the same time as Stamenopiles.

Red algae are primarily warm, tropical, marine inhabitants, although some are found in fresh water and soil. Other features include:

- Chlorophyll *a*, carotenoids, phycobilins, and chlorophyll *d* in some
- Red color of plastids due to the accessory pigment, phycoerythrin
  ⇒ Phycoerythrin is a phycobilin, a pigment found only in red algae and cyanobacteria.
- Color of the thallus may vary (even in a single species) with depth, as pigmentation changes to optimize photosynthesis.
  - Deep water forms are almost black, moderate depth forms are red, and shallow water forms are green.
  - One species has been discovered near the Bahamas at a depth of 260 meters.
  - Some tropical species lack pigmentation and survive as parasites on other red algae.

- Carbohydrate food reserves stored as floridean starch (similar to glycogen).
- Cell walls are cellulose with agar and carageenan.
- Most red algae are multicellular and the largest are designated as seaweeds.
- Most thalli are filamentous and are often branched forming an interwoven lacy network (see Campbell, Figure 28.21)

All red algae reproduce sexually.

- Have no flagellated stages, unlike other algal protists
- Alternation of generations is common

K. Green algae and plants probably had a common photoautotrophic ancestor

Green algae (Chlorophyta) are named for their grass-green chloroplasts, which are similar in ultrastructure and pigment composition to the organelles of organism s traditionally referred to as plants.

Molecular and structural features suggest that green algae and plants are closely related and were derived from a common ancestor different from that giving rise to stramenopiles and red algae.

- Some systematists argue for the inclusion of green algae in the plant kingdom.

At least 7000 species of green algae are known; most are freshwater, some are marine.

- Many unicellular types live as plankton, inhabit damp soil, coat snow surfaces, or are symbionts with protozoa or invertebrates.
- When living mutualistically with fungi they form the association known as *lichens*.
- Colonial forms are often filamentous ("pond scum").
- Multicellular forms may have large, complex structures resembling true plants and comprise a group of seaweeds.

Evolutionary trends probably produced colonial and multicellular forms from flagellated unicellular ancestors.

1. Formation of colonies of individual cells, as seen in *Volvox* (see Campbell, Figure 28.22a)
2. Repeated division of nuclei with no cytoplasmic division, as in *Caulerpa* (see Campbell, Figure 28.22b)
3. Formation of true multicellular forms, as in *Ulva* (see Campbell, Figure 28.22c)

Most green algae have complex life histories involving sexual and asexual reproductive stages.

- Nearly all reproduce sexually by way of biflagellated gametes.
- Some are conjugating algae (e.g., *Spirogyra*), which produce amoeboid gametes (see Campbell, Figure 28.23).

The life cycle of *Chlamydomonas* is a good example of the life history of a unicellular chlorophyte. Note: a mature *Chlamydomonas* is a single haploid cell (see Campbell, Figure 28.24) .

- During asexual reproduction, the flagella are resorbed and the cell divides twice by mitosis to form four cells (more in some species).
  ⇒ The daughter cells develop and emerge as swimming zoospores. Zoospore development includes formation of flagella and cell walls.
  ⇒ Zoospores grow into mature cells, thus completing asexual reproduction.

Sexual reproduction is stimulated by environmental stress from such things as a shortage of nutrients, drying of the pond, or some other factor.

⇒ During sexual reproduction, many gametes are produced by mitotic division within the wall of the parent cell. The gametes escape the parent cell wall.

⇒ Gametes of opposite mating strains (+ and −) pair off and cling together by the tips of their flagella.

⇒ The gametes are morphologically indistinguishable and their fusion is known as *isogamy*.

⇒ The slow fusion of the gametes forms a diploid zygote which secretes a resistant coat that protects it from harsh environmental conditions.

⇒ When dormancy of the zygote is broken, four haploid individuals (two of each mating type) are produced by meiosis.

⇒ These haploid cells emerge from the coat and develop into mature cells, thus completing the sexual life cycle.

Many features of *Chlamydomonas* sex are believed to have evolved early in the chlorophyte lineage. Using this basic life cycle, many refinements that evolved among the chlorophytes have been identified.

• Some green algae produce gametes that differ from vegetative cells and, in some species, the male gamete differs in size or morphology from the female gamete (*anisogamy*).

• Many species exhibit *oogamy*, a type of anisogamy in which a flagellated sperm fertilizes a nonmotile egg.

• Some multicellular species also exhibit alternation of generations.

⇒ *Ulva* produces isomorphic thalli for its diploid sphorophyte and haploid gametophyte (see Campbell, Figure 28.25).

L. Multicellularity originated independently many times

Early eukaryotes were more complex than prokaryotes and this increase in complexity allowed for greater morphological variations to evolve.

• Extant protists are more complex in structure and show a greater diversity of morphology than the simpler prokaryotes.

• The ancestral stock which gave rise to new waves of adaptive radiations were the protists with multicellular bodies.

Multicellularity evolved several times among the early eukaryotes and gave rise to the multicellular algae, plants, fungi, and animals.

Most researchers believe that the earliest multicellular forms arose from unicellular ancestors as colonies or loose aggregates of interconnected cells (see Campbell, Figure 28.26).

• Multicellular algae, plants, fungi, and animals probably evolved from several lineages of protists that formed by amalgamations of individual cells.

• Evolution of multicellularity from colonial aggregates involved cellular specialization and division of labor.

⇒ The earliest specialization may have been locomotor capabilities provided by flagella.

⇒ As cells became more interdependent, some lost their flagella and performed other functions.

• Further division of labor may have separated sex cells from somatic cells.

⇒ This type of specialization and cooperation is seen today in colonial species such as *Volvox* (a green alga).

⇒ Gametes specialized for reproduction are dependent on somatic cells while developing.

⇒ The evolution of telomerase enzymes, which add nucleotides to the ends of DNA (telomeres; see chapter 19) and protects genes from degradation during DNA replication, may have been involved in gamete formation.

- Many additional steps were involved in the evolution of specialized somatic cells capable of performing all the nonreproductive function in a multicellular organism.

    ⇒ Extensive division of labor exists among the different tissues that comprise the thalli of seaweeds.

- Multicellular forms more complex than filamentous algae appeared approximately 700 million years ago.

- A variety of animal fossils has been found in late Precambrian strata and many new forms evolved in the Cambrian period (about 570 million years ago).

- Seaweeds and other complex algae were also abundant during the Cambrian period.

- Primitive plants are believed to have evolved from certain green algae living in shallow waters about 400 million years ago.

REFERENCES

Bold, Harold C., and Michael J. Wynne. *Introduction to the Algae: Structure and Reproduction.* 2nd ed. Englewood Cliffs, New Jersey: Prentice-Hall, Inc., 1985.

Campbell, N., et al. *Biology.* 5th ed. Menlo Park, California: Benjamin/Cummings, 1999.

Lee, John J., Seymour H. Hutner and Eugene C. Bovee. *An Illustrated Guide to the Protozoa.* Lawrence, Kansas: Allen Press, Inc., 1985.

CHAPTER 29
PLANT DIVERSITY I:
THE COLONIZATION OF LAND

OUTLINE

I. An Overview of Plant Evolution
   A. Structural, chemical, and reproductive adaptations enabled plants to colonize land
   B. The history of terrestrial adaptation is the key to modern plant diversity

II. The Origins of Plants
   A. Plants probably evolved from green algae called charophytes
   B. Alternation of generations in plants may have originated by delayed meiosis
   C. Adaptations to shallow water as preadapted plants for living on land

III. Bryophytes
   A. The embryophyte adaptation evolved in bryophytes
   B. The gametophyte is the dominant generation in the life cycles of byrophytes
   C. The three divisions of bryophytes are mosses, liverworts, and hornworts

IV. The Origin of Vascular Plants
   A. Additional terrestrial adaptations evolved as vascular plants descended from bryophyte-like ancestors
   B. The branched sporophytes of vascular plants amplified the production of spores and made complex bodies possible

V. Seedless Vascular Plants
   A. A sporophyte-dominant life cycle evolved in seedless vascular plants
   B. The three divisions of seedless vascular plants are lycophytes, horsetails, and ferns
   C. Seedless vascular plants formed vast "coal forests" during the carboniferous period

OBJECTIVES

After reading this chapter and attending lecture, the student should be able to:

1. List characteristics that distinguish plants from organisms in the other kingdoms.
2. Diagram a generalized plant life cycle indicating which generation is the sporophyte/gametophyte, which individuals are haploid/diploid, where meiosis occurs and where mitosis occurs.
3. Describe four major periods of plant evolution that opened new adaptive zones on land.
4. Distinguish between the categories division and phylum.
5. Using the classification scheme presented in the text, list the plant divisions; give the common name for each; and categorize them into nonvascular, vascular seedless and vascular seed plants.
6. Provide evidence to defend the position that plants evolved from green algae.
7. Describe two adaptations that made the bryophytes' move onto land possible.

8. Explain how bryophytes are still tied to water.
9. List and distinguish among three division of Bryophyta.
10. Diagram the life cycle of a moss including gamete production, fertilization, and spore production.
11. Compare environmental conditions faced by algae in an aquatic environment and plants in a terrestrial environment.
12. Provide evidence that suggests the division Bryophyta is a phylogenetic branch separate from vascular plants.
13. Describe adaptations of vascular plants, including modifications of the life cycle and modifications of the sporophyte, that have contributed to their success on land.
14. List and distinguish among the four extant divisions of seedless vascular plants.
15. Distinguish between homosporous and heterosporous.
16. Distinguish among spore, sporophyte, sporophyll and sporangium.
17. Diagram the life cycle of a fern including spore production, gamete production and fertilization.
18. Point out the major life cycle differences between mosses and ferns.
19. Describe how coal is formed and during which geological period the most extensive coal beds were produced.

KEY TERMS

stomata	sporophyte	sporangium	megaspores
cuticle	vascular tissue	mosses	microspores
secondary products	gymnosperm	liverworts	lycophytes
lignin	angiosperm	hornworts	epiphytes
sporopollenin	division	xylem	sporophylls
gametangia	charophyte	phloem	horsetails
embryophyte	antheridium	homosporous	ferns
gametophyte	archegonium	heterosporous	

LECTURE NOTES

Plants appeared on land about 475 million years ago, and the evolutionary history of the plant kingdom reflects increasing adaptation to the terrestrial environment. The colonization of land by plants transformed the biosphere. This transformation created new adaptive zones and paved the way for other organisms.

I. An Overview of Plant Evolution

A. Structural, chemical, and reproductive adaptations enabled plants to colonize land

1. Some characteristics of plants

Plants are multicellular eukaryotes that are photosynthetic autotrophs; however, not all organisms with these characteristics are plants. Plants share the following characteristics with their green algal ancestors:

- Chloroplasts with the photosynthetic pigments: chlorophyll *a*, chlorophyll *b*, and carotenoids
- Cell walls containing cellulose
- Food reserve that is starch stored in plastids

It is the set of structural, chemical, and reproductive adaptations associated with terrestrial life that distinguishes plants from algae. Plants have evolved complex bodies with cell specialization for different functions.

- Plants have developed structural specializations in order to extract the resources needed for photosynthesis (water, minerals, carbon dioxide, light) from the terrestrial environment (above and below ground).
  - In most plants, gas exchange occurs via *stomata*, special pores on the surfaces of leaves
- Chemical adaptation includes the secretion of a waxy *cuticle*, a coating on the surface of plants that helps prevent desiccation.
  - Cuticle waxes are *secondary products*, so named because they arise through metabolic pathways not common to all plants. (Cellulose is an example of a primary product).
  - Other secondary products include *lignin* (cell wall component of "woody" plants) and *sporopollenin* (a resilient polymer in the walls of spores and pollen grains).

2. Plants as embryophytes

With the move from an aquatic to terrestrial environment, a new mode of reproduction was necessary to solve two problems:

1. Gametes must be dispersed in a nonaquatic environment. Plants produce gametes within *gametangia*, organs with protective jackets of sterile (nonreproductive) cells that prevent gametes from drying out (see Campbell, Figure 29.1a). The egg is fertilized within the female organ.
2. Embryos must be protected against desiccation. The zygote develops into an embryo that is retained for awhile within the female gametangia's jacket of protective cells (see Campbell, Figure 29.1b). Emphasizing this terrestrial adaptation, plants are often referred to as *embryophytes*.

3. Alternation of generations: *a review*

Most plants reproduce sexually, and most are also capable of asexual propagation. All plants have life cycles with an *alternation of generations* (also occurs in some groups of algae).

- A haploid *gametophyte* generation produces and alternates with a diploid *sporophyte* generation. The sporophyte, in turn, produces gametophytes (see Campbell, Figure 29.2).
- The life cycles are heteromorphic; that is, sporophytes and gametophytes differ in morphology.
- The sporophyte is larger and more noticeable than the gametophyte in all plants but mosses and their relatives.

A comparison of life cycles among plant divisions is instructive because:

- It points to an important trend in plant evolution: reduction of the haploid gametophyte generation and dominance of the diploid sporophyte.
- Certain life cycle features are adaptations to a terrestrial environment; for example the replacement of flagellated sperm by pollen.

4. Some highlights of plant phylogeny

There are four major periods of plant evolution that opened new adaptive zones on land (see Campbell, Figure 29.3):

1. Origin of plants from aquatic ancestors (probably green algae) in the Ordovician about 475 million years ago (mya).
   - Cuticle and jacketed gametangia evolved which protected gametes and embryos.

- *Vascular tissue* evolved with conducting cells that transport water and nutrients throughout the plant.

2. Diversification of seedless vascular plants, such as ferns, during the early Devonian about 400 mya.

3. Origin of the seed near the end of the Devonian about 360 mya.
   - *Seed* = Plant embryo packaged with a store of food within a resistant coat
   - Early seed plants bore seeds as naked structures and evolved into *gymnosperms,* including conifers.
   - Conifers and ferns coexisted in the landscape for more than 200 million years.

4. Emergence of flowering plants during the early Cretaceous, about 130 mya.
   - Unlike gymnosperms, flowering plants bear seeds within the flower's protective ovaries.
   - Most contemporary plants are flowering plants or *angiosperms*.

5. Classification of plants

The major taxonomic category of plants is the *division*; it is comparable to phylum, the highest category in the animal kingdom.

- Divisions are subdivided into lower taxa (class, order, family, genus).
- Currently, eleven divisions of plants are recognized (see also Campbell, Table 29.1).

A CLASSIFICATION OF PLANTS

	Common Name	Approximate Number of Extant Species
Nonvascular Plants (byrophytes)		
Division Bryophyta	Mosses	12,000
Division Hepatophyta	Liverworts	6,500
Division Anthocerophyta	Hornworts	100
Vascular Plants		
Seedless Vascular Plants		
Division Lycophyta	Lycophytes	1,000
Division Sphenophyta	Horsetails	15
Division Pterophyta	Ferns	12,000
Seed Plants		
Gymnosperms		
Division Coniferophyta	Conifers	550
Division Cycadophyta	Cycads	100
Division Ginkgophyta	Ginkgo	1
Division Gnetophyta	Gnetae	70
Angiosperms		
Division Anthophyta	Flowering plants	250,000

II.   The Origin of Plants

A.   Plants probably evolved from green algae called charophytes

The green algae are likely the photosynthetic protists most closely related to plants. This conclusion is based on homologies in:

- Cell wall composition
- Structure and pigmentation of chloroplasts

Available evidence supports the hypothesis that plants and green algae called *charophytes* both evolved from a common ancestor (see Campbell, Figure 29.4). Researchers have found the following homologies between charophytes and plants:

1. Homologous chloroplasts
    - Green algae and plants both have the accessory pigments, chlorophyll *b* and beta-carotene.
    - Green algae and plants both have chloroplasts with thylakoid membranes stacked as grana.
    - Compared to chloroplast DNA of various green algae, plant chloroplast DNA most closely matches that of charophytes.
2. Biochemical similarity
    - Most green algae and plants contain cellulose in their cell walls. Charophytes are the most plantlike in wall composition with cellulose making up 20% - 26% to 26% of the wall material.
    - Charophyte peroxisomes are the only algal peroxisomes with the same enzyme composition as plant peroxisomes.
3. Similarity in mitosis and cytokinesis. During cell division in charophytes and plants:
    - The nuclear envelope completely disperses during late prophase.
    - The mitotic spindle persists until cytokinesis begins.
    - Cell plate formation during cytokinesis involves cooperation of microtubules, actin microfilaments, and vesicles.
4. Similarity in sperm ultrastructure. Charophyte sperm ultrastructure is more similar to certain plants than to other green algae.
5. Genetic relationship. DNA and rRNA similarities in charophytes and plants provides additional evidence for the hypothesis that charophytes are the closest relatives of plants.

B.   Alternation of generations in plants may have originated by delayed meiosis

The alternation of haploid and diploid generations apparently evolved independently among various groups of algae.

- Since alternation of generations does not occur among modern charophytes, it is presumed that alternation of generations in plants has had a separate origin from alternation of generations in other algal groups.
- Its appearance in plants is thus *analogous*, not homologous, to the alternation of generations observed in various groups of algae.

How did alternation of generations evolve in plant ancestors?

Coleochaete, a modern charophyte, holds some clues:

- The Coleochaete thallus is haploid.
    - ⇒ In contrast to most algae, the parental thallus of *Coleochaete* retains the eggs, and after fertilization, the zygotes remain attached to the parent.

⇒ Nonreproductive cells of the thallus grow around each zygote, which enlarges, undergoes meiosis, and releases haploid swimming spores.

⇒ Haploid spores develop into new individuals.

⇒ The only diploid stage is the zygote; there is no alternation of multicellular diploid and haploid generations.

- If an ancestral charophyte delayed meiosis until after the zygote divided mitotically, there would be a multicellular diploid generation (sporophyte) still attached to the haploid parent (gametophyte). Such a life cycle would be an alternation of generations.

- If specialized gametophyte cells formed protective layers around a tiny sporophyte, this hypothetical ancestor would also be a primitive embryophyte (see Campbell, Figure 29.5).

What would be the adaptive advantage of delaying meiosis and forming a mass of diploid cells?

It may maximize the production of haploid spores.

- If the zygote undergoes meiosis directly, each fertilization event results in only a few haploid spores.

- Mitotic division of the zygote to form a multicellular sporophyte amplifies the sexual product. Many diploid cells can undergo meiosis producing a large number of haploid spores, enhancing the chances of survival in unfavorable environments.

C. Adaptations to shallow water preadapted plants for living on land

Some adaptations for life in shallow water could also have been adaptive for life on land.

- Many modern charophytes live in shallow water, and some ancient charophytes may have also lived in shallow-water habitats subject to occasional drying.

- About 440 million years ago, during the transition from Ordivician to Silurian, repeated glaciation and climatic changes caused fluctuations in the water levels of lakes and ponds.

- Natural selection may have favored shallow-water plants tolerant to periodic drying. Adaptations to shallow water may also have been preadaptive for terrestrial life.

> Examples:
> - Waxy cuticles
> - Protection of gametes
> - Protection of developing embryos

- Eventually, accumulated adaptations made it possible for ancestral plants to live permanently above the water line, opening a new adaptive zone with:
  ⇒ Sunlight unfiltered by water and algae
  ⇒ Soil rich in minerals
  ⇒ Absence of terrestrial herbivores

III. Bryophytes

A. The embryophyte adaptation evolved in bryophytes

The *bryophytes* include plants found in three divisions:

- Bryophyta (mosses)
- Hepatophyta (liverworts)

- Anthocerophyta (hornworts)

Bryophytes display a pivotal adaptation that made the move onto land possible: the embryophyte condition.

- Gametangia protect developing gametes.
    a. *Antheridium,* or male gametangium, produces flagellated sperm cells.
    b. *Archegonium*, or female gametangium, produces a single egg; fertilization occurs within the archegonium, and the zygote develops into an embryo within the protective jacket of the female organ (embryophyte condition).

Bryophytes are not totally free from their ancestral aquatic habitat.

- They need water to reproduce. Their flagellated sperm cells must swim from the antheridium to the archegonium to fertilize the egg.
- Most have no vascular tissue to carry water from the soil to aerial plant parts; they imbibe water and distribute it throughout the plant by the relatively slow processes of diffusion, capillary action, and cytoplasmic streaming.

Bryophytes lack woody tissue and cannot support tall plants on land; they may sprawl horizontally as mats, but always have a low profile (see Campbell, Figure 29.6).

B. The gametophyte is the dominant generation in the life cycles of bryophytes

The life cycle of a bryophyte alternates between haploid and diploid generations (see Campbell, Figure 29.7)

- The sporophyte (2n) produces haploid spores by meiosis in a *sporangium*; the spores divide by mitosis to form new gametophytes.
- Contrary to the life cycles of vascular plants, the haploid gametophyte is the dominant generation in mosses and other bryophytes. Sporophytes are generally smaller and depend on the gametophyte for water and nutrients.

C. The three divisions of bryophytes are mosses, liverworts, and hornworts

1. Mosses (Division Bryophyta)

A tight pack of many moss plants forms a spongy mat that can absorb and retain water.

Each plant grips the substratum with *rhizoids*, elongate cells or cellular filaments.

Photosynthesis occurs mostly in the small stemlike and leaflike structures found in upper parts of the plant; these structures are not homologous with stems and leaves in vascular plants.

Mosses cover about 3% of the land surface, and they contain vast amounts of organic carbon (see Campbell, Figure 29.6).

2. Liverworts (Division Hepatophyta)

Liverworts are less conspicuous than mosses.

They sometimes have bodies divided into lobes.

They have a life cycle similar to mosses. Their sporangia have elaters, coil-shaped cells, that spring out of the capsule and disperse spores.

They can also reproduce asexually from gemmae (small bundles of cells that can bounce out of cups on the surface of the gametophyte when hit by rainwater) (see Campbell, Figure 29.8).

They display their greatest diversity in tropical forests.

3. Hornworts (Division Anthocerophyta)

Hornworts resemble liverworts, but sporophytes are horn-shaped, elongated capsules that grow from the matlike gametophyte (see Campbell, Figure 29.9).

Their photosynthetic cells have only one large chloroplast, unlike the many smaller ones of other plants.

Recent molecular evidence suggest that they are most closely related to vascular plants.

IV.   The Origin of Vascular Plants

A. Additional terrestrial adaptations evolved as vascular plants descended from bryophyte-like ancestors

In addition to cuticles and jacketed sex organs, other adaptations for terrestrial life evolved in vascular plants as they colonized land:

1. *Regional specialization of the plant body.* Unlike aquatic environments, terrestrial environments spatially segregate the resources of water and light. This problem was solved as plants evolved subterranean roots that absorb water and minerals from the soil and an aerial shoot system of stems and leaves to make food.

2. *Structural support.* In aquatic environments, the denser medium of water buoys plants up toward the light, but in terrestrial environments plants must have structural support to stand upright in air. Such support was provided as the hard material lignin was embedded into the cellulose matrix of cell walls.

3. *Vascular system.* Regional specialization of the plant body presented the problem of transporting substances between the root and shoot systems. This problem was solved as a vascular system evolved with two types of conducting tissues:

    Xylem = Complex, plant vascular tissue that conducts water and minerals from the roots to the rest of the plant

    • Composed of dead, tube-shaped cells that form a microscopic water-pipe system

    • Cell walls are usually lignified, giving the plant structural support

    Phloem = Plant vascular tissue that conducts food throughout the plant

    • Composed of living cells arranged into tubules

    • Distributes sugars, amino acids, and other organic nutrients

4. *Pollen.* Pollination eliminated the need for water to transport gametes.

5. *Seeds*

6. *Increased dominance of the diploid sporophyte*

B. The branched sporophytes of vascular plants amplified the production of spores and made more complex bodies possible

Oldest fossilized vascular plant is *Cooksonia* (late Silurian):

• Discovered in both European and North American Silurian rocks; North America and Europe were probably connected during the late Silurian, about 400 million years ago

• Simple plant that displayed dichotomous branching and bulbous terminal sporangia on sporophyte (see Campbell, Figure 29.10)

• True roots and leaves were absent; the largest species was about 50 cm tall

• Grew in dense stands around marshes

• As vascular plants became more widespread, new species appeared

V.   Seedless Vascular Plants

The earliest vascular plants were seedless and they dominated the Carboniferous forests. Modern flora includes three divisions of seedless vascular plants.

A. A sporophyte-dominant life cycle evolved in seedless vascular plants

The sporophyte (diploid) generation emerged as the larger and more complex plant from the time of *Cooksonia* and other early vascular plants. It is the dominant stage in the life cycle in all extant vascular plants.

The sporophyte-dominant life cycle is exemplified by ferns, one group of the seedless vascular plants (see Campbell, Figure 29.11).

- The familiar leafy plant is the sporophyte.
- Gametophytes are quite small and grow on or below the surface of the soil.

Vascular plants display two distinct reproductive strategies:

- The sporophyte of a *homosporous* plant produces a single type of spore (e.g., ferns); each spore develops into a bisexual gametophyte with both male (antheridia) and female (archegonia) sex organs.
- The sporophyte of a *heterosporous* plant produces two kinds of spores:
1. *Megaspores* develop into female gametophytes possessing archegonia.
2. *Microspores* develop into male gametophytes possessing antheridia.

B. The three divisions of seedless vascular plants are lycophytes, horsetails, and ferns

1. Lycophytes (division Lycophyta)

The division Lycophyta includes the club mosses and ground pines.

Lycophytes survived through the Devonian period and dominated land during the Carboniferous Period (340–280 million years ago).

Some are temperate, low-growing plants with rhizomes and true leaves.

Some species of *Lycopodium* are *epiphytes*, plants that use another organism as a substratum, but are not parasites.

- The sporangia of *Lycopodium* are borne on *sporophylls*, leaves specialized for reproduction. In some, sporophylls are clustered at branch tips into club-shaped strobili—hence the name club moss.
- Spores develop into inconspicuous gametophytes. The non-photosynthetic gametophytes are nurtured by symbiotic fungi.

Most are homosporous.

Genus *Selaginella* is heterosporous.

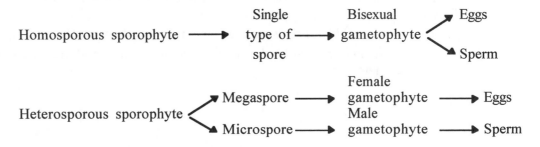

2. Horsetails (division Sphenophyta)

The division Sphenophyta includes the horsetails; it survived through the Devonian and reached its zenith during the Carboniferous period.

The only existing genus is *Equisetum*, which (see Campbell, Figure 29.13):

- Lives in damp locations and has flagellated sperm
- Is homosporous
- Has a conspicuous sporophyte generation

- Has photosynthetic, free-living gametophytes (not dependent on the sporophyte for food)

3. Ferns (division Pterophyta)

Appearing in the Devonian, ferns radiated into diverse species that coexisted with tree lycopods and horsetails in the great Carboniferous forests.

- Ferns are the most well represented seedless plants in modern floras. There are more than 12,000 existing species of ferns; most diverse in the tropics.
- Fern leaves are generally much larger than those of lycopods and probably evolved in a different way.
    ⇒ Lycopods have microphylls, small leaves that probably evolved as emergences from the stem that contained a single strand of vascular tissue.
    ⇒ Ferns have megaphylls, leaves with a branched system of veins. Megaphylls probably evolved from webbing formed between separate branches growing close together.

Most ferns have fronds, compound leaves that are divided into several leaflets (see Campbell, Figure 29.11).

- The emerging frond is coiled into a fiddlehead that unfurls as it grows.
- Leaves may sprout directly from a prostrate stem (bracken and sword ferns) or from upright stems many meters tall (tropical tree ferns).

Ferns are *homosporous* and the conspicuous leafy fern plant is the sporophyte.

- Specialized sporophylls bear sporangia on their undersides; many ferns have sporangia arranged in clusters called sori and are equipped with springlike devices that catapult spores into the air, where they can be blown by the wind far from their origin.
- The spore is the dispersal stage.
- The free-living gametophyte is small and fragile, requiring a moist habitat.
- Water is necessary for fertilization, since flagellated sperm cells must swim from the antheridium to the archegonium, where fertilization takes place.
- The sporophyte embryo develops protected within the archegonium.

C. Seedless vascular plants formed vast "coal forests" during the Carboniferous period

During the Carboniferous period, the landscape was dominated by extensive swamp forests

- Organic rubble of the seedless plants mentioned above accumulated as peat.
- When later covered by the sea and sediments, heat and pressure transformed the peat into coal.

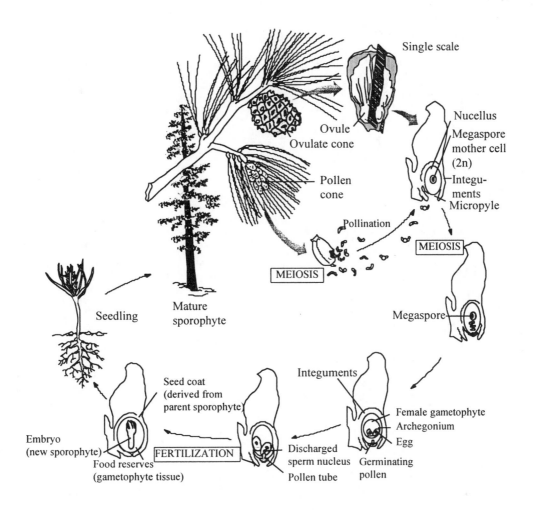

Single scale

Ovule
Ovulate cone

Pollen
cone

Nucellus
Megaspore
mother cell
(2n)
Integu-
ments
Micropyle

Pollination

MEIOSIS

MEIOSIS

Seedling

Mature
sporophyte

Megaspore

Integuments

Female gametophyte
Archegonium
Egg

Seed coat
(derived from
parent sporophyte)

Embryo
(new sporophyte)

FERTILIZATION

Food reserves
(gametophyte tissue)

Discharged
sperm nucleus
Pollen tube

Germinating
pollen

REFERENCES

Campbell, N., et al. *Biology*. 5th ed. Menlo Park, California: Benjamin/Cummings, 1999.

Raven, P.H., R.F. Evert and S.E. Eichhorn. *Biology of Plants*. 6th ed. New York: Worth Publishers, Inc., 1998.

OUTLINE

I. Overview of Reproductive Adaptations of Seed Plants
   A. The gametophytes of seed plants became even more reduced than the gametophytes of seedless vascular plants
   B. In seed plants, the seed replaced the spore as the main means of dispersing offspring
   C. Pollen became the vehicles for sperm cells in seed plants

II. Gymnosperms
   A. The Mesozoic era was the age of gymnosperms
   B. The four divisions of extant gymnosperms are the cycads, the ginkgo, the gnetophytes, and the conifers
   C. The life cycle of a pine demonstrates the key reproductive adaptations of seed plants

III. Angiosperms (Flowering Plants)
   A. Terrestrial adaptation continued with the refinement of vascular tissue in angiosperms
   B. The flower is the defining reproductive adaptation of angiosperms
   C. Fruits help disperse the seeds of angiosperms
   D. The life cycle of an angiosperm is a highly refined version of the alternation of generations common to all plants
   E. The radiation of angiosperms marks the transition from the Mesozoic to the Cenozoic era
   F. Angiosperms and animals have shaped one another's evolution
   G. Agriculture is based almost entirely on angiosperms

IV. The Global Impact of Plants
   A. Plants transformed the atmosphere and climate
   B. Plant diversity is a nonrenewable resource

OBJECTIVES

After reading this chapter and attending lecture, the student should be able to:

1. Describe the adaptations of seed plants that have contributed to their success on land.
2. List the four divisions of gymnosperms.
3. Describe the structures of ovulate and pollen cones of a pine and distinguish between the two.

4.   Describe the life history of a pine and indicate which structures are part of the gametophyte generation and which are part of the sporophyte generation.

5.   Point out the major life cycle differences in ferns and pines.

6.   Distinguish between pollination and fertilization.

7.   Describe a pine seed and indicate which structures are old sporophyte, gametophyte, and new sporophyte.

8.   Describe how the needle-shaped leaves of pines and firs are adapted to dry conditions.

9.   List and give examples of the two classes of Anthophyta.

10.   Compare the life cycles of mosses, ferns, conifers, and flowering plants in terms of:
      a. Dominant life cycle stage (gametophyte/sporophyte)
      b. Whether they are homosporous or heterosporous
      c. Mechanism of gamete transfer

11.   Describe some refinements in vascular tissue that occurred during angiosperm evolution.

12.   Explain how evolution of the flower enhanced the reproductive efficiency of angiosperms.

13.   Identify the following floral structures and describe a function for each:
      a. Sepals   f. Anther
      b. Petals   g. Stigma
      c. Stamens   h. Style
      d. Carpels   i. Ovary
      e. Filament

14.   Describe four commonly recognized evolutionary trends in floral structure found in various angiosperm lineages.

15.   Define fruit and explain how fruits are modified in ways that help disperse seeds.

16.   Diagram the generalized life cycle of an angiosperm, identify which structures are haploid, and explain how it differs from the life cycle of a pine.

17.   Explain the process of double fertilization and describe the fate of the polyploid nucleus.

18.   Explain how an angiosperm seed differs from that of a pine.

19.   Explain why paleobotanists have difficulty piecing together the origin of angiosperms and describe some current theories on how flowering plants may have evolved.

20.   Explain how animals may have influenced the evolution of terrestrial plants, and vice versa.

KEY TERMS

nucellus	fiber	anther	cross-pollination
integuments	flower	stigma	double fertilization
ovule	sepal	style	cotyledons
seed	petal	ovary	endosperm
conifer	stamen	fruit	coevolution
tracheids	carpel	pollen grains	
vessel elements	filament	embryo sac	

LECTURE NOTES

The emergence of seed plants further transformed the Earth. Seeds and other adaptations of gymnosperms and angiosperms heightened the ability of plants to survive and reproduce in diverse terrestrial environments; these plants became the principal producers in the food webs of most ecosystems on land.

I.   **Overview of Reproductive Adaptations of Seed Plants**

   Three life cycle modifications contributed to terrestrial seed plant success:
   1. Reduction of the gametophyte. They were retained in the moist reproductive tissue of the sporophyte generation (not independent).
   2. Origin of the seed
      - Zygotes developed into embryos packaged with a food supply within a protective seed coat.
      - Seeds replaced spores as main means of dispersal.
   3. Evolution of pollen. Plants were no longer tied to water for fertilization.

A. **The gametophytes of seed plants became even more reduced than the gametophytes of seedless vascular plants**

   While the gametophytes of seedless vascular plants develop in the soil as an independent generation, those of seed plants are reduced in size and retained within the moist reproductive tissue of the sporophyte generation.
   - This evolutionary trend reverses the gametophyte-sporophyte relationship observed in bryophytes (see Campbell, Figure 30.1).
   - Dominance of the diploid generation may afford protection from solar radiation-induced mutations of the genome (damaging radiation is more extensive on land than in aquatic habitats).

B. **In seed plants, the seed replaced the spore as the main means of dispersing offspring**

   The relatively harsh terrestrial environment led to the development of resistant structures for the dispersal of offspring.
   - Bryophytes and seedless vascular plants produce and release hardy single-celled spores.
   - Seeds are more hardy because of their multicellular nature.

   A seed consists of a sporophyte embryo together with a food supply surrounded by a protective coat.
   - The sporophytes do not release their spores, but retain them in their sporangia, as a result, the sporophyte also contains a gametophyte.

   All seed plants are heterosporous in that they possess two different kinds of sporangia, each producing a different type of spore.
   - Megasporangia produce megaspores that give rise to female egg-containing gametophytes.
   - Microsporangia produce microspores that give rise to male sperm-containing gametophytes.

   The development of the seed is associated with the megasporangia.
   - The megasporangium of seed plants is not a chamber, but a fleshy structure called a *nucellus*.
   - Additional tissues called *integuments* surround the megasporangium (contribute to the protective coat).

- The resulting structure—megaspore, megasporangium, and integuments—is called an *ovule* (see Campbell, Figure 30.2a).
- The female gametophyte develops within the wall of the megaspore and is nourished by the nucellus.
- If the egg cell of a female gametophyte is fertilized by a sperm cell (see Campbell, Figure 30.2b), the zygote develops into a sporophyte embryo.
- The resulting sporophyte-containing ovule develops into a seed (see Campbell, Figure 30.2c).

C. Pollen became the vehicles for sperm cells in seed plants

The microspores develop into pollen grains, which in turn, mature to form the male gametophores of seed plants.

- Pollen grains are coated with a resilient polymer, sporopollenin (see Chapter 29).
- Pollen grains can be carried away by wind or animals (e.g., bees) following release from microsporangia.

A pollen grain near an ovule will extend a tube through sperm cells into the female gametophyte within the ovule.

- In some gymnosperms, the sperm cells are flagellated (ancestral condition).
- Other gymnosperms (including conifers) and angiosperms do not have flagellated sperm cells.

II. Gymnosperms

A. The Mesozoic era was the age of gymnosperms

Gymnosperms appear in the fossil record much earlier than flowering plants. Gymnosperms most likely descended from Devonian progymnosperm and were seedless.

- Seeds evolved by the end of the Devonian.
- Adaptive radiation during the Carboniferous and Permian periods led to today's divisions.
- During the Permian, Earth became warmer and drier; therefore, lycopods, horsetails, and ferns (previously dominant) were largely replaced by conifers and their relatives, the cycads (two divisions of gymnosperms).
- This large change marks the end of the Paleozoic era and the beginning of the Mesozoic era.

Gymnosperms lack enclosed chambers (ovaries) in which seeds develop.

B. The four divisions of extant gymnosperms are the cycads, the ginkgo, the gnetophytes, and the conifers

Campbell, Figure 30.3, shows the four divisions of living gymnosperms.

The *conifers* are the largest division of gymnosperms.

- Most are evergreens: pines, firs, spruces, larches, yews, junipers, cedars, cypresses, and redwoods all belong to this division.
- Includes some of the tallest (redwoods and some eucalyptus); largest (giant sequoias); and oldest (bristle cone pine) living organisms.
- Most lumber and paper pulp is from conifer wood.

Needle-shaped conifer leaves are adapted to dry conditions.

- Thick cuticle covers the leaf
- Stomata are in pits, reducing water loss
- Despite the shape, needles are megaphylls, as are leaves of all seed plants.

C. The life cycle of a pine demonstrates the key reproductive adaptations of seed plants

The life cycle of pine, a representative conifer, is characterized by the following:

- The multicellular sporophyte is the most conspicuous stage; the pine tree is a sporophyte, with its sporangia located on cones.
- The multicellular gametophyte generation is reduced and develops from haploid spores that are retained within sporangia.
    ⇒ The male gametophyte is the pollen grain; there is no antheridium.
    ⇒ The female gametophyte consists of multicellular nutritive tissue and an archegonium that develops within an ovule.

Conifer life cycles are heterosporous; male and female gametophytes develop from different types of spores produced by separate cones.

- Trees of most pine species bear both pollen cones and ovulate cones, which develop on different branches.
- Pollen cones have microsporangia; cells in these sporangia undergo meiosis producing haploid microspores, small spores that develop into pollen grains—the male gametophytes.
- Ovulate cones have megasporangia; cells in these sporangia undergo meiosis producing large megaspores that develop into the female gametophyte (see Campbell, Figure 30.4). Each ovule initially includes a megasporangium (nucellus) enclosed in protective integuments with a single opening, the micropyle.

It takes nearly three years to complete the pine life cycle, which progresses through a complicated series of events to produce mature seeds.

- Windblown pollen falls onto the ovulate cone and is drawn into the ovule through the micropyle.
- The pollen grain germinates in the ovule, forming a pollen tube that begins to digest its way through the nucellus.
- A megaspore mother cell in the nucellus undergoes meiosis producing four haploid megaspores, one of which will survive; it grows and divides repeatedly by mitosis producing the immature female gametophyte.
- Two or three archegonia, each with an egg, then develop within the multicellular gametophyte.
- More than a year after pollination, the eggs are ready to be fertilized; two sperm cells have developed and the pollen tube has grown through the nucellus to the female gametophyte.
- Fertilization occurs when one of the sperm nuclei unites with the egg nucleus. All eggs in an ovule may be fertilized, but usually only one zygote develops into an embryo.
- The pine embryo, or new sporophyte, has a rudimentary root and several embryonic leaves. It is embedded in the female gametophyte, which nourishes the embryo until it is capable of photosynthesis. The ovule has developed into a pine seed, which consists of an embryo (2n), its food source (n), and a surrounding seed coat (2n) derived from the integuments of the parent tree.
- Scales of the ovulate cone separate, and the winged seeds are carried by the wind to new locations. Note, that with the seed plants, the seed has replaced the spore as the mode of dispersal.
- A seed that lands in a habitable place germinates, its embryo emerging as a pine seedling.

III. Angiosperms (Flowering Plants)

Flowering plants are the most widespread and diverse; 250,000 species are now known.

- There is only one division, Anthophyta, with two classes, Monocotyledones (monocots) and Dicotyledones (dicots).
- Most use insects and animals for transferring pollen, and therefore, are less dependent on wind and have less random pollination.

A. Terrestrial adaptation continued with refinement of vascular tissue in angiosperms

Vascular tissue became more refined during angiosperm evolution.

- Conifers have *tracheids* (see Campbell, Figure 30.5), water-conducting cells that are:
  - ⇒ An early type of xylem cell
  - ⇒ Elongated, tapered cells that function both in mechanical support and water movement up the plant
- Most angiosperms also have *vessel elements* that are:
  - ⇒ Shorter, wider cells than the more primitive tracheids
  - ⇒ Arranged end to end forming continuous tubes
  - ⇒ Compared to tracheids, vessel elements are more specialized for conducting water, but less specialized for support
- Angiosperm xylem is reinforced by other cell types called *fibers*, which are:
  - ⇒ Specialized for support with a thick lignified wall
  - ⇒ Evolved in conifers. (Conifer xylem contains both fibers and tracheids, but not vessel elements.)

B. The flower is the defining reproductive adaptation of angiosperms

Flower = The reproductive structure of an angiosperm which is a compressed shoot with four whorls of modified leaves (see also Campbell, Figure 30.6)

Parts of the flower:

Sepals - Sterile, enclose the bud

Petals – Sterile, aid in attracting pollinators

Stamen – Produces the pollen

Carpel – Evolved from a seed-bearing leaf that became rolled into a tube

Stigma – Part of the carpel that is a sticky structure that receives the pollen

Ovary – Part of the carpel that protects the ovules, which develop into seeds after fertilization

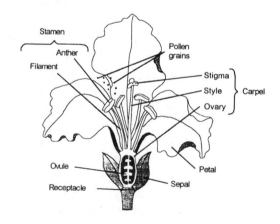

There are four evolutionary trends in various angiosperm lineages:

1. The number of floral parts have become reduced.
2. Floral parts have become fused.
3. Symmetry has changed from radial to bilateral.
4. The ovary has dropped below the petals and sepals, where the ovules are better protected.

C. Fruits help disperse the seeds of angiosperms

Fruit = A ripened ovary that protects dormant seeds and aids in their dispersal; some fruits (like apples) incorporate other floral parts along with the ovary (see Campbell, Figure 30.7)

Aggregate fruits = Several ovaries that are part of the same flower (e.g., raspberry)

Multiple fruit = One that develops from several separate flowers (e.g., pineapple)

Modifications of fruits that help disperse seeds include:

- Seeds within fruits that are shaped like kites or propellers to aid in wind dispersal
- Burr-like fruit that cling to animal fur
- Edible fruit with tough seeds which pass through the digestive tract of herbivores unharmed, dispersing seeds miles away

D. The life cycle of an angiosperm is a highly refined version of the alternation of generations common to all plants

Life cycles of angiosperms are heterosporous (in common with all seed plants) and the two types of sporangia are found in the flower (see Campbell, Figure 30.8):

- Microsporangia in anthers produce microspores that form male gametophytes.
- Megasporangia in ovules produce megaspores that develop into female gametophytes.

Immature male gametophytes:

- Are *pollen grains*, which develop within the anthers of stamens
- Each pollen grain has two haploid nuclei that will participate in *double fertilization* characteristic of angiosperms.

Female gametophytes:

- Do *not* produce an archegonium
- Are located within an ovule
- Consist of only a few cells: an *embryo sac* with eight haploid nuclei in seven cells (a large central cell has two haploid nuclei)
- One of the cells is the egg

An outline of the angiosperm life cycle follows:

- Pollen from the anther lands on the sticky stigma at the carpel's tip; most flowers do not self-pollinate, but have mechanisms to ensure *cross-pollination.*
- The pollen grain germinates on the stigma by growing a pollen tube down the style of the carpel.
- When it reaches the ovary, the pollen tube grows through its micropyle and discharges two sperm cells into the embryo sac.
- Double fertilization occurs as one sperm nucleus unites with the egg to form a diploid zygote; the other sperm nucleus fuses with two nuclei in the embryo sac's central cell to form triploid ($3n$) endosperm.
- After double fertilization, the ovule matures into a seed.

The seed is a mature ovule, consisting of:

1. Embryo. The zygote develops into an embryo with a rudimentary root and one (in monocots) or two (in dicots) *cotyledons* or seed leaves.
2. Endosperm. The triploid nucleus in the embryo sac divides repeatedly forming triploid endosperm, rich in starch and other food reserves.
3. Seed coat. This is derived from the integuments (outer layers of the ovules).

Monocots and dicots use endosperm differently.

- Monocot seeds store most food in the endosperm.
- Dicots generally restock most of the nutrients in the developing cotyledons.

In a suitable environment the seed coat ruptures and the embryo emerges as a seedling, using the food stored in the endosperm and cotyledons.

E. The radiation of angiosperms marks the transition from the Mesozoic to the Cenozoic era

Angiosperms showed a relatively sudden appearance in the fossil record with no clear transitional links to ancestors.

- Earliest fossils are early Cretaceous (approximately 130 million years ago)
- dominant, as they are today.

There are two theories about their sudden appearance:

1. Angiosperms originated where fossilization was unlikely (they are an artifact of an imperfect fossil record).
2. Angiosperms evolved and radiated relatively abruptly (*punctuated equilibrium*).

Perhaps angiosperms evolved from seed ferns, an extinct group of unspecialized gymnosperms.

F. Angiosperms and animals shaped one another's evolution

Terrestrial plants and animals have coevolved, a consequence of their interdependence.

Coevolution = Reciprocal evolutionary responses among two or more interacting species; adaptive change in one species is in response to evolutionary change in the other species.

Coevolution between angiosperms and their pollinators led to diversity of flowers.

- Some pollinators are specific for a particular flower. The pollinator has a monopoly on a food source and guarantees the flower's pollen will pollinate a flower of the same species (see Campbell, Figure 30.9).
- Often, the relationship between angiosperms and their pollinators is not species specific; a pollinator may not depend exclusively on one flower species, or a flower species may not depend exclusively on one species of pollinator. However, flower color, fragrance, and structure are usually adaptations for *types* of pollinators, such as various species of bees or hummingbirds.

Edible fruits of angiosperms have coevolved with animals that can disperse seeds. Animals become attracted to ripening fruits as they:

- Become softer, more fragrant, and higher in sugar
- Change to a color that attracts birds and mammals, animals which are large enough to disperse the seeds

G. Agriculture is based almost entirely on angiosperms

Angiosperms provide nearly all our food: fruit, vegetable crops, and grains, such as corn, rice, wheat.

Flowering plants are also used for other purposes, such as:

- Fiber
- Medication
- Perfume
- Decoration

Through agriculture, humans have influenced plant evolution by artificially selecting for plants that improved the quantity and quality of foods and other crops.

- Many of our agricultural plants are so genetically removed from their origins that they probably could not survive in the wild.

- As a consequence, cultivated crops that require human intervention to water, fertilize, provide protection from insects and disease, and even to plant their seeds, are vulnerable to natural and human-caused disasters.

V.   The Global Impact of Plants

A. Plants transformed the atmosphere and the climate

In addition to being the primary producers of the terrestrial environment, plants also changed the physical environment of Earth.

- They decreased atmospheric carbon dioxide, resulting in global cooling.
- The cooler environment may have made terrestrial life more habitable for other organisms.

B. Plant diversity is a nonrenewable resource

Plant diversity is a nonrenewable resource, and the irrevocable extinction of plant species is occurring at an unprecedented rate.

- The exploding human population demands space and natural resources.
- The toll of habitat destruction is greatest in the tropics because this is where:
  ⇒ Most species live
  ⇒ More than half the human population lives and human population growth is fastest
  ⇒ Most deforestation is caused by slash-and-burn clearing for agriculture

As the forest disappears, so do thousands of plant and animal species.

- Habitat destruction also endangers animal species that depend on plants in the tropical rainforest.
- Habitat destruction by humans has not been limited to the tropics. Europeans eliminated most of their forests centuries ago, and in North America, destruction of habitat is endangering many species.

There are many reasons to value plant diversity and to find ways to protect it.

- Ecosystems are living treasures that can regenerate only slowly.
- Humans depend on plants for products such as medicines, food and building materials.
- We still know so little of the 250,000 known plant species. (Food agriculture is based on only about two dozen species.)

REFERENCES

Campbell, N., et al. *Biology*. 5th ed. Menlo Park, California: Benjamin/Cummings, 1999.

Raven, P.H., R.F. Evert and S.E. Eichhorn. *Biology of Plants*. 6th ed. New York: Worth Publishers, Inc., 1988.

CHAPTER **31**
FUNGI

OUTLINE

I.   Introduction to Fungi
    A. Absorptive nutrition enables fungi to live as decomposers and symbionts
    B. Extensive surface area and rapid growth adapt fungi for absorptive nutrition
    C. Fungi reproduce by releasing spores that are produced either sexually or asexually

II. Diversity of Fungi
    A. Division Chytridiomycota: chytrids may provide clues about fungal origins
    B. Division Zygomycota: zygote fungi form resistant dikaryotic structures during sexual reproduction
    C. Division Ascomycota: sac fungi produce sexual spores in saclike asci
    D. Division Basidiomycota: club fungi have long-lived dikaryotic mycelia and a transient diploid stage
    E. Molds, yeasts, lichens, and mycorrhizae represent unique lifestyles that evolved independently in three fungal divisions

III. Ecological Impacts of Fungi
    A. Ecosystems depend on fungi as decomposers and symbionts
    B. Some fungi are pathogens
    C. Many animals, including humans, eat fungi

IV. Phylogenetic Relationships of Fungi
    A. Fungi and animals probably evolved from a common protistan ancestor

OBJECTIVES

After reading this chapter and attending lecture, the student should be able to:

1. List characteristics that distinguish fungi from organisms in other kingdoms.
2. Explain how fungi acquire their nutrients.
3. Explain how non-motile fungi seek new food sources and how they disperse.
4. Describe the basic body plan of a fungus.
5. Distinguish between septate and aseptate (coenocytic) fungi.
6. Describe some advantages to the dikaryotic state.
7. Distinguish among fungi and list some common examples of each.
8. Describe asexual and sexual reproduction in Zygomycota, Ascomycota, and Basidiomycota, and the sexual structure that characterizes each group.
9. Explain the difference between conidia and ascospores.
10. Explain why ascomycetes can be useful to geneticists studying genetic recombination.
11. Explain why the Deuteromycota are called imperfect fungi.

12. Describe the anatomy of lichens and explain how they reproduce.

13. Provide evidence for both sides of the debate on whether symbiosis in lichens is parasitic or mutualistic.

14. Describe the ecological importance of lichens.

15. Explain why fungi are ecologically and commercially important.

16. Describe how the mutualistic relationship in mycorrhizae is beneficial to both the fungus and the plant, and explain its importance to natural ecosystems and agriculture.

17. Describe a scenario for fungal phylogeny and list two possible ancestors of Zygomycota.

KEY TERMS

absorption	plasmogamy	asci	imperfect fungi
hyphae	karyogamy	ascocarp	yeast
mycelium	dikaryon	conidia	lichen
septa	chytrids	basidium	soredia
chitin	zygote fungi	club fungus	
coenocytic	mycorrhizae	basidiocarps	
haustoria	sac fungi	mold	

LECTURE NOTES

I. **Introduction to the Fungi**

Fungi are eukaryotes, and nearly all are multicellular (although yeasts are unicellular). Their nutrition, structural organization, growth, and reproduction distinguish them from organisms in other kingdoms.

A. **Absorptive nutrition enables fungi to live as decomposers and symbionts**

Fungi are heterotrophs that acquire nutrients by *absorption*.

- They secrete hydrolytic enzymes and acids to decompose complex molecules into simpler ones that can be absorbed.
- Fungi are specialized into three main types:
  1. Saprobes, which absorb nutrients from dead organic material
  2. Parasitic fungi, which absorb nutrients from the cells of living hosts; some are pathogenic
  3. Mutualistic fungi, which absorb nutrients from a host, but reciprocate to benefit the host.

Fungi exist in diverse habitats and form symbioses with many organisms.

For example, fungi are found in:
- Terrestrial habitats
- Aquatic habitats, both freshwater and marine
- Symbiotic relationships with algae to form lichens

B. Extensive surface area and rapid growth adapt fungi for absorptive nutrition

The basic structural unit of a fungal vegetative body (*mycelium*) is the *hypha*. Except for yeasts, fungal bodies are diffuse, intertwining mats of hyphae that are organized around and within their food source (see Campbell, Figure 31.1).

These hyphae:

- Are composed of tubular walls containing *chitin*, a strong, flexible nitrogen-containing polysaccharide similar to that found in arthropod exoskeletons
- Provide enormous surface area for the absorptive mode of nutrition. Parasitic fungi have modified hyphae called *haustoria*, which are nutrient absorbing hyphae that penetrate host tissue, but remain outside the host cell membrane (see Campbell, Figure 31.2c).

Fungal hyphae may be aseptate or septate.

- Hyphae of aseptate fungi lack cross-walls and are *coenocytic*, formed from repeated nuclear division without cytokinesis (see Campbell, Figure 31.2b).
- Hyphae of septate fungi are divided into cells by crosswalls called *septa*. Pores in the septa allow organelles to move from cell to cell.

True fungi have no flagellated stages in their life cycle. This characteristic is partly why the Chytridiomycota and Oomycota have been moved to the Protista.

C. Fungi reproduce by releasing pores that are produced either sexually or asexually

Mycelial growth is adapted to the absorptive node of nutrition.

- Mycelia grow in length, not girth, which maximizes the surface area for absorption.
- Mycelia grow rapidly, as much as a kilometer of hyphae each day. Fast growth can occur because cytoplasmic streaming carries molecules synthesized by the mycelium to the growing hyphal tips.
- Since fungi are nonmotile, they cannot search for food or mates. Instead, they grow in hyphal length to reach new food sources and territory.

Fungal chromosomes and nuclei are relatively small, and the nuclei divide differently from most other eukaryotes.

- During mitosis, the nuclear envelope remains intact from prophase to anaphase; the spindle is inside the nuclear envelope.
- After anaphase, the nuclear envelope pinches in two, and the spindle disappears.

Fungi reproduce by releasing spores that are:

- Usually unicellular, haploid, and of various shapes and sizes.
- Produced either sexually (by meiosis) or asexually (by mitosis). In favorable conditions, fungi generally produce enormous numbers of spores asexually. For many fungi (not all), sexual reproduction occurs only as a contingency for stressful environmental conditions.
- The agent of dispersal responsible for the wide geographic distribution of fungi. Carried by wind or water, spores germinate if they land in a moist place with an appropriate substratum.

Except for transient diploid stages in sexual life cycles, fungal hyphae and spores are haploid. Some mycelia may, however, be genetically heterogeneous resulting from fusion of hyphae with different nuclei.

- The different nuclei may stay in separate parts of the same mycelium.
- Alternatively, the different nuclei may mingle and even exchange genes in a process similar to crossing over.

The sexual cycle in fungi differs from other eukaryotic organisms in that syngamy occurs in two stages that are separated in time.

Syngamy = The sexual union of haploid cells from two individuals. In fungi, syngamy occurs in two stages:

1. *Plasmogamy,* the fusion of cytoplasm
2. *Karyogamy,* the fusion of nuclei

After plasmogamy, haploid nuclei from each parent pair up, forming a *dikaryon,* but they do not fuse.

- Nuclear pairs in dikaryons may exist and divide synchronously for months or years.
- The dikaryotic condition has some advantages of diploidy; one haploid genome may compensate for harmful mutations in the other nucleus.
- Eventually, the haploid nuclei fuse forming a diploid cell that immediately undergoes meiosis.

II.   Diversity of Fungi

There are four divisions of fungi (see Campbell, Figure 31.4). They differ in the:

- Structures involved in plasmogamy
- Length of time spent as a dikaryon
- Location of karyogamy; the fungal divisions are named for the sexual structures in which karyogamy occurs.

A. Division Chytridiomycota: chytrids may provide clues about fungal origins

The Chytridiomycota and Fungi may share a protistan ancestor.

- Chytrids were placed in the Kingdom Protista because they form flagellated zoospores and gametes—a protistan characteristic.
- However, chytrids and fungi share many characteristics, such as:
  ⇒ An absorptive mode of nutrition
  ⇒ Cell walls of chitin
  ⇒ Most form hyphae
  ⇒ Key enzymes and metabolic pathways that are not found in the other fungus-like protists (slime molds and water molds)
  ⇒ Similar sequences of proteins and nucleic acids
- This evidence lends support for
  ⇒ Combining the chytrids with fungi as a monophyletic group
  ⇒ The hypothesis that chytrids are the most primitive fungi, diverging earliest in fungal phylogeny.
  ⇒ The hypothesis that fungi evolved from protists with flagella, a feature retained by the chytrids.

B. Division Zygomycota: zygote fungi form resistant dikaryotic structures during sexual reproduction

Fungi in the division Zygomycota are characterized by the presence of dikaryotic *zygosporangia*, resistant structures formed during sexual reproduction.

- Zygomycetes are mostly terrestrial and live in soil or on decaying organic material.
- Some form *mycorrhizae*, mutualistic associations with plant roots (see Campbell, Figure 31.16).
- Zygomycete hyphae are coenocytic; septa are found only in reproductive cells.

See Campbell, Figure 31.6 for the life cycle of the zygomycete, *Rhizopus stolonifer*, a common bread mold.

- The mycelium consists of horizontal hyphae that spread out and penetrate the food source.
- Under favorable environmental conditions, *Rhizopus* reproduces asexually:
  ⇒ Sporangia develop at the tips of upright hyphae.
  ⇒ Mitosis produces hundreds of haploid spores that are dispersed through the air.
  ⇒ If they land in a moist, favorable environment, spores germinate into new mycelia.
- In unfavorable conditions, *Rhizopus* begins its sexual cycle of reproduction:
  ⇒ Mycelia of opposite mating types (+ and −) form gametangia that contain several haploid nuclei walled off by the septum.
  ⇒ Plasmogamy of the + and − gametangia occurs, and the haploid nuclei pair up forming a dikaryotic zygosporangium that is metabolically inactive and resistant to desiccation and freezing.
  ⇒ When conditions become favorable, karyogamy occurs between paired nuclei; the resulting diploid nuclei immediately undergo meiosis producing genetically diverse haploid spores.
  ⇒ The zygosporangium germinates a sporangium that releases the genetically recombined haploid spores.
  ⇒ If they land in a moist, favorable environment, spores germinate into new mycelia.

Even though air currents are not a very precise way to disperse spores, *Rhizopus* releases so many that enough land in hospitable places. Some zygomycetes, however, can actually aim their spores.

- For example, *Pilobolus*, a fungus that decomposes animal dung, bends sporangium-bearing hyphae toward light, where grass is likely to be growing.
- The sporangium is shot out of the hypha, dispersing spores away from the dung and onto surrounding grass. If an herbivore eats the grass and consumes the spores, the asexual life cycle is completed when the animal disperses the spores in its feces.

C. Division Ascomycota: sac fungi produce sexual spores in saclike asci

Ascomycetes include unicellular yeasts and complex multicellular cup fungi (see Campbell, Figure 31.7).

- Hyphae are septate.

- In asexual reproduction, the tips of specialized hyphae form *conidia*, which are chains of haploid, asexual spores that are usually wind dispersed.

- In sexual reproduction, haploid mycelia of opposite mating strains fuse. One acts as "female" and produces an ascogonium which receives haploid nuclei from the antheridium of the "male" (see Campbell, Figure 31.8).

The ascogonium grows hyphae with dikaryotic cells. Syngamy is delayed.

↓

In terminal cells of dikaryotic hyphae, syngamy occurs.

↓

Meiosis forms four haploid nuclei which undergo mitotic division to yield eight haploid nuclei.

↓

The nuclei form walls and become ascospores within an *ascus*, the sac of sexually produced spores. Multiple asci may form an ascocarp.

Ascocarps = Fruiting structures of many asci packed together

- The ascospores of each ascus are lined up in a row in the order in which they formed from a single zygote, allowing geneticists to study genetic recombination.

- Unicellular yeasts appear dissimilar, but produce the equivalent of an ascus during sexual reproduction and bud during asexual reproduction in a manner similar to the formation of conidia. Thus, they are classified as ascomycetes.

- Includes important decomposers and both mutualistic and parasitic symbionts

- Many live symbiotically with algae as lichens.

D. Division Basidiomycota: club fungi have long-lived dikaryotic mycelia and a transient diploid stage

The division Basidiomycota, or *club fungi*, includes mushrooms, shelf fungi, puffballs, and stinkhorns (see Campbell, Figure 31.9).

Basidiomycetes:

- Are named for a transient diploid stage called the *basidium*, a club-shaped spore-producing structure.

- Are important decomposers of wood and other plant material. Saprobic basidiomycetes can decompose the complex polymer lignin, an abundant component of wood.

- Include mycorrhiza-forming mutualists and plant parasites. Many shelf fungi are tree parasites that function later as saprobes after the trees die.

- Include mushroom-forming fungi, only a few of which are strictly parasitic. About half are saprobic and the other half form mycorrhizae.
- Include the rusts and smuts, which are plant parasites.

Basidiomycete life cycles are characterized by a long-lived *dikaryotic* mycelium that reproduces sexually by producing fruiting bodies called *basidiocarps*. Refer to Campbell, Figure 31.10 for the life cycle of a mushroom-forming basidiomycete.

- Haploid basidiospores grow into short-lived haploid mycelia. Under certain environmental conditions, *plasmogamy* occurs between two haploid mycelia of opposite mating types (+ and −).
- The resulting dikaryotic mycelium grows; depending upon the species, it may form mycorrhizae with trees. Certain environmental cues stimulate the mycelium to produce mushrooms (basidiocarps). A "fairy ring" is an expanding ring of living mycelium that produces mushrooms above it; it slowly increases in diameter, about 30 cm per year.
- The mushroom cap supports and protects a large surface area of gills; karyogamy in the terminal, dikaryotic cells lining the gills produces diploid basidia.
- Each basidium immediately undergoes meiosis producing four haploid basidiospores. When mature, these sexual spores drop from the cap and are dispersed by wind.

Asexual reproduction occurs less often than in ascomycetes, but also results in conidia formation.

E. Molds, yeasts, lichens, and mycorrhizae represent unique lifestyles that evolved independently in three fungal divisions

1. Molds

Mold = A rapidly growing, asexually reproducing fungus

Molds may be saprobes or parasites on a great variety of substrates.

Molds only include asexual stages; they may be zygomycetes, ascomycetes, basidiomycetes or fungi with no known sexual stage.

Since molds are classified by their sexual stages (zygosporangium, ascogonium or basidium), molds with no known sexual stage cannot be classified as zygomycetes, ascomycetes, or basidiomycetes.

Molds with no known sexual stages are classified as Deuteromycota or *imperfect fungi*.

- Imperfect fungi reproduce asexually by producing spores.
- Deuteromycetes are sources of antibiotics. Penicillin is produced by some species of *Penicillium,* which are ascomycetes.
- Other commercial uses of imperfect fungi include flavoring for cheeses, such as blue cheese, Brie, Camembert and Roquefort; fermenting food products such as soybeans; and providing pharmaceuticals such as cyclosporine.
- Some deuteromycetes are predatory soil fungi that kill small animals such as soil nematodes (see Campbell, Figure 31.12).

2. Yeasts

Yeasts are unicellular fungi that inhabit liquid or moist habitats; some can alternate between mycelium or yeast, depending on the amount of liquid in the environment.

Yeasts reproduce:

- Asexually by simple cell division or by budding off from a parent cell; some are classified as Deuteromycota, if no sexual stages are known.
- Sexually by forming asci (Ascomycota) or basidia (Basidiomycota)

Though humans have used yeasts to raise bread and ferment alcoholic beverages for thousands of years, only recently have they been separated into pure culture for more controlled human use.

- *Saccharomyces cerevisiae* is the most important of all domesticated fungi (see Campbell, Figure 31.13). Highly active metabolically, this ascomycete is available as baker's and brewer's yeast.
- In an aerobic environment, baker's yeast respires, releasing small bubbles of carbon dioxide that leaven dough; cultured anaerobically, *Saccharomyces* ferments sugars to alcohol.
- Researchers use *Saccharomyces* to study eukaryotic molecular genetics because it is easy to culture and manipulate.

Some yeasts cause problems for humans.

- *Rhodotorula*, a pink yeast, grows on shower curtains and other moist surfaces.
- *Candida*, a normal inhabitant of moist human tissues, can become pathogenic when there is a change in pH or other environmental factor; or when an individual's immune system is compromised.

3. Lichens

Lichen = Highly integrated symbiotic association of algal cells (usually filamentous green algae or blue-green algae) with fungal hyphae (usually ascomycetes)

Though lichens vary in shape and physiology, some shared general features characterize the symbiotic relationship.

The alga, which is below the lichen's surface (see Campbell, Figure 31.15),

- Always provides the fungus with food
- May fix nitrogen (e.g.,. cyanobacteria)

The fungus provides a suitable environment for algal growth:

The hyphal mass:

- Absorbs needed minerals from airborne dust or rain
- Retains water and minerals
- Allows gas exchange
- Protects the algae

The fungus produces unique organic compounds with several functions.

- Fungal pigments shade the algae from intense sunlight.
- Toxic fungal compounds prevent lichens from being eaten by consumers.
- Fungal acid secretion aids the uptake of minerals.

Most of the lichen's mass is hyphal tissue which gives the lichen its shape and structure. Named for their fungal component, lichens are informally categorized as:

- Foliose (leafy)
- Fruticose (shrubby)

- Crustose (crusty)

Lichen reproduction occurs as a combined unit or as independent reproduction of the symbionts.

- Many lichen fungi reproduce sexually by forming ascocarps or rarely, basidiocarps.
- Lichen algae reproduce independently by asexual cell division.
- Symbiotic units commonly reproduce asexually by:
  ⇒ Fragmentation of the parental lichen
  ⇒ Formation of *soredia*, specialized reproductive structures that are small clusters of hyphae with embedded algae.

Though most evidence points to a mutualistic symbiosis, some debate that the relationship may actually be parasitic.

- The argument for mutualism is that fungi benefit the algae and that lichens can survive in habitats that are inhospitable to either organism alone.
- The argument for "controlled parasitism" is based on the fact that the fungus actually kills some algal cells, though not as fast as the algae replenishes itself.

Lichens are important pioneers, breaking down rock and allowing for colonization by other plants.

- Some can tolerate severe cold.
- Photosynthesis occurs when lichen water content is 65-75%.

Lichens are sensitive to air pollution due to their mode of mineral uptake.

4. Mycorrhizae

Mycorrhizae are specific, mutualistic associations of plant roots and fungi (see Campbell, Figure 31.16).

- The fungi increase the absorptive surface of roots and exchanges soil minerals.

Mycorrhizae are seen in 95% of all vascular plants.

They are necessary for optimal plant growth.

III.   Ecological Impacts of Fungi

A. Ecosystems depend on fungi as decomposers and symbionts

Fungi and bacteria are the principal decomposers in ecosystems. Decomposition allows for the recycling of nutrients between biotic and abiotic components.

Fungi decompose food, wood, and even certain plastics.

Between 10% - 50% of the world's fruit harvest is lost each year to fungal attack.

B. Some fungi are pathogens

Many fungi are pathogenic (e.g., athletes foot, ringworm, and yeast infections).

Plants are particularly susceptible. For example, Dutch elm disease, caused by an ascomycete, drastically changed the landscape of northeastern United States.

Ergots = Purple structure on rye caused by an ascomycete

- Causes gangrene, hallucinations, and burning sensations (St. Anthony's fire).
- Produces lysergic acid, from which LSD is made.

Toxins from fungi may be used in weak doses for medical purposes such as treating high blood pressure.

C. Many animals, including humans, eat fungi

Fungi are consumed as food by a variety of animals, including humans.

- In the U.S., mushroom (basidiomycete) consumption is usually restricted to one species of *Agaricus*, which is cultivated commercially on compost in the dark.

- In many other countries, however, people eat a variety of cultivated and wild mushrooms.

- Truffles prized by gourmets are underground ascocarps of mycelia that are mycorrhizal on tree roots (see Campbell, Figure 31.17). The fruiting bodies (ascocarps) release strong odors that attract mammals and insects—consumers that excavate the truffles and disperse their spores.

- Since it is difficult for novices to distinguish between poisonous and edible mushrooms, only qualified experts at identification should collect wild mushrooms for eating.

IV. Phylogenetic Relationships of Fungi

A. Fungi and animals probably evolved from a common protistan ancestor

The presence of flagella in the most primitive group of fungi, the chytrids, suggests that the ancestors of fungi were flagellated and that the lack of flagella in the other fungi divisions is a secondary condition.

There is compelling evidence that animals and fungi diverged from a common protistan ancestor.

- Animals also probably evolved from flagellated protists.

- Proteins and rRNA comparisons indicate that fungi and animals are more closely related to each other than either is to plants. Molecular systematists believe the most likely protistan ancestor common to fungi and animals was a choanoflagellate.

Perhaps the fungi are a consequence of adaptive radiation when life began to colonize land.

- The oldest undisputed fossils are 450-500 million years old.

- All major groups of fungi evolved by the end of the Carboniferous period (approximately 300 million years ago).

- Plants and fungi moved from water to land together. Fossils of the first vascular plants have mycorrhizae.

REFERENCES

Alexopoulos, C.J. and C.W. Mims. *Introductory Mycology*. 3rd ed. New York: John Wiley and Sons, Inc., 1979.

Campbell, N., et al. *Biology*. 5th ed. Menlo Park, California: Benjamin/Cummings, 1999.

CHAPTER 32
INTRODUCTION TO ANIMAL EVOLUTION

OUTLINE
I. What Is an Animal?
II. An Overview of Animal Phylogeny and Diversity
   A. Parazoans lack true tissues
   B. Evolution of body cavities led to more complex animals
   C. Coelomates branched into protostomes and deuterostomes
III. The Origins of Animal Diversity
   A. Most animal phyla originated in a relatively brief span of geological time
   B. Developmental genetics may clarify our understanding of the Cambrian diversification

OBJECTIVES
After reading this chapter and attending lecture, the student should be able to:
1. List characteristics that distinguish animals from organisms in the other four kingdoms.
2. Distinguish between radial and bilateral symmetry.
3. Outline the major phylogenetic branches of the animal kingdom, which are based upon grade of organization; symmetry and embryonic germ layers; absence or presence of a body cavity; and protostome-deuterostome dichotomy.
4. Distinguish among acoelomate, pseudocoelomate and coelomate.
5. Distinguish between spiral and radial cleavage; determinant and indeterminate cleavage; schizocoelous and enterocoelous.
6. Compare developmental differences between protostomes and deuterostomes including:
   a. Plane of cleavage
   b. Determination
   c. Fate of the blastopore
   d. Coelom formation
7. Compare and contrast two hypotheses about animal origins from unicellular ancestors: syncytial hypothesis and colonial hypothesis.
8. Explain why it is difficult to resolve what the first animals looked like.
9. Describe two views about discontinuities between Ediacaran and Cambrian fauna.

KEY TERMS
ingestion	bilateral symmetry	archenteron	deuterostomes
cleavage	dorsal	mesoderm	spiral cleavage
blastula	ventral	diploblastic	determinate cleavage
gastrulation	anterior	triploblastic	radial cleavage
larva	posterior	acoelomates	indeterminate cleavage
metamorphosis	bilateria	pseudocoelom	blastopore

parazoa	cephalization	pseudocoelomates	schizocoelous
eumetazoa	germ layers	coelomates	enterocoelous
radial symmetry	ectoderm	coelom	Ediacaran period
radiata	endoderm	protostomes	Cambrian explosion

LECTURE NOTES

Over one million species of animals are living today; 95% of these are invertebrates.

- Grouped into about 35 phyla depending on the taxonomic view followed.
- Most are aquatic.
- The most familiar belong to the subphylum Vertebrata of the phylum Chordata. This is only about 5% of the total.

I.   What Is an Animal?

Although there is great animal diversity, most animals share the following characteristics:

- Multicellular, eukaryotic organisms
- Heterotrophy is by ingestion.
    - *Ingestion* = Eating other organisms or decomposing organic matter (detritus). This mode of nutrition distinguishes animals from the plants and fungi.
    - Carbohydrate reserves generally are stored as glycogen.
- No cell walls are present, but animals do have intercellular junctions: desmosomes, gap junctions, and tight junctions.
- Highly differentiated body cells which are organized into tissues, organs and organ systems for such specialized functions as digestion, internal transport, gas exchange, movement, coordination, excretion, and reproduction.
- Nervous tissue (impulse conduction) and muscle tissue (movement) are unique to animals.
- Reproduction is typically sexual with flagellated sperm fertilizing nonmotile eggs to form diploid zygotes. A diploid stage dominates the life cycle.
    - The zygote undergoes a series of mitotic divisions known as *cleavage* which produces a *blastula* in most animals.
    - *Gastrulation* occurs after the blastula has formed; during this process, the embryonic forms of adult body tissues are produced.
    - Development in some animals is direct to maturation while the life cycles of others include *larvae* which undergo *metamorphosis* into a sexually mature adults.
    - *Larva* = Free-living, sexually immature forms

The seas contain the greatest diversity of animal phyla, although many groups live in fresh water and terrestrial habitats.

II.   An Overview of Animal Phylogeny and Diversity

Animals diversified so rapidly during the late Precambrian and early Cambrian periods that it is difficult to determine the exact sequence of branching from the fossil record.

- To reconstruct the evolutionary history of the animal phyla, zoologists use information from comparative anatomy, embryology of living animals, and molecular systematics.

- Most zoologists agree that the animal kingdom is monophyletic and that the ancestral organism was probably a colonial flagellated protist related to choanoflagellates (see Campbell, Figure 32.2).

A. Parazoans lack true tissues

Sponges (Phylum Porifera) represent an early branch of the animal kingdom (see Campbell, Figure 32.3:

- Have unique development and simple anatomy that separates them from other animals
- Lack true tissues, therefore, they are called parazoa ("beside the animals")

The presence of true tissues is characteristic of nearly all the other groups of animals, collectively known as *eumetazoa*. True tissues permitted the evolution of a more complex anatomy.

B. Radiata and bilateria are the major branches of eumetazoans

The division of eumetazoans into two branches is based partly on body symmetry (see Campbell, Figure 32.3).

- *Radiata* exhibit *radial symmetry* (see also Campbell, Figure 32.4).
  - These animals have an oral (top) and aboral (bottom) side, but no front, back, left, or right sides.
- *Bilateria* exhibit *bilateral symmetry* (see also Campbell, Figure 32.5).
  - ⇒ Bilaterally symmetrical animals have *dorsal* (top), *ventral* (bottom), *anterior* (head), *posterior* (tail), left and right body surfaces.
  - ⇒ These animals exhibit *cephalization* (an evolutionary trend toward concentration of sensory structures at the anterior end).

Care must be taken when assigning an animal to an evolutionary line as symmetry may change between the larval and adult forms. The phylum Echinodermata shows a secondary radial symmetry in adults, which evolved as an adaptation to their sedentary lifestyle. They are actually in the bilateria.

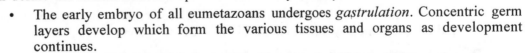

Examination of development and body plan can define the radiata-bilateria split better than symmetry.

- The early embryo of all eumetazoans undergoes *gastrulation*. Concentric germ layers develop which form the various tissues and organs as development continues.
  - ⇒ The radiata (e.g., Phylum Cnidaria, Phylum Ctenophores) develop only two germ layers (*ectoderm* and *endoderm*) and are termed *diploblastic-*.
  - ⇒ The bilateria (e.g., all eumetazoan phyla except Phylum Cnidaria and Phylum Ctenophores) develop three germ layers (ectoderm, endoderm, and *mesoderm*) and are termed *triploblastic*.

The germ layers of an early embryo include:

1. Ectoderm
   - Covers the surface of the embryo
   - Forms the animal's outer covering and the central nervous system in some phyla

2. Endoderm
   - Innermost germ layer which lines the *archenteron* (primitive gut)
   - Forms the lining of the digestive tract, and outpocketings give rise to the liver and lungs of vertebrates

3. Mesoderm
   - Located between the ectoderm and endoderm in triploblastic animals
   - Forms the muscles and most organs located between the digestive tract and outer covering of the animal

C. Evolution of body cavities led to more complex animals

Triploblastic animals can also be grouped on the basis of whether a body cavity develops and how that cavity develops.

Animals in which no body cavity develops are termed acoelomate.

- *Acoelomate* = An animal body plan characterized by no body cavity present between the digestive tract and the outer body wall (see Campbell, Figure 32.5a)
- The area between the digestive tract and outer wall is filled with cells, producing a solid body (e.g., Phylum Platyhelminthes).

Animals in which a body cavity develops may be termed pseudocoelomate or coelomate, depending on how the cavity develops.

- *Pseudocoelomate* = Animal body plan characterized by a fluid-filled body cavity that separates the digestive tract and the outer body wall. (See Campbell, Figure 32.5b.)
- This cavity (the *pseudocoelom*) is not completely lined with tissue derived from mesoderm (e.g., Phylum Nematoda).
- *Coelomate* = Animal body plan characterized by a fluid-filled body cavity completely lined with tissue derived from mesoderm (the *coelom*) that separates the digestive tract from the outer body wall (see Campbell, Figure 32.5c)
- Mesenteries connect the inner and outer mesoderm layers and suspend the internal organs in the coelom (e.g., Annelida).
- The fluid-filled body cavities:
  - Cushion the organs, thus preventing injury
  - Allow internal organs can grow and move independently of the outer body wall.
  - Serve as a hydrostatic skeleton in soft bodied coelomates such as earthworms.

In addition to the presence of a body cavity, acoelomates differ from pseudocoelomates and eucoelomates by not having a blood vascular system.

D. Coelomates branched into protostomes and deuterostomes

Distinguished by differences in development, the coelomate phyla can be divided into two distinct evolutionary lines (see Campbell, Figure 32.3):

1. *Protostomes* (e.g., mollusks, annelids, arthropods)
2. *Deuterostomes* (e.g., echinoderms, chordates)

Developmental differences between protostomes and deuterostomes include: cleavage patterns, coelom formation, and fate of the blastopore.

1. Cleavage

Most protostomes undergo spiral cleavage and determinate cleavage during their development.

- *Spiral cleavage* = Cleavage in which the planes of cell division are diagonal to the vertical axis of the embryo (see Campbell, Figure 32.6)
- *Determinate cleavage* = Cleavage in which the developmental fate of each embryonic cell is established very early; a cell isolated from the four-cell stage of an embryo will not develop fully.

Deuterostomes undergo radial cleavage and indeterminate cleavage during their development.

- *Radial cleavage* = Cleavage during which the cleavage planes are either parallel or perpendicular to the vertical axis of the embryo
- *Indeterminate cleavage* = Cleavage in which each early embryonic cell retains the capacity to develop into a complete embryo if isolated from other cells; this type of cleavage in the human zygote results in identical twins.

2. Coelom formation

Schizocoelous = Descriptive term for coelom development during which, as the archenteron forms, the coelom begins as splits within the solid mesodermal mass; coelom formation found in protostomes (see Campbell, Figure 32.6b)

Enterocoelous = Coelom development during which the mesoderm arises as lateral outpocketings of the archenteron with hollows that become the coelomic cavities; coelom formation found in deuterostomes (see Campbell, Figure 32.6c)

3. Blastopore fate

Blastopore = The first opening of the archenteron which forms during gastrulation

- In protostomes, the blastopore forms the mouth.
- In deuterostomes, the blastopore forms the anus.

SUMMARY OF PROTOSTOME - DEUTEROSTOME SPLIT	
PROTOSTOMES	**DEUTEROSTOMES**
Spiral cleavage	Radial cleavage
Determinate cleavage	Indeterminate cleavage
Blastopore forms the mouth	Blastopore forms the anus
Schizocoelous coelom formation	Enterocoelous coelom formation

III. The Origins of Animal Diversity

A. Most animal phyla originated in a relatively brief span of geological time

The animal kingdom probably originated from colonial protists related to choanoflagellates. The diversification that produced many phyla occurred in a relatively short time on the geological scale. This evolutionary episode is called the *Cambrian explosion.*

The Cambrian explosion encompassed a 20-million-year time span at the beginning of the Cambrian period (ca. 545 to 525 million years ago).

- Nearly all of the major animal body plans seen today evolved during this time.
- New taxa appeared later but were variations on the basic plans already evolved. For example, mammals evolved about 220 million years ago, but are only a variation of the chordate body plan which evolved during the Cambrian explosion.

A much less diverse fauna preceded the Cambrian explosion.

- This Precambrian fauna dated back to the *Ediacaran period* (700 million years ago).
  - ⇒ This period is named for the Ediacara Hills of Australia where Precambrian animal fossils were first discovered.
  - ⇒ Fossils similar in age to these have since been discovered on other continents.
- Most Ediacaran fossils appear to be cnidarians although bilaterial animals are also indicated by fossilized burrows probably left by worms.

The diversity of Cambrian animals is represented in three fossil beds:
- The Burgess Shale in British Columbia is the best known (see Campbell, Figure 32.7).
- A fossil bed in Greenland and one in the Yunnan region of China predate the Burgess Shale by 10 million years.

Two contrasting interpretations of Burgess Shale fossils have been proposed:
1. The Cambrian explosion resulted in a large number of phyla which included the current phyla, many of which are now extinct.
   - ⇒ During the mass extinction at the end of the Cambrian, only the base stock of 35 or so extant phyla survived.
2. The diversity of the Cambrian fossils represents ancient variations within the taxonomic boundaries of extant phyla.
   - ⇒ As these fossils undergo continued study, many are classified into extant phyla. Thus, the number of exclusively Cambrian fossils is decreasing.

B. Developmental genetics may clarify our understanding of the Cambrian diversification

Several hypotheses about external factors have been proposed as explanations for the Cambrian explosion and the lack of subsequent major diversification.
1. The Cambrian explosion was an adaptive radiation resulting from the origin of the first animals.
   - ⇒ These early animals diversified as they adapted to the various, previously unoccupied, ecological niches.
2. Predator-prey relationships emerged and triggered diverse evolutionary adaptations.
   - ⇒ Various kinds of shells and different forms of locomotion evolved as defense mechanisms against predation.
   - ⇒ Predators also evolved new mechanisms to capture prey.
3. Major environmental change provided an opportunity for diversification during the Cambrian explosion.
   - ⇒ The accumulation of atmospheric oxygen may have finally reached a concentration to support the more active metabolism needed for feeding and other activities by mobile animals.

Other hypotheses for the Cambrian explosion have emphasized internal changes in the organisms.

1. The origin of mesoderm may have stimulated diversification of the body plan.
   - ⇒ This third tissue layer permits development of more complex anatomical structure.
2. Variation in genes that control pattern formation during animal development may have played a role in diversification.

⇒ Some of the genes that determine features such as segmentation and placement of appendages and other structures are common to diverse animal phyla.

⇒ Variation in expression of these genes during development results in morphological differences that distinguish the phyla.

⇒ This same kind of variation in expression may have resulted in the relatively rapid origin of diverse animal types during the Cambrian explosion.

⇒ The phyla, once developed, may have become locked into developmental patterns that permitted subtle variation to allow speciation and the origin of lower taxa, but prevented large scale morphological evolution resulting in new phyla.

The hypotheses presented for external and internal factors are not mutually exclusive. A combination of factors may have combined to produce the Cambrian explosion.

REFERENCES

Brusca, R and Brusca, G. Invertebrates. Sunderland, Massachusetts: Sinauer Associates, 1990.

Campbell, N., et al. *Biology*. 5th ed. Menlo Park, California: Benjamin/Cummings, 1999.

Harris, C.L. Concepts on Zoology, 2nd ed. New York, NY: HarperCollins, 1996.

CHAPTER 33
INVERTEBRATES

OUTLINE

I. The Parazoa
    A. Phylum Porifera: sponges are sessile with porous bodies and choanocytes

II. The Radiata
    A. Phylum Cnidaria: cnidarians have radial symmetry, a gastrovascular cavity, and cnidocytes
    B. Phylum Ctenophora: comb jellies possess rows of ciliary plates and adhesive colloblasts

III. The Acoelomates
    A. Phylum Platyhelminthes: flatworms are dorsoventrally flattened acoelomates

IV. The Pseudocoelomates
    A. Phylum Rotifera: rotifers have jaws and a crown of cilia
    B. Phylum Nematoda: roundworms are unsegmented and cylindrical with tapered ends

V. The Coelomates: Protostomes
    A. Phylum Nemertea: The phylogenetic position of proboscis worms is uncertain
    B. The lophophorate phyla: bryozoans, phoronids, and brachiopods have ciliated tentacles around their mouths
    C. Phylum Mollusca: mollusks have a muscular foot, a visceral mass, and a mantle
    D. Phylum Annelida: annelids are segmented worms
    E. Phylum Arthropoda: arthropods have regional segmentation, jointed appendages, and an exoskeleton

VI. The Coelomates: Deuterostomes
    A. Phylum Echinodermata: Echinoderms have a water vascular system and secondary radial symmetry
    B. Phylum Chordata: the chordates include two invertebrate subphyla and all vertebrates

OBJECTIVES

After reading this chapter and attending lecture, the student should be able to:

1. From a diagram, identify the parts of a sponge and describe the function of each including the spongocoel, porocyte, epidermis, choanocyte, mesohyl, amoebocyte, osculum, and spicule.

2. List characteristics of the phylum Cnidaria that distinguish it from the other animal phyla.

3. Describe the two basic body plans in Cnidaria and their role in Cnidarian life cycles.

4. List the three classes of Cnidaria and distinguish among them based upon life cycle and morphological characteristics.

5. List characteristics of the phylum Ctenophora that distinguish it from the other animal phyla.

6. List characteristics that are shared by all bilaterally symmetrical animals.

7. List characteristics of the phylum Platyhelminthes that distinguish it from the other animal phyla.

8. Distinguish among the four classes of Platyhelminthes and give examples of each.

9. Describe the generalized life cycle of a trematode and give an example of one fluke that parasitizes humans.

10. Describe the anatomy and generalized life cycle of a tapeworm.

11. List distinguishing characteristics descriptive of the phylum Nemertea.

12. Explain why biologists believe proboscis worms evolved from flatworms.

13. Describe features of digestive and circulatory systems that have evolved in the Nemertea and are not found in other acoelomate phyla.

14. Describe unique features of rotifers that distinguish them from other pseudocoelomates.

15. Define parthenogenesis and describe alternative forms of rotifer reproduction.

16. List characteristics of the phylum Nematoda that distinguish it from other pseudocoelomates.

17. Give examples of both parasitic and free-living species of nematodes.

18. List characteristics that distinguish the phylum Mollusca from the other animal phyla.

19. Describe the basic body plan of a mollusk and explain how it has been modified in the Polyplacophora, Gastropoda, Bivalvia, and Cephalopoda.

20. Distinguish among the following four Molluscan classes and give examples of each:

    a. Polyplacophora   c. Bivalvia
    b. Gastropoda   d. Cephalopoda

21. Explain why some zoologists believe the mollusks evolved from ancestral annelids while others propose that mollusks arose from flatworm-like ancestors.

22. List characteristics that distinguish the phylum Annelida from the other animal phyla.

23. Explain how a fluid-filled septate coelom is used by annelids for burrowing.

24. Distinguish among the classes of annelids and give examples of each.

25. List characteristics of arthropods that distinguish them from the other animal phyla.

26. Describe advantages and disadvantages of an exoskeleton.

27. Distinguish between hemocoel and coelom.

28. Provide evidence for an evolutionary link between the Annelida and Arthropoda.

29. Describe major independent arthropod lines of evolution represented by the subphyla:

    a. Trilobitomorpha   c. Crustacea
    b. Cheliceriformes   d. Uniramia

30. Explain what arthropod structure was a preadaptation for living on land.

31. Distinguish among the following arthropod classes and give an example of each:

    a. Arachnida   d. Chilopoda
    b. Crustacea   e. Insecta
    c. Diplopoda

32. Distinguish between incomplete metamorphosis and complete metamorphosis.

33. Define lophophore and list three lophophorate phyla.

34. Explain why lophophorates are difficult to assign as protostomes or deuterostomes.

35. List at least four characteristics shared by the deuterostome phyla that distinguish them from protostomes.

36. List characteristics of echinoderms that distinguish them from other animal phyla.

37. Describe the structures and function of a water vascular system, including ring canal, radial canal, tube feet and ampulla.
38. Distinguish among the classes of echinoderms and give examples of each.

KEY TERMS

invertebrates	parthenogenesis	cuticle	Class Insecta
spongocoel	closed circulatory	exoskeleton	entomology
osculum	system	molting	Malpighian tubules
choanocyte	lophophorate animals	open circulatory system	tracheal system
mesohyl	lophophore	trilobite	incomplete
amoebocyte	bryozoans	Chelicerates	metamorphosis
hermaphrodites	phoronids	Uniramians	complete
gastrovascular cavity	brachiopods	Crustaceans	metamorphosis
polyp	foot	chelicerae	echinoderms
medusa	visceral mass	mandibles	water vascular system
cnidocytes	mantle	antennae	tube feet
cnidae	mantle cavity	compound eyes	nematocysts
radula	eurypterids	colloblasts	trochophore
Class Arachnida	planarian	torsion	book lungs
complete digestive	ammonites	Class Diplodia	metanephridia
tract	Class Chilopodia		

LECTURE NOTES

Over one million species of animals are living today; 95% of these are invertebrates.

• Most are aquatic.
• The most familiar belong to the subphylum Vertebrata of the phylum Chordata. This is only about 5% of the total.

I. **The Parazoa**

A. **Phylum Porifera: sponges are sessile with porous bodies and choanocytes**

The sponges, in the phylum Porifera, are the only members of the subkingdom Parazoa due to their unique development and simple anatomy (see Campbell, Figure 33.1).

• Approximately 9000 species, mostly marine with only about 100 in fresh water
• Lack true tissues and organs, and contain only two layers of loosely associated unspecialized cells
• No nerves or muscles, but individual cells detect and react to environmental changes
• Size ranges from 1 cm to 2 m
• All are suspension-feeders (= filter-feeders)
• Possibly evolved from colonial choanoflagellates

Parts of the sponge include (see Campbell, Figure 33.2):

• *Spongocoel* = Central cavity of sponge
• *Osculum* = Larger excurrent opening of the spongocoel
• *Epidermis* = Single layer of flattened cells which forms outer surface of the sponge

- *Porocyte* = Cells which form pores; possess a hollow channel through the center which extends from the outer surface (incurrent pore) to spongocoel
- *Choanocyte* = Collar cell, majority of cells which line the spongocoel; possess a flagellum which is ringed by a collar of fingerlike projections. Flagellar movement moves water and food particles which are trapped on the collar and later phagocytized.
- *Mesohyl* = The gelatinous layer located between the two layers of the sponge body wall (epidermis and choanocytes)
- *Amoebocyte* = Wandering, pseudopod bearing cells in the mesohyl; function in food uptake from choanocytes, food digestion, nutrient distribution to other cells, formation of skeletal fibers, gamete formation
- *Spicule* = Sharp, calcium carbonate or silica structures in the mesohyl which form the skeletal fibers of many sponges
- *Spongin* = Flexible, proteinaceous skeletal fibers in the mesohyl of some sponges

Most sponges are hermaphrodites, but usually cross-fertilize.

- Eggs and sperm form in the mesohyl from differentiated amoebocytes or choanocytes.
- Eggs remain in the mesohyl.
- Sperm are released into excurrent flow of the spongocoel and are then drawn in with incurrent flow of another sponge.
- Sperm penetrate into mesohyl and fertilize the eggs.
- The zygote develops into a flagellated larva which is released into the spongocoel and escapes with the excurrent water through the osculum.
- Surviving larvae settle on the substratum and develop. In most cases the larva turns inside-out during metamorphosis, moving the flagellated cells to the inside.

Sponges possess extensive regeneration abilities for repair and asexual reproduction.

II.   The Radiata

The branch radiata is composed of phylum Cnidaria and phylum Ctenophora

A. Phylum Cnidaria: Cnidarians have radial symmetry, a gastrovascular cavity, and cnidocytes

There are more than 10,000 species in the phylum Cnidaria, most of which are marine. The phylum contains hydras, jellyfish, sea anemones and coral animals.

Some characteristics of cnidarians include:

- Radial symmetry
- Diploblastic
- Simple, sac-like body
- *Gastrovascular cavity,* a central digestive cavity with only one opening (functions as mouth and anus)

There are two possible cnidarian body plans: sessile polyp and motile, floating medusa (see Campbell, Figure 33.3). Some species of cnidarians exist only as polyps, some only as medusae, and others are dimorphic (both polyp and medusa stages in their life cycles).

Polyp = Cylindrical form which adheres to the substratum by the aboral end of the body stalk and extends tentacles around the oral end to contact prey

Medusa = Flattened, oral opening down, bell-shaped form; moves freely in water by passive drifting and weak bell contractions; tentacles dangle from the oral surface which points downward.

Cnidarians are carnivorous.

- Tentacles around the mouth/anus capture prey animals and push them through the mouth/anus into the gastrovascular cavity.

- Digestion begins in the gastrovascular cavity with the undigested remains being expelled through the mouth/anus.

- Tentacles are armed with stinging cells, called cnidocytes—after which the Cnidaria are named.

Cnidocytes = Specialized cells of cnidarian epidermis that contain eversible capsule-like organelles, or *cnidae*, used in defense and capture of prey (see Campbell, Figure 33.4). Nematocysts are stinging capsules.

The simplest forms of muscles and nerves occur in the phylum Cnidaria.

- Epidermal and gastrodermal cells have bundles of microfilaments arranged into contractile fibers.

  ⇒ The gastrovascular cavity, when filled with water, acts as a hydrostatic skeleton against which the contractile fibers can work to change the animal's shape.

- A simple nerve net coordinates movement; no brain is present.

  ⇒ The nerve net is associated with simple sensory receptors radially distributed on the body. This permits stimuli to be detected and responded to from all directions.

There are three major classes of cnidarians (see Campbell, Figure 33.5 and Table 33.1):

1. **Class Hydrozoa**

   Most hydrozoans alternate polyp and medusa forms in the life cycle although the polyp is the dominant stage. Some are colonial (e.g., *Obelia*, Campbell, Figure 33.6), while others are solitary (e.g., *Hydra*).

   Hydra is unique in that only the polyp stage is present.

   - They usually reproduce asexually by budding; however, in unfavorable conditions they reproduce sexually. In this case a resistant zygote is formed and remains dormant until environmental conditions improve.

2. **Class Scyphozoa**

   The planktonic medusa (jellyfish) is the most prominent stage of the life cycle.

   - Coastal species usually pass through a small polyp stage during the life cycle.

   - Open ocean species have eliminated the polyp entirely.

3. **Class Anthozoa**

   This class contains sea anemones and coral animals.

   They only occur as polyps.

   Coral animals may be solitary or colonial and secrete external skeletons of calcium carbonate.

   - Each polyp generation builds on the skeletal remains of earlier generations. In this way, coral reefs are formed.

   - Coral is the rock-like external skeletons.

B. Phylum Ctenophora: combjellies possess rows of ciliary plates and adhesive colloblasts

This phylum contains the comb jellies. There are about 100 species, all of which are marine.

Some characteristics of ctenophores include:

- A resemblance to the medusa of Cnidarians in that the body of most is spherical or ovoid; a few are elongate and ribbonlike.
- Transparent body, 1 - 10 cm in diameter (spherical/ovoid forms) or up to 1 m long (ribbonlike forms) (see Campbell, Figure 33.7)
- Eight rows of comblike plates composed of fused cilia which are used for locomotion
- One pair of long retractable tentacles that function in capturing food; these tentacles have adhesive structures called *colloblasts*.
- A sensory organ containing calcareous particles is present.
    ⇒ The particles settle to the low point of the organ which then acts as an orientation cue.
    ⇒ Nerves extending from the sensory organ to the combs of cilia coordinate movement.

III. The Acoelomates

A. Phylum Platyhelminthes: flatworms are dorsoventrally flattened acoelomates

The members of the phylum Platyhelminthes differ from the phylum Cnidaria in that they:

- Exhibit bilateral symmetry with moderate cephalization
- Are triploblastic (develop from three-layered embryos: ectoderm, endoderm and mesoderm)
- Possess several distinct organs, organ systems, and true muscles

Although more advanced than cnidarians, two things point to the early evolution of platyhelminths in bilateria history.

- A gastrovascular cavity is present.
- They have an acoelomate body plan.

There are more than 20,000 species of Platyhelminthes which are divided into four classes (see Campbell, Table 33.2):

- Class Turbellaria
- Classes Trematoda and Monogenea
- Class Cestoda

1. Class Turbellaria

Mostly free-living, marine species; a few species are found in freshwater and moist terrestrial habitats (see Campbell, Figure 33.8).

Planarians are familiar and common freshwater forms (see Campbell, Figure 33.9).

- Carnivorous, they feed on small animals and carrion
- Lack specialized organs for gas exchange or circulation
    ⇒ Gas exchange is by diffusion (flattened body form places all cells close to water).
    ⇒ Fine branching gastrovascular cavity distributes food throughout the animal.
- Flame cell excretory apparatus present which functions primarily to maintain osmotic balance of the animal.

⇒ Nitrogenous waste (ammonia) diffuses directly from cells to the water.
- Move by using cilia on the ventral dermis to glide along a film of mucus. Muscular contractions produce undulations which allow some to swim.
- On the head are a pair of eyespots which detect light and a pair of lateral auricles that are olfactory sensors.
    ⇒ Possess a rudimentary brain which is capable of simple learning.
- Reproduce either asexually or sexually.
    ⇒ Asexually by regeneration: mid-body constriction separates the parent into two halves, each of which regenerates the missing portion
    ⇒ Sexually by cross-fertilization of these hermaphroditic forms

2. Classes Monogenea and Trematoda

All members of these two classes are parasitic.

Flukes are members of the class Trematoda.
- Suckers are usually present for attaching to host internal organs.
- Primary organ system is the reproductive system; a majority are hermaphroditic.
- Life cycles include alternations of sexual and asexual stages with asexual development taking place in an intermediate host.
    ⇒ Larvae produced by asexual development infect the final hosts where maturation and sexual reproduction occurs (see Campbell, Figure 33.10)
- *Schistosoma* spp. (blood flukes) infect 200 million people worldwide.

Members of the class Monogenea are mostly external parasites of fish.
- Structures with large and small hooks are used for attaching to the host animal.
- All are hermaphroditic and reproduce sexually.

3. Class Cestoidea

Adult tapeworms parasitize the digestive system of vertebrates.
- Possess a scolex (head) which may be armed with suckers and/or hooks that help maintain position by attaching to the intestinal lining (see Campbell, Figure 33.11).
- Posterior to the scolex is a long ribbon of units called proglottids.
    ⇒ A proglottid is filled with reproductive organs.
- No digestive system is present.

The life cycle of a tapeworm includes an intermediate host.
- Mature proglottids filled with eggs are released from the posterior end of the worm and pass from the body with the feces.
- Eggs are eaten by an intermediate host and a larva develops, usually in muscle tissue.
    ⇒ The final host becomes infected when it eats an intermediate host containing larvae.
- Humans can become infected with some species of tapeworms by eating undercooked beef or pork containing larvae.

IV. The Pseudocoelomates

The pseudocoelomate body plan probably arose independently several times.

A. Phylum Rotifera: rotifers have jaws and a crown of cilia

There are approximately 1800 species of rotifers. They are small, mainly freshwater organisms, although some are marine and others are found in damp soil.

* Size ranges from 0.05-2.0 mm
* Pseudocoelomate with the pseudocoelomic fluid serving as a hydrostatic skeleton and as a medium which transports nutrients and wastes when the body moves
* *Complete digestive system* is present.
    ⇒ Rotifer refers to the crown of cilia that draws a vortex of water into the mouth.
    ⇒ Posterior to the mouth, a jawlike organ grinds the microscopic food organisms suspended in the water.

Reproduction in rotifers may be by *parthenogenesis* or sexual.

* Some species consist only of females with new females developing by parthenogenesis from unfertilized eggs.
* Other species produce two types of eggs, one that develops into females, the other into degenerate males.
    ⇒ Males produce sperm that fertilize eggs which develop into resistant zygotes that survive desiccation.
    ⇒ When conditions improve, the zygotes break dormancy and develop into a new female generation that reproduces by parthenogenesis until unfavorable conditions return.
    * Rotifers have no regeneration or repair abilities.

Rotifers contain a certain and consistent number of cells as adults. The zygotes undergo a specific number of divisions and the adult contains a fixed number of cells.

B. Phylum Nematoda: roundworms are unsegmented and cylindrical with tapered ends

There are about 90,000 species of roundworms, ranging in size from less than 1.0 mm to more than 1 m.

* Bodies are cylindrical with tapered ends
* Very numerous in both species and individuals
* Found in fresh water, marine, moist soil, tissues of plants, and tissues and body fluids of animals
* A complete digestive tract is present and nutrients are transported through the body in the pseudocoelomic fluid
* A tough, transparent cuticle forms the outer body covering (see Campbell, Figure 33.12a)
* Longitudinal muscles are present and provide for whip-like movements
* Dioecious with females larger than males
* Sexual reproduction only, with internal fertilization
* Female may produce 100,000 or more resistant eggs per day
* Like rotifers, nematodes have a fixed number of cells as adults

Nematodes fill various roles in the community.

* Free-living forms are important in decomposition and nutrient cycling.
* Plant parasitic forms are important agricultural pests.
* Animal parasitic forms can be hazardous to health (*Trichinella spiralis* in humans via undercooked infected pork) (see Campbell, Figure 33.13b).

- One species, *Caenorhabitis elegans*, is cultured extensively and is a model species for the study of development.

V. The Coelomates: Protostomes

The protostome lineage of coelomate animals gave rise to many phyla. In many, the coelom functions as a hydrostatic skeleton (e.g., mollusks, annelids).

A. Phylum Nemertea: the phylogenetic position of proboscis worms is uncertain

There are about 900 species, most are marine with a few in fresh water and damp soil.

The phylum Nemertea contains the proboscis worms (see Campbell, Figure 33.14)

- Sizes range from 1 mm to more than 30 m.
- Some active swimmers, others burrow in sand.
- Possess a long, retractable hollow tube (proboscis) which is used to probe the environment, capture prey, and as defense against predators.
- Excretory, sensory, and nervous systems are similar to planarians.
- Structurally acoelomate, like flatworms.

There are some important differences between the Nemertea and Platyhelminthes:

- Nemertea possess a closed circulatory system, which consists of vessels through which blood flows. Some species have red blood cells containing a form of hemoglobin which transports oxygen. No heart is present, but body muscle contractions move the blood through vessels.
- Nemertea possess a complete digestive system with a mouth and an anus.

The phylogenetic position of the Nemertea is uncertain.

- Although the body is structurally acoelomate, the fluid-filled proboscis sac is considered a true coelom by some researchers.
- A simple blood vascular system and a complete digestive system are characteristics shared with more advanced phyla.

B. The lophophorate phyla: bryozoans, phoronids, and brachiopods have ciliated tentacles around their mouths

The lophophorate animals contain three phyla: Phoronida, Bryozoa and Brachiopoda.

- These three phyla are grouped together due to presence of a lophophore (see Campbell, Figure 33.15)

Lophophore = Horseshoe-shaped or circular fold of the body wall bearing ciliated tentacles that surround the mouth at the anterior end of the animal.

- Cilia direct water toward the mouth between the tentacles which trap food particles for these suspension-feeders.
- The presence of a lophophore in all three groups suggests a relationship among these phyla.

The three phyla also possess a U-shaped digestive tract (the anus lies outside of the tentacles) and have no distinct head—both adaptations for a sessile existence.

Lophophorates are difficult to assign as protostomes or deuterostomes.

- Their embryonic development more closely resembles deuterostomes; however, in the Phoronida, the blastopore develops into the adult mouth.
- Molecular systematics places the lophophorate phyla closer to the protostomes than the deuterostomes.

1. Bryozoans

This phylum contains the moss animals. There are about 5000 species which are mostly marine and are widespread.

Bryozoans are small, colonial forms (see Campbell, Figure 33.15a)

- In most, the colony is enclosed within a hard exoskeleton and the lophophores are extended through pores when feeding.
- Some are important reef builders.

2. Phoronids

This phylum contains about 15 species of tube-dwelling marine worms.

- Length from 1 mm to 50 cm
- Phoronids live buried in sand in chitinous tubes with the lophophore extended from the tube when feeding.

3. Brachiopods

The phylum Brachiopoda contains the lamp shells. There are approximately 330 extant species, all marine.

- More than 30,000 fossil species of the Paleozoic and Mesozoic have been identified.

The body of a brachiopod is enclosed by dorsal and ventral shell halves (see Campbell, Figure 33.15b)

- Attach to the substratum by a stalk
- Open the shell slightly to allow water to flow through the lophophore

C. Phylum Mollusca: mollusks have a muscular foot, visceral mass, and a mantle

There are more than 50,000 species of snails, slugs, oysters, clams, octopuses, and squids.

Mollusks are mainly marine, though some inhabit fresh water and many snails and slugs are terrestrial.

Mollusks are soft-bodied, but most are protected by a hard calcium carbonate shell.

- Squids and octopuses have reduced, internalized shells or no shell.

The molluscan body consists of three primary parts: muscular *foot* for locomotion, a *visceral mass* containing most of the internal organs, and a *mantle,* which is a heavy fold of tissue that surrounds the visceral mass and secretes the shell (see Campbell, Figure 33.16).

A *radula* is present in many and functions as a rasping tongue to scrap food from surfaces.

Some species are monoecious while most are dioecious.

- Gonads are located in the visceral mass.

Some zoologists believe the mollusks evolved from annelid-like ancestors (although true segmentation is absent in mollusca) because the life cycle of many mollusks includes a ciliated larva, called a *trochophore,* which also is characteristic of annelids, while others believe that mollusks arose earlier in the protostome lineage before segmentation evolved.

1. Class Polyplacophora

The class Polyplacophora contains the marine species known as chitons.

- They have an oval shape with the shell divided into eight dorsal plates (see Campbell, Figure 33.17).
- Cling to rocks along the shore at low tide using the foot as a suction cup to grip the rock. This muscular foot also allows it to creep slowly over the rock surface.
- A radula is used to cut and ingest ("graze") algae.

2. Class Gastropoda

The class Gastropoda contains the snails and slugs.

- Largest molluscan class with more than 40,000 species
- Mostly marine, but many species are freshwater or terrestrial
- *Torsion* during embryonic development is a distinctive characteristic:
  ⇒ Uneven growth in the visceral mass causes the visceral mass to rotate 180°, placing the anus above the head in adults (see Campbell, Figure 33.18).
- Body protected by a shell (absent in slugs and nudibranchs) which may be conical or flattened (see Campbell, Figure 33.19).
- Many species have distinct heads with eyes at the tips of tentacles.
- Movement results from a rippling motion along the elongated foot.
- Most gastropods are herbivorous, using the radula to graze on plant material; several groups are predatory and possess modified radulae.
- Most aquatic gastropods exchange gases via gills; terrestrial forms have lost the gills and utilize a vascularized lining of the mantle cavity for gas exchange.

3. Class Bivalvia

The class Bivalvia contains the clams, oysters, mussels and scallops.

Possess a shell divided into two halves (see Campbell, Figure 33.20).

- The shell halves are hinged at the mid-dorsal line and are drawn together by two adductor muscles to protect the animal.
- Bivalves may extend the foot for motility or anchorage when the shell is open.
- The mantle cavity (between shells) contains gills which function in gas exchange and feeding.

Most are suspension-feeders and they trap small food particles in the mucus coating of the gills and then use cilia to move the particles to the mouth.

  ⇒ Water enters the mantle cavity through an incurrent siphon, passes over the gills, and then exits through an excurrent siphon.
  ⇒ No radula or distinct head is present.

Bivalves lead sedentary lives. They use the foot as an anchor in sand or mud. Sessile mussels secrete threads that anchor them to rocks, docks or other hard surfaces.

Scallops can propel themselves along the sea floor by flapping their shells.

4. Class Cephalopoda

The Class Cephalopoda contains the squids and octopuses.

Cephalopods are agile carnivores (see Campbell, Figure 33.22).

- Use beak-like jaws to crush prey
- The mouth is at the center of several long tentacles

A mantle covers the visceral mass, but the shell is either reduced and internal (squids) or totally absent (octopuses).

- The chambered nautilus is the only shelled cephalopod alive today.
- Squids swim backwards in open water by drawing water into the mantle cavity, and then firing a jetstream of water through the excurrent siphon which points anteriorly.
  ⇒ Directional changes can be made by pointing the siphon in different directions.
  ⇒ Most squid are less than 75 cm long but the giant squid may reach 17 m and weigh 2 tons.

- Octopuses usually don't swim in open water, but move along the sea floor in search of food.

Cephalopods are the only mollusks with a *closed circulatory system* in which the blood is always contained in vessels.

Cephalopods have well developed nervous systems with complex brains capable of learning. They also have well developed sense organs.

The cephalopod ancestors were probably shelled, carnivorous forms - the *ammonites*.

- These cephalopods were the dominant invertebrate predators in the oceans until they became extinct at the end of the Cretaceous.

D. Phylum Annelida: annelids are segmented worms

The presence of a true coelom and segmentation are two important evolutionary advances present in the annelids.

- The coelom serves as a hydrostatic skeleton, permits development of complex organ systems, protects internal structures, and permits the internal organs to function separately from the body wall muscles.
- Segmentation also provided for the specialization of different body regions.

There are more than 15,000 species of annelids.

- They have segmented bodies and range in size from less than 1 mm to 3 m.
- There are marine, freshwater, and terrestrial (in damp soil) annelids.

Annelids have a coelom partitioned by septa. The digestive tract, longitudinal blood vessels, and nerve cords penetrate the septa and extend the length of the animal (see Campbell, Figure 33.23.)

The complete digestive system is divided into several parts, each specialized for a specific function in digestion:

pharynx ⟶ esophagus ⟶ crop ⟶ gizzard ⟶ intestine

Annelids have a closed circulatory system.

- Hemoglobin is present in blood cells.
- Dorsal and ventral longitudinal vessels are connected by segmental pairs of vessels.
- Five pairs of hearts circle the esophagus.
- Numerous tiny vessels in the skin permit gas exchange is across the body surface.

An excretory system of paired *metanephridia* is found in each segment; each metanephridium has a nephrostome (which removes wastes from the coelomic fluid and blood) and exits the body through an exterior pore.

The annelid nervous system is composed of a pair of cerebral ganglia lying above and anterior to the pharynx.

- A nerve ring around the pharynx connects these ganglia to a subpharyngeal ganglion, from which a pair of fused nerve cords run posteriorly.
- Along the ventral nerve cords are fused segmental ganglia.

Annelids are hermaphroditic but cross-fertilize during sexual reproduction.

- Two earthworms exchange sperm and store it temporarily.
- A special organ, the clitellum, secretes a mucous cocoon which slides along the worm, picking up its eggs and then the stored sperm.
- The cocoon slips off the worm into the soil and protects the embryos while they develop.

- Asexual reproduction occurs in some species by fragmentation followed by regeneration.

Movement involves coordinating longitudinal and circular muscles in each segment with the fluid-filled coelom functioning as a hydrostatic skeleton.

- Circular muscle contraction makes each segment thinner and longer; longitudinal muscle contraction makes the segment shorter and thicker.
- Waves of alternating contractions pass down the body.
- Most aquatic annelids are bottom-dwellers that burrow, although some swim in pursuit of food.

1. Class Oligochaeta

The class Oligochaeta contains earthworms and a variety of aquatic species.

Earthworms ingest soil, extract nutrients in the digestive system and deposit undigested material (mixed with mucus from the digestive tract) as casts through the anus.

- Important to farmers as they till the soil and castings improve soil texture.
- Darwin estimated that one acre of British farmland had about 50,000 earthworms that produced 18 tons of castings per year.

2. Class Polychaeta

The class Polychaeta contains mostly marine species (see Campbell, Figure 33.24.)

A few drift and swim in the plankton, some crawl along the sea floor, and many live in tubes they construct by mixing sand and shell bits with mucus.

- Tube-dwellers include the fanworms that feed by trapping suspended food particles in their feathery filters which are extended from the tubes.

Each segment has a pair of parapodia which are highly vascularized paddle-like structures that function in gas exchange and locomotion.

- Traction for locomotion is proved by several chitinous setae present on each parapodium.

3. Class Hirudinea

The class Hirudinea contains the leeches.

- A majority of species are freshwater but some are terrestrial in moist vegetation.
- Many are carnivorous and feed on small invertebrates, while some attach temporarily to animals to feed on blood.
- Size ranges from 1 – 30 cm in length.

Some blood-feeding forms have a pair of blade-like jaws that slit the host's skin while others secrete enzymes that digest a hole in the skin.

- An anesthetic is secreted by the leech to prevent detection of the incision by the host.
- Leeches also secrete hirudin which prevents blood coagulation during feeding.
- Leeches may ingest up to ten times their weight in blood at a single meal and may not feed again for several months.
- Leeches are currently used to treat bruised tissues and for stimulating circulation of blood to fingers and toes reattached after being severed in accidents.

E. Phylum Arthropoda: arthropods have regional segmentation, jointed appendages, and exoskeletons

The phylum Arthropoda is the largest phylum of animals with approximately one million described species.

Arthropods are the most successful phylum based on species diversity, distribution, and numbers of individuals.

1. General characteristics of arthropods

The success and great diversity of arthropods is related to their segmentation, jointed appendages, and hard exoskeleton.

The segmentation in this group is much more advanced than that found in annelids.

- In the arthropods, different segments of the body and their associated appendages have become specialized to perform specialized functions.
- Jointed appendages are modified for walking, feeding, sensory reception, copulation and defense.
- Campbell, Figure 33.25 illustrates the diverse appendages and other arthropod characteristics of a lobster.

The arthropod body is completely covered by the *cuticle*, an *exoskeleton* (external skeleton) constructed of layers of protein and chitin.

- The cuticle is thin and flexible in some locations (joints) and thick and hard in others.
- The exoskeleton provides protection and points of attachment for muscles that move the appendages.
- The exoskeleton is also relatively impermeable to water.
- The old exoskeleton must be shed for an arthropod to grow (molting) and a new one secreted.

Arthropods show extensive cephalization with many sensory structures clustered at the anterior end. Well-developed sense organs including eyes, olfactory receptors, and tactile receptors are present.

An *open circulatory system* containing hemolymph is present.

- Hemolymph leaves the heart through short arteries and passes into the sinuses (open spaces) which surround the tissues and organs.
- The hemolymph reenters the heart through pores equipped with valves.
- The blood sinuses comprise the hemocoel. Though the hemocoel is the main body cavity, it is not part of the coelom.
- The true coelom is reduced in adult arthropods.

Gas exchange structures are varied and include:

- Feathery gills in aquatic species
- Tracheal systems in insects
- Book lungs in other terrestrial forms (e.g., spiders)

2. Arthropod phylogeny and classification

Arthropods are segmented protostomes which probably evolved from annelids or a segmented protostome common ancestor.

- Early arthropods may have resembled onychophorans which have unjointed appendages.
  - ⇒ However, many fossils of jointed-legged animals resembling segmented worms support the evolutionary link between the Annelida and Arthropoda (see Campbell, Figure 33.26).

⇒ Such comparisons also indicate that annelids and arthropods are *not* closely related.

- Parapodia may have been forerunners of appendages.

- Some systematists suggest that comparisons of ribosomal RNA and other macromolecules indicate that onychophorans are arthropods and not transitional forms.

- This evidence presents an alternative hypothesis that segmentation evolved independently in annelids and arthropods.

  ⇒ Thus, the most recent common ancestor of these two phyla would have been an unsegmented protostome.

Although the origin of arthropods is unclear, most zoologists agree that four main evolutionary lines can be identified in the arthropods. Their divergence is represented by the subgroups: Trilobites (all extinct), Chelicerates, Uniramians, and Crustaceans (see Campbell, Table 33.5).

3. Trilobites

Early arthropods, called trilobites, were very numerous, but became extinct approximately 250 million years ago (see Campbell, Figure 33.27).

- Trilobites had extensive segmentation, but little appendage specialization.

- As evolution continued, the segments tended to fuse and appendages became specialized for a variety of functions.

4. Spiders and other chelicerates

Other early arthropods included chelicerates, such as the *eurypterids* (sea scorpions) which were predaceous and up to 3 m in length.

The chelicerate body is divided into an anterior cephalothorax and a posterior abdomen.

Their appendages were more specialized than those of trilobites, with the most anterior ones being either pincers or fangs.

Chelicerates are named for their feeding appendages, the *chelicerae*.

Only four marine species remain; one is the horseshoe crab (see Campbell, Figure 33.28).

The bulk of the modern chelicerates are found on land in class Arachnida.

- Includes terrestrial spiders, scorpions, ticks, and mites (see Campbell, Figure 33.29)

- Arachnids possess a cephalothorax with six pairs of appendages: chelicerae, pedipalps (used in sensing and feeding), and four pairs of walking legs (see Campbell, Figure 33.30).

In spiders,

- Fang-like chelicerae, equipped with poison glands, are used to attack prey.

- Chelicerae and pedipalps masticate the prey while digestive juices are added to the tissues. This softens the food and the spider sucks up the liquid.

- Gas exchange is by book lungs (stacked plates in an internal chamber), whose structure provides an extensive surface area for exchange (see Campbell, Figure 33.30b).

- Spiders weave silken webs to capture prey.

  - The proteinaceous silk is produced as a liquid by abdominal glands and spun into fibers by spinnerets. The fibers harden on contact with air.

  - Web production is apparently an inherited complex behavior.

  - Silk fibers are also used for escape, egg covers, and wrapped around food presented to females during courtship.

5. **Millipedes and centipedes**

The class Diplopoda includes the millipedes.

- Wormlike with a large number of walking legs (two pairs per segment) (see Campbell, Figure 33.31a)
- Eat decaying leaves and other plant matter
- Probably among the earliest land animals

The class Chilopoda includes the centipedes.

- They are carnivorous.
- One pair of antennae and three pairs of appendages modified as mouthparts (including mandibles) are located on the head.
- Each trunk segment has one pair of walking legs (see Campbell, Figure 33.31b).
- Poison claws on the most anterior trunk segment are used to paralyze prey and for defense.

6. **Insects**

The class Insecta has a greater species diversity than all other forms of life combined.

- There are about 26 orders of insects (see Campbell, Table 33.6)
- They inhabit terrestrial and freshwater environments, but only a few marine forms exist.

Entomology = The study of insects

The oldest insect fossils are from the Devonian period (about 400 million years ago), and an increase in insect diversity can be attributed to:

- The evolution of flight during the Carboniferous and Permian
- The evolution of specialized mouth parts for feeding on gymnosperms and other Carboniferous plants
    ⇒ The fossil record holds examples of a diverse array of specialized mouth parts.

A second major radiation of insects, which occurred during the Cretaceous period, was once thought to have paralleled radiation of flowering plants.

- Current research indicates the major diversification of insects preceded angiosperm radiation during the Cretaceous period (65 million years ago).
- If this is true, insect diversity played a major role in angiosperm radiation, the reverse of the original hypothesis.

Flight is the key to the success of insects, enabling them to escape predators, find food and mates, and disperse more easily than nonflying forms.

- One or two pairs of wings emerge from the dorsal side of the thorax in most species (see Campbell, Figure 33.32).
- Wings are extensions of the cuticle and not modified appendages.
- Wings may have first evolved to help absorb heat, then developed further for flight.
- Other views suggest wings may have initially served for gliding, as gills in aquatic forms, or even as structures for swimming.

Dragonflies were among the first to fly and have two coordinated pairs of wings. Modifications are found in groups which evolved later.

- Bees and wasps hook their wings together (act as one pair).
- Butterflies have overlapping anterior and posterior wings.

- Beetles have anterior wings modified to cover and protect the posterior (flying) wings.

Insects have several complex internal organ systems (see Campbell, Figure 33.33).

- Complete digestive system with specialized regions
- Open circulatory system with hemolymph
- Excretory organs are the *Malpighian tubules,* which are outpocketings of the gut
- Gas exchange is by a tracheal system, which opens to the outside via spiracles that can open or close to regulate air and limit water loss
- Nervous system is composed of a pair of ventral nerve cords (with several segmental ganglia) which meet in the head where the anterior ganglia are fused into a dorsal brain close to the sense organs.
    - ⇒ Insects show complex behavior which is apparently inherited (e.g., social behavior of bees and ants).

Many insects undergo metamorphosis during their development.

Incomplete metamorphosis = A type of development during which young resemble adults but are smaller and have different body proportions

- For example, in grasshoppers a series of molts occur with each stage looking more like an adult until full size is reached.

Complete metamorphosis = A type of development characterized by larval stages (e.g., maggot, grub, caterpillar) which are very different in appearance from adults.

- Larva eat and grow before becoming adults.
- Adults find mates and reproduce with the females laying eggs on the appropriate food source for the larval forms.

Insects are dioecious and usually reproduce sexually with internal fertilization.

- In most, sperm are deposited directly into the female's vagina during copulation. Some males produce spermatophores which are picked up by the female.
- Inside the female, sperm are stored in the spermatheca.
- Most insects produce eggs although some flies are viviparous.
- Many insects mate only once in a lifetime with stored sperm capable of fertilizing many batches of eggs.

Insects impact terrestrial organisms in a number of ways by:

- Competing for food
- Serving as disease vectors
- Pollinating many crops and orchards

7. Crustaceans

There are more than 40,000 species of crustaceans in marine and fresh waters.

The crustaceans have extensive specialization of their appendages.

- Two pairs of antennae, three or more pairs of mouthparts including mandibles, walking legs on the thorax, appendages are present on the abdomen.
- Lost appendages can be regenerated.

Characteristics of their physiology:

- Gas exchange may take place across thin areas of the cuticle (small forms) or by gills (large forms).
- An open circulatory system is present with hemolymph.

- Nitrogenous wastes are excreted by diffusion across thin areas of the cuticle.
- Salt balance of the hemolymph is regulated by a pair of specialized antennal or maxillary glands.

Most are dioecious and some males (e.g., lobsters) have a specialized pair of appendages to transfer sperm to the female's reproductive pore during copulation.

- Most aquatic crustaceans have at least one swimming larval stage.

The decapods are relatively large crustaceans that have a carapace (calcium carbonate hardened exoskeleton over the cephalothorax).

> Examples:
> - Freshwater crayfish
> - Marine lobsters, crabs and shrimp
> - Tropical land crabs

The isopods are mostly small marine crustaceans but include terrestrial sow bugs and pill bugs.

- Terrestrial forms live in moist soil and damp areas.

Copepods are numerous small marine and freshwater planktonic crustaceans.

- The larvae of larger crustaceans may also be planktonic.

Barnacles are sessile crustaceans with parts of their cuticle hardened into shells by calcium carbonate.

- Barnacles feed by directing suspended particles toward the mouth with specialized appendages.

VI. The Coelomates: Deuterostomes

The deuterostomes, while a very diverse group, share characteristics which indicate their association: radial cleavage, enterocoelous coelom formation, and the blastopore forms the anus.

A. Phylum Echinodermata: echinoderms have water vascular systems and secondary radial symmetry

Most *echinoderms* are sessile or sedentary marine forms with radial symmetry as adults.

- Internal and external parts radiate from the center, often as five spokes.
- A thin skin covers a hard calcareous platelike exoskeleton.
- Most have bumps and spines which serve various functions.

A unique feature of echinoderms is the *water vascular system*, a network of hydraulic canals which branch into extensions called *tube feet* that function for locomotion, feeding, and gas exchange.

Echinoderms are dioecious with sexual reproduction and external fertilization.

- Bilaterally symmetrical larvae metamorphose into radial adults.
- Early embryonic development exhibits the characteristics of deuterostomes.

There are about 7000 species of echinoderms, all of which are marine. The six recognized classes are:

1. Class Asteroidea

This class includes the sea stars which have five or more arms extending from a central disc (see Campbell, Figure 33.36)

- Tube feet on the undersurface of the arms are extended by fluid forced into them by contraction of their ampulla.
- Suction cups at the end of each tube foot attach to the substratum and muscles in the tube foot wall contract and shorten the foot.

Coordination of extension, attaching, contraction and release allow slow movement and attachment to prey (see Campbell, Figure 33.37).

- Prey are obtained by attaching tube feet to the shells of clams and oysters; the arms of the sea star wrap around the prey and hold tightly using the tube feet.
- The muscles of the mollusk fatigue and the shell is pulled open.
- The sea stars evert their stomachs between the shell halves and secrete digestive juices onto the soft tissues of the mollusk.

Sea stars have a strong ability to regenerate. One species can even regrow an entire body from a single arm. Fishermen chopping up sea stars may actually increase their numbers.

2. Class Ophiuroidea

This class contains the brittle stars which differ from sea stars in that they have:

- Smaller central discs than sea stars
- Longer, more flexible arms than sea stars (see Campbell, Figure 33.37c)
- No suckers on their tube feet
- Locomotion is by serpentine lashing of flexible arms
- Varying feeding mechanisms

3. Class Echinoidea

The class Echinoidea contains the sea urchins and sand dollars.

Sea urchins are spherical in shape, while sand dollars are flattened in the oral-aboral axis.

Echinoideans lack arms but have:

- Five rows of tube feet present that provide slow movement (see Campbell, Figure 33.37d)
- Muscles that pivot their spines to aid in locomotion

Echinoideans have a complex jaw-like structure present around the mouth which is used for feeding on seaweeds and other food.

4. Class Crinoidea

The class Crinoidea contains the sea lilies.

Most sea lilies are sessile, living attached to substratum by stalks.

Motile sea lilies use their arms for a crawling form of locomotion as well as for feeding.

Arms circle the mouth (which points upward) and are used in suspension-feeding (see Campbell, Figure 33.37e).

Crinoids have exhibited a very conservative evolution. Extant forms are very similar to fossilized forms from Ordovician period (500 million years ago).

5. Class Holothuroidea

The class Holothuroidea contains the sea cucumbers which have little resemblance to other echinoderms (see Campbell, Figure 33.37f).

- They lack spines.
- The hard endoskeleton is reduced.
- The body is elongated in the oral-aboral axis.

Species in the Holothuroidea do possess five rows of tube feet, a part of the unique water vascular system.

- Some tube feet around the mouth have developed into feeding tentacles.

6. **Class Concentricycloidea**

This class contains the sea daisies which are small (less than 1 cm), disc-shaped marine animals.

- They live in deep water.
- They do not possess arms.
- Tube feet are located around the disc margin.
- Possess a rudimentary digestive system or an absorptive velum on the oral surface.
- Water vascular system consists of two concentric ring canals.

B. **Phylum Chordata: the chordates include two invertebrate subphyla and all vertebrates**

The Chordata diverged from a common deuterostome ancestor with echinoderms at least 500 million years ago.

- The two phyla are grouped together due to similarities in early embryonic development.
- This phylum contains three subphyla: Urochordata, Cephalochordata, and Vertebrata.

REFERENCES

Brusca, R and Brusca, G. Invertebrates. Sunderland, Massachusetts: Sinauer Associates, 1990.

Campbell, N., et al. *Biology*. 5th ed. Menlo Park, California: Benjamin/Cummings, 1999.

Harris, C.L. Concepts on Zoology, 2nd ed. New York, NY: HarperCollins, 1996.

CHAPTER 34
VERTEBRATE EVOLUTION AND DIVERSITY

OUTLINE

I. Invertebrate Chordates and the Origin of Vertebrates
    A. Four anatomical features characterize phylum Chordata
    B. Invertebrate chordates provide clues to the origin of vertebrates

II. Introduction to the Vertebrates
    A. Neural crest, cephalization, a vertebral column, and a closed circulatory system characterize subphylum Vertebrata
    B. Overview of vertebrate diversity

III. Superclass Agnatha: Jawless Vertebrates
    A. Lampreys and hagfishes are the only extant agnathans

IV. Superclass Gnathostomata I: The Fishes
    A. Vertebrate jaws evolved from skeletal supports of the pharyngeal slits
    B. A cartilaginous endoskeleton reinforced by calcified granules is diagnostic of class Chondrichthyes
    C. A bony endoskeleton, operculum, and swim bladder are hallmarks of class Osteichthyes

V. Superclass Gnathastomata II: The Tetrapods
    A. Amphibians are the oldest class of tetrapods
    B. Evolution of the amniotic egg expanded the success of vertebrates on land
    C. A reptilian heritage is evident in all amniotes
    D. Birds began as flying reptiles
    E. Mammals diversified extensively in the wake of the Cretaceous extinctions

VI. Primates and the Phylogeny of *Homo sapiens*
    A. Primate evolution provides a context for understanding human origins
    B. Humanity is one very young twig on the vertebrate tree

OBJECTIVES

After reading this chapter and attending lecture, the student should be able to:

1. Describe the four unique characteristics of chordates.
2. Distinguish between the three subphyla of the phylum Chordata and give examples of each.
3. Describe the specialized characteristics found in the subphylum Vertebrata and explain how each is beneficial to survival.
4. Compare and contrast members of Agnatha, Placadermi, and Chondrichthyes.

5. Explain how members of the class Osteichthyes have becomes so diversified.
6. Summarize the evidence supporting the fact that amphibians evolved from crossopterygians.
7. Distinguish between the three orders of living amphibians.
8. List the distinguishing characteristics of members of the class Reptilia and explain any special adaptations to the terrestrial environment.
9. Explain how environmental changes during the Cretaceous Period may have affected the dinosaurs.
10. List the distinguishing characteristics of members of the class Aves and explain any special adaptations for flight.
11. Summarize the evidence supporting the fact that birds evolved from reptilian ancestors.
12. Explain why mammals underwent an adaptive radiation during the Cenozoic.
13. Distinguish between monotreme, marsupial, and placental mammals.
14. Explain how convergent evolution produced marsupial and placental ecological counterparts on different continents.
15. Compare and contrast the four main evolutionary lines of placental mammals.
16. Describe the characteristics found in early primates which indicate an arboreal existence.
17. Appraise the significance of the three most prominent misconceptions about human evolution.
18. Diagram an evolutionary tree for humans.
19. Explain how humans have influenced the extinction rates of other organisms.

KEY TERMS

vertebrates	Superclass	lungfishes	Chelonia
chordates	Gnathostomata	Subclass Sarcopterygii	Squamata
notochord	placoderms	Class Amphibia	Crocodilia
urochordates	Class Chondrichthyes	urodeles	Class Aves
tunicates	spiral valve	anurans	ratites
lancelets	lateral line system	apodans	carinates
cephalochordates	oviparous	extraembryonic membranes	Class Mammalia
somites	ovoviviparous	Class Reptilia	placenta
paedogenesis	viviparous	ectotherms	therapsids
neural crest	cloaca	synapsids	monotremes
tetrapods	Class Osteichthyes	sauropods	eutherian mammals
amniotic egg	operculum	anapsids	prosimians
amniotes	swim bladder	diapsids	anthropoids
Superclass Agnatha	ray-finned fishes	therapsids	paleoanthropology
ostracoderms	lobe-finned fishes	endothermic	mosaic evolution
passeriforms			

LECTURE NOTES

I. **Invertebrate Chordates and the Origin of Vertebrates**

The phylum Chordata includes three subphyla: two invertebrate subphyla, Urochordata, Cephalochordata; and the subphylum Vertebrata.

A. Four anatomical features characterize phylum Chordata

Chordates are deuterostomes with four unique characteristics which appear at some time during the animal's life. These characteristics are the notochord, a dorsal, hollow nerve cord, pharyngeal slits, and a muscular postanal tail (see Campbell, Figure 34.1).

1. Notochord

Notochord = A longitudinal, flexible rod located between the gut and nerve cord

- Present in all chordate embryos
- Composed of large, fluid-filled cells encased in a stiff, fibrous tissue
- Extends through most of the length of the animal as a simple skeleton

In some invertebrate chordates and primitive vertebrates it persists to support the adult.

In most vertebrates, a more complex, jointed skeleton develops and the notochord is retained in adults as the gelatinous material of the discs between the vertebrae.

2. Dorsal, hollow nerve cord

Develops in the embryo from a plate of dorsal ectoderm that rolls into a tube located dorsal to the notochord.

Unique to chordates; other animal phyla have solid, usually ventral, nerve cords.

The brain and spinal cord (central nervous system) develops from this nerve cord.

3. Pharyngeal slits

Chordates have a complete digestive system (mouth and anus). The pharynx is the region just posterior to the mouth and it opens to the outside through several pairs of slits.

- The presence of these pharyngeal slits permits water entering the mouth to exit without passing through the entire digestive system.
- These pharyngeal gill slits function for suspension-feeding in invertebrate chordates.
- They have become modified for gas exchange and other functions during the evolution of vertebrates.

4. Muscular postanal tail

A tail extending beyond the anus, it is found in most chordates and contains skeletal elements and muscles.

- Provides much of the propulsive force in many aquatic species.

The digestive tract in most nonchordates extends nearly the whole length of the body.

II. Invertebrate Chordates Provide Clues to the Origin of Vertebrates

A. Subphylum Urochordata

Species in the subphylum Urochordata (*urochordates*) are commonly called *tunicates*.

Entire animal is cloaked in a tunic made of a celluloselike carbohydrate.

Most are sessile marine animals that adhere to rocks, docks, and boats (see Campbell, Figure 34.2a).

Some species are planktonic, while others are colonial.

The tunicates are filter feeders.

- Seawater enters through an incurrent siphon, passes through the slits of the pharynx into a chamber called the atrium, and exits via an excurrent siphon, the atriopore (see Campbell, Figure 34.2b).
- Food filtered from the water by a mucus net of the pharynx is moved by cilia into the intestine.

- The anus empties into the excurrent siphon.
- When disturbed, tunicates eject a jet of water through the excurrent siphon, so they are commonly called sea squirts.

Adult tunicates bear little resemblance to other chordates.

- They lack a notochord, a nerve cord and tail.
- They possess only pharyngeal slits.

Larval tunicates are free swimmers and possess all four chordate characteristics (see Campbell, Figure 34.2c).

- Larva attach by the head on a surface and undergo metamorphosis to adult form.
- In some species, if the *Manx* gene (named after the tailless cat) is turned off during development, the larvae will be tailless. *Manx* expression also is required for notochord and nerve cord development.
  - These observations remind us that a relatively small number of genes that regulate development may influence the evolution of some basic aspect of an animal body plan.

B. Subphylum Cephalochordata

Animals in the subphylum Cephalochordata (*cephalochordates*) are known as *lancelets* due to their bladelike shape (see Campbell, Figure 34.3). Chordate characteristics are prominent and persist in the adult. These include:

- Notochord
- Dorsal nerve cord
- Numerous gill slits
- Postanal tail

Cephalochordates are marine filter feeders.

- They burrow tail first into the sand with only the anterior exposed.
- Water is drawn into mouth by ciliary action and food is trapped on a mucous net secreted across the pharyngeal slits.
- Water exits through the slits and trapped food passes down the digestive tube.

Cephalochordates are feeble swimmers with fishlike motions.

- Frequently move to new locations
- Muscle segments are serially arranged in chevronlike rows, and coordinated contraction flexes the notochord from side to side in a sinusoidal pattern.
- Muscle segments develop from blocks of mesoderm called *somites* that are arranged along each side of the notochord in the embryo.
- The serial musculature is evidence of the lancelet's segmentation (see Campbell, Figure 34.4).
  ⇒ Whether segmentation evolved independently in annelids, arthropods, and chordates, or from a common ancestor of all bilateral animals, is currently under debate.

C. Relationship of invertebrate chordates to the vertebrates

Vertebrates first appeared in the fossil records in Cambrian rocks.

- Fossilized invertebrates (about 550 million years old) resembling cephalochordates were found in Burgess Shale of British Columbia.
- This is about 50 million years older than the oldest known vertebrates.

Most zoologists feel the vertebrate ancestors possessed all four chordate characteristics and were suspension-feeders.

- They may have resembled lancelets but were less specialized.
- Information provided by molecular systematics supports the idea that cephalochordates are the closest relatives of vertebrates.

Cephalochordates and vertebrates may have evolved from a sessile ancestor by paedogenesis.

Paedogenesis = Precocious attainment of sexual maturity in a larva

- Cephalochordates more closely resemble urochordate larvae than adult urochordates.
- Changes in the developmental control genes can alter the timing of developmental events (e.g., gonad maturation).
- Zoologists postulate that some early urochordatelike larval forms became sexually mature and reproduced before undergoing metamorphosis.
- If reproducing larvae were successful, natural selection may have reinforced the absence of metamorphosis and a vertebrate life cycle may have evolved.

The divergence of cephalochordates from vertebrates occurred about 500 million years ago.

- The differences between the two groups can be viewed as vertebrate adaptations to larger size and a more mobile life style.

III.   Introduction to the Vertebrates

Vertebrates have retained the primitive chordate features while adding other specializations. These synapomorphies distinguish the vertebrates from urochordates and cephalochordates.

A. Neural crest, pronounced cephalization, a vertebral column, and a closed circulatory system characterize subphylum Vertebrata

The unique vertebrate structures probably evolved in association with increased size and more active foraging for food. The unique vertebrate adaptations include:

- The *neural crest*, a group of embryonic cells found only in vertebrates, contributes to the formation of certain skeletal components and many other structures distinguishing vertebrates from other chordates.
   - ⇒ The dorsal, hollow nerve cord develops from an infolding of the edges of an ectodermal plate on the surface of the embryo.
   - ⇒ The neural crest forms near the dorsal margins of the tube resulting from this infolding (see Campbell, Figure 34.5).
   - ⇒ Cells from the neural crest then migrate to various specific areas of the embryo and help form a variety of structures including some of the bones and cartilage of the cranium.
- Vertebrates show a much greater degree of *cephalization* than cephalochordates.
   - ⇒ The brain and sense organs are located at the anterior end which is the portion of the body which is in contact with the most environmental stimuli.
- A skeleton including a cranium and vertebral column is the main axis of the body, replacing the notochord as the basic skeleton.
   - ⇒ The cranium protects the brain.
   - ⇒ The vertebral column provides support and a strong, jointed anchor that provides leverage to the segmental swimming muscles.
   - ⇒ The axial skeleton made larger size and stronger, faster movement possible.
   - ⇒ Most vertebrates also have ribs (anchor muscles and protect internal organs) and an appendicular skeleton supporting two pairs of appendages.

- The vertebrate skeleton may be composed of bone, cartilage, or a combination of the two.
  - ⇒ A majority of the skeleton is a non-living matrix which is secreted and maintained by living skeletal cells.
  - ⇒ The living vertebrate endoskeleton can grow with the animal unlike the exoskeleton of arthropods.

Some anatomical adaptations also support the greater metabolic demands of increased activity.

- The generation of ATP by cellular respiration, to replace the energy used by vertebrates in obtaining food or escaping predators, consumes oxygen.
  - ⇒ The respiratory and circulatory systems of vertebrates show adaptations which support the mitochondria of muscles and other active tissues.
- Vertebrates have a closed circulatory system composed of a ventral chambered (two to four) heart, arteries, capillaries, and veins.
  - ⇒ The heart pumps the blood through the system.
  - ⇒ The blood becomes oxygenated as it passes through the capillaries of the gills or lungs.
- The more active the lifestyle, the larger the amounts of organic molecules necessary to produce energy.
  - ⇒ Vertebrates have several adaptations for feeding, digestion and nutrient absorption.
  - ⇒ For example, muscles in the walls of the digestive tract move food from organ to organ along the tract.

B. Overview of Vertebrate Diversity

The vertebrates are divided into two major groups, or superclasses: Superclass Agnatha, whose members lack jaws, and Superclass Gnathastomata, whose members possess jaws (see Campbell, Table 34.1)

The gnathostomes are divided into six classes: Chondrichthyes, Osteichthyes, Amphibia, Reptilia, Aves, and Mammalia. The last four classes are collectively known as the tetrapod vertebrates.

- *Tetrapod* = An animal possessing two pairs of limbs that support it on land

In addition to being tetrapods, the reptiles, birds and mammals have other adaptations for a terrestrial lifestyle which are not found in the amphibians.

- The *amniotic egg* (a shelled, water resistant egg) allows completion of the life cycle on land.
- Most mammals do not lay eggs but retain other features of the amniotic condition, consequently, they are also considered *amniotes* along with the birds and reptiles.

> Review Figure 34.6 to establish the relationships among the classes of the subphylum Vertebrata.

IV. Superclass Agnatha: Jawless Vertebrates

Vertebrates probably arose in the late Precambrian and early Cambrian. The oldest fossils of vertebrates are of jawless animals found in rock strata 400 to 500 million years old.

Early agnathans were small, less than 50 cm in length.

- They were jawless with oval or slitlike mouths; most lacked paired fins and were bottom-dwellers.

- Some were active and had paired fins.
- Were probably bottom- or suspension-feeders that trapped organic debris in their gill slits.

Ostracoderms and most other agnathans declined and disappeared during the Devonian.

A. Lampreys and hagfishes are the only extant agnathans

Extant forms include about 60 species of lampreys and hagfishes which lack paired appendages and external armor (see Campbell, Figure 34.7).

Lampreys are eel-shaped and feed by clamping their round mouths onto live fish.

- Once attached, they use a rasping tongue to penetrate the skin and feed on the prey's blood.
- Sea lampreys spend their larval development in freshwater streams and migrate to the sea or lakes as they mature.
- Larva are suspension-feeders that resemble lancelets (cephalochordates).
- Some lamprey species feed only as larvae. Once they mature and reproduce, they die within a few days.

Hagfishes superficially resemble lampreys.

- They are scavengers without rasping mouthparts.
- Some species will feed on sick or dead fish while others feed on marine worms.
- Lack a larval stage and are entirely marine (Physiologically, they are the only extant vertebrate that is a true osmoconformer; all other vertebrates are osmoregulators).

V.   Superclass Gnathostomata I: The Fishes

The agnathans were gradually replaced by vertebrates with jaws (Superclass Gnathostomata) during the late Silurian and early Devonian. Early gnathostomes included ancestors of class Chondrichthyes and class Osteichthyes and a now extinct group of armored fishes called *placoderms* appeared.

Most placoderms were less than 1 m in length, but some were up to 10 m long.

- Differed from agnathans in that they possessed paired fins and hinged jaws.
- Paired fins enhanced swimming ability and hinged jaws allowed more varied feeding habits including predation.

A. Vertebrate jaws evolved from skeletal supports of the pharyngeal slits

Hinged jaws evolved as modifications of the skeletal rods which previously supported the anterior pharyngeal (gill) slits (see Campbell, Figure 34.8).

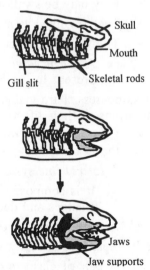

- Remaining gill slits retained function as major gas exchange sites.
- Hinged jaws of vertebrates work in an up and down direction; those in arthropods work from side to side.

Placoderms and another group of jawed fishes, the acanthodians, radiated during the Devonian period (the Age of Fishes) and many new forms evolved in fresh and salt waters.

- Placoderms and acanthodians disappeared by the start of the Carboniferous period (350 million years ago).
- Ancestors of the placoderms and acanthodians also gave rise to early sharks (class Chondrichthyes) and bony fishes (class Osteichthyes).

B. A cartilaginous endoskeleton reinforced by calcified granules is diagnostic of class Chondrichthyes

The class Chondrichthyes contains about 750 extant species of cartilaginous fishes (e.g., sharks, skates, rays) (see Campbell, Figure 34.9).

Species in the class Chondrichthyes have flexible skeletons composed of cartilage, well-developed jaws and paired fins.

- The ancestors of members of this class had bony skeletons.
- The characteristic cartilaginous skeleton is thus a derived characteristic, having evolved secondarily.
- The developmental sequence in cartilaginous fishes differs from other vertebrates in that the initial (first) cartilaginous skeleton does not become ossified.
- In most species, however, parts of the skeleton are strengthened by calcified granules
- The cartilaginous skeleton is more elastic and lighter than a bone skeleton.

Sharks have streamlined bodies and are swift swimmers.

- The tail provides propulsion.
- The dorsal fins serve as stabilizers.
- Pectoral and pelvic fins produce lift.
- Some buoyancy is provided by large amounts of oil stored in liver, but most must swim continuously to remain in the water column.
  ⇒ Continual swimming also produces water flow through mouth and over gills for gas exchange.
  ⇒ Some sharks are known to rest on the sea floor and in caves; they use jaw and pharynx muscles to pump water over their gills while resting.

Most sharks are carnivorous, although the largest sharks and rays are suspension-feeders.

- Prey may be swallowed whole or pieces may be torn from large prey.
- Teeth evolved as modified scales.
- The digestive tract is proportionately shorter than in other vertebrates.
  ⇒ A *spiral valve*, which increases surface area and slows food movement, is present in the intestine.

Sharks possess sharp vision (cannot distinguish color) and olfactory senses that are adaptations to their lifestyle.

- Electric sensory regions that detect muscle contractions of prey are located on the head.
- A *lateral line system* is present along the flanks.
  ⇒ It is composed of rows of microscopic organs sensitive to water pressure changes and detects vibrations.
- A pair of auditory organs also detect sound waves passing through the water.

Sharks reproduce sexually with internal fertilization.

- A pair of claspers on the pelvic fins of males transfers sperm into the female reproductive tract.
- Some species are *oviparous*, some are *ovoviviparous*, and a few are *viviparous*.

- A *cloaca* (common chamber for reproductive, digestive and excretory systems) is present.

Rays are adapted to a bottom-dwelling lifestyle.
- They have dorsoventrally flattened bodies.
- Their jaws are used to crush mollusks and crustaceans.
- Enlarged pectoral fins provide propulsion for swimming.
- The tail in many species is whiplike and, in some, bears venomous barbs.

C. A bony endoskeleton, operculum, and a swim bladder are hallmarks of class Osteichthyes

The class Osteichthyes contains the bony fishes, which are represented by more than 30,000 extant species.
- Abundant in marine and fresh waters (see Campbell, Figure 34.10)
- Range from 1 cm to 6 m in length
- Skeleton is bony, reinforced with a matrix of calcium phosphate
- Skin is covered with flattened bony scales
- Skin glands produce mucus that reduces drag when swimming
- A lateral line system is present as a row of tiny pits in the skin on both sides of the body (see Campbell, Figure 34.11).

Gas exchange occurs by drawing water over the four or five pairs of gills located in chambers covered by an operculum.
- Water is drawn into the mouth, through the pharynx and out between the gills by movement of the operculum and contraction of muscles within the gill chambers.
  ⇒ Allows bony fishes to breath while stationary.

A *swim bladder*, located dorsal to the digestive tract, provides buoyancy.
- Transfer of gases between blood and swim bladder varies bladder inflation and adjusts the density of the fish.

Bony fishes are very maneuverable swimmers. The flexible fins provide better steering and propulsion than the stiff fins of sharks.
- The fastest bony fish can swim to 80 km per hour in short bursts.
- A fusiform body shape is common to all fast fishes and aquatic mammals.
  ⇒ This body shape reduces drag produced by the density of water (convergent evolution).

Most bony fish are oviparous and utilize external fertilization.
- Some are ovoviviparous or viviparous and utilize internal fertilization.
- Some display complex mating behavior.

The cartilaginous and bony fishes diversified during the Devonian and Carboniferous periods.
- Sharks arose in the sea, bony fishes probably originated in fresh water.
- The swim bladder was modified from lungs of ancestral fishes which supplemented the gills for gas exchange in stagnant waters.

The two extant subclasses of bony fishes had diverged by the end of the Devonian: Subclass Actinopterygii (ray-finned fishes) and Subclass Sarcopterygii

The *subclass Actinopterygii* includes fish with fins supported mainly by flexible rays.
- These are the most familiar fishes.

- They spread from fresh water to the seas and many returned to fresh water during evolution of the taxon.
- Some bony fish (e.g., salmon and sea-run trout) reproduce in fresh water and mature in the sea.

The *subclass Sarcopterygii* includes lobe-finned fishes (coelocanths and rhipidistans) and lungfishes that evolved in fresh water The ancestors of these fishes continued to use their lungs to aid the gills in gas exchange.

Coelocanths and rhipidistians are referred to as lobe-finned fishes.

- Their fins were fleshy, muscular, and supported by extensions of the bony skeleton.
- Many were large, bottom-dwelling forms that used their paired fins to walk on the substratum.
- The only extant species, the coelocanth, is marine and lungless. It belongs to a lineage that became marine at some point in its evolution.
- All rhipidistians are extinct.

Three genera of lungfishes exist in the Southern Hemisphere.

- They live in stagnant ponds and swamps where they surface to gulp air into lungs connected to the pharynx.
- When ponds dry, lungfishes burrow in the mud and aestivate.

Lobe-finned fishes of the Devonian were numerous and important in vertebrate genealogy because they probably gave rise to amphibians (see Campbell, Figure 34.13).

VI. Superclass Gnathostomata II: The Tetrapods

A. Amphibians are the oldest class of tetrapods

The first vertebrates to move onto land were members of *Class Amphibia*.

Today there are about 4000 extant species of frogs, salamanders, and caecilians.

1. Early amphibians

Some scientists suggest that early amphibians evolved from lobe-finned fishes that adapted to environmental variations (drought and flooding) of the Devonian.

- The skeletal structure of the lobed fins suggest they could have assisted in movement on land.
- Fossil lobe-fins, such as the rhipidistian *Eusthenopteron*, exhibited many anatomical similarities to early amphibians (see Campbell, Figure, 34.14a)

Recent molecular and other data, however, suggest that amphibians are more closely related to lungfishes than they are to lobe-fin fishes.

The oldest amphibian fossils are from the late Devonian (365 million years ago) (see Campbell, Figure 34.14b).

- Early amphibians probably were mostly aquatic and periodically came on land to eat insects and other invertebrates that had moved previously onto land.
- Amphibians were the only vertebrates on land in the late Devonian and early Carboniferous.

Radiation of forms occurred during the early Carboniferous period.

- Some forms reached 4 m in length and some resembled reptiles.
- Amphibians began to decline during the late Carboniferous. At the beginning of the Triassic period (245 million years ago), most of the survivors resembled modern species.

2. Modern amphibians

There are three extant orders of amphibians: Urodela (salamanders), Anura (frogs and toads) and Apoda (caecilians) (see Campbell, Figure 34.15).

The order Urodela (urodeles; e.g., salamanders) contains about 400 species.

- Some are aquatic and some are terrestrial.
- Terrestrial forms walk with a side-to-side bending of the body. Aquatic forms swim sinusoidally or walk along the bottom of streams or ponds.

The order Anura (*anurans*; e.g., frogs and toads) contains about 3500 species, which are better adapted to the terrestrial habitat than urodeles.

- Enlarged hindlegs provide better movement (hopping) than in urodeles.
- They capture prey by flicking the sticky tongue which is attached anteriorly.
- Predator avoidance is aided by camouflage color patterns and distasteful or poisonous mucus secreted by skin glands.
    ⇒ Bright coloration is common in poisonous species.

The order Apoda (*apodans*; e.g., caecilians) contains about 150 species.

- They are legless and almost blind.
- Most species burrow in moist tropical soils; a few species inhabit freshwater ponds and streams.

Many frogs exhibit a metamorphosis from the larval to adult stage (see Campbell, Figure 34.16):

- The tadpole (larval stage) is usually an aquatic herbivore. It possesses internal gills, a lateral line system, and a long, finned tail.
- The tadpole lacks legs and swims by undulating the tail.
- During metamorphosis, legs develop and the gills and lateral line system disappear.
- A young frog is tetrapod. It has air-breathing lungs, a pair of external eardrums, and a digestive system that can digest animal protein.
- The adult is usually terrestrial or semiaquatic and a predator.

Many amphibians, including some frogs, do not have a tadpole stage.

- Some species in each order are strictly aquatic while others are strictly terrestrial.
- Urodeles and apodans have larva that more closely resemble adults and both larva and adults are carnivorous.
- Paedogenesis is common in some groups of urodeles.

Most amphibians maintain close ties with water and are most abundant in damp habitats.

- Terrestrial forms in arid habitats spend much of their time in burrows where humidity is high.
- Gas exchange is primarily cutaneous and terrestrial forms must keep the skin moist.
    ⇒ Lungs can aid in gas exchange although most are small and inefficient. Some forms lack lungs.
    ⇒ Many species also exchange gases across moist surfaces of the mouth.

Amphibians are dioecious.

- The reproduce sexually usually with external fertilization in water (e.g., ponds, streams, temporary pools).

- In frogs, the male grasps the female and sperm are released as the female sheds her eggs (see Campbell, Figure 34.16a)
- Eggs are unshelled and produced in large numbers by most species.

Some species exhibit parental behavior and produce small numbers of eggs.

- Males or females (species dependent) incubate eggs on their back, in the mouth or in the stomach.
- Some tropical species lay eggs in a moist foamy nest that prevents drying.
- Some species are ovoviviparous and a few are viviparous with the eggs developing in the female's reproductive tract.

Amphibians exhibit complex and diverse social behavior especially during breeding season (e.g., vocalization by male anurans, migrations, navigation or chemical signaling).

B. Evolution of the amniotic egg expanded the success of vertebrates on land

Many specialized adaptations for living in a terrestrial habitat were necessary for reptiles to evolve from their amphibian ancestor.

- The amniotic egg was important. Its development broke the last ties with the aquatic environment by allowing life cycles to be completed on land.
  - ⇒ The shell of the amniotic egg helps prevent desiccation, therefore, it can be laid in a dry place.
  - ⇒ Most mammals have dispensed with the shell, instead, the embryo implants in the wall of the uterus and obtains its nutrients from the mother.
- The *extraembryonic membranes* within the egg develop from tissue layers that grow out from the embryo.
  - ⇒ These specialized membranes function in gas exchange, transfer of stored nutrients to the embryo, and waste storage.
  - ⇒ One of these membranes, the amnion, encloses a compartment filled with amniotic fluid that bathes the embryo and provides a cushion against shocks (see Campbell, Figure 34.17).

Reptiles , birds, and mammals make up a monophyletic group, the amniotes.

C. A reptilian heritage is evident in all amniotes

The *class Reptilia* is a diverse group with about 7000 extant species and a wide array of extinct forms.

- This grouping is based on the apparent similarity of the tetrapods (lizards, snakes, turtles, and crocodilians), but cladistic analysis indicates that grouping these vertebrates in a class that does not include birds is inconsistent with phylogeny.
- Birds appear to be more closely related to crocodiles than are turtles.
- Class Reptilia can only be defined by the absence of features that distinguish birds (feathers) and mammals (hair and mammary glands)

1. Reptilian characteristics

Reptiles possess several adaptations for terrestrial living not found in amphibians.

- Scales contain the protein keratin which helps prevent dehydration.
- Gas exchange occurs via lungs although many turtles also use moist cloacal surfaces.
- They are dioecious with sexual reproduction and internal fertilization.
  - ⇒ Most are oviparous and produce an amniotic egg (see Campbell, Figure 34.18).

⇒ Some species of snakes and lizards are viviparous with the young obtaining nutrients from the mother across a "placenta" which forms from the extraembryonic membranes.

Reptiles are ectotherms and use behavioral adaptations to regulate their body temperature.

- *Ectotherm* = An animal that uses behavioral adaptations to absorb solar energy and regulate its body temperature.
- Due to ectothermy, reptiles can survive on less than 10% of the calories required by mammals of comparable size.

Reptiles were abundant and diverse in the Mesozoic era.

2. The Age of Reptiles

a. Origin and early evolutionary radiation of reptiles

The oldest reptilian fossils are found in late Carboniferous rock (300 million years old).

- Ancestors were probably Devonian amphibians.
- Two waves of adaptive radiation resulted in reptiles being the dominant terrestrial vertebrates for 200 million years.

The first reptilian radiation was in the early Permian period and gave rise to two main evolutionary branches: the synapsids and sauropsids (see Campbell, Figure 34.19).

- Synapsids were terrestrial predators and gave rise to the therapsid lineage which were mammal-like reptiles.
  - ⇒ Therapsids were large, dog-sized predators from which mammals are believed to have evolved.
- Sauropsids gave rise to the modern amniote groups other than mammals. They split into two lineages early in their history: the anapsids and the diapsids.
  - ⇒ The anapsids are presently represented only by the turtles.
  - ⇒ The extant diapsids are represented by the lizards, snakes, and crocodilians; dinosaurs and some other extinct groups were also diapsids. Cladistic analysis suggests that birds are the closest living relatives of extinct dinosaurs.

b. Dinosaurs and pterosaurs

The second reptilian radiation began in the late Triassic (about 200 million years ago) and several lineages evolved during this event.

- Two groups are most important: the dinosaurs and pterosaurs (flying reptiles).
- Dinosaurs varied in body shape, size, and habitat.
  - ⇒ Some fossilized forms measure 45 meters in length.
- Pterosaurs had wings formed from skin stretched from the body wall, along the forelimb to the tip of an elongate finger and supported by stiff fibers.

Evidence indicates that dinosaurs were agile, fast moving, and social, some may have even exhibited parental care of the young (see Campbell, Figure 34.20).

There is also some anatomical evidence supporting the hypothesis that dinosaurs were endothermic.

- *Endothermy* = The ability to keep the body warm through an animal's own metabolism

- Skeptics of this hypothesis feel that the Mesozoic climate was warm and consistent, and that basking may have been sufficient for maintaining body temperature.
- Low surface-to-volume ratios of large forms reduced fluctuations of body temperature vs. air temperature; thus, dinosaurs may not have been endothermic.

c. The Cretaceous crisis

In the Cretaceous (last period of the Mesozoic), the climate became cooler and more variable, and mass extinctions occurred.

A few dinosaurs survived into the early Cenozoic, but all these reptiles were gone by the end of the Cretaceous (65 million years ago).

3. Modern reptiles

The largest and most diverse extant orders are the: *Chelonia* (turtles), *Squamata* (lizards and snakes), and *Crocodilia* (alligators and crocodiles).

Turtles of the order Chelonia evolved from anapsids during the Mesozoic.

- They show little change from the earliest forms (see Campbell, Figure 34.21a).
- They are protected from predators by a hard shell.
- All turtles, even aquatic species, lay their eggs on land.

Lizards and snakes are classified in the order Squamata.

Lizards are the most numerous and diverse group of extant reptiles (see Campbell, Figure 34.21b).

- They evolved from the diapsid lineage.
- Most are small.
- Many nest in crevices and decrease activity during cold periods.

Snakes probably descended from burrowing lizards (see Campbell, Figure 34.21c).

- They are limbless and most live above ground.
- Vestigial pelvic and limb bones present in primitive snakes (boas) are evidence of a limbed ancestor.
- Snakes are carnivorous and have a number of adaptations for hunting prey.
  - They have acute chemical sensors.
  - They are sensitive to ground vibrations (although lacking eardrums).
  - Pit vipers have sensitive heat-detecting organs between their eyes and nostrils.
  - Flicking tongue helps transmit odors toward olfactory organs on roof of mouth.
  - Poisonous snakes inject a toxin through a pair of sharp, hollow teeth and loosely articulated jaws allow them to swallow large prey.

Crocodiles and alligators are among the largest living reptiles (see Campbell, Figure 34.21d).

- Crocodilians also evolved from the diapsid lineage.
- The spend most of their time in the water, breathing air through upturned nostrils.
- They are confined to warm regions of Africa, China, Indonesia, India, Australia, South America, and the southeastern United States.
- They are the living reptiles most closely related to dinosaurs.

D. Birds began as flying reptiles

The *class Aves* (birds) evolved during the great reptilian radiation of the Mesozoic era (see Campbell, Figure 34.19).

- They possess distinct reptilian characteristics such as the amniotic egg and scales on the legs, but modern birds also look quite different because of their feathers and other flight equipment.

1. Characteristics of birds

Each part of the bird's anatomy is modified in some way that enhances flight.

- The bones have a honeycombed internal structure that provides strength while reducing weight (see Campbell, Figure 34.22).
- Some organ systems are reduced (only one ovary in females).
- Birds have no teeth (reduces weight) and food is ground in the gizzard.
    ⇒ The beak is made of keratin and evolution has produced many shapes in relation to the bird's diet.

Flying requires much energy production from an active metabolism.

- Birds are endothermic with insulation provided by feathers and a fat layer.
- Birds have efficient circulatory system with a four-chambered heart that segregates oxygenated blood from unoxygenated blood.
- They have efficient lungs with tubes connecting to elastic air sacs that help dissipate heat and reduce the body density.

Birds also have a very well developed nervous system.

- Acute vision and well-developed visual and coordinating areas of the brain aid in flying.
- They show complex behavior especially during breeding season when elaborate courtship rituals are performed.

Birds are dioecious with sexual reproduction and internal fertilization.

- Sperm are transferred from the cloaca of the male to the cloaca of the female (males of most species lack a penis) during copulation.
- Eggs are laid and must be kept warm through brooding by the female, male or both depending on the species.

Wings are airfoils, formed by the shape and arrangement of the feathers, that illustrate the same aerodynamic principles as airplane wings (see Campbell, Figure 34.23).

- Power is supplied to the wings by contraction of the large pectoral (breast) muscles which are anchored to a keel on the sternum (breastbone).
- Some birds have wings adapted for soaring (hawks) while others must beat their wings continuously to stay aloft (hummingbirds).

Feathers are made of keratin and are extremely light and strong.

- Feathers evolved from the scales of reptiles and may have first functioned as insulation.
- Feathers also function to control air movements around the wing.

Radical alteration of body form was necessary for evolution of flight, but flight provides many benefits.

- Allows aerial reconnaissance that enhances hunting and scavenging.
- Birds can exploit flying insects as an abundant, highly nutritious food resource.
- Flight provides an escape mechanism from land-bound predators.

- Flight also allows migration to utilize different food resources and seasonal breeding areas.

2. The origin of birds

Birds shared a common ancestor with *Archaeopteryx lithographica*.

- Fossils of *Archaeopteryx* have been recovered from limestone dating to the Jurassic period (150 million years ago).
- *Archaeopteryx* had clawed forelimbs, teeth, a long tail containing vertebrae and feathers (see Campbell, Figure 34.24).
- *Archaeopteryx* is not considered the ancestor to modern birds, but a side branch of the avian lineage.
    ⇒ The skeleton indicates a weak flyer that may have been a tree-dwelling glider.

Cladistic analysis suggests that birds arose from a theropod dinosaur. Some researchers, however, believe that birds arose from early Mesozoic reptiles (thecodonts), a group that also was ancestral to the dinosaurs.

E. Modern birds

There are about 8600 extant species in 28 orders. Most birds can fly but several are flightless (ostrich, kiwi, and emu) (see Campbell, Figure 34.25a).

- Flightless birds are called *ratites* because the breastbone lacks a keel and large breast muscles used for flying are absent.
- Flying birds are referred to as *carinates* due to the presence of a sternal keel (carina) that supports the large breast muscles used in flying.
- Carinate birds exhibit a variety of feather colors, beak and foot shape, behavior and flight ability (see Campbell, Figure 34.25 b and c).
    ⇒ Penguins are carinate birds that do not fly, but use powerful breast muscles in swimming.
- Almost 60% of extant species belong to one order of carinate birds (the *passeriforms*, or perching birds), which includes the jays, swallows, sparrows, warblers and many others (see Campbell, Figure 34.25d).

F. Mammals diversified extensively in the wake of the Cretaceous extinctions

There are about 4500 species of extant mammals.

Extinction of the dinosaurs and the fragmentation of continents opened new adaptive zones at the end of the Mesozoic era.

Mammals underwent a massive radiation to fill these vacant zones.

1. Mammalian characteristics

Species in *class Mammalia* have the following characteristics:

- Hair that is composed of keratin, but is not believed to have evolved from reptilian scales; it provides insulation
- Endothermic with an active metabolism
    ⇒ An efficient respiratory system that utilizes a diaphragm for ventilation supports the metabolism.
    ⇒ A four-chambered heart segregates oxygenated from unoxygenated blood.
- Mammary glands that produce milk to nourish the young
- Teeth that are differentiated into various sizes and shapes, which are adapted to chewing many types of food.

⇒ The jaw apparatus of the ancestral reptiles was also modified during evolution with two of the jaw bones becoming incorporated into the middle ear.

Mammals are dioecious with sexual reproduction and internal fertilization.

- Most are viviparous with the developing embryo receiving nutrients from the female across the *placenta*.
- A few are oviparous.

Mammals have large brains in comparison to other vertebrate groups and are capable of learning.

- Parental care of long duration helps young learn from the parents.

2. The evolution of mammals

Mammals evolved from therapsid ancestors (part of the synapsid branch) during the Triassic period.

- The oldest fossil mammals are dated to 220 million years ago.
- Early mammals coexisted with dinosaurs throughout the Mesozoic era.
- Most Mesozoic mammals were small, probably insectivorous and nocturnal.

Mammals continued to diversify during the Cenozoic.

- During the Cretaceous period (last of the Mesozoic), mass extinctions and mass radiations transformed the flora and fauna of Earth.
- By the beginning of the Cenozoic, mammals were undergoing an adaptive radiation, and their diversity is represented today by three major groups: monotremes, marsupials, and eutherian (placental) mammals.

3. Monotremes

The *monotremes* include the platypuses and echidnas, which are characterized by:

- Oviparity
- A reptilian-like egg with large amounts of yolk that nourishes the developing embryos (see Campbell, Figure 34.27a)
- Hair
- Milk production from specialized glands on the belly of females
    - ⇒ After hatching, young suck milk from the fur of the mother who lacks nipples.

The mixture of ancestral reptilian and derived mammalian traits suggests that monotremes descended from an early branch of the mammalian lineage.

- Extant monotremes are found in Australia and New Guinea.

4. Marsupials

The *marsupials* include opossums, kangaroos, koalas, and other mammals that complete their development in a marsupium (maternal pouch).

Marsupial eggs contain a moderate amount of yolk that nourishes the embryo during early development in the mother's reproductive tract.

- Young are born in an early stage of development and are small (about the size of a honeybee in kangaroos).
- The hindlegs are simple buds, but the forelimbs are strong enough for the young to climb from the female reproductive tract exit to the marsupium.
- In the marsupium, the young attaches to a teat and completes its development while nursing.

Convergent evolution in Australian marsupials has produced a diversity of forms which resemble eutherian (placental) counterparts in all ecological roles (see Campbell, Figure 34.28).

- Opossums are the only extant marsupials outside of the Australian region.
- South America had an extensive marsupial fauna during the Tertiary period, as seen in the fossil record.

Plate tectonics and continental drift provide a mechanism which explains the distribution of fossil and modern marsupials.

- Fossil evidence indicates marsupials probably originated in what is now North America and spread southward while the land masses were joined.
- The breakup of Pangaea produced two island continents: South America and Australia.
  ⇒ With isolation, their marsupial faunas diversified away from the placental mammals that began adaptive radiation on the northern continents.
- Australia has remained isolated from other continents for about 65 million years, thus isolating its developing fauna.
- When North and South America joined at the isthmus of Panama, extensive migrations took place over the land bridge in both directions.
  ⇒ The most important migrations occurred about 12 million years ago and again about 3 million years ago.

5. Eutherian (placental) mammals

In placental mammals, embryonic development is completed within the uterus where the embryo is joined to the mother by the placenta.

Adaptive radiation during the late Cretaceous and early Tertiary periods (about 70 to 45 million years ago) produced the orders of extant placental mammals (see Campbell, Table 34.2)

- Fossil evidence indicates that placentals and marsupials diverged from a common ancestor about 80 to 100 million years ago; thus, they are more closely related than either is to the monotremes.

Most mammalogists favor a genealogy that recognizes at least four main evolutionary lines of placental mammals.

- One lineage consists of the orders Chiroptera (bats) and Insectivora (shrews) which resemble early mammals.
  ⇒ The modified forelimbs which serve a wings in bats probably evolved from insectivores that fed on flying insects.
  ⇒ Some bats feed on fruits while others bite mammals and lap the blood.
  ⇒ Most bats are nocturnal.
- A second lineage consists of medium-sized herbivores that underwent a massive adaptive radiation during the Tertiary period.
  ⇒ This led to such modern orders as the Lagomorpha (rabbits), Perissodactyla (odd-toed ungulates), Artiodactyla (even-toed ungulates), Sirenia (sea cows), Proboscidea (elephants) and Cetacea (whales, porpoises).
- The third evolutionary lineage produced the order Carnivora which probably first appeared during the Cenozoic.
  ⇒ Included in this order are the cats, dogs, raccoons, skunks, and pinnipeds (seals, sea lions, walruses).
  ⇒ Seals and their relatives evolved from middle Cenozoic carnivores that became adapted for swimming.
- The fourth lineage had the greatest adaptive radiation and produced the primate-rodent complex.

⇒ Includes the orders Rodentia (rats, squirrels, beavers) and Primates (monkeys, apes, humans).

VII. Primates and the Phylogeny of *Homo sapiens*

A. Primate evolution provides a context for understanding human origins

The first primates were small arboreal mammals.

- Dental structure suggests they descended from insectivores in the late Cretaceous.
- *Purgatorius unio*, found in Montana, is considered to be the oldest primate.

Primates have been present for 65 million years (end of Mesozoic era) and are defined by characteristics shaped by natural selection for living in trees. These characteristics include:

- Limber shoulder joints which make it possible to *brachiate* (swing from one hold to the next).
- Dexterous hands for hanging on branches and manipulating food.
- Sensitive fingers with nails, not claws.
- Eyes are close together on the front of the face, giving overlapping fields of vision for enhanced depth perception (necessary for brachiating).
- Excellent eye-hand coordination.
- Parental care with usually single births and long nurturing of offspring.

B. Modern primates

Modern primates are divided into two suborders: Prosimii (premonkeys) and Anthropoidea (monkeys, apes, humans).

- Prosimians (lemurs, lorises, pottos, tarsiers) probably resemble early arboreal primates (see Campbell, Figure 34.29).

There is a question as to which early *prosimian* lineage is ancestral to the *anthropoids*.

- Two groups of prosimian fossils are recognized by paleontologists.
  - ⇒ One ancestral to the tarsiers, the other to lemurs, lorises, and pottos (see Campbell, Figure 34.30).
- The divergence of these two groups occurred at least 50 million years ago and it has been debated as to which of these two groups was also ancestral to the anthropoids.
- Recently discovered fossils raise another possibility.
  - ⇒ Fossils found in Asia and Africa, which date at least 50 million years ago, appear to be more similar to anthropoids than to either groups of prosimian fossils.
  - ⇒ These fossils indicate an early divergence of prosimians into three lineages with the third being ancestral to the anthropoids.

Fossils of monkeylike primates indicate anthropoids were established in Africa and Asia by 40 million years ago in Africa or Asia (South America and Africa had already separated).

- Ancestors of New World monkeys may have reached South America by rafting from Africa or migration southward from North America.

New World monkeys and Old World monkeys have evolved along separate pathways for many millions of years (see Campbell, Figure 34.31)

- All New World monkeys are arboreal.
- Old World monkeys include arboreal and ground-dwelling forms.

- Most monkeys, both New and Old World, are diurnal and usually live in social bands.

There are also four genera of apes included in the anthropoid suborder: *Hylobates* (gibbons), *Pongo* (orangutans), *Gorilla* (gorillas) and *Pan* (chimpanzees) (see Campbell, Figure 34.32).

- Apes are confined to the tropical regions of the Old World.
- They are larger than monkeys (except the gibbons) with relatively long legs, short arms, and no tails.
- Only gibbons and orangutans are primarily arboreal although all are capable of brachiation.
- Social organization varies with the gorillas and chimpanzees being highly social.
- Apes have larger brains than monkeys and thus exhibit more adaptable behavior.

C. Humanity is one very young twig on the vertebrate tree

Paleoanthropology concentrates on the small span of geological time during which humans and chimpanzees diverged from a common ancestor.

Paleoanthropology = The study of human origins and evolution

Competition between researchers has often clouded the field of paleoanthropology.

- Researchers gave new names to fossil forms which were actually the same species recovered by others. (This practice ended about 20 years ago.)
- Theories were proposed on insufficient evidence; often a few teeth or a jawbone fragment.
- Such actions resulted in many persistent misconceptions about human evolution even though fossil discoveries have disproved many of the myths.

1. Some common misconceptions

Our ancestors were chimpanzees or other modern apes.

Humans and chimpanzees represent two divergent branches of the anthropoid lineage which evolved from a common, less specialized ancestor.

Human evolution represents a ladder with a series of steps leading directly from an ancestral anthropoid to Homo sapiens.

- This progression is usually shown as a line of fossil hominids becoming progressively more modern.
- Human evolution included many branches which led to dead ends with several different human species coexisting at times (see Campbell, Figure 34.33).
- If punctuated equilibrium applies to humans, most evolutionary change occurred with the appearance of new hominid species, not phyletic (anagenic) change in an unbranched lineage.

Various human characteristics like upright posture and an enlarged brain evolved in unison.

- *Mosaic evolution* occurred with different features evolving at different times.
- Some ancestral forms walked upright but had small brains.

Present understanding of our ancestry remains unclear even after dismissing many of these myths.

2. Early anthropoids

The oldest known fossils of apes are of *Aegyptopithecus*, the "dawn ape," which was a cat-sized tree-dweller from about 35 million years ago.

About 23 million years ago (during the Miocene epoch), descendants of the first apes diversified and spread to Eurasia.

About 20 million years ago, the Indian plate collided with Asia and the Himalayan range formed.

- The climate became drier and the African and Asian forests contracted.
    - ⇒ This isolated these regions of anthropoid evolution from each other.

Most anthropologists believe that humans and apes diverged from a common African anthropoid ancestor 6 to 8 million years ago.

- Evidence from the fossil record and DNA comparisons between humans and chimpanzees supports this conclusion.

3.   **The first humans**

Australopithecus africanus was discovered by Raymond Dart in 1924.

- Additional fossils proved that *Australopithecus* was a hominid that walked fully erect and had humanlike teeth and hands.
- The brain was about one-third the size of modern humans.
- Various species of *Australopithecus* began appearing about 4 million years ago and existed for over 3 million years.

"Lucy", an *Australopithecus* skeleton, was discovered in 1974 in the Afar region of Ethiopia by paleoanthropologists.

- Lucy is 3.2 million years old.
- The skeleton was 40% complete and small, about one meter tall with a head the size of a softball.
- The structure indicates an upright posture.
- It was different enough to be placed in a different species, *Australopithecus afarensis*.
- Similar fossils have been discovered which indicate the species existed for about one million years.

Since 1994, other hominid bone fragments have been recovered in east Africa areas near the site of Lucy's discovery. These fragments are so different from *A. afarensis* that new hominid species have been named:

- *Australopithecus anamensis*, from about 4 million years ago
- *Australopithecus ramidus.*, from about 4.4 million years ago

A. ramidus, which represents the oldest known hominid, exhibited some interesting characteristics.

- Skull fragments indicated the head balanced on top of the spinal column— evidence of the early evolution of upright posture.
- The skeletons of forest-dwelling animals were found among the bones—this challenges the view that bipedalism evolved when humans began living on the savanna.

The discovery of A. ramidus and A. afarensis has raised several questions.

- Is *A. ramidus* the ancestor of *A. afarensis* or an extinct evolutionary branch?
- Is *A. afarensis* ancestral to other hominids or did it share a common ancestor with *Homo*?
    - *A. afarensis* underwent little change during its one million year span.
    - Several new hominid species resulted from an adaptive radiation which began about 3 million years ago.

- The new species included *A. africanus*, several heavy-boned species of *Australopithecus*, and *Homo habilis* (appeared about 2.5 million years ago).

While the phylogeny of early hominids is uncertain, one fact is clear: hominids walked upright for two million years without a substantial increase in brain size.

- This posture may have freed the hands for other things such as gathering food or caring for infants.

4. Homo habilis

Enlargement of the human brain is first evident in fossils dating to about 2.5 million years ago.

- Skulls with brain capacities of about 650 cubic centimeters have been found compared with the 500 cc capacity of *A. africanus*.
- Simple stone tools have been found at times with the larger-brained fossils.
- Most paleoanthropologists believe these advances warrant placing the larger-brained fossils in the genus *Homo* and naming them *Homo habilis*.
- It is clear from the fossil record that after walking upright for more than two million years, hominids began to use their brains and hands to fashion tools.

Homo habilis and other new hominids were a part of a larger speciation event among African mammals.

- About 2.5 million years ago, Africa's climate began to become drier and savannas started to replace forests.
- The fauna began to adapt to these new conditions.
- *Homo habilis* coexisted with the smaller-brained *Australopithecus* for nearly one million years.
- One hypothesis is that *Australopithecus africanus* (and other australopithecines) and *Homo habilis* were two distinct lines of hominids.
  ⇒ *Australopithecus africanus* was an evolutionary dead end, while *Homo habilis* may have been on the line to modern humans, leading first to *Homo erectus* which later gave rise to *Homo sapiens*.

5. *Homo erectus* and descendants

Homo erectus was the first hominid to migrate out of Africa into Europe and Asia.

- Fossils known as Java Man and Beijing Man are examples if *H. erectus*.

Homo erectus lived from about 1.8 million years ago until 250,000 years ago.

- Fossils found in Africa cover the entire span of *H. erectus*' existence.
- These populations existed during the same period as *H. erectus* populations on other continents.
- The spread to new continents may have resulted from a gradual range expansion associated with a shift in diet to include a larger portion of meat.
  ⇒ In general, carnivores need a larger range than herbivores.

Homo erectus was taller and had a larger brain than *H. habilis*.

- The *H. erectus* brain capacity increased to as large as 1200 cc during the 1.5 million years of its existence.
  ⇒ This overlaps the normal range of modern humans.

The intelligence that evolved in *H. habilis* allowed early humans to survive in the colder climates to which they migrated.

- *Homo erectus* lived in huts or caves, built fires, wore clothes of skins, and designed more refined stone tools than *H. habilis*.

- *Homo erectus* was poorly equipped in a physical sense to live outside of the tropics but made up for the deficiencies with intelligence and social cooperation.

Some descendants of *H. erectus* developed larger brain capacities and exhibited regional diversity in populations.

- The Neanderthals are the best known descendants of *H. erectus*.
- Neanderthals lived in Europe, the Middle East, and Asia from 130,000 to 35,000 years ago.
- They had heavier brows, less pronounced chins, and slightly larger brain capacities than modern man.
- They were skilled tool makers who participated in burials and other rituals requiring abstract thought.

Many paleoanthropologists group the African post-*Homo erectus* fossils with Neanderthals and other descendants from Asia and Australasia. They believe these fossils represent the earliest forms of *Homo sapiens*.

- Some post-*H. erectus* fossils date to 300,000 years ago.

6. **The emergence of *Homo sapiens*: Out of Africa...but when?** *science as a Process*

The debate over the origin of modern humans continues unabated with two widely divergent models currently being discussed. These are the multiregional model and the monogenesis model.

The *multiregional model* proposes: 1) Neanderthals and other post-*Homo erectus* hominids were ancestors to modern humans; and 2) modern humans evolved along the same lines in different parts of the world (see also, Campbell, Figure 34.35a).

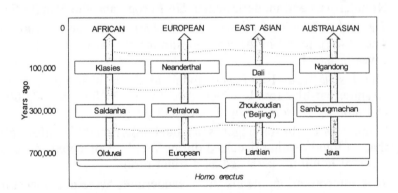

- If this model is correct, the geographic diversity of humans originated between one and two million years ago when *Homo erectus* spread from Africa to other continents.
- Supporters feel that interbreeding among neighboring populations provided opportunities for gene flow over the entire range and resulted in the genetic similarity of modern humans.

During the 1980s, some paleoanthropologists who interpreted the fossil record in a different way began to develop an alternative to the multiregional model. This alternative became known as the "Out of Africa" or *monogenesis model*.

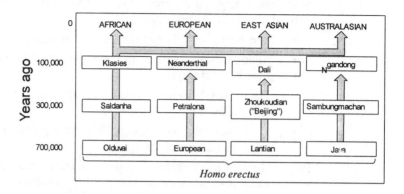

The *monogenesis model* proposes: 1) *Homo erectus* was the ancestor to modern humans who evolved in Africa; and 2) modern humans dispersed from Africa, displacing the Neanderthals and other post-*H. erectus* hominids (see also Campbell, Figure 34.35b).

- If this model is correct, the diversity of modern humans has developed from geographic diversification within the last 100,000 years.

- To supporters of this model, an exclusively African genesis for modern humans is strongly indicated by the fact that the complete transition from archaic *Homo sapiens* to modern humans is found only in African fossils.

- The focus of their interpretations is on the relationship between Neanderthals and modern humans in Europe and the Middle East.

- The oldest fossils of modern *Homo sapiens* are about 100,000 years old. These were found in Africa and similar fossils have been recovered from caves in Israel.

- The fossils from Israel were found in caves near other caves containing Neanderthal-like fossils which date from 120,000 to 60,000 years ago—overlapping *H. sapiens* by about 40,000 years.

- Supporters of the monogenesis model interpret this information to mean that no interbreeding occurred during the time of coexistence since the two types of hominids persisted as distinct forms.

- This interpretation means the Neanderthals were not ancestors of modern humans since they coexisted and were probably evolutionary dead ends along with other dispersed post-*H. erectus* hominids.

Modern molecular techniques are being used to examine the question of human origins.

- In the late 1980s, a group of geneticists compared the mitochondrial DNA (mtDNA) from a multiethnic group of more than 100 people from four different continents.

- The premise for this analysis is that the greater the differences in the mtDNA of two people, the longer the time of divergence from a common source.

- Analysis of the mtDNA comparisons resulted in the tracing of the source of all human mtDNA back to Africa with the divergence from that common source beginning about 200,000 years ago.

- These results appeared to support the monogenesis model in that the divergence from the common source was too late to represent dispersal of *Homo erectus*, but supported a later dispersal of modern humans.

Several researchers have challenged the interpretation of the mtDNA study, especially the methods used to construct evolutionary trees from this type of data and the reliability of mtDNA as a biological clock.

- This criticism has encouraged advocates of the multiregional model to argue that the fossil evidence supports the multiregional evolution of humans more strongly than African monogenesis.
- These scientists also consider certain fossils from different geographical regions to be links between that region's archaic *Homo sapiens* and the modern humans currently on that continent.

New evidence from comparisons of nuclear DNA and new methods for using the mtDNA data to trace the relationships of populations are stimulating further debate.

- This new information strengthens the monogenesis model.
- Greater genetic diversity is found in African populations south of the Sahara than in other parts of the world.
    ⇒ If modern humans evolved in southern Africa, these populations would have the longest history of genetic diversification.
    ⇒ If populations of humans in other parts of the world resulted from migrations out of Africa, the smaller genetic diversity could be the result of the founder effect and genetic drift.

Debate continues about whether the multiregional model or the monogenesis model is more accurate.

- New evidence obtained by attempts to extract DNA from fossilized Neanderthal and other archaic *Homo sapiens* skulls may shed new light on this question.

7.   Cultural evolution: a new force in the history of life

Erect stance was a very radical anatomical change in our evolution and required major changes in the foot, pelvis and vertebral column.

Enlargement of the brain was a secondary alteration made possible by prolonging the growth period of the skull and its contents.

The brains of nonprimate mammalian fetuses grow rapidly, but growth slows and stops soon after birth.

- The brains of primates continue to grow after birth and the period is longer for a human than other primates.
- Parental care is lengthened due to this extended development and this contributes to the child's learning.

Learning from the experiences of earlier generations is the basis of culture (transmission of accumulated knowledge over generations); the transmission is by written and spoken language.

A cultural evolution is a continuum, but three stages are recognized:

1. Nomads of the African grasslands made tools, organized communal activities and divided labors about 2 million years ago.
2. The development of agriculture in Africa, Eurasia, and the Americas about 10,000 to 15,000 years ago encouraged permanent settlements.
3. The Industrial Revolution began in the eighteenth century. Since then, new technology and the human population have escalated exponentially.

No significant biological change in humans has occurred from the beginning to now.

Evolution of the human brain may have been anatomically simpler than acquiring an upright stance, but the consequences of cerebral growth have been enormous.

- Cultural evolution resulted in *Homo sapiens* becoming a species that could change the environment to meet its needs and not have to adapt to an environment through natural selection.

  ⇒ Humans are the most numerous and widespread of large animals.

- Cultural evolution outpaces biological evolution and we may be changing the world faster than many species can adapt.

  ⇒ The rate of extinctions this century is 50 times greater than the average for the past 100,000 years.

  ⇒ The overwhelming rate of extinction is due primarily to habitat destruction and chemical pollution, both functions of human cultural changes and overpopulation.

  ⇒ Global temperature increase and alteration of world climates are a result of escalating fossil fuel consumption.

  ⇒ Destruction of tropical rain forests, which play a role in maintenance of atmospheric gas balance and moderating global weather, is startling.

The effect of *Homo sapiens* is the latest and may be the most devastating crisis in the history of life.

REFERENCES

Campbell, N., et al. *Biology*. 5th ed. Menlo Park, California: Benjamin/Cummings, 1999.

Carroll, R.L. Vertebrate Paleontology and Evolution. New York, NY: Freeman, 1988.

Coppens, Y. "The East Side Story: The Origin of Humankind." Scientific American, May, 1994.

Dean, D., and E. Delson. "Homo at the Gates of Europe." Nature, February 9, 1995.

CHAPTER 35
PLANT STRUCTURE AND GROWTH

OUTLINE

I. Introduction to Modern Plant Biology
   A. Molecular biology is revolutionizing the study of plants
   B. Plant biology reflects the major themes in the study of life
II. The Angiosperm Body
   A. A plant's root and shoot systems are evolutionary adaptations to living on land
   B. Structural adaptations of protoplasts and walls equip plant cells for their specialized functions
   C. The cells of a plant are organized into dermal, vascular, and ground tissue systems
III. Plant Growth
   A. Meristems generate cells for new organs throughout the lifetime of a plant: *an overview of plant growth*
   B. Primary growth: apical meristems extend roots and shoots by giving rise to the primary plant body
   C. Secondary growth: lateral meristems add girth by producing secondary vascular tissue and periderm

OBJECTIVES

After reading this chapter and attending lecture, the student should be able to:

1. List the characteristics of an angiosperm.
2. Explain the differences between monocots and dicots.
3. Describe the importance of root systems and shoot systems to plants and explain how they work together.
4. Explain how taproot systems and fibrous root systems differ.
5. Explain the differences between stolons and rhizomes.
6. Describe how plant cells grow.
7. Distinguish between parenchyma and collenchyma cells with regards to structure and function.
8. Describe the differences in structure and function of the two types of sclerenchyma cells.
9. Explain the importance of tracheids and vessel elements to plants.
10. Distinguish between water-conducting cells and sieve-tube members with regards to structure and function.
11. Explain the differences between simple tissues and complex tissues.
12. Explain the importance of a cuticle on the aerial parts of a plant and its absence on roots.

13. Describe the functions of the dermal tissue system, vascular tissue system and ground tissue system.
14. Distinguish among annual, biennial, and perennial plants.
15. Explain the importance of the zones of cell division, cell elongation, and cell differentiation in primary growth of roots.
16. Explain the importance of the endodermis to a plant.
17. Describe the importance of an apical meristem to the primary growth of shoots.
18. Distinguish between the arrangement of vascular tissues in roots and shoots.
19. Describe how "wood" forms due to secondary growth of stems.
20. Using a diagram, describe the basic structure of a root, a stem, and a leaf.

KEY TERMS

monocots	petiole	meristem	endodermis
dicots	protoplast	apical meristem	lateral roots
root system	parenchyma cell	primary growth	pericycle
shoot system	collenchyma cell	secondary growth	vascular bundle
xylem	sclerenchyma cell	lateral meristem	stomata
phloem	fiber	primary plant body	guard cells
taproot	sclereids	root cap	transpiration
fibrous root	tracheids	zone of cell division	mesophyll
root hairs	vessel elements	quiescent center	secondary plant body
adventitious	pits	protoderm	vascular cambium
stem	xylem vessels	pericycle	cork cambium
node	sieve-tube members	procambium	ray initials
internode	sieve plates	ground meristem	fusiform initials
axillary bud	companion cell	zone of elongation	periderm
terminal bud	dermal tissue system	zone of maturation	bark
apical dominance	epidermis	stele	lenticel
leaves	annuals	pith	
blade	perennials	cortex	

LECTURE NOTES

Plants form the foundation on which most terrestrial ecosystems are built.

- As the primary producers in most systems (through the process of photosynthesis), plants serve as the first link in the food chain which affects all species of animals in the system.

Plants were first studied by early humans who had to distinguish between edible and poisonous plants.

- These early humans later began to use plant products to make useful tools and other items.
- Modern plant biology continues to center on how to use plants and plant products to benefit humans.

I. Introduction to Modern Plant Biology

A. Molecular biology is revolutionizing the study of plants

New methods and the discovery of unique, interesting experimental organisms has expanded the knowledge base in many areas of plant biology.

Studying organisms such as *Arabidopsis thaliana*, which has a short generation time and a tiny genome, has enabled scientists to gain new insight into the genetic control of plant development.

- For example, researchers have identified the genes responsible for flower development.
- Eventually, researchers may be able to understand the mechanisms that control several basic process, such as how a plant causes water to flow upward, how plants resist disease, and what makes roots grow down and shoots grow up.

B. Plant biology reflects the major themes in the study of life

Plants, as do all living systems, have a hierarchy of structural levels. As research into the biology of plants progresses, the relationships of the molecular and cellular mechanisms to the function of the whole organism will emerge.

Correlating the structure and functions of plants is a second focus of plant biology.

- Structure and function of plants are the result of interactions with the environment over both evolutionary and short-term time scales.
  ⇒ In their evolutionary journey, plants adapted to the problems of a terrestrial existence as they moved from water to land (e.g., woody tissues provide support).
  ⇒ Over the short-term, environmental stimuli cause individual plants to exhibit structural and physiological responses (e.g., stomata close during the hottest part of the day, thus conserving water) (see Campbell, Figure 35.2).

While plants and animals faced many of the same problems (e.g., support, internal transport) they solved such problems in different ways.

- The two groups evolved independently from unicellular ancestors which had different modes of nutrition; this alone established a different evolutionary direction.

II. The Angiosperm Body

A. A plant's root and shoot systems are evolutionary adaptations to living on land

Plant biologists study two levels of plant architecture: morphology and anatomy.

Plant morphology = The study of the external structure of plants (e.g., arrangement of the parts of a flower)

Plant anatomy = The study of the internal structure of plants (e.g., arrangement of the cells and tissues in a leaf)

Angiosperms (flowering plants) are the most diverse and widespread of the plants with about 275,000 extant species.

- Characterized by flowers and fruits, which are believed to be adaptations for reproduction and seed dispersal
- Taxonomists divide angiosperms into two taxonomic classes: *monocots* and *dicots* which possess either one or two seed leaves, respectively, in combination with other characteristics (see Campbell, Figure 35.3).

A plant can be divided into two basic systems, a subterranean *root system* and an *aerial shoot system* (stems, leaves, flowers).

- This two system arrangement reflects the evolutionary history of plants as terrestrial organisms (see Campbell, Figure 35.4).
- Unlike the algal ancestors which were completely surrounded by nutrient rich water, terrestrial plants face a divided habitat:
  ⇒ Air is the source of CO_2 for photosynthesis, and sunlight cannot penetrate into the soil.
  ⇒ Soil provides water and dissolved minerals to the plant.
- Each system depends on the other for survival of the whole plant.
  ⇒ Roots depend on shoots for sugar and other organic nutrients.
  ⇒ Shoots depend on roots for minerals, water and support.
- Materials are transported through the plant by vascular tissues.
  ⇒ *Xylem* conveys water and dissolved minerals to the shoots.
  ⇒ *Phloem* conveys food from shoots to roots and other nonphotosynthetic parts, and from storage roots to actively growing shoots.

1. The root system

Root structure is well adapted to:

- Anchor plants
- Absorb and conduct water and nutrients
- Store food

There are two major types of root systems:

1. The *taproot* system is seen in many dicots.
   - One large, vertical root (the taproot) produces many smaller secondary roots.
   - Provides firm anchorage
   - Some taproots, such as carrots, turnips, and sweet potatoes, are modified to store a large amount of reserve food.
2. A *fibrous root* system is found primarily in monocots (palms, bamboo, grasses).
   - A mat of threadlike roots spreads out below the soil surface.
   - Provides extensive exposure to soil water and minerals
   - Roots are concentrated in the upper few centimeters of soil, preventing soil erosion.

Absorption of water is greatly enhanced by *root hairs*, which increase the surface area of the root (see Campbell, Figure 35.5).

- Root hairs are normally most numerous near the root tips.
- Water and mineral absorption are also enhanced by mycorrhizae, symbiotic associations between roots and fungi.

Many plants possess root nodules, which contain symbiotic bacteria capable of converting atmospheric nitrogen to nitrogenous compounds that the plant can use for the synthesis of proteins and other organic molecules.

Adventitious roots = Roots arising above ground from stems or leaves

- Form in addition to the normal root system
- Some, such as prop roots of corn, help support the plant stem

2. The shoot system

Shoot systems are comprised of vegetative shoots and floral shoots.

- Vegetative shoots consist of a stem and attached leaves; may be the main shoot or a vegetative branch.

- Floral shoots terminate in flowers.

a. Stems

Stem morphology includes:(see Campbell, Figure 35.4):

- *Nodes* = The points where leaves are attached to stems
- *Internodes* = The stem segments between the nodes
- *Axillary bud* = An embryonic side shoot found in the angle formed by each leaf and the stem; usually dormant
- *Terminal bud* = The bud on a shoot tip; it usually has developing leaves and a compact series of nodes and internodes

Growth of a shoot is usually concentrated at the apex of the shoot where the terminal bud is located.

- The presence of a terminal bud inhibits development of axillary buds, a condition called *apical dominance*.
- Apical dominance appears to be an evolutionary adaptation to increase exposure of plant parts to light by concentrating resources on increasing plant height.

Axillary buds begin to grow under certain conditions and after damage or removal of the terminal bud.

- Some may develop into floral shoots.
- Others develop into vegetative shoots with terminal buds, leaves, and axillary buds.
    - ⇒ This development results in branching which also increases exposure of plant parts to light.

Some plants have modified stems which are often mistaken for roots. There are several types of modified stems, each of which performs a specific function (see Campbell, Figure 35.6)

- Stolons are horizontal stems growing along the surface of the ground (e.g., strawberry plant runners).
- Rhizomes are horizontal stems growing underground (e.g., irises).
    - ⇒ Some end in enlarged tubers where food is stored (e.g., potatoes)
- Bulbs are vertical, underground shoots with leaf bases modified for food storage (e.g., onions).

b. Leaves

Leaves are the main photosynthetic organs of a plant.

A leaf usually exists in the shape of a flattened blade which is joined to the node of a stem by a petiole.

- Most monocots lack petioles; instead, the leaf base forms a sheath surrounding the stem.

Leaves of monocots and dicots vary in the arrangement of their major veins.

- Monocot leaves have parallel major veins running the length of the blade.
- Dicot leaves have a multi-branched network of major veins; can be palmate or pinnate.
- All leaves have numerous minor cross-veins.

Plant taxonomists use a variety of leaf characteristics to classify plants, including:

- Leaf shape

- Spatial arrangement of leaves on a stem
- Pattern of a leaf's veins

Campbell, Figure 35.7 illustrates simple vs. compound leaves.

Some plants have leaves that have become adapted for functions other than photosynthesis (see Campbell, Figure 35.8).

- Tendrils are modified leaflets that cling to supports.
- Spines of cacti function in protection.
- Many succulents have leaves modified for storing water.
- Some plants have brightly colored leaves that help attract pollinators to the flower.

B. Structural adaptations of protoplasts and walls equip plant cells for their specialized functions

Each type of plant cell has structural adaptations that make it possible to perform that cell's function. Some are coupled with specific characteristics of the protoplast (see Campbell, Figure 35.9)

Protoplast = Contents of a plant cell exclusive of the cell wall

1. Parenchyma cells

Parenchyma cells are the least specialized plant cells (see Campbell, Figure 35.10a).

Primary walls are thin and flexible.

Most lack secondary walls.

The protoplast usually has a large central vacuole.

They function in synthesizing and storing organic products.

⇒ Photosynthesis occurs in the chloroplasts of mesophyll cells.

Some parenchyma cells in stems and roots have colorless plastids that store starch.

Most mature cells do not divide, but retain the ability to divide and differentiate into other cell types under special conditions (e.g., repair and replacement after injury).

2. Collenchyma cells

Collenchyma cells usually lack secondary walls.

The primary cell wall is thicker than in parenchyma cells but is of an uneven thickness (see Campbell, Figure 35.10b).

They are usually grouped in strands or cylinders to support young parts of plants without restraining growth.

They are living cells which elongate as the stems and leaves they support grow.

3. Sclerenchyma cells

Sclerenchyma cells function in support.

They have very rigid, thick secondary walls strengthened by lignin.

Many lack protoplasts at functional maturity, so they cannot elongate. In fact, they may be dead, functioning only as support.

There are two forms: *fibers* (long, slender, tapered cells occurring in bundles) and *sclerids* (shorter, irregularly-shaped cells) (see Campbell, Figure 35.10c).

4. Water-conducting cells of xylem: tracheids and vessel elements

Xylem consists of two cell types, both with secondary walls and both dead at functional maturity.

- Before the protoplast dies, secondary walls are deposited in spiral or ring patterns (which allows them to stretch) in parts of the plant that are still growing.

Tracheids are long, thin, tapered cells having lignin-hardened secondary walls with *pits* (thinner regions where only primary walls are present) (see Campbell, Figure 35.10a)

- Water flows from cell to cell through pits.
- They also function in support.

Vessel elements are wider, shorter, thinner-walled, and less tapered than tracheids.

- Vessel elements are aligned end to end.
- The end walls are perforated, permitting the free flow of water through long chains of vessel elements called xylem vessels.

5. **Food-conducting cells of phloem: sieve-tube members**

Sieve-tube members are chains of phloem cells that transport sucrose, other organic compounds, and some minerals.

- The cells are alive at functional maturity.
- Protoplasts lack a nucleus, ribosomes, and a distinct vacuole.

In angiosperms, the end walls of sieve-tube members have pores and are called *sieve plates*.

- The pores probably facilitate the movement of fluid between cells.
- At least one *companion cell* is connected to each sieve-tube member by many plasmodesmata; the companion's nucleus and ribosomes may also serve the sieve-tube member which lacks these organelles.
  - Companion cells also help load sugar produced in the mesophyll into sieve-tubes of leaves of some plants.

> It is useful to emphasize how close inspection of cell structure often reveals its function, and vice versa.

C. **The cells of a plant are organized into dermal, vascular, and ground tissue systems**

Each organ of a plant (leaf, stem, root) has three tissue systems: dermal, vascular, and ground tissue systems.

Each of the three systems is continuous throughout the plant, although specific characteristics and spatial relationships vary in different plant organs (see Campbell, Figure 35.12).

1. *Dermal tissue system* (or *epidermis*) = Single layer of tightly packed cells covering and protecting the young parts of the plant
   - Functions in protection and has special characteristics consistent with the function of the organ it covers.
     - Root hairs specialized for water and mineral absorption are extensions of epidermal cells near root tips.
     - The waxy cuticle that helps the plant retain water is secreted by epidermal cells of leaves and most stems.
2. *Vascular tissue system* = The xylem and phloem that functions in transport and support; is continuous throughout the plant
3. *Ground tissue system* = Predominantly parenchyma, with some collenchyma and sclerenchyma present; occupies the space between dermal and vascular tissue systems; functions in photosynthesis, storage, and support

III. Plant Growth

A. Meristems generate cells for new organs throughout the lifetime of a plant: *an overview of plant growth*

Plant growth begins with germination of the seed and continues for the lifespan of the plants.

Plants do not live indefinitely, they have finite life spans.

- Most have genetically determined lifespans.
- Some are environmentally determined.
- *Annuals* complete their life cycle in one year or growing season.
- Biennials typically have life spans of two years.
- *Perennials*, such as trees and some grasses, live many years.

Indeterminate growth is made possible by *meristems* (perpetually embryonic tissues).

- *Indeterminate growth* = Continued growth as long as the plant lives
  - In contrast, most animals cease growing after reaching a certain size (determinate growth).
  - Certain plant organs, such as flower parts, show determinate growth.
- Meristematic cells are unspecialized and divide to generate new cells near the growing point.
  - New meristematic cells formed by division that remain in the region and produce new cells are called initials.
  - New cells are displaced (derivatives), and become incorporated and specialized into tissues.
- *Apical meristems*, located in root tips and shoot buds, supply cells for plants to grow in length.
  - *Primary growth* (elongation) is initiated by apical meristems and forms primary tissues organized into the 3 tissue systems.
  - *Secondary growth* (increased girth) is the thickening of roots and shoots which occurs in woody plants due to development of lateral meristems.
- *Lateral meristems* = Cylinders of dividing cells extending along the lengths of roots and shoots.
  - Cell division in the lateral meristems produces secondary dermal tissues which are thicker and tougher than the epidermis it replaces.
  - Also adds new layers of vascular tissues.

B. Primary growth: apical meristems extend roots and shoots by giving rise to the primary plant body

Primary growth produces the *primary plant body*, which consists of three tissue systems (see Campbell, Figure 35.12).

- The youngest portions of woody plants and herbaceous plants are examples of primary plant bodies.
- Apical meristems are responsible for the primary growth of roots and shoots.

1. Primary growth of roots

Root growth is concentrated near its tip and results in roots extending through the soil.

- The root tip is covered by a *root cap*, which protects the meristem and secretes a polysaccharide coating that lubricates the soil ahead of the growing root.

- The root tip contains three zones of cells in successive stages of primary growth. Although described separately, these zones blend into a continuum (see Campbell, Figure 35.15).

1. Zone of cell division
   - Located near the tip of the root; includes the apical meristem and its derivatives, the primary meristems.
   - The apical meristem is centrally located; it produces the primary meristems and replacement cells of the root cap.
   - A *quiescent center* is located near the center of the apical meristem. It is composed of resistant, slowly dividing cells which may serve as reserve replacement cells in case of damage to the meristem.
   - The primary meristems form as three concentric cylinders of cells (the *protoderm*, *procambium*, and *ground meristem*) that will produce the tissue systems of the roots.

2. Zone of cell elongation
   - In this region, cells elongate to at least ten times their original length.
   - The elongation of cells in this region pushes the root tip through the soil.
   - Continued growth is sustained by the meristem's constant addition of new cells to the youngest end of the elongation zone.

3. Zone of maturation
   - Is located farthest from the root tip
   - Region where the new cells become specialized in structure and function and where the three tissue systems complete their differentiation.

a. **Primary tissues of roots**

The apical meristem produces three primary meristems, which in turn give rise to the three primary tissues of roots (see Campbell, Figure 35.16).

1. The *protoderm* is the outermost primary meristem.
   - Gives rise to the epidermis over which water and minerals must cross.
   - Root hairs are epidermal extensions which increase surface area, thus enhancing uptake.

2. The *procambium* forms a *stele* (central cylinder) where xylem and phloem develop.
   - In dicots, xylem radiates from the stele's center in two or more spokes, with phloem in between the spokes.
   - In monocots, a central *pith* (core of parenchyma cells) is ringed by vascular tissue in an alternating pattern of xylem and phloem.

3. The ground meristem is located between the protoderm and procambium; it gives rise to the ground tissue system. The ground tissue:
   - Is mostly parenchyma and fills the *cortex* (root area between the stele and epidermis)
   - Stores food; the cell membranes are active in mineral uptake
   - Has *endodermis*, the single-cell thick, innermost layer of the cortex that forms the boundary between the cortex and the stele; it selectively regulates passage of substances from soil to the vascular tissue of the stele

Lateral roots may sprout from the outermost layer of the stele of a root.

- The *pericycle*, just inside the endodermis, is a layer of cells that may become meristematic and divide to form the lateral root (see Campbell, Figure 35.17).
- A lateral root forms as a clump of cells in the pericycle, then elongates and pushes through the cortex until it emerges from the primary root.
- The lateral root maintains its vascular connection to the stele of the main root.

2. Primary growth of shoots

A shoot's apical meristem is a dome-shaped mass of dividing cells at the tip of the terminal bud (see Campbell, Figure 35.18).

- Forms the primary meristems that differentiate into the three tissue systems.
- On the flanks of the apical meristem dome are *leaf primordia* which form leaves.
- Meristematic cells left by the apical meristem at the base of the leaf primordia develop into axillary buds.

Most shoot elongation actually occurs due to growth of slightly older internodes below the shoot apex.

- Growth is a result of both cell division and elongation within the internode.
- Intercalary meristems are present at the base of each internode in grasses and some other plants.
    - ⇒ These tissues permit prolonged internode elongation along the length of the shoot.

Axillary buds may form branches later in the life of the plant.

- Branches originate at the surface of the shoot and are connected to the vascular system which lies near the surface.
- This is a direct contrast to development of lateral roots which form deep in the pericycle of the root.

a. Primary tissues of stems

The vascular tissue of the stem is organized into strands of *vascular bundles* that run the length of the stem (see Campbell, Figure 35.19).

- They converge at the transition zone (shoot → root) to join the root stele.
- Each bundle is surrounded by ground tissue, including pith and cortex.

In dicots, bundles are arranged in a ring with pith inside and cortex outside.

- Xylem faces the pith, phloem faces the cortex.
- Pith and cortex are connected by pith rays, thin layers of ground tissue.

In monocots, vascular bundles are scattered throughout the ground tissue of the stem.

- The stem ground tissue is mostly parenchyma.
- Ground tissue of stems is mostly parenchyma, strengthened in many plants by collenchyma located beneath the epidermis.

The protoderm of the terminal bud gives rise to the epidermal portion of the dermal tissue system.

b. Tissue organization of leaves

Leaves are cloaked by an epidermis of tightly interlocked cells (see also Campbell, Figure 35.20).

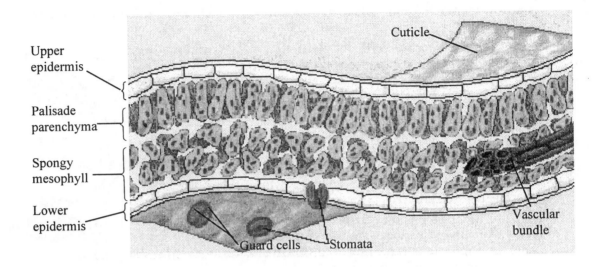

- It protects against physical damage and pathogens.
- A waxy cuticle prevents water loss.
- *Stomata* are pores flanked by *guard cells* which regulate gas exchange with the surrounding air and photosynthetic cells inside the leaf.
- Stomata also allow *transpiration* (water loss from plant by evaporation).
- Stomata are usually more numerous on the bottom surface of the leaf. This location minimizes water loss.

The ground tissue of a leaf is *mesophyll*.

- Consists mainly of parenchyma cells equipped with chloroplasts for photosynthesis
- Dicots usually have two distinct regions of mesophyll:
  1. *Palisade parenchyma* = One or more layers of columnar cells of the upper half of a leaf
  2. *Spongy parenchyma* = Irregularly shaped cells surrounded by air spaces through which oxygen and carbon dioxide circulate. Located in the lower half of the leaf

The leaf vascular tissue is continuous with that of the stem through leaf traces which are branches from the stem vascular bundles.

- Leaf traces continue in the petiole as a vein, which branches repeatedly throughout the mesophyll of the leaf blade.
- This arrangement brings the photosynthetic tissue of the leaf into close contact with xylem and phloem.
- Functions also as a skeleton to support the shape of the leaf.

c. Modular shoot construction and phase changes during development

Shoots are constructed of a series of modules produced by the serial development of nodes and internodes within the shoot apex (see also Campbell, Figure 35.21).

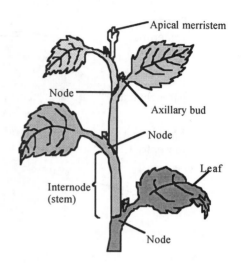

- Each module consists of a stem, one or more leaves, and an axillary bud associated with each leaf.
- Elongation of the internode provides for primary growth of the plant.

Modules found in the body of a plant represent various ages of tissue. The age of a module is proportional to its distance from the apical meristem.

- The developmental phase of each module changes over time.
- There is a gradual change from a juvenile vegetative state to a mature vegetative state.
  ⇒ Leaf morphology usually changes during this transition (see Campbell, Figure 35.22).
- In some plants, the mature shoot apex may undergo a second phase change to a flower-producing (reproductive) state.
  ⇒ Primary growth of these shoot tips is terminated since the apical meristem is consumed in production of the flower.

C. Secondary growth: lateral meristems add girth by producing secondary vascular tissue and periderm

A *secondary plant body* results from secondary growth.

- The secondary plant body is comprised of the secondary tissues produced during growth in diameter.
- Secondary growth results from two lateral meristems: the *vascular cambium* and the *cork cambium*.
  ⇒ Vascular cambium produces secondary xylem and phloem.
  ⇒ Cork cambium produces a tough, thick covering for roots and stems that replaces the epidermis.
- Secondary growth occurs in all gymnosperms and most dicot angiosperms.
  ⇒ It is rare in monocots.

1. Secondary growth of stems

a. Vascular cambium and the production of secondary vascular tissue

Vascular cambium forms when meristemic parenchyma cells develop between the primary xylem and primary phloem of each vascular bundle and in the rays of ground tissue between the bundles (see Campbell, Figures 35.23 and 35.24)

Fasicular cambium = Cambium within the vascular bundle

Interfasicular cambium = Cambium in the rays between vascular bundles

- Together, meristematic bands in the fasicular and interfasicular regions form a continuous cylinder of dividing cells around the xylem and pith of the stem.

- *Ray initials* (meristematic cells of the interfasicular cambium) produce radial files of parenchyma cells (= xylem and phloem rays) which permit lateral transport of water and nutrients as well as storage of starch and other reserves.
- *Fusiform initials* (cells of fasicular cambium) produce new vascular tissues; secondary xylem to the inside of the vascular cambium and secondary phloem to the outside.

Accumulated layers of secondary xylem produces wood that consists mostly of tracheids, vessel elements and fiber.

- The hardness and strength of wood results from these cells which, while dead at maturity, have thick, lignified walls.
- Forms annual growth rings due to yearly activity: cambium dormancy, spring wood production and summer wood production.

The secondary phloem does not accumulate extensively. The secondary phloem, and all tissues external to it, develop into bark which eventually sloughs off the tree trunk.

It is important to note that primary and secondary growth occur simultaneously, but in different regions of the stem.

b. Cork cambium and the production of periderm

Cork cambium is a cylinder of meristematic tissue that forms protective layers of the secondary plant body.

- It first forms in the outer cortex of the stem.
- Phelloderm (parenchyma cells) forms to the inside of cork cambium initials as they divide.
- Cork cells form to the outside of cork cambium initials as they divide.
- As cork cells mature, they deposit suberin (a waxy material) in their walls and die.
- These dead cork tissues protect the stem from damage and pathogens while reducing water loss.
- The combination of cork cambium, layers of cork, and phelloderm form the *periderm* (the protective coat of the secondary plant body that replaces primary epidermis).
- The term *bark* refers to all the tissues external to the vascular cambium (phloem, phelloderm, cork cambium and cork) (see Campbell, Figure 35.25).

Cork cambium is a cylinder of fixed size and does not grow in diameter.

- As continued secondary growth splits it, it is replaced by new cork cambium formed deeper in the cortex.
- When no cortex is left, it develops from parenchyma cells in the secondary phloem (only the youngest secondary phloem, internal to cork cambium, functions in sugar transport).
- Lenticels are present in the bark.

Lenticel = Spongy region in the bark which permit gas exchange by living cells within the trunk

2. Secondary growth of roots

Vascular cambium and cork cambium also function in secondary growth of roots.

- The vascular cambium produces secondary xylem to its inside and secondary phloem to its outside.
    ⇒ It is first located between xylem and phloem of the stele.

⇒ The cortex and epidermis split and are shed as the stele grows in diameter.

- Cork cambium forms from the pericycle of the stele and produces the periderm, which becomes secondary dermal tissue.

⇒ Periderm is impermeable to water, consequently, the roots with secondary growth function to anchor the plant and transport water and solutes between the younger roots and the shoot system.

Older roots become woody, and annual rings appear in the secondary xylem.

- Old roots and old stems may look very similar.

REFERENCES

Campbell, N., et al. *Biology*. 5th ed. Menlo Park, California: Benjamin/Cummings, 1999.

Raven, P.H., R.F. Evert and S.E. Eichhorn. *Biology of Plants*. 6th ed. New York: Worth Publishers, Inc., 1998.

CHAPTER 36
TRANSPORT IN PLANTS

OUTLINE

I. An Overview of Transport Mechanisms in Plants
   A. Transport at the cellular level depends on the selective permeability of membranes
   B. Proton pumps play a central role in transport across plant membranes
   C. Differences in water potential drive water transport
   D. Vacuolated cells have three major compartments
   E. The symplast and apoplast both function in transport within tissues and organs
   F. Bulk flow functions in long-distance transport

II. Absorption of Water and Mineral by Roots
   A. Root hairs, mycorrhizae, and a large surface area of cortical cells enhance water and mineral absorption
   B. The endodermis functions as a selective sentry between the root cortex and vascular tissue

III. Transport of Xylem Sap
   A. The ascent of xylem sap depends mainly on transpiration and the physical properties of water
   B. Review: xylem sap ascends by solar-powered bulk flow

IV. The Control of Transpiration
   A. Guard cells mediate the photosynthesis-transpiration compromise
   B. Xerophytes have evolutionary adaptations that reduce transpiration

V. Translocation of Phloem Sap
   A. Phloem translocates its sap from sugar sources to sugar sinks
   B. Pressure flow is the mechanism of translocation in angiosperms

OBJECTIVES

After reading this chapter and attending lecture, the student should be able to:

1. List three levels in which transport in plants occurs and describe the role of aquaporins.
2. Trace the path of water and minerals from outside the root to the shoot system.
3. Provide experimental evidence that links plant cellular respiration to mineral accumulation.
4. Explain how a proton pump may affect mineral transport in plants.
5. Describe the symplast and apoplast routes for the transit of water and minerals across the root cortex from the epidermis to the stele.
6. Explain the function of the Casparian strip.
7. Explain how solutes are transferred between the symplast and apoplast.

8. Define water potential.

9. Explain how solute concentration and pressure affects water potential.

10. Predict the direction of net water movement based upon differences in water potential between a plant cell and a hypoosmotic environment, a hyperosmotic environment or an isosmotic environment.

11. Explain how root pressure is created by some plants and how it causes guttation.

12. According to the transpiration-cohesion-adhesion theory, describe how xylem sap can be pulled upward in xylem vessels.

13. Explain why a water potential gradient is required for the passive flow of water through a plant, from soil.

14. Compare the transpiration-to-photosynthesis ratio between C_3 and C_4 plants.

15. Describe both the disadvantages and benefits of transpiration.

16. Explain how guard cells control the stomatal aperture and how this, in turn, can affect photosynthetic rate and transpiration.

17. Explain how K^+ fluxes across the guard cell membrane affects guard cell function.

18. List three cues that contribute to stomatal opening at dawn.

19. Describe environmental stresses that can cause stomata to close during the daytime.

20. Explain how xerophytes can be adapted to arid climates.

21. Explain how crassulacean acid metabolism allows CAM plants to reduce the transpiration rate.

22. Describe source-to-sink transport in phloem and explain what determines the direction of phloem sap flow.

23. Compare the process of phloem loading between plants such as corn and squash.

24. Give one explanation for how a proton pump can allow for selective accumulation of sucrose in the symplast.

25. Explain what causes phloem sap to flow from source to sink and describe how a scientist can study pressure-flow in phloem.

KEY TERMS

transport proteins	turgor pressure	endodermis	circadian rhythms
selective channels	turgid	Casparian strip	translocation
chemiosmosis	aquaporins	transpiration	sugar source
osmosis	tonoplast	cohesion	sugar sink
water potential	symplast	root pressure	transfer cells
megapascals	apoplast	guttation	
tension	bulk flow	transpiration-to-	
plasmolyze	mycorrhizae	photosynthesis ratio	

LECTURE NOTES

Land plants require a transport system, because unlike their aquatic ancestors, photosynthetic plant organs have no direct access to water and minerals.

I.   **An Overview of Transport Mechanisms in Plants**

Three levels of transport occur in plants:

1. Uptake and release of water and solutes by individual cells
2. Short-distance cell-to-cell transport at the level of tissues and organs
3. Long-distance transport of sap in xylem and phloem at the whole-plant level

Campbell, Figure 36.1 provides an overview of these transport functions.

A. Transport at the cellular level depends on the selective permeability of membranes

> Water and solute transport is covered in detail in Chapter 8. This section highlights a few of the transport processes in the context of plant cells.

The plasma membrane's selective permeability controls the movement of solutes between a plant cell and the extracellular fluids. Solutes may move by passive or active transport.

Passive transport occurs when a solute molecule diffuses across a membrane down a concentration gradient.

- Requires no direct expenditure of energy by the cell
- Transport proteins embedded in the membrane may increase the speed at which solutes cross.
  - *Transport proteins* may facilitate diffusion by serving as carrier proteins or forming selective channels.
  - Carrier proteins bind selectively to a solute molecule on one side of the membrane, undergo a conformational change, and release the solute molecule on the opposite side of the membrane.
  - *Selective channels* are simply passageways by which selective molecules may enter and leave a cell; some gated selective channels are stimulated to open or close by environmental conditions.

Active transport occurs when a solute molecule is moved across a membrane against the concentration gradient.

- Energy requiring process
- Active transport is not accomplished by the use of transport proteins, such as those involved in facilitated diffusion, but is conducted by active transporters, a special class of membrane proteins (e.g., the proton pump is an active transporter important to plants).

B. Proton pumps play a central role in transport across plant membranes

A *proton pump* hydrolyzes ATP and uses the energy to pump hydrogen ions (H^+) out of the cell.

- Produces a proton gradient with a higher concentration outside of the cell
- Produces a membrane potential, since the inside of the plant cell is negative in relation to the outside

This membrane potential and the stored energy of the proton gradient are used by the plant to transport many different molecules (see Campbell, Figure 36.2).

- Potassium ions (K^+) are pulled into the cells because of the electrochemical gradient.
- Nitrate (NO_3^-) enters plant cells against the electrochemical gradient by in exchange for hydrogen ions by a process known as *cotransport*.

The involvement of proton pumps in the transport processes of plant cells is a specific application of *chemiosmosis*.

C. Differences in water potential drive water transport in plant cells

Osmosis results in the net uptake or loss of water by the cell and depends on which component, the cell or extracellular fluids, has the highest water potential.

Water potential (Ψ) = The free energy of water that is a consequence of solute concentration and applied pressure; physical property predicting the direction water will flow

- Water will always move across the membrane from the solution with the higher water potential to the one with lower water potential.
- Water potential is measured in units of pressure called *megapascals* (MPa); one MPa is equal to ten atmospheres of pressure.

1. How solutes and pressure affect water potential

Pure water in an open container has a water potential of zero megapascals ($\Psi = 0$ MPa).

- Addition of solutes to water lowers the Ψ into the negative range.
- Increased pressure raises the Ψ into the positive range.
- A negative pressure, or *tension*, may also move water across a membrane; this *bulk flow* (movement of water due to pressure differences) is usually faster than the movement caused by different solute concentrations.

Campbell, Figure 36.3 illustrates water potential and water movement.

2. Quantitative analysis of water potential

The effects of pressure and solute concentration on water potential are represented by:

$$\Psi = \Psi_P + \Psi_S$$

- Ψ_P = Pressure potential
- Ψ_S = Solute potential or osmotic potential

> Example: A 0.1M solution has a Ψ_S of -0.23 MPa; in the absence of physical pressure, the water potential is -0.23 MPa ($\Psi = 0 + (-0.23) = -0.23$ MPa).
>
> The addition of pressure to the solution could counter the effects of osmotic pressure by stopping net water movement (if P = 0.23) or by forcing water from the solution back into the pure water (if $P \geq 0.3$).
>
> Similar changes would result if a negative pressure were applied to the pure water side of the membrane.

Plant cells will gain or lose water to intercellular fluids depending upon their water potential.

- A flaccid cell (P = 0) placed in a hyperosmotic solution will lose water by osmosis; the cell will *plasmolyze* (protoplast pulls away from the cell wall) in response (see Campbell, Figure 36.4a).
- A flaccid cell placed in a hypoosmotic solution will gain water by osmosis; the cell will swell and a *turgor pressure* develops; when pressure from the cell wall is equal to the osmotic pressure, an equilibrium is reached and no net water movement occurs ($\Psi = 0$) (see Campbell, Figure 36.4b).

3. Aquaporins

Until recently, most biologists accepted the hypothesis that leakage of water through the lipid bilayer accounted for the flux of water across plasma membranes.

Recent experimental data suggests that water transport is too specific and too rapid to be explained entirely by diffusion through the lipid bilayer,

- Water-specific channels made up of transport proteins, called *aquaporins*, have been discovered in plant and animal cells.
- Aquaporins do not actively transport water, but rather facilitate its diffusion (osmosis).

D. Vacuolated cells have three major compartments

The three major compartments of a plant cell are the 1) cell wall, 2) cytosol of the protoplast, and 3) tonoplast (see Campbell, Figure 36.5).

Tonoplast = Membrane surrounding the large central vacuole found in plant cells; important in regulating intracellular conditions

- Contains integral transport proteins that control the movement of solutes between the cytosol and the vacuole
- Has a membrane potential; proton pumps in the tonoplast help the plasma membrane maintain a low H^+ concentration in the cytosol by moving H^+ into the vacuole
- Several solutes are transported between the cytosol and vacuole due to this membrane potential and proton gradient.

Two of the three cellular compartments of plants are continuous between cells.

- Plasmodesmata connect the cytosolic compartments of neighboring cells; this cytoplasmic continuum is called the *symplast*.
- The walls of adjacent cells are connected to forming a continuum of cell walls called the *apoplast*.

E. The symplast and apoplast both function in transport within tissues and organs

Lateral transport is short-distance transport, usually along the radial axis of plant organs.

It can occur by three routes in plant tissues and organs (see Campbell, Figure 36.5):

1. Across the plasma membranes and cell walls. Solutes move from one cell to the next by repeatedly crossing plasma membranes and cell walls.
2. The symplast route. A symplast is the continuum of cytoplasm within a plant tissue formed by the plasmodesmata which passes through pores in the cell walls. Once water or a solute enters a cell by crossing a plasma membrane, the molecules can enter other cells by traveling through the plasmodesmata.
3. The apoplast route. An apoplast is the continuum between plant cells which is formed by the continuous matrix of cell walls. Water and solute molecules can move from one area of a root or other organ via the apoplast without entering a cell.

Water and solute molecules can move laterally in a plant organ by any one of these routes or by switching from one route to another.

F. Bulk flow function in long-distance transport

This type of transport is usually along the vertical axis of the plant (up and down) from the roots to the leaves and vice versa.

Vascular tissues are involved in this type of transport as diffusion would be too slow.

- *Bulk flow* (movement due to pressure differences) moves water and solutes through xylem vessels and sieve tubes.
- Transpiration reduces pressure in the leaf xylem; this creates a tension which pulls sap up through the xylem from the roots.
- Hydrostatic pressure develops at one end of the sieve tubes in the phloem; this forces the sap to the other end of the tube.

II. Absorption of Water and Minerals by Roots

Water and minerals enter plants through the following transport pathway (see Campbell, Figure 36.6):

$$Soil \longrightarrow Epidermis \longrightarrow Root\ cortex \longrightarrow Xylem$$

A. Root hairs, mycorrhizae, and a large surface area of cortical cells enhance water and mineral absorption

Soil ——▶ Epidermis:

- Most absorption occurs near root tips where the epidermis is permeable to water.

- Root hairs, extensions of epidermal cells, increase the surface area available for absorption.

- Most plants form symbiotic relationships with fungi; the "infected" roots form *mycorrhizae*, a structure made from the plant roots and the hyphae of fungi. Water and minerals absorbed by the hyphae are transferred to the host plant.

Epidermis ——▶ Root cortex:

Lateral transport of minerals and water through the root is usually by a combination of apoplastic and symplastic routes.

- The apoplastic route exposes parenchymal cortex cells to soil solution.

  ⇒ Soil solution, containing soil particles, water and dissolved minerals, flows into hydrophilic walls of epidermal cells and passes freely along the apoplast into root cortex.

  ⇒ Compared to the epidermis, the apoplastic route exposes greater membrane surface area for water and mineral uptake into cytoplasm.

- The symplastic route makes selective mineral absorption possible.

  ⇒ As soil solution moves along cell walls, some water and solutes cross the plasma membrane of epidermal and cortex cells.

  ⇒ Cells cannot absorb a sufficient supply of mineral ions by diffusion alone—the soil solution is too dilute. Active transport permits root cells to accumulate essential minerals in very high concentrations.

  ⇒ For example, transport proteins of the plasma membrane and tonoplast actively transport K^+ into root cells as Na^+ is pumped out.

B. The endodermis functions as a selective sentry between the root cortex and vascular tissue

Root Cortex ——▶ Xylem:

- Only minerals using the symplastic route may move directly into the vascular tissues. They have been previously selected by a membrane.

- Minerals and water passing through apoplasts are blocked at the *endodermis* (the innermost layer of cells in the root cortex) by a *Casparian strip* (a ring of suberin around each cell in the endodermis) and must enter an endodermal cell (see Campbell, Figure 36.6).

Water and minerals enter into the stele through the cells of the endodermis.

- Casparian strips ensure that all substances entering the stele pass through at least one membrane, allowing only selected ions to pass into the stele. Also prevents stele contents from leaking back into the apoplast and out into the soil.

- Water and minerals enter the stele via symplast, but tracheids and xylem vessels are part of the apoplast.

- Endodermal and parenchymal cells selectively discharge minerals into the apoplast so they may enter the xylem. This action probably involves diffusion and active transport.

- Those minerals and water that move into the apoplast are free to enter the tracheids and xylem vessels.

III.   Transport of Xylem Sap

A.   The ascent of xylem sap depends mainly on transpiration and the physical properties of water

The shoot depends upon an efficient delivery of its water supply.

- Xylem sap flows upward at 15m per hour or faster.
- Xylem vessels are close to each leaf cell, because veins branch throughout the leaves.

Water transported up from roots must replace that lost by *transpiration*.

- Transpiration is the evaporation of water from the aerial parts of a plant.
- The upward flow of xylem sap also provides nutrients (minerals) to the shoot system.

1.   Pushing xylem sap: root pressure

When transpiration is low, active transport of ions into the xylem decreases the stele's water potential and causes water flow into the stele. This osmotic water uptake increases pressure which forces fluid up the xylem (= *root pressure*).

- Root pressure causes *guttation* (exudation of water droplets at leaf margins).
- The water droplets escape through specialized structures called hydrathodes, which relieve the pressure caused by more water entering the leaves than is lost by transpiration.

Root pressure is not the major mechanism driving the ascent of xylem sap.

- Cannot keep pace with transpiration
- Can only force water up a few meters

2.   Pulling xylem sap: the transpiration-cohesion-tension mechanism

Transpiration pulls xylem sap upward, and cohesion of water transmits the upward-pull along the entire length of xylem.

a.   Transpirational pull

Transpirational pull depends upon the creation of negative pressure.

Gaseous water in damp intercellular leaf spaces diffuses into the drier atmosphere through stomata (see Campbell, Figure 36.8).

The lost water vapor is replaced by evaporation from mesophyll cells bordering the airspaces.

The remaining water film, adhering to the hydrophilic cell walls, retreats into the cell wall pores (see Campbell, Figure 36.9).

Cohesion in this surface film of water resists an increase in the surface area of the film—a surface tension effect.

The water film forms a meniscus due to the negative pressure caused by the adhesion and cohesion.

This negative pressure pulls water from the xylem, through the mesophyll, toward the surface film on cells bordering the stomata.

Water moves through symplasts and apoplasts to a region of low water potential.

Mechanism results in water from the xylem replacing water transpired through the stomata.

b.   Cohesion and adhesion of water

The transpirational pull on the xylem sap is transmitted to the soil solution (see Campbell, Figure 36.10). Cohesion of water due to H bonds allows for the pulling of water from the top of the plant without breaking the "chain."

The adhesion of water (by H bonds) to the hydrophilic walls of xylem cells also helps pull against gravity.

The small diameter of vessels and tracheids is important to the adhesion effect.

The upward pull of sap causes tension (negative pressure) in xylem, which decreases water potential and allows passive flow of water from soil into stele.

Transpirational pull can extend down to the roots only through an unbroken chain of water molecules.

Cavitation (formation of a water vapor pocket in xylem) breaks the chain of water molecules and the pull is stopped.

- Vessels cannot function again unless refilled with water by root pressure. (This can only occur in small plants.)
- Pits between adjacent xylem vessels allow for detours around a cavitated area.
- Secondary growth also adds new xylem vessels each year.

B. Review: xylem sap ascends by solar-powered bulk flow

Bulk flow is the movement of fluid due to pressure differences at opposite ends of a conduit.

The ascent of xylem sap is ultimately powered by the sun, which causes evaporative water loss, and thus, negative pressure.

- Xylem vessels or chains of tracheids serve as the conduits in plants.
- Transpirational pull lowers the pressure at the upper (leaf) end of the conduit.
- Osmotic movement of water from cell to cell in roots and leaves are due to small gradients in water potential caused by both solute and pressure gradients.

In contrast, bulk flow through the xylem vessels depends only on pressure.

IV. The Control of Transpiration

A. Guard cells mediate the photosynthesis-transpiration compromise

Transpiration results in a tremendous water loss from the plant. This water is replaced by the upward movement of water through the xylem. Guard cells surrounding stomata balance the requirements for photosynthesis with the need to conserve water.

1. The photosynthesis-transpiration compromise

Large surface areas along a leaf's airspaces are needed for CO_2 intake for photosynthesis, but also results in greater surface area for evaporative water loss.

- Internal surface area of a leaf may be 10 to 30 times the external surface area.
- Stomata are more concentrated on the bottom of leaves, away from the sun; this reduces evaporative loss.
- The waxy leaf cuticle prevents water loss from the rest of the leaf surface.

Transpiration-to-photosynthesis ratios measure efficiency of water use. This ratio is g H_2O lost/g CO_2 assimilated into organic material.

- Ratio of 600:1 is common in C_3 plants; 300:1 in C_4 plants.
- C_4 plants can assimilate CO_2 at greater rates than C_3 plants.

Benefits of transpiration:

- It assists in mineral transfer from roots to shoots.
- Evaporative cooling reduces risk of leaf temperatures becoming too high for enzymes to function.

If transpiration exceeds delivery of water by xylem, plants wilt.

- Plants can adjust to reduce risk of wilting.
- Regulating the size of stomatal openings also reduces transpiration.

2. How stomata open and close

Guard cells = Cells that flank stomata and control stomatal diameter by changing shape

- When turgid, guard cells "buckle" due to radially-arranged microfibrils and stomata open
- When flaccid, guard cells sag and stomatal openings close

The change in turgor pressure that regulates stomatal opening results from reversible uptake and loss of K^+ by guard cells.

- Uptake of K^+ decreases guard cell water potential so H_2O is taken up, cells become turgid, and stomata open. The tonoplast plays a role as most of the K^+ and water are stored in the vacuole.
- The increase in positive charge is countered by the uptake of chloride (Cl^-), export of H^+ ions released from organic acids, and the negative charges acquired by organic acids as they lose their protons.
- Closing of the stomata results when K^+ exits the guard cells and creates an osmotic loss of water.

Evidence from studies using patch clamping techniques indicates that K^+ fluxes across the guard cell membrane are likely coupled to membrane potentials created by proton pumps.

- Stomatal opening correlates with active transport of H^+ out of the guard cell.
- The resulting membrane potential drives K^+ into the cell through specific membrane channels.

By integrating internal and external environmental cues, guard cells open and close, balancing the requirements for photosynthesis with the need to conserve water from transpirational loss.

Stomata open at dawn in response to three cues:

1. Light. Induces guard cells to take up K^+ by:
   - Activating a blue-light receptor which stimulates proton pumps in the plasma membrane
   - Driving photosynthesis in guard cell chloroplasts, making ATP available for the ATP-driven proton pumps
2. Decrease of CO_2 in leaf air spaces due to photosynthesis in the mesophyll
3. An internal clock in the guard cells. This will make them open even if plant is kept in dark (a *circadian rhythm* approximates a 24-hour cycle)

Guard cells may close stomata during the daytime if:

- There is a water deficiency resulting in flaccid guard cells.
- Mesophyll production of abscisic acid (a hormone) in response to water deficiency signals guard cells to close.
- High temperature increases CO_2 in leaf air spaces due to increased respiration, closing guard cells.

B. Xerophytes have adaptations that reduce transpiration

Xerophytes, plants adapted to arid climates, have some of the following evolutionary adaptations that reduce transpiration:

- Small, thick leaves (reduced surface area:volume, so less H_2O loss)
- A thick cuticle
- Stomata are in depressions on the underside of leaves to protect from water loss due to drying winds (see Campbell, Figure 36.12)
- Some shed leaves in the driest time of the year.
- Cacti and others store water in stems during the wet season.

Plants of family Crassulaceae use CAM (crassulacean acid metabolism) to assimilate CO_2.

- At night, mesophyll cells assimilate CO_2 into organic acids.
- During the day, the acids are broken down, releasing CO_2, which is used to synthesize sugars by the conventional C_3 pathway.
- Thus, stomata can close during the day, the plant conserves water, and there is still an ample supply of CO_2 for photosynthesis.

V.   Translocation of Phloem Sap

Translocation = The transport of the products of photosynthesis by phloem to the rest of the plant

- In angiosperms, sieve-tube members are the specialized cells of phloem that function in translocation.
   ⇒ Sieve-tube members are arranged end-to-end forming long sieve tubes.
   ⇒ Porous cross walls called sieve plates are in between the members and allow phloem to move freely along the sieve tubes.

Phloem sap contains primarily sucrose, but also minerals, amino acids, and hormones.

A.   Phloem translocates its sap from sugar sources to sugar sinks

Phloem sap movement is not unidirectional; it moves through the sieve tubes from a source (production area) to a sink (use or storage area).

Sugar source = Organ where sugar is produced by photosynthesis or by the breakdown of starch (usually leaves).

Sugar sink = Organ that consumes or stores sugar (growing parts of plants, fruits, non-green stems and trunks, and others)

Sugar flows from source to sink.

- Source and sink depend on season. A tuber is the sink when stockpiling in the summer, but is the source in the spring.
- Minerals may also be transported to sinks.
- The sink is usually supplied by the nearest source.
- Direction of flow within a phloem element can change, depending on locations of the source and sink.

B.   Phloem loading and unloading

Sugar produced at a source must be loaded into sieve-tube members before it can be translocated to a sink.

- In some plant species, the sugar may move through the symplast from mesophyll cells to the sieve members.
- In other species, the sugar uses a combination of symplastic and apoplastic routes (see Campbell, Figure 36.13a)
- Some plants have *transfer cells*. These are modified companion cells which have numerous ingrowths of their walls. These structures increase the cells' surface area and enhances solute transfer between apoplast and symplast.

- In plants such as corn, active transport accumulates sucrose in sieve-tube members to two to three the concentration in mesophyll cells.
    - ⇒ Proton pumps power this transport by creating a H⁺ gradient (see Campbell, Figure 36.13b).
    - ⇒ A membrane protein uses the potential energy stored in the gradient to drive the cotransport of sucrose by coupling sugar transport to the diffusion of H⁺ back into the cell.

Sucrose is unloaded at the sink end of sieve tubes.

- In some plants, sucrose is unloaded from the phloem by active transport.
- In other species diffusion moves the sucrose from the phloem into the cells of the sink.
- Both symplastic and apoplastic routes may be involved.

C. Pressure flow is the mechanism of translocation in angiosperms

Phloem sap flows up to 1 m per hour, too fast for just diffusion or cytoplasmic streaming.

The flow is by a bulk flow (pressure-flow) mechanism; buildup of pressure at the source and release of pressure at the sink causes source-to-sink flow (see Campbell, Figure 36.14).

- At the source end, phloem loading causes high solute concentrations.
    - ⇒ Water potential decreases, so water flows into tubes creating hydrostatic pressure.
    - ⇒ Hydrostatic pressure is greatest at the source end of the tube.
- At the sink end, the water potential is lower outside the tube due to the unloading of sugar; osmotic loss of water releases hydrostatic pressure.
- Xylem vessels recycle water from the sink to the source.

Aphids have been used to study flow in phloem (see Campbell, Figure 36.15).

- The aphid stylet punctures phloem and the aphid is "force-fed" by the pressure.
- The aphid is severed from its stylet and flow is measured through the stylet.
- The flow exerts greater pressure and has a higher sugar concentration the closer to the source.

REFERENCES

Campbell, N., et al. *Biology*. 5th ed. Menlo Park, California: Benjamin/Cummings, 1999.

Raven, P.H., R.F. Evert and S.E. Eichhorn. *Biology of Plants*. 6th ed. New York: Worth Publishers, Inc., 1998.

OUTLINE

I.   Nutritional Requirements of Plants
    A. The chemical composition of plants provides clues to nutritional requirements
    B. Plants require nine macronutrients and at least eight micronutrients
    C. The symptoms of mineral deficiency depend on the function and mobility of the element
II. Soil
    A. Soil characteristics are key environmental factors of terrestrial ecosystems
    B. Soil conservation is one step toward sustainable agriculture
III. The Special Case of Nitrogen As a Plant Nutrient
    A. The metabolism of soil bacteria makes nitrogen available to plants
    B. Improving the protein yield of crops is a major goal of agricultural research
IV. Nutritional Adaptations: Symbiosis of Plants and Soil Microbes
    A. Symbiotic nitrogen fixation results from intricate interactions between roots and bacteria
    B. Mycorrhizae are symbiotic associations of roots and fungi that enhance plant nutrition
    C. Mycorrhizae and root nodules may have an evolutionary relationship
V.   Nutritional adaptations: Parasitism and Predation By Plants
    A. Parasitic plants extract nutrients from other plants
    B. Carnivorous plants supplement their mineral nutrition by preying on animals

OBJECTIVES

After reading this chapter and attending lecture, the student should be able to:

1. Describe the chemical composition of plants including:
    a. Percent of wet weight as water
    b. Percent of dry weight as organic substances
    c. Percent of dry weight as inorganic minerals
2. Explain how hydroponic culture is used to determine which minerals are essential nutrients.
3. Distinguish between macronutrient and micronutrient.
4. Recall the nine macronutrients required by plants and describe their importance in normal plant structure and metabolism.
5. List seven micronutrients required by plants and explain why plants need only minute quantities of these elements.

6. Explain how a nutrient's role and mobility determine the symptoms of a mineral deficiency.

7. Explain how soil is formed.

8. Explain what determines the texture of topsoil and list the type of soil particles from coarsest to smallest.

9. Describe the composition of loams and explain why they are the most fertile soils.

10. Explain how humus contributes to the texture and composition of soil.

11. Explain why plants cannot extract all of the water in soil.

12. Explain how the presence of clay in soil helps prevent the leaching of mineral cations.

13. Define cation exchange, explain why it is necessary for plant nutrition, and describe how plants can stimulate the process.

14. Explain why soil management is necessary in agricultural systems but not in natural ecosystems such as forests and grasslands.

15. List the three mineral elements that are most commonly deficient in farm soils.

16. Describe the environmental consequence of overusing commercial fertilizers.

17. Explain how soil pH determines the effectiveness of fertilizers and a plant's ability to absorb specific mineral nutrients.

18. Describe problems resulting from farm irrigation in arid regions and list several current approaches to solving these problems.

19. Describe precautions that can reduce wind and water erosion.

20. Define nitrogen fixation and write the overall equation representing the conversion of gaseous nitrogen to ammonia.

21. Distinguish between nitrogen-fixing bacteria and nitrifying bacteria.

22. Recall the forms of nitrogen that plants can absorb and describe how they are used by plants.

23. Beginning with free-living rhizobial bacteria, describe the development of a root nodule.

24. Explain why the symbiosis between a legume and its nitrogen-fixing bacteria is considered to be mutualistic.

25. Recall two functions of leghemoglobin and explain why its synthesis is evidence for coevolution.

26. Describe the basis for crop rotation.

27. Describe agricultural research methods used to improve the quality and quantity of proteins in plant crops.

28. Discuss the relationships between root nodule formation and mycorrhizae development

29. Describe modifications for nutrition that have evolved among plants including parasitic plants, carnivorous plants, and mycorrhizae.

KEY TERMS

mineral nutrients	topsoil	cation exchange	nodules
essential nutrient	horizons	nitrogen-fixing bacteria	bacteroids
macronutrients	loams	nitrogen fixation	mycorrhizae
micronutrients	humus	nitrogenase	ectomycorrhizae

LECTURE NOTES

I. Nutritional Requirements of Plants

Plants and other photosynthetic autotrophs play a critical role in the energy flow and chemical cycling of ecosystems by transforming:

- Light energy into chemical bond energy
- Inorganic compounds into organic compounds

To accomplish these tasks, plants require:

- Sunlight as the energy source for photosynthesis
- Inorganic, raw materials, such as:
  - Carbon dioxide
  - Water
  - Variety of inorganic, mineral ions in the soil

To acquire the resources of light and inorganic nutrients, plants have highly ramified root and shoot systems with large surface areas for exchange with their environment—soil and air.

A. The chemical composition of plants provides clues to nutritional requirements

Early ideas about plant nutrition:

- In the fourth century B.C., Aristotle thought that soil was the substance for plant growth and that leaves only provided shade for fruit.
- In the 1600s, Jean-Baptiste Van Helmont wanted to discover if plants grew by absorbing soil, so he:
  - Planted a seed in a measured amount of soil
  - Weighed the grown plant
  - Measured how much soil was left.

The soil did not decrease proportionately. Van Helmont concluded plants grew mainly from the water they absorbed

- In the 1700s, Stephen Hales postulated plants were nourished mostly by air.

Although certain minerals taken up from the soil are essential to the growth of plants (e.g., nitrogen taken up in the form of nitrate ions), these mineral nutrients make up only a tiny fraction of a plant's mass

Plants grow mainly by accumulating water in their cells' central vacuoles.

- Water is a nutrient since it supplies most of the hydrogen and some oxygen incorporated into organic compounds by photosynthesis.
- 80–85% of an herbaceous plant is water.
- More than 90% of the water absorbed is lost by transpiration.
- Retained water functions as a solvent, allows cell elongation, and keeps cells turgid.

By weight, the bulk of a plant's organic material is from assimilated CO_2 (taken from air) (see Campbell, Figure 37.1). The composition of the dry weight of plants is:

- 95% organic substances, mostly carbohydrates
- 5% minerals, to some extent determined by soil composition

B. Plants require nine macronutrients and at least eight micronutrients

An *essential nutrient* is one that is required for a plant to grow from a seed and complete its life cycle (see Campbell, Table 37.1) .

- Determined by hydroponic culture, in which roots are bathed in an aerated aqueous solution of known mineral content. If a mineral is omitted and plant growth is abnormal, it is essential (see Campbell, Figure 37.2).

*Macronutrient*s = Elements required by plants in large amounts

Micronutrients = Elements required by plants in small amounts

- These function primarily as cofactors of enzymatic reactions.
- Optimal concentrations vary for different plant species.

C. The symptoms of a mineral deficiency depend on the function and mobility of the element

Symptoms of mineral deficiencies depend on:

1. The role of the nutrient in the plant
2. Its mobility within the plant
   - Deficiencies of nutrients mobile in the plant appear in older organs first since some are preferentially shunted to growing parts.
   - Deficiencies of immobile nutrients affect young parts of plant first because older tissues may have adequate reserves.

Deficiencies of N, K, and P are the most common.

- Shortages of micronutrients are less common and often localized.
- Overdoses of some micronutrients can be toxic.

II.   Soil

A. Soil characteristics are key environmental factors in terrestrial ecosystems

Plants growing in an area are adapted to the texture and chemical composition of the soil.

1. Texture and composition of soils

Soil is produced by the weathering of solid rock. Living organisms may accelerate the process once they become established.

Horizons = Distinct soil layers

Topsoil = Mixture of decomposed rock of varying texture, living organisms, and humus.

The texture of a topsoil depends on the particle size.

- Coarse sand has diameter of 0.2 – 2 mm.
- Sand is 20 – 200 mm.
- Silt is 2 – 20 mm.
- Clay is less than 2 mm.

The most fertile soils are *loams*, a mixture of sand, silt and clay.

- Fine particles retain water and minerals.
- Coarse particles provide air spaces with oxygen for cellular respiration.

Soil contains many bacteria, fungi, algae, protists, insects, earthworms, nematodes, and plant roots. These affect the soil composition. For example, earthworms aerate soil; bacteria alter soil mineral composition.

Humus = Decomposing organic material

- Prevents clay from packing together
- Builds a crumbly soil that retains water but is still porous for good root aeration
- Acts as a reservoir of mineral nutrients

Soil composition determines which plants may grow in it, and plant growth, in turn, affects soil characteristics.

2. **The availability of soil water and minerals**

Some water is bound so tightly to hydrophilic soil that it cannot be extracted by plants.

Water bound less tightly is generally available to the plant as a soil solution containing minerals. This solution is absorbed into the root hairs and passes via the apoplast to the endodermis (see Campbell, Figure 37.6a)

Positively charged minerals (K^+, Ca^+, Mg^+) adhere by electrical attraction to negatively charged clay particles.

- Clay provides much surface area for binding
- Prevents leaching of mineral nutrients

Cation exchange = H ions in soil displace positively charged mineral ions from clay, making them available to plants (see Campbell, Figure 37.6b).

- Stimulated by roots which release acids to add H^+ to the soil solution.

Negatively charged minerals (NO_3^-, $H_2PO_4^-$, SO_4^-) are not tightly bound to soil particles.

- Tend to leach away more quickly.

B. **Soil conservation is one step toward sustainable agriculture**

Good soil management is necessary to maintain soil fertility, which may have taken centuries to develop through decomposition and accumulation of organic matter.

Agriculture is unnatural and depletes the mineral content of the soil, making soil less fertile. Crops also use more water than natural vegetation.

Three important aspects of soil management are:

1. Fertilizers
2. Irrigation
3. Erosion prevention

1. **Fertilizers**

Fertilizers may be mined, chemically produced, or organic.

- Usually enriched in N, P, K
- Marked with three numbers corresponding to percentage of nitrogen (as ammonium or nitrate), percentage of phosphorus (as phosphoric acid), and percentage of potassium (as potash).
- Organic fertilizers are manure, fishmeal, and compost. Release minerals more gradually than chemical fertilizers, thus, soil retains minerals longer. Excess minerals from chemical fertilizers may be leached from soil and may pollute streams and lakes.

To properly use fertilizers, one must consider the soil pH.

- Acidity affects cation exchange and the chemical form of the minerals.
- A change in soil pH may make one essential element more available while causing another to adhere so tightly to soil particles that it is unavailable.

2. **Irrigation**

Availability of water limits plant growth in many environments.

Problems of irrigation arid land:

- Huge drain of water resources.
- Can gradually make soil salty and infertile.

Solutions to these problems:

- Use of drip irrigation. Perforated pipes slowly drip water close to plant roots, which reduces water evaporation and drainage.
- Development of plant varieties that require less water or can tolerate more salinity.

3. Erosion

Wind and water erode away much of the topsoil each year. Measures to prevent these losses include:

- Rows of trees to divide fields act as windbreaks.
- Terracing hillsides helps prevent water erosion.
- Planting alfalfa and wheat provide good ground cover and protection.

Proper management makes soil a renewable resource that will remain fertile for many generations.

III.   The Special Case of Nitrogen as a Plant Nutrient

A.   The metabolism of soil bacteria makes nitrogen available to plants

Plants require nitrogen to produce proteins, nucleic acids and other organic molecules.

- Plants can not use nitrogen in gaseous form (N_2).
- To be assimilated by plants, nitrogen must be in the form of ammonium (NH_4^+) or nitrate (NO_3^-).

There is a complex cycling of nitrogen in ecosystems:

- Over the short term, the main source of nitrogenous minerals is the decomposition of humus by microbes.
  - For example, ammonifying bacteria (see Campbell, Figure 37.8)
  - Nitrogen in organic compounds is repackaged into inorganic compounds that can be absorbed as minerals by roots.
- Nitrogen is lost from this local cycle when soil denitrifying bacteria convert NO_3^- to N_2, which diffuses form the soil to the atmosphere.
- *Nitrogen–fixing bacteria* restock nitrogenous minerals in the soil by converting N_2 to NH_3 (ammonia), a metabolic process called nitrogen fixation.
  - *Nitrogen fixation* = The process of converting atmospheric nitrogen (gaseous state) to nitrogenous compounds that can be directly used by plants (nitrate or ammonia) (see Campbell, Chapter 49, for more details)•
    The process is catalyzed by the enzyme nitrogenase:

$$N_2 + 8e^- + 8H^+ + 16ATP \xrightarrow{\text{nitrogenase}} 2NH_3 + H_2 + 16ADP + 16\,P_i$$

  - Some soil bacteria possess nitrogenase.
  - Very energy consuming process, costing the bacteria at least 8 ATPs for each ammonia molecule synthesized.
  - In the soil, ammonia is converted to the ammonium ion which plants can absorb:

$$NH_3 + H^+ \longrightarrow NH_4^+$$

  - Plants acquire most of their nitrogen in the form of nitrate (NO_3^-), which is produced in soil by nitrifying bacteria that oxidize ammonium (see Campbell, Figure 37.8). Other species of nitrogen-fixing bacteria live in plant roots in symbiotic relationships (see below).
  - Nitrogen absorbed by the plant is incorporated into organic compounds.

- Most plant species export nitrogen from the roots to the shoots (through xylem) in the form of nitrate or of organic compounds (e.g., amino acids) that were synthesized in the roots.

B. Improving the protein yield of crops is a major goal of agricultural research

A majority of the world's population depends mainly on plants for protein. Some food plants have a low protein content while others may be deficient in certain amino acids.

Ways in which to increase the protein content of plants include:

1. Plant breeding to create new varieties enriched in protein
   - Unfortunately these "super" varieties require large quantities of nitrogen in the form of commercial fertilizer, which is too expensive for many countries to afford.

2. Improving the productivity of symbiotic nitrogen fixation
   - Mutant strains of *Rhizobium* have been isolated that continue to produce nitrogenase even after fixed nitrogen accumulates, thus releasing the excess into the soil.
   - *Rhizobium* varieties may be selected that fix N_2 at a lower cost in photosynthetic energy, yielding a higher total food content in the plant.

IV. Nutritional Adaptations: Symbiosis of Plants and Soil Microbes

A. Symbiotic nitrogen fixation results from intricate interactions between roots and bacteria

Legumes (e.g., peas, beans, soybeans, peanuts, alfalfa, clover) have a built-in source of fixed nitrogen because they possess root nodules.

Nodules = Root swellings composed of plant cells that contain nitrogen-fixing bacteria of the genus *Rhizobium*. Each species of legume is associated with a particular species of *Rhizobium* (see Campbell, Figure 37.10).

Nodules form as follows:

- Roots secrete chemicals that attract nearby bacteria.
- Attracted bacteria emit chemicals that stimulate root hairs to elongate and curl to prepare for bacterial infection.
- Bacteria enter the root through an "infection thread" that carries them to the root cortex.
- Bacteria become enclosed in vesicles and assume a form called *bacteroids* (see Campbell, Figure 37.9).
  - Bacteroids produce a chemical that induces the host's cells to divide and form a nodule.
- Nodules continue to grow and a connection with the xylem and phloem develops.

This association is mutualistic; the bacteria supplies fixed nitrogen, and the plant supplies carbohydrates and other organic compounds.

Leghemoglobin = An iron-containing protein that binds oxygen

- The plant and the bacteria each make a part of the molecule.
- Releases oxygen for the intense respiration needed to produce ATP for nitrogen fixation.
- Keeps the free oxygen concentration low in root nodules so that the oxygen cannot inhibit the function of nitrogenase.

Most of the ammonium produced is used by the nodules to make amino acids for export to the shoots and leaves.

1. **Symbiotic nitrogen fixation and agriculture**

   The basis for crop rotation is that, under favorable conditions, root nodules fix more nitrogen than the legume uses. The excess is secreted as ammonium into the soil.

   - One year a nonlegume crop is planted, and the next year a legume is planted to restore the fixed nitrogen content of the soil.
   - Legumes may be plowed under to further increase the fixed nitrogen content of the soil.

   Some nonlegumes host nitrogen-fixing symbionts.

   - Alders and tropical grasses may host nitrogen-fixing actinomycetes.
   - Rice farmers culture a fern (*Azolla*), containing symbiotic nitrogen-fixing cyanobacteria, with the rice.

2. **The molecular biology of root nodule formation in legumes**

   Chemical signals between plant roots and bacteria direct their association and the formation of nodules (see Campbell, Figure 37.11).

   The specificity of the interaction between a plant and a particular bacterial species (e.g., *Rhizobium*) in the soil results from the unique chemical structure of the signal molecules.

   - The initial signal molecule is produced by the plant.
   - In response to the plant signal, bacteria produce an "answering" signal.
   - The signals alter gene expression in cell of the recipient that results in the production of enzymes and other signal molecules (e.g., bacterial Nod factors; named for their action on nodulation and similar in structure to chitins).
     - Recent evidence reveals that plants produce chitin-like growth factors and that the bacterial Nod factors "cross-talk" with signal systems in the plant.

   Through genetic engineering, scientists may:

   - Create varieties of *Rhizobium* that can infect nonlegumes
   - Transfer the genes required for nitrogen-fixation directly into plant genomes, using bacterial plasmids as vectors (see Campbell, Chapter 20).

B. **Mycorrhizae are symbiotic associations of roots and fungi that enhance plant nutrition**

   Mycorrhizae = Symbiotic associations (mutualistic) between plant roots and fungi; the fungus either forms a sheath around the root or penetrates root tissue

   - Help the plant absorb water
   - Absorb minerals, and may secrete acid that increases mineral solubility and converts minerals to forms easily used by the plant
   - May help protect the plant against certain soil pathogens
   - The plant nourishes the fungus with photosynthetic products.

   Almost all plants are capable of forming mycorrhizae if exposed to the proper species of fungi. Plants grow more vigorously when mycorrhizae are present.

   Mycorrhizae may have permitted early plants to colonize land.

   - Fossils indicate the earliest land plants possessed mycorrhizae.
   - This mutualistic association may have allowed the early plants to obtain enough nutrients to survive colonization.

1. **Two main types of mycorrhizae**

   In *ectomycorrhizae*, the mycelium forms a sheath over the root, but does not penetrate it (see Campbell, Figure 37.12a).

   - Hyphae increase absorptive surface
   - Common in woody plants (e.g., pine, oak, walnut)

   Endomycorrhizae do not form a sheath surrounding the root and hyphae extend into root cell walls (but do not penetrate plasma membrane) (see Campbell, Figure 37.12b).

   - More common that ectomycorrhizae; found in over 90% of plant species, including crop plants (e.g., wheat, corn)

2. **Agricultural importance of mycorrhizae**

   Mycorrhizae can only form if the plant is exposed to the appropriate species of fungus.

   - In agriculture, seeds are often collected in one environment to be planted in foreign soil devoid of the correct fungus and the resulting plant may display symptoms of malnutrition.
   - Inoculating seeds with the spores of mycorrhizal fungi promotes the formation of mycorrhizae and helps to assure good plant health.

C. **Mycorrhizae and root nodules may have an evolutionary relationship**

   Recent research indicates that the molecular biology of root nodule formation is closely related to the mechanisms involved with mycorrhizae formation.

   - The same plant genes that are activated in the early stages of nodule formation are the same genes activated during the early development of endomycorrhizae.
   - The chemical cues produced by the microbes appear structurally related and activate plant gene expression through the same signal transduction pathway.

   Mycorrhizae probably evolved over 400 million years ago in the earliest vascular plants, whereas root nodules most likely emerged some 65 to 130 million years ago, during the evolution of angiosperms

   The recent observations concerning the molecular mechanisms associated with the symbiotic relationships of roots suggests that root nodule development is partially adapted from a signaling pathway already in place in mycorrhizae (e.g., an exaptation).

V. **Nutrition Adaptations: Parasitism and Predation By Plants**

   Some plant adaptations enhance nutrition through interactions with other organisms.

A. **Parasitic plants extract nutrients from other plants**

   Some parasitic plants:

   - Are photosynthetic, and only supplement nutrition by using haustoria (*not* homologous to those of parasitic fungi) to obtain xylem sap from their host plant (e.g., mistletoe)
   - Have lost photosynthesis entirely, drawing all nutrients from the host plant by tapping into the phloem (e.g., dodder) (see Campbell, Figure 37.13)

   Epiphyte*s* are plants that:

   - Grow on the surface of other plants, anchored by roots, but are *not parasitic*
   - Nourish themselves from the water and minerals absorbed from rain
   - Examples include Spanish moss and staghorn ferns

B. Carnivorous plants supplement their mineral nutrition by digesting animals

Carnivorous plants:

- Live in habitats with poor (usually nitrogen deficient) soil conditions
- Are photosynthetic, but obtain some nitrogen and minerals by killing and digesting insects

Most insect traps evolved by modification of leaves and are usually equipped with glands that secrete digestive juices.

REFERENCES

Campbell, N., et al. *Biology*. 5th ed. Menlo Park, California: Benjamin/Cummings, 1999.

Raven, P.H., R.F. Evert and S.E. Eichhorn. *Biology of Plants*. 6th ed. New York: Worth Publishers, Inc., 1998.

Reganold, J.P., R.I. Papendick and J.F. Parr. "Sustainable Agriculture." *Scientific American*, June 1990.

CHAPTER 38
PLANT REPRODUCTION
AND DEVELOPMENT

OUTLINE

I. Sexual Reproduction
    A. Sporophyte and gametophyte generations alternate in the life cycles of plants: *a review*
    B. Male and female gametophytes develop within anthers and ovaries, respectively
    C. Pollination brings female and male gametes together
    D. Researchers are unraveling the molecular mechanisms of self-incompatibility
    E. Double fertilization gives rise to the zygote and endosperm
    F. The ovule develops into a seed containing an embryo and a supply of nutrients
    G. The ovary develops into a fruit adapted for seed dispersal
    H. Evolutionary adaptations of seed germination contribute to seedling survival

II. Asexual Reproduction
    A. Many plants can clone themselves by asexual reproduction
    B. Vegetative propagation of plants is common in agriculture
    C. Sexual and asexual reproduction are complementary in the life histories of many plants: *a review*

III. Cellular Mechanisms of Plant Development
    A. Growth, morphogenesis, and differentiation produce the plant body: *an overview*
    B. The cytoskeleton guides cell division and expansion
    C. Cell differentiation depends on gene regulation
    D. Pattern formation determines the location and tissue organization of plant organs

OBJECTIVES

After reading this chapter and attending lecture, the student should be able to:
1. Outline the angiosperm life cycle.
2. List the four floral parts in their order from outside to inside of the flower.
3. From a diagram of an idealized flower, correctly label the following structures and describe their function:
    a. Sepals   c. Stamen: filament and anther
    b. Petals   d. Carpel: style, ovary, ovule and stigma
4. Distinguish between complete and incomplete flowers.
5. Distinguish between a perfect and imperfect flower.
6. Distinguish between monoecious and dioecious.

7. Explain by which generation, structure, and process spores are produced.
8. Explain by which generation, structures, and process gametes are produced.
9. Explain why it is technically incorrect to refer to stamens and carpels as male and female sex organs.
10. Describe the formation of a pollen grain in angiosperms.
11. With reference to the developing pollen grain, distinguish among generative nucleus, tube nucleus, and sperm nucleus.
12. Describe the development of an embryo sac, and explain what happens to each of its cells.
13. Distinguish between pollination and fertilization.
14. Describe how pollen can be transferred between flowers.
15. Describe mechanisms that prevent self-pollination, and explain how this contributes to genetic variation.
16. Outline the process of double fertilization, and describe the function of endosperm.
17. Describe the development of a plant embryo from the first mitotic division to an embryonic plant with rudimentary organs.
18. From a diagram, identify the following structures of a seed and recall a function for each:

a.	Seed coat	d.	Radicle	g.	Endosperm
b.	Embryo	e.	Epicotyl	h.	Cotyledons
c.	Hypocotyl	f.	Plumule	i.	Shoot apex

19. Explain how a monocot and dicot seed differ.
20. Describe several functions of fruit and explain how fruits form.
21. Distinguish among simple, aggregate, and multiple fruits and give examples of each.
22. Explain how seed dormancy can be advantageous to a plant and describe some conditions for breaking dormancy.
23. Using a cereal as an example, explain how a seed mobilizes its food reserves and describe the function of aleurone, α-amylase, and gibberellic acid.
24. Describe variations in the process of germination including the fate of the radicle, shoot tip, hypocotyl, epicotyl, and cotyledons.
25. Distinguish between sexual reproduction and vegetative reproduction.
26. Describe natural mechanisms of vegetative reproduction in plants including fragmentation and apomixes.
27. Describe various methods horticulturists use to vegetatively propagate plants from cuttings.
28. Explain how the technique of plant tissue culture can be used to clone and genetically engineer plants.
29. Describe the process of protoplast fusion and its potential agricultural impact.
30. Define monoculture and list its benefits and risks.
31. Compare sexual and asexual reproduction in plants and explain their adaptive roles in plant populations.

KEY TERMS

alternation of generations	monoecious	scutellum	stock
sporophyte	dioecious	coleorhiza	scion
gametophyte	microspore	coleoptile	protoplast fusion
	megaspore	fruit	monoculture

sepal	embryo sac	pericarp	development
petal	pollination	simple fruit	growth
stamen	self-incompatible	aggregate fruit	morphogenesis
ovules	endosperm	multiple fruit	cellular differentiation
complete flower	double fertilization	imbibition	preprophase band
incomplete flower	seed coat	vegetative reproduction	pattern formation
perfect flower	hypocotyl	fragmentation	positional information
imperfect flower	radicle	apomixis	
organ-identity genes	epicotyl	callus	

LECTURE NOTES

Modifications in reproduction were key adaptations enabling plants to spread into a variety of terrestrial habitats.

- Water has been replaced by wind and animals as a means for spreading gametes.
- Embryos are protected in seeds.
- Vegetative reproduction is an asexual mechanism for propagation in many environments.

I. Sexual Reproduction

A. Sporophyte and gametophyte generations alternate in the life cycles of plants: *a review*

The angiosperm (flowering plant) life cycle includes *alternation of generations* during which multicellular haploid gametophyte generations alternate with diploid sporophyte generations (see Campbell, Figure 38.1).

- The *sporophyte* is the recognizable "plant" most familiar to us. It produces haploid spores by meiosis in sporangia.
- Spores will undergo mitotic division and develop into a multicellular male or female *gametophyte*.
- Gametophytes produce gametes (sperm and egg) by mitosis. The gametes fuse to form a zygote that develops into a multicellular sporophyte.
- The sporophyte is dominant in the angiosperm life cycle with the gametophyte stages being reduced and totally dependent on the sporophyte.

B. Male and female gametophytes develop within anthers and ovaries, respectively

Flowers are the reproductive structure of angiosperm sporophytes.

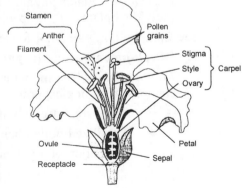

- Develop from compressed shoots with four whorls of modified leaves separated by very short internodes.
- The four sets of modified leaves are the: *sepals*, *petals*, *stamens*, and *carpels* (see also Campbell, Figure 38.2).

- Stamens and carpels contain the sporangia and are the reproductive parts of the flower.
- Female gametophytes develop in carpel sporangia as *embryo sacs*, which contain the eggs. This occurs inside the *ovules,* which are at the base of the carpel and surrounded by ovaries.
- Male gametophytes develop in the stamen sporangia as pollen grains. These form at the stamen tips within chambers of the anthers.

Pollination occurs when wind- or animal-born pollen released from anthers lands on the stigma at the tip of a carpel.

- A pollen tube grows from the pollen grain, down the carpel, into the embryo sac.
- Sperm are discharged resulting in fertilization of the eggs.
- The zygote will develop into an embryo; as the embryo grows, the ovule surrounding it develops into a seed.
- While seed formation is taking place, the entire ovary is developing into a *fruit*, which will contain one or more seeds.

Seeds are dispersed from the source plant when fruits are moved about by the wind or animals.

- Seeds deposited in soil of the proper conditions (moisture, nutrients) will germinate.
- The embryo starts growing and develops into a new sporophyte.
- After flowers are produced by the sporophyte, a new generation of gametophytes develop and the life cycle continues.

Several variations on the basic flower structure have evolved during the angiosperm evolutionary history.

Complete flower = A flower with sepals, petals, stamens and carpels

Incomplete flower = A flower that is missing one or more of the parts listed for a complete flower (e.g., most grasses do not have petals on their flowers)

Perfect flower = A flower having both stamens and carpels (may be incomplete by lacking either sepals or petals)

Imperfect flower = A flower that is either *staminate* (having stamens but no carpels) or carpellate (having carpels but no stamens); a unisex flower

Monoecious = Plants having both staminate flowers and carpellate flowers on the same individual plant

Dioecious = Plants having staminate flowers and carpellate flowers on separate individual plants of the species

See Campbell, Figure 34.3, for examples of floral diversity.

1. **Development of male gametophyte (pollen)**

Pollen grain = The immature male gametophyte that develops within the anthers of stamens in an angiosperm

- They are extremely durable; their tough coats are resistant to biodegradation.
- Fossilized pollen has provided many important evolutionary clues.
- Formation of a pollen grain is as follows (see also Campbell, Figure 38.4a):

Within the sporangial chamber of an anther,
diploid microsporocytes undergo meiosis to
form four haploid microspores.

↓

The haploid microspore nucleus undergoes
mitotic division to give rise to a generative
cell and a tube cell.

↓

The wall of the microspore then thickens
and becomes sculptured into a species-
specific pattern.

↓

These two cells and the thickened wall
are the pollen grain, an immature
male gametophyte.

2. Development of female gametophyte (embryo sac)

Ovule = Structure which forms within the chambers of the plant ovary and contains the female sporangium

The female gametophyte is the *embryo sac*, and it generally develops as follows (see also Campbell, Figure 38.4b):

A megasporocyte in the sporangium of each ovule
grows and goes through meiosis to form four
haploid *megaspores* (only one usually survives).
(Details of next steps vary extensively, depending on species.)

↓

The remaining megaspore grows and its
nucleus undergoes three mitotic divisions,
forming one large cell with eight haploid nuclei.

↓

Membranes partition this into
a multicellular *embryo sac*.

Within the embryo sac:
- The egg cell is located at one end and is flanked by two other cells (synergids).
- At the opposite end are three antipodal cells.
- The other two nuclei (polar nuclei) share the cytoplasm of the large central cell.
- At the end containing the egg is the micropyle (an opening through the integuments surrounding the embryo sac).

C. Pollination brings female and male gametophytes together

1. Pollination

Pollination = The placement of pollen onto the stigma of a carpel

- Some plants use wind to disperse pollen.
- Other plants interact with animals that transfer pollen directly between flowers.
- Some plants self-pollinate, but most cross-pollinate.

Most monoecious angiosperms have mechanisms to prevent self-pollination. These mechanisms thus contribute to genetic variation in the species by ensuring sperm and eggs are from different plants.

- The stamens and carpels mature at different times in some species.
- Structural arrangement of the flower in many species pollinated by animals reduces the chance that pollinators will transfer pollen from anthers to the stigma of the same flower.
- Other species are *self-incompatible*. If a pollen grain lands on the stigma of the same flower, a biochemical block prevents the pollen grain from developing and fertilizing the egg.

D. Researchers are unraveling the molecular mechanisms of self-incompatibility

Self-incompatibility = The rejection of pollen from the same, or closely related, plant by the stigma

The recognition of "self" pollen is based on S-genes (named for self-incompatibility).

- Many alleles for the S-locus are found in a plant population's gene pool (see Campbell, Figure 38.5).
- A pollen grain that lands on a stigma with matching alleles at the S-locus is self-incompatible.
  - The pollen grain will either not initiate or complete formation of the pollen tube.
  - This prevents self-fertilization and fertilization between plants with a common S-locus (usually closely related plants).

Although all self-compatibility genes are described as being S-loci, such genes appear to have evolved independently in numerous plant families. As a consequence, the mechanism underlying inhibition of pollen tube formation varies.

- In some cases, the block occurs in the pollen grain (e.g., gametophyte self-incompatibility). RNAases from the carpel enter the pollen grain (only if the pollen is identified as a "self") and destroy RNA (e.g., legumes).
- In other cases, the block is caused by cells in the stigma of the carpel (e.g., sporophyte incompatibility). For example, in members of the mustard family, self recognition activates a signal transduction system in epidermal cells of the carpel that inhibits germination of the pollen grain (see Campbell, Figure 38.6).

Studies on self-incompatibility may lead to benefits for agricultural production.

- Many important agricultural plants are self-compatible.
  ⇒ Different varieties of these crop plants are hybridized to combine the best traits of the varieties and prevent loss of vigor from excessive inbreeding.
  ⇒ To maximize the numbers of hybrids, plant breeders prevent self-fertilization by laboriously removing the anthers from parent plants that provide the seeds.
- If the molecular mechanism responsible for self-incompatibility can be imposed on normally self-compatible crop species, production of hybrids would be simplified.

E. Double fertilization gives rise to the zygote and endosperm

When a compatible pollen grain (different S-locus alleles) lands on a stigma of an angiosperm, double fertilization occurs.

- *Double fertilization* = The union of two sperm cells with two cells of the embryo sac
- After adhering to a stigma, the pollen grain germinates and extends a pollen tube between the cells of the style toward the ovary (see Campbell, Figure 38.7).
- The generative cell divides (mitosis) to form two sperm. (A pollen grain with a tube enclosing two sperm = mature male gametophyte.)
- Directed by a chemical attractant (usually calcium), the tip of the pollen tube enters through the micropyle and discharges its two sperm nuclei into the embryo sac.
- One sperm unites with the egg to form the zygote.
- The other sperm combines with the two polar nuclei to form a $3n$ nucleus in the large central cell of the embryo sac.
  - ⇒ This central cell will give rise to the *endosperm,* which is a food storing tissue.

After double fertilization is completed, each ovule will develop into a seed and the ovary will develop into a fruit surrounding the seed(s).

F. The ovule develops into a seed containing an embryo and a supply of nutrients

1. Endosperm development

Endosperm development begins before embryo development.

The triploid nucleus divides to form a milky, multinucleate "supercell" after double fertilization.

This endosperm undergoes cytokinesis to form membranes and cell walls between the nuclei, thus becoming multicellular.

- ⇒ Endosperm is rich in nutrients, which it provides to the developing embryo.
- ⇒ In most monocots, the endosperm stocks nutrients that can be used by the seedling after germination.
- ⇒ In many dicots, food reserves of the endosperm are exported to the cotyledons, thus mature seeds have no endosperm.

2. Embryo development (embryogenesis)

The zygote's first mitotic division is transverse, creating a larger basal cell and a smaller terminal cell (see Campbell, Figure 38.8).

- The basal cell divides transversely to form the *suspensor*, which anchors the embryo and transfers nutrients to it from the parent plant.
- The terminal cell divides several time to form a spherical proembryo attached to the suspensor.

Cotyledons appear as bumps on the proembryo and the embryo elongates.

- The apical meristem of the embryonic shoot is located between the cotyledons.

The suspensor (the opposite end of the axis) attaches at the apex of the embryonic root with its meristem.

- The basal cell gives rise to part of the root meristem in some species.

After germination, the apical meristems at the root and shoot tips will sustain primary growth.

- The embryo also contains protoderm, ground meristem, and procambium.

Two features of plant form are established during embryogenesis.

1. The root-shoot axis with meristems at opposite ends
2. A radial pattern of protoderm, ground meristem, and procambium ready to produce the dermal, ground, and vascular tissue systems

As the embryo develops, proteins, oil, and starch accumulate and are stored until the seed germinates.

3. **Structure of the mature seed**

In mature seeds, the embryo is quiescent until germination.

- The seed dehydrates until its water content is only 5% to 15% by weight.
- The embryo is surrounded by endosperm, enlarged cotyledons, or both.
- The seed coat is formed from the integuments of the ovule.

The arrangement within the seed of a dicot is shown in Campbell, Figure 38.9a.

- Below the cotyledon attachment point, the embryonic axis is called the *hypocotyl*, which terminates in the *radicle*, or embryonic root.
- Above the cotyledons, the embryonic axis is called the *epicotyl*, which terminates in the *plumule* (shoot tip with a pair of tiny leaves).
- Fleshy cotyledons are present in some dicots before germination due to their absorption of nutrients from the endosperm.
- In other dicots, thin cotyledons are found, and nutrient absorption and transfer occur only after germination (see Campbell, Figure 38.9b).

A monocot seed has a single cotyledon (see Campbell, Figure 38.9c). Members of the grass family, including wheat and corn, have a specialized cotyledon called the *scutellum*.

- The scutellum has a large surface area and absorbs nutrients from the endosperm during germination.
- The embryo is enclosed in a sheath comprised of the *coleorhiza* (covers the root) and the *coleoptile* (covers the shoot).

G. **The ovary develops into a fruit adapted for seed dispersal**

A fruit develops from the ovary of the flower while seeds are developing from the ovules.

- A *fruit* protects the seeds and aids in their dispersal by wind or animals.
- In some angiosperms, other floral parts also contribute to formation of what we call fruit:
    ⇒ The core of an apple is the true fruit.
    ⇒ The fleshy part of the apple is mainly derived from the fusion of flower parts located at the base of the flower.

A true fruit is a ripened ovary.

- Pollination triggers hormonal changes that cause the ovary to grow (see Campbell, Figure 38.10).
- The wall of the ovary thickens to become the *pericarp*.
- Transformation of a flower into a fruit parallels seed development.
- In most plants, fruit does not develop without fertilization of ovules. (In parthenocarpic plants, fruit does develop without fertilization.)

Depending upon their origin, fruits can be classified as (see Campbell, Table 38.1):

1. *Simple fruits*. Fruit derived from a single ovary; for example, cherries (fleshy) or soybeans (dry).
2. *Aggregate fruits*. Fruit derived from a single flower with several separate carpels; for example, strawberries.
3. *Multiple fruits*. Fruit derived from an *inflorescence* or separate tightly clustered flowers; for example, pineapple.

Fruits ripen about the time seeds are becoming fully developed.

- In dry fruits, such as soybean pods, the fruit tissues age and the fruit (pod) opens and releases the seeds.
- Fleshy fruits ripen through a series of steps guided by hormonal interactions.
  ⇒ The fruit becomes softer as a result of enzymes digesting the cell wall components.
  ⇒ Colors usually change and the fruit becomes sweeter as organic acids or starch are converted to sugar.
  ⇒ These changes produce an edible fruit which entices animals to feed, thus dispersing the seeds.

H. Evolutionary adaptations contribute to seedling survival

Seed germination represents the continuation of growth and development, which was interrupted when the embryo became quiescent at seed maturation.

- Some seeds germinate as soon as they reach a suitable environment.
- Other seeds require a specific environmental cue before they will break dormancy.

1. Seed dormancy

The evolution of the seed was an important adaptation by plants to living in terrestrial habitats.

- The environmental conditions in terrestrial habitats fluctuate more often than conditions in aquatic habitats.

Seed dormancy prevents germination when conditions for seedling growth are unfavorable.

- It increases the chance that germination will occur at a time and place most advantageous to the success of the seedling.

Conditions for breaking dormancy vary depending on the type of environment the plant inhabits.

- Seeds of desert plants may not germinate unless there has been heavy rainfall (not after a light shower).
- In chaparral regions where brushfires are common, seeds may not germinate unless exposed to intense heat, after a fire has cleared away older, competing vegetation.
- Other seeds may require exposure to cold, sunlight, or passage through an animal's digestive system before germination will occur.

A dormant seed may remain viable for a few days to a few decades (most are viable for at least a year or two). This provides a pool of ungerminated seeds in the soil, which is one reason vegetation appears so rapidly after environmental disruptions.

2. From seed to seedling

The first step in seed germination in many plants is *imbibition* (absorption of water).

- Hydration causes the seed to swell and rupture the seed coat.
- Hydration also triggers metabolic changes in the embryo that cause it to resume growth.
- Storage materials of the endosperm or cotyledons are digested by enzymes and the nutrients transferred to the growing regions of the embryo.

> Example: The embryo of a cereal grain releases a hormone (a gibberellin) as a messenger to the aleurone (outer layer of endosperm) to initiate production of α-amylase and other enzymes that digest starch stored in the endosperm (see Campbell, Figure 38.11).

- The radicle (embryonic root) then emerges from the seed.

The next step in the change from a seed to a seedling is the shoot tip breaking through the soil surface.

- In many dicots, a hook forms in the hypocotyl.
  ⇒ Growth pushes the hypocotyl above ground.
- Light stimulates the hypocotyl to straighten, raising the cotyledons and epicotyl.
- The epicotyl then spreads the first leaves which become green and begin photosynthesis.

Germination may follow different methods depending on the plant species (see Campbell, Figure 38.12).

- In peas, a hook forms in the epicotyl and the shoot tip is lifted by elongation of the epicotyl and straightening of the hook. The cotyledons remain in the ground.
- In monocots, the coleoptile pushes through the soil and the shoot tip grows up through the tunnel of the tubular coleoptile.

Only a small fraction of the seedlings will survive to the adult plant stage.

- Large numbers of seeds and fruits are produced to compensate for this loss.
- This utilizes a large proportion of the plant's available energy.

II.   Asexual Reproduction

A. Many plants can clone themselves by asexual reproduction

Asexual reproduction (or *vegetative reproduction*) = The production of offspring from a single parent; occurs without genetic recombination, resulting in a clone

- Meristematic tissues composed of dividing, undifferentiated cells can sustain or renew growth indefinitely.
- Parenchyma cells can also divide and differentiate into various types of specialized cells.

There are two major natural mechanisms of vegetative reproduction:

1. *Fragmentation* = Separation of a parent plant into parts that re-form whole plants (see Campbell, Figure 38.13a)
   - The most common form of vegetative reproduction
   - Some species of dicots exhibit a variation of fragmentation during which the parental root system develops adventitious shoots that become separate shoot systems.
2. *Apomixis* = The production of seeds without meiosis and fertilization
   - A diploid cell in the ovule gives rise to an embryo.
   - The ovules mature into seeds which are dispersed.
   - An example would be a dandelion.

B. Vegetative propagation of plants is common in agriculture

Most methods of vegetative propagation in agriculture are based on the ability of plants to form adventitious roots or shoots.

The objective is to improve crops, orchards, and ornamental plants.

1. Clones from cuttings

Clones may be obtained from either shoot or stem cuttings (plant fragments).

- At the cut end of the shoot, a mass of dividing, undifferentiated cells form (called a *callus*).
- If the shoot fragment includes a node, then adventitious roots can form without a callus stage.
- Cuttings may come from stems, leaves (African violets), or specialized storage stems (potatoes).

It is possible to combine the best qualities of different varieties or species by grafting a twig of one plant onto a closely related species or different variety of the same species.

- The plant providing the root system is the *stock*.
- The twig grafted onto the stock is the *scion*.
- The quality of a fruit is usually determined by the scion, although sometimes the stock can alter the characteristics of the shoot system that develops from the scion.

2. Test-tube cloning and related techniques

Test-tube cloning makes it possible to culture small explants (pieces of parental tissue) or single parenchyma cells on an artificial medium containing nutrients and hormones (see Campbell, Figure 38.14)

- The cultured cells divide to form an undifferentiated callus.
- The callus sprouts fully differentiated roots and shoots when the hormone balance of the culture media is manipulated.
- A single plant can be cloned into thousands of copies by subdividing calluses as they grow.

Tissue culture is often used to regenerate genetically engineered plants.

- Foreign genes are typically introduced into small pieces of plant tissue or into single plant cells.
- The use of test-tube culture techniques permits the regeneration of genetically altered plants from a single plant cell that received foreign DNA.
    ⇒ The protein quality of sunflower seeds has been improved in transgenic plants that received a gene for bean protein as cultured cells.

Protoplast fusion, coupled with tissue culture methods, can produce new plant varieties that can be cloned.

- Protoplasts are plant cells which have had their cell walls removed.
- Protoplasts may be fused to form hybrid protoplasts.
- Protoplast can be screened for mutations that will improve the agricultural value of the plant.
- Protoplasts regenerate cell walls and become hybrid plantlets.

3. Benefits and risks of monoculture

Monoculture = The cultivation of large areas of land with a single plant variety

Genetic variability in many crops has been purposefully reduced by plant breeders who have selected for self-pollinating varieties or used vegetative reproduction to clone exceptional plants.

- Benefits of such genetic unity are: plant growth is uniform; fruits ripen in unison; crop yields are dependable.
- A great disadvantage is that little genetic variability means little adaptability. One disease could destroy a whole plant variety.
- "Gene banks," where seeds of many plant varieties are stored, are maintained to retain diverse varieties of crop plants.

C. Sexual and asexual reproduction are complementary in the life histories of many plants: *a review*

Both sexual and asexual reproduction have had featured roles in the adaptation of plant populations to their environments.

Benefits of sexual reproduction:

- Generates variation, an asset when the environment (biotic and abiotic) changes
- Produces seeds, which can disperse to new locations and wait until hostile environments become favorable

Benefits of asexual reproduction:

- In a stable environment, plants can clone many copies of themselves in a short period.
- Progeny are mature fragments of the parent plant, and not as fragile as seedlings produced by sexual reproduction.

III.   Cellular Mechanisms of Plant Development

Regardless of whether a plant is sexually produced or results from vegetative reproduction, the initial individual will go through a series of changes that will produce a whole plant.

Development = The sum of all changes that progressively elaborate an organism's body.

- These changes include a number of mechanisms that shape the leaves, roots, and other organs into functional structures.

A. Growth, morphogenesis, and differentiation produce the plant body: *an overview*

The change from a fertilized egg to a plant involves growth, morphogenesis, and cellular differentiation.

Growth is an irreversible increase in size resulting from cell division and cell enlargement.

- The zygote divides mitotically to produce a multicellular embryo in the seed.
- Mitosis resumes in the root and shoot apical meristems after germination.
- Enlargement of the newly produced cells results in most of the actual size increase.

Morphogenesis is the development of body shape and organization.

- Begins in the early divisions of the embryo to produce the cotyledons and rudimentary roots and shoots
- Continues to shape the root and shoot systems as the plant grows
  - The meristems, which remain embryonic, continue growth and morphogenesis throughout the life of the plant.

Cellular differentiation is the divergence in structure and function of cells as they become specialized during the plant's development.

- Every organ of the plant body has a diversity of cells within its total structure.
- Each cell of each organ is fixed in a certain location and performs a specific function (e.g., guard cells, xylem).

B. The cytoskeleton guides cell division and expansion

Plant shape depends on the spatial orientations of cell divisions and cell expansions.

- Plant cells cannot move about as individuals within a developing organ due to their cell walls being cemented to those of neighboring cells.
- Since movement is eliminated, when the cell elongates, its growth is perpendicular to the plane of division.

1. Orienting the plane of cell division

During late interphase (G_2), the cytoskeleton of the cell is rearranged and the microtubules of the cortex become concentrated into the *preprophase band* (see Campbell, Figure 38.17).

The microtubules of the preprophase band disperse leaving behind an array of actin microfilaments.

- These microfilaments hold the nucleus in a fixed orientation until the spindle forms and then direct movement of the vesicles that produce the cell plate.

The walls that develop at the end of cell division form along the plane established by the preprophase band.

2. Orienting the direction of cell expansion

Plant cells expand (elongate) when the cell wall yields to the turgor pressure of the cell.

- Crosslinks between cellulose microfibrils in the cell wall are weakened (via broken hydrogen bonds) by acid-inducible enzymes in the cell wall upon secretion of acids from the cell.
- The loosened wall permits uptake of water by the hypertonic cell; water uptake causes the cell to expand.
- Growth continues until the crosslinks become re-established firmly enough to offset the turgor pressure.
- About 90% of the cell's expansion is due to water uptake, although some cytoplasm is also produced by the cell.
- Most of the water entering the cell is stored in the large central vacuole which forms due to coalescence of small vacuoles as the cell grows.

Plant cells show very little increase in width as they elongate.

- Cellulose microfibrils in the innermost cell wall layers stretch very little; consequently, the cell expands in the direction perpendicular to the orientation of the microfibrils.
- Alignment of microfibrils in the wall mirrors the microtubule orientation found in the cortex (see Campbell, Figure 38.15). This is believed to result from microtubular control of the flow of cellulose-producing enzymes in a specific direction along the membrane.

C. Cell differentiation depends on gene regulation

The progressive development of specialized structures and functions in plant cells reflects the different types of proteins synthesized by different types of cells. It should be noted that differentiative processes continue throughout the life of a plant because meristems sustain indeterminate growth (see Campbell, Figure 38.20).

Xylem cells function in both transport within the plant and structural support.

- Cell walls are hardened by lignin, which is produced by enzymes made by the cell.
- The final stage of differentiation includes the production of hydrolytic enzymes which destroy the protoplast.
    ⇒ This leaves only the cell walls intact and permits the movement of xylem sap through the cells.

Guard cells regulate the size of the stomatal opening.

- Must have flexible walls, thus the enzymes that produce lignin are not produced.
- The protoplast remains intact and regulates ion exchange necessary to increase and decrease turgor.

All cells in a plant possess a common genome. This has been proven by cloning whole plants from single somatic cells.

- All the genes necessary are present since these cells dedifferentiate in tissue cultures and then redifferentiate to produce the diversity of cells found in the plant.
- This ability indicates that cellular differentiation is controlled by gene expression leading to the production of specific proteins.
- Different cell types (like xylem and guard cells) selectively express certain genes at different times during their differentiation; this results in the different developmental pathways that gives rise to the diverse cell types.

D. Pattern formation determines the location and tissue organization of plant organs

The organization in a plant can be seen in the characteristic pattern of cells in each tissue, the pattern of tissues in each organ, and the spatial organization of the organs on the plant.

Pattern formation = The development of specific structures in specific locations

1. Positional information

Pattern formation depends on positional information.

Positional information = Signals indicating a cell's location relative to other cells in an embryonic mass

- Genes respond to these signals and their response affects the localized rates and planes of cell division and expansion.
- This signal detection continues in each cell as the organs develop and cells respond by differentiating into particular cell types.

Several hypotheses have been proposed as to how embryonic cells detect their positions. One hypothesis about positional information transmission is that it relies on gradients of chemical signals.

- A chemical signal might diffuse from a shoot's apical meristem and the decreasing gradient farther from the source would indicate to cells their relative position from the tip.
- A second chemical signal released from the outermost lateral cells would diffuse inward indicating the radial position to each cell.

- Each cell could thus determine its relative longitudinal and radial position from the gradients of these two chemical signals.

2.   Clonal analysis of the shoot apex

Positional information is the basis for the processes involved in plant development: growth, morphogenesis, and differentiation. Plant developmental biologists have developed the technique of clonal analysis to study the relationships of these processes.

- Clonal analysis involves mapping the cell lineages derived from each cell of the apical meristem, noting their position as the plant organs develop.
- Mapping is possible due to induced somatic mutations in each cell which can be used to distinguish that cell and its derived cells from neighboring cells.

This technique has been used to determine that the developmental fates of cells in the apex are somewhat predictable.

- For example, almost all cells developing from the outermost meristematic cells become part of the dermal tissues of leaves and stems.

It is not currently possible to predict what meristematic cells will develop into specific tissues and organs.

- The outermost cells usually divide on a plane perpendicular to the shoot surfaces, thus adding cells to the surface layer.
- Random changes can result in one of these cells dividing on a plane parallel to the surface; this indicates meristematic cells are not dedicated early in their development to forming specific tissues and organs.
- Consequently, it is the cell's final position in a developing organ which determines what type of cell it becomes, not the particular cell lineage to which it belongs.

3.   The genetic basis of pattern formation in flower development

The shoot tip in flowering plants shifts from indeterminate growth to determinate growth when the flower is produced.

- The meristem is consumed during the formation of primordia for sepals, petals, stamens, and carpels.

Positional information commits each primordium to develop into an organ of specific structure and function.

- Some *organ-identity genes* that function in development of the floral pattern and are regulated by positional information have been identified.
- Mutations in these organ-identity genes can cause abnormal floral patterns.
  - ⇒ For example, an extra whorl of sepals may develop instead of petals.
  - ⇒ Such abnormal patterns indicate wild-type alleles are responsible for normal floral pattern development.

Arabidopsis thaliana is the experimental organism many plant biologists are now using to study plant development.

- It is a small plant with a relatively rapid life cycle.
- It also has a small genome that simplifies the search for specific genes.
- Several organ-identity genes affecting floral pattern development have been identified and a few have been cloned.
- Similar organ-identity genes have been found in a distantly related plant, the snapdragon.

⇒ This finding suggests a conservative evolution of the genes controlling basic angiosperm body plan development.

One hypothesis about how positional information influences a particular floral-organ primordium is based on the overall genetic basis of pattern formation (see Campbell, Figure 38.21).

- Organ-identity genes code for transcription factors that are regulatory proteins that help control expression of other genes.
- This control involves binding of the transcription factor to specific sites on the DNA, which affects transcription.
- It is believed that positional information determines which organ-identity gene is expressed, and the resulting transcription factor induces expression of these genes controlling development of specific organs.

REFERENCES

Campbell, N., et al. *Biology.* 5th ed. Menlo Park, California: Benjamin/Cummings, 1999.

Raven, P.H., R.F. Evert and S.E. Eichhorn. *Biology of Plants.* 6th ed. New York: Worth Publishers, Inc., 1998.

CHAPTER 39
CONTROL SYSTEMS IN PLANTS

OUTLINE

I. Plant Hormones
   A. Research on how plants grow toward light led to the discovery of plant hormones: *science as a process*
   B. Plant hormones help coordinate growth, development, and responses to environmental stimuli
   C. Analysis of mutant plants is enhancing plant research
   D. Signal-transduction pathways link cellular responses to hormone signals and environmental stimuli

II. Plant Movements as Models for Studying Control Systems
   A. Tropisms orient the growth of plant organs
   B. Turgor movements are relatively rapid, reversible plant responses

III. Control of Daily and Seasonal Responses
   A. Biological clocks control circadian rhythms
   B. Photoperiodism synchronizes many plant responses to changes of season

IV. Phytochromes
   A. Phytochromes function as photoreceptors in many plant responses to light and photoperiod
   B. Phytochromes may help entrain the biological clock

V. Plant Responses to Environmental Stress
   A. Plants cope with environmental stress through a combination of developmental and physiological responses

VI. Defense Against Pathogens
   A. Resistance to disease depends on a gene-for-gene recognition between plant and pathogen
   B. The hypersensitive response (HR) contains an infection
   C. Systemic acquired resistance (SAR) extends protection against pathogens to the whole plant

OBJECTIVES

After reading this chapter and attending lecture, the student should be able to:

1. For each of the following scientists, describe their hypothesis, experiments, and conclusions about the mechanism of phototropism:
    a. Charles Darwin   c. Peter Boysen Jensen
    b. Francis Darwin   d. F.W. Went

2.   List five classes of plant hormones, describe their major functions, and recall where they are produced in the plant.

3.   Explain how a hormone may cause its effect on plant growth and development.

4.   Describe a possible mechanism for polar transport of auxin.

5.   According to the acid-growth hypothesis, explain how auxin can initiate cell elongation.

6.   Explain why 2,4-D is widely used as a weed killer.

7.   Explain how the ratio of cytokinin to auxin affects cell division and cell differentiation.

8.   Define apical dominance and describe the check-and-balance control of lateral branching by auxins and cytokinins.

9.   List several factors besides auxin from the terminal bud that may control apical dominance.

10.   Describe how stem elongation and fruit growth depend upon a synergism between auxin and gibberellins.

11.   Explain the probable mechanism by which gibberellins trigger seed germination.

12.   Describe how abscisic acid (ABA) helps prepare a plant for winter.

13.   Explain the antagonistic relationship between ABA and gibberellins, and how it is possible for growing buds to have a higher concentration of ABA than dormant buds.

14.   Give an example of how ABA can act as a "stress hormone".

15.   Describe the role of ethylene in plant senescence, fruit ripening and leaf abscission.

16.   Discuss how the study of mutant varieties of plants has heightened our understanding of plant hormones

17.   Describe the components of a signal-transduction pathway.

18.   List two environmental stimuli for leaf abscission.

19.   Define tropism and list three stimuli that induce tropisms and a consequent change of body shape.

20.   Explain how light causes a phototropic response.

21.   Describe how plants apparently tell up from down, and explain why roots display positive gravitropism and shoots exhibit negative gravitropism.

22.   Distinguish between thigmotropism and thigmomorphogenesis.

23.   Describe how motor organs within pulvini can cause rapid leaf movements and sleep movements.

24.   Provide a plausible explanation for how a stimulus that causes rapid leaf movement can be transmitted through the plant.

25.   Define circadian rhythm and explain what happens when an organism is artificially maintained in a constant environment.

26.   List some common factors that entrain biological clocks.

27.   Define photoperiodism.

28.   Distinguish among short-day plants, long-day plants, and day-neutral plants; give common examples of each; and explain how they depend upon critical night length.

29.   Provide evidence for the existence of a florigen.

30.   Explain how the interconversion of phytochrome can act as a switching mechanism to help plants detect sunlight and trigger many plant responses to light.

31.   Using photoperiodism as an example, explain how an integrated control system can regulate a plant process such as flowering.

32.   Explain the molecular basis of resistance to nonvirulent pathogens.

33.   Describe the local and systemic response to virulent pathogens.

KEY TERMS

hormone	oligosaccharins	circadian rhythm	phytoalexins
phototropism	brassinosteroids	photoperiodism	PR proteins
auxin	tropisms	short-day plant	hypersensitive response
cytokinins	gravitropism	long-day plants	systemic acquired
gibberellin	statoliths	day-neutral plants	resistance (SAR)
abscisic acid (ABA)	thigmomorphogenesis	phytochrome	
ethylene	action potentials	heat-shock proteins	
senescence	sleep movements	gene-for-gene recognition	

LECTURE NOTES

Control systems in plants are adaptations that evolved over time in response to interactions with their environment. Plants respond to environmental stimuli by:

- Sending signals between different parts of the plant.
- Tracking the time of day and the time of year.
- Sensing and responding to gravity and direction of light, etc.
- Adjusting their growth pattern and development.

I.   Plant Hormones

Hormone = A compound produced by one part of an organism that is transported to other parts where it triggers a response in target cells and tissues

A.   Research on how plants grow toward light led to the discovery of plant hormones: *science as a process*

Phototropism = Growth toward or away from light

- Growth of a shoot toward light is positive phototropism; growth away from the light is negative phototropism.
- Results form differential growth of cells on opposite sides of a shoot, or in the case of a grass seedling, coleoptile.
- Cells on the darker side elongate faster than those on the light side (see Campbell, Figure 39.1).

Experiments on phototropism led to the discovery of a plant hormone.

Charles and Francis Darwin removed the tip of the coleoptile from a grass seedling (or covered it with an opaque cap) and it failed to grow toward light (see Campbell, Figure 39.2). They concluded that:

- The coleoptile tip was responsible for sensing light.
- Since the curvature occurs some distance below the tip, the tip sends a signal to the elongating region.

Peter Boysen-Jensen separated the tip from the remainder of the coleoptile by a block of gelatin, preventing cellular contact, but allowing chemical diffusion.

- Seedlings behaved normally.
- If an impenetrable barrier was substituted, no phototropic response occurred.
- These experiments demonstrated that the signal was a mobile substance.

In 1926, F.W. Went removed the coleoptile tip, placed it on an agar block, and then put the agar (without the tip) on decapitated coleoptiles kept in the dark (see Campbell, Figure 39.3).

- A block centered on the coleoptile caused the stem to grow straight up.
- If the block was placed off-center, the plant curved away from the side with the block.
- Went concluded the agar block contained a chemical that diffused into it from the coleoptile tip, and that this chemical stimulated growth.
- Went called this chemical an *auxin*.

Kenneth Thimann later purified and characterized auxin.

B. Plant hormones help coordinate growth, development, and responses to environmental stimuli

Plant hormones control plant growth and development by affecting division, elongation, and differentiation of cells.

- Effects depend on site of action, stage of plant growth and hormone concentration.
- The hormonal signal is amplified, perhaps by affecting gene expression, enzyme activity, or membrane properties.
- Reaction to hormones depends on hormonal balance (relative concentration of one hormone compared with others).

Major classes of plant hormones include (see also Campbell, Table 39.1):

1. Auxin (such as IAA)
2. Cytokinins (such as zeatin)
3. Gibberellins (such as GA_3)
4. Abscisic acid
5. Ethylene

1. Auxin

Auxin = A hormone that promotes elongation of young developing shoots or coleoptiles

The natural auxin found in plants is a compound named indoleacetic acid (IAA).

a. Auxin and cell elongation

The apical meristem is a major site of auxin production.

Auxin stimulates cell growth only at concentrations between 10^{-8} to 10^{-3} M.

Auxin moves from the apex down to the zone of cell elongation at a rate of about 10 mm per hour.

- This is faster than would be found in diffusion but much slower than in phloem translocation.
- Polar transport of auxin is unidirectional and requires metabolic energy (see Campbell, Figure 39.4).
- Energy for auxin transport is provided by chemiosmosis.
- IAA is actively transported down a stem by auxin carriers located on the basal ends of cells (carriers are absent on the apical ends).
- Movement of auxin is aided by the differences in pH between the acidic cell wall and the neutral cytoplasm.
  - ATP-driven pumps maintain a proton gradient across the plasma membrane (see Campbell, Figure 39.4).

- As auxin passes through the acidic cell wall, it picks up a proton to become electrically neutral, which allows it to pass through the plasma membrane.
- Auxin is ionized in the neutral intracellular environment which temporarily traps it within the cell since the plasma membrane is less permeable to ions.
- Auxin can only exit the cell by the basal end, where specific carrier proteins are built into the membrane. The proton gradient contributes to auxin efflux by favoring the transport of anions out of the cell.

The acid-growth hypothesis states that cell elongation is due to stimulation of a proton pump that acidifies the cell wall (see Campbell, Figure 39.5).

- Acidification causes the crosslinks between the cellulose myofibrils of the cell walls to break (via disruption of hydrogen bonds).
- This loosens the wall, allowing water uptake, which results in elongation of the cell.

b. Other effects of auxin

Affects secondary growth by inducing vascular cambium cell division and differentiation of secondary xylem

Promotes formation of adventitious roots

Promotes fruit growth in many plants

Auxins are used as herbicides. 2,4-D is a synthetic auxin which affects dicots selectively, allowing removal of broadleaf weeds from a lawn or grain field.

2. Cytokinins

Cytokinins = Modified forms of adenine that stimulate cytokinesis

Cytokinins function in several areas of plant growth:

- Cell division and differentiation
- Apical dominance
- Anti-aging hormones

a. Control of cell division and differentiation

Move from the roots to target tissues by moving up in the xylem sap.

Stimulate RNA and protein synthesis. The new proteins produced by stimulation of RNA appear to be involved in cell division.

Cytokinins, in conjunction with auxin, control cell division and differentiation.

- Stem parenchyma cells cultured without cytokinins grow very large and do not divide.
- Cytokinins added alone have no effect on cells grown in tissue culture.
- Equal concentrations of cytokinins and auxins stimulate cells to grow and divide, but they remain an undifferentiated callus.
- More cytokinin than auxin causes shoot buds to develop from the callus.
- More auxin than cytokinin causes roots to form.

b. Control of apical dominance

Cytokinins and auxin contribute to apical dominance through an antagonistic mechanism.

- Auxin from the terminal bud restrains axillary bud growth, causing the shoot to lengthen.
- Cytokinins (from the roots) stimulate axillary bud growth.

- Auxin cannot suppress axillary bud growth once it has begun.
- Lower buds thus grow before higher ones since they are closer to the cytokinin source than the auxin source.

Auxin stimulates lateral root formation while cytokinins restrain it.

This stimulation-inhibition action may help balance plant growth since an increase in the root system would signal the plant to produce more shoots.

c. Cytokinins as anti-aging hormones

Cytokinins can retard aging of some plant organs, perhaps by inhibiting protein breakdown, stimulating RNA and protein synthesis, and mobilizing nutrients.

May slow leaf deterioration on plants since detached leaves dipped in a cytokinin solution stay green longer.

3. Gibberellins

More than 80 different gibberellins, many naturally occurring, have been identified.

a. Stem elongation

Gibberellins are produced primarily in roots and young leaves. They:

- Stimulate growth in leaves and stems but show little effect on roots
- Stimulate cell division and elongation in stems, possibly in conjunction with auxin
- Cause bolting (rapid growth of floral stems, which elevates flowers)

b. Fruit growth

Fruit development is controlled by both gibberellins and auxin.

- In some plants, both must be present for fruit set.

The most important commercial application of gibberellins is in the spraying of Thompson seedless grapes (see Campbell, Figure 39.8). The hormones cause the grapes to grow larger and farther apart after treatment.

c. Germination

The release of gibberellins signals seeds to break dormancy and germinate.

- A high concentration of gibberellins is found in many seeds, especially in the embryo.
- Imbibed water appears to stimulate gibberellin release.
- Environmental cues may also cause gibberellin release in seeds which require special conditions to germinate.

In cereal grains, gibberellins stimulate germination and support growth by stimulating synthesis of mRNA coding for α-amylase. The α-amylase then digests the stored nutrients, making them available to the embryo and seedling.

In breaking both seed dormancy and apical bud dormancy, gibberellins act antagonistically with abscisic acid, which inhibits plant growth.

4. Abscisic acid (ABA)

Abscisic acid is produced in the terminal bud and helps prepare plants for winter by suspending both primary and secondary growth.

- Directs leaf primordia to develop scales that protect dormant buds.
- Inhibits cell division in vascular cambium.

The onset of seed dormancy is another time it is advantageous to suspend growth.

- In most cases, the ratio of ABA:gibberellins determines whether seeds remain dormant or germinate.
- In other plants, seeds germinate when ABA is washed out of the seeds (desert plants) or degraded by some other stimulus such as sunlight.

ABA also acts as a stress hormone, closing stomata in times of water-stress thus reducing transpirational water loss.

5. Ethylene

Ethylene = A gaseous hormone that diffuses through air spaces between plant cells

- Ethylene can also move in the cytosol, traveling from cell to cell in the phloem or symplast.

High auxin concentrations induce release of ethylene, which acts as a growth inhibitor.

a. Senescence in plants

Senescence (aging) is a natural process in plants that may occur at the cellular, organ, or whole plant level. Ethylene probably plays an important role at each level.

> Examples:
> - Xylem vessel elements and cork cells that die before becoming fully functional
> - Leaf fall in the autumn
> - Withering of flowers
> - Death of annuals after flowering

The best studied forms of senescence are fruit ripening and leaf abscission.

b. Fruit ripening

During fruit ripening, ethylene triggers senescence, and then the aging cells release more ethylene.

- The breakdown of cell walls and loss of chlorophyll are considered aging processes.
- The signal to ripen spreads from fruit to fruit since ethylene is a gas.

c. Leaf abscission

Leaf abscission is an adaptation that prevents deciduous trees from desiccating during winter when roots cannot absorb water from the frozen ground.

- Before abscission, the leaf's essential elements are shunted to storage tissues in the stem from which they are recycled to new leaves in the spring.
- Environmental stimuli are shortening days and cooler temperatures.

When a leaf falls, the breakpoint is an abscission layer near the petiole base (see Campbell, Figure 39.10).

- Weak area since the small parenchyma cells have very thin walls and there are no fiber cells around the vascular tissue.

Mechanics of abscission are controlled by a change in the balance of ethylene and auxin.

- Auxin decrease makes cells in the abscission layer more sensitive to ethylene. Cells then produce more ethylene which inhibits auxin production.
- Ethylene induces synthesis of enzymes that digest the polysaccharides in the cell walls, further weakening the abscission layer.

Wind and weight cause the leaf to fall by causing a separation in the abscission layer.

Even before the leaf falls, a layer of cork forms a protective scar on the twig's side of the abscission layer. The cork prevents pathogens from entering the plant.

D. Analysis of mutant plants is extending the list of hormones and their functions

Until recently, plant hormone research was conducted mainly by applying compounds to whole plants or tissue cultures and measuring their effects on growth and development.

Recently, researchers have gained new insight into hormone synthesis and action by studying mutant varieties that grow or develop abnormally.

Studies with mutant also has led to the discovery of new plant hormones.

- *Oligosaccharides* = Short chains of sugars released from cell walls by the hydrolytic action of enzymes on cell wall polysaccharide; these compounds function in pathogen defense, cell growth and differentiation, and flower development
- *Brassinosteroids* = Steroids that are critical for normal growth

E. Signal-transduction pathways link cellular responses to hormonal signals and environmental stimuli

> Chemical signaling was covered in detail in Chapter 11. It may be useful to review highlights in the special context of plant systems, which are discussed at various points in the remainder of this chapter.

Plant cell responses to hormones and environmental stimuli are mediated by intracellular signals (signal-transduction pathways).

- *Signal-transduction pathway* = A mechanism linking a mechanical or chemical stimulus to a cellular response
- Three steps are involved in each pathway: reception, transduction, and induction (see Campbell, Figure 39.11).

Reception is the detection of a hormone or environmental stimulus by the cell.

- May take various forms depending on the stimulus.

> Examples:
> - Absorption of a particular wavelength of light by a pigment within a cell
> - The binding of a hormone to a specific protein receptor in the cell or on its membrane

- Reception of a hormone only occurs in *target cells* for that hormone.
  - Target cells possess the specific protein receptor to which the hormone must bind; other cells do not possess the receptor.

Transduction in the pathway results in an amplification of the stimulus and its conversion into a chemical form that can activate the cell's responses.

- The hormone (first messenger) binds to a specific receptor and the hormone-receptor combination stimulates the second messenger (a substance that increases in concentration within a cell stimulated by the first messenger).
- The receptor may be bound to the cell membrane and its activation results in a chemical change to the cell.
- Amplification of the signal results from a single first messenger molecule binding to its receptor giving rise to many second messengers, which activate an even larger number of proteins and other molecules.
  - Calcium ions appear to be important second messengers in many plant responses. Calcium ion concentration increases in the cell and the ions bind to the protein calmodulin.
  - The calmodulin-calcium complex then activates other target molecules within the cell.

- A second part of transduction that is important is the specificity of the responses.
  - Two cell types both may have receptors for a hormone but respond differently, because each contains different target proteins for the second messenger.

Induction is the pathway step in which the amplified signal induces the cell's specific response to the stimulus.

- Some responses occur rapidly. For example,
  - ABA stimulation of stomatal closing
  - Auxin-induced acidification of cell walls during cell elongation
- Other responses take longer, especially if they require changes in gene expression (thigmomorphogenesis).

II.   Plant Movements As Models for Studying Control Systems

A.   Tropisms orient the growth of plant organs toward or away from stimuli

Tropisms = Growth responses that result in curvatures of whole plant organs toward or away from stimuli.

- Mechanism is a differential rate of cell elongation on opposite sides of the organ.

Three primary stimuli that result in tropisms are light (phototropism), gravity (gravitropism), and touch (thigmotropism).

1.   Phototropism

Phototropism = Growth either toward or away from light

It is generally accepted that cells on the darker side of a grass coleoptile elongate faster than cells on the bright side due to asymmetric distribution of auxins moving down from the shoot tip.

- For organs, other than grass coleoptiles, the mechanism may be different.

No evidence exists that unilateral light causes an asymmetric distribution of auxins in the stems of many dicots.

There is evidence that other substances that act as growth inhibitors do have an asymmetric distribution toward the lighted side of the stem.

Regardless of the mechanism, the shoot tip is the site of the photoreception that triggers the growth response.

- A photoreceptor sensitive to blue light is present in the shoot tip; this receptor is believed to be a yellow pigment related to riboflavin.
- The same receptor may be involved in other plant responses to light.

2.   Gravitropism

Gravitropism = Orientation of a plant in response to gravity

Roots display positive gravitropism (curve downward).

Shoots display negative gravitropism (bend upward).

The possible mechanisms of gravitropism in roots:

- Specialized plastids containing dense starch grains (*statoliths*) aggregate in the low points of plant cells (see Campbell, Figure 39.12).
- In roots, statoliths occur in certain root cap cells.
  - Aggregating statoliths trigger calcium redistribution, which results in lateral transport of auxin in the root.
  - Calcium and auxin accumulate on the lower side of the elongation zone.

- Roots curve down, because at high concentrations, auxin *inhibits* root cell elongation, so cells on the upper side elongate faster than those on the lower side.

Researchers are challenging the falling statolith hypothesis for positive gravitropism in root growth.

- Insufficient energy is released by starch grains settling to the bottom of cells to account for gravitational detection.
- Many plants lacking starch grains distinguish up from down.
- Studies on *Chara*, a green alga closely related to plants, indicate the settling of the entire protoplasm provides a cell with its up-down orientation.
  - The protoplast is attached to the inside of the cell wall by proteins.
  - When the protoplast settles, the protein tethers at the top of the cell are stretched and those at the bottom are compressed.
  - The sense of up and down is related to this stretching and compressing of the proteins.
  - Experiments where *Chara* was placed in a solution more dense than the protoplast resulted in the protoplast floating upward and an upside down growth pattern.
  - Whether this mechanism is at work in true plants is currently under investigation.

3. Thigmotropism

Thigmotropism = Directional growth in response to touch

- Contact of tendrils stimulates a coiling response caused by differential growth of cells on opposite sides of the tendril.

Thigmomorphogenesis = Developmental response to mechanical perturbation

- Usually results from increased ethylene production in response to chronic mechanical stimulation.
- Stem lengthening decreases while stem thickening increases.

B. Turgor movements are relatively rapid, reversible plant responses

Turgor movements = Reversible movements caused by changes in turgor pressure of specialized cells in response to stimuli

1. Rapid leaf movements

Rapid leaf movements occur in plants such as *Mimosa*.

- When the compound leaf is touched, it collapses and folds together (see Campbell, Figure 39.14).
- Results from rapid a loss of turgor within pulvini (special motor organs located in leaf joints).
- Motor cells lose potassium, which causes water loss by osmosis.
- Turgor pressure is regained and natural leaf form restored in about ten minutes.

Rapid leaf movements travel from the leaf that was stimulated to adjacent leaves along the stem.

- This may be a response to reduce water loss or protect against herbivores.
- The stimulus and response travel wavelike through the plant at 1 cm/sec.
- This transmission is correlated with *action potentials* (electrical impulses) resembling those in animals, but thousands of times slower.
- Action potentials may be widely used as a form of internal communication since they have been found in many algae and plants.

2. Sleep movements

Sleep movements = Lowering of leaves to a vertical position in evening and raising of leaves to a horizontal position in morning (see Campbell, Figure 39.15)

Occurs in many legumes.

Due to daily changes in turgor pressure of motor cells of pulvini.

Cells on one side of the pulvinus are turgid while those on the other side are flaccid.

Migration of potassium ions from one side of the pulvinus to the other is the osmotic agent leading to reversible uptake and loss of water by motor cells.

III. Control of Daily and Seasonal Responses

A. Biological clocks control circadian rhythms in plants and other eukaryotes

Biological clocks (internal oscillators that keep accurate time) are common in all eukaryotes and control many rhythmic phenomena.

- Many human features (e.g., blood pressure, temperature, metabolic rate) fluctuate with the time of the day.
- Certain fungi produce spores for only certain hours during the day.
- Plants display sleep movements and a rhythmic pattern of opening and closing stomata.

Circadian rhythm = A physiological cycle with a frequency of about 24 hours

- Persists even when an organism is sheltered from environmental cues.
- The oscillator is probably endogenous and is set to a 24-hour period by daily signals from the environment.
- When the organism is sheltered from environmental cues, rhythm may deviate from 24 hours (called free-running periods) and can vary from 21 to 27 hours.

Deviation of a free-running period from 24 hours does not indicate erratic drift of a biological clock, just absence of a synchronizing cue.

- Most biological clocks are cued to the light-dark cycle resulting from the Earth's rotation.
- The clock may take days to reset once the cues change.
- Jet lag is a human condition resulting from a lack of synchronization of the internal clock to the time zone.

The nature of the internal oscillator is still currently of great research interest.

- Recent research suggests that the clock is a molecular mechanism common to all eukaryotes. Timekeeping appears to be related to the synthesis of a protein that regulates its own production through feedback.
  - The protein is a transcription factor that inhibits the gene that encodes for the transcription factor; it has been suggested that cyclic changes in the level of the protein form the basis for the internal clock.
  - Such genes have been found in a wide range of organisms, including fruit flies and bread mold, but not plants as of yet.

B. Photoperiodism synchronizes many plant responses to changes of season

Photoperiodism = A physiological response to day length

Seasonal events (seed germination, flowering) are important in plant life cycles.

Plants detect the time of year by the photoperiod (relative lengths of night and day).

1. Photoperiodism and the control of flowering

W.W. Garner and H.A. Allard (1920) postulated that the amount of day length controls flowering. Based on their studies, they classified plants into three categories:

1. *Short-day plants* require a light period shorter than a critical length and generally flower in late summer, fall and winter.

2. *Long-day plants* flower only when the light period is longer than a certain number of hours, generally in late spring and summer.

3. *Day-neutral plants* are unaffected by photoperiod and flower when they reach a certain stage of maturity.

a. Critical night length

It was discovered in the 1940s that night length, not day length, actually controls flowering and other responses to photoperiod.

- If the daytime period is broken by a brief exposure to darkness, there is no effect on flowering.

- If the nighttime period is interrupted by short exposure to light, photoperiodic responses are disrupted and the plants do not flower (see Campbell, Figure 39.16).

- Therefore, short-day plants flower if night is longer than a critical length, and long-day plants need a night shorter than a critical length.

Some plants flower after a single exposure to the proper photoperiod.

Some require several successive days of the proper photoperiod to bloom.

Others respond to photoperiod only if they have been previously exposed to another stimulus. For example, vernalization is a requirement for pretreatment with cold before flowering.

b. Is there a flowering hormone?

There is evidence that a "flowering hormone" is present in plants since leaves detect the photoperiod while buds produce flowers.

- Only requires one leaf for a plant to detect photoperiod and for floral buds develop.

- If all leaves are removed, no photoperiod detection occurs.

Most plant physiologists believe an unidentified hormone is produced in the leaves and moves to the buds or there is a change in the relative concentrations of two more hormones (see Campbell, Figure 39.17).

The hormone (or mixture of hormones) appears to be the same in both long-day and short-day plants.

c. Meristem transition from vegetative growth to flowering

A combination of environmental cues (e.g., photoperiod) and internal signals (e.g., hormones) induces the transition of a bud's meristem from a vegetal state into a flowering state.

The transition requires the coordinate expression of genes that control pattern formation.

IV. Phytochromes

A. Phytochromes function as a photoreceptors in many plant responses to light and photoperiod

Pigments named phytochromes help plants measure the length of darkness in a photoperiod.

- *Phytochrome* = A protein containing a chromophore (light-absorbing component) responsible for a plant's response to photoperiod (see Campbell, Figures 39.18 and 39.19)

- Discovered during studies on how different colors of light affect responses to photoperiod.

Red light (λ of 660 nm) is most effective in interrupting night length.

- Brief exposure of short-day plants to red light prevents flowering even if the plant is kept at critical night length conditions.
- A long-day plant is induced to flower by a brief exposure to red light even if kept at a night length exceeding the critical number of hours.
- If a flash of red light (R flash) is followed by a flash of far-red (FR) light (λ of 730 nm), the plant perceives no interruption of night length.
- Only the wavelength of the last flash affects the plant's measurement of night length, regardless of the number of alternating flashes (see Campbell, Figure 39.18).

Phytochromes alternate between two photoreversible forms: P_r (red absorbing) and P_{fr} (far-red absorbing). The P_r and P_{fr} interconversion is a switching mechanism controlling various plant events (see Campbell, Figure 39.20).

1. **The ecological significance of phytochrome as a photoreceptor**

Phytochrome functions as a photodetector that tells the plant if light is present.

- Plants synthesize phytochrome as P_r, and if kept in dark, it remains as P_r, but if the phytochrome is illuminated, some P_r is converted to P_{fr}.
- P_{fr} triggers many plant responses to light (e.g., seed germination).
- A shift in the P_r and P_{fr} equilibrium indicates the relative amounts of red and far-red light present in the sunlight.
- Shifts in the P_r and P_{fr} ratio may cause changes (e.g., increased growth) which would adjust a plant's growth and development in response to some environmental changes.

Photoreception by phytochrome has a large effect on the whole plant even though very little of the pigment is present in plant cells.

- This fact implies the photoconversion from P_r to P_{fr} produces a signal that is amplified in some way.
- The amplification may be by either an alteration of membrane permeability and/or by affecting gene expression.
  - ⇒ Photoconversion of phytochrome triggers the potassium fluxes in cells of the pulvini that produces the sleep movements in legumes.
  - ⇒ Light induces the synthesis of starch digesting α-amylase required for seed germination in some species.

Complementing phytochrome's effect, other photoreceptors help coordinate a plant's growth and development with its environment.

B. **Phytochromes may help entrain the biological clock**

P_{fr} gradually reverts to P_r.

- This occurs every day after sunset.
- The pigment is synthesized as P_r and degradative enzymes destroy more P_{fr} than P_r.
- At sunrise, the P_{fr} level increases due to photoconversion of P_r.

Plants do not use the disappearance of P_{fr} to measure night length since:

- The conversion is complete within a few hours after sunset.
- Temperature affects the conversion rate, thus, it would not be reliable.

Night length is measured by the biological clock, not by phytochrome.

- Perhaps phytochrome synchronizes the clock to the environment.

- The clock measures night length very accurately (some short-day plants will not flower if night is even one minute shorter than the critical length).

V.   Plant Responses to Environmental Stress

A.   Plants cope with environmental stress through a combination of developmental and physiological responses

A plant must adjust to environmental fluctuations every day of its life. Severe fluctuations may put plants under stress.

- *Stress* = An environmental condition that can have an adverse effect on a plant's growth, reproduction, and survival

Some plants have evolutionary adaptations that enable them to live in environments that are stressful to other plants.

- For example, halophytes have special anatomical and physiological adaptations that permit them to grow best in salty soils.
  ⇒ Salt glands on the leaves eliminate excess salt from the plants, thus the saline environment is not an environmental stress.

1.   Responses to water deficit

Plants have control systems in both the leaves and roots that help them cope with water deficits.

Most control systems in leaves help the plant conserve water by reducing transpirational water loss.

- Guard cells lose turgor and the stomata close when a leaf faces a water deficit.
- Mesophyll cells in the leaf are also stimulated to increase synthesis and release abscisic acid which acts on guard cell membranes to help keep the stomata closed.
- Growth of young leaves is inhibited by a water deficit since cell expansion is a turgor-dependent process.
  ⇒ This reduces transpiration by slowing the increase in leaf surface area.
- The leaves of many grasses and other plants wilt; they roll into a shape that reduces the surface area exposed to the sun, thus reducing transpiration.

Roots respond to water deficits by reducing growth.

- Drying of the soil from the surface down inhibits the growth of shallow roots.
  ⇒ The cells cannot retain the turgor necessary for elongation.
- Deeper roots surrounded by moist soil continue to grow.
  ⇒ This maximizes root exposure to soil moisture.

2.   Responses to oxygen deprivation

Waterlogged soil lacks the air spaces that provide oxygen for cellular respiration in the roots.

- Some plants form air tubes that extend from submerged roots to the surface, thus oxygen can reach the roots (see Campbell, Figure 39.21).
- Mangroves are structurally adapted to their coastal marsh environments in that their submerged roots are continuous with aereal roots that provide access to oxygen.

3.   Responses to salt stress

Excess salts (sodium chloride or others) in the soil may:

1. Lower the water potential of the soil solution causing a water deficit even though sufficient water is present
    - A water potential in the soil that is more negative than that of the root tissue will cause roots to lose water instead of absorb it.
2. Have a toxic effect on the plant at relatively high concentrations
    - The uptake of most harmful ions is impeded by the selectively permeable membranes of root cells.
    - This causes a problem with acquiring water from solute rich soils.

Many plants produce compatible solutes in response to moderately saline soils.
- *Compatible solute* = An organic compound that keeps the water potential of cells more negative than the soil solution without admitting toxic quantities of salt

With the exception of halophytes, most plants cannot survive extended periods of salt stress.

4. Responses to heat stress

Transpiration is one mechanism that helps plants respond to excessive heat and prevent the denaturing of enzymes and damage to metabolism.
- The evaporative cooling associated with transpiration keeps the temperature of the leaf 3° to 10°C lower than ambient temperature.
- Cooling via transpiration will continue while stomata remain open; however, if a water deficit occurs, the stomata close and the cooling function is lost in order to conserve water.

Most plants will begin producing *heat-shock proteins* when exposed to excessive temperatures (40°C or above for temperate zone plants).
- This is a back-up system to transpiration.
- Some heat-shock proteins are identical to chaperone proteins found in unstressed cells.
    ⇒ Chaperone proteins serve as temporary supports which help other proteins fold into their functional conformations.
- Heat-shock proteins may help enzymes and other proteins maintain their conformation, thus preventing denaturation.

5. Responses to cold stress

Chilling of a plant (reduction of ambient temperature to a non-freezing level) causes a change in the fluidity of cell membranes.
- *Fluidity* = The lateral drifting of proteins and lipids in the plane of the membrane; a result of the fluid mosaic structure of membranes
- At a critical point, lipids become locked into crystalline structures causing a loss of fluidity.
- Solute transport and membrane protein function are adversely affected by the loss of fluidity.

Plants respond to the cold stress of chilling by altering the lipid composition of their membranes.
- The proportion of saturated fatty acids in the membrane is increased.
    ⇒ The shape of the fatty acids reduces crystal formation, thus maintaining fluidity at lower temperatures.
- This modification works best for gradual temperature changes as it takes several hours to days to occur.

Subfreezing temperatures are the most severe form of cold stress because ice crystals begin to form in the plant.

- Less threat to plant survival occurs if the ice crystals form only in the cell walls and intracellular spaces.
- When ice crystals form in the protoplasts, the cell usually dies.
  ⇒ The ice crystals perforate the membranes and organelles.
- Wood plants which are native to regions where cold winters occur have adaptations to cope with the stress of freezing.
  ⇒ The solute composition of live cells is changed in a way that prevents ice crystal formation even when the cytosol is supercooled.
  ⇒ This is effective even when ice crystals form in the cell walls.

6. Responses to herbivores

Plants counter excessive grazing by herbivores with both physical and chemical defense measures.

- Physical defenses include structures such as thorns and spines.
- Chemical defenses take the form of distasteful or toxic compounds such as canavanine.
  ⇒ Similar to arginine, canavanine is an unusual amino acid produced by some plants.
  ⇒ When ingested by insects, it is incorporated in place of arginine in the insect's proteins.
  ⇒ Incorporation of canavanine disrupts protein conformation and the insect dies.

Some plants even recruit predatory animals to help defend against herbivores.

- Certain wasps inject their eggs into their prey (e.g., caterpillars). Upon hatching, the larvae feed upon, and eventually kill, the caterpillar (this benefits the plant).
- The plant attracts wasps via volatile chemical signals released from caterpillar-induced wounded plant tissue.

VI. Defense Against Pathogens

Plants, like animals, are subject to infection from various pathogens. Plant structure (epidermis and periderm) offers the first line of defense. Plants also posses a chemical-based second line of defense akin to the immune system of animals.

A. Resistance to disease depends on a gene-for-gene recognition between plant and pathogen

Most pathogen-plant interactions are nonvirulent, meaning that the pathogen gains access to the host to perpetuate itself without causing severe damage to the plant.

Specific resistance to disease is based on a *gene-for-gene recognition* in which a precise match-up occurs between an allele in the plant and an allele in the pathogen (see Campbell, Figure 39.23).

- A plant is resistant if one of its *R* (resistance) genes is a dominant allele that corresponds to a dominant *Avr* (avirulent) allele in the pathogen.
- The molecular mechanisms actually involve the products of the *R* and *Avr* genes, a receptor and a signal molecule, respectively. (The *Avr* product most likely mimics the action of an endogenous signal molecule of the plant.)

B. The hypersensitive response (HR) contains the infection

Plants infected by a virulent pathogen are capable of resisting infection through a localized chemical signaling system. The response involves the following factors:

- *Phytoalexins* = Antimicrobial compounds released from wounded cells
- Activation of genes encoding *PR proteins* (for pathogenesis-related), some of which are antimicrobial, and others of which act as signals to adjacent cells
- Lignin synthesis and cross-linking of cell wall components, actions aimed at isolating the infection

If the pathogen is a virulent based on an R-Avr match, the local response is more vigorous and is referred to as a *hypersensitive response* (HR).

C. Systemic acquired resistance (SAR) helps prevent infection throughout the plant

Although an HR response is local and specific, signals produced from an HR are conveyed throughout the plant.

Such signals, or "alarm hormones," initiate a non-specific *systemic acquired resistance* (SAR) response to help protect uninfected tissue from a pathogen that might spread from its point of invasion.

- The SAR includes production of phytoalexin and PR proteins.
- A possible alarm hormone is salicylic acid. A modified form of this compound, acetylsalicylic acid, is the active component of aspirin.

REFERENCES

Campbell, N., et al. *Biology*. 5th ed. Menlo Park, California: Benjamin/Cummings, 1999.

Raven, P.H., R.F. Evert and S.E. Eichhorn. *Biology of Plants*. 5th ed. New York: Worth Publishers, Inc., 1992.

CHAPTER 40
AN INTRODUCTION TO ANIMAL STRUCTURE AND FUNCTION

OUTLINE

I. Levels of Structural Organization
    A. Function correlates with structure in the tissues of animals
    B. The organ systems of an animal are interdependent

II. Introduction to the Bioenergetics of Animals
    A. Animals are heterotrophs that harvest chemical energy from the food they ingest
    B. Metabolic rate provides clues to an animal's bioenergetic "strategy"
    C. Metabolic rate per gram is inversely related to body size among similar animals

III. Body Plans and the External Environment
    A. Physical support on land depends on adaptations of body proportions and posture
    B. Body size and shape affect interactions with the environment

IV. Regulating the Internal Environment
    A. Mechanisms of homeostasis moderate changes in the internal environment
    B. Homeostasis depends on feedback circuits

OBJECTIVES

After reading this chapter and attending lecture, the student should be able to:

1. Define tissue and explain where it falls in the hierarchy of structural organization.

2. From micrographs or diagrams, correctly identify the following animal tissues, explain how their structure relates to function and give examples of each.
    a. Epithelial tissue: cuboidal, columnar, squamous
    b. Connective tissue: adipose, cartilage, bone
    c. Muscle: skeletal (striated), cardiac, smooth, nervous

3. Describe how metabolic rate can be determined for animals, and distinguish between basal metabolic rate and standard metabolic rate.

4. Describe several body shapes that maximize external surface area in contact with the environment.

5. Explain how animals with complex internal organization and relatively small surface area to volume ratio can have adequate surface area for materials exchange with the environment.

6. Describe the location and function of interstitial fluid.

7. Define homeostasis.

8. Distinguish between negative and positive feedback.

KEY TERMS

tissues	elastic fibers	osteoblasts	abdominal cavity
epithelial tissue	reticular fibers	Haversian systems	organ systems
basement membrane	fibroblasts	blood	metabolic rate
simple epithelium	macrophages	nervous tissue	calories (cal)
stratified epithelium	adipose tissue	neuron	kilocalories (kcal)
cuboidal	tendons	muscle tissue	basal metabolic rate
columnar	fibrous connective	skeletal muscle	standard metabolic rate
squamous	tissue	striated muscle	interstitial fluid
mucous membrane	ligaments	cardiac muscle	homeostasis
connective tissue	cartilage	organs	negative feedback
loose connective tissue	chondrocytes	mesenteries	positive feedback
collagenous fibers	bone	thoracic cavity	

LECTURE NOTES

There are several unifying themes in the study of animal anatomy and physiology.

- There is a correlation between form and function; functions are properties that emerge from the specific shape and order of body parts.
- A comparative approach allows us to see how species of diverse evolutionary history and varying complexity solve problems common to all.
- Animals, as all living organisms, have the capacity to respond and adjust to environmental change in two temporal scales:
  ⇒ Over the long term by adaptation due to natural selection
  ⇒ Over the short term by physiological responses

The objectives of this chapter are to:

- Illustrate the hierarchy of structural order characterizing animals
- Emphasize the importance of energetics in animal life (how animals obtain, process, and use chemical energy)
- Examine how animal body forms affect their interactions with the environment
- Preview how regulatory systems maintain favorable internal environments

I. Levels of Structural Organization

There is a structural hierarchy of life:

- Atoms → molecules → supramolecular structures → cell
- The cell is the lowest level of organization that can live as an organism.

The hierarchy of multicellular organisms is: cell → tissues → organs → organ systems.

A. Function correlates with structure in the tissues of animals

Tissues = Groups of cells with common structure and function

- Cells may be held together by a sticky coating or woven together in a fabric of extracellular fibers.

There are four main categories of tissues: epithelial tissue, connective tissue, muscle tissue, and nervous tissue.

1. Epithelial Tissue

Formed from sheets of tightly packed cells, *epithelial tissue* covers the outside of the body and lines organs and body cavities. Characteristics of epithelium include:

- Cells are closely joined and are riveted by tight junctions in some tissue types (see Campbell, Chapter 7).
- It functions as a barrier against mechanical injury, invading microbes, and fluid loss.
- Its free surface is exposed to air or fluid. Cells at the base are attached to a *basement membrane,* which is a dense layer of extracellular material.

Epithelial tissue cells are categorized by the number of layers and shape of the free surface cells:

- *Simple epithelium* is one layer of cells.
- *Stratified epithelium* has multiple tiers of cells.
- Pseudostratified epithelium is one layer of cells that appear to be multiple because the cells vary in length.
- Cell shapes are *cuboidal* (like dice), *columnar* (bricks on end), or *squamous* (like flat floor tiles).
- A tissue may be described by a combination of terms such as stratified squamous epithelium (see Campbell, Figure 40.1).

Some epithelia are specialized for absorption or secretion of chemical solutions, in addition to their protective role.

- Some epithelia are ciliated (e.g., the lining of the respiratory system).
- The *mucous membranes* lining the oral cavity and nasal passageways secrete mucus which moistens and lubricates the surfaces.
- The structure fits function. For example, simple squamous epithelium is leaky and is specialized for exchange of materials by diffusion. It is found in blood vessel linings and air sacs in the lungs.

2. Connective tissue

Connective tissue is characterized by a sparse cell population scattered through an extensive extracellular matrix.

- Functions to bind and support other tissues.
- Matrix is a web of fibers embedded in a homogenous ground substance.
- Consists of a loose weave of three types of proteinaceous fibers:
  - *Collagenous fibers* are bundles of fibers containing three collagen molecules each. They have great tensile strength and resist stretching (see Campbell, Figure 40.2).
  - *Elastic fibers* are long threads of the protein elastin. Elastic properties lend tissue a resilience to quickly return to the original shape.
  - *Reticular fibers* are branched and form a tightly woven fabric joining connective tissue to adjacent tissues.

Major types of connective tissue include (see Campbell, Figure 40.3):

- Loose connective tissue
- Adipose tissue
- Fibrous connective tissue
- Cartilage
- Bone
- Blood

Loose connective tissue holds organs in place and attaches epithelia to underlying tissues.

- Consists of two types of cells
  1. *Fibroblasts secrete the proteins of the extracellular fibers.*

2. *Macrophages are phagocytic amoeboid cells that function in immune defense of the body.*

- Has all three fiber types

Adipose tissue is loose connective tissue specialized to store fat in adipose cells distributed throughout its matrix.

- Insulates the body and stores fuel molecules.
- Each adipose cell has one large fat droplet which can vary in size as fats are stored or utilized.

Fibrous connective tissue is dense due to the arrangement of a large number of collagenous fibers in parallel bundles, which impart great tensile strength.

- Found in *tendons* (attach muscles to bones) and *ligaments* (attach bones together at joints).

Cartilage is composed of collagenous fibers embedded in chondroitin sulfate, a protein-carbohydrate ground substance.

- *Chondrocytes* secrete both collagen and chondroitin sulfate, which make cartilage both strong and flexible:
  ⇒ Chondrocytes are confined to lacunae, scattered spaces within the ground substance.
- Cartilage comprises the skeleton of all vertebrate embryos.
  ⇒ Some vertebrates (e.g., sharks) retain the cartilaginous skeleton in adults.
  ⇒ Most vertebrates eventually replace most of the cartilage with bone. Cartilage is retained in areas such as the nose, ears, trachea, intervertebral discs and ends of some bones.

Bone is a mineralized connective tissue.

- *Osteoblasts*, bone-forming cells, deposit a matrix of collagen and calcium phosphate, which hardens into the mineral hydroxyapatite. The combination of collagen and hydroxyapatite makes the bone harder than cartilage, but not brittle.
- Bone consists of repeating *Haversian systems* (concentric layers or lamellae deposited around a central canal containing blood vessels and nerves).
- Once osteoblasts are trapped in their secretions, they are called osteocytes. Osteocytes are located in spaces called *lacunae* surrounded by a hard matrix and are connected to each other by cell extensions called canaliculi.
- In long bones, only the outer area is hard and compact; the inner area is filled with spongy bone tissue called marrow.

Blood is a connective tissue composed of:

- Liquid extracellular matrix of plasma that contains water, salts, and proteins.
- Cellular component that contains:
  ⇒ Leukocytes, white blood cells that function in immune defense
  ⇒ Erythrocytes, red blood cells that transport oxygen
  ⇒ Platelets, cell fragments that function in blood clotting
- Blood cells are made in red marrow near the ends of long bones.

3. **Nervous tissue**

Nervous tissue senses stimuli and transmits signals from one part of the animal to another.

Neuron = Nerve cell specialized to conduct an impulse or bioelectric signal (see Figure 40.4); consists of:

- Cell body
- Dendrites, extensions that conduct impulses to the cell body
- Axons, extensions that transmit impulses away from the cell body (see Campbell, Chapter 48)

4. Muscle tissue

Muscle tissue consists of long, excitable cells capable of contraction.

In the muscle cell cytoplasm are parallel bundles of microfilaments made of the contractile proteins, actin and myosin.

Muscle is the most abundant tissue in most animals.

There are three types of vertebrate muscle tissue (see Campbell, Figure 40.5):

1. *Skeletal muscle* is responsible for voluntary movements.
   ⇒ Attached to bones by tendons
   ⇒ Microfilaments are aligned to form a banded or striated appearance (hence, skeletal muscle is also called *striated muscle*).
2. *Cardiac muscle* forms the contractile wall of the heart.
   ⇒ Cells are striated and branched.
   ⇒ Ends of cells are joined by *intercalated disks*, which relay the contractile impulse from cell to cell.
3. *Smooth muscle* is so named because it is unstriated.
   - Found in the walls of internal organs (e.g., digestive tract, bladder) and arteries
   - Spindle-shaped cells contract slowly, but can retain contracted condition longer than skeletal muscle
   - Responsible for involuntary movements (e.g., churning of the stomach)

B. The organ systems of an animal are interdependent

Tissues are organized into *organs* in all but the simplest animals.

In some organs, the tissues may be layered; for example, the vertebrate stomach (see Campbell, Figure 40.6)

Many organs are suspended by sheets of connective tissue called *mesenteries*.

- In mammals, for example, the heart and lungs are suspended in the *thoracic cavity*; the thoracic cavity is separated from the *abdominal cavity* by the diaphragm.

Organs may be organized into organ systems.

- *Organ systems* = Several organs with separate functions that act in a coordinated manner (e.g., digestive, circulatory, and respiratory systems) (see Campbell, Table 40.1)
- Systems are interdependent: an organism is a living whole greater than the sum of its parts.

II.   Introduction to the Bioenergetics of Animals

A. Animals are heterotrophs that harvest chemical energy from the food they ingest

Animals, as living organisms, exchange energy with the environment. Since they are heterotrophic, animals acquire energy from organic molecules synthesized by other organisms (see also Campbell, Figure 40.7).

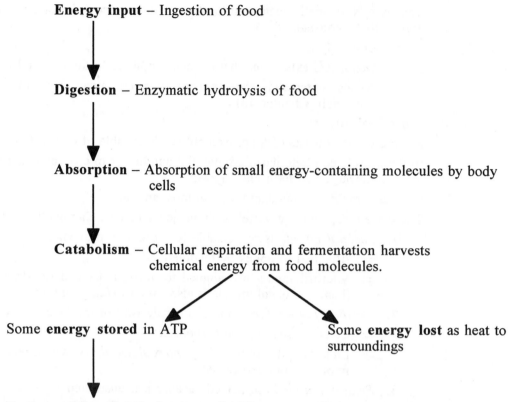

Energy input – Ingestion of food

Digestion – Enzymatic hydrolysis of food

Absorption – Absorption of small energy-containing molecules by body cells

Catabolism – Cellular respiration and fermentation harvests chemical energy from food molecules.

Some **energy stored** in ATP

Some **energy lost** as heat to surroundings

Energy used – Chemical energy of ATP powers cellular work. After the needs of staying alive are met, leftover chemical energy and carbon skeletons from food molecules can be used in biosynthesis.

Energy lost – Cellular work generates heat, which is lost to the surroundings.

B. Metabolic rate provides clues to an animal's bioenergetic "strategy"

Bioenergetics, the study of the dynamic balance between energy intake and loss in an organism, gives clues to how an animal adapts to its environment. By measuring the rate of energy use, physiologists can determine:

- How much food energy an animal needs just to stay alive
- The energy costs for specific activities such as walking or running

Metabolic rate = Total amount of energy an animal uses per unit of time; usually measured in *calories* or *kilocalories* (kcal or CAL = 1000 calories).

- Can be determined by measuring:
  - The amount of oxygen used for an animal's cellular respiration
  - An animal's heat loss per unit of time
    - ⇒ Heat loss, a byproduct of cellular work, is measured with a calorimeter (a closed, insulated chamber with a device that records heat production).
    - ⇒ Calorimeters are effectively used with small animals that have high metabolic rates, but are less precise with small animals that have low metabolic rates and with large animals.

Every animal has a range of metabolic rates.

- Minimal rates support basic life functions, such as breathing.
- Maximal rates occur during peak activity, such as all-out running.

- Between these extremes, metabolic rates can be influenced by many factors, including:
  ⇒ Age, sex, and size
  ⇒ Body temperature
  ⇒ Environmental temperature
  ⇒ Food quality and quantity
  ⇒ Activity level
  ⇒ Amount of available oxygen
  ⇒ Hormonal balance
  ⇒ Time of day

Endotherms = Animals that generate their own body heat metabolically

- Examples include birds and mammals
- Require more kilocalories to sustain minimal life functions that ectotherms
- Many are also homeothermic, that is, their body temperature must be maintained within narrow limits

Basal metabolic rate (BMR) = An endothermic animal's metabolic rate measured under resting, fasting and stress-free conditions

- Average human BMR is 1600-1800 kcal/day for adult males; 1300-1500 kcal/day for adult females.

Ectotherms = Animals that acquire most of their body heat from the environment

- Include most fish, amphibians, reptiles, and invertebrates
- Are energetically different from endotherms; body temperature and metabolic rate changes with environmental temperature
- Because it is influenced by temperature, an ectotherm's minimal metabolic rate (SMR) must be determined at a specific temperature.
  - *Standard metabolic rate (SMR)* = An ectotherm's metabolic rate measured under controlled temperature and under resting, fasting, and stress-free conditions

C. Metabolic rate per gram is inversely related to body size among similar animals

There is an inverse relationship between metabolic rate and size.

- Smaller animals consume more calories per gram than larger animals
- Correlated with a higher metabolic rate and need for faster O_2 delivery to the tissues, small animals also have higher:
  ⇒ Breathing rates
  ⇒ Blood volume
  ⇒ Heart rates
- This inverse relationship between metabolic rate and body size holds true for both endotherms and ectotherms, and is not simply a function of surface area to volume ratio.

III.   Body Plans and the External Environment

An animal's body plan results from a developmental pattern programmed by its genome—a product of millions of years of evolution due to natural selection.

A. Physical support on land depends on adaptations of body proportions and posture

Body proportions and size-weight relationships change in animal bodies as they become larger.

- Body design must accommodate the greater demand for support that comes with increasing size. (The strain on body supports depends on an animal's weight, which increases as the cube of its height or other linear dimension.)

- In mammals and birds, the most important design feature in supporting body weight is posture—leg position relative to the main body—rather than leg bone size (see Campbell, Figure 40.8).

> Examples:
> - The legs of an elephant are in a more upright position than those of small mammals.
> - Large mammals run with legs nearly extended, which reduces strain; small mammals run with legs bent and crouch when standing.

Bioenergetics also plays an important role in load-bearing, since crouched posture is partly a function of muscle contraction, powered by chemical energy.

B. Body size and shape affect interactions with the environment

Animal cells must have enough surface area in contact with an aqueous medium to allow adequate environmental exchange of dissolved oxygen, nutrients, and wastes. This requirement imposes constraints on animal size and shape.

- Single-celled organisms, such as protozoans, must have sufficient surface area of plasma membrane to service the entire volume of cytoplasm and are thus limited in size (see Campbell, Figure 40.9a). Recall that:
  ⇒ The upper limits of cells size are imposed by the surface area to volume ratio.
  ⇒ As cell size increases, volume increases proportionately more than surface area.

- Some multicellular animals have a body plan that places all cells in direct contact with their aqueous environments. Two such body plans include:
  1. *Two-layered sac.* A body wall only two cell layers thick (see Campbell, Figure 40.9b); for example, the body cavity of *Hydra* opens to the exterior, so both outer and inner layers of cells are bathed in water.
  2. *Flat-shaped body with maximum surface area exposed to the aqueous environment.* For example, tapeworms are thin and flat, so most cells are bathed in the intestinal fluid of the worm's vertebrate host.

- Most complex animals have a smaller surface area to volume ratio and thus lack adequate exchange area on the outer surface.
  ⇒ Instead, highly folded, moist, internal surfaces exchange materials with the environment (see Campbell, Figure 40.10).
  ⇒ The circulatory system shuttles materials between these specialized exchange surfaces.

Though logistical problems exist with environmental exchange, there are some distinct advantages to a complex body form.

- Environmental exchange surfaces are internal and protected from desiccation, so the animal can live on land.

- Cells are bathed with internal body fluid, so the animal can control the quality of the cells' immediate environment.

IV. Regulating the Internal Environment

A. Mechanisms of homeostasis moderate changes in the internal environment

Interstitial fluid = The internal environment of vertebrates, composed of fluid between the cells

- Fills spaces between cells
- Exchanges nutrients and wastes with blood carried in capillaries

Homeostasis = Dynamic state of equilibrium in which internal conditions remain relatively stable; steady state (see Campbell, Figure 40.11)

- French physiologist Claude Bernard first described the "constant internal milieu" in animals; he recognized many animals can maintain constant conditions in their internal environment—even when the external environment changes.

B. Homeostasis depends on feedback circuits

Homeostatic control systems have three functional components:

1. *Receptor* detects internal change.
2. *Control center* processes information from the receptor and directs the effector to respond.
3. *Effector* provides the response.

As a control system operates, the effector's response feeds back and influences the magnitude of the stimulus by either depressing it (negative feedback) or enhancing it (*positive feedback*).

Negative feedback = Homeostatic mechanism that stops or reduces the intensity of the original stimulus and consequently causes a change in a variable that is opposite in direction to the initial change

- Most common homeostatic mechanism in animals
- There is a lag time between sensation and response, so the variable drifts slightly above and below the *set point*.
- A nonbiological example is the thermostatic control of room temperature (see Campbell, Figure 40.12.)
- Human examples include hormonal control of blood glucose levels and the regulation of body temperature by the hypothalamus.
  - If the human hypothalamus detects a high blood temperature, it sends nerve impulses to sweat glands, which increase sweat output and cause evaporative cooling.
  - When the body temperature returns to normal, no additional signals are sent.

Set point = A variable's range of values that must be maintained to preserve homeostasis

Positive feedback = Homeostatic mechanism that enhances the initial change in a variable

- Rarer than negative feedback and usually controls only episodic events
- Examples include blood clotting and the heightening of labor contractions during childbirth.
  - ⇒ During childbirth, the baby's head against the uterine opening stimulates contractions which cause greater pressure of the head against the uterine opening.
  - ⇒ The greater pressure, in turn, further enhances uterine contractions.

REFERENCES

Campbell, N. *Biology*. 5th ed. Menlo Park, California: Benjamin/Cummings, 1999.

Randall, D., Burggren, W., and French, K. *Animal Physiology: Mechanisms and Adaptations*, 4th ed. New York, NY: W.H. Freeman and Company, 1997

Schmidt-Nielsen, Knut. *Animal Physiology: Adaptation and Environment*. 4th ed. New York: Cambridge University Press, 1990.

CHAPTER 41
ANIMAL NUTRITION

OUTLINE

I.   Nutritional Requirements
   A. Animals are heterotrophs that require food for fuel, carbon skeletons, and essential nutrients: *an overview*
   B. Homeostatic mechanisms manage an animal's fuel
   C. An animal's diet must supply essential nutrients and carbon skeletons for biosynthesis
II. Food Types and Feeding Mechanisms
   A. Most animals are opportunistic feeders
   B. Diverse feeding adaptations have evolved among animals
III. Overview of Food Processing
   A. The four main stages of food processing are ingestion, digestion, absorption, and elimination
   B. Digestion occurs in specialized compartments
IV. The Mammalian Digestive System
   A. The oral cavity, pharynx, and esophagus initiate food processing
   B. The stomach stores food and performs preliminary digestion
   C. The small intestine is the major organ of digestion and absorption
   D. Hormones help regulate digestion
   E. Reclaiming water is a major function of the large intestine
V.   Evolutionary Adaptations of Vertebrate Digestive Systems
   A. Structural adaptations of the digestive system are often associated with diet
   B. Symbiotic microorganisms help nourish many vertebrates

OBJECTIVES

After reading this chapter and attending lecture, the student should be able to:

1. Distinguish among herbivores, carnivores, and omnivores.
2. Describe the following feeding mechanisms and give examples of animals that use each:
    a. Filter-feeding   b. Substrate-feeding
    c. Deposit-feeding   d. Fluid-feeding
3. Define digestion and describe why it is a necessary process.
4. Explain how anhydro bonds are formed and describe the role of hydrolysis in digestion.
5. Distinguish between intracellular and extracellular digestion.
6. Explain why intracellular digestion must be sequestered in a food vacuole, and give examples of organisms which digest their food in vacuoles.

7. Define gastrovascular cavity and explain why extracellular digestive cavities are advantageous.

8. Using *Hydra* as an example, describe how a gastrovascular cavity functions in both digestion and distribution of nutrients.

9. List major animal phyla which use gastrovascular cavities for digestion.

10. Describe some distinct advantages that complete digestive tracts have over gastrovascular cavities, and list the major animal phyla with alimentary tracts.

11. Define peristalsis and describe its role in the digestive tract.

12. Describe how salivation is controlled and list the functions of saliva.

13. Describe the role of salivary amylase in digestion.

14. Describe the sequence of events which occur as a result of the swallowing reflex.

15. Describe the function of the esophagus, and explain how peristalsis in the esophagus is controlled.

16. Describe the role of the cardiac and pyloric sphincters.

17. List the three types of secretory cells found in stomach epithelium and what substances they secrete.

18. Recall the normal pH of the stomach, and explain the function of stomach acid.

19. Describe the function of pepsin.

20. Explain why the stomach normally does not digest itself.

21. Explain how pepsin and acid secretion are regulated and describe the roles of the hormones gastrin and enterogastrone.

22. Describe the cause of ulcers, and explain why they are frequently found in the duodenum.

23. Explain how chyme is moved through the small intestine.

24. Describe the sequence of events which occur in response to acid chyme entering the duodenum and include the roles of:

    a. Secretin   e. Bile
    b. Bicarbonate   f. Pancreatic enzymes
    c. Cholecystokinin (CCK) g. Enterogastrone
    d. Gall bladder

25. Describe how pancreatic zymogens for proteolytic enzymes are activated in the duodenum and include the role of the intestinal enzyme enterokinase.

26. Describe enzymatic digestion of carbohydrates, proteins, lipids and nucleic acids, including the reactants and products for each enzymatic reaction and whether they occur in the:

    a. Oral cavity   c. Lumen of small intestine
    b. Stomach   d. Brush border of small intestine

27. Explain the function of bile, describe where it is produced and stored, and describe its composition.

28. State whether the lumen of the digestive tract is technically inside or outside the body.

29. Explain where most nutrient absorption occurs.

30. Explain why the many folds, villi, and microvilli are important in the small intestine.

31. Describe how specific nutrients are absorbed across the intestinal epithelium and across the capillary or lacteal wall, and indicate whether the transport is with or against the concentration gradient.

32. Explain what happens to glycerol and fatty acids after they are absorbed into the intestinal epithelium, and describe the fate of chylomicrons and lipoproteins.

33. Explain the function of the hepatic portal vein.

34. Explain where in the digestive tract that most reabsorption of water occurs.

35. Describe the composition of feces, and explain what the main source of vitamin K isfor humans.

36. Give examples of vertebrates with the following digestive adaptations and explain how these adaptations are related to diet:
    a. Variation in dentition
    b. Variation in length of the digestive tract
    c. Fermentation chambers

37. Explain why animals need a nutritionally adequate diet

38. Describe the effects of undernourishment or starvation.

39. List some of the risks of obesity.

40. Distinguish between malnourished and undernourished.

41. List four classes of essential nutrients, and describe what happens if they are deficient in the diet.

42. List and distinguish between water-soluble and fat-soluble vitamins, and explain how they are used by the body.

43. Describe the dietary sources, major body functions and effects of deficiency for the following required minerals in the human diet: calcium, phosphorus, sulfur, potassium, chlorine, sodium, magnesium, and iron.

KEY TERMS

undernourished	bulk-feeders	gastric juice	villi
essential nutrients	intracellular digestion	pepsin	microvilli
malnourished	extracellular digestion	pepsinogen	brush border
essential amino acids	gastrovascular cavity	acid chyme	lacteal
essential fatty acids	complete digestive tract	pyloric sphincter	chylomicrons
vitamins	alimentary canal	small intestine	hepatic portal vessel
minerals	peristalsis	duodenum	gastrin
herbivores	sphincters	bile	secretin
carnivores	salivary gland	trypsin	cholecystokinin (CCK)
omnivores	pancreas	chymotrypsin	enterogastrone
ingestion	liver	carboxypeptidase	large intestine
digestion	gallbladder	aminopeptidase	colon
enzymatic hydrolysis	oral cavity	dipeptidase	cecum
absorption	salivary amylase	enteropeptidase	appendix
elimination	bolus	nucleases	feces
suspension-feeders	pharynx	emulsification	rectum
substrate-feeders	epiglottis	lipase	ruminant
deposit-feeders	esophagus	jejunum	anus
fluid-feeders	stomach	ileum	

LECTURE NOTES

I. **Nutritional Requirements**

A. **Animals are heterotrophs that require food for fuel, carbon skeletons, and essential nutrients:** *an overview*

Like all heterotrophs, animals must rely on organic compounds in their food to supply energy and the raw materials for growth and repair.

A nutritionally adequate diet provides an animal with:
- Fuel (chemical energy) for cellular respiration
- Raw organic materials for biosynthesis
- Essential nutrients which must be obtained in prefabricated form

B. **Homeostatic mechanisms manage an animal's fuel**

Chemical energy is obtained from the oxidation of complex organic molecules.
- Monomers from any of the complex organic molecules can be used to produce energy, although those from carbohydrates and fats are used first.
- Oxidation of a gram of fat liberates 9.5 kcal, twice that of a gram of carbohydrate or protein.

The basal energy requirements of an animal must be met to sustain their metabolic functions.
- When an animal takes in more calories than it consumes, the liver and muscles store the excess in the form of glycogen; further excess is stored in adipose tissue in the form of fat.
- When the diet is deficient in calories, glycogen stored in the liver and muscles is utilized first and fat is then withdrawn from adipose tissues.
  - An *undernourished* person or animal is one whose diet is deficient in calories.
    - If starvation persists, the body begins to breakdown its own proteins as a source of energy.
    - The breakdown of the body's own proteins can cause muscles to atrophy and can result in the consumption of the brain's proteins.
  - Obesity (overnourishment) is a greater problem in the United States and other developed countries than undernourishment.
    - It increases the risk of heart attack, diabetes, and other disorders.

C. **An animal's diet must supply essential nutrients and carbon skeletons for biosynthesis**

Heterotrophs cannot use inorganic materials to make organic molecules; they must obtain organic precursors for these molecules from the food they ingest.

Given a source of carbon and nitrogen, heterotrophs can fabricate a great variety of organic molecules by using enzymes to rearrange the molecular skeletons of precursors acquired from food.
- A single type of amino acid can supply the nitrogen necessary to build other amino acids.
- Fats can be synthesized from carbohydrates.
- The liver is responsible for most of the conversion of nutrients from one type of organic molecule to another.

An animal's diet also must include essential nutrients, in addition to providing fuel and carbon skeletons.

Essential nutrients = Chemicals an animal requires but cannot synthesize
  • Vary from species to species
  • An animal is malnourished if its diet is missing one or more essential nutrients (see Campbell, Figure 41.2).
  • Includes essential amino acids, essential fatty acids, vitamins, and minerals

Essential amino acids are those that must be obtained in the diet in a prefabricated form.
  • Most animals can synthesize about half of the 20 kinds of amino acids needed to make proteins.
  • Human adults can produce 12, leaving eight as essential in the diet. (Human infants can only produce 11.)
  • Protein deficiency results when the diet lacks one or more essential amino acids.
      ⇒ The syndrome known as kwashiorkor is a form of protein deficiency in some parts of Africa.
  • The human body cannot store essential amino acids, thus a deficiency retards protein synthesis.
      ⇒ This is most frequent in individuals, who for economic or other reasons, have unbalanced diets.

Essential fatty acids are those unsaturated fatty acids that cannot be produced by the body.
  • An example in humans is linoleic acid, which is required to produce some of the phospholipids necessary for membranes.
  • Fatty acid deficiencies are rare, as most diets include sufficient quantities.

Vitamins are organic molecules required in the diet in much smaller quantities (0.01 to 100 mg/day) than essential amino acids or fatty acids.
  • Many serve a catalytic function as coenzymes or parts of coenzymes.
  • Vitamin deficiencies can cause very severe effects as shown in Table 41.1.
  • Water-soluble vitamins are not stockpiled in the body tissues; amounts ingested in excess of body needs are excreted in the urine.
  • Fat-soluble vitamins (vitamins A, D, E, and K) can be held in the body; excess amounts are stored in body fat and may accumulate over time to toxic levels.
  • If the body of an animal can synthesize a certain compound, it is not a vitamin.
      ⇒ A compound such as ascorbic acid is a vitamin for humans (vitamin C) and must be included in our diets; it is not a vitamin in rabbits where the normal intestinal bacteria produce all that is needed.

Minerals are inorganic nutrients required in the diet in small quantities ranging from 1 mg to 2500 mg per day, depending on the mineral.
  • Some minerals serve structural and maintenance roles in the body (calcium, phosphorous) while others serve as parts of enzymes (copper) or other molecules (iron).
  • Refer to Table 41.2 for the mineral requirements of humans.

II.   **Food Types and Feeding Mechanisms**

A. **Most animals are opportunistic feeders**
Animals usually ingest other organisms.
  • The food organism may be either dead or alive.
  • The food organism may be ingested whole or in pieces.

- Some parasitic animals (e.g., tapeworms) are exceptions.

Animals are categorized based on the kinds of food they usually eat and their adaptations for obtaining and processing food items.

- Herbivores eat autotrophic organisms (plants, algae, and autotrophic bacteria).
- Carnivores eat other animals.
- Omnivores eat other animals and autotrophs.

Animals are opportunistic and may eat foods outside their principal dietary category.

- For example, most carnivores obtain some nutrients from plants that remain in the digestive tract of their prey.
- All animals consume bacteria along with their other food items.

B. Diverse feeding adaptations have evolved among animals

The varied diets exhibited by animals are accompanied by a variety of mechanisms used to obtain food.

- *Suspension-feeders* sift small food particles from the water.
  ⇒ Many are aquatic animals such as clams and oysters (trap food on gills) and baleen whales (strain food from water forced through the screen-like plates on their jaws) (see Campbell, Figure 41.4).
- *Substrate-feeders* live on or in their food source and eat their way through the food.
  ⇒ Leaf miners (larvae of various insects) tunnel through the interior of leaves (see Campbell, Figure 41.5).
- *Deposit-feeders* are a type of substrate-feeder that ingests partially decayed organic materials along with their substrate.
  ⇒ Earthworms ingest soil and their digestive systems extract the organic materials.
- *Fluid-feeders* suck nutrient-rich fluids from a living host (see Campbell, Figure 41.6).
  ⇒ Aphids ingest the phloem sap from plants; leeches and mosquitoes suck blood from animals; hummingbirds and bees ingest nectar from flowers.
- *Bulk-feeders* eat relatively large pieces of food (see Campbell, Figure 41.7).
  ⇒ Most animals; they possess various adaptations to kill prey or tear off pieces of meat or vegetation.

III. Overview of Food Processing

A. The four main stages of food processing are ingestion, digestion, absorption, and elimination

Ingestion, the act of eating, is the first stage.

Digestion, the process of breaking down food into small molecules the body can absorb, is the second stage.

- Organic food material is composed of macromolecules (proteins, fats, carbohydrates) that are too large to cross the membranes and enter an animal's cells.
- Digestion enzymatically cleaves these macromolecules into component monomers that can be used by the animal (polysaccharides and disaccharides to simple sugars; proteins to amino acids; fats to glycerol and fatty acids).
- Digestion uses *enzymatic hydrolysis* to break bonds in macromolecules.
  ⇒ Hydrolytic enzymes catalyze the digestion of each class of macromolecule by adding water.

⇒ The chemical digestion is usually preceded by mechanical fragmentation (e.g., chewing) that increases the surface area exposed to digestive juices.

⇒ Occurs in a specialized compartment where the enzymes are contained so they don't damage the animal's own cells.

Absorption is the third stage, and it involves the uptake of the small molecules resulting from digestion.

Elimination is the fourth and final stage in which undigested material passes out of the digestive compartment.

B. Digestion occurs in specialized compartments

1. Intracellular digestion

Food vacuoles are the simplest digestive compartments.

- They are organelles in which a single cell digests its food without hydrolytic enzymes mixing with the cell's cytoplasm.
- Protozoa have food vacuoles that form around food by endocytosis.
- Hydrolytic enzymes are secreted into the food vacuole and digestion occurs; this is referred to as *intracellular digestion* (see Campbell, Figure 41.8)
- Sponges differ from other animals in that all digestion is by the intracellular mechanism.

2. Extracellular digestion

Extracellular digestion occurs within compartments that are continuous, via passages, with the outside of the body.

- At least some hydrolysis occurs in most animals by this mechanism.

Gastrovascular cavity = Digestive sac with a single opening; functions in both digestion and nutrient distribution

- Most animals that have simple body plans possess a gastrovascular cavity.

Digestion in *Hydra* involves both extracellular and intracellular digestion (see Campbell, Figure 41.9).

- *Hydra* is a carnivore that captures prey.
- Food items are immobilized by stings from nematocysts on the tentacles.
- Tentacles then force prey through the mouth into the gastrovascular cavity.
- Specialized gastrodermal cells secrete digestive enzymes that fragment the soft tissues of the prey into tiny pieces.
- Some gastrodermal cells also possess flagella, whose movement prevents settling of food particles and distributes them through the cavity.
- The small pieces are phagocytized by nutritive gastrodermal cells and surrounded by food vacuoles.
- Hydrolysis is completed by intracellular digestion.
- Undigested materials are expelled from the gastrovascular cavity through the single opening.

The combination of extracellular and intracellular digestion that occurs in most animals permits these organisms to feed on larger prey items.

- Phagocytosis is limited to microscopic food.
- Extracellular digestion begins the digestive process by breaking down large food items into smaller particles.

Animals with body plans more complex than cnidarians and platyhelminths have *complete digestive tracts* or *alimentary canals*.

- *Alimentary canal* = Digestive tube running between two openings: the mouth (where food is initially ingested) and the *anus* (where undigested wastes are eliminated) (see Campbell, Figure 41.10)
- Since food moves in one direction along the tube, the tube can be organized into specialized regions that carry out digestion and absorption of nutrients in a stepwise fashion.
- The unidirectional passage of food and the specialization of function for different regions makes the alimentary canal more efficient.

IV. The Mammalian Digestive System

The digestive system in mammals includes the alimentary canal and accessory glands that secrete digestive juices into the canal through ducts.

- *Peristalsis* (rhythmic smooth muscle contractions) pushes food along the tract.
- *Sphincters* (modifications of the muscle layer into ringlike valves) occur at some junctions between compartments and regulate passage of materials through the system.
- The accessory glands are: three pairs of *salivary glands*, the *pancreas*, the *liver*, and the *gallbladder*.

Refer to Campbell, Figure 41.11, for an orientation to the specialized compartments in the human system.

A. The oral cavity, pharynx, and esophagus initiate food processing

1. The oral cavity

Physical and chemical digestion begin in the oral cavity.

- Chewing breaks down large pieces of food into smaller pieces.
- This makes food easier to swallow and increases the surface area available for enzyme action.

The presence of food in the oral cavity stimulates the salivary glands to secrete saliva into the oral cavity.

- Saliva contains mucin (protects the mouth from abrasion and lubricates food); buffers that neutralize acids; antibacterial agents; and salivary amylase, an enzyme that hydrolyzes starch and glycogen to the disaccharide maltose or small polysaccharides.

The tongue tastes and manipulates food during chewing and forms it into a *bolus*, which it pushes to the back of the oral cavity and into the pharynx.

2. The pharynx

The pharynx is an intersection for both the digestive and respiratory systems.

- The movement of swallowing moves the *epiglottis* to block the entrance of the windpipe (the glottis).
- This directs food through the pharynx and into the esophagus (see Campbell, Figure 41.12 a and b)

3. The esophagus

The esophagus is a muscular tube that conducts food from the pharynx to the stomach.

- Peristalsis moves the bolus along the esophagus to the stomach (see Campbell, Figure 41.12 c).
- The initial entrance of the bolus into the esophagus is voluntary (swallowing); once in, the peristalsis results from involuntary contraction of the smooth muscles.
- Salivary amylase remains active as the bolus moves through the esophagus.

B. The stomach stores food and performs preliminary digestion

The stomach is a large, saclike structure located just below the diaphragm on the left side of the abdominal cavity. It functions in:

Food storage

⇒ The stomach has an elastic wall with rugae, folds that can expand to accommodate up to 2 L of food.

⇒ Storage capacity permits periodic feeding (meals).

Churning

⇒ The stomach's longitudinal, vertical, and diagonal muscles contract in churning movements that mix the food.

⇒ Stomach contents are mixed about every 20 minutes.

⇒ Churning and enzyme action convert food to a nutrient broth called *acid chyme*.

◊ The passage of acid chyme into the small intestine is regulated by the *pyloric sphincter* at the bottom of the stomach.

◊ The pyloric sphincter relaxes at intervals and permits small quantities of chyme to pass.

Secretion

• Gastric secretion is controlled by nerve impulses and the hormone *gastrin*. The stomach epithelium contains three types of secretory cells.

1. Mucous cells, which secrete:

• Mucin, a thin mucus that protects the stomach lining from being digested

• *Gastrin*, a hormone produced by the stomach; gastrin is released into the bloodstream and its action is to stimulate further secretion of gastric juice (HCl and pepsin).

2. Chief cells, which secrete:

• *Pepsinogen*, an inactive protease or zymogen that is the precursor to pepsin

• *Zymogen*, an inactive form of a protein-digesting enzyme

3. Parietal cells, which secrete HCl

• Protein digestion. Both components of *gastric juice*, HCl and pepsin, are involved with protein digestion:

⇒ HCl provides acidity (pH 1 - 4) which:

◊ Kills bacteria

◊ Denatures protein

◊ Starts the conversion of pepsinogen to pepsin; newly formed pepsin can also catalyze this reaction:

⇒ Pepsin splits peptide bonds next to some amino acids.

◊ Does not hydrolyze protein completely

◊ Is an endopeptidase that splits peptide bonds located within the polypeptide chain

C. The small intestine is the major organ of digestion and absorption

The human *small intestine* is about 6 m in length and is the site of most enzymatic hydrolysis of food and absorption of nutrients.

- Remember, only limited digestion of carbohydrates occurs in the oral cavity and esophagus (by salivary amylase) and of proteins in the stomach (by pepsin).

The pancreas, liver, gall bladder, and small intestine all contribute to the digestion that occurs in the small intestine. Their products are released into the *duodenum*, the first 25 cm of the small intestine.

- The (exocrine) *pancreas* produces:
    ⇒ Hydrolytic enzymes that break down all major classes of macromolecules—carbohydrates, lipids, proteins, and nucleic acids.
    ⇒ Bicarbonate buffer that helps neutralize the acid chyme coming from the stomach.

> NOTE: The typical vertebrate pancreas is a compound gland, having both an exocrine, ducted component as noted above and an endocrine, ductless component that produces and secretes hormones, such as insulin and glucagon.

- The *liver* performs many functions including the production of bile, which:
    ⇒ Is stored in the *gallbladder*
    ⇒ Does not contain digestive enzymes
    ⇒ Contains bile salts which emulsify fat
    ⇒ Contains pigments that are byproducts of destroyed red blood cells

1. **Enzymatic action in the small intestine**
   a. **Carbohydrate digestion**
   Begins with the action of salivary amylase in the mouth.

   Begins again in the duodenum where pancreatic amylases hydrolyze starch and glycogen into disaccharides.

   Disaccharidases attached to the surface of the duodenal epithelium hydrolyze disaccharides into monosaccharides.
    ⇒ Each disaccharide has its own disaccharidase. For example, maltose is hydrolyzed by maltase, sucrose by sucrase, lactose by lactase, etc.
    ⇒ Since the disaccharidases are on the surface of the epithelium, the final breakdown of carbohydrates occurs where the sugars will be absorbed.

   b. **Protein digestion**
   Protein digestion involves the efforts of teams of enzymes (see Campbell, Figure 41.13b):
    - Pepsin, an endopeptidase, begins protein digestion in the stomach.
    - The pancreas secretes proteases in the form of zymogens that will be activated only in the lumen of the duodenum by the intestinal enzyme, *enteropeptidase*.
        - Enteropeptidase converts *trypsinogen* to *trypsin*.

   Trypsinogen $\xrightarrow{\text{enteropeptidase}}$ Trypsin

        - Trypsin the catalyzes the conversion of more trypsinogen to trypsin.
        - Trypsin catalyzes conversion of the other zymogens.

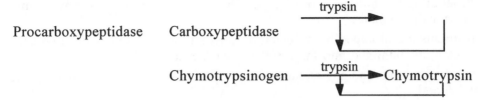

- Trypsin and chymotrypsin (endopeptidases) digest large polypeptides into shorter chains by breaking internal peptide bonds adjacent to certain amino acids.
- *Carboxypeptidase* (exopeptidase) splits amino acids, one at a time, off the end of a polypeptide that has a free carboxyl group.
- The lining of the small intestine also secretes protein-digesting enzymes, *aminopeptidase* and *dipeptidases*.
  ⇒ Aminopeptidase begins at the end of a polypeptide that has a free amino group and splits off one amino acid at a time.
  ⇒ Dipeptidases attached to the intestinal lining split small polypeptides.
- Since the protein-digesting enzymes from the pancreas and small intestine break bonds in specific areas of the polypeptide, protein digestion to amino acids is a combined effort from all of these enzymes.

c. Nucleic acid digestion

Nucleic acid digestion also involves teams of enzymes.
- *Nucleases* hydrolyze DNA and RNA into nucleotides (see Campbell, Figure 41.13c).
- Other hydrolytic enzymes (nucleotidases and nucleosidases) break nucleotides into nucleosides and nitrogenous bases, sugars and phosphates.

d. Fat digestion

Fat digestion occurs only in the small intestine, so most fat in food is undigested when it reaches the duodenum. If fat is present in chyme,
- *Emulsification* produces many small fat droplets that collectively have a large surface area exposed for digestion.
- Pancreatic *lipase*, secreted into the duodenum, hydrolyzes fats into the building blocks, glycerol and fatty acids(see Campbell, Figure 41.13d).

Macromolecules are completely hydrolyzed as peristalsis moves the digestive juice-chyme mixture through the duodenum. See Campbell, Figures 41.11 and 41.12 for a summary of digestion.

The remaining areas of the small intestine, the *jejunum* and *ileum*, are specialized for absorption of nutrients.

2. Absorption of nutrients

Nutrients resulting from digestion must cross the digestive tract lining to enter the body. While a small number of nutrients are absorbed by the stomach and large intestine, most absorption occurs in the small intestine.
- Large folds in the walls are covered with projections called villi, which in turn have many microscopic microvilli; this results in a surface area for absorption of about 300 m^2 (see Campbell, Figure 41.15).
  ⇒ This *brush border* (microvillar surface) is exposed to the lumen of the intestine.

- Penetrating the hollow core of each villus are capillaries and a tiny lymph vessel called a *lacteal*.
- Nutrients are absorbed by diffusion or active transport across the two cell-thick epithelium and into the capillaries or lacteals.
    ⇒ Amino acids and sugars enter the capillaries and are transported by the blood.
    ⇒ Absorbed glycerol and fatty acids are recombined in epithelial cells to form fats; most are coated with proteins to form *chylomicrons* which enter the lacteals.
- Capillaries and veins draining nutrients away from the villi converge into the *hepatic portal vessel*, which leads directly to the liver.
    ⇒ Here various organic molecules are used, stored, or converted to different forms.
    ⇒ Blood flows at a rate of about 1 L per minute through the hepatic portal vessel.

D. Hormones help regulate digestion

Hormonal control of digestion involves many different factors; chief among them are the four following regulatory hormones (see Campbell, Table 41.3):

1. *Gastrin.* Released from the stomach in response to presence of food; stimulates the stomach to release gastric juice (HCl and pepsin); stimulates mitosis and development of new mucosa cells.
2. *Secretin.* Released from the duodenum in response to acid chyme entering from the stomach; signals the pancreas to release bicarbonate buffer to neutralize acid chyme.
3. *Cholecystokinin (CCK).* Released from the duodenum in response to chyme entering from the stomach; signals the gall bladder to release bile and the pancreas to release pancreatic enzymes into the duodenum; also may be involved with the satiety reflex of the brain.
4. *Enterogastrone.* Released from the duodenum in response to the presence of fat in the chyme; inhibits peristalsis in the stomach, and slows digestion.

E. Reclaiming water is a major function of the large intestine

The *large intestine*, or colon, connects to the small intestine at a T-shaped junction containing a sphincter; the blind end of the T is called the *cecum* (see Campbell, Figure 41.11).

- The *appendix* is a fingerlike extension of the cecum and is composed of lymphoid tissue.
- The *colon* is about 1.5 m long and is in the shape of an inverted "U". Its major function is water reabsorption.

Feces (wastes of the digestive tract) are moved through the colon by peristalsis.

- Intestinal bacteria live on organic material in the feces, and some produce vitamin K which is absorbed by the host.
- Feces may also contain an abundance of salts.
- Feces are stored in the rectum and pass through the two sphincters (one involuntary, one voluntary) to the anus for elimination.

V. Evolutionary Adaptations of Vertebrate Digestive Systems

A. Structural adaptations of the digestive system are often associated with diet

The digestive systems of vertebrates are variations on a common plan, and many adaptations associated with diet are found.

Variation in the dentition (assortment of teeth) of mammals reflects the animal's diet (see Campbell, Figure 41.16).

- Carnivores (e.g., dogs and cats) generally have pointed canines and incisors.
    - ⇒ These teeth are used to kill prey and rip away pieces of flesh.
    - ⇒ Premolars and molars are jagged and used to crush and shred the food.
- Herbivores (e.g., cows) have teeth with broad, ridged surfaces that are used for grinding vegetation.
    - ⇒ Incisors and canines are modified for biting off pieces of vegetation.
- Omnivores (e.g., humans) have relatively unspecialized dentition.
    - ⇒ They are adapted to eat both vegetation and meat and teeth similar to those of both herbivores and carnivores are found.

Nonmammalian vertebrates typically have less specialized dentition, although there are exceptions.

- Poisonous snakes have fangs which are teeth modified to inject venom into prey.
    - ⇒ Some fangs are hollow, others are grooved.
    - ⇒ Snakes can also swallow very large prey items due to the loosely hinged lower jaw-skull articulation.

A correlation is also found between length of the vertebrate digestive system and diet.

- Herbivores and omnivores have longer alimentary canals than carnivores relative to size (see Campbell, Figure 41.17).
    - The cell walls in vegetation make it more difficult to digest than meat, and nutrients are less concentrated.
    - The longer tract allows for more time for digestion and provides a greater surface area for absorption.
- The functional length of an alimentary canal may be longer than superficial appearance reveals, as it is with the spiral valve structure in sharks.

B. Symbiotic microorganisms help nourish many vertebrates

Special fermentation chambers are present in the alimentary canals of many herbivores.

- Symbiotic bacteria and protozoa present in these chambers produce cellulase which can digest the cellulose. (Animals do not produce cellulase.)
- The microorganisms digest cellulose to simple sugars and convert the sugars to nutrients essential to the animal.
- The microorganisms may be housed in the cecum (e.g., horses), cecum and colon (e.g., rabbit), or the much more elaborate structure found in ruminants (see Campbell, Figure 41.18).

REFERENCES

Campbell, N., et al. *Biology*. 5th ed. Menlo Park, California: Benjamin/Cummings, 1999.

Marieb, Elaine N. *Human Anatomy and Physiology*. 4th ed. Redwood City, California: Benjamin/Cummings, 1997.

CHAPTER 42
CIRCULATION AND GAS EXCHANGE

OUTLINE

I. Circulation in Animals
   A. Transport systems functionally connect the organs of exchange with the body cells: an overview
   B. Most invertebrates have a gastrovascular cavity or a circulatory system for internal transport
   C. Closed cardiovascular systems accommodate gill breathing or lung breathing in the vertebrates
   D. Rhythmic pumping of the mammalian heart drives blood through pulmonary and systemic circuits
   E. Structural differences among the blood vessels correlate with regional functions of the circulatory system
   F. Natural laws governing the movement of fluids in pipes effect blood flow and blood pressure
   G. Transfer of substances between the blood and the interstitial fluid occurs across the thin walls of capillaries
   H. The lymphatic system returns fluid to the blood and aids in body defense
   I. Blood is a connective tissue with cells suspended in plasma
   J. Cardiovascular diseases are the leading cause of death in the United States and many other developed nations
II. Gas Exchange in Animals
   A. Gas exchange supplies oxygen for cellular respiration and disposes of carbon dioxide: *an overview*
   B. Gills are respiratory adaptations of most aquatic animals
   C. Tracheal systems and lungs are respiratory adaptations of terrestrial animals
   D. Control centers in the brain regulate the rate and depth of breathing
   E. Gases diffuse down pressure gradients in the lungs and other organs
   F. Respiratory pigments transport gases and help buffer the blood
   G. Deep-diving mammals stockpile oxygen and consume it slowly

OBJECTIVES

After reading this chapter and attending lecture, the student should be able to:
   1. List the major animal phyla with gastrovascular cavities, and explain why they do not need a circulatory system.
   2. Distinguish between open and closed circulatory systems.
   3. Using an arthropod as an example, describe the circulation of hemolymph.

4. Explain how hemolymph differs from blood.

5. Using an earthworm as an example, describe circulation of blood, and explain how it exchanges materials with interstitial fluid.

6. List the components of a vertebrate cardiovascular system.

7. Distinguish between an artery and a vein.

8. Using diagrams, compare and contrast the circulatory schemes of birds, amphibians, and mammals.

9. Distinguish between pulmonary and systemic circuits, and explain the function of each.

10. Explain the advantage of double circulation over a single circuit.

11. Trace a drop of blood through the human heart, listing the structures it passes through en route.

12. List the four heart valves, describe their location, and explain their function.

13. Distinguish between systole and diastole.

14. Describe the events of the cardiac cycle, and explain what causes the first and second heart sounds.

15. Define heart murmur, and explain its cause.

16. Define pulse, and describe the relationship between size and pulse rate among different mammals.

17. Define cardiac output, and explain how it is affected by a change in heart rate or stroke volume.

18. Define myogenic, and describe some unique properties of cardiac muscle which allows it to contract in a coordinated manner.

19. Define pacemaker, and describe the location of two patches of nodal tissue in the human heart.

20. Describe the origin and pathway of the action potential (cardiac impulse) in the normal human heart.

21. Explain why it is important that the cardiac impulse be delayed at the AV node.

22. Explain how the pace of the SA node can be modulated by sympathetic and parasympathetic nerves, changes in temperature, physical conditioning, and exercise.

23. Compare the structures of arteries and veins, and explain how differences in their structures are related to differences in their functions.

24. Describe how capillary structure differs from other vessels, and explain how this structure relates to its function.

25. Recall the law of continuity, and explain why blood flow through capillaries is substantially slower than it is through arteries and veins.

26. Define blood pressure and describe how it is measured.

27. Explain how peripheral resistance and cardiac output affect blood pressure.

28. Explain how blood returns to the heart, even though it must travel from the lower extremities against gravity.

29. Define microcirculation and explain how blood flow through capillary beds is regulated.

30. Explain how osmotic pressure and hydrostatic pressure regulate the exchange of fluid and solutes across capillaries.

31. Describe the composition of lymph, and explain how the lymphatic system helps the normal functioning of the circulatory system.

32. Explain why protein deficiency can cause edema.

33. Explain how the lymphatic system helps defend the body against infection.

34. Explain why vertebrate blood is classified as connective tissue.

35. List the components of blood and describe a function for each.

36. Outline the formation of erythrocytes from stem cells to destruction by phagocytic cells.

37. Outline the sequence of events that occur during blood clotting, and explain what prevents spontaneous clotting in the absence of injury.

38. Explain how atherosclerosis affects the arteries.

39. Distinguish between thrombus and embolus; atherosclerosis and arteriosclerosis; low-density lipoproteins (LDLs) and high-density lipoproteins (HDLs).

40. List the factors that have been correlated with an increased risk of cardiovascular disease.

41. Describe the general requirements for a respiratory surface and list the variety of respiratory organs adapted for this purpose.

42. Describe respiratory adaptations of aquatic animals.

43. Describe countercurrent exchange, and explain why it is more efficient than concurrent flow of water and blood.

44. Describe the advantages and disadvantages of air as a respiratory medium, and explain how insect tracheal systems are adapted for efficient gas exchange in a terrestrial environment.

45. For the human respiratory system, describe the movement of air through air passageways to the alveolus, listing the structures it must pass through on the journey.

46. Define negative pressure breathing, and explain how respiratory movements in humans ventilate the lungs.

47. Define the following lung volumes, and give a normal range of capacities for the human male:

    a. Tidal volume

    b. Vital capacity

    c. Residual volume

48. Explain how breathing is controlled.

49. List three barriers oxygen must cross from the alveolus into the capillaries, and explain the advantage of having millions of alveoli in the lungs.

50. Describe how oxygen moves from the alveolus into the capillary, and explain why a pressure gradient is necessary.

51. Distinguish between hemocyanin and hemoglobin.

52. Describe the structure of hemoglobin, explain the result of cooperative binding, and state how many oxygen molecules a saturated hemoglobin molecule can carry.

53. Draw the Hb-oxygen dissociation curve, explain the significance of its shape, and explain how the affinity of hemoglobin for oxygen changes with oxygen concentration.

54. Describe the Bohr effect, and explain how the oxygen dissociation curve shifts with changes in carbon dioxide concentration and changes in pH.

55. Explain the advantage of the Bohr shift.

56. Describe how carbon dioxide is picked up at the tissues and deposited in the lungs, describe the role of carbonic anhydrase, and state the form most of the carbon dioxide is in as it is transported.

57. Explain how hemoglobin acts as a buffer.

58. Describe respiratory adaptations of diving mammals including the role of myoglobin.

KEY TERMS

open circulatory system
hemolymph
sinuses
closed circulatory system
cardiovascular system
atrium
ventricles
arteries
arterioles
capillaries
capillary bed
venules
veins
systemic circuit
double circulation
pulmonary circuit
atrioventricular valve
semilunar valves
pulse
heart rate
cardiac cycle
systole
hemocyanin

diastole
cardiac output
stroke volume
sinoatrial (SA) node
pacemaker
atrioventricular (AV)
   node
electrocardiogram
endothelium
blood pressure
peripheral resistance
lymphatic system
lymph
lymph nodes
plasma
red blood cells
erythrocytes
hemoglobin
white blood cells
leukocytes
platelets
pluripotent stem cells
dissociation curve

erythropoietin
fibrinogen
fibrin
hemophilia
thrombus
cardiovascular disease
heart attack
stroke
artherosclerosis
arteriosclerosis
hypertension
low-density lipo-
   proteins (LDLs)
high-density lipo-
   proteins (HDLs)
gas exchange
respiratory medium
respiratory surface
gills
ventilation
countercurrent
   exchange
myoglobin

tracheal system
lungs
vocal cords
larynx
trachea
bronchi
bronchioles
alveoli
breathing
positive pressure
   breathing
negative pressure
   breathing
diaphragm
tidal volume
vital capacity
residual volume
parabronchi
breathing control
   centers
partial pressure
respiratory pigments

LECTURE NOTES

The exchange of materials (whether nutrients, gases, or waste products) between an organism and its environment must take place across a moist cell membrane.

- The molecules must be dissolved in water in order to diffuse or be transported across the membrane.
- In protozoans, the entire external surface may be used for this exchange.
- The simple, multicellular animals (sponges, cnidarians) have body structures such that each cell is exposed to the surrounding waters.

Three-dimensional animals face the problem that some of their cells are isolated from the surrounding environment.

- These animals have specialized organs, where exchange with the environment occurs, coupled with special systems for the internal transport through body fluids to the cells.
- The association of specialized organs with an internal transport system not only reduces the distance over which molecules must diffuse to enter or leave a cell, but permits regulation of internal fluid composition.

I.   Circulation in Animals

   A. **Transport systems functionally connect body cells with the organs of exchange:** *an overview*

      Diffusion is too slow of a process to transport chemicals through the body of an animal.

- Time of diffusion is proportional to the square of the distance the chemical must travel.
- If a glucose molecule takes 1 second to diffuse 100 μm, it will take 100 seconds to diffuse 1 mm.

The presence of a circulatory system reduces the distance a substance must diffuse to enter or leave a cell.

- The distance is reduced because the circulatory system connects the aqueous environment of the cell with organs specialized for exchange.
- Molecules will diffuse or be transported into the blood, which carries these molecules to the cells or to the exchange surface, depending on whether the molecules are used (nutrients) or produced (wastes) by the cells.
  ⇒ Oxygen diffuses from air in the lungs across the thin lung epithelium into the blood.
  ⇒ This oxygenated blood is carried by the circulatory system to all parts of the body.
  ⇒ As the blood passes through capillaries in the tissues, oxygen diffuses from the blood into the cells across the cells' plasma membranes.
  ⇒ Carbon dioxide is produced by the cells and moves in the opposite direction through the same system.

The circulatory system does more than move gases, it is a critical component to maintaining homeostasis of the body.

- The chemical and physical properties of the immediate surroundings of the cells can be controlled using the carrying capacity of the blood.
  ⇒ The blood passes from the cells through organs (liver, kidneys) that regulate the nutrient and waste content of the blood.

B. Most invertebrates have a gastrovascular cavity or a circulatory system for internal transport

1. Gastrovascular cavities

The cnidarian body plan does not require a specialized internal transport system. The body wall is only two cells thick and encloses a central gastrovascular cavity.

- The water inside the gastrovascular cavity is continuous with the surrounding water.
- The gastrovascular cavity functions in digestion and distribution of nutrients.
- Gastrodermal cells lining the cavity have direct access to the nutrients produced by digestion, however, the structure of the body is such that nutrients only have a short distance to diffuse to the outer cell layer (see Campbell, Figure 42.1).

Planarians and other flatworms also have a gastrovascular cavity.

- The highly ramified structure of the cavity and the flattened body shape ensure all cells are exposed to cavity contents.

A gastrovascular cavity cannot perform the necessary internal transport required by more complex animals, especially if they are terrestrial.

- These animals have some type of circulatory system.

C. Open and closed circulatory systems

Open and closed circulatory systems are alternative solutions for moving materials efficiently through the bodies of animals.

They usually function in combination with other organ systems, which increases the efficiency of the overall body.

1. *Open circulatory system* = Circulatory system in which hemolymph bathes internal organs directly while moving through sinuses (see Campbell, Figure 42.2a)

   - *Hemolymph* is a body fluid which acts as both blood and interstitial fluid.

   - Circulation results from contractions of the dorsal vessel (heart) and body movements.

   - Relaxation of the "heart" draws blood through the ostia (pores) into the vessel.

   - Chemical exchange between the hemolymph and cells occurs in the *sinuses*, which are an interconnected system of spaces surrounding the organs.

   - Insects, other arthropods, and mollusks have an open circulatory system.

2. *Closed circulatory system* = Circulatory system in which blood is confined to vessels and a distinct interstitial fluid is present

   - The heart (or hearts) pumps blood into large vessels.

   - The major vessels branch into smaller vessels, and they supply blood to organs (see Campbell, Figure 42.2b)

   - In the organs, materials are exchanged between the blood and the interstitial fluid bathing the cells.

   - A closed circulatory system is found in annelids, squids, octopuses, and vertebrates.

D. Closed cardiovascular systems accommodate gill-breathing or lung-breathing in the vertebrates

A *cardiovascular system* consists of a heart, blood vessels, and blood.

- The heart has of one *atrium* or two *atria*, chambers that receive blood, and one or two *ventricles*, chambers that pump blood out.

- *Arteries* carry blood away from the heart to organs where they branch into smaller *arterioles* that give rise to microscopic *capillaries* (the site of chemical exchange between blood and interstitial fluid).

  ⇒ Capillaries have thin, porous walls and are usually arranged into networks called *capillary beds* that infiltrate each tissue.

- Capillaries rejoin to form *venules*, which converge to form the veins that return blood to the heart.

An examination of vertebrate circulatory systems shows various adaptations have evolved within this taxon.

Fish have a two-chambered heart with one atrium and one ventricle (see Campbell, Figure 42.3a)

- Blood pumped from the ventricle goes to the gills. Here, oxygen diffuses into the gill capillaries and CO_2 diffuses out. Gill capillaries converge into arteries that carry blood to capillary beds in other organs. Blood from the organs travel through veins to the atrium of the heart, then into the ventricle.

- Blood flows through two capillary beds during each complete circuit: one in the gills and a second in the organ systems (systemic capillaries).

  - As blood flows through a capillary bed, blood pressure drops substantially.

- Blood flow to the tissues and back to the heart is aided by swimming motions.

Amphibians have a three-chambered heart with two atria and one ventricle (see Campbell, Figure 42.3b)

- Blood flows through a *pulmocutaneous circuit* (to the lungs and skin) and a *systemic circuit* (to all other organs) in a scheme called *double circulation*.
- Blood flow pattern: ventricle → lungs and skin to become oxygenated → left atrium → ventricle → all other organs → right atrium.
  - The second passage through the ventricle ensures sufficient blood pressure for the systemic circulation.
- There is some mixing of oxygen-rich and oxygen-poor blood in the single ventricle although a ridge present in the ventricle diverts most of the oxygenated blood to the systemic circuit and most of the deoxygenated blood to the pulmonary circuit.

Most reptiles (excluding crocodilians) have a three-chambered heart; although, the ventricle is partially divided.

- The anatomical arrangement provides for double circulation: a systemic circuit and a *pulmonary circuit*.
- The partially divided ventricle reduces mixing of oxygenated and deoxygenated blood (even more than in amphibians).

Birds and mammals have a four-chambered heart with two atria and two ventricles (see Campbell, Figure 42.3c). (Crocodiles have a four-chambered heart.)

- Double circulation is similar to that of amphibians except that oxygenated and deoxygenated blood do not mix due to the presence of two separate ventricles.
- The complete separation of oxygenated and deoxygenated blood increases the efficiency of oxygen delivery to the cells.

E. Rhythmic pumping of the mammalian heart drives blood through pulmonary and systemic circuits

1. The mammalian heart

To fully understand the double circulation, consider the human heart (see Campbell Figure 42.5)

- Located beneath the sternum
- Cone-shaped and about the size of a clenched fist
- Surrounded by a sac with a two-layered wall
- Comprised mostly of cardiac muscle tissue
- The two atria have thin walls and function as collection chambers for blood returning to the heart
- The ventricles have thick, powerful walls that pump blood to the organs

There are four valves in the heart which prevent backflow of blood during systole.

- The valves consist of flaps of connective tissue.
- *Atrioventricular valves* are found between each atrium and ventricle and keep blood from flowing back into the atria during ventricular contraction.
- *Semilunar valves* are located where the aorta leaves the left ventricle and where the pulmonary artery leaves the right ventricle; these prevent blood from flowing back into ventricles when they relax.
- A heart murmur is a defect in one or more of the valves that allows backflow of blood. Serious defects are usually corrected by surgical replacement of the valve.

Heart rate = The number of heartbeats per minute

- Usually measured by taking the *pulse* (the sensation you feel is the rhythmic stretching of the arteries caused by the increased pressure of blood during systole).
- In humans, the average resting heart rate of a young adult is 60-70 beats per minute; rates vary depending on the individual's activity level.
- It is not uncommon for heart rates to change during the day.
- There is an inverse relationship between animal body size and pulse; elephants have a rate of 25 beats per minute while some shrews have 600 beats per minute.

The heart chambers alternately contract and relax in a rhythmic cycle (see Campbell, Figure 42.6).

- A complete sequence of contraction and relaxation is the *cardiac cycle*.
- During *systole*, heart muscle contracts and the chambers pump blood.
- During *diastole*, the heart muscles relax and the chambers fill with blood.

Cardiac output = The volume of blood per minute that the left ventricle pumps into the systemic circuit; depends on heart rate and stroke volume

- *Stroke volume* is the amount of blood pumped by the left ventricle each time it contracts. The average human stroke volume is about 75 mL per beat.
- The cardiac output of an average human is 5.25 L/ min.

2. Maintaining the heart's rhythmic beat

Cardiac muscle cells are myogenic (self-excitable) and can contract without input from the nervous system. The tempo of contraction is controlled by the *sinoatrial (SA) node*, a specialized region of the heart, sometimes called the *pacemaker*.

- Located in the right atrium wall near the entrance of the superior vena cava (see Campbell, Figure 42.7).
- The SA node is composed of specialized muscle tissue, which has characteristics of both muscle and nerve tissue.
- Contraction of the SA node initiates a wave of excitation that spreads rapidly from the node and causes the two atria to contract in unison.
- This wave of contraction will pass down the atria until it reaches the *atrioventricular (AV) node*; a second mass of specialized muscle tissue located near the base of the wall separating the atria.
- The impulse is delayed at the AV node for 0.1 second to ensure the atria are completely empty before the ventricles contract.
- The impulse is then carried by a mass of specialized muscle fibers throughout the walls of the ventricle.
- The impulses produce electrical currents as they pass through the cardiac muscle.
  - The currents can be detected by electrodes placed on the skin and recorded as an *electrocardiogram (ECG or EKG)*.

Although the SA node controls the rate of heartbeat, it is influenced by several factors:

- Two antagonistic sets of nerves influence the heart rate; one speeds up contractions in the SA node, the other slows contractions.
- Hormones influence the SA node; for example, epinephrine increases heart rate.
- Others factors, including body temperature changes and exercise, also have an influence.

- Exercise creates a greater demand for oxygen in the muscles and an increase in heart rate is an adaptation to meet this demand.

F. Structural differences among the blood vessels correlate with regional functions of the circulatory system

The walls of arteries and veins have three layers:

- An outer layer of connective tissue with elastic fibers that permits stretching and recoil of the vessel
- A middle layer of smooth muscle and elastic fibers
- An inner *endothelium* of simple squamous epithelium (see Campbell, Figure 42.8)

The middle and outer layers of arteries are thicker than those in veins.

- The thicker walls of arteries provide the strength necessary to accommodate the high force with which the heart pumps the blood; the elasticity helps to even out the flow of blood so that it is not pulsatile.
- The thinner-walled veins accommodate the low pressure and low velocity of blood following emergence from capillary beds; flow is facilitated by muscular activity.

Capillaries are comprised only of the endothelial lining, which permits the exchange of chemicals with the interstitial fluids.

G. Natural laws governing the movement of fluids in pipes effect blood flow and blood pressure

1. Blood flow velocity

There is a great difference in the speed at which blood flows through the various parts of the circulatory system. Blood travels about 30 cm/sec in the aorta and about 0.026 cm/sec in capillaries.

- The velocity decreases in accordance with the *law of continuity* which states that a fluid will flow faster through narrow portions of a pipe than wider portions if the volume of flow remains constant.
- An artery gives rise to so many arterioles and then capillaries that the total diameter of vessels is much greater in capillary beds than in the artery, thus blood flows more slowly in capillaries.
- Resistance to blood flow is greater in the smaller vessels since the blood contacts more epithelial surface area.
- Blood flows faster as it enters the venules and veins since the cross-sectional area is decreased (see Campbell, Figure 42.10).

2. Blood pressure

Fluids are driven through pipes by hydrostatic pressure, which is the force exerted by fluids against the surfaces they contact.

Blood pressure = The hydrostatic force that blood exerts against a vessel wall

- Pressure is greater in arteries than in veins and greatest during ventricular systole.
  - ⇒ This is the main force propelling blood from the heart through the vessels.
- *Peripheral resistance* results from impedance by arterioles; blood enters the arteries faster than it can leave.
  - ⇒ Thus, there is a pressure even during diastole, driving blood into capillaries continuously.
- Determined by cardiac output and degree of peripheral resistance

⇒ Stress may trigger neural and hormonal responses that cause the smooth muscles of vessel walls to contract, constricting blood vessels and increasing resistance.

- In veins, blood pressure is near zero; blood returns to the heart by the action of skeletal muscles around the veins.

⇒ Veins have valves that allow blood to flow only toward the heart.

- Breathing also helps return blood to the heart, since the pressure change in the thoracic cavity during inhalation cause the vena cava and large veins near the heart to expand and fill.

H. Transfer of substances between the blood and the interstitial fluid occurs across the thin walls of capillaries

1. Blood flow through capillary beds

All tissues and organs receive a sufficient supply of blood even though only 5% to 10% of the capillaries are carrying blood at any one time; the supply is adequate due to the vast number of capillaries present in each tissue.

- Capillaries in the brain, heart, kidneys, and liver usually carry a full load of blood.

Two mechanisms regulate the distribution of blood in capillary beds. Both are controlled by nerve signals and hormones.

1. One mechanism involves the contraction and relaxation of the smooth muscle layer in the walls of arterioles.

⇒ Contraction of the muscle layer constricts the arteriole and reduces the blood flow from it into the capillary bed.

⇒ Relaxation of the muscle layer dilates the arteriole and increases the blood flow into the capillary bed (see Campbell, Figure 42.11).

2. The second mechanism involves the contraction and relaxation of precapillary sphincters, which are rings of smooth muscle located at the entrance to capillary beds.

⇒ Contraction of these sphincters reduces blood flow into the capillary bed; relaxation increases blood flow into the capillary bed.

The diversion of blood from one area of the body to another changes the blood supply to capillary beds.

- After ingesting food, the digestive tract receives a large supply of blood due to dilation of arterioles and the opening of precapillary sphincters associated with the system.

- During exercise, blood is diverted to the skeletal muscles.

2. Capillary exchange

The exchange of materials between the blood and interstitial fluids that are in direct contact with the cells occurs across the thin walls of capillaries.

- The capillary wall is a single, "leaky" layer of flattened endothelial cells that overlap at their edges.

- Materials may cross these cells in vesicles (via endocytosis and exocytosis), by diffusion through the cell, or by bulk flow between the cells due to hydrostatic pressure.

Direction of fluid movement at any point along a capillary depends on the relative forces of hydrostatic pressure and osmotic pressure (see Campbell, Figure 42.12).

- Fluid flows out of a capillary at the upstream end near an arteriole and into a capillary at the downstream end near a venule.

- About 85% of the fluid which leaves the blood at the arteriole end of the capillary, re-enters from the interstitial fluid at the venous end.
- The remaining 15% of the fluid is eventually returned by the lymphatic system.

I.   The lymphatic system returns fluid to the blood and aids in body defense

Capillary walls leak fluid and some blood proteins, which return to the blood via the *lymphatic system*. This is the 15% (4 L/day) of fluid that does not re-enter the capillaries.

- The fluid, *lymph*, is similar in composition to interstitial fluid.
- The lymph enters the system by diffusing into lymph capillaries which intermingle with the blood capillaries.
- The lymphatic system drains into the circulatory system at two locations near the shoulders.

Lymph vessels have valves that prevent backflow and depend mainly on movement of skeletal muscles to squeeze the fluid along.

- Rhythmic contractions of vessel walls also help draw fluid into the lymphatic capillaries.

Lymph nodes are specialized swellings along the system that filter the lymph and attack viruses and bacteria.

⇒ This defense is conducted by specialized white blood cells inhabiting the lymph nodes.

⇒ The nodes become swollen and tender during an infection due to rapid multiplication of the white blood cells present.

Lymph capillaries penetrate small intestine villi and absorb fats, thus transporting them from the digestive system to the circulatory system.

J.   Blood is a connective tissue with cells suspended in plasma

Vertebrate blood is connective tissue with several cell types suspended in a liquid matrix called *plasma*.

- The average human has 4 to 6 L of whole blood (plasma + cellular elements).
- Cellular elements are the cells and cell fragments of the blood and represent about 45% of the blood volume (see Figure 42.13).

1.   Plasma

Water accounts for 90% of plasma, which also contains electrolytes and plasma proteins.

- *Electrolytes* = Inorganic salts in the form of dissolved ions that help maintain osmotic balance of the blood; some also help buffer the blood
- Electrolyte balance in the blood is maintained by the kidneys.

Plasma proteins help buffer blood, help maintain the osmotic balance between blood and interstitial fluids, and contribute to its viscosity.

- Some escort lipids through blood, some are immunoglobulins, some (fibrinogens) are clotting factors.
- Serum is blood plasma that has had the clotting factors removed.

Plasma also contains substances in transit through the body such as nutrients, metabolic wastes, respiratory gases, and hormones.

2.   Cellular elements

Erythrocytes = *Red blood cells (RBC)*; biconcave discs that function in transport of oxygen.

- Each cubic millimeter of human blood contains 5 to 6 million erythrocytes.

- In mammals, RBCs lack nuclei and mitochondria; they generate ATP exclusively by anaerobic metabolism.
- They contain *hemoglobin*, an iron-containing protein that reversibly binds oxygen. About 250 million molecules of hemoglobin are in each erythrocyte.

Leukocytes = *White blood cells* that function in defense and immunity.

- There are five major types of leukocytes: basophils, eosinophils, neutrophils, lymphocytes and monocytes.
- There are usually 5000 to10,000 leukocytes per cubic millimeter of blood, although this number increases during an infection.
- Actually spend most of their time outside the circulatory system in the interstitial fluid and lymphatic system; large numbers are found in the lymph nodes.
- Lymphocytes become specialized during an infection and produce the body's immune response.

Platelets = Fragments of cells 2 to 3 μm in diameter

- Originate as pinched-off cytoplasmic fragments of large cells in the bone marrow
- Lack nuclei
- Function in blood clotting

3. Replacement of cellular elements

The cellular elements of the blood must be replaced as they wear out.

- The average erythrocyte circulates in the blood for 3 to 4 months before being destroyed by phagocytic cells in the liver and spleen.
- The components are usually recycled with new molecules being constructed from the macromolecule components of the old cells.

The *pluripotent stem cells* give rise to all three of the cellular elements.

- These are found in the red marrow of bones, especially the ribs, vertebrae, breastbone, and pelvis.
- Form in the early embryo and are renewed by mitosis (see Campbell, Figure 42.14).
- Produce a number of new blood cells equivalent to the number of dying cells.

Erythrocyte production is controlled by a negative-feedback mechanism. If tissues are not receiving enough oxygen, the kidneys convert a plasma protein to the hormone *erythropoietin*, which stimulates production of erythrocytes in the bone marrow.

Using DNA technology to correct genetic defects in pluripotent stem cells may provide a treatment for such diseases as leukemia and sickle-cell anemia.

4. Blood clotting

A clot forms when platelets clump together to form a temporary plug and release clotting factors (some are also released from damaged cells) that initiate a complex reaction resulting in conversion of inactive *fibrinogen* to active *fibrin* (see Campbell, Figure 42.15)

- Fibrin aggregates into threads that form the clot.

An inherited defect in any of the steps involved in clot formation results in *hemophilia*, a disorder characterized by excessive bleeding from minor injuries.

Anticlotting factors normally prevent spontaneous clotting in the absence of injury.

- On occasion, spontaneous clots called a *thrombus* form from platelets aggregating with coagulated fibrin. Such clots are more likely to form in individuals with cardiovascular disease.

K. Cardiovascular diseases are the leading cause of death in the United States and many other developed nations

Diseases of the heart and blood vessels are referred to as *cardiovascular diseases*.

- Account for more than 50% of all deaths in the United States
- May culminate in a heart attack or stroke
  - *Heart attack* = Death of the cardiac muscle resulting from prolonged blockage of one or more coronary arteries
  - *Stroke* = Death of nervous tissue in the brain often resulting from blockage of arteries in the brain

A thrombus is often associated with a heart attack or stroke.

- *Thrombus* = A blood clot that blocks a key blood vessel
- If it blocks the coronary arteries, a heart attack occurs.
- An *embolus* is a moving clot.
- If the thrombus or embolus blocks an artery in the brain, a stroke results.

Arteries may gradually become impaired by atherosclerosis (see Campbell, Figure 42.16).

- *Atherosclerosis* = Chronic cardiovascular disease characterized by plaques that develop on the inner walls of arteries and narrow the bore of the vessels
- Decreases blood flow through the vessels.
- Increase the risk of clot formation and heart attack.

Arteriosclerosis = Degenerative condition of arteries (form of atherosclerosis) in which plaques become hardened by calcium deposits

Angina pectoris = Chest pains that occur when the heart receives insufficient oxygen due to the build up of plaques in the arteries

- An indicator that the heart is not receiving sufficient oxygen

Hypertension = High blood pressure; may promote atherosclerosis

- Increases risk of heart attack and stroke

Smoking, lack of exercise, and a diet rich in animal fats correlate with increased risk of cardiovascular disease.

Abnormally high concentrations of low density lipoproteins (LDLs) in the blood correlate with atherosclerosis; high density lipoproteins(HDLs) may reduce deposition of cholesterol in arterial plaques.

- Exercise tends to increase HDL concentration.
- Smoking increases the LDL to HDL ratio.

II.   Gas Exchange in Animals

A. Gas exchange supplies oxygen for cellular respiration and disposes of carbon dioxide: *an overview*

Circulatory systems transport oxygen and carbon dioxide between respiratory organs and other parts of the body.

Gas exchange = The movement of O_2 and CO_2 between the animal and its environment (see Campbell, Figure 42.17)

- Supports cellular respiration by supplying O_2 and removing CO_2.

The *respiratory medium* (source of oxygen) is air for terrestrial animals and water for aquatic animals.

- Air is 21% oxygen, while the amount of dissolved oxygen in water varies due to temperature, solute concentrations, and other factors.

Respiratory surface = Portion of the animal surface where gas exchange with the respiratory medium occurs. Oxygen diffuses in; carbon dioxide diffuses out

- O_2 and CO_2 can only diffuse through membranes if they are first dissolved in the water that coats the respiratory surface.
- This surface must be large enough to provide O_2 and expel CO_2 for the entire body.

A variety of respiratory surfaces have evolved that are adaptive for organism size and environment.

- Protozoa and other unicellular organisms exchange gases over their entire surface area.
- Animals such as sponges, cnidarians, and flatworms have body structures such that the plasma membrane of each body cell contacts the outside environment and can function in exchange.
- In animals with a more three-dimensional body plan, most of the body cells are isolated from the environment and the respiratory surfaces are generally thin, moist epithelium with a rich blood supply.
  - ⇒ Usually only a single cell layer separates the respiratory medium from the blood or capillaries.

Animals that are relatively small or have a shape (long, thin) that results in a high surface area to volume ratio may use their outer skin as a respiratory organ.

- Earthworms have moist skin which overlays a dense network of capillaries.
- Gases diffuse across the entire surface (O_2 in; CO_2 out) and into the circulatory system.
- Earthworms and other animals (some amphibians) that use their skin for gas exchange must live in water or damp places in order to keep the exchange surface moist.

Most other animals lack sufficient body surface area to exchange gases for the entire body. These animals possess a region of the body surface that is extensively branched or folded, thus providing a large enough respiratory surface area for gas exchange.

- In most aquatic animals, external *gills* are present and are in direct contact with the water.
- Terrestrial animals have internal respiratory surfaces that open to the atmosphere through narrow tubes.
  - ⇒ Air lacks the supportive density and moisturizing qualities of water.
  - ⇒ Lungs and insect tracheae are two variations.

B. Gills are respiratory adaptations of most aquatic animals

Gills are outfoldings of the body surface specialized for gas exchange (see Campbell, Figure 42.18).

- In some invertebrates (e.g., echinoderms), gills have simple shapes and are distributed over the entire body.
- In other invertebrates (e.g., annelids), gills may be flap-like and extend from each body segment or be clustered at one end and be long and feathery.
- Other animals (mollusks, fishes) have gills that are localized on a body region where the surface is finely subdivided to provide a large amount of surface area.

- The gills must be efficient; although water keeps the respiratory surface wet, it has a lower oxygen concentration than air.

Ventilation = Any method of increasing the flow of the respiratory medium over the respiratory surface; brings in a fresh supply of O_2 and removes CO_2

Due to the density and low oxygen concentration of water, fish must expend a large amount of energy to ventilate water.

- Fish have a unique arrangement of blood vessels in their gills, which maximizes O_2 uptake from H_2O.
- Called *countercurrent exchange*, blood flows opposite to the direction in which water passes over gills, maintaining a constant concentration gradient for oxygen between the blood and the water passing over the gill surface (see Campbell, Figures 42.19 and 42.20)

C. Tracheal systems and lungs are respiratory adaptations of terrestrial animals

Air has several advantages over water as a respiratory medium:

- A higher oxygen concentration
- Oxygen and carbon dioxide diffuse faster through air than water.
- Respiratory surfaces do not have to be ventilated as thoroughly.

The major disadvantage is that respiratory surfaces are continually desiccated.

- The evolution of respiratory surfaces within the body (tracheae, lungs) helped solve this problem.

1. Tracheal systems

Tracheae are tiny air tubes that branch throughout the insect body; air enters the system through pores called spiracles and diffuses through the small branches which extend to the surface of nearly every cell (see Campbell, Figure 42.21).

- Some small insects rely on diffusion alone to move O_2 into the system and CO_2 out; others use rhythmic body movements for ventilation.
- Cells are exposed directly to the respiratory medium so insects do not use their circulatory systems to transport O_2 and CO_2.
  - This is a major reason why the open circulatory system works so well in insects.

2. Lungs

Lungs are highly vascularized invaginations of the body surface that are restricted to one location.

- The circulatory system must transport oxygen from the lungs to the rest of the body.

Land snails use an internal mantle as a lung.

Spiders possess booklungs.

Various degrees of lung development are found in terrestrial vertebrates: frogs have simple balloonlike lungs with limited surface area; mammals have highly subdivided lungs with a large surface area.

3. Mammalian respiratory systems: *a closer look*

Located in the thoracic cavity, mammalian lungs are enclosed in a sac consisting of two layers held together by the surface tension of fluid between the layers.

Air entering the nostrils is filtered by hairs, warmed and moistened.

The air then travels through the pharynx, then through the glottis, and into *larynx* (which possesses vocal cords and functions as a voice box).

The flow then enters the cartilage-lined *trachea* that forks into two *bronchi* which further branch into finer *bronchioles* that dead-end in *alveoli*.

Alveoli are lined with a thin layer of epithelium which serves as the respiratory surface.

Oxygen dissolves in the moist film covering the epithelium and diffuses across to the capillaries covering each alveolus; carbon dioxide moves in the opposite direction by diffusion.

4. **Ventilating the lungs**

Vertebrates ventilate lungs by *breathing* (alternate inhalation and exhalation of air).

Maintains a maximum O_2 concentration and minimum CO_2 concentration in the alveoli.

Frogs ventilate the lungs by *positive pressure breathing*.

- Air is pushed down the windpipe into the lungs.
- Air is pulled into the mouth by lowering the floor region; this enlarges the oral cavity.
- The nostrils and mouth are closed and the floor of the mouth is raised, forcing air down the trachea.
- Air is exhaled by elastic recoil of lungs and by compression of the lungs by the muscular body wall.

Mammals ventilate their lungs by *negative pressure breathing*.

- During inhalation, air is pulled into the lungs by the negative pressure created as the thoracic cavity enlarges by two possible mechanisms:
- When a mammal is at rest, most of the shallow inhalation results from contraction of the *diaphragm*.
  ⇒ The diaphragm is a dome-shaped, thin sheet of muscle that forms the bottom wall of the thoracic cavity (see Campbell, Figure 42.23). When it contracts, it pushes downward towards the abdomen, enlarging the thoracic cavity.
  ⇒ This lowers the air pressure in the lungs below atmospheric pressure and causes inhalation.
- Action of the rib muscles in increasing lung volume is important during vigorous exercise.
  ⇒ Contraction of the rib muscles pulls the ribs upwards, which expands the rib cage.
  ⇒ As the thoracic cavity enlarges, the lungs also expand, since the surface tension of the fluid between the layers of the lung sac causes the lungs to follow.
  ⇒ Lung volume increases, resulting in negative pressure within the alveoli, causing air to rush in.

Exhalation occurs when the diaphragm and/or the rib muscles relax, decreasing the volume of the thoracic cavity.

The amount of air inhaled and exhaled depends upon size, activity level, and state of health.

- *Tidal volume* is the volume of air an animal inhales and exhales with each breath during normal quiet breathing. Averages about 500 mL in humans.
- *Vital capacity* is the maximum air volume that can be inhaled and exhaled during forced breathing; averages 3400 mL to 4800 mL in college-age females and males, respectively.
- *Residual volume* is the amount of air that remains in the lungs even after forced exhalation.

Birds have a more complex process for ventilation (see Campbell, Figure 42.24):

- Besides lungs, birds have 8 or 9 air sacs in their abdomen, neck and wings that serve to trim the density of the body and act as sinks for the heat dissipation by metabolism of flight muscles.
- The air sacs do not exchange gases, they serve as bellows to keep the air moving.
- The lungs also have tiny channels called *parabronchi*; air flows through the entire system, lungs and air sacs, in only one direction regardless of whether it is inhaling or exhaling.
- The continuous flow of air through the parabronchi provides a constant supply of oxygen to the blood whether the bird is inhaling or exhaling.

D. Control centers in the brain regulate the rate and depth of breathing

Breathing is an automatic action. We inhale when nerves in the *breathing control centers* of the medulla oblongata and pons send impulses to the rib muscles or diaphragm, stimulating the muscles to contract (see Campbell, Figure 42.25).

- This occurs about 10 to 14 times per minute and the degree of lung expansion is controlled by a negative-feedback mechanism involving stretch receptors in the lungs.
- The medulla's control center also monitors blood and cerebrospinal fluid pH, which drops as blood CO_2 concentrations increase. When it senses a drop in pH (increased CO_2 level), the tempo and depth of breathing are increased and the excess CO_2 is removed in exhaled air.

O_2 concentration in the blood only affects the breathing control centers when it becomes severely low.

- ⇒ O_2 sensors in the aorta and carotids send signals to the breathing control centers and the centers respond by increasing the breathing rate.

The breathing control centers thus respond to a variety of neural and chemical signals.

- The response is an adjustment in the rate and depth of breathing to meet the demands of the body.
- The control is effective since it is coordinated with control of the circulatory system.

E. Gases diffuse down pressure gradients in the lungs and other organs

The diffusion of gases, whether in air or water, depends on differences in *partial pressure* (e.g., P_{O_2} = partial pressure of oxygen; note similarity to the diffusion of solutes being dependent on solute concentration). Oxygen comprises about 21% of the atmosphere and carbon dioxide about 0.03%.

- The partial pressure of a gas is the proportion of the total atmospheric pressure (760 mm) contributed by the gas; P_{O_2} = 160 mm Hg (0.21 × 760) and P_{CO_2} = 0.23 mm Hg.

Gases always diffuse from areas of high partial pressure to those of low partial pressure (see Campbell, Figure 42.26).

- Blood arriving at the lungs from the systemic circulation has a lower P_{O_2} and a higher P_{CO_2} than air in the alveoli.
  - ⇒ Thus, the blood exchanges gases with air in the alveoli and the P_{O_2} of the blood increases while the P_{CO_2} decreases.
- In systemic capillaries gradients of partial pressure favor diffusion of oxygen out of the blood and diffusion of CO_2 into it, since cellular respiration rapidly depletes interstitial fluid of O_2 and adds CO_2.

F. Respiratory pigments transport gases and help buffer the plasma

1. Oxygen transport

Oxygen is carried by *respiratory pigments* in the blood of most animals since oxygen is not very soluble in water.

- The pigments are proteins which contain metal atoms; the metal atoms are responsible for the color of the protein.

In arthropods and mollusks, *hemocyanin* is the O_2 carrying pigment.

- The oxygen-binding component is copper and results in a blue color.
- It is dissolved directly in plasma, not confined to cells.

Hemoglobin is the oxygen-transporting pigment in almost all vertebrates.

- Consists of four subunits, each containing a heme group that bonds O_2; an iron atom is at the center of each heme group and actually binds to oxygen.
- The binding of oxygen to hemoglobin is reversible. (Binding occurs in the lungs; release occurs in the tissues.)
- Binding of O_2 to one subunit induces a shape change that increases the affinity of the other three subunits for oxygen (cooperativity).
- The unloading of oxygen from one heme group results in a conformational change that stimulates unloading from the other three.
- The cooperative nature of this mechanism is evident in the *dissociation curve* depicted in Figure 42.27.
- Bohr shift is the lowering of hemoglobins' affinity for oxygen upon a drop in pH (see Campbell, Figure 42.27b) This occurs in active tissues due to the entrance of CO_2 into the blood.

2. Carbon dioxide transport

Hemoglobin not only transports oxygen, but it also helps the blood transport carbon dioxide and assists in buffering the blood against harmful pH changes (see Campbell, Figure 42.28).

Carbon dioxide is transported by the blood in three forms:

- Dissolved CO_2 in the plasma (7%)
- Bound to the amino groups of hemoglobin (23%)
- As bicarbonate ions in the plasma (70%)

Carbon dioxide from cells diffuses into the blood plasma and then into erythrocytes. In the erythrocytes, carbonic anhydrase catalyzes a reversible reaction wherein CO_2 is converted into bicarbonate.

- The CO_2 reacts with water to form carbonic acid.
- The carbonic acid quickly dissociates to bicarbonate and hydrogen ions.
- Bicarbonate then diffuses out of the erythrocyte and into the blood plasma (in exchange for a chloride ion = chloride shift).
- The hydrogen ions attach to hemoglobin and other proteins which results in only a slight change in the pH and the dissociation of oxygen from hemoglobin.
- The process is reversed in the lungs.

G. Deep-diving mammals stockpile oxygen and consume it slowly

Diving mammals such as seals, dolphins, and whales have special adaptations which allow them to make long underwater dives.

Weddell seals, which make dives to 200 to 500 m depths for 20 minutes or more, have been extensively studied (see Campbell, Figure 42.29).

- These mammals can store large amounts of oxygen, about twice as much per kilogram body weight as humans.
  - ⇒ 70% of the oxygen load is found in the blood (51% in humans) and 5% in the lungs (36% in humans).
  - ⇒ They also possess a higher *myoglobin* (a respiratory pigment similar to hemoglobin except that it is comprised of a single subunit contains one heme unit) concentration in their muscles.
- The high oxygen load in the blood may be due to the higher blood volume in these seals (twice the amount of blood per kilogram of body weight as humans).
- The seals also have a very large spleen which can store about 24 L of blood.
  - ⇒ The spleen probably contracts after a dive and forces more erythrocytes loaded with oxygen into the blood.
- The high concentration of myoglobin permits these seals to store about 25% of their oxygen in muscles (compared to 13% in humans).

These adaptations provide diving mammals with a large oxygen reservoir at the beginning of a dive, but they also have several adaptations to conserve oxygen during the dive.

- A diving reflex slows the pulse, and oxygen consumption declines as the cardiac output slows.
- Most of the blood is routed to the brain, spinal cord, eyes, adrenal glands, and placenta due to regulatory mechanisms that reduce blood flow to the muscles.
- The reduced blood flow to the muscles results in a depletion of the myoglobin oxygen reserves after about 20 minutes, the muscles then shift to anaerobic production of ATP.

The special adaptations found in diving mammals emphasizes that animals can respond to environmental pressures in the short term through physiological adaptation as well as in the long term by natural selection.

REFERENCES

Campbell, N., et al. *Biology*. 5th ed. Menlo Park, California: Benjamin/Cummings, 1999.

Marieb, E.N. *Human Anatomy and Physiology*. 4th ed. Redwood City, California: Benjamin/Cummings, 1997.

CHAPTER 43
THE BODY'S DEFENSES

OUTLINE

OBJECTIVES

After reading this chapter and attending lecture, the student should be able to:

1. Explain what is meant by nonspecific defense, and list the nonspecific lines of defense in the vertebrate body.
2. Explain how the physical barrier of skin is reinforced by chemical defenses.
3. Define phagocytosis, and list two types of phagocytic cells derived from white blood cells.
4. Explain how the function of natural killer cells differs from the function of phagocytes.
5. Describe the inflammatory response including how it is triggered.
6. Explain how the inflammatory response prevents the spread of infection to surrounding tissue.
7. List several chemical signals that initiate and mediate the inflammatory response.

8. Describe several systemic reactions to infection, and explain how they contribute to defense.

9. Describe a plausible mechanism for how interferons can fight viral infections and might act against cancer.

10. Explain how complement proteins may be activated and how they function in cooperation with other defense mechanisms.

11. Explain how the immune response differs from nonspecific defenses.

12. Distinguish between active and passive immunity.

13. Explain how humoral immunity and cell-mediated immunity differ in their defensive activities.

14. Outline the development of B and T lymphocytes from stem cells in red bone marrow.

15. Describe where T and B cells migrate, and explain what happens when they are activated by antigens.

16. Characterize antigen molecules, in general, and explain how a single antigen molecule may stimulate the immune system to produce several different antibodies.

17. Describe the mechanism of clonal selection.

18. Distinguish between primary and secondary immune response.

19. Describe the cellular basis for immunological memory.

20. Describe the cellular basis for self-tolerance.

21. Explain how the humoral response is provoked.

22. Explain how B cells are activated.

23. Diagram and label the structure of an antibody, and explain how this structure allows antibodies to perform the following functions:

    a. Recognize and bind to antigens

    b. Assist in destruction and elimination of antigens

24. Distinguish between variable (V) regions and constant (C) regions of an antibody molecule.

25. Compare and contrast the structure and function of an enzyme's active site and an antibody's antigen-binding site.

26. List the five major classes of antibodies in mammals and distinguish among them.

27. Describe the following effector mechanisms of humoral immunity triggered by the formation of antigen-antibody complexes:

    a. Neutralization   c. Precipitation

    b. Agglutination   d. Activation of complement system

28. Explain how monoclonal antibodies are produced and give examples of current and potential medical uses.

29. Explain how T-cell receptors recognize "self" and how macrophages, B cells, and some T cells recognize one another in interactions.

30. Describe an antigen-presenting cell (APC).

31. Design a flow chart describing the sequence of events that follows the interaction between antigen presenting macrophages and helper T cells, including both cell-mediated and humoral immunity.

32. Define cytokine, and distinguish between interleukin I and interleukin II.

33. Distinguish between T-independent antigens and T-dependent antigens.

34. Describe how cytotoxic T cells recognize and kill their targets.

35. Explain how the function of cytotoxic T cells differs from that of complement and natural killer cells.

36. Describe the function of suppressor T cells.
37. Distinguish between complement's classical and alternative activation pathways.
38. Describe the process of opsonization.
39. For ABO blood groups, list all possible combinations for donor and recipient in blood transfusions; indicate which combinations would cause an immune response in the recipient; and state which blood type is the universal donor.
40. Explain how the immune response to Rh factor differs from the response to A and B blood antigens.
41. Describe the potential problem of Rh incompatibility between a mother and her unborn fetus, and explain what precautionary measures may be taken.
42. Explain why, other than with identical twins, it is virtually impossible for two people to have identical MHC markers.
43. Describe the rejection process of transplanted tissue in terms of normal cell-mediated immune response, and describe how the immune system can be suppressed in transplant patients.
44. List some known autoimmune disorders, and describe possible mechanisms of autoimmunity.
45. Explain why immunodeficient individuals are more susceptible to cancer than normal individuals.
46. Describe an allergic reaction including the role of IgE, mast cells, and histamine.
47. Explain what causes anaphylactic shock and how it can be treated.
48. Recall the infectious agent that causes AIDS and explain how it weakens the immune system.
49. Explain how AIDS is transmitted and why it is difficult to produce vaccines to protect uninfected individuals.
50. Describe what it means to be HIV-positive.
51. Explain how general health and mental well being might affect the immune system.

KEY TERMS

lysozyme
phagocytosis
macrophages
eosinophils
natural killer cells
inflammatory response
histamine
basophils
mast cells
prostaglandins
chemokines
pyrogens
complement system
interferon
B lymphocytes (B cells)
T lymphocytes (T cells)
antigens
antibodies

effector cells
memory cells
clonal selection
primary immune response
plasma cells
secondary immune response
major histocompatibility complex (MHC)
Class I MHC
Class II MHC
antigen presentation
cytotoxic T cells (T_C)
helper T cells (T_H)
humoral immunity
cell-mediated immunity
antigen-presenting cells (APCs)

interleukin-1 (IL-1)
suppressor T cells (T_S)
CD8
CD4
target cell
perforin
tumor antigen
T-dependent antigens
T-independent antigens
epitope
immunoglobulins (Ig)
heavy chains
light chains
monoclonal antibodies
neutralization
opsonization
agglutination
complement fixation

immune adherence
active immunity
immunization
vaccination
passive immunity
transfusion reaction
Rh factor
graft versus host reaction
anaphylactic shock
acquired immunodeficiency syndrome (AIDS)
opportunistic diseases
human immunodeficiency

| antigen receptors | cytokines | membrane attack | virus (HIV) |
| T cell receptors | interleukin-2 (IL-2) | complex (MAC) | HIV-positive |

LECTURE NOTES

The vertebrate body possesses two mechanisms which protect it from potentially dangerous viruses, bacteria, other pathogens, and abnormal cells which could develop into cancer (see Campbell, Figure 43.1 for an overview diagram).

1. One of these mechanisms is *nonspecific*, that is, it does not distinguish between infective agents.

2. The second mechanism is *specific* in that it responds in a very specific manner (e.g., production of antibodies) to the particular type of infective agent.

I.   Nonspecific Defense Against Infection

Nonspecific defense mechanisms help prevent entry and spread of invading microbes in an animal's body.

- An invading microbe must cross the external barrier formed by the skin and mucous membranes.

- If the external barrier is penetrated, the microbe encounters a second line of defense: interacting mechanisms of phagocytic white blood cells, antimicrobial proteins, and the inflammatory response.

A. The skin and mucous membranes provide first-line barriers to infection

The skin and mucous membranes act as physical barriers preventing entry of pathogens, and as chemical barriers of anti-pathogen secretions.

- In humans, oil and sweat gland secretions acidify the skin (pH 3 to 5), which discourages microbial growth.

- The normal bacterial flora of the skin (adapted to the acidity) may release acids and other metabolic wastes to further inhibit pathogen growth.

- Saliva, tears, and mucous secretions contain antimicrobial proteins and wash away potential invading microbes.

- An enzyme (*lysozyme*) in perspiration, tears, and saliva attacks the cell walls of many bacteria and destroys other microbes entering the respiratory system and eyes.

- In the respiratory tract, nostril hairs filter inhaled particles and mucus traps microorganisms that are then swept out of the upper respiratory tract by cilia, thus preventing their entrance into the lungs (see Campbell, Figure 43.2).

- In the digestive tract, stomach acid kills many bacteria that enter with foods or those trapped in swallowed mucus from the upper respiratory system.

B. Phagocytic cells, inflammation, and antimicrobial proteins function early in infection

Microbes that penetrate the skin or mucous membranes encounter amoeboid white blood cells capable of *phagocytosis* or cell lysis.

1. Phagocytic and natural killer cells

Neutrophils are cells that become phagocytic in infected tissue.

- Comprise 60% – 70% of total white cells

- Attracted by chemical signals, they enter infected tissues by amoeboid movement

- Only live a few days as they destroy themselves when destroying pathogens

Monocytes comprise only about 5% of the white blood cells, but they provide an even more effective phagocytic defense. They mature, circulate for a few hours, then migrate to the tissues where they enlarge and become macrophages.

Macrophages are large amoeboid cells that use pseudopodia to phagocytize microbe, which is then destroyed by digestive enzymes and reactive forms of oxygen within the cell (see Campbell, Figure 43.3).

- Most wander through interstitial fluid phagocytosing bacteria, viruses, and cell debris.
- Some reside permanently in organs and connective tissues. They are fixed in place, but are located where they will have contact with infectious agents circulating in the blood and lymph.
- Fixed macrophages are especially numerous in the lymph nodes and spleen (see Campbell, Figure 43.4)..

Eosinophils represent about 1.5% of the total white cell count and have limited phagocytic activity.

- Contain destructive enzymes in cytoplasmic granules, which are discharged against the outer covering of the invading pathogen.
- Main contribution is defense against larger invaders such as parasitic worms.

Natural killer (NK) cells destroy the body's own infected cells, especially those harboring viruses.

- Also assault abnormal cells that could form tumors
- Are not phagocytic, but attack the membrane, causing cell lysis

2. The inflammatory response

A localized *inflammatory response* occurs when there is damage to a tissue due to physical injury or entry of microorganisms.

- Vasodilation of small vessels near the injury increases the blood supply to the area, which produces the characteristic redness.
- The dilated vessels become more permeable allowing fluids to move into surrounding tissues resulting in localized edema.

Chemical signals are important in initiating an inflammatory response (see Campbell, Figure 43.5)

- *Histamine* is released from injured circulating *basophils* and *mast cells* in the connective tissue.
  - Released histamine causes localized vasodilation, and the capillaries in the area become leakier.
- Prostaglandins are also released from white blood cells and damaged tissues.
  - These and other substances promote increased blood flow to the injured area.
  - Increased blood flow to the site of injury delivers clotting elements that help block the spread of pathogenic microbes and begin the repair process.

Migration of phagocytic cells into the injured area is also a result of increased blood flow and increased leakage from the capillaries.

- Phagocytes are attracted to the damaged tissues by several chemical mediators including chemotactic factors called *chemokines*.
- Neutrophils arrive first, followed closely by monocytes which develop into macrophages.
- The neutrophils eliminate microorganisms and then die.

- Macrophages destroy pathogens and clean up the remains of damaged tissue cells and dead neutrophils.
- Dead cells and fluid leaked from the capillaries may accumulate as pus in the area before it is absorbed by the body.

More widespread (systemic) inflammatory responses may also occur in cases of severe infections (meningitis, appendicitis).

- The bone marrow may be stimulated to release more neutrophils by molecules emitted by injured cells.
- There may also be a severalfold increase in the number of leukocytes within a few hours of response onset.
- A fever may develop in response to toxins produced by pathogens or due to *pyrogens* released by leukocytes.
- While a high fever is dangerous, moderate fevers inhibit the growth of some microorganisms.
- Moderate fevers may also facilitate phagocytosis and speed up tissue repairs.

3. Antimicrobial proteins

A number of proteins function in nonspecific defense by either directly attacking microorganisms or impeding their reproduction.

The two most important nonspecific protein groups are *complement proteins* and the interferons.

The *complement system* is a group of at least 20 proteins that interact with other defense mechanisms.

- These proteins interact in a series of steps that results in lysis of the invading microbes.
- Some components of the system function along with chemotaxin to help attract phagocytes to the infected site.

The *interferons* are substances produced by virus-infected cells that help other cells resist infection by the virus.

- They are secreted by infected cells as a nonspecific defense earlier than specific antibodies appear.
- Cannot save the infected cell, but their diffusion to neighboring cells stimulates production of proteins in those cells that inhibit viral replication.
- Not a virus-specific defense; interferon produced to infection by one strain of virus produces resistance in cells to other unrelated viruses.
- Most effective against short-term infections (colds and influenza).
- One type of interferon activates phagocytes, which enhances their ability to ingest and kill microorganisms.
- Interferons are now being mass produced using recombinant DNA technology and are being tested as treatments for viral infections and cancer.

II.   How Immunity Arises

The *immune system* is the body's third line of defense and is very specific in its response. It is distinguished from nonspecific defenses by:

- Specificity
- Diversity

- Self/nonself recognition
- Memory

A. Lymphocytes provide the specificity and diversity of the immune system

Lymphocytes are responsible for both humoral and cell-mediated immunity

The different responses are due to the two main classes of lymphocytes in the body: *B cells* and *T cells*.

- Early B and T cells (as well as other lymphocytes) develop from multipotent stem cells in the bone marrow and are very much alike. They only differentiate after reaching their site of maturation.
- B cells (B lymphocytes) are responsible for the humoral immune response.
  - They form in the bone marrow and remain there to complete their maturation.
- T cells (T lymphocytes) are responsible for the cell-mediated immune response.
  - They also form in the bone marrow, then migrate to the thymus gland to mature.

Specificity refers to the immune system's ability to recognize and eliminate particular microorganisms and foreign molecules.

Antigen = A foreign substance that elicits a specific response by lymphocytes

Antibody = An antigen-binding immunoglobulin (protein), produced by B cells, that functions as the effector in an immune response

- Antigens may be molecules exhibited on the surface of, produced by, or released from bacteria, viruses, fungi, protozoans, parasitic worms, pollen, insect venom, transplanted organs, or worn-out cells.
- Each antigen has a unique molecular shape and stimulates production of an antibody that defends specifically against that particular antigen.
- *Antigen receptors* on the plasma membranes of lymphocytes recognize and distinguish among antigens.
  - Antigen receptors on B cells are actually transmembrane forms of antibodies and are sometimes referred to as *membrane antibodies*.
  - Antigen receptors on T cells, called *T cell receptors*, are similar in structure to those of B cells, but are never produced in a secreted form.
- The immune response is thus very specific and distinguishes between even closely related invaders.

Diversity refers to the immune system's ability to respond to numerous kinds (millions) of invaders, which are recognized by their antigenic markers.

- Based on the wide variety of lymphocyte populations in the immune system.
- Each population of antibody-producing lymphocytes is stimulated by a specific antigen; the stimulated lymphocytes synthesize and secrete the appropriate antibody.

B. Antigens interact with specific lymphocytes, inducing immune responses and immunological memory

The ability of the immune system to respond to the wide variety of antigens which enter the body is based in the enormous diversity of antigen-specific lymphocytes present in the system.

- Each lymphocyte will recognize and respond to only one antigen.

- This specificity is determined during embryonic development before any antigens are encountered, and is the consequence of the antigen receptor on the lymphocyte's surface.

When an antigen enters the body and binds to receptors on the specific lymphocytes, those lymphocytes are activated and begin to divide and to differentiate.

- The divisions produce a large number of identical effector cells (clones), which bind to the antigen that stimulated the response.
- If, for example, a B cell is activated, it will proliferate to produce a large number of plasma cells that will each secrete an antibody that functions as an antigen receptor for the specific antigen that activated the original B cell.

Thus, each antigen activates only a small number of the diverse group of lymphocytes. The activated cells proliferate to produce a clone of millions of effector cells which are specific for the original antigen (clonal selection) (see Campbell, Figure 43.6).

- *Clonal selection* = Antigenic-specific selection of a lymphocyte that activates it to produce clones of effector cells dedicated to eliminating the antigen that provoked the initial immune response. One clone of cells consists of effector cells. Another clone of cells consists of memory cells.
- *Effector cells* are the short-lived cells that actually defend the body during an immune response.
  - Effector cells are populations of cells resulting from division of lymphocytes that were activated by the binding of antigens to their antigen receptors.

The *primary immune response* is the proliferation of lymphocytes to form clones of effector cells specific to an antigen during the body's first exposure to the antigen.

- There is a 10- to 17-day lag period between initial exposure and maximum production of effector cells.
- The lymphocytes selected by the antigen are differentiating into effector B cells and T cells during the lag period.
  - Activated B cells give rise to effector cells called *plasma cells*, which secrete antibodies (humoral response) that eliminate the activating antigen.

A *secondary immune response* occurs when the body is exposed to a previously encountered antigen.

- The response is faster (2 to 7 days) and more prolonged than a primary response.
- The antibodies produced are more numerous, and they are more effective at binding the antigen (see Campbell, Figure 43.7).

This ability to recognize a previously encountered antigen is known as immunological memory.

- Based on memory cells, which are produced during clonal selection for effectors in a primary immune response.
- Memory cells are not active during the primary response and survive in the system for long periods. (Effector cells produced in the primary response are active, and thus, short-lived.)
- When the same antigen that caused a primary immune response again enters the body, the memory cells are activated and rapidly proliferate to form a new clone of effector cells and memory cells.
- These new clones of effector and memory cells are the secondary immune response.
- This acquired immunity has long been recognized as a resistance to some infections encountered earlier in life (e.g., chicken pox).

C. Lymphocyte development gives rise to an immune system that distinguishes self from nonself

Like all blood cells, lymphocytes derive from pluripotent stem cells in bone marrow or liver of a developing embryo. Initially, all lymphocytes are alike, but depending on the site of maturation, they develop into T cells or B cells (see Campbell, Figure 43.8).

- Cells that remain in the bone marrow develop into B cells
- Cells that migrate to the thymus develop into T cells

1. Immune tolerance for self

Self/nonself recognition is the ability of the immune system to distinguish between the body's own molecules and foreign molecules (antigens).

- Develops before birth when T and B lymphocytes begin to mature in the thymus and bone marrow of the embryo.

Antigen receptors on the surfaces of lymphocytes are responsible for detecting foreign molecules that enter the body. There are no lymphocytes reactive against the body's own molecules under normal conditions.

Self-tolerance = The lack of a destructive immune response to the body's own cells.

- Failure of this system leads to autoimmune disorders that destroy the body's own tissues.

2. The role of cell surface markers in T cell function and development

The *major histocompatibility complex* (MHC; sometimes referred to as human leukocyte antigens or HLA in humans) is a group of glycoproteins embedded in the plasma membranes of cells.

- Important "self-markers" coded for by a family of genes
- There are at least 20 MHC genes and at least 100 alleles for each gene.
- The probability that two individuals will have matching MHC sets is virtually zero unless they are identical twins.

There are two main classes of MHC molecules in the body:

- *Class I MHC* molecules are located on all nucleated cells of the body.
- *Class II MHC* molecules are found only on specialized cells, such as macrophages, B cells, and activated T cells.

MHCs function in antigen presentation by binding to an antigen, thereby facilitating antigen binding to a T cell

- Class MHC I molecules facilitate antigen binding to cytotoxic T cells (see Campbell, Figure 43.9a).
- Class MHC II molecules facilitate antigen binding to helper T cells (see Campbell, Figure 43.9b).

III. Immune Responses

The body will mount either a *humoral* response or a *cell-mediated* response, depending on the antigen which stimulates the system (see Campbell, Figure 43.10).

Humoral immunity produces antibodies in response to toxins, free bacteria, and viruses present in the body fluids.

- Antibodies to these types of antigens are synthesized by certain lymphocytes and then secreted as soluble proteins which circulate through the body in blood plasma and lymph.

Cell-mediated immunity is the response to intracellular bacteria and viruses, fungi, protozoans, worms, transplanted tissues, and cancer cells.

- Depends on the direct action of certain types of lymphocytes rather than antibodies.

A. Helper T lymphocytes function in both humoral and cell-mediated immunity

Cells that take up antigens, such as B cells and macrophages, are known as *antigen-presenting cells (APC)*. They alert the immune system, via helper T cells, of the presence of a foreign antigen (see Campbell Figure 43.11).

- Such cells engulf foreign material (e.g., bacteria)
- Foreign protein (e.g., from broken down bacteria) binds to a newly synthesized class II MHC molecule and is conveyed to the outside of the APC.
- The class II MHC bound foreign antigen is recognized by a helper T cell.
- The interaction between the APC and the helper T cell is enhanced by the presence of a T cell membrane protein, *CD4*.

Helper T cells that bind to a class II MHC-antigen complex on an APC are induced to differentiate into either of two clones of cells:

1. *Activated helper T cells.* Secrete *cytokines*, factors that stimulate other lymphocytes (e.g., interleukin-2 (IL-2) stimulates differentiation of B cells into antibody-secreting plasma cells and induces cytotoxic T cells to become active killers).

2. *Memory helper T cells*
   - Interleukin-1 (IL-1), secreted from APCs, promotes activation of the helper T cell and the subsequent secretion of IL-2 from the activated helper T cell.
   - Helper T cells themselves are regulated by cytokines

Another type of T lymphocyte, *suppressor T cells*, may function to suppress the immune system when an antigen is no longer present.

- Action is not well understood and some immunologists feel T_S cells are actually a form of T_H cells.

B. In the cell-mediated response, cytotoxic T cells defend against intracellular pathogens

Antigen-activated cytotoxic T cells kills cells that are infected by pathogens (e.g., viruses, bacteria).

- Host cells infected by viruses and other pathogens display antigens complexed with class I MHC molecules on their surfaces.
- T_C cells have specific receptors that recognize and bind to antigen-class I MHC markers. (Note that this differs from T_H cells which bind to antigen-class II MHC complexes.)
- The T_C receptor can bind to any cell in the body displaying the antigen-class I MHC marker since class I MHC is present on all nucleated cells.
  ⇒ T_C cells carry a surface molecule called CD8, which has an affinity for class I MHC molecules, and facilitates the interaction between an APC and the cytotoxic T cell (see Campbell, Figure 43.12).
- When a T_C cell binds to an infected cell (*target cell*), it releases *perforin* which is a protein that forms a lesion in the infected cell's membrane.
- Cytoplasm escapes through the lesion and eventually cell lysis occurs.
- Destruction of the host cell not only removes the site where pathogens can reproduce, but also exposes the pathogens to circulating antibodies from the humoral response.
- T_C cells continue to live after destroying the infected cell and may kill many others displaying the same antigen-class I MHC marker.

Cytotoxic T cells also function to destroy cancer cells, which develop periodically in the body.

- Cancer cells possess distinctive markers not found on normal cells, known as *tumor antigen*.
- T_C cells recognize these markers as nonself and attach and lyse the cancer cells.
- Cancers develop primarily in individuals with defective or declining immune systems.
- Certain types of cancers and viruses have diminish the amounts of class I MHC proteins on affected cells, thereby reducing the ability of cytotoxic T cell to recognize and destroy them.

C. In the humoral response, B cells produce antibodies against extracellular pathogens

The humoral response occurs when an antigen binds to B cell receptors that are specific for the antigen epitopes.

- The B cells differentiate into a clone of plasma cells which begin to secrete antibodies that are most effective against pathogens circulating in the blood or lymph.
- Memory cells are also produced and form the basis for secondary immune responses.

The selective activation of a B cell results from one of two mechanisms:

1. *T-dependent antigens* = Antigens that evoke the cooperative response involving macrophages, helper T cells, and B cells
   - These antigens cannot stimulate antibody production without T_H cell involvement (via cytokines such as IL-2).
   - Most antigens are T-dependent.
   - Memory cells are produced in T-dependent responses.

2. *T-independent antigens* = Antigens that trigger humoral immune responses without macrophages or T cell involvement
   - These antigens usually are long chains of repeating units, such as polysaccharides or protein subunits often found in bacterial capsules and flagella.
   - B cells are stimulated directly by the antigen, which probably binds simultaneously to several antigen receptors on the B cell surface.
   - The antibody production (humoral response) is usually much weaker than that of T-dependent antigens.
   - No memory cells are generated in T-independent responses.

Whether activated by T-dependent or T-independent antigens, a B cell gives rise to a clone of plasma cells.

- Each of these effector cells secretes up to 2000 antibodies per second into the body fluids for its 4- to 5-day lifespan.
- The specific antibodies help eliminate the foreign invader from the body.

1. Antibody structure and function

Antigens are usually proteins or large polysaccharides that make up a portion of the outer covering of pathogens or transplanted cells.

- May be components of the coats of viruses, capsules and cell walls of bacteria, or surface molecules of other cell types.
- Molecules on the cell surface of transplanted tissues and organs or blood cells from other individuals are also recognized as foreign.

Antibodies recognize a localized region on the surface of an antigen (epitope), not the entire antigen molecule (see Campbell, Figure 43.14).

- *Epitope* = On an antigen's surface, a localized region that is chemically recognized by antibodies; also called an antigenic determinant
- Several types of antibodies from several different B cells may be produced by a single bacterial cell since it may have different antigens on different areas and each bacterial antigen may possess more than one recognizable epitope.

Antibodies comprise a specific class of proteins called *immunoglobulins* (Igs) (see Campbell, Figure 43.15a).

The structure of the immunoglobulin is associated with its function.

- Antibodies are Y-shaped molecules comprised of four polypeptide chains: two identical *light chains* and two identical *heavy chains*.
- All four chains have constant (C) regions that vary little in amino acid sequence among antibodies that perform a particular type of defense.
- At the tips of the Y are found variable (V) regions in all four chains; the amino acid sequences in the variable region show extensive variation from antibody to antibody.
- The variable regions function as antigen-binding sites and their amino acid sequences result in specific shapes that fit and bind to specific antigen epitopes.

The antigen-binding site is responsible for the antibody's ability to identify its specific antigen epitope and the stem (constant) regions are responsible for the mechanism by which the antibody inactivates or destroys the antigenic invader.

There are five types of heavy-chain constant regions, and these determine the five major classes of antibodies (see Table 43.1 for a summary):

1. IgM. Consists of five Y-shaped monomers arranged in a pentamer structure; they are circulating antibodies that appear in response to an initial exposure to an antigen.

2. IgG. A Y-shaped monomer; most abundant circulating antibody; readily crosses blood vessels and enters tissue fluids; protects against bacteria, viruses, and toxins circulating in blood and lymph; triggers complement system action.

3. IgA. A dimer consisting of two Y-shaped monomers; produced primarily by cells abundant in mucous membranes; prevents attachment of bacteria and viruses to epithelial surfaces; also found in saliva, tears, perspiration, and colostrum.

4. IgD. A Y-shaped monomer; found primarily on external membranes of B cells; probably functions as an antigen-receptor that initiates differentiation of B cell

5. IgE. A Y-shaped monomer; stem regions attach to receptors on mast cells and basophils; stimulates these cells to release histamine and other chemicals that cause allergic reactions when triggered by an antigen

The specificity of antigen-antibody interactions has formed the bases of laboratory technologies used in research and clinical diagnosis.

- The production of monoclonal antibodies has made significant contributions to biomedical science (see Campbell, Methods box).

2. Antibody-mediated disposal of antigen

Antibodies do not directly destroy an antigenic pathogen. The antibody binds to the antigen to form an antigen-antibody complex, which tags the invader for destruction by one of several effector mechanisms (see Campbell, Figure 43.16).

- *Neutralization* is the simplest mechanism.

- The antibody blocks viral attachment sites or coats bacterial toxins, making them ineffective. Phagocytic cells eventually destroy the complex.
- In *opsonization*, bound antibodies enhance macrophage attachment to and phagocytosis of microbes.
- Antibody-mediated *agglutination* neutralizes and opsonizes the microbes. Each antibody has two or more antigen-binding sites and can cross-link adjacent antigens. The cross-linking can result in clumps of a bacteria being held together by the antibodies, making it easier for phagocytes to engulf the mass.
- Precipitation is similar to agglutination but involves the cross-linking of soluble antigen molecules instead of cells; these immobile precipitates are easily engulfed by phagocytes.
- In *complement fixation*, antibodies combine with complement proteins; this combination activates the complement proteins, which produce lesions in the foreign cell's membrane that result in cell lysis.

The 20 or so complement proteins circulate in the blood in inactive forms.

- In an infection, these proteins become activated in a cascading fashion, with each component activating the next in the series.
- Completion of the complement cascade results in lysis of many types of viruses and pathogenic cells. Lysis by complement can be achieved two ways:

1. The classical pathway describes complement's activation in the specific defense mechanism (see Campbell, Figure 43.17)
   - Initiated when IgM or IgG antibodies bind to a specific pathogen; this targets the cell for destruction.
   - A complement protein attaches to and bridges the gap between two adjacent antibody molecules.
   - The antibody-complement association activates complement proteins to form in a step-by-step sequence, generating a *membrane attack complex*, which forms a 7 to 10 nm-diameter pore in the bacterial membrane.
   - The MAC pore allows ions and water to rush into the cell, causing it to swell and lyse.
2. The *alternative pathway* is how complement is activated in nonspecific defense mechanisms.
   - Does not require cooperation with antibodies.
   - Complement proteins are activated by substances that are naturally present on many pathogens (yeasts, viruses, virus-infected cells, protozoans) to form a membrane attack complex.
   - The complex lyses the pathogen without the aid of antibodies.

Complement proteins also contribute to inflammation by binding to histamine-containing cells; this association triggers the release of histamine from those cells.

Several complement proteins also attract phagocytic cells to infected sites.

Complement and phagocytes also work together in two ways to destroy pathogens.

1. *Opsonization* is a cooperative mechanism in which complement proteins attach to a foreign cell and stimulate phagocytes to engulf the cell.
2. In *immune adherence*, complement proteins and antibodies coat a microbe, which causes it to adhere to blood vessel walls and other surfaces; this makes the cell easy prey for circulating phagocytes.

IV. **Immunity and Health and Disease**

A. **Immunity can be achieved naturally or artificially**

Active immunity is the immunity conferred by recovery from an infectious disease.

- Depends on response by the person's own immune system
- May be acquired naturally from an infection to the body or artificially by *immunization*, also known as *vaccination*
  - Vaccines may be inactivated bacterial toxins, killed microbes, parts of microbes, or viable but weakened microbes.
  - In all cases the organisms can no longer cause the disease but can act as antigens and stimulate an immune response.
  - A person vaccinated against an infectious agent who encounters the pathogen will show the same rapid, memory-based secondary response as someone who has had the disease.

Passive immunity is immunity which has been transferred from one individual to another by the transfer of antibodies.

- Naturally occurs when IgG antibodies cross the placenta from a pregnant woman to her fetus.
- Some antibodies are transferred to nursing infants though breast milk.
- Provides temporary protection to newborns whose immune systems are not fully operational at birth.
- Persists as long as the antibodies last (a few weeks or months), but it can provide protection from infections until a baby's own immune system has matured.
- May also be transferred artificially from an animal or human already immune to the disease.
  - Rabies is treated by injecting antibodies from people vaccinated against rabies; produces an immediate immunity important to quickly progressing infections.
  - Artificial passive immunity is of short duration but permits the body's own immune system to begin to produce antibodies against the rabies virus.

B. **The immune system's capacity to distinguish self from nonself limits blood transfusion and tissue transplantation**

The body's immune system distinguishes between self (the body's own cells) and nonself (foreign cells).

- Nonself includes pathogens and cells from other individuals of the same species.

1. **Blood groups and blood transfusion**

The human ABO blood groups provide a good example of nonself recognition. The antigen present on the surface of the erythrocytes is not antigenic to that person but may be recognized as foreign if placed in the body of another individual.

- Individuals of blood type A have the A antigen and make anti-B antibodies.
- Individuals of blood type B have the B antigen and make anti-A antibodies.
- Individuals of blood type AB have the A and B antigen and make no antibodies.
- Individuals of blood type O have neither the A nor B antigen and make anti-A and anti-B antibodies.

Blood group antibodies can cause blood of a different antigenic type to agglutinate, resulting in a life-threatening *transfusion reaction*.

- Type O individuals are universal donors since their blood has neither antigen.
- Type AB individuals are universal recipients since they produce neither antibody A or antibody B.

The blood group antibodies are present in the body before a transfusion occurs because they form in response to the body's normal bacterial flora that have epitopes very similar to blood group antigens.

⇒ Usually IgM class antibodies do not cross the placenta, thus they present no harm to a developing fetus with a blood type different from the mother.

The *Rh factor* is another blood group antigen. Rh factor causes problems when a mother is Rh negative and her fetus is Rh positive (inherited from the father).

- When small amounts of fetal blood cross the placenta and come into contact with the mother's lymphocytes, the mother develops antibodies against the Rh factor.
- Usually only a problem in the second child since the response will be quick due to sensitization and formation of memory cells during the first baby's gestation.
    ⇒ Unlike blood group antibodies, Rh antibodies are IgG class, which can cross the placenta.
    ⇒ The mother's antibodies cross the placenta and destroy the red blood cells of the Rh positive fetus.
- Can be prevented by injection of anti-Rh antibodies which destroy Rh-positive red cells before the mother develops immunological memory.

2. Tissue grafts and organ transplantation

The MHC is a biochemical fingerprint unique to each individual.

- Complicates tissue grafts and organ transplants since foreign MHC molecules are antigens and cause cytotoxic T cells to mount a cell-mediated response.
- Cyclosporin A and FK506 suppress cell-mediated immunity without crippling humoral immunity, thus increasing the chance of successful grafts and transplants.

In the case of bone marrow transplants, which are used to treat leukemia and other cancers, the graft rather than the host is the source of immune rejection

- The donated bone marrow contains lymphocytes that will react against the recipient. This graft versus host rejection is limited if the MHC molecules of the donor and recipient are well matched.

The reactions of the immune system to transfusions, tissue grafts, and organ transplants are normal reactions of a healthy immune system, not disorders of the system.

C. Abnormal immune function can lead to disease

1. Allergies

Allergy = A hypersensitivity of the body's defense system to an environmental antigen called an allergen

Some believe these reactions to be evolutionary remnants to infection by parasitic worms due to similarities in the responses.

IgE class antibodies are commonly involved in allergic reactions; these antibodies recognize pollen as allergens (see Campbell, Table 43.1)

- IgE antibodies attach by their tails to noncirculating mast cells found in connective tissues.

- When a pollen grain bridges the gap between two adjacent IgE monomers, the mast cell responds with a reaction called *degranulation* (see Campbell, Figure 43.18).
- Degranulation involves the release of histamine and other inflammatory agents.
- Histamine causes dilation and increased permeability of small blood vessels, which results in the common symptoms of an allergy.
- Antihistamines are drugs used to treat allergies since they interfere with the action of histamine.

Anaphylactic shock is a life-threatening reaction to injected or ingested antigens; it is the most serious type of acute allergic response.

- Occurs when mast cell degranulation causes a sudden dilation of peripheral blood vessels and a drastic drop in blood pressure.
- Death may occur in a few minutes.
- This hypersensitivity may be associated with foods (peanuts, fish) or insect venoms (wasp or bee stings).
- Epinephrine may be injected to counteract the allergic response.

2. Autoimmune diseases

Autoimmune disease = An immune system reaction against self

Examples:
- Some cases involve immune reactions against components of the body's own cells, which are released by the normal breakdown of skin and other tissues, especially nucleic acids in lupus erythematous.
- Rheumatoid arthritis is an autoimmune disease in which inflammation damages cartilage and bones in joints.
- Destruction of insulin-producing pancreas cells by an autoimmune reaction appears to cause insulin-dependent diabetes.
- In multiple sclerosis, T cells reactive against myelin infiltrate the central nervous system and destroy the myelin of neurons.
- Antibodies produced to repeated streptococcal infections may react with heart tissues and cause valve damage in some people.
- Other autoimmune diseases are Grave's disease and rheumatic fever.

3. Immunodeficiency diseases

Immunodeficiency refers to a condition where an individual is inherently deficient in either humoral or cell-mediated immune defenses.

- Severe combined immunodeficiency (SCID) is a congenital disorder in which both the humoral and cell-mediated immune defenses fail to function.
- Gene therapy has had some success in the treatment of a type of SCID where there is a deficiency of the enzyme adenosine deaminase.

Not all cases of immunodeficiency are inborn conditions.

- Some cancers, like Hodgkin's disease, damage the lymphatic system and make the individual susceptible to infection.
- Some viral infections cause depression of the immune system (e.g., AIDS).

Physical and emotional stress may compromise the system.

- Adrenal hormones secreted by stressed individuals affect the number of leukocytes and may suppress the system in other ways.
- Physiological evidence suggests direct links between the nervous system and the immune system.

- There is a network of nerve fibers that penetrates deep into the thymus.
- Lymphocytes have also been found to possess surface receptors for chemical signals secreted by nerve cells.

4. Acquired immunodeficiency syndrome (AIDS)

Acquired immunodeficiency syndrome is a severe immune system disorder caused by infection with the *human immunodeficiency virus (HIV)*.

Individuals with AIDS are highly susceptible to *opportunistic diseases*, infections, and cancers that take advantage of a deficient immune system.

Mortality rate approaches 100%.

HIV probably evolved from another virus in central Africa and may have gone unrecognized for many years.

There are two major strains: HIV-1 and HIV-2.

HIV infects cells, including T_H cells, which carry the CD4 receptor on their surface.

- Entry by the virus requires CD4 and a coreceptor, such as fusin or CCR5.
- After entry, HIV RNA is reverse-transcribed and the product DNA is integrated into the host cell genome.
- In this provirus form, the viral genome directs the production of new virus particles.

HIV is not eliminated from the body by antibodies for several reasons:

- The latent provirus is invisible to the immune system.
- The virus undergoes rapid mutational changes in antigens during replication which eventually overwhelms the immune system.
- The population of helper T-cells eventually declines to the point where cell-mediated immunity collapses.
- Secondary infections characteristic of HIV infection develop (*Pneumocystis pneumonia* and Kaposi's sarcoma).

A person who tests positive for the presence of antibodies to the virus is *HIV-positive*.

AIDS is the late stage of HIV infection and is defined by a reduced T cell population and the appearance of secondary infections.

- Takes an average of about ten years to reach this stage of infection.
- During most of this time, only moderate symptoms are shown.
- Changes in the level of T cells can be monitored as an indication of disease progression, however, recently it has been shown that measures of viral load may be a better indicator of disease progress and the effectiveness of anti-HIV treatment.

HIV is only transmitted through the transfer of body fluids, such as blood or semen, containing infected cells.

- Most commonly transmitted in the U.S. and Europe through unprotected sex between male homosexuals and unsterilized needles in intravenous drug users.
- In Africa and Asia, transmission through unprotected heterosexual sex is rapidly increasing; especially in areas with a high incidence of other sexually transmitted diseases.
- Transmission during fetal development occurs in 25% of HIV-infected mothers.
- Transmission through blood transfusions has also been reported, but the incidence has declined greatly with implementation of screening procedures.

HIV infection is currently considered an incurable disease.

New drug combinations designed to slow the progression of the disease to AIDS are expensive and not available to all HIV-positive people.

- Drugs that slow viral replication (when used in combinations) include DNA-synthesis inhibitors, reverse transcriptase inhibitors (e.g., AZT, ddI), and protease inhibitors.
- Other drugs are used to fight opportunistic infections common in AIDS patients.

The best way to prevent additional infections is to educate people on how the disease is transmitted and how to protect themselves.

D. Invertebrates have a rudimentary immune system

How invertebrates react against pathogens that enter their bodies is poorly understood, although it is known that they have a well developed ability to distinguish self from nonself.

- Experiments have shown that if the cells from two sponges of the same species are mixed, the cells from each individual will aggregate in separate groups, excluding cells from the other individual.
- Coelomocytes, amoeboid cells that destroy foreign materials, have been found in many invertebrates.

A memory response has also been identified in earthworms.

- A body wall graft from one worm to another will survive for about eight months before rejection if the worms are from the same population.
- A graft involving worms from different populations is rejected in two weeks.
- A second graft from the same donor to the same recipient is rejected in less than one week due to coelomocyte activity.

REFERENCES

Beardsley, T. "Better Than a Cure." *Scientific American*, 272(1). 1995.

Campbell, N., et al. *Biology*. 5th ed. Menlo Park, California: Benjamin/Cummings, 1999.

Coleman, R.M., M.F. Lombard, and R.E. Sicard. *Fundamental Immunology*. 2nd ed. Dubuque, IA: Wm. C. Brown, Publishers, 1992.

Gallo, R.C. "The AIDS Virus." *Scientific American*. 256(1). 1987.

Laurence, J. "The Immune System in AIDS." *Scientific American*. 253(6). 1985.

Marrack, P. and J. Kappler. "The T Cell and Its Receptor." *Scientific American*. 254(2). 1986.

Newman, J. "How Breast Milk Protects Newborns." *Scientific American*, 273(6). 1995.

Tonegawa, S. "The Molecules of the Immune System." *Scientific American*. 253(4). 1985.

CHAPTER **44**
CONTROLLING THE INTERNAL ENVIRONMENT

OUTLINE

I. Regulation of Body Temperature
   A. Four physical processes account for heat gain or loss
   B. Ectotherms derive body heat mainly from their surroundings; endotherms derive it mainly from metabolism
   C. Thermoregulation involves physiological and behavioral adjustments
   D. Most animals are ectothermic, but endothermy is widespread
   E. Torpor conserves energy during environmental extremes

II. Water Balance and Waste Disposal
   A. Water balance and waste disposal depend on transport epithelia
   B. An animal's nitrogenous wastes are correlated with its phylogeny and habitat
   C. Cells require a balance between osmotic gain and loss of water
   D. Osmoregulators expend energy to control their internal osmolarity; osmoconformers are isoosmotic with their surroundings

III. Excretory Systems
   A. Most excretory systems produce urine by refining a filtrate derived from body fluids: *an overview*
   B. Diverse excretory systems are variations on a tubular theme
   C. Nephrons and associated blood vessels are the functional units of the mammalian kidney
   D. From blood filtrate to urine: a closer look
   E. The mammalian kidney's ability to conserve water is a key terrestrial adaptation
   F. Nervous and hormonal feedback circuits regulate kidney functions
   G. Diverse adaptations of the vertebrate kidney have evolved
   H. Interacting regulatory systems maintain homeostasis

OBJECTIVES

After reading this chapter and attending lecture, the student should be able to:

1. Distinguish between osmoregulators and osmoconformers.
2. Discuss the problems that marine organisms, freshwater organisms, and terrestrial organisms face in maintaining homeostasis, and explain what osmoregulatory adaptations serve as solutions to these problems.
3. Explain the role of transport epithelia in osmoregulation.
4. Describe how a flame-bulb (protonephridial) excretory system functions.

5. Explain how the metanephridial excretory tubule of annelids functions, and describe any structural advances over a protonephridial system.

6. Explain how the Malpighian tubule excretory system contributed to the success of insects in the terrestrial environment.

7. Using a diagram, identify and give the function of each structure in the mammalian excretory system.

8. Using a diagram, identify and give the function of each part of the nephron.

9. Describe and show the relationship among the processes of filtration, secretion, and reabsorption.

10. Explain the significance of the fact that juxtamedullary nephrons are only found in birds and mammals.

11. Explain how the loop of Henle enhances water conservation by the kidney.

12. Describe the mechanisms involved in the hormonal regulation of the kidney.

13. Describe structural and physiological adaptations in the kidneys of non-mammalian species that allow them to osmoregulate in different environments.

14. Explain the correlation between the type of nitrogenous waste produced (ammonia, urea, or uric acid) by an organism and its habitat.

15. Describe the adaptive advantages of endothermy.

16. Discuss the four general categories of physiological and behavioral adjustments used by land mammals to maintain relatively constant body temperatures.

17. Distinguish between the two thermoregulatory centers of the hypothalamus.

18. Describe the thermoregulatory adaptations found in animals other than terrestrial mammals.

19. Describe several mechanisms for physiological acclimatization to new temperature ranges.

20. Distinguish between hibernation and aestivation.

KEY TERMS

thermoregulation	stress-induced proteins	protonephridium	collecting duct
osmoregulation	heat-shock proteins	metanephridium	cortical nephrons
excretion	torpor	Malpighian tubules	juxtamedullary nephrons
conduction	hibernation	renal artery	afferent arteriole
convection	estivation	renal vein	efferent arteriole
radiation	transport epithelium	ureter	peritubular capillaries
evaporation	ammonia	urinary bladder	vasa recta
ectotherm	uric acid	urethra	antidiuretic hormone
endotherm	osmolarity	renal cortex	juxtaglomerular apparatus
vasodilation	osmoconformer	renal medulla	angiotensin II
vasoconstriction	osmoregulator	nephron	aldosterone
countercurrent heat	stenohaline	glomerulus	renin
exchanger	euryhaline	Bowman's capsule	renin-angiotensin-
nonshivering	anhydrobiosis	podocytes	aldosterone system
thermogenesis	filtration	proximal tubule	atrial natriuretic factor
brown fat	secretion	loop of Henle	
acclimatization	reabsorption	distal tubule	

LECTURE NOTES

Most animals can survive environmental fluctuations more extreme than any of their individual cells could tolerate. This is possible because mechanisms of homeostasis maintain internal environments within ranges tolerable to body cells.

Homeostatic mechanisms include:

- Adaptation to the thermal environment (*thermoregulation*)
- Adaptation to the osmotic environment (*osmoregulation*)
- Strategies for the elimination of waste products of protein catabolism (*excretion*)

They are long-term adaptations that evolved in populations facing environmental problems.

They include cellular mechanisms and short-term physiological adjustments.

I.   Regulation of Body Temperature

Metabolism and membrane properties are very sensitive to changes in an animal's internal temperature.

- Each animal lives in, and is adapted to, an optimal temperature range in which it can maintain a constant internal temperature when external temperatures fluctuate.
- Maintaining the body temperature within a range that permits cells to function efficiently is known as *thermoregulation*.

A.   Four physical processes account for heat gain or loss

An organism exchanges heat with its environment by four physical processes (see Campbell, Figure 44.1):

1. *Conduction* is the direct transfer of thermal motion (heat) between molecules of the environment and a body surface.
   - Heat is always conducted from a body of higher temperature to one of lower temperature.
   - Water is 50 to 100 times more effective than air in conducting heat.
   - For example, on a hot day, an animal in cold water cools more rapidly than one on land.
2. *Convection* is the transfer of heat by the movement of air or liquid past a body surface.
   - For example, breezes contribute to heat loss from an animal with dry skin.
3. *Radiation* is the emission of electromagnetic waves produced by all objects warmer than absolute zero.
   - It can transfer heat between objects not in direct contact.
   - For example, an animal can be warmed by the heat radiating from the sun.
4. *Evaporation* is the loss of heat from a liquid's surface that is losing some molecules as gas.
   - Production of sweat greatly increases evaporative cooling.
   - Can only occur if surrounding air is not saturated with water molecules

Evaporation and convection are the most variable causes of heat loss.

Terrestrial animals are affected by all four of the above processes; aquatic animals are impacted minimally by radiation and not at all by change of state.

B.   Ectotherms derive body heat mainly from their surroundings; endotherms derive it mainly from metabolism

Animals may be classified as either ectotherms or endotherms depending on their major source of body heat.

Ectotherm = An animal that warms its body mainly by absorbing heat from surroundings

- Includes most invertebrates, fishes, reptiles and amphibians
- May derive a small amount of body heat from metabolism.

Endotherm = An animal that derives most or all of its body heat from its own metabolism

- Includes mammals, birds, some fishes, and numerous insects.
- Many maintain a consistent internal temperature even as the environmental temperature fluctuates (see Campbell, Figure 44.2).

The main source of body heat distinguishes endotherms from ectotherms, not the body temperature.

- Distinction is not absolute.
  - Most ectothermic insects and some fishes are partial endotherms. These organisms retain metabolic heat to warm only certain body parts (e.g., locomotor muscles).
  - Some birds and mammals acquire additional body heat by basking in the sun.

A terrestrial lifestyle presents certain problems that have been solved by the evolution of endothermy.

Environmental temperatures fluctuate more in terrestrial habitats than in aquatic habitats.

Endothermic vertebrates are usually warmer than their surroundings, but also have mechanisms to cool their bodies.

- Maintaining a warm body temperature requires an active metabolism which contributes to a high level of cellular respiration
- This permits endotherms to be more physically active for a longer period of time in comparison to most ectotherms.

Endothermy requires more energy than ectothermy.

- At 20°C, a human has a resting metabolic rate of 1300 to 1800 kcal/day while an alligator (ectotherm) of similar weight has a resting metabolic rate of about 60 kcal/day.
- Endotherms usually consume more food than ectotherms of similar size to offset the energy requirements.

The bioenergetic connections between body temperature, an active metabolism, and mobility were important to the evolution of endothermy.

- The evolution of endothermy and a high metabolic rate were accompanied by an increase in efficiency in circulatory and respiratory systems as seen in the birds and mammals.
- Terrestrial ectotherms exhibit their own adaptations for adjusting to temperature changes in the terrestrial environment.

C. Thermoregulation involves physiological and behavioral adjustments

Endotherms and ectotherms use a combination of strategies to thermoregulate.

1. *Adjusting the rate of heat exchange between the animal and its surroundings*

   Heat loss is reduced by the presence of hair, feathers, and fat just below the skin.

   Adaptations found in the circulatory system also help regulate heat exchange.

   - The amount of blood flowing to the skin can be changed by many endotherms and some ectotherms.

- *Vasodilation* increases the blood flow to the skin due to an increase in the diameter of blood vessels near the body surface.
- Nerve signals cause muscles in the walls of the blood vessels to relax which permits more blood to flow through the vessel.
- As blood flow increases, more heat is transferred to the environment by conduction, convection, and radiation.
- *Vasoconstriction* reduces blood flow to the skin due to a decrease in the diameter of blood vessels near the body surface.
- Reduced blood flow decreases the amount of heat transferred to the environment.
- Heat exchange with the environment is also altered by a *countercurrent heat exchanger* (see Campbell, Figure 44.3).
  - This is a special arrangement of arteries and veins found in the extremities of many endothermic animals.
  - Arteries carrying warm blood from the body to the legs of a bird or flipper of a dolphin are in close contact with veins carrying blood from the appendage into the body.
  - This vessel arrangement enhances heat transfer from arteries to veins along the entire length of the vessel.
  - This is possible since venous blood returning from the tip of the appendage is always cooler than the arterial blood.
  - The venous blood entering the body has been warmed to almost core temperature by the exchange.
  - In some species, an alternative set of vessels which bypass the exchanger is present.
  - The rate of heat loss is controlled by the relative amount of blood that enters the appendage by the two paths.

2. *Cooling by evaporative heat loss*

Water is lost by terrestrial endotherms and ectotherms through breathing and across the skin.
- In low humidity, water evaporates and heat is lost by evaporative cooling.
- Panting increases evaporation from the respiratory system.
- Sweating or bathing in mammals increases evaporative cooling across the skin (see Campbell, Figure 44.4).

3. *Behavioral responses*

Relocating allows animals to increase or decrease heat loss from the body.
- In winter, many animals bask in the sun or on warm rocks.
- In summer, many animals burrow or move to damp areas.
- Some animals migrate to more suitable climates.

4. *Changing the rate of metabolic heat production*

This is found only in birds and mammals.

Increased skeletal muscle activity and non-shivering thermogenesis can greatly increase the amount of metabolic heat produced.

D. Most animals are ectothermic, but endothermy is widespread

1. Invertebrates

Most invertebrates have little control over body temperature, but some do adjust temperature by behavioral or physiological mechanisms.

- The desert locust orients the body to maximize heat absorption from the sun.
- Some large flying insects (e.g., bees) can generate internal heat by contracting all flight muscles in synchrony (functionally analogous to shivering).
  ⇒ Little wing movement occurs but large amounts of heat are produced.
  ⇒ Allows activity even on cold days and at night (see Campbell, Figure 44.5a).
- Endothermic insects such as honeybees, bumblebees, and noctuid moths have a countercurrent heat exchanger that maintains a high temperature in the thorax where flight muscles are located.

Honeybees also use social organization to regulate temperature.

- They increase movements and huddle to retain heat in cold weather.
  ⇒ Maintain constant temperature by changing the density of huddling; heat is distributed by the movement of individuals from the core to the margins of the huddle.
- In warm weather, they cool hives by transporting water to it and fanning with their wings to promote evaporation and convection.

2. Amphibians and reptiles

Amphibians produce little heat and most lose heat rapidly by evaporative cooling from their body surface, making thermoregulation difficult.

- The optimal temperature range varies greatly depending on the species.
- Seek cooler or warmer microenvironments as necessary (behavioral adaptation).
- Some can vary the amount of mucus they secrete to regulate evaporative cooling.

Reptiles are generally ectotherms that warm themselves mainly by behavioral adaptations.

- Orient the body toward the heat sources to increase uptake and maximize the body surface exposed to the heat source.
- They regulate temperature by alternately seeking sun or shade (for example, by turning direction to reduce surface area exposed to the sun).

Some reptiles have physiological adaptations that help regulate heat loss.

- When swimming in cold water, the Galapagos iguana utilizes superficial vasoconstriction to reduce heat loss.
- Female pythons that are incubating eggs increase their metabolic rate by shivering.
  ⇒ This maintains their body temperature 5° to 7°C above ambient temperature.

3. Fishes

Fish are generally ectotherms although some are endothermic.

  ⇒ The body temperature of most is within 1° to 2°C of the water temperature.
- Metabolic heat produced by the swimming muscles is lost to the water as blood passes through the gills.

⇒ The dorsal aorta carries blood directly from the gills to the tissues, cooling the body core.

Endothermic fishes have an adaptation to reduce heat loss (see Campbell, Figure 44.6).

- Includes large active species such as the bluefin tuna, swordfish, and great white shark.
- Large arteries carry most of the blood from the gills to tissues just under the skin.
- Branches carry blood to the deep muscles where smaller vessels are arranged into a countercurrent heat exchanger.
- The swimming muscles produce enough metabolic heat to raise temperatures at the body core.
- Endothermy enhances the activity level of these fishes by keeping the swimming muscles several degrees warmer than tissues closer to the animal's surface.

4. **Mammals and birds**

These organisms maintain high body temperatures within a narrow range.

- 36° – 38°C for most mammals
- 40° – 42°C for most birds

Maintaining these narrow ranges requires the ability to closely balance the rate of metabolic heat production with the rate of heat loss or gain from the outside environment.

The rate of heat production may be increased by:

1. The increased contraction of muscles

   ⇒ Moving or shivering produces metabolic heat

2. The action of hormones that increase the metabolic rate and the production of heat instead of ATP.

   - *Nonshivering thermogenesis* is the hormonal triggering of heat production.
     - Found in numerous mammals and a few birds
     - Occurs throughout the body
   - Some mammals have *brown fat* between the shoulders and in the neck that is specialized for rapid heat production.

Birds and mammals also use other mechanisms to regulate environmental heat exchange.

- Vasodilation and vasoconstriction permit the maintenance of a proper core temperature even when the extremities are cooler.
- Fur and feathers trap a layer of insulating air next to the body.

   ⇒ Most land mammals and birds raise their hair or feathers, thereby trapping a thicker layer of air, in response to cold.
- A layer of fat just below the human skin provides insulation (see Campbell, Figure 44.7).

Marine mammals live in water colder than their body temperature, but their adaptations to conserve heat are more effective than those of land mammals.

- They are insulated by a very thick layer of fat called blubber under the skin.

   ⇒ Helps them maintain a body temperature of 36° – 38°C even in Arctic and Antarctic waters.

- Countercurrent heat exchangers reduce heat loss in the extremities where no blubber is found.

Thermoregulation by endotherms also involves cooling the body. Many birds and mammals have behavioral and physiological adaptations to cool the body.

- When whales and other marine mammals move into warm waters, excess heat is eliminated by vasodilation.
  ⇒ A large number of blood vessels in the outer skin dilate which permits increased blood flow (and heat loss by conduction) to occur.
- Terrestrial birds and mammals depend on evaporative cooling.
  ⇒ Panting increases heat loss.
  ⇒ The fluttering of vascularized pouches in the floor of the mouth in some birds increases evaporative cooling.
- Many terrestrial mammals have sweat glands.
- Some mammals spread saliva (some kangaroos and rodents) or a combination of saliva and urine (some bats) over the body to increase evaporative heat loss.

5. Feedback mechanisms in thermoregulation

Thermoregulation in humans and other terrestrial mammals involves a complex homeostatic system and feedback mechanisms (see Campbell, Figure 44.8).

- The hypothalamus contains nerve cells that control thermoregulation and many other aspects of homeostasis.
- The hypothalamus responds to changes in body temperature that are above or below the normal range ($36.1°$ – $37.8°C$).

The body's temperature is monitored by nerve cells in the skin, hypothalamus, and other parts of the body.

- Some nerve cells function as warm receptors that signal the hypothalamus when the skin or blood temperature increases.
- Others function as cold receptors that signal the hypothalamus when there is a decrease in temperature of the skin or blood.

When the body's temperature increases above the normal range, cooling mechanisms are activated.

- Vasodilation of skin vessels occurs and capillaries fill with blood; the heat radiates from the skin's surface.
- Sweat glands are activated which increases evaporative cooling.
- These changes result in a decrease in body temperature to within the normal range.
- The cooling mechanisms are "turned off" by the hypothalamus when normal body temperature is achieved.

When the body's temperature decreases below the normal range due to a cold environment, warming mechanisms are activated.

- Vasoconstriction of skin vessels occurs.
  ⇒ A smaller volume of blood passes to the skin, thus reducing heat loss from the body surface.
  ⇒ The warm blood is diverted from the skin to deeper tissues.
- Skeletal muscles are activated and shivering occurs, which generates metabolic heat.
- Nonshivering thermogenesis increases heat production.

- Changes resulting in an increase in body temperature to within the normal range.
- The warming mechanisms are "turned off" by the hypothalamus when normal body temperature is achieved.

6. **Temperature range adjustments**

A physiological response called *acclimatization* allows many animals to adjust to a new range of environmental temperature.

- The adjustment takes place over a period of days or weeks.
- The process is important to animals that must adjust to seasonal changes in temperature.

Acclimatization of an animal to a new external temperature range may involve several physiological changes.

- May involve changes in the cooling and warming mechanisms which maintain the internal temperature.
- May involve adjustments at the cellular level.
  ⇒ Cells may increase production of certain enzymes to compensate for the lowered activity of each molecule at lower temperatures.
  ⇒ Cells may produce enzyme variants having the same function but different temperature optima.
- Membranes may remain fluid by changing the proportions of saturated and unsaturated lipid in their composition.

Cells can also make rapid adjustments to temperature changes.

- Shock due to a large temperature increase (or other stress) stimulates the accumulation of a class of factors called *stress-induced proteins*, including *heat-shock proteins*.
  ⇒ Found in animal cells, yeast, and bacteria
  ⇒ Help prevent denaturation of other proteins by high temperatures
  ⇒ Help prevent cell death and may help maintain homeostasis while the organisms is adjusting to the external environment

E. **Torpor conserves energy during environmental extremes**

When food supply is low and/or environmental temperatures are extreme, many endotherms will enter a state of torpor.

- *Torpor* = Alternative physiological state in which metabolism decreases and the heart and respiratory systems slow down
- This is a mechanism to conserve energy.

Torpor may be a long-term state (hibernation, estivation) or short-term as in the daily period of torpor seen in many small mammals and birds.

Hibernation and estivation are often triggered by seasonal changes in day length.

- In *hibernation*, the body temperature is lowered; this allows an animal to survive long periods of cold and diminished food supplies.
  ⇒ Some animals will begin to eat huge quantities of food as the amount of daylight decreases.
- *Estivation* is characterized by slow metabolism and inactivity.
  ⇒ Allows the animal to survive long periods of high temperature and diminished water supply.
  ⇒ Also known as summer torpor.

Daily periods of torpor appear to be adapted to feeding patterns.

- They allow many relatively small endotherms to survive on stored energy during hours when they cannot feed.
  ⇒ These animals have a very high metabolic rate and rate of energy consumption when active.
- Most bats and shrews feed at night and enter torpor during daylight hours.
- Chickadees and hummingbirds feed during the day and undergo torpor on cold nights.
  ⇒ The body temperature of chickadees in the cold, northern forests may drop 10°C during winter nights.

An animal's biological clock appears to control its daily cycle and torpor.

- For example, shrews continue to undergo a daily torpor even when food is continually available.
- Human sleep periods and the corresponding drop in body temperature may be a remnant of a daily torpor in ancestral mammals.

Many animals can tolerate wide fluctuations in their body temperature, but no animal can withstand high concentrations of its own metabolic wastes or much change in the relative amounts of dissolved solutes and water in its cellular environment.

II.   Water Balance and Waste Disposal

The majority of cells in most animals (all but sponges and cnidarians) are not exposed to the external environment, but are bathed by an extracellular fluid.

- Animals with open circulatory systems have an extracellular compartment containing hemolymph which bathes the cells.
- Animals with closed circulatory systems have two extracellular compartments—interstitial fluid and blood plasma.

By balancing water gain and loss and disposing of metabolic wastes, an animal's homeostatic mechanisms prevent environmental fluctuations from having a harmful impact.

A.   Water balance and waste disposal depend on transport epithelia

The maintenance of water and ion balance as well as the excretion of metabolic wastes is managed by *transport epithelia*, which regulate the movement of solutes between their internal fluids and the external environment (see Campbell, Figure 44.9).

- Usually a single layer of cells, joined by impermeable tight junctions, facing the external environment
- May face a channel that leads to the exterior through an opening on the body surface

In most animals, transport epithelia are arranged into tubular networks with extensive surface areas.

- The transport epithelia in the nasal glands of marine birds are very efficient at eliminating the excess salts obtained from drinking seawater.

The molecular composition of the epithelium's plasma membrane determines the specific osmoregulatory functions.

- For example, the gill epithelium pumps salt out of marine fishes and pumps salts into freshwater fishes.
- May also function in excretion of metabolic wastes in some animals.

B.   An animal's nitrogenous wastes are correlated with its phylogeny and habitat

The metabolism of proteins and nucleic acids produces ammonia, a small and very toxic waste product. Some animals excrete the ammonia directly, while others first convert it to urea or uric acid, which are less toxic, but require ATP to produce.

The kind of nitrogenous wastes an animal excretes depends on its evolutionary history and habitat (see Campbell, Figure 44.10).

1. Ammonia

Most aquatic animals excrete wastes as ammonia.

Easily permeates membranes since molecules are very water soluble.

In soft-bodied invertebrates, ammonia just diffuses across the body surface and into the surrounding water.

In fishes, ammonia is excreted as ammonium ions across gill epithelium.

⇒ The gill epithelium of freshwater fishes exchanges Na^+ from the water for NH_4^+, thus maintaining a higher Na^+ concentration in the blood.

2. Urea

Ammonia excretion is unsuitable for animals in a terrestrial habitat.

- It requires large amounts of water and is so toxic it must be eliminated quickly.

Urea is the nitrogenous waste excreted by mammals and most adult amphibians.

- Can be more concentrated in the body since it is about 100,000 times less toxic than ammonia; reduces water loss for terrestrial animals.
- Produced in liver by a metabolic cycle combining ammonia with CO_2. It is transported to kidneys via the circulatory system.
- Sharks produce and retain urea in the blood as an osmoregulatory agent.
- Amphibians that undergo metamorphosis and move to land as adults switch from excreting ammonia to excreting urea.

3. Uric acid

Uric acid is the primary form of nitrogenous waste excreted by land snails, insects, birds, and many reptiles.

- Much less soluble in water than ammonia or urea, it can be excreted as a precipitate after reabsorption of nearly all the water from the urine.
- Eliminated in a pastelike form through the cloaca (mixed with feces) in birds and reptiles.

The mode of reproduction is an important factor in determining whether uric acid or urea excretion evolved in a particular group.

- If an embryo released ammonia or urea within a shelled egg, the soluble waste would accumulate to toxic concentrations: uric acid precipitates out of solution and can be stored as a solid within the egg.

The animal's habitat, along with the phylogenetic position, influences the type of nitrogenous waste produced.

- Terrestrial reptiles excrete mostly uric acid; crocodiles excrete ammonia and uric acid; aquatic turtles excrete urea and ammonia.
- Some animals can modify their nitrogenous wastes when the temperature or water availability changes.

C. Cells require a balance between osmotic gain and loss of water

Animals cells cannot survive a net gain (swell and burst) or loss (shrivel and die) of water.

Osmosis = Diffusion of water across a selectively permeable membrane

⇒ Occurs when two solutions separated by a membrane differ in *osmolarity* (total solute concentration).

⇒ If a selectively permeable membrane separates two solutions of differing osmolarities, water flows from the hypoosmotic solution to the hyperosmotic solution.

Hyperosmotic solution = When comparing two solutions, the solution with a greater solute concentration; net water movement occurs into the solution.

Hypoosmotic solution = When comparing two solutions, the solution with a lower solute concentration; *net* water movement occurs out of the solution.

Isoosmotic solution = When comparing two solutions, a solution with a solute concentration equal to the other solution; no net water movement occurs between the solutions.

D. Osmoregulators expend energy to control their internal osmolarity; osmoconformers are isoosmotic with their surroundings

Water may enter the body of a terrestrial animal through food, drinking, and oxidative metabolism; water exits the body via evaporation and excretion. Aquatic animals are not affected by evaporation, but face the problem of osmosis where water may enter (fresh water) or leave (marine) the body.

- Even animals with specialized body coverings that retard water gain or loss have some unprotected structures exposed to the environment for gas exchange (lungs, gills).

An animal may be an osmoconformer or osmoregulator depending on their strategy for adaptation to the osmotic environment.

Osmoconformers = Animals that do not actively adjust their internal osmolarity

- Most marine invertebrates
- Body fluids are isoosmotic with surroundings

> Although these animals have an internal osmolarity that is equal to that of their surroundings, they do adjust the specific ion composition of their body fluids. For example, the osmolarity of starfish hemolymph is ca. 1000 mosm, identical to that of the seawater in which it lives. The magnesium ion concentration in the hemolymph of the starfish, however, is substantially lower than that in seawater.

Osmoregulators = Animals that regulate internal osmolarity by discharging excess water or taking in additional water

- Many marine animals, all freshwater animals, terrestrial animals, and humans.
- Net movement of water requires an osmotic gradient, the maintenance of which requires energy.
- Osmoregulation permits animals to live in a variety of habitats, but the tradeoff is that it requires an energy expenditure by the animal.

A large change in external osmolarity is fatal to most animals, although some can survive radical fluctuations.

- *Stenohaline animal* = Animal that cannot survive a wide fluctuation in external osmolarity
- *Euryhaline animal* = Animal that can survive wide fluctuations in external osmolarity
  - Can be osmoconformers or osmoregulators

1. Maintaining water balance in the sea

Most marine invertebrates are osmoconformers.

- Body fluids are isoosmotic to the environment.

- Body fluid composition usually differs from the external medium due to internal regulation of specific ions.

Some vertebrates of the Class Agnatha (hagfishes) are also osmoconformers; however, all other marine vertebrates are osmoregulators.

Most cartilaginous fishes, including sharks, maintain internal salt concentrations lower than sea water by pumping salt out through rectal glands and through the kidneys, yet their osmolarity is slightly hyperosmotic to seawater.

- Sharks retain urea as a dissolved solute in the body fluids.
- Sharks also produce and retain trimethylamine oxide (TMAO), which protects their proteins from denaturation by urea.
- Retention of these organic solutes (urea, TMAO) in the body fluids actually makes them slightly hyperosmotic to seawater.
- Water enters the shark's body by osmosis rather than by drinking, and they balance osmotic uptake of water by excreting urine.

Marine bony fishes are hypoosmotic to seawater.

- Compensate for osmotic water loss by drinking large amounts of seawater and pumping excess salt out with their gill epithelium.
- Excrete only a small amount of urine.

2. Maintaining osmotic balance in fresh water

Freshwater animals are hyperosmotic to their environment and constantly take in water by osmosis.

- Freshwater protists compensate with contractile vacuoles that pump out excess water.
- Many freshwater animals, including fish, compensate by excreting large amounts of very dilute urine.
    ⇒ Since salts are lost in this process, salt is replenished either by eating substances with a higher salt content, or in the case of some fish, by active uptake of sodium and chloride ions from the surrounding water by gill epithelium.

Anadromous fishes such as salmon are euryhaline and migrate between seawater and fresh water.

- While in the ocean, they osmoregulate like other marine fishes.
- When in fresh water, they cease drinking and their gills start taking up salt from the dilute environment, just like other freshwater fishes.

3. Special problems of living in temporary waters

Anhydrobiosis is an adaptation found in a small number of aquatic invertebrates that permits them to survive in a dormant state when their habitat dries up.

    ⇒ Best exemplified by the tardigrades (see Campbell, Figure 44.12)
- Hydrated animals are about 1 mm long and are about 85% water.
- They can dehydrate to less than 2% water and survive in an inactive state.
- Tardigrades can survive many years in this state and will rehydrate and become active when water returns.

Dehydrated and frozen animals face the problem of keeping their cell membranes intact.

    ⇒ Researchers have found that dehydrated anhydrobiotic animals contain large amounts of the disaccharide trehalose along with other sugars.
    ⇒ Trehalose appears to replace the water associated with membranes and proteins.

$\Rightarrow$ Trehalose is also found in insects that survive freezing.

4. Maintaining osmotic balance on land

Terrestrial animals live in a dehydrating environment and cannot survive desiccation.

- Humans die if 12% of their body water is lost.

Osmoregulatory mechanisms in terrestrial animals include protective outer layers, drinking and eating moist foods, behavioral adaptations, and excretory organ adaptations that conserve water.

- Arthropods have waxy cuticles, land snails possess shells, and vertebrates are covered by a multi-layer skin comprised of dead, keratinized cells.
- Drinking and eating moist foods replaces much of the water lost during gas exchange.
- Some desert animals are nocturnal; being active only at night reduces dehydration and some like the kangaroo rat produce large amounts of metabolic water (see Campbell, Figure 44.13).
- The excretory organs of terrestrial animals are adapted to conserve water while eliminating wastes.

III.   Excretory Systems

A. Most excretory systems produce urine by refining a filtrate derived from body fluids: an overview

Although the excretory systems of animals are structurally diverse, they share functional similarities

In general, urine is produced in two steps:

1. Filtration of body fluids
2. Modification of the filtrate

Modification of the filtrate can occur by two means:

1. Selective secretion of solutes (e.g., salts, toxins, etc.) from body fluids into the filtrate
2. Selective reabsorption of solutes from the filtrate back into body fluids (e.g., glucose)

B. Diverse excretory systems are variations on a tubular theme

1. Protonephridia: flame-bulb systems

Flatworms, which have neither circulatory systems nor coeloms, have a simple tubular excretory system called a protonephridium (see Campbell, Figure 44.15).

- *Protonephridium* = A network of closed tubules lacking internal openings that branch throughout the body; the smallest branches are capped by a cellular flame bulb
- Interstitial fluid passes through a flame bulb and is propelled by a tuft of cilia (in the flame bulb) along the branched system of tubules.
- Urine from the system empties into the external environment through numerous openings called nephridiopores.
- Transport epithelium lining the tubules function in osmoregulation by absorbing salts before the fluid exits the body.

Some parasitic flatworms are isoosmotic to their hosts and this closed system is used mainly to excrete nitrogenous wastes.

Protonephridia are also found in rotifers, some annelids, the larvae of mollusks, and lancelets.

2. Metanephridia

Each segment of most annelids, including earthworms, contains a pair of metanephridia, excretory tubules that have internal openings to collect body fluids (see Campbell, Figure 44.16)

- Coelomic fluid enters the funnel-shaped nephrostome, which is surrounded with cilia.
- The fluid passes through the metanephridium and empties into a storage bladder that empties outside the body through the nephridiopore.
- The nephrostome collects coelomic fluid from the body segment just anterior.
- A network of capillaries envelops each metanephridium.
    - ⇒ These capillaries reabsorb essential salts pumped out of the collecting tubules by transport epithelium bordering the lumen.
- Excretion of hypoosmotic, dilute urine offsets the continual osmosis of water from damp soil across the skin.

3. Malpighian tubules

Malpighian tubules = Excretory organs of insects and other terrestrial arthropods that remove nitrogenous wastes from the hemolymph and function in osmoregulation (see Campbell, Figure 44.17)

They are outpocketings of the gut that open into the digestive tract at the midgut-hindgut juncture.

The tubules dead-end at the tips away from the gut and are bathed in the hemolymph.

Transport epithelium lining each tubule moves solutes (salts and nitrogenous wastes) from the hemolymph into the tubule's lumen.

Accumulates nitrogenous wastes from the hemolymph and water follows by osmosis.

The fluid in the tubule then passes through the hindgut to the rectum.

Salts and water are reabsorbed across the epithelium of the rectum and dry nitrogenous wastes are excreted with feces.

4. Vertebrate kidneys

The invertebrate chordate ancestors of vertebrates probably possessed segmentally arranged excretory structures arranged throughout the body.

- Extant hagfishes have segmentally arranged excretory tubules associated with their kidneys.

The kidneys of vertebrates (other than hagfishes) are compact organs and contain large numbers of non-segmentally arranged tubules.

- Kidney structure also includes a dense capillary network intimately associated with the tubules.
- The tubules function in both excretion and osmoregulation in vertebrates that osmoregulate.

The vertebrate excretory system is comprised of the kidneys, blood vessels serving the kidneys, and the structures that carry urine from the kidneys out of the body (see Campbell, Figure 44.18).

Variations of the basic system are found among the vertebrate classes.

C. Nephrons and associated blood vessels are the functional units of the mammalian kidney

The human kidneys are a pair of bean-shaped organs about 10 cm long (see Campbell, Figure 44.18a).

Blood enters each kidney via the *renal artery* and exits via the *renal vein*.

About 20% of the blood pumped by each heartbeat passes through the kidneys.

Urine exits each kidney through a ureter and both ureters drain into a common urinary bladder.

Urine leaves the body from the urinary bladder through the urethra.

Sphincter muscles near the junction of the urethra and bladder control urination.

1. **Structure and function of the nephron and associated structures**

   The two distinct regions of the kidney are the outer *renal cortex* and inner *renal medulla* (see Campbell, Figure 44.18b).

   - Each region contains many microscopic *nephrons* and *collecting ducts* (see Campbell, Figure 44.18c).
   - Associated with each excretory tubule is a network of capillaries.

   Nephron = Functional unit of the kidney, consisting of a single long tubule and its associated capillaries

   The blind end of the renal tubule that receives filtrate from the blood forms a cup-shaped *Bowman's capsule*, which embraces a ball of capillaries called the *glomerulus* (see Campbell, Figure 44.18d).

   The Bowman's capsules and the proximal and distal convoluted tubules are located in the cortex.

   The composition of blood is regulated by transport epithelia of the nephrons and collecting ducts through three processes: filtration, secretion, and reabsorption.

   a. **Filtration of the blood**

      Blood pressure forces fluid (containing water, salts, urea and other small molecules) from the glomerulus into the lumen of the Bowman's capsule.

      - Porous capillaries and *podocytes* (specialized cells of the capsule) nonselectively filter out blood cells and large molecules; any molecule small enough to be forced through the capillary wall enters the nephron tubule.
      - Filtrate at this point contains a mixture of glucose, salts, vitamins, nitrogenous wastes, and small molecules in concentrations similar to that in blood plasma.

   b. **Pathway of the filtrate**

      Filtrate then passes through the *proximal tubule*, the *loop of Henle* (a long hairpin turn with a descending limb and an ascending limb) and the *distal tubule*, which empties into a *collecting duct*.

      - The collecting duct receives filtrate from many nephrons.
      - Filtrate, now called presumptive urine (further modification can occur in the bladder), flows from the collecting ducts into the renal pelvis. The presumptive urine then drains from the chamberlike pelvis into the ureter.

      Two types of nephrons are found in mammals and birds: cortical nephrons and juxtamedullary nephrons.

      - *Cortical nephrons* = Nephrons that have reduced loops of Henle and are confined to the renal cortex; 80% of the nephrons in humans are cortical nephrons
      - *Juxtamedullary nephrons* = Nephrons that have long loops that extend into the renal medulla; 20% of the nephrons are juxtamedullary nephrons (see Campbell, Figure 44.18c).
      - The nephrons in other vertebrates lack loops of Henle.

The nephron and collecting duct are lined by transport epithelium that processes the filtrate into urine.

- About 1100 to 1200 L of blood flow through the human kidneys each day.
- The nephrons process 180 L of filtrate per day, and the transport epithelium processes this filtrate to form the approximately 1.5 L urine excreted daily.
- The rest of the filtrate is reabsorbed into the blood.

The loops of Henle and collecting tubules extend into the medulla.

c. **Blood vessels associated with the nephrons**

Each nephron is closely associated with blood vessels:

- *Afferent arteriole* is a branch of the renal artery that divides to form the capillaries of the glomerulus.
- *Efferent arteriole* forms from the converging capillaries as they leave the glomerulus. This subdivides to form the *peritubular capillaries* which intermingle with the proximal and distal tubules.
- *Vasa recta* is the capillary system branching downward from the peritubular capillaries that serves the loop of Henle.

Materials are exchanged between capillaries and nephrons through interstitial fluids.

d. **Secretion**

Filtrate is joined by substances transported across the tubule epithelium from the surrounding interstitial fluid as it moves through the nephron tubule.

- Adds plasma solutes to the filtrate.

The proximal and distal tubules are the most common sites of secretion.

Secretion is a very selective process involving both passive and active transport.

- For example, controlled secretion of H^+ ions helps maintain constant body fluid pH.

e. **Reabsorption**

Reabsorption is the selective transport of filtrate substances across the excretory tubule epithelium from the filtrate back to the interstitial fluid.

- Reclaims small molecules essential to the body
- Occurs in the proximal tubule, distal tubule, loop of Henle, and collecting duct
- Nearly all sugar, vitamins, organic nutrients are reabsorbed. In mammals and birds, water is also reabsorbed.

The composition of the filtrate is modified by selective secretion and reabsorption.

- The concentration of beneficial substances in the filtrate is reduced as they are returned to the body.
- The concentration of wastes and nonuseful substances is increased and excreted from the body.

The kidneys are central to the process of homeostasis as they clear metabolic wastes from the blood and respond to body fluid imbalances by selectively secreting ions.

D. **From blood to filtrate:** *a closer look*

Reclamation of small molecules and water from the filtrate as it flows through the nephron and collecting duct converts the filtrate into urine (see Campbell, Figure 44.19).

1. The proximal tubule alters the volume and composition of filtrate by reabsorption and secretion.

   - In this area, ammonia, drugs, and poisons processed in the liver are secreted to join the filtrate.
   - Helps maintain a constant body fluid pH by controlled secretion of H^+ and reabsorption of bicarbonate.
   - Nutrients such as glucose and amino acids are reabsorbed (by active transport) from the filtrate and returned to the interstitial fluid from which they enter the blood.

   The reabsorption of NaCl and water is also an important function of the proximal tubule.

   - Salt diffuses into the transport epithelium cells; the membranes facing the interstitial fluid then actively pump Na^+ out of the cells, which is balanced by passive transport of Cl^-. Water follows passively by osmosis.
   - Cells facing the interstitial fluid (outside the tubule) have a small surface area to minimize leakage of salt and water back into the filtrate.

2. In the descending limb of the loop of Henle, the transport epithelium is freely permeable to water but not to salt and other small solutes.

   - Filtrate moving down the tubule from the cortex to the medulla continues to lose water by osmosis since the interstitial fluid in this region increases in osmolarity.
   - The NaCl concentration of the filtrate increases due to the water loss.

3. In the ascending limb of the loop of Henle, the transport epithelium is very permeable to salt, but not to water.

   - As the filtrate ascends through the thin segment near the loop tip, NaCl diffuses out and contributes to the high osmolarity of interstitial fluids of the medulla.
   - In the thick segment leading to the distal convoluted tubule, NaCl is actively transported out into the interstitial fluid.
   - The filtrate becomes more dilute due to the removal of salts without loss of water.

4. The distal tubule is an important site of selective secretion and absorption.

   - It regulates K^+ and NaCl concentrations of body fluids by regulating K^+ secretion into the filtrate and NaCl reabsorption from the filtrate.
   - This region also contributes to pH regulation by controlled secretion of H^+ and reabsorption of bicarbonate.

5. The collecting duct carries filtrate back through the medulla into the renal pelvis.

   - The transport epithelium here is permeable to water but not to salt.
     ⇒ The filtrate loses water by osmosis to the hyperosmotic fluid outside the duct which results in a concentration of urea in the filtrate.
   - The lower portion of the duct is permeable to urea, some of which diffuses out.

⇒ This contributes to the high osmolarity of the interstitial fluid of the renal medulla, which enables the kidney to conserve water by excreting a hyperosmotic urine.

E. The mammalian kidney's ability to conserve water is a key terrestrial adaptation

Cooperative action between the loop of Henle and the collecting duct maintains the osmolarity gradient in the tissues of the kidney.

- The two solutes responsible for the gradient are NaCl (deposited by the loop of Henle) and urea which leaks across the epithelium of the collecting ducts.
- The urine formed is up to four times as concentrated as the blood (about 1200 mosm/L).

1. Conservation of water by two solute gradients

The juxtamedullary nephron, with its urine-concentrating features, is a key adaptation to terrestrial life that enables mammals to excrete nitrogenous waste without squandering water.

Filtrate passing from the Bowman's capsule to the proximal tubule has about the same osmolarity as blood (300 mosm/L) (see Campbell, Figure 44.20)

A large amount of water and salt is reabsorbed as filtrate passes through the proximal tubule, which is located in the renal cortex.

⇒ The osmolarity remains about the same during the decrease in volume.

Water moves out of the filtrate by osmosis as it flows from the cortex into the medulla through the descending loop of Henle.

⇒ The osmolarity steadily increases due to loss of water until it peaks at the apex.

As filtrate moves up the ascending loop of Henle back to the cortex, salt leaves the filtrate first by passive transport and then by active transport.

⇒ Osmolarity decreases to about 100 mosm/L at this point.

Osmolarity changes very little as filtrate flows through the distal tubule as it is hypoosmotic to the interstitial fluids of the cortex.

After entering the collecting duct, the filtrate passes back through the medulla.

⇒ The filtrate loses water which increases osmolarity.

⇒ Some urea also leaks out of the lower portion of the collecting duct.

The passage of the filtrate back through the hyperosmotic medulla causes a gradual increase in osmolarity to about 1200 mosm/L; the remaining molecules are excreted in a minimal amount of water as urine which passes to the renal pelvis to the ureter.

> NOTE: The loss of salt from the filtrate passing through the ascending limb of the loop of Henle to the interstitial fluid of the medulla contributes to the high osmolarity of the medulla. This, in turn, helps conserve water.

The *vasa recta* (capillary network of the renal medulla) maintains the crucial osmolarity gradient in the kidney.

- As the descending vessel conveys blood toward the inner medulla, water is lost from the blood and NaCl diffuses into the blood.
- These fluxes are reversed as blood flows back toward the cortex in the ascending vessel.
- This countercurrent system allows the vasa recta to supply the tissues with necessary substances without interfering with the osmolarity gradient.

Urine, at its most concentrated, is isoosmotic to the interstitial fluid of the inner medulla, but is hyperosmotic to body fluids elsewhere.

F. Nervous and hormonal feedback circuits regulate kidney functions

The kidney is a versatile osmoregulatory organ subject to a combination of nervous and hormonal controls.

- It excretes hyper- or hypoosmotic urine as necessary.

Three mechanisms regulate the kidney's ability to change the osmolarity, salt concentration, volume, and blood pressure: 1) antidiuretic hormone, 2) juxtaglomerular apparatus, and 3) atrial natriuretic factor.

1. *Antidiuretic hormone (ADH)* enhances fluid retention by increasing the water permeability of epithelium of the distal tubules and the collecting duct (see Campbell, Figure 44.21a)

   - ADH is produced in the hypothalamus and stored and released from the posterior pituitary gland.

   - ADH release is triggered when osmoreceptor cells in the hypothalamus detect increased blood osmolarity due to an excessive loss of water from the body.

   - Increased water reabsorption reduces blood osmolarity and reduces stimulation of the osmoreceptor cells.

     ⇒ Results in less ADH being secreted

     ⇒ Ingestion of water returns blood osmolarity to normal

   - When a large volume of water has been ingested, little ADH is released and the kidneys produce a dilute urine since little water is absorbed.

   - Alcohol can inhibit ADH release, causing dehydration.

2. The *juxtaglomerular apparatus (JGA)* is a specialized tissue near the afferent arterioles, which carries blood to the glomeruli (see Campbell, Figure 44.21b).

   - It responds to a decrease in blood pressure or blood volume, as well as to a decrease in blood sodium ion concentration by releasing the enzyme renin into the blood.

   - Renin leads to the conversion of inactive angiotensinogen to active *angiotensin II*, which functions as a hormone.

   - Angiotensin II directly increases blood pressure by causing arteriole constriction.

   - Angiotensin II acts indirectly by:

     ⇒ Signaling adrenal glands to release *aldosterone*, which stimulates Na^+ reabsorption by the distal tubules (water follows by osmosis)

     ⇒ Stimulating thirst centers in brain to induce drinking behavior

     ⇒ Increasing blood pressure and blood volume.

   - The increased blood pressure and blood volume suppresses further release of renin.

   - The *renin-angiotensin-aldosterone system (RAAS)* is part of a complex feedback circuit that functions in homeostasis.

ADH and the RAAS cooperate in homeostasis.

- ADH is released in response to increased blood osmolarity, but this does not compensate for excessive loss of salts and body fluids if blood osmolarity does not change.

- RAAS responds to a decrease in blood volume caused by fluid loss.

- ADH alone would lower blood Na⁺ concentration by increasing water reabsorption, but the RAAS maintains balance by stimulating Na⁺ reabsorption.

3. The hormone *atrial natriuretic factor (ANF)* opposes the RAAS.

- Released by the heart's atrial walls in response to increased blood volume and pressure
- Inhibits release of renin from the JGA, inhibits NaCl absorption by the collecting ducts, and reduces aldosterone release from the adrenal glands; this decreases blood volume and lowers blood pressure

G. Diverse adaptations of the vertebrate kidney have evolved in different habitats

Nephrons vary in structure and physiology and help different vertebrates to osmoregulate in their various habitats.

- Desert mammals have very long loops of Henle to maintain steep osmotic gradients that conserve water by allowing urine to become very concentrated.
- Mammals living in aquatic environments (e.g., beavers) have nephrons with very short loops of Henle, which result in production of dilute urine.
- Birds have shorter loops of Henle and produce a more dilute urine than mammals.
- Reptiles have only cortical nephrons and produce isoosmotic urine, but the epithelium of their cloaca conserves fluid by reabsorbing water from urine and feces.
  ⇒ Also, most excrete nitrogenous wastes as uric acid which conserves water.
- Freshwater fish (hyperosmotic to surroundings) nephrons use cilia to sweep the large volume of very dilute urine from the body.
  ⇒ Salts are conserved by efficient ion reabsorption from filtrate.
- Amphibians excrete dilute urine and accumulate certain salts from the water by active transport across the skin. On land, body fluid is conserved by water reabsorption across urinary bladder epithelium.
- Bony marine fishes (hypoosmotic to surroundings) excrete very little concentrated urine (many lack glomeruli and capsules).
  ⇒ The kidneys function mainly to rid the body of divalent ions such as Ca^{2+}, Mg^{2+}, and SO_4^{2-} taken in by drinking seawater.
  ⇒ Monovalent ions like Na⁺ and Cl⁻, and the majority of nitrogenous waste (in the form of ammonium) is excreted mainly by the gills.

H. Interacting regulatory systems maintain homeostasis

Maintaining homeostasis in the internal environment involves a number of regulatory systems within an animal's body.

- Regulation of body temperature involves mechanisms that control osmolarity, metabolic rate, blood pressure, tissue oxygenation, and body weight.

While the regulatory systems normally work in concert to maintain homeostasis, under extreme physiological stress, demands of one regulatory system may conflict with those of other systems.

- Water conservation takes precedence over evaporative heat loss in very warm, dry climates.
- Many desert animals must tolerate occasional hyperthermia (abnormally high body temperature) in order to maintain body water.

The functions of the vertebrate liver are critical to maintaining homeostasis and involve interactions with most of the body's organ systems.

1. Excretion

   Excretion by the kidneys is supported by the liver, which synthesizes ammonia, urea, or uric acid from the nitrogen of amino acids.

   The liver also detoxifies many chemical poisons.

2. Bioenergetics

   The liver plays a role in regulating basal metabolic rate, a thyroid hormone-sensitive process, by converting thyroxin to the bioactive triiodothyronine.

   The liver also plays a central role in nutrient metabolism, especially the metabolism of carbohydrates.

   - Liver cells take up glucose from the blood and store excess amounts as glycogen.
   - When glucose is needed by the body, the liver converts glycogen back to glucose and releases it into the blood.
   - Blood glucose levels are closely controlled by feedback circuits which regulate the homeostatic mechanisms.

3. Osmoregulation. The liver plays a role in osmoregulation through its production of angiotensinogen, the precursor to angiotensin II.

4. Growth. The liver plays a role in growth regulation through its production and release of the proximate regulator of growth, insulin-like growth factor.

Thus, while the liver performs the essential functions, feedback circuits are in turn controlled by the nervous and endocrine systems.

REFERENCES

Campbell, N., et al., *Biology*. 5th ed. Menlo Park, California: Benjamin/Cummings, 1999.

Randall, D., Burggren, W., and French, K *Animal Physiology: Mechanisms and Adaptations*. 4th ed. San Francisco: W.H. Freeman and Company, 1997.

Marieb, E.N. *Human Anatomy and Physiology*. 4th ed. Menlo Park, California: Benjamin/Cummings, 1997.

CHAPTER 45
CHEMICAL SIGNALS
IN ANIMALS

OUTLINE

I.   An Introduction to Regulatory Systems
   A. The endocrine system and the nervous system are structurally, chemically, and functionally related
   B. Invertebrate regulatory systems clearly illustrate endocrine and nervous system interactions

II. Chemical Signals and Their Modes of Action
   A. A variety of local regulators affect neighboring target cells
   B. Chemical signals bind to specific receptor proteins within target cells or on their surface
   C. Most chemical signals bind to plasma-membrane proteins, initiating signal-transduction pathways
   D. Steroid hormones, thyroid hormones, and some local regulators enter target cells and bind with intracellular receptors

III. The Vertebrate Endocrine System
   A. The hypothalamus and pituitary integrate many functions of the vertebrate endocrine system
   B. The pineal gland is involved in biorhythms
   C. Thyroid hormones function in development, bioenergetics, and homeostasis
   D. Parathyroid hormone and calcitonin balance blood calcium
   E. Endocrine tissues of the pancreas secrete insulin and glucagon, antagonistic hormones that regulate blood glucose
   F. The adrenal medulla and adrenal cortex help the body manage stress
   G. Gonadal steroids regulate growth, development, reproductive cycles, and sexual behavior.

OBJECTIVES

After reading this chapter and attending lecture, the student should be able to:

1. Compare the response times of the two major systems of internal communication: the nervous system and the endocrine system.
2. On the basis of structure and function, distinguish among types of chemical messengers.
3. Distinguish between endocrine and exocrine glands.
4. Describe the relationships among endocrine system components: hormones, endocrine glands, target cells, and target cell receptors.
5. List the general chemical classes of hormones and give examples of each.
6. Explain how pheromone function differs from hormone function.

7. Provide indirect evidence that humans may communicate with pheromones.

8. State which of the two classes of hormones is lipid soluble, and explain how this property affects hormone function.

9. Describe the mechanism of steroid hormone action, and explain the location and role of steroid hormone receptors.

10. Explain how to account for specificity in target cell response to hormonal signals.

11. Compare and contrast the two general modes of hormone action.

12. Describe hormonal regulation of insect development including the roles of ecdysone, brain hormone, and juvenile hormone.

13. Describe the location of the hypothalamus, and explain how its hormone-releasing cells differ from both endocrine gland secretory cells and other neurons.

14. Describe the location of the pituitary, and explain the functions of the posterior and anterior lobes.

15. List the posterior pituitary hormones, and describe their effects on target organs.

16. Using antidiuretic hormone as an example, explain how a hormone contributes to homeostasis and how negative feedback can control hormone levels.

17. Define tropic hormone, and describe the functions of tropic hormones produced by the anterior pituitary.

18. Explain how the anterior pituitary is controlled.

19. List hormones of the thyroid gland, and explain their role in development and metabolism.

20. Diagram the negative feedback loop which regulates the secretion of thyroid hormones.

21. State the location of the parathyroid glands, and describe hormonal control of calcium homeostasis.

22. Distinguish between α and β cells in the pancreas and explain how their antagonistic hormones (insulin and glucagon) regulate carbohydrate metabolism.

23. List hormones of the adrenal medulla, describe their function, and explain how their secretion is controlled.

24. List hormones of the adrenal cortex, describe their function, and explain how their secretion is controlled.

25. Describe both the short-term and long-term endocrine responses to stress.

26. Identify male and female gonads, and list the three categories of gonadal steroids.

27. Define gonadotropin, and explain how estrogen and androgen synthesis is controlled.

28. Describe the location of the pineal and thymus glands, list their hormone products, and describe their functions.

29. Explain how the endocrine and nervous systems are structurally, chemically, and functionally related.

KEY TERMS

hormone
target cells
endocrine system
endocrine glands
neurosecretory cells
ecdysone
brain hormone (BH)
juvenile hormone (JH)

releasing hormones
inhibiting hormones
posterior pituitary
   (neurohypophysis)
growth hormone (GH)
insulinlike growth
   factors (IGFs)
prolactin (PRL)

endorphins
pineal gland
melatonin
thyroid gland
triiodothyronine (T_3)
thyroxine (T_4)
calcitonin
parathyroid glands

adrenal glands
adrenal cortex
adrenal medulla
epinephrine
norepinephrine
catecholamines
corticosteroids
glucocorticoids

nitric oxide	follicle-stimulating	parathyroid hormone	mineralocorticoids
growth factors	hormone (FSH)	(PTH)	androgen
prostaglandins (PGs)	luteinizing hormone (LH)	pancreas	testosterone
signal-transduction	thyroid-stimulating	islets of Langerhans	estrogen
pathways	hormone (TSH)	alpha (α) cells	progestins
tropic hormones	gonadotropins	glucagon	
hypothalamus	adrenocorticotropic	beta (β) cells	
pituitary gland	hormone (ACTH)	insulin	
anterior pituitary	melanocyte-stimulating	type I diabetes mellitus	
(adenohypophysis)	hormone (MSH)	type II diabetes mellitus	

LECTURE NOTES

I.   An Introduction to Regulatory Systems

The activities of the various specialized parts of an animal are coordinated by the two major systems of internal communication: the nervous system and the endocrine system.

- The nervous system is involved with high-speed messages.
- The endocrine system is slower and involves the production, release, and movement of chemical messages.

Endocrine glands = Ductless glands that secrete hormones into the body fluids for distribution throughout the body

Exocrine glands = Secrete chemicals, such as sweat, mucus, and digestive enzymes, into ducts which convey the products to the appropriate locations

As we learn more about the regulatory processes of animals, it becomes increasingly clear that the endocrine system and the nervous system are interrelated; homeostasis depends heavily on their overlap.

A.   The endocrine system and the nervous system are structurally, chemically, and functionally related

Many endocrine organs and tissues contain specialized nerve cells called *neurosecretory cells* that secrete hormones.

Several chemicals serve both as hormones of the endocrine system and as signals in the nervous system.

- Norepinephrine functions as both an adrenal hormone and as a neurotransmitter.

The regulation of several physiological processes involves structural and functional overlap between the two systems.

Positive and negative feedback regulate mechanisms of both systems (see Campbell, Figure 45.1).

B.   Invertebrate regulatory systems clearly illustrate endocrine and nervous system interactions

Invertebrates possess a diversity of hormones which function in homeostasis, reproduction, development, and behavior.

In many cases, the hormone may stimulate one activity while inhibiting another.

- In *Hydra*, one hormone stimulates growth and budding while inhibiting sexual reproduction.
- In the sea hare *Aplysia*, a peptide hormone secreted by specialized neurons stimulates egg laying while inhibiting feeding and locomotion.

All arthropods have extensive endocrine systems that regulate growth and reproduction, water balance, pigment movement in the integument and eyes, and metabolism.

Hormones may act together, such as in the case of insect development and molting.

- *Ecdysone* is secreted by a pair of prothoracic glands; it triggers molting and favors development of adult characteristics and metamorphosis (see Campbell, Figure 45.2).
  - Ecdysone stimulates transcription of specific genes, the same mechanism of action as vertebrate steroid hormones.
- *Brain hormone* promotes development by stimulating production of ecdysone.
  - Brain hormone and ecdysone are balanced by *juvenile hormone (JH)*, which actively promotes retention of larval characteristics; JH is secreted by the corpora allata.
  - When JH levels decrease, ecdysone-induced molting produces a pupa; in the pupa, adult anatomy replaces larval anatomy during metamorphosis.

II.   Chemical Signals and Their Mode of Action

Chemical signals operate at virtually all levels of organization

- Intracellular (includes elements of signal transduction systems)
- Cell to cell
- Tissue to tissue
- Organ to organ (this is level that was considered the subject of classic endocrinology, with hormones acting as the chemical signals)
- Organism to organism (includes pheromones, chemical signals that function between organisms of the same species; classified according to function e.g., mate attractant, territorial marker, alarm substance).

Compounds called local regulators operate at the cell to cell and tissue to tissue levels of organization (see Campbell, Figure 11.3a).

> NOTE: It is important to note that chemical mediators may act without regard to the conceptual boundaries that we place on them. A single compound may act in one instance as a cell to cell chemical mediator and in another instance as hormones or pheromones. Prostaglandins, for example, have been shown to operate at every level of biological organization.

A.   A variety of local regulators affect neighboring target cells

Examples of local regulators include the following:

- Histamine (involved with various immune and regulatory responses; see Campbell, Chapters 41 and 43)
- Interleukins (involve with various immune responses; see Campbell, Chapter 43)
- Retinoic acid (involved with vertebrate development; see Campbell, Chapter 47)
- Nitric oxide (NO)
  - NO released by endothelial cells of blood vessels makes the adjacent smooth muscle cells relax, dilating the vessel.
  - NO released by white blood cells kills certain cancer cells and bacteria in the body fluids.

Growth factors = Peptides and proteins that regulate the behavior of cells in growing and developing tissues

- Must be present in the extracellular environment for certain cell types to grow and develop normally.
    - Studied mainly in cultures of mammalian cells, but also regulate development within the animal body.
    - It is likely that the interaction of numerous growth factors regulates cell behavior in developing tissues and organs.

Prostaglandins (PGs) = Modified fatty acids released into interstitial fluid that function as local regulators

- Often derived from lipids of the plasma membrane
- Very subtle differences in their molecular structure profoundly affect how these signals affect target cells (e.g., antagonistic actions of PGE and PGF).
- PGs secreted by the placenta help induce labor during childbirth by causing chemical changes in the nearby uterine muscles.
- Other PGs help defend the body by inducing fever and inflammation.

B. Chemical signals bind to specific receptor proteins within target cells or on their surface

A chemical signal can affect different target cells within an animal differently, or it may affect different species differently (e.g., thyroxine).

The action of a particular chemical signal depends upon:

- Recognition of the signal by a specific receptor in or on the target cell (see Campbell, Figure 45.3)
- Activation of a signal-transduction pathway that leads to a specific cellular response

Binding of a chemical signal to a specific receptor protein triggers chemical events within the target cell that result in a change in its behavior.

The nature of the response to a chemical signal depends on the number and affinity of the receptor proteins (see Campbell, Figure 45.4).

C. Most chemical signals bind to plasma-membrane proteins, initiating signal-transduction pathways

Because of their chemical nature, most signal molecules (e.g., peptides, proteins, glycoproteins) are unable to diffuse through the plasma membrane.

- The biological action of these factors begins at the plasma membrane, where the signal molecule binds to a specific plasma membrane receptor.
- Binding of the signal molecule to a plasma membrane receptor initiates a *signal transduction pathway*, the series of events that converts the signal into a specific cellular response.

> A specific example is the binding on the polypeptide hormone insulin to the insulin receptor. Hormone binding initiates a chain of events that leads to the activation of glucose transporters, plasma membrane transport molecules, and the subsequent uptake of glucose that accounts for the blood sugar lowering effects of insulin. Refer to Chapter 11 for specific details of signal transduction.

D. Steroid hormones, thyroid hormone, and some local regulators enter target cells and bind with intracellular receptors

The chemical nature of some regulators allows them to pass through the plasma membrane (e.g., steroids, thyroid hormones, NO). The receptors for these factors are located within target cells (see Campbell, Figure 45.5).

The binding of the signal molecule with a specific receptor initiates the signal transduction process. In many cases, the signal-receptor complex binds to DNA to modify gene expression.

III.   The Vertebrate Endocrine System

The vertebrate endocrine system, through its production of numerous hormones, coordinates various aspects of metabolism, growth, development, and reproduction.

- Campbell, Figure 45.6 shows where the major endocrine glands in humans are located.
- Campbell, Table 45.1 summarizes the functions of the major vertebrate hormones.

Some of the hormones in vertebrates have a single action while others have multiple actions.

Tropic hormones act on other endocrine glands.

A.   The hypothalamus and pituitary integrate many functions of the vertebrate endocrine system

The *hypothalamus* is a region of the lower brain that receives information from nerves throughout the body and brain and initiates endocrine signals appropriate to the environmental conditions.

- Contains two sets of neurosecretory cells whose secretions are stored in or regulate activity of the pituitary gland.

The *pituitary gland* is an extension of the brain located at the base of the hypothalamus. It consists of two lobes and has numerous endocrine functions (see Campbell, Figure 45.7).

The *adenohypophysis*, or *anterior pituitary*, consists of endocrine cells that synthesize and secrete several hormones directly into the blood.

- Controlled by two kinds of hormones secreted by neurosecretory cells in the hypothalamus: releasing hormones and inhibiting hormones.
  - *Releasing hormones* stimulate the anterior pituitary to secrete its hormones.
  - *Inhibiting hormones* stop the anterior pituitary from secreting its hormones.

The *neurohypophysis*, or *posterior pituitary*, stores and secretes peptide hormones that are made by the hypothalamus (e.g., oxytocin and antidiuretic hormone).

1.   Posterior pituitary hormones

The hypothalamic peptide hormones, oxytocin and antidiuretic hormone (ADH), are stored in and released from the posterior pituitary.

- They are synthesized in neurosecretory cell bodies located in the hypothalamus and are secreted from the neurosecretory cell axons located in the posterior pituitary.
- Oxytocin induces uterine muscle contraction and causes the mammary glands to eject milk during nursing.
- ADH acts on the kidneys to increase water retention, which results in a decrease in urine volume.

ADH is part of the feedback mechanism that helps regulate blood osmolarity.

- Osmoreceptors (specialized nerve cells) in the hypothalamus monitor blood osmolarity.

- When plasma osmolarity increases, the osmoreceptors shrink slightly (lose water by osmosis) and transmit nerve impulses to certain hypothalamic neurosecretory cells.
- These neurosecretory cells respond by releasing ADH into the general circulation from their tips in the posterior pituitary.
- The target cells for ADH are cells lining the collecting ducts of nephrons in the kidneys.
- ADH binds to receptors on the target cells and activates a signal-transduction pathway that increases the water permeability of the collecting duct.
- Water retention is increased as water exits the collecting ducts and enters nearby capillaries.
- The osmoreceptors also stimulate a thirst drive.
  ⇒ When an individual drinks water it reduces blood osmolarity to the set point.
- As more dilute blood (lower osmolarity) arrives at the brain, the hypothalamus responds by reducing ADH secretion and lowering thirst sensations.
  ⇒ This prevents overcompensation by stopping hormone secretion and quenching thirst.
- Note that this negative feedback scheme includes a hormonal action and a behavioral response (drinking).

2. Anterior pituitary hormones

The anterior pituitary produces many different hormones and is regulated by releasing factors and release-inhibiting factors from the hypothalamus.

Four of the hormones (TSH, ACTH, FSH, LH) secreted from the anterior pituitary are tropic hormones that stimulate other endocrine glands to synthesize and release their hormones.

> Keep in mind that hormones may have multiple actions. The effect of GH on the liver to secrete IGFs is a tropic action, one of many actions of GH in vertebrates.

Growth hormone (GH) is a protein hormone that affects a wide variety of tissues.
- It promotes growth directly and stimulates the production of growth factors.
- For example, GH stimulates the liver to secrete *insulinlike growth factors (IGFs,)* which stimulate bone and cartilage growth.

Prolactin (PRL) is a protein hormone similar in structure to GH, although their physiological roles are very different.
- PRL produces a diversity of effects in different vertebrates, including:
  - Stimulation of mammary gland development and milk synthesis in mammals
  - Regulation of fat metabolism and reproduction in birds
  - Delay of metamorphosis; it may also function as a larval growth hormone in amphibians
  - Regulation of salt and water balance in freshwater fish

Follicle-stimulating hormone (FSH) is a tropic hormone that affects the gonads (*gonadotropin*).
- In males, it is necessary for spermatogenesis.
- In females, it stimulates ovarian follicle growth.

Luteinizing hormone (LH) is another gonadotropin, which stimulates ovulation and corpus luteum formation in females and spermatogenesis in males.

Thyroid-stimulating hormone (TSH) is a tropic hormone that stimulates the thyroid gland to produce and secrete its own hormones.

The remaining hormones from the anterior pituitary are formed by the cleaving of a single large protein, pro-opiomelanocortin, into short fragments.

- At least three of these fragments become active peptide hormones:
    - *Adrenocorticotropin (ACTH)* stimulates the adrenal cortex to produce and secrete its steroid hormones.
    - *Melanocyte-stimulating hormone (MSH)* regulates the activity of pigment-containing skin cells in some vertebrates.
    - *Endorphins* inhibit pain perception (see Campbell, Chapter 48).

B. The pineal gland is involved in biorhythms

The *pineal gland* is a small mass of tissue near the center of the mammalian brain (it is closer to the surface in some other vertebrates).

The pineal contains light-sensitive cells or has nervous connections with the eyes.

It secretes *melatonin*, a modified amino acid.

- Melatonin modulates skin pigmentation.
- Melatonin regulates functions related to light and to seasons marked by changes in day length, such as biological rhythms associated with reproduction.
- Melatonin is secreted only at night; larger amounts are secreted during winter.
- The role of melatonin in regulating biological rhythms is not yet understood.

C. Thyroid hormones function in development, bioenergetics, and homeostasis

The *thyroid gland* consists of two lobes located on the ventral surface of the trachea in mammals, and on the two sides of the pharynx in other vertebrates.

- Produces two modified amino acid hormones, T_3 (*triiodothyronine*) and T_4 (*thyroxine* or tetraiodothyronine) derived from the amino acid tyrosine.
    - ⇒ These differ in structure by only one iodine atom.
- Both have the same effects on their target, although T_3 is usually more active in mammals than T_4.

The thyroid gland plays a major role in vertebrate development and maturation.

- Thyroid hormones control metamorphosis in amphibians.
- Normal function of bone-forming cells and the branching of nerve cells during embryonic brain development of nonhuman animals also requires the presence of the thyroid hormones.

The thyroid gland is critical for maintaining homeostasis in mammals.

- It helps maintain normal blood pressure, heart rate, muscle tone, digestion, and reproductive functions.
- T_3 and T_4 tend to increase the rate of oxygen consumption and cellular metabolism.
- Serious metabolic disorders can result from a deficiency or excess of thyroid hormones.
    - ⇒ Hyperthyroidism, excessive secretion of thyroid hormones, causes high body temperature, sweating, weight loss, irritability, and high blood pressure.
    - ⇒ Hypothyroidism, low secretion of thyroid hormones, can cause cretinism in infants and weight gain, lethargy, and cold-intolerance in adults.
    - ⇒ Goiter (enlarged thyroid) is caused by a dietary iodine deficiency.

Thyroid hormone secretion is regulated by the hypothalamus and pituitary through a negative feedback system (see Campbell, Figure 45.9).

- The hypothalamus secretes TRH (TSH-releasing hormone), which stimulates TSH (thyroid-stimulating hormone) secretion by the anterior pituitary.
- When TSH binds to receptors in the thyroid, cAMP is generated and triggers release of T_3 and T_4.
- High levels of T_3, T_4, and TSH inhibit TRH secretion.

The thyroid gland in mammals also produces and secretes *calcitonin*, a peptide hormone that lowers blood calcium levels.

D. Parathyroid hormone and calcitonin balance blood calcium

Four *parathyroid glands* are embedded in the surface of the thyroid and function in the homeostasis of calcium ions.

- They secrete *PTH (parathyroid hormone)*, which raises blood Ca^{2+} levels (antagonistic to calcitonin) and needs vitamin D to function.
- PTH stimulates Ca^{2+} reabsorption in the kidney and induces osteoclasts to decompose bone and release Ca^{2+} into the blood.

PTH and calcitonin (which have opposite affects) work in an antagonistic manner and their balance in the body maintains the proper blood calcium levels (see Campbell, Figure 45.10).

E. Endocrine tissues of the pancreas secrete insulin and glucagon, antagonistic hormones that regulate blood glucose

The pancreas of many vertebrates is a compound gland that performs both exocrine and endocrine functions.

- In addition to secreting two hormones, it produces the pancreatic enzymes associated with digestion, which are carried to the small intestine via ducts.

The endocrine cells are typically clustered together is solid balls called the *islets of Langerhans*.

- Each islet is composed of *alpha (α) cells*, which secrete the peptide hormone *glucagon*, and *beta (β) cells* which secrete the hormone *insulin*.

Glucagon and insulin work together in an antagonistic manner to regulate the concentration of glucose in the blood (see Campbell, Figure 45.11)

- Blood glucose levels must remain near 90 mg/100 mL in humans for proper body function.
- At glucose levels above the set point, insulin is secreted and lowers blood glucose concentration by stimulating body cells to take up glucose from the blood.
  ⇒ It also slows glycogen breakdown in the liver and inhibits the conversion of amino acids and fatty acids to sugar.
- When blood glucose levels drop below the set point, glucagon is secreted and increases blood glucose concentrations by stimulating the liver to increase the hydrolysis of glycogen, convert amino acids and fatty acids to glucose, and slowly release glucose into the blood.

Glucose homeostasis is critical due to its function as the major fuel for cellular respiration and a key source of carbon for the synthesis of other organic compounds.

Serious conditions can result when glucose homeostasis is unbalanced. Diabetes mellitus is caused by a deficiency of insulin or a loss of response to insulin in target tissues. Diabetes occurs in two forms:

- *Type I diabetes mellitus* (insulin-dependent diabetes) is an autoimmune disorder, in which the immune system attacks the cells of the pancreas.

⇒ Usually occurs suddenly during childhood and destroys the ability of the pancreas to produce insulin

⇒ Treated by insulin injections several times each day.

- *Type II diabetes mellitus* (non-insulin-dependent diabetes) occurs most frequently in adults over 40.

   ⇒ May be due to an insulin deficiency but more commonly results from reduced responsiveness in target cells because of changes in insulin receptors

   ⇒ Heredity is a major factor

   ⇒ Is non-insulin-dependent and can be treated with exercise and dietary controls

Both types of diabetes mellitus will result in high blood sugar concentrations if untreated.

- Kidneys excrete glucose, resulting in higher concentrations in the urine.
- More water is excreted due to the high concentration of glucose (results in the symptoms of copious urine production accompanied by thirst).
- Fat must serve as the major fuel source for cellular respiration since glucose does not enter the cells.
- In severe cases, acidic metabolites formed during fat metabolism may lower the blood pH to a life-threatening level.

F. The adrenal medulla and adrenal cortex help the body manage stress

Adrenal glands are located adjacent to kidneys.

In mammals, each gland has an outer *adrenal cortex* and inner *adrenal medulla*, which are composed of different cell types, have different functions, and are of different embryonic origin.

- Different arrangements of these same tissues are found in other vertebrates.
- The adrenal medulla has close developmental and functional ties with the nervous system.

The adrenal medulla synthesizes and secretes catecholamines (epinephrine and norepinephrine) (see Campbell, Figure 45.13).

- *Catecholamines* are secreted in times of stress when nerve cells excited by stressful stimuli release the neurotransmitter acetylcholine in the medulla. Acetylcholine combines with cell receptors, stimulating release of epinephrine.

   ⇒ Norepinephrine is released independently of epinephrine.

- Epinephrine and norepinephrine released into the blood results in rapid and dramatic effects on several targets:

   ⇒ Glucose is mobilized in skeletal muscle and liver cells.

   ⇒ Fatty acid release from fat cells is stimulated (may serve as extra energy sources).

   ⇒ The rate and stroke volume of the heartbeat is increased.

   ⇒ Blood is shunted away from the skin, gut, and kidneys to the heart, brain, and skeletal muscles by stimulation of smooth muscle contraction in some blood vessels and relaxation in others.

   ⇒ Oxygen delivery to the body cells is increased by dilation of bronchioles in the lungs.

The adrenal cortex synthesizes and secretes corticosteroids.

- Stressful stimuli cause the hypothalamus to secrete a releasing hormone that stimulates release of ACTH from the anterior pituitary.
- ACTH stimulates release of corticosteroids from the adrenal cortex.

In humans, the two primary types are glucocorticoids (e.g., cortisol) and mineralocorticoids (e.g., aldosterone).

Glucocorticoids promote glucose synthesis from noncarbohydrate substances such as proteins.

⇒ Skeletal muscle proteins are broken down and the carbon skeletons transported to the liver and kidneys.

⇒ The liver and kidneys convert the carbon to glucose which is released into the blood to increase the fuel supply.

⇒ Also have immunosuppressive effects and are used to treat inflammation.

Mineralocorticoids affect salt and water balance.

- Aldosterone stimulates kidney cells to reabsorb sodium ions and water from the filtrate.
  - This raises blood volume and blood pressure.
- Aldosterone from the RAAS (renin-angiotensin-aldosterone system), ADH, and ANF (atrial natriuretic factor) from the heart form a regulatory complex that influences the kidneys' ability to maintain the blood's ion and water concentrations.
  - ⇒ The RAAS regulates aldosterone secretion in response to changes in plasma ion concentrations.
  - ⇒ Severe stress also stimulates aldosterone secretion by causing the hypothalamus to secrete releasing hormones that increase ACTH secretion from the anterior pituitary, which increases aldosterone secretion by the adrenal cortex.

Current evidence indicates glucocorticoids and mineralocorticoids are important to maintaining body homeostasis during extended periods of stress (see Campbell, Figure 45.15).

G. Gonadal steroids regulate growth, development, reproductive cycles, and sexual behavior

The testes of males and ovaries of females produce steroid hormones that affect growth and development as well as regulate reproductive cycles and behaviors.

The gonads of both males and females produce all three categories of gonadal steroids (androgens, estrogens, and progestins), although the proportions differ. (see Campbell, Figure 45.14).

Androgens generally stimulate the development and maintenance of the male reproductive system.

- They are produced in greater quantities in males than females.
- Primary androgen is *testosterone*.
- Androgens produced during early embryonic development determine whether the fetus will be male or female.
- High androgen concentrations at puberty stimulate development of male secondary sex characteristics.

Estrogens perform the same functions in females as androgens do in males.

- Estradiol is the primary estrogen produced.
- They maintain the female system and stimulate development of female secondary sex characteristics.

Progestins are primarily involved with preparing and maintaining the uterus for reproduction in mammals.

- Include progesterone

The gonadotropins from the anterior pituitary (FSH and LH) control the synthesis of both androgens and estrogens.

- FSH and LH are in turn controlled by gonadotropin-releasing hormone (GnRH) from the hypothalamus.

REFERENCES

Campbell, N., et al. *Biology*. 5th ed. Menlo Park, California: Benjamin/Cummings, 1999.

Marieb, E.N. *Human Anatomy and Physiology*. 4th ed. Menlo Park, California: Benjamin/Cummings, 1997.

Norris, D. *Vertebrate Endocrinology*. 3rd ed. New York, NY: Academic Press, 1997.

CHAPTER 46
ANIMAL REPRODUCTION

OUTLINE

I. Overview of Animal Reproduction
    A. Both asexual and sexual reproduction occur in the animal kingdom
    B. Diverse means of asexual reproduction enable animals to produce identical offspring rapidly
    C. Reproductive cycles and patterns vary extensively among animals

II. Mechanisms of Sexual Reproduction
    A. Internal and external fertilization both depend on mechanisms ensuring that mature sperm encounter fertile eggs of the same species
    B. Species with internal fertilization usually produce fewer zygotes but provide more parental protection than species with external fertilization
    C. Complex reproductive systems have evolved in many animal phyla

III. Mammalian Reproduction
    A. Human reproduction involves intricate anatomy and complex behavior
    B. Spermatogenesis and oogenesis both involve meiosis but differ in three significant ways
    C. A complex interplay of hormones regulates reproduction
    D. Embryonic and fetal development occur during pregnancy in humans and other eutherian (placental) mammals
    E. Modern technology offers solutions for some reproductive problems

OBJECTIVES

After reading this chapter and attending lecture, the student should be able to:

1. Distinguish between asexual and sexual reproduction.
2. List and describe four forms of asexual reproduction.
3. Explain how asexual reproduction may be advantageous for a population of organisms living in a stable, favorable environment.
4. Explain the advantages of sexual reproduction.
5. Explain the importance of reproductive cycles.
6. Distinguish among parthenogenesis, hermaphroditism, and sequential hermaphroditism.
7. Describe three mechanisms which increase the probability of successful fertilization that are found in organisms that use external fertilization.
8. List and describe the various methods of parental care by animals.
9. Using a diagram, identify and give the function of each part of the reproductive systems of an insect and a platyhelminth.

10. Using a diagram, identify and give the function of each component of the reproductive system of the human male.

11. Using a diagram, identify and give the function of each component of the human female reproductive system.

12. Discuss the hormonal control of reproduction in male mammals.

13. Explain the differences between menstrual cycles and estrous cycles.

14. Discuss the hormonal control of reproduction in female mammals.

15. Explain how the menstrual cycle and ovarian cycle are synchronized in female mammals.

16. Describe spermatogenesis.

17. Describe oogenesis.

18. Compare and contrast oogenesis and spermatogenesis.

19. Describe the hormonal changes which occur at puberty in humans.

20. Describe the four phases of the sexual response cycle.

21. Describe the changes which occur in the developing embryo and the mother during each trimester of a human pregnancy.

22. Describe the hormonal control of a pregnancy in a human female.

23. Explain the possible mechanisms which prevent the mother's immune system from rejecting the developing embryo.

24. List the various methods of contraception and explain how they work.

25. Explain how technological advancements are used to study human reproductive problems.

KEY TERMS

asexual reproduction
sexual reproduction
gametes
zygote
ovum
spermatozoon
fission
budding
gemmules
fragmentation
regeneration
parthenogenesis
hermaphroditism
sequential hermaphro-
   ditism
protogynous
protandrous
fertilization
external fertilization
internal fertilization
pheromones
gonads
spermatheca

seminiferous tubules
Leydig cells
scrotum
epididymis
ejaculation
 vas deferens
  ejaculatory duct
  urethra
semen
seminal vesicles
prostate gland
bulbourethral glands
penis
baculum
glans penis
prepuce
ovaries
follicles
ovulation
corpus luteum
oviduct
uterus
endometrium

hymen
vestibule
labia minora
labia majora
clitoris
Bartholin's glands
mammary glands
vasocongestion
myotonia
coitus
orgasm
spermatogenesis
acrosome
oogenesis
menstrual cycle
estrous cycles
menstruation
estrus
menstrual flow phase
proliferative phase
secretory phase
ovarian cycle
follicular phase

pregnancy (gestation)
embryos
conception
trimesters
cleavage
blastocyst
placenta
organogenesis
fetus
human chorionic
gonadotropin (hCG)
parturition
labor
lactation
contraception
rhythm method
natural family planning
barrier methods
condom
tubal ligation
vasectomy
in vitro fertilization

cloaca   cervix   luteal phase
testes   vagina   menopause

LECTURE NOTES

An individual organism exists for only a length of time (lifespan).

For a species to remain viable, its members must reproduce.

Only through reproduction can extinction be avoided.

I. Overview of Animal Reproduction

A. Both asexual and sexual reproduction can occur in the animal kingdom

Two principal modes of reproduction are found in animals: asexual reproduction and sexual reproduction.

Asexual reproduction is the production of offspring whose genes come from one parent without the fusion of egg and sperm.

- Relies completely on mitotic cell division

Sexual reproduction is the production of offspring by the fusion of *gametes* to form a diploid *zygote*.

- The gametes are formed by meiosis.
- The female gamete, or *ovum*, is usually a large, nonmotile cell, while the male gamete, or *spermatozoan*, is usually a small, motile cell.
- Usually involves two parents, each contributing genes to the offspring.

In sexual reproduction, the offspring have a combination of genes inherited from both parents. This mode of reproduction increases genetic variability and is, therefore, advantageous in a fluctuating environment.

B. Diverse means of asexual reproduction enable animals to produce identical offspring rapidly

There are various types of asexual reproduction found in invertebrates: fission, budding, the release of specialized cells, and fragmentation and regeneration.

Fission involves the separation of a parent into two or more individuals of approximately equal size (see Campbell, Figure 46.1).

Budding occurs when a new individual splits off from an existing one.

⇒ In cnidarians, new individuals grow out from the parental body.

⇒ Offspring may detach from the parent or remain joined to form extensive colonies.

Some invertebrates release specialized groups of cells that grow into new adults.

⇒ Freshwater sponges produce *gemmules*, which form from aggregates of several types of cells surrounded by a protective coat.

Fragmentation is the breaking of the body into several pieces, each of which develops into a complete adult.

⇒ Must be accompanied by *regeneration,* the regrowth of lost body parts

⇒ Found in sponges, cnidarians, polychaete annelids, and tunicates

⇒ Regeneration allows many animals to replace lost parts following an injury. (When the arm of a sea star is removed, it will grow a new arm.)

Several advantages are associated with asexual reproduction:

- Allows animals living in isolation to produce offspring without finding mates (e.g., sessile animals or animals in conditions of low population density)
- Allows production of many offspring in a short time

- Most advantageous in stable, favorable environments because it perpetuates successful genotypes precisely.

C. Reproductive cycles and patterns vary extensively among animals

Most animals have reproductive cycles, often related to changing seasons.

- It allows for conservation of resources; animals reproduce when energy is above that needed for maintenance.
- It allows for reproduction when environmental conditions favor offspring survival.
- Cycles are controlled by a combination of hormonal and environmental cues.

Animals employ various patterns of reproduction and may use either sexual or asexual reproduction exclusively or alternate between the two.

- In animals that employ both mechanisms, asexual reproduction often occurs under favorable conditions and sexual reproduction during times of environmental stress.
- Environmental conditions will stimulate parthenogenesis in aphids, rotifers, some freshwater crustaceans, and other animals.
  - *Parthenogenesis* = Development of an egg without fertilization
  - *Daphnia*, a freshwater crustacean, will switch from sexual to asexual reproduction depending on the season and environmental conditions.
    - ⇒ Each female can produce two types of eggs, one type is fertilized, the other develops by parthenogenesis.
    - ⇒ Adults developing by parthenogenesis are often haploid and their cells do not undergo meiosis when forming new eggs.
  - Parthenogenesis plays a role in the social organization of some insects; male honey bees (drones) are produced parthenogenetically, females develop from fertilized eggs.
  - Some fish, amphibians and lizards reproduce via a more complex form of parthenogenesis which involves a doubling of chromosomes after meiosis to create diploid "zygotes" (see Campbell, Figure 46.2).
- *Hermaphroditism* is a solution found in many animals that may have difficulty finding a member of the opposite sex.
  - Each individual has both functional male and female reproductive parts.
  - This solves the problem of finding a mate of the opposite sex for some sessile, burrowing, or parasitic animals.
  - Some self-fertilize, but most mate with another; each donates and receives sperm, thus potentially producing twice as many offspring from one mating.
- *Sequential hermaphroditism* occurs in some species.
  - An individual reverses its sex during its lifetime.
  - Some species are *protogynous* (female first); others are *protandrous* (male first).
  - Reversal is often associated with age and size.
    - ⇒ In protogynous species, such as the bluehead wrasse, the largest (usually the oldest) fish in the harem becomes a male and defends the harem (see Campbell, Figure 46.3).
    - ⇒ Some oysters are protandrous; the largest individuals become females and can produce more egg cells.

II. Mechanisms of Sexual Reproduction

Two major mechanisms of *fertilization* (the union of sperm and egg) have evolved in animals: external fertilization (see Campbell, Figure 46.4) and internal fertilization. Each has specific environmental and behavioral requirements.

A. Internal and external fertilization both depend on mechanisms ensuring that mature sperm encounter fertile eggs of the same species

Internal fertilization occurs when sperm are deposited in or near the female reproductive tract and fertilization occurs within the female's body.

- Usually requires more sophisticated reproductive systems and cooperative mating behaviors.
- Copulatory organs for sperm delivery and receptacles for sperm storage and transport must be present.
- Mating behaviors must include specific reproductive signals for copulation to occur.

In *external fertilization*, eggs are shed by a female and fertilized by a male's sperm in the environment.

- Occurs almost exclusively in moist habitats where development can occur without desiccation or heat stress.
- Some aquatic invertebrates release their eggs and sperm into the surrounding water with no contact occurring between the parents.
- Environmental cues and pheromones trigger release of mature gametes in close proximity, which increases the probability of successful fertilization.
- In vertebrates exhibiting external fertilization (fishes and amphibians), courtship behavior increases the probability of successful fertilization and permits mate selection.

Pheromones, chemical signals between organisms of the same species, may be operative in organisms that use either internal or external fertilization.

- Pheromones are easily dispersed in the environment and are active in minute amounts.
- Sex attractants in insects may be effective at distances of up to one mile.
- The role of pheromones in social behavior is further explored in Campbell, Chapter 51.

B. Species with internal fertilization usually produce fewer zygotes but provide more parental protection than species with external fertilization

Once fertilization occurs, the zygote undergoes embryonic development. The degree to which these developing embryos are protected, varies with the mechanism of fertilization and the organisms involved.

Organisms using external fertilization do not have protective coverings for the embryos.

- Fish and amphibian eggs are covered with a gelatinous coat, which permits the free exchange of gases and water.
- The water or moist habitat into which the gametes are released prevent desiccation and temperature stress.
- Very large numbers of zygotes are usually formed although only a small proportion survive to complete development.

Many types of protection are found in those animals that use internal fertilization.

- Eggs resistant to harsh environments are produced by many reptiles, all birds, and monotremes. The amniotic egg of these animals has a protective shell of protein (reptiles and monotremes) or calcium (birds), which makes them resistant to water loss and physical damage.
- Mammals (other than monotremes) do not produce a shelled egg, but possess reproductive tracts that permit embryonic development within the female parent.
    ⇒ In marsupials, the embryo develops for a short time in the uterus, then moves to the mother's pouch to complete development.
    ⇒ Eutherian (placental) mammals retain the embryo in the uterus until development is completed.
- Internal fertilization usually produces fewer zygotes than external fertilization, but survival through development is much greater due to the protection and care of the developing offspring.

Parental protection is important to survival of the developing embryo (see Campbell, Figure 46.5).

    ⇒ While some organisms using external fertilization show forms of parental care (nesting fish will guard the eggs against predators), parental care is most highly developed in animals using internal fertilization.

C. Complex reproductive systems have evolved in many animal phyla

The only prerequisite for sexual reproduction is that the animal possesses the capability to produce and deliver gametes to the gametes of the opposite sex.

- The simplest systems do not contain distinct *gonads*, organs that produce gametes in most animals.
- The most complex forms involve many sets of accessory structures for the transport and protection of gametes and developing embryos.

The complexity of the reproductive system is *not* necessarily related to the phylogenetic position of the animal.

- For example, the reproductive systems of parasitic flatworms is among the most complex of all animals (see Campbell, Figure 46.6).

Most polychaetes have separate sexes without distinct gonads.

- Gametes develop from undifferentiated cells lining the coelom and fill the coelom as they mature.
- In some species they are released through excretory openings; in others, the parental body splits open to release the eggs to the environment as the parent dies.

Most insects have separate sexes and complex reproductive structures (see Campbell, Figure 46.7).

- In males, sperm develop in a pair of testes, pass through a coiled duct to seminal vesicles, where they are stored.
    ⇒ During mating, the sperm are ejaculated into the female's system.
- In females, eggs develop in a pair of ovaries and pass through ducts to the vagina where they are fertilized.
    ⇒ May have a *spermatheca*, a blind-ended sac in which sperm may be stored for a year or more.

The reproductive systems of vertebrates are all similar, but some important differences are found.

- Most mammals have separate openings for digestive, excretory, and reproductive tracts, while many nonmammalian vertebrates have only a common opening, the *cloaca*.
- Some mammals, birds and snakes have a uterus with only one branch, while most other vertebrates have a uterus that is partly or completely divided into two chambers.
- Nonmammalian vertebrates do not have well-developed *penises* and use other mechanisms to transfer spermatozoa.

III. Mammalian Reproduction

A. Human reproduction involves intricate anatomy and complex behavior

1. Reproductive anatomy of the human male

In most mammalian species, including humans, the reproductive system includes the external genitalia and the internal reproductive organs.

- The external genitalia includes the *scrotum* and *penis*.
- The internal reproductive organs consist of the gonads (*testes*), accessory glands, and associated ducts.

The male genitalia (*scrotum* and *penis*) aid the reproductive process in different ways.

- Testes develop in the abdomen and descend into the scrotum just before birth. This is important since sperm cannot develop at normal body temperature.
- By having the testes hang outside the abdominal cavity in the scrotum, the temperature is 2°C lower and sperm production can occur.
- The penis serves as the male copulatory organ. The ejaculatory duct joins the urethra (from the excretory system) which opens at the tip of the penis.
- The movement of semen through the urethra during copulation results in the sperm being deposited directly in the female system.

Internal male reproductive organs are the gonads and associated ducts (see Campbell, Figure 46.8):

- The *testes* are comprised of highly coiled tubes surrounded by layers of connective tissue.
  ⇒ The tubules, or *seminiferous tubules*, are where sperm form.
  ⇒ *Interstitial cells*, or *Leydig cells*, are scattered between the tubules and produce testosterone and other androgens.
- Sperm pass from the seminiferous tubules into the tubules of the *epididymis*.
  ⇒ These coiled tubules are where sperm are stored and mature (gain motility and fertilizing power).
- At *ejaculation*, sperm are forced through the *vas deferens*, which is a muscular duct running from epididymis to the *ejaculatory duct*.
  ⇒ The ejaculatory duct forms by the joining of the two vas deferens ducts (one from each testis) with the duct from the seminal vesicles.
  ⇒ The ejaculatory duct opens into the *urethra*, the tube that runs through the penis and drains both the excretory and reproductive systems.

There are three sets of accessory glands associated with the male system. These glands add their secretions to the *semen*.

  1. A pair of *seminal vesicles* is located below and behind the bladder and empty into the ejaculatory duct.

⇒ They secrete a fluid containing mucus, amino acids (causes semen to coagulate after deposited in female), fructose (provides energy for sperm) and prostaglandins (stimulate female uterine contractions to help move semen to the uterus).

⇒ Seminal vesicle secretions make up about 60% of the total semen volume.

2. The *prostate gland* is a large gland that surrounds the upper portion of, and empties directly into, the urethra.

⇒ It secretes a thin, milky alkaline fluid that contains several enzymes.

⇒ Prostatic fluid balances the acidity of residual urine in the male system and the acidity of the vagina and helps activate sperm.

3. The *bulbourethral glands* are a pair of small glands below the prostate that empty into the urethra at the base of the penis.

⇒ Secrete a clear mucus before sperm ejaculation

⇒ The fluid neutralizes any acidic urine remaining in the urethra.

Prostaglandins in the semen thin the mucus at the opening of the uterus and stimulate contractions of the uterine muscle to help the semen move up the uterus.

The human penis is composed of three cylinders of spongy erectile tissue that fill with blood during sexual arousal.

- Rodents, raccoons, walruses and some other mammals also possess a *baculum* (a bone that helps stiffen the penis).
- The head of the penis, the *glans penis*, is covered by *prepuce*, or foreskin.

2. Reproductive anatomy of the human female

The *human female reproductive system* is more complicated than that of the male; it possesses structures not only for the production of female gametes, but also to house the embryo and fetus (see Campbell, Figure 46.9).

- The internal reproductive organs are the gonads (*ovaries*) and associated ducts and chambers, which are involved with gamete movement and embryo development.
- The external genitalia include the *clitoris* and the two sets of *labia* that surround the clitoris and vaginal opening.

The *ovaries* are located in the abdominal cavity and enclosed in a tough protective capsule.

- A mesentery flanks and attaches each ovary to the uterus.
- Each ovary contains many *follicles* (one egg cell surrounded by follicle cells, which nourish and protect the developing egg).

  ⇒ All are formed at birth.
- Follicle cells also produce estrogens.
- Starting at puberty, and continuing to menopause, one follicle matures and releases its egg cell during each menstrual cycle.

During *ovulation*, the egg is expelled from the follicle (see Campbell, Figure 46.10). The remaining follicular tissue forms the *corpus luteum*, which secretes progesterone (maintains the uterine lining) and additional estrogen.

- If the egg is not fertilized, the corpus luteum degenerates.
- The egg cell is expelled into the abdominal cavity near the opening of the *oviduct*.
- Cilia lining the oviduct draw in the egg cell and convey it to the uterus.
- The *uterus* (or womb) is a thick muscular organ that can expand to accommodate a 4-kg. fetus.

- The inner uterine lining, the *endometrium*, is richly supplied with blood vessels.

The remaining female reproductive structures are:

Cervix = The neck of the uterus which opens into the vagina

Vagina = Thin-walled chamber that is the repository for semen during copulation; also forms the birth canal

The *hymen*, a vascularized membrane, usually covers the vaginal opening from birth until ruptured by vigorous physical activity or sexual intercourse.

Vestibule = Chamberlike area formed by the two pairs of skin folds covering the vaginal orifice and urethral opening

Labia minora = The slender skin folds bordering the vestibule

Labia majora = A pair of thick, fatty ridges enclosing and protecting the labia minora and vestibule

Clitoris = Bulb of erectile tissue at the front edge of the vestibule which is covered by a prepuce (small hood of skin)

Bartholin's glands = Small glands located near the vaginal opening that secrete mucus into the vestibule during sexual arousal, which facilitates intercourse by lubricating the vagina

Mammary glands are important to mammalian reproduction, although not actually a part of the reproductive system.

- They consist of small sacs of epithelial tissue that secrete milk.
- The milk drains into a series of ducts that open at the nipple.
- In a nonlactating female mammal, the mammary glands are composed primarily of adipose tissue.

3. **Human sexual response**

Human sexuality includes a diversity of stimuli and responses.

- Although variable, human sexual behavior is based on a common physiological pattern, the *sexual response cycle*.
- The sexual responses of males and females show similarities and differences.

Physiological reactions that predominate in both sexes can be divided into two types: vasocongestion and myotonia.

- *Vasocongestion* is the filling of a tissue (e.g., penis, clitoris) with blood due to an increased in blood flow through the arteries of that tissue.
- *Myotonia* is increased muscle tension, and both skeletal and smooth muscles may show sustained or rhythmic contractions.

There are four phases in the sexual response cycle:

1. In the *excitement* phase, the vagina and penis are prepared for *coitus* (sexual intercourse).
   ⇒ Vasocongestion of the penis and clitoris occurs along with enlargement of testes, labia, and breasts. Vaginal lubrication and myotonia also occur.
2. The *plateau* phase continues the responses of the excitement phase.
   ⇒ The vagina forms a depression to receive the sperm: the outer third becomes vasocongested, the inner two-thirds slightly expand, and the uterus elevates.
   ⇒ In both sexes, the heart rate rises and breathing rates increase in response to stimulation of the autonomic nervous system.
3. *Orgasm* is the third and shortest phase, it is characterized by rhythmic, involuntary contractions in the reproductive systems of both sexes.

⇒ Two stages occur in males:
  ◊ *Emission* is the forcing of semen into the urethra due to contraction of the glands and ducts of the reproductive system.
  ◊ *Expulsion* (ejaculation) expels the semen due to contraction of the urethra.
⇒ In females, the uterus and outer vagina contract but the inner two-thirds of the vagina do not.

4. The *resolution* phase reverses the responses of earlier phases and completes the cycle.
   ⇒ Vasocongested organs return to normal size and color.
   ⇒ Muscles relax.

B. Spermatogenesis and oogenesis both involve meiosis but differ in three significant ways

Spermatogenesis is the production of mature sperm cells in adult males.

* A continuous process in adult males, which can result in 100 to 650 million sperm cells per ejaculation on a daily basis.
* Occurs in seminiferous tubules of the testes (see Campbell, Figure 46.12).

The thick head of a sperm cell contains the haploid nucleus tipped with the *acrosome*, which contains enzymes to aid in egg penetration (see Campbell, Figure 46.11).

Behind the head, the sperm cell contains many mitochondria that provide ATP for movement of the flagellum.

Oogenesis is the development of ova (mature, unfertilized egg cells) (see Campbell, Figure 46.13).

* Begins in the embryo when primordial germ cells undergo mitotic divisions to produce diploid oogonia.
* Each oogonium will develop into a primary oocyte by the time of birth of the female, resulting in all potential ova being present in the ovaries at birth.
* Between birth and puberty, primary oocytes enlarge and their surrounding follicles grow.
  ⇒ They replicate their DNA and enter prophase I and remain there until activated by hormones.
* After puberty, during each ovarian cycle, FSH stimulates a follicle to enlarge and the primary oocyte within completes meiosis I to produce a haploid secondary oocyte and the first polar body. Meiosis then stops again.
* LH triggers ovulation and the secondary oocyte is released from the follicle.
* If a sperm cell penetrates the secondary oocyte's membrane, meiosis II will occur and the second polar body will separate from the ovum; this completes oogenesis.

The three important differences between spermatogenesis and oogenesis are:

1. In spermatogenesis, all four products of meiosis I and II become mature spermatozoa. In oogenesis, the unequal cytokinesis, which occurs during meiosis I and II, results in most of the cytoplasm being in one daughter cell, which will form the single ovum—the other cells (polar bodies) will degenerate.
2. Spermatogenesis is a continuous process throughout the reproductive life of the male. All potential ova that can be produced via oogenesis will be present as primary oocytes in the ovaries at the time of the female's birth.

3. Spermatogenesis occurs as an uninterrupted sequence. In oogenesis, long "resting" periods occur between the formation of the initial steps and final production of the ovum.

C. A complex interplay of hormones regulates reproduction

1. The male pattern

In males, androgens are directly responsible for formation of primary sex characteristics (reproductive organs) and secondary sex characteristics (deepening of voice, hair growth pattern, muscle growth).

- Androgens are steroid hormones produced primarily by the Leydig cells of the testes.
- Testosterone is the most important androgen produced.
- Androgens are also potent determinants of sexual and aggressive behaviors.
- GnRH from the hypothalamus stimulates the anterior pituitary to release LH (stimulates androgen production) and FSH (acts on seminiferous tubules to increase sperm production).

2. The female pattern

Hormonal control in females is more complicated and reflects the cyclic nature of female reproduction. Female mammals display two different types of cycles: estrous cycles and menstrual cycles.

Estrous cycles occur in non-primate mammals.

- Ovulation occurs after the endometrium thickens and vascularizes.
- If pregnancy does not occur, the endometrium is reabsorbed by the uterus.
- Pronounced behavioral changes occur, and seasonal and climatic changes affect the estrous cycle more than the menstrual cycle.
- *Estrus* is the period of sexual activity surrounding ovulation and is the only time most non-primate mammals will copulate. The length and frequency varies widely among species.

Menstrual cycles occur in humans and many other primates.

- Ovulation occurs after the endometrium thickens and vascularizes (as in the estrous cycle).
- If pregnancy does not occur, the endometrium is shed from the uterus through the cervix and vagina during menstruation.

The term *menstrual cycle* refers to the changes that occur in the uterus during the reproductive cycle.

- In human females, the cycle varies from one woman to another.
- Usually ranges from 20 – 40 days with an average of 28 days.
- Some women have very regular cycles; others vary from one cycle to the next.

There are three phases of the menstrual cycle. (These refer specifically to changes in the uterus; see Campbell, Figure 46.15)

1. The *menstrual flow phase* is the time during which most of the endometrium is being lost from the uterus (menstruation).
   ⇒ Persists only a few days
   ⇒ The first day of this phase is usually designated day 1 of the cycle.
2. The *proliferative phase* lasts for one to two weeks and involves the regeneration and thickening of the endometrium.
   ⇒ The endometrium is thin at the beginning since most of its structure was lost during the preceding phase.

3. The *secretory phase* lasts about two weeks and is a time when the endometrium continues to develop.

⇒ The endometrium continues to thicken, becomes more vascularized, and develops glands which secrete a glycogen-rich fluid.

⇒ If an embryo does not implant in the uterine lining by the end of this phase, a new menstrual flow phase begins.

An *ovarian cycle* parallels the menstrual (uterine) cycle.

- The *follicular phase* begins this cycle and is a time during which several follicles in the ovaries begin to grow.

  ⇒ The egg cells within the follicles enlarge and the follicle cell coat becomes multilayered.

  ⇒ Only one of the growing follicles will continue to mature while the others degenerate.

  ⇒ A fluid-filled cavity develops in the maturing follicle and grows large enough to form a bulge on the surface of the ovary.

  ⇒ The ovulatory phase ends with ovulation, an event marked by eruption of the follicle and expulsion of the egg from the ovary.

- The *luteal phase* begins after ovulation.

  ⇒ Follicular tissue remaining in the ovary after ovulation forms a *corpus luteum*.

  ⇒ The corpus luteum is endocrine tissue that secretes female hormones.

Five hormones work together (by positive and negative feedback mechanisms) to coordinate the menstrual and ovarian cycles. This coordination synchronizes follicle growth and ovulation with preparation of the uterine lining for embryo implantation (see Campbell, Figure 46.15).

- During the follicular phase of the ovarian cycle, *GnRH* secreted by the hypothalamus stimulates the anterior pituitary to secrete small quantities of *FSH* and *LH*. Note that GnRH is secreted in a phasic manner; this secretory pattern, linked with higher brain centers (biological clock), drives the cyclic nature of the female reproductive pattern.

- FSH stimulates the immature follicles in the ovary to grow and these follicle cells secrete small amounts of *estrogen*.

- As the follicle continues to grow, the amount of estrogen secreted increases. The mature follicle secretes a large amount of estrogen.

- The high concentration of estrogen stimulates the hypothalamus to increase secretion of GnRH, which results in a sudden increase in FSH and LH secretion.

  ⇒ LH increases more than FSH because high estrogen levels increases the sensitivity of LH-releasing mechanisms (in the pituitary) to GnRH.

  ⇒ The follicles have LH receptors and can respond directly to the hormone.

- The sudden surge in LH concentration stimulates final maturation of the follicle and ovulation (after about 24 hours). The high concentration of LH stimulates the ruptured follicular tissue to transform into the corpus luteum.

- The presence of LH during the luteal phase of the ovarian cycle stimulates the corpus luteum to continue to secrete estrogen, but to also secrete increasing amounts of *progesterone*.

- Increasing concentrations of estrogen and progesterone inhibits GnRH secretion by the hypothalamus resulting in a decrease in FSH and LH.

- As the LH concentration declines, the corpus luteum begins to atrophy; this results in a sudden drop in estrogen and progesterone concentrations.
- Decreasing levels of estrogen and progesterone removes the inhibition exerted on the hypothalamus, which begins to secrete small amounts of GnRH that stimulates the anterior pituitary to secrete low levels of FSH and LH.
- A new follicular phase begins at this point.

Coordination of the menstrual cycle with the ovarian cycle depends primarily on the levels of estrogen and progesterone.

- Increasing amounts of estrogen secreted by growing follicles during the follicular phase stimulate the endometrium lining the uterus to thicken in preparation for the embryo.
    ⇒ Coordination of the follicular phase of ovarian cycle with proliferative phase of the menstrual cycle.
- After ovulation, the estrogen and progesterone secreted by the corpus luteum stimulate continued development and maintenance of the endometrium.
    ⇒ Arteries supplying blood to the endometrium enlarge and the endometrial glands that supply the nutritional fluid to the early embryo grow and mature.
    ⇒ Coordination of the luteal phase of the ovarian cycle with the secretory phase of the menstrual cycle.
- Decreasing concentrations of estrogen and progesterone due to the disintegration of the corpus luteum reduce blood flow to the endometrium.
    ⇒ The endometrium breaks down and passes out of the uterus as the menstrual flow.
- A new menstrual cycle begins with a new ovarian cycle.
- Estrogens are also responsible for development of the female secondary sex characteristics.

a. Menopause

Human females stop menstruation and ovulation at *menopause*, usually between the ages of 46 to 54.

- Ovaries lose their responsiveness to LH and FSH.
- Some scientists suggest that menopause is adaptive so that females can provide better care to their children and grandchildren.

D. Embryonic and fetal development occur during pregnancy in humans and other eutherian (placental) mammals

1. From conception to birth

In placental mammals, *pregnancy (gestation)* is the condition of carrying one or more developing *embryos* in the uterus.

- It is preceded by *conception* (fertilization of an egg by a sperm cell) and ends with birth of the offspring.
- Human pregnancy averages 266 days from conception.
- Duration in other species correlates with body size and extent of development of young at birth.

Human gestation is divided into three *trimesters*, each of about 3 months duration.

The *first trimester* is when the most radical changes occur for both the mother and baby.

- Fertilization occurs in the oviduct and *cleavage* (cell division) begins in about 24 hours (see Campbell, Figure 46.16).
- As cleavage continues, the zygote develops into a ball of cells passing down the oviduct to the uterus.
- The embryo reaches the uterus in 3 to 4 days and develops into a hollow ball of cells called a *blastocyst*.
  - This stage develops about one week after fertilization.
- The blastocyst will implant into the endometrium in the next 5 days.
- During implantation, the blastocyst bores into the endometrium, which grows over the blastocyst.
  - For the first 2 to 4 weeks of development, nutrients are obtained directly from the endometrium.
- Embryonic tissues begin to mingle with the endometrium to form the *placenta*, which functions in respiratory gas exchange, nutrient transfer, and waste removal for the embryo.
  - Blood from the embryo passes through the umbilical arteries to the placenta and returns through the umbilical vein (see Campbell, Figure 46.17).
- The first trimester is also the main period of *organogenesis* (development of organs) (see Campbell, Figure 46.18).
  - After eight weeks, the embryo develops into a *fetus* and possesses all organs of the adult in rudimentary form.
  - The fetus is about 5 cm in length by the end of the first trimester.
- During this trimester, the embryo secretes hormones that signal its presence and controls the mother's reproductive system.
- *Human chorionic gonadotropin* (hCG) is an embryonic hormone that maintains progesterone and estrogen secretion by the corpus luteum to prevent menstruation, which would end the pregnancy.
- High progesterone levels also stimulate formation of a protective mucous plug in the cervix, growth of the maternal part of the placenta, uterus enlargement, and cessation of ovulation and menstrual cycling.

In the *second trimester*, rapid growth occurs and the fetus is very active.

- The fetus grows rapidly to about 30 cm in length.
- The mother may feel movement during the early part of this trimester.
- Hormone levels stabilize as hCG declines, the corpus luteum degenerates, and the placenta secretes its own progesterone to maintain the pregnancy.
- The uterus grows sufficiently for the pregnancy to become obvious.

In the *third trimester*, growth is rapid and fetal activity decreases.

- The fetus grows to about 50 cm in length and 3 to 3.5 kg in weight.
- The maternal abdominal organs become compressed and displaced.
- Labor is induced and regulated by interplay among estrogen, oxytocin, and prostaglandins (see Campbell, Figure 46.19).
  - High estrogen levels during the last weeks of pregnancy trigger the formation of oxytocin receptors on the uterus.
  - Oxytocin (from the fetus and maternal posterior pituitary) stimulate the smooth muscles of the uterus to contract.

- Oxytocin also stimulates prostaglandin secretion by the placenta (the prostaglandins enhance the muscle contractions).
- The physical and emotional stresses caused by the uterine muscle contractions stimulate secretion of additional oxytocin and prostaglandins.

Parturition (birth) occurs through a series of strong, rhythmic contractions of the uterus usually called *labor* (see Campbell, Figure 46.20).

- The first stage of labor involves the opening and thinning of the cervix until it is completely dilated.
- The second stage is the expulsion of the baby from the uterus.
  ⇒ During this time there are continuous strong uterine contractions.
- The last stage is expulsion of the placenta from the uterus.

Mammals are unique among animals due to the *lactation* component of their postnatal care.

- Decreasing levels of progesterone after birth remove the inhibition from the anterior pituitary, which allows *prolactin* secretion.
- Prolactin stimulates milk production after 2 to 3 days' delay.
- Oxytocin controls the release of milk from the mammary glands.

2. Reproductive immunology

One function of the immune system is to protect the body from bodies identified as foreign (nonself).

- Although the embryo possesses many chemical markers that are "foreign" to the mother, an immune response is not mounted against the embryo.
  ⇒ Foreign markers result from the half of the genome inherited from the father.
- Reasons why the mother does not reject the developing embryo are only partly known but include the presence of the trophoblast and a number of hypotheses.

The *trophoblast* is a protective layer that forms a physical barrier preventing the embryo from actually contacting maternal tissues.

- Develops from embryonic cells of the blastocyst and penetrates the endometrium.
- May not be detected as foreign due to production of a chemical signal that stimulates production of specialized type of white blood cells in the uterus.
  ⇒ These special white blood cells act to suppress other white blood cells by secreting a chemical that blocks the interleukin-2 action required for normal immune responses.
- Some researchers believe this interference with localized immune responses may occur only after the trophoblast has been identified as foreign by uterine white blood cells which have begun an immune response.
  ⇒ No suppressor cells are produced if the paternal cellular markers are very similar to the maternal markers.

Many women who have multiple miscarriages may be rejecting the embryo as foreign tissue.

- Some researchers have speculated that in some cases, spontaneous abortions are due to a failure to suppress the immune response in the uterus.

- The hypothesis is that the mother's immune response is too weak to trigger suppression and that the continued immune response is sufficient to reject the embryo.

3. Contraception

There are three major ways to achieve contraception: prevent fertilization by keeping sperm and eggs apart; prevent implantation of the embryo; or prevent the release of mature eggs and sperm from the gonads (see Campbell, Figure 46.21).

Preventing the egg and sperm from meeting in the female reproductive tract prevents fertilization.

- The *rhythm method* (or *natural family planning*) involves refraining from intercourse when conception is most likely (failure rate = 10-20%).
    - Temporary abstinence during the days before and after ovulation is necessary since the egg can survive for 24 to 48 hours in the oviduct and sperm up to 72 hours.
    - The time of ovulation may be hard to predict even in women with regular menstrual cycles.
- Condoms, diaphragms, cervical caps, and contraceptive sponges are physical barriers that block the sperm from meeting the egg.
    - Failure rate < 10%; most effective when used in conjunction with spermicidal foam or jelly.
- Preventing implantation of a blastocyst in the uterus can be accomplished by using an *intrauterine device* (*IUD*).
    - IUDs are small plastic devices fitted into the uterine cavity.
    - Probably works by irritating the endometrium but precise mechanism is unknown.
    - May have serious side effects such as persistent bleeding, uterine infections and other complications.
- Withdrawal is unreliable since sperm may be present in secretions preceding ejaculation.
- Chemical contraception prevents the release of mature gametes from the gonads.
    - Failure rates < 1%.
    - The commonly used birth control pills are combination of synthetic estrogen and progestin (similar to progesterone).
    - These hormones stop release of GnRH, LH and FSH through a negative feedback mechanism.
        - By blocking LH release, progestin prevents ovulation.
        - Estrogen inhibits FSH so no follicles develop.
    - The *minipill* contains only progestin and may be in the form of an oral pill or an implant.
        - Prevents sperm from entering the uterus by altering the woman's cervical mucus.
- Sterilization by *tubal ligation* (cutting the oviduct in women) or *vasectomy* (cutting the vas deferens of men) is nearly 100% effective, safe, but difficult to reverse.

If contraceptive methods are unsuccessful, a pregnancy will occur. The pregnancy may be terminated in several ways.

- A *miscarriage* (spontaneous abortion) occurs in as many as one-third of all pregnancies.

⇒ May occur so early that the woman isn't aware of the pregnancy.

- Physician assisted abortions are chosen by about 1.5 million women annually in the United States.
- A drug-induced abortion using RU-486 is a nonsurgical method, which is effective in the first few weeks of the pregnancy.
  ⇒ RU-486 is a progesterone analog which blocks progesterone receptors in the uterus.

E. Modern technology offers solutions for reproductive problems

Technological advances have made it possible to detect problems in the developing fetus (see Campbell, Figure 46.22).

Ultrasound imaging is a noninvasive procedure in which a scanner emits high-frequency sound waves that are reflected back by tissues of varying densities to form an image of the fetus.

Sampling of maternal blood also provides a means of diagnosing fetal status. A few fetal blood cells (which are nucleated as compared to the non-nucleated red blood cells of adults) cross the placenta a can serve as source material for genetic screening.

Amniocentesis is an invasive procedure in which a needle is inserted into the amnion and a sample of fluid is withdrawn.

- Fluid contains fetal cells that are cultured and analyzed for chromosomal defects and genetic disorders
- Often takes weeks to obtain the results
- 1% risk of spontaneous abortion following procedure

Chorionic villus sampling is another invasive procedure where a small portion of the chorion (a fetal part of the placenta) is removed, cultured and analyzed for genetic and metabolic disorders.

- 5% to 20% risk of spontaneous abortion following procedure
- Can be performed earlier in pregnancy and results are obtained in days

Scientific discovery has solved some fertility problems.

- For cases of male infertility, sperm from anonymous donors are widely available from sperm banks.
- *In vitro fertilization* can permit women with blocked oviducts to become pregnant.
  - Ova are surgically removed from hormonally stimulated follicles, fertilized in petri dishes and the embryo placed in the uterus for implantation.
  - Difficult and costly procedure with a success rate of about 1 in every 6 attempts.
  - Embryos may be frozen for use after unsuccessful attempts.

REFERENCES

Campbell, N., et al. *Biology*. 5th ed. Menlo Park, California: Benjamin/Cummings, 1999.

Marieb, E.N. *Human Anatomy and Physiology*. 4th ed. Menlo Park, California: Benjamin/Cummings, 1997.

CHAPTER 47
ANIMAL DEVELOPMENT

OUTLINE

I.   The Stages of Early Embryonic Development
    A. From egg to organism, an animal's form develops gradually: *the concept of epigenesis*
    B. Fertilization activates the egg and brings together the nuclei of sperm and egg
    C. Cleavage partitions the zygote into many smaller cells
    D. Gastrulation rearranges the blastula to form a three-layered embryo with a primitive gut
    E. In organogenesis, the organs of the animal body from the three embryonic germ layers
    F. Amniote embryos develop in a fluid-filled sac within a shell or uterus

II.   The Cellular and Molecular Basis of Morphogenesis and Differentiation
    A. Morphogenesis in animals involves specific changes in cell shape, position, and adhesion
    B. The developmental fate of cells depends on cytoplasmic determinants and cell-cell induction: *a review*
    C. Fate mapping can reveal cell geneologies in chordate embryos
    D. The eggs of vertebrates contain cytoplasmic determinants that help establish the body axes and differences among cells of the early embryo
    E. Inductive signals drive differentiation and pattern formation in vertebrates

OBJECTIVES

After reading this chapter and attending lecture, the student should be able to:

1. List the two functions of fertilization.
2. Describe the acrosomal reaction, and explain how it ensures the gametes are conspecific.
3. Describe the cortical reaction.
4. Explain how the acrosomal and cortical reactions function sequentially to prevent polyspermy.
5. Describe the changes that occur in an activated egg and explain the importance of cytoplasmic materials to egg activation.
6. Explain the importance of embryo polarity during cleavage.
7. Describe the process of gastrulation and explain its importance.
8. List adult structures derived from each of the primary tissue layers.
9. Using diagrams, identify the various stages of embryonic development of an amphibian.
10. Distinguish between meroblastic cleavage and holoblastic cleavage.
11. List and explain the functions of the extraembryonic membranes in bird and reptile eggs.

12. Compare and contrast development in birds and mammals.
13. Explain the relationships among polarity, cytoplasmic determinants, and development.
14. Describe how cell extension, contraction and adhesion are involved in shaping the embryo.
15. Explain how interactions among the three primary tissue layers influence organogenesis.
16. Explain the relationship between cytoplasmic cues and cell determination.
17. Describe the importance of cell location and orientation along the three body axes with respect to polarity in the embryo, morphogenetic movements, and pattern formation.
18. Explain how positional cues influence pattern formation.

KEY TERMS

preformation	morula	yolk plug	inner cell mass
epigenesis	blastocoel	organogenesis	trophoblast
acrosomal reaction	blastula	notochord	convergent extension
fast block to polyspermy	meroblastic cleavage	neural tube	cell adhesion molecules
cortical reaction	holoblastic cleavage	somites	(CAMs)
cortical granules	gastrulation	amniotes	cadherins
fertilization membrane	gastrula	blastodisc	fate map
slow block to polyspermy	ectoderm	primitive streak	pattern formation
zona pellucida	endoderm	extraembryonic	positional information
cleavage	mesoderm	membranes	apical epidermal ridge
blastomeres	invagination	yolk sac	(AER)
yolk	archenteron	amnion	zone of polarizing
vegetal pole	blastopore	chorion	activity (ZPA)
animal pole	dorsal lip	allantois	
gray crescent	involution	blastocyst	

LECTURE NOTES

Animals develop throughout their lifetime. Development begins with the changes that form a complete animal from the zygote and continue as progressive changes in form and function.

I.   **The Stages of Early Embryonic Development**

   A. **From egg to organism, an animal's form develops gradually:** *the concept of epigenesis*

   Two early views of how animals developed from an egg competed for supporters until modern techniques were developed. One view was termed preformation and the other epigenesis.

   Preformation suggested that the embryo contained all of its descendants as a series of successively smaller embryos within embryos (see Campbell, Figure 47.1).

   • Based on this idea, by dissecting an egg an individual would be able to find smaller and smaller individuals as the dissection continued.

   • This was a popular idea as recently as the eighteenth century.

   Epigenesis proposed that the form of an embryo gradually emerged from a formless egg.

   • Originally proposed by Aristotle

   • Gained support in the nineteenth century as improved microscopy permitted biologists to view embryos as they gradually developed.

Modern biology has found that an organism's development is mostly determined by the zygote's genome and the organization of the egg cell's cytoplasm.

- The heterogenous distribution of messenger RNA, proteins, and other components in the unfertilized egg greatly impacts the development of the embryo in most animals.
- After fertilization, cell division partitions the heterogenous cytoplasm in such a way that nuclei of different embryonic cells are exposed to different cytoplasmic environments.
- These different cytoplasmic environments result in the expression of different genes in different cells.
- This leads to an emergence of inherited traits that is ordered in space and time by mechanisms controlling gene expression.

B. Fertilization activates the egg and brings together the nuclei of sperm and egg

Fertilization is important because:

- It forms a diploid zygote from the haploid sets of chromosomes from two individuals.
- It triggers onset of embryonic development.

1. The acrosomal reaction

The *acrosomal reaction* is the discharge of hydrolytic enzymes from a vesicle in the acrosome of a sperm cell (based on studies with sea urchins):

- Upon contacting the egg's jelly coat, the acrosomal vesicle in the head of the sperm releases hydrolytic enzymes via exocytosis.
- These enzymes enable an acrosomal process to elongate and penetrate the jelly coat (see Campbell, Figure 47.2).
- A protein coating the tip of the process attaches to specific receptors on the egg's vitelline layer (just external to the plasma membrane).
- This provides species specificity for fertilization.
- Enzymes of the acrosomal process probably digest vitelline layer materials allowing the tip of the process to contact the egg's plasma membrane.
- The sperm and egg's plasma membranes fuse, allowing the sperm nucleus to enter the egg and causing a depolarization of the plasma membrane that prevents other sperm cells from also uniting with the egg (*fast block to polyspermy*).

2. The cortical reaction

The fusion of the egg and sperm membranes stimulates a series of changes in the egg's cortex known as a *cortical reaction* (see Campbell, Figure 47.2, step 6).

- The fusion of sperm and egg stimulates a signal transduction pathway that causes the release of calcium (Ca^{2+}) from the egg cell's endoplasmic reticulum (ER) (see Campbell, Figure 47.3).
- The signaling pathway also leads to the production of IP_3, which in turn, opens ligand-gated calcium channels on the ER.
- The high Ca^{2+} concentration results in a change in the egg's *cortical granules* (special vesicles).
- The increase in Ca^{2+} causes the cortical granules to fuse with the plasma membrane and release their contents into the perivitelline space outside the plasma membrane.
  ⇒ The contents of the cortical granules include enzymes that separate the vitelline layer from the plasma membrane.

> ⇒ Mucopolysaccharides produce an osmotic gradient which stimulates osmosis into the perivitelline space causing the area to swell.
>
> ⇒ The swelling elevates the vitelline layer and other granule enzymes harden it to form the *fertilization* membrane.

- The fertilization membrane prevents entry of additional sperm.
- The fast block to polyspermy no longer functions and is replaced by a *slow block to polyspermy* consisting of the fertilization membrane and other changes to the egg's surface.

3. Activation of the egg

The sharp rise in cytoplasmic Ca^{2+} concentration also incites metabolic changes that *activates* the egg cell.

- Cellular respiration and protein synthesis rates increase.
- Cytoplasmic pH changes from slightly acidic to slightly alkaline due to H^+ extrusion.
- Activation can be artificially induced by injection of Ca^{2+}.
- The sperm nucleus within the egg swells and merges with the egg nucleus to form the zygote (actual fertilization).
- DNA replication begins and the first division occurs in about 90 minutes.

The events of fertilization in sea urchins are illustrated in Campbell, Figure 47.4.

4. Fertilization in mammals

The events of internal fertilization in mammals are similar to those of external fertilization discussed for sea urchins but include some important differences.

Fertilization in terrestrial animals is generally internal (see Campbell, Figure 47.5).

Capacitation (enhanced sperm function) results from secretion in the female's reproductive tract.

- Alters certain molecules on the surface of sperm cells and increases sperm motility.

The capacitated sperm cell must reach the *zona pellucida* for the process to continue.

- The secondary oocyte (egg) released at ovulation is surrounded by follicle cells released at the same time.

The sperm cell must migrate through this layer of cells.

- The zona pellucida (extracellular matrix of the egg) is a three-dimensional network of cross-linked filaments formed by three different glycoproteins.
- The glycoprotein ZP3 acts as a sperm receptor by binding to a complementary molecule on the surface of the sperm head.
- The binding of ZP3 and its complementary molecule stimulates an acrosomal reaction similar to that described earlier.
  - Protein-digesting enzymes and other hydrolases from the acrosome allow the sperm cell to penetrate the zona pellucida and reach the plasma membrane of the egg.
  - The acrosomal reaction also exposes a sperm membrane protein that binds and fuses with the egg membrane.

Binding of the sperm to the egg depolarizes the egg's plasma membrane (a fast block to polyspermy).

- A cortical reaction occurs, however, cortical granule contents stimulate a hardening of the zona pellucida (slow block to polyspermy) instead of raising a fertilization membrane.

- Microvilli from the egg pull the whole sperm cell into the egg cell.
  - ⇒ The sperm cell's flagellar basal body divides and forms the zygote's centrosomes.
  - ⇒ Nuclear envelopes disperse and the chromosomes from the gametes share a common spindle apparatus for the first mitotic division of the zygote.
  - ⇒ The chromosomes from the two parents form a common nucleus (offspring's genome) as diploid nuclei form in the daughter cells after the first division.

C. Cleavage partitions the zygote into many smaller cells

The basic body plan of an animal is established in three successive stages following fertilization: cleavage, gastrulation, and organogenesis. The following information is based upon studies of the sea urchin, frogs, and *Drosophila*.

Cleavage is a succession of rapid mitotic cell divisions following fertilization that produce a multicellular embryo, the blastula (see Campbell, Figure 47.6).

- During cleavage, the cells undergo the S and M phases of the cell cycle but the G_1 and G_2 phases are virtually skipped.
- Very little gene transcription occurs during cleavage and the embryo does not grow.
- The cytoplasm of the zygote is simply divided into many smaller cells called *blastomeres*, each of which has a nucleus.
- The heterogenous nature of the zygote's cytoplasm results in blastomeres with differing cytoplasmic components.

A definite polarity is shown by the eggs of most animals and the planes of division during cleavage follow a specific pattern relative to the poles of the zygote.

- The polarity results from concentration gradients in the egg of such cellular components as mRNA, proteins, and yolk (stored nutrients).
- The yolk gradient is a key factor in determining polarity and influencing the cleavage pattern in frogs and other animals.
- The *vegetal pole* of the egg has the highest concentration of yolk.
- The *animal pole*, opposite the vegetal pole, has the lowest concentration of yolk and is the site where polar bodies are budded from the cell.
- The animal pole also marks the area where the most anterior part of the embryo will form in most animals.

The zygote is composed of two hemispheres named for the respective poles: vegetal and animal.

- The hemispheres in the egg of many frogs have different coloration due to the heterogeneous distribution of cytoplasmic substances (see Campbell, Figure 47.7).
  - ⇒ The animal hemisphere has a gray hue due to the presence of melanin granules in the outer cytoplasm.
  - ⇒ The vegetal hemisphere has a light yellow hue due to the yellow yolk.
- The cytoplasm in amphibian eggs is rearranged at fertilization.
  - ⇒ The plasma membrane and the outer cytoplasm rotate toward the point of sperm entry.
  - ⇒ This rotation may be due to a reorganization of the cytoskeleton around the centriole introduced by the sperm.
  - ⇒ A narrow *gray crescent* is seen due to the exposure of a light-colored region of cytoplasm by the rotation.

⇒ The gray crescent is located near the equator of the egg on the side opposite of sperm entry.

⇒ The presence of the gray crescent is an early marker of the egg's polarity.

Cleavage in the animal hemisphere of a frog's zygote is more rapid than in the vegetal hemisphere.

- Large amounts of yolk impedes cell division.
- This discrepancy in the rate of cleavage divisions results in a frog embryo with different size cells.
  ⇒ Animal hemisphere cells are smaller than those in the vegetal pole.
- In sea urchins and many other animals, the blastomeres are about equal in size due to small amounts of yolk.
  ⇒ Animal-vegetal pole axes are present but are due to concentration gradients of cytoplasmic components other than yolk.
  ⇒ The absence of yolk permits cleavage divisions to occur at about the same rate.

The first two cleavage divisions in sea urchins and frogs are vertical and divide the embryo into four cells that extend from the animal pole to the vegetal pole (see Campbell, Figures 47.7 an 47.8)

- The third cleavage plane is horizontal and produces an eight cell embryo with two tiers (animal and vegetal) of four cells each.
- In deuterostomes, which have radial cleavage, the upper tier of cells is aligned directly over the lower tier.
- In protostomes, which have spiral cleavage, the upper tier of cells align with the grooves between cells of the lower tier.

A continuation of cleavage produces a solid ball of cells called a *morula* (see Campbell, Figure 47.8b).

- The *blastocoel*, a fluid-filled cavity, develops within the morula as cleavage continues which changes the embryo from the solid morula to a hollow ball of cells, the *blastula* (see Campbell, Figure 47.8c).
- In sea urchins, the blastocoel is centrally located in the blastula due to equal cell divisions.
- Unequal cell divisions in the frog embryo produces a blastocoel in the animal hemisphere (see Campbell, Figure 47.8d).

The amount of yolk present in an egg greatly effects the cleavage.

- In eggs with little yolk (sea urchins) or moderate amounts of yolk (frogs), a complete division of the egg occurs and is called *holoblastic cleavage*.
- In eggs which contain large amounts of yolk (birds, reptiles), cleavage is incomplete and confined to a small disc of yolk-free cytoplasm at the animal pole of the egg; this is termed *meroblastic cleavage*.

A unique type of meroblastic cleavage occurs in the yolk-rich eggs of *Drosophila* and other insects.

- The zygote's nucleus is located within a mass of yolk.
- Cleavage begins with the nucleus undergoing a series of mitotic division *without* cytoplasmic divisions.
- The several hundred nuclei produced by these division migrate to the outer margin of the egg.
- Cytoskeletal proteins surround each nucleus and separate them from each other without membranes.

⇒ At this point the embryo is a syncytium, a cell containing many nuclei within a common cytoplasm.

- Several more mitotic divisions occur before plasma membranes form around the nuclei.
- The formation of the plasma membranes forms a blastula consisting of about 6000 cells in a single layer surrounding a yolk mass.

D. Gastrulation rearranges the blastula to form a three-layered embryo with a primitive gut

Gastrulation involves an extensive rearrangement of cells which transforms the blastula, a hollow ball of cells, into a three-layered embryo called the *gastrula*.

- A set of common cellular changes is involved in all animals: changes in cell motility; changes in cell shape; and changes in cellular adhesion to other cells and to molecules of the extracellular matrix.
  ⇒ Some cells at or near the surface move to a more interior location.
- Specific details of processes may differ from one animal group to the next.

The three layers produced by gastrulation are embryonic tissues called embryonic germ layers. These three cell layers (the primary germ layers) will eventually develop into all parts of the adult animal.

- The *ectoderm* is the outermost layer of the gastrula. The nervous system and outer layer of skin in adult animals develop from ectoderm.
- The *endoderm* lines the archenteron. The lining of the digestive tract and associated organs (i.e. liver, pancreas) develop from endoderm.
- The *mesoderm* partly fills the space between the ectoderm and endoderm. The kidneys, heart, muscles, inner layer of the skin, and most other organs develop from mesoderm.

Gastrulation in sea urchins begins at the vegetal pole (see Campbell, Figure 47.9).

- The sea urchin blastula consists of a single layer of cells.
- Vegetal pole cells form a flattened plate that buckles inward (*invagination*).
- Cells near the plate detach and enter the blastocoel as migratory mesenchyme cells.
- The invaginated plate undergoes rearrangement to form a deep, narrow pouch, the *archenteron* or primitive gut.
  ⇒ The archenteron opens to the surface through the *blastopore* which will become the anus.
- A second opening forms at the other end of the archenteron, forming the mouth end of the rudimentary digestive tube.
- At this point, gastrulation has produced an embryo with a primitive gut and three germ layers.
- The sea urchin embryo will develop into a ciliated larva that feeds on bacteria and unicellular algae while drifting near the ocean's surface as plankton.

Gastrulation during frog development also results in an embryo with the three embryonic germ layers and an archenteron that opens through a blastopore.

- The mechanics of gastrulation are more complicated than in the sea urchin because of the large, yolk-laden cells in the vegetal hemisphere and the presence of more than one cell layer in the blastula wall (see Campbell, Figure 47.10).
- A small crease forms on one side of the blastula where the blastopore will eventually form.
  ⇒ Invagination is due to a cluster of cells burrowing inward.

712   Unit VII   Animal Form and Function

- The invagination produces an external tuck that becomes the dorsal lip (upper edge) of the blastopore.

  ⇒ The dorsal lip forms where the gray crescent was located in the zygote.

- *Involution* then occurs. This is a process in which cells on the surface of the embryo roll over the dorsal lip and move into the embryo's interior away from the blastopore.

  ⇒ These cells continue to migrate along the dorsal wall of the blastocoel.

- As involution continues, migrating internal cells become organized into layered mesoderm and endoderm.

  ⇒ The archenteron forms within the endoderm.

- The cell movements of gastrulation produce a three-layered embryo.

- As gastrulation continues, more and more lateral cells involute with the dorsal cells resulting in a wider blastopore lip which eventually forms a complete circle.

- The circular blastopore lip surrounds a group of large, food-laden cells from the vegetal pole called the *yolk plug*.

- Gastrulation is complete with the formation of the three germ layers.

- Ectoderm is formed from those cells remaining on the surface (other than the yolk plug).

  ⇒ Surrounds the layers of mesoderm and endoderm.

E. In organogenesis, the organs of the animal body form from the three embryonic layers

The three germ layers that develop during gastrulation will give rise to rudimentary organs through the process of *organogenesis* (see Campbell, Table 47.1).

- The first evidence of organ development is morphogenetic changes (folds, splits, condensation of cells) that occur in the layered embryonic tissues.

The neural tube and notochord are the first organs to develop in frogs and other chordates.

- The dorsal mesoderm above the archenteron condenses to form the *notochord* in chordates.

- Ectoderm above the rudimentary notochord thickens to form a neural plate that sinks below the embryo's surface and rolls itself into a *neural tube*, which will become the brain and spinal cord (see Campbell, Figure 47.11).

- The notochord elongates and stretches the embryo lengthwise; it functions as the core around which mesoderm cells that form the vertebrae gather.

- Strips of mesoderm lateral to the notochord condense into blocks of mesodermal cells called *somites* from which will develop the vertebrae and muscles associated with the axial skeleton.

- As organogenesis continues, other organs and tissues develop from the embryonic germ layers.

  ⇒ Ectoderm also gives rise to epidermis, epidermal glands, inner ear, and eye lens.

  ⇒ Mesoderm also gives rise to the notochord, coelom lining, muscles, skeleton, gonads, kidneys and most of the circulatory system.

  ⇒ Endoderm forms the digestive tract linings, liver, pancreas and lungs.

- The neural crest forms from ectodermal cells which develop along the border where the neural tube breaks off from the ectoderm.

⇒ These cells migrate to other parts of the body and form pigment cells in the skin, some bones and muscles of the skull, teeth, adrenal medulla, and parts of the peripheral nervous system.

The end result of embryonic development in the frog is an aquatic, herbivorous tadpole.

- Metamorphosis will later change this larval stage into a terrestrial, carnivorous adult.

F. Amniote embryos develop in a fluid-filled sac within a shell or uterus

All vertebrate embryos require an aqueous environment for development.

- Fish and amphibians lay their eggs in water.
- Terrestrial animals live in dry environments and have evolved two solutions to this problem: the shelled egg in birds and reptiles and the *uterus* in placental mammals.
- The embryos of reptiles, birds, and mammals develop in a fluid-filled sac, the *amnion*.

These three classes of animals are referred to as *amniotes* due to the presence of the amnion around the embryo.

Important differences between cleavage and gastrulation occur in the development of birds and mammals, as well as differences between these amniotes and nonamniotes, such as the frog previously discussed.

1. Avian development

The larger yellow "yolk" of a bird egg is actually the ovum containing a large food reserve properly called yolk.

- Surrounding this large cell is a protein-rich solution (the egg white) that provides additional nutrients during development.
- After fertilization, meroblastic cleavage will be restricted to a small disc of yolk-free cytoplasm at the animal pole.
- Cell division partitions the yolk-free cytoplasm into a cap of cells called the *blastodisc* which rests on large undivided yolk mass.
- The blastomeres sort into an upper layer (epiblast) and lower layer (hypoblast) with a cavity (blastocoel) forming between them (see Campbell, Figure 47.12).
- Although different in appearance, this blastula is equivalent to the hollow ball stage in the frog.

Gastrulation begins with the movement of some epiblast cells toward the midline of the blastodisc. These cells then detach and move inward toward the yolk.

- The medial and inward movement of these cells produce a groove called the *primitive streak*, in the upper cell layer along the anterior-posterior axis.

Some cells in the epiblast migrate through the primitive streak and into the blastocoel to form mesoderm; others invade the hypoplast and contribute to endoderm.

- The ectoderm forms from cells remaining in the epiblast.
- Hypoblast cells appear to direct formation of the primitive streak and are necessary for development even though they contribute no cells to the embryo.
  ⇒ Hypoblast cells segregate from the endoderm and form portions of a sac surrounding the yolk and a stalk which connects the embryo and yolk.
- Embryonic disc borders fold downward, come together below the embryo, and form a three-layered embryo attached to the yolk below by a stalk at midbody.

Subsequent organogenesis occurs as in the frog, except that primary germ layers also form four extraembryonic membranes: the *yolk sac, amnion, chorion* and *allantois* (see Campbell, Figure 47.14).

2. Mammalian development

Fertilization occurs in the oviducts of most mammals and the early development occurs while the embryo travels down the oviduct to the uterus.

- The egg of placental mammals stores little nutrients and the zygote displays holoblastic cleavage.
- Gastrulation and organogenesis follow a similar pattern to that in birds.

Cleavage is relatively slow in mammals and the zygote has no apparent polarity.

- Cleavage planes appear randomly oriented and blastomeres are of equal size.
- The process of compaction, which occurs at the 8-cell stage, produces cadherins, which help the cells to tightly adhere to one another.

The development of a human embryo can represent mammalian development (see Campbell, Figure 47.15).

- Cleavage is relatively slow with the first, second and third divisions being completed at 36, 60, and 72 hours, respectively.
- At 7 days post-fertilization the embryo consists of about 100 cells arranged around a central cavity forming the *blastocyst*.
- The *inner cell mass* protrudes into one end of the cavity and will develop into the embryo and some of its extraembryonic membranes.
- The *trophoblast* is the outer epithelium surrounding the cavity which will, along with mesodermal tissue, form the fetal part of the placenta.
- During implantation, the inner cell mass forms a flat disc with an epiblast and hypoblast similar to those in birds; the embryo develops from epiblast cells and the yolk sac from hypoblast cells.

The blastocyst stage reaches the uterus and begins to implant.

- The trophoblast layer:
  ⇒ Secretes enzymes that enable blastocyst implantation in the uterus
  ⇒ Thickens and extends fingerlike projections into the endometrium
- Gastrulation occurs by inward movement of mesoderm and endoderm through a primitive streak.

Four extraembryonic membranes homologous to those in birds and reptiles form in mammals.

- The chorion forms from the trophoblast and surrounds the embryo and all extraembryonic membranes.
- The amnion forms as a dome above the epiblast and encloses the embryo in a fluid-filled cavity.
- The yolk sac encloses a fluid-filled cavity but no yolk; its membrane is the site of early blood cell formation.
  ⇒ These cells later migrate into the embryo.
- The allantois develops from an outpocketing of the rudimentary gut and is incorporated into the umbilical cord where it forms blood vessels that transport oxygen and nutrients from the placenta to the embryo and waste products from the embryo to the placenta.

Organogenesis begins with formation of the neural tube, notochord, and somites.

- Rudiments of all major organs have developed from the three germ layers by the end of the first trimester in humans.

II. The Cellular and Molecular Basis of Morphogenesis and Differentiation

Early in the embryonic development of an animal, a sequence of changes takes place that establishes the basic body plan of that animal.

- These include not only morphogenetic changes, which result in characteristic shapes, but also differentiation of many kinds of cells in specific locations.

A. Morphogenesis in animals involves specific changes in cell shape, position, and adhesion

The changes in cell shape and the cell migrations during cleavage, gastrulation and organogenesis are morphogenetic movements. These various morphogenetic movements help shape an embryo.

- Cell extension, contraction and adhesion are involved in these movements.
- Changes in shape usually involve reorganization of the cytoskeleton.
- Amoeboid or "cell crawling" movement based on extension and contraction is an important factor, especially during gastrulation.
    - Cell crawling is involved in *convergent extension*, a morphogenic movement in which the cells of a tissue layer rearrange themselves such that the sheet of cells becomes narrower as it lengthens.

Morphogenetic movements are partially guided by the extracellular matrix.

- The matrix contains adhesive substances and fibers that may direct migrating cells. (Laminin and fibronectins are extracellular glycoproteins that help cells adhere to their substratum as they migrate.)
- Nonmigrant cells along the pathways may promote or inhibit migration depending on the substances they secrete.
- Migrating cells receive direction cues, by way of surface receptor proteins, as they migrate along specific pathways.
    - Signals from the receptors direct the cytoskeletal elements to propel the cell in the proper direction.
- Other substances in the extracellular matrix direct cells by preventing migration along certain paths.

Cell adhesion molecules (CAMs) are substances on a cell surface that contribute to the selective association of certain cells with each other.

- CAMs vary either in amount or chemical identity from one type of cell to another which helps regulate morphogenetic movements and tissue building.
- *Cadherins* are a class of CAM that require the presence of calcium in order to function. Various cadherin genes have been found to be expressed in a tissue-specific manner as development proceeds.

B. The developmental fate of cells depends on cytoplasmic determinants and cell-cell induction: *a review*

Two general principles appear to integrate the genetic and cellular mechanisms underlying differentiation during embryonic development.

1. The heterogeneous distribution of cytoplasmic determinants in the unfertilized egg leads to regional differences in the early embryos of many animal species.
    - Different blastomeres receive different substances (mRNA, proteins, etc.) during cleavage due to the partitioning of the heterogenous cytoplasm of the ovum.
    - Gene expression in, and the developmental fate of, cells in the early embryo are influenced by these local differences in the distribution cytoplasmic determinants.

2. In induction, interactions among the embryonic cells themselves induce changes in gene expression.

- Interactions among embryonic cells induce development of many specialized cell types.
- Induction may be mediated by diffusible chemical signals.
- Membrane interaction between cells that are in contact may also induce cytoplasmic changes.

C. Fate mapping can reveal cell geneologies in chordate embryos

It is often possible to develop a fate map for embryos whose axes are defined early in development.

- The *fate map* shows which parts of the embryo are derived from each region of the zygote or blastula.
- First accomplished by W. Vogt in the 1920s, using nontoxic dyes to color different regions of the amphibian blastula surface and subsequently sectioning the embryo to see where each color turned up.

Cell lineage analysis is a more detailed form of fate mapping, which is possible due to the development of new techniques.

- Modern techniques permit the marking of an individual blastomere during cleavage and following the marker as it is distributed to the mitotic descendants of that cell.
- Using such techniques, it has been possible to map the developmental fate of every cell in the nematode *Caenorhabditis elegans* (see Campbell, Figures 21.4 and 47.20)

Events crucial to the normal development and growth of *C. elegans* and other animals have been determined through cell lineage analyses combined with experimental manipulation of parts of organisms. Two conclusions have emerged:

- In most animals, certain early "founder" cells generate specific tissues of the embryo.
- The developmental potential of a cell, that is the range of structures to which it can give rise, becomes limited as development proceeds.

D. The eggs of vertebrates contain cytoplasmic determinants that help establish the body axes and differences among cells of the early embryo

1. Polarity and the basic body plan

Bilaterally symmetrical animals have an anterior-posterior axis, a dorsal-ventral axis, and left and right sides.

- The first step in morphogenesis is establishment of this body plan, which is prerequisite to tissue and organ development.
- In humans and other mammals, the basic polarities do not appear to be established until after cleavage.
- In most animals polarity of the embryo is established in the unfertilized egg or during early cleavage.

In frogs and many other animals, all three axes of the embryo are defined before cleavage begins.

- The animal-vegetal axis of the egg is due to concentrations gradients of cytoplasmic contents and marks the anterior-posterior axis.
- The dorsal-ventral and left-right axes are marked by the gray crescent which forms opposite of the sperm entry point.
- The embryo continues development along all three axes.

2. Restriction of cellular potency

Two blastomeres with equal developmental potential can develop from a zygote with asymmetrically distributed cytoplasmic determinants.

- Requires that the axis of the first cleavage equally distribute these substances.
- In amphibians, a normal first cleavage divides the gray crescent between the two blastomeres.
    ⇒ If the two blastomeres are separated, each will develop into a normal tadpole.
    ⇒ The cells are totipotent in that each cell retains the zygote's potential to form all parts of the animal.
    ⇒ Experimental manipulation of the first cleavage plane so that it bypasses the gray crescent produces one cell that will develop into a normal tadpole and one that fails to develop properly (see Campbell, Figure 47.21).

The developmental fate of different regions of the embryos of some animals are affected by the distribution of cytoplasmic determinants and the zygote's characteristic pattern of cleavage.

- Cytoplasmic determinants are substances localized within specific regions of the egg's cytoplasm, which leads them to be included in specific blastomeres.
    ⇒ May control gene expression

Only the zygote is totipotent in many species, while in others, there is a progressive restriction in potency of the cells.

- Only the zygote is totipotent in those where the first cleavage plane divides the cytoplasmic determinants in a way that each blastomere will give rise to only certain parts of the embryo.
- In mammals, cells of the embryo remain totipotent until they become arranged into the trophoblast and inner cell mass of the blastocyst.
- The cells of the early gastrula of some species can still give rise to more than one type of cell even though they have lost their totipotency.
- By the late gastrula stage, the developmental fate of all cells is fixed.

Determination is the progressive restriction of a cell's developmental potential.

- A determined cell is one whose developmental fate can not be changed by moving it to a different location in the embryo.
- Daughter cells receive a developmental commitment from the original cell.
- Involves the cytoplasmic environment's control of the genome of the cell.
- The partitioning of heterogeneous cytoplasm of an egg during cleavage exposes the nuclei of the cells to cytoplasmic determinants that will affect which genes are expressed as the cells begin to differentiate.

E. Inductive signals drive differentiation and pattern formation in vertebrates

Induction = The ability of one cell group to influence the development of another

1. The "organizer" of Spemann and Mangold

In the 1920s, Spemann and Mangold performed a series of transplantation experiments in which they discovered that the dorsal lip of the blastopore acts as a primary organizer, setting up the interaction between chordamesoderm and the overlying ectoderm.

- Transplanting the chordamesoderm, which forms the notochord, to an abnormal site in the embryo will cause the neural plate to develop in an abnormal location.
- The rudimentary notochord induces the dorsal ectoderm of a gastrula to form the neural plate.

Recent experiments provide clues to the molecular basis of induction by Spemann's organizer

- In amphibians, bone morphogenic protein 4 (BMP-4) is exclusively active on the ventral side of the gastrula.
- Organizer cells on the dorsal side release factors that inactivate BMP-4.

Morphogenetic movements are important to cellular determination and differentiation since cells move to areas with differing physical and chemical environments.

Interactions among cells and cell layers are crucial during and after gastrulation.

Many inductions appear to involve a sequence of inductive steps that progressively determine the fate of the cells, as in the development of an amphibian's eye.

2. **Pattern formation in the vertebrate limb**

Pattern formation is the development of an animal's spatial organization with organs and tissues in their characteristic places in the three dimensions of the animal.

- Occurs in addition to the determination and differentiation of cells.

Pattern formation is controlled by *positional information*, which is a set of molecular cues that indicate a cell's location relative to other cells in an embryonic structure and that help to determine how the cell and its descendants respond to future molecular signals.

- Vertebrate limbs all develop from undifferentiated limb buds (see Campbell, Figure 47.23).
- A specific pattern of tissues (e.g., bone, muscle) emerges as the limb develops.
- Every component has a precise location and orientation relative to three axes.
    ⇒ The proximal-distal axis (shoulder to fingertip)
    ⇒ The anterior-posterior axis (thumb to little finger)
    ⇒ The dorsal-ventral axis (knuckle to palm)
- For proper development to occur, embryonic cells in the limb bud must receive positional information indicating location along all three axes.

Experiments on the development of chick limbs have provided information about positional information.

- A region of mesoderm located beneath an *apical epidermal ridge (AER)* at the wing bud tip is believed to assign embryonic tissue position along the proximal-distal axis. Transplantation experiments reversing the position of this tissue have produced limbs with reversed orientation. The AER produces several proteins of the fibroblast growth factor (FGF) family which serve as the inducing signals.
- A *zone of polarized activity (ZPA)* has been found to be the point of reference for the anterior-posterior axis.
    ⇒ The ZPA is located where the posterior part of the bud is attached to the body.
    ⇒ ZPA cells release a protein growth factor called Sonic hedgehog.

- Cells nearest the ZPA give rise to posterior structures, while cells farthest from the ZPA give rise to anterior structures (see Campbell, Figure 47.24)

Such experiments have led to the conclusion that pattern formation requires cells to receive and interpret environmental cues that vary between locations.

- Certain polypeptides (e.g., FGF, Sonic hedgehog) are believed to be the cues which function as positional signals for vertebrate limb development.
- Regional gradients of these polypeptides along the three orientation axes would provide a cell positional information needed to determine its position in a three-dimensional organ.
  - The regional variation in production of these polypeptides result from differential gene expression in different locations of the embryo.

A hierarchy of gene expression events affects the expression of the homeobox-containing (Hox) genes (see Chapter 21 for a review of homeoboxes). These genes appear to be involved in specifying the identity of various regions of the limb as well as of the entire body.

REFERENCES

Browder, L.W. *Developmental Biology*. 2nd ed. Philadelphia: Saunders College/Holt, Rinehart and Winston, 1984.

Campbell, N., et al. *Biology*. 5th ed. Menlo Park, California: Benjamin/Cummings, 1999.

Fjose, A. "Spatial Expression of Homeotic Genes in *Drosophila*." *Bioscience*, September 1986.

Gehring, Walter J. "The Molecular Basis of Development." *Scientific American*, October 1985.

Gilbert, S.F. *Developmental Biology*. 2nd ed. Sunderland, MA: Sinauer Associates, 1988.

Walbot, V., and N. Holder. *Developmental Biology*. New York: Random House, Inc., 1987.

CHAPTER 48
NERVOUS SYSTEMS

OUTLINE

I.   An Overview of Nervous Systems
     A. Nervous systems perform the three overlapping functions of sensory input, integration, and motor output
     B. The nervous system is composed of neurons and supporting cells

II. The Nature of Neural Signals
     A. Membrane potentials arise from differences in ion concentration between a cell's contents and the extracellular fluid
     B. An action potential is an all-or-none change in the membrane potential
     C. Action potentials "travel" along an axon because they are self-propagating
     D. Chemical or electrical communication between cells occurs at synapses
     E. Neural integration occurs at the cellular level
     F. The same neurotransmitter can produce different effects on different types of cells

III. Organization of Nervous Systems
     A. Nervous system organization tends to correlate with body symmetry
     B. Vertebrate nervous systems are highly centralized and cephalized
     C. The vertebrate PNS has several components differing in organization and function

IV. Structure and Function of the Vertebrate Brain
     A. The vertebrate brain develops from three anterior bulges of the spinal cord
     B. The brain stem conducts data and controls automatic activities essential for survival
     C. The cerebellum controls movement and balance
     D. The thalamus and hypothalamus are prominent integrating centers of the diencephalon
     E. The cerebrum contains the most sophisticated integrating centers
     F. The human brain is a major research frontier

OBJECTIVES

After reading this chapter and attending lecture, the student should be able to:

1. Compare the two coordinating systems in animals.
2. Describe the three major functions of the nervous system.
3. List and describe the three major parts of a neuron, and explain the function of each.
4. Explain how neurons can be classified by function.
5. Describe the function and location of each type of supporting cell.
6. Explain what a resting potential is, and list four factors that contribute to the maintenance of the resting potential.

7. Define equilibrium potential, and explain why the K^+ equilibrium potential is more negative than the resting potential.

8. Define graded potential, and explain how it is different from a resting potential or action potential.

9. Describe the characteristics of an action potential, and explain the role membrane permeability changes and ion gates play in the generation of an action potential.

10. Explain how the action potential is propagated along a neuron.

11. Describe two ways to increase the effectiveness of nerve transmission.

12. Describe synaptic transmission across an electrical synapse and a chemical synapse.

13. Describe the role of cholinesterase, and explain what would happen if acetylcholine was not destroyed.

14. List some other possible neurotransmitters.

15. Define neuromodulator, and describe how it may affect nerve transmission.

16. Explain how excitatory postsynaptic potentials (EPSP) and inhibitory postsynaptic potentials (IPSP) affect the postsynaptic membrane potential.

17. Explain how a neuron integrates incoming information, including a description of summation.

18. List three criteria for a compound to be considered a neurotransmitter.

19. List two classes of neuropeptides, and explain how they illustrate overlap between endocrine and nervous control.

20. Describe two mechanisms by which a neurotransmitter affects the postsynaptic cell.

21. Diagram or describe the three major patterns of neural circuits.

22. Compare and contrast the nervous systems of the following invertebrates and explain how variation in design and complexity correlate with phylogeny, natural history, and habitat:

    a. *Hydra*   d. Annelids and arthropods
    b. Jellyfish, ctenophores, and echinoderms   e. Mollusks
    c. Flatworms

23. Outline the divisions of the vertebrate nervous system.

24. Distinguish between sensory (afferent) nerves and motor (efferent) nerves.

25. Define reflex and describe the pathway of a simple spinal reflex.

26. Distinguish between the functions of the autonomic nervous system and the somatic nervous system.

27. List the major components of the central nervous system.

28. Distinguish between white matter and gray matter.

29. Describe three major trends in the evolution of the vertebrate brain.

30. From a diagram, identify and describe the functions of the major structures of the human brain:

    a. Medulla oblongata   f. Diencephalon
    b. Pons   g. Thalamus
    c. Cerebellum   h. Hypothalamus
    d. Superior and inferior colliculi   I. Cerebral cortex
    e. Telencephalon   j. Corpus callosum

31. Explain how electrical activity of the brain can be measured, and distinguish among alpha, beta, theta, and delta waves.

32. Describe the sleep-wakefulness cycle, the associated EEG changes, and the parts of the brain that control sleep and arousal.

33. Define lateralization and describe the role of the corpus callosum.

34. Distinguish between short-term and long-term memory.

35. Using a flowchart, outline a possible memory pathway in the brain.

KEY TERMS

central nervous system (CNS)

effector cells

nerves

peripheral nervous system (PNS)

neurons

cell body

dendrites

axons

myelin sheath

Schwann cells

oligodendrocytes

synaptic terminals

synapse

sensory neurons

interneurons

motor neurons

reflex

ganglion (pl., ganglia)

nuclei

supporting cells (glia)

blood-brain barrier

membrane potential

excitable cells

resting potential

gated ion channels

hyperpolarization

depolarization

graded potentials

threshold potential

action potential

voltage-gated ion channels

refractory period

saltatory conduction

presynaptic cell

postsynaptic cell

synaptic cleft

synaptic vesicles

neurotransmitter

white matter

presynaptic membrane

postsynaptic membrane

excitatory postsynaptic potential (EPSP)

inhibitory postsynaptic potential (IPSP)

summation

biogenic amines

epinephrine

norepinephrine

dopamine

serotonin

gamma aminobutyric acid (GABA)

glycine

glutamate

aspartate

neuropeptides

substance P

endorphins

cephalization

nerve cord

metencephalon

gray matter

central canal

ventricles

cerebrospinal fluid

meninges

cranial nerves

spinal nerves

sensory division

motor division

somatic nervous system

autonomic nervous system

parasympathetic division

sympathetic division

forebrain

midbrain

hindbrain

telencephalon

diencephalon

mesencephalon

myencephalon

brain stem

medulla oblongata

pons

inferior colliculi

superior colliculi

cerebellum

epithalamus

choroid plexus

thalamus

hypothalamus

suprachiasmatic nuclei

cerebral hemispheres

basal nuclei

cerebral cortex

corpus callosum

electroencephalogram (EEG)

reticular formation

limbic system

amygdala

short-term memory

long-term memory

hippocampus

long-term depression

long-term-potentiation

consciousness

nerve net

LECTURE NOTES

The *endocrine* and *nervous* systems of animals often cooperate and interact to maintain homeostasis and control behavior. Though structurally and functionally linked, these two systems play different roles.

	Nervous System	Endocrine System
Complexity	More structurally complex; can integrate vast amounts of information and stimulate a wide range of responses	Less structurally complex
Structure	System of *neurons* that branch throughout the body	*Endocrine glands* secrete *hormones* into the bloodstream where they are carried to the *target organ*
Communication	Neurons conduct electrical signals directly to and from specific targets; allows fine pinpoint control.	Hormones circulate as chemical messengers throughout the body via the bloodstream; most body cells are exposed to the hormone and only target cells with receptors respond
Response Time	Fast transmission of nerve impulses up to 100m/sec	May take minutes, hours or days for hormones to be produced, carried by blood to target organ, and for response to occur

I. An Overview of Nervous Systems

A. Nervous systems perform the three overlapping functions of sensory input, integration, and motor output

The nervous system has three overlapping functions (see Campbell, Figure 48.1):

1. Sensory input is the conduction of signals from sensory receptors to integration centers of the nervous system.

2. Integration is a process by which information from sensory receptors is interpreted and associated with appropriate responses of the body.

3. Motor output is the conduction of signals from the processing center to *effector cells* (muscle cells, gland cells) that actually carry out the body's response to stimuli.

The signals are conducted by *nerves*, threadlike extensions of nerve cells wrapped in connective tissue.

These functions involve both parts of the nervous system:

1. *Central nervous system (CNS)* = Comprised of the brain and spinal cord; responsible for integration of sensory input and associating stimuli with appropriate motor output

2. *Peripheral nervous system (PNS)* = Consists of the network of nerves extending into different parts of the body that carry sensory input to the CNS and motor output away from the CNS

B. The nervous system is composed of neurons and supporting cells

The nervous system includes two main types of cells: neurons, which actually conduct messages; and supporting cells (glia), which provide structural reinforcement as well as protect, insulate, and assist the neuron.

1. Neurons

Neurons = Cells specialized for transmitting chemical and electrical signals from one location in the body to another

They have a large *cell body*.

⇒ Contains most of the cytoplasm, the nucleus, and other organelles (see Campbell, Figure 48.2)

⇒ The cell bodies of most neurons are located in the CNS, although certain types of neurons have their cell bodies located in ganglia outside of the CNS.

Have *two* types of fiberlike extensions (processes) that increase the distance over which the cells can conduct messages:

- *Dendrites* convey signals to the cell body. They are short, numerous, and extensively branched to increase the surface area where the cell is most likely to be stimulated.

- *Axons* conduct impulses away from the cell body. They are long, single processes.

  - Vertebrate axons in PNS are wrapped in concentric layers of *Schwann cells*, which form an insulating *myelin sheath*. In the CNS, the myelin sheath is formed by *oligodendrocytes*.

  - Axons extend from the axon hillock (where impulses are generated) to many branches, which are tipped with *synaptic terminals* that release neurotransmitters.

    - *Synapse* = Gap between a synaptic terminal and a target cell—either dendrites of another neuron or an effector cell

    - *Neurotransmitters* = Chemicals that cross the synapse to relay the impulse

2. Functional organization of neurons

There are three major classes of neurons (see Campbell, Figure 48.3):

1. *Sensory neurons* convey information about the external and internal environments from sensory receptors to the central nervous system. Most sensory neurons synapse with interneurons.

2. *Interneurons* integrate sensory input and motor output; located within the CNS, they synapse only with other neurons..

3. *Motor neurons* convey impulses from the CNS to effector cells.

Neurons are arranged in groups, or circuits, of two or more kinds of neurons

- The simplest circuits involve synapses between sensory neurons and motor neurons, resulting in a simple *reflex* arc (e.g., knee-jerk reflex) (see Campbell, Figure 48.4).

- Complex circuits, such as those associated with most behaviors, involve integration by interneurons in the central nervous system.

  ⇒*Convergent circuits* = Neural circuit in which information from several presynaptic neurons come together at a single postsynaptic neuron; permits integration of information from several sources.

  ⇒*Divergent circuits* = Neural circuit in which information from a single neuron spreads out to several postsynaptic neurons; permits transmission of information from a single source to several parts of the brain.

  ⇒*Reverberating circuits* = Circular circuits in which the signal returns to its source; believed to play a role in memory storage.

Nerve cell bodies are often arranged into clusters; these clusters allow coordination of activities by only part of the nervous system.

- A *nucleus* is a cluster of nerve cell bodies within the brain.

- A *ganglion* is a cluster of nerve cell bodies in the peripheral nervous system.

3. Supporting cells

Supporting cells, or *glia*, structurally reinforce, protect, insulate, and generally assist neurons.

- Do not conduct impulses
- Outnumber neurons 10- to 50-fold

Several types of glia are present.

- Astrocytes encircle capillaries in the brain.
  - ⇒ Contribute to the *blood-brain barrier*, which restricts passage of most substances into the CNS.
  - ⇒ Probably communicate with one another and with neurons via chemical signals.
- Oligodendrocytes form myelin sheaths that insulate CNS nerve processes.
- Schwann cells form the insulating myelin sheath around axons in the PNS.

Myelination of neurons in a developing nervous system occurs when Schwann cells or oligodendrocytes grow around an axon so their plasma membranes form concentric layers.

- Provides electrical insulation since membranes are mostly lipid, which is a poor current conductor.
- Insulating myelin sheath increases the speed of nerve impulse propagation.
- In multiple sclerosis, myelin sheaths deteriorate causing a disruption of nerve impulse transmission and consequent loss of coordination.

II.   The Nature of Neural Signals

Signal transmission along the length of a neuron depends on voltages created by ionic fluxes *across* neuron plasma membranes.

A. Membrane potentials arise from differences in ion concentrations between a cell's contents and the extracellular fluid

All cells have an electrical *membrane potential* or voltage across their plasma membranes.

- Ranges from -50 to -100 mV in animal cells
- The charge outside the cell is designated as zero, so the minus sign indicates that the cytoplasm inside is negatively charged compared to the extracellular fluid.
- Is about -70mV in a resting neuron

The membrane potential arises from differences in the ionic composition of the intracellular and extracellular fluids (see Campbell, Figure 48.5). For example, the approximate concentrations in millimoles per liter (mM) are listed below for ions in mammalian cells:

Ion	Inside Cell	Outside Cell
[Na$^+$]	15 mM	150mM
[K$^+$]	150mM	5mM
[Cl$^-$]	10mM	120mM
[A-]	100mM	---------------

Note: [A-] symbolizes all anions within the cell including negatively charged proteins, amino acids, sulfate and phosphate.

- Principal cation inside the cell is K$^+$, while the principal cation outside the cell is Na$^+$.

- Principal anions inside the cell are proteins, amino acids, sulfate, phosphate, and other negatively charged ions (A-); principal anion outside the cell is Cl$^-$.

- Because internal anions (A-) are primarily large organic molecules, they cannot cross the membrane and remain in the cell as a reservoir of internal negative charge.

The selective permeability of the plasma membrane maintains ionic differences.

- As charged molecules, ions cannot readily diffuse through the hydrophobic core of the plasma membrane's phospholipid bilayer.

- Ions can only cross membranes by carrier-mediated transport or by passing through ion channels.

Ion channel = Integral transmembrane protein that allows a specific ion to cross the membrane

- May be passive and open all the time; or may be gated, requiring a stimulus to change into an open conformation

- Are selective for a specific ion, such as Na$^+$, K$^+$ and Cl$^-$

- Membrane permeability to each ion is a function of the type and number of ion channels. For example, membranes are usually more permeable to K$^+$ than to Na$^+$, suggesting that there are more potassium channels than sodium channels.

How does the cell create and maintain the membrane potential?

- K$^+$ diffuses out of the cell down its concentration gradient, since the K$^+$ concentration is greater inside cell and the membrane has a high permeability to potassium.

- K$^+$ diffusion out of the cell transfers positive charges from inside of the cell to outside.

- The cell's interior becomes progressively more negative as K$^+$ leaves because the molecules of the anion pool (A$^-$) are too large to cross the membrane.

- As the electrical gradient increases, the negatively charged interior attracts K$^+$ back into the cell.

- If K$^+$ was the only ion to cross the membrane, an *equilibrium potential* for potassium ions (= -85 mV) would be reached; this is the potential at which no net movement of K$^+$ would occur since the electrical gradient attraction of K$^+$ would balance the K$^+$ loss due to the concentration gradient.

- However, K⁺ is not the only ion to cross the membrane; although the membrane is less permeable to Na⁺ than K⁺, some Na⁺ diffuses into the cell down both its concentration gradient and the electrical gradient.

- The Na⁺ trickle into the cell transfers positive charge to the inside resulting in a slightly more positive charge (-70 mV) than if the membrane were permeable only to K⁺ (-85 mV).

- If left unchecked, the Na⁺ trickle into the cell would cause a progressive increase in Na⁺ concentration and a decrease in K⁺ concentration (the electrical gradient would be weakened by the positive sodium ions and the potassium ions would diffuse down its concentration gradient).

- This shift in ionic gradients is prevented by sodium-potassium pumps, special transmembrane proteins which use energy from ATP to:

  ⇒ Pump sodium back out of the cell against concentration and electrical gradients.

  ⇒ Pump potassium into the cell, restoring its concentration gradient.

B. An action potential is an all-or-none change in the membrane potential

All cells in the body exhibit the properties of membrane potential described above. However, only neurons and muscle cells can change their membrane potentials in response to stimuli.

- Such cells are called *excitable cells*.

- The membrane potential of an excitable cell at rest (unexcited state) is called a *resting potential*.

The presence of *gated ion channels* in neurons permits these cells to change the plasma membrane's permeability and alter the membrane potential in response to stimuli received by the cell.

- Sensory neurons are stimulated by receptors that are triggered by environmental stimuli.

- Interneurons normally receive stimuli produced by activation of other neurons.

The effect on the neuron depends on the type of gated ion channel the stimulus opens (see Campbell, Figure 48.6).

- Stimuli that open potassium channels *hyperpolarize* the neuron. K⁺ effluxes from the cell; this increases the electrical gradient making the interior of the cell more negative.

- Stimuli that open sodium channels *depolarize* the neuron. Na⁺ influxes into the cell; this reduces the electrical gradient and membrane potential because the inside of the cell becomes more positive.

Voltage changes caused by stimulation are called *graded potentials* because the magnitude of the change depends on the strength of the stimulus.

- Each excitable cell has a *threshold* to which depolarizing stimuli are graded. This threshold potential is usually slightly more positive (−50 to −55 mV) than the resting potential.

- If depolarization reaches the threshold, the cell responds differently by triggering an action potential.

- Hyperpolarizing stimuli do not produce action potentials since they cause the potential to become more negative; actually reduces the probability an action potential will occur by making it more difficult for depolarizing stimuli to reach the threshold.

An *action potential* is the rapid change in the membrane potential of an excitable cell, caused by stimulus-triggered selective opening and closing of *voltage-gated ion channels*.

- Voltage-gated ion channels open and close in response to changes in membrane potential.
- Voltage-gated sodium channels have two gates: the activation gate opens *rapidly* at depolarization, the inactivation gate closes slowly at depolarization.
- The voltage-gated potassium channel has one gate that opens *slowly* in response to depolarization.

An action potential has four phases (see Campbell, Figure 48.7):

1. Resting state; no channels are open.
2. Large depolarizing phase during which the membrane briefly reverses polarity (cell interior becomes positive to the exterior). The Na^+ activation gates open, allowing an influx of Na^+, while potassium gates remain closed.
3. The steep repolarizing p*hase* follows quickly and returns the membrane potential to its resting level; inactivation gates close the sodium channels and the potassium channels open.
4. The undershoot phase is a time when the membrane potential is temporarily more negative than the resting state (hyperpolarized); sodium channels remain closed but potassium channels remain open since the inactivation gates have not had time to respond to repolarization of the membrane.

A *refractory period* occurs during the undershoot phase; during this period, the neuron is insensitive to depolarizing stimuli. The refractory period limits the maximum rate at which action potentials can be stimulated in a neuron.

Action potentials are all-or-none events and their amplitudes are not affected by stimulus intensity. The nervous system distinguishes between strong and weak stimuli based on the frequency of action potentials generated.

- Strong stimuli produce action potentials more rapidly than weak stimuli.
- Maximum frequency is limited by the refractory period of the neuron.

C. Action potentials "travel" along an axon because they are self-propagating

A neuron is stimulated at its dendrites or cell body, and the action potential travels along the axon to the other end of the neuron.

- Action potentials in the axon are usually generated at the axon hillock.
- Strong depolarization in one area results in depolarization above the threshold in neighboring areas.
- The action potential does not travel down the axon, but is regenerated at each position along the membrane.

The signal travels in a perpendicular direction along the axon regenerating the action potential (see Campbell, Figure 48.8).

- Na^+ influx in the area of the action potential results in depolarization of the membrane just ahead of the impulse, surpassing the threshold.
- The voltage-sensitive channels in the new location will go through the same sequence previously described regenerating the action potential.
- Subsequent portions of the axons are depolarized in the same manner.
- The action potential moves in only one direction (down the axon) since each action potential is followed by a refractory period when sodium channel inactivation gates are closed and no action potential can be generated.

Factors affecting the speed of action potential propagation include:

1. The larger the diameter of the axon, the faster the rate of transmission since resistance to the flow of electrical current is inversely proportional to the cross-sectional area of the "wire" conducting the current.

2. *Saltatory conduction.* The action potential "jumps" from one node of Ranvier to the next, skipping the myelinated regions of membrane (see Campbell, Figure 48.9)

- *Nodes of Ranvier* are gaps in the myelin sheath between successive glial cells.
- Voltage-sensitive ion channels are concentrated in node regions of the axon.
- Extracellular fluid only contacts the axon membranes at the nodes; restricts the area for ion exchange to these regions.
- This results in faster transmission of the nerve impulse.

D. Chemical or electrical communication between cells occurs at synapses

Synapse = Tiny gap between a synaptic terminal of an axon and a signal-receiving portion of another neuron or effector cell.

- Also found between sensory receptors and sensory neurons, and between motor neurons and muscle cells.
- *Presynaptic cell* is the transmitting cell; *postsynaptic cell* is the receiving cell.
- There are two types of synapses: electrical and chemical.

1. Electrical synapses

Electrical synapses allow action potentials to spread directly from pre- to postsynaptic cells via gap junctions (intercellular channels).

- Allows impulses to travel from one cell to the next without delay or loss of signal strength
- Much less common than chemical synapses
- Example: the giant neuron processes of crustaceans

2. Chemical synapses

At a chemical synapse a *synaptic cleft* separates the pre- and postsynaptic cells so they are not electrically coupled (see Campbell, Figure 48.10).

- Within the cytoplasm of the synaptic terminal of a presynaptic cell are numerous *synaptic vesicles* containing thousands of neurotransmitter molecules.
- An action potential arriving at the synaptic terminal depolarizes the *presynaptic membrane* causing Ca^{2+} to rush through voltage-sensitive channels.
- The sudden rise in Ca^{2+} concentration stimulates synaptic vesicles to fuse with the presynaptic membrane and release *neurotransmitter* into the synaptic cleft by exocytosis.
- The neurotransmitter diffuses to the *postsynaptic membrane* where it binds to specific receptors, causing ion gates to open.
- Depending on the type of receptors and the ion gates they control, the neurotransmitter may either excite the membrane by depolarization or inhibit the postsynaptic cell by hyperpolarization.
- The neurotransmitter molecules are quickly degraded by enzymes and the components recycled to the presynaptic cell.

Chemical synapses allow transmission of nerve impulses in only one direction.

- Synaptic vesicles and their neurotransmitter molecules are found only in the synaptic terminals at the tip of axons.
- Receptors for neurotransmitters are located only on postsynaptic membranes.

E. Neural integration occurs at the cellular level

One neuron may receive information from thousands of synapses. Some synapses are excitatory, others are inhibitory (see Campbell, Figure 48.11).

- *Excitatory postsynaptic potentials (EPSP)* occur when excitatory synapses release a neurotransmitter that opens gated channels allowing Na^+ to enter the cell and K^+ to leave (depolarization).
- *Inhibitory postsynaptic potentials (IPSP)* occur when neurotransmitters released from inhibitory synapses bind to receptors that open ion gates, which make the membrane more permeable to K^+ (which leaves the cell) and/or to Cl⁻ (which enters the cell) causing hyperpolarization.

EPSPs and IPSPs are graded potentials; they vary in magnitude with the number of neurotransmitter molecules binding to postsynaptic receptors.

- Change in voltage lasts only a few milliseconds since neurotransmitters are inactivated by enzymes soon after release.
- The electrical impact on the postsynaptic cell also decreases with distance from the synapse.

A single EPSP is rarely strong enough to trigger an action potential, although an additive effect (*summation*) from several terminals or repeated firing of terminals can change membrane potential (see Campbell, Figure 48.12).

- Temporal summation is when chemical transmissions from one or more synaptic terminals occur so close in time that each affects the membrane while it is partially depolarized and before it has returned to resting potential.
- Spatial summation is when several different synaptic terminals, usually from different presynaptic neurons, stimulate the postsynaptic cell at the same time and have an additive effect on membrane potential.
- EPSPs and IPSPs can summate, each countering the effects of the other.

At any instant, the axon hillock's membrane potential is an average of the summated depolarization due to all EPSPs and the summated hyperpolarization due to all IPSPs.

- An action potential is generated when the EPSP summation exceeds the IPSP summation to the point where the membrane potential of the axon hillock reaches threshold voltage.

F. The same neurotransmitter can produce different effects on different types of cells

Dozens of different molecules are known to be neurotransmitters and many others are suspected to function as such.

Criteria for neurotransmitters include:

- Must be present in and discharged from synaptic vesicles in the presynaptic cell when stimulated and affect the postsynaptic cell's membrane potential
- Must cause an IPSP or EPSP when experimentally injected into the synapse
- Must be rapidly removed from the synapse by an enzyme or uptake by a cell permitting the postsynaptic membrane to return to resting potential

Types of neurotransmitters (see Campbell, Table 48.2):

1. *Acetylcholine* may be excitatory or inhibitory depending on the receptor; functions in the vertebrate neuromuscular junction (between a motor neuron and muscle cell) and in the central nervous system.
   - This is the most common neurotransmitter in both vertebrates and invertebrates.
2. *Biogenic amines* are derived from amino acids.

- *Epinephrine*, *norepinephrine*, and *dopamine* are produced from tyrosine; *serotonin* is synthesized from tryptophan.
- Commonly function in the central nervous system. Norepinephrine also functions in the peripheral nervous system.
- Imbalances in dopamine and serotonin are associated with mental illness.

3. Amino acids *glycine*, *glutamate*, *aspartate*, and *gamma aminobutyric acid (GABA)* function as neurotransmitters in the central nervous system.
   - GABA is the most abundant inhibitory transmitter in the brain.
4. *Neuropeptides* are short chains of amino acids.
   - *Substance P* is an excitatory signal that mediates pain perception.
   - *Endorphins* (or enkephalins) function as natural analgesics in the brain.

Neurotransmitters may affect the postsynaptic cell by:

- Altering the permeability of the postsynaptic membrane to specific ions (e.g., acetylcholine, amino acid transmitters)
- Affecting postsynaptic cell metabolism (e.g., biogenic amines, neuropeptides) by triggering a signal transduction pathway

A single neurotransmitter can elicit different responses in different postsynaptic cells by

- Binding to different receptor (e.g., acetylcholine)
- Binding to the same receptor, but the receptor is linked to different signal-transduction pathways

1. Gaseous signals of the nervous system

Some neurons of the vertebrate PNS and CNS release gas molecules, such as nitric oxide (NO) and carbon monoxide (CO), as local regulators. For example,

- During sexual arousal, neurons release NO into erectile tissue of the penis, which causes blood vessels to dilate and fill with blood, producing an erection.
- Acetylcholine released by neurons into blood vessel walls stimulates their endothelial cells to produce and release NO. In response, neighboring smooth muscle cells relax, dilating the vessels.
- Similarly, nitroglycerin is effective in treating angina because it is converted to NO, which dilates the heart's blood vessels.

Cells do not store gaseous messengers, so they must be produced on demand.

- Within a few seconds, they diffuse into target cells, produce a change, and are broken down.
- NO often works by a signal-transduction pathway; it stimulates a membrane-bound enzyme to synthesize a second messenger that directly affects cellular metabolism.

III.   Organization of Nervous Systems

There is great diversity in nervous system organization among animals; some even lack nervous systems (e.g., sponges).

A. Nervous system organization tends to correlate with body symmetry

The *Hydra*, a cnidarian, has a *nerve net*—a loosely organized system of nerves with no central control.

- Impulses are conducted in both directions causing movements of the entire body.

- Some cnidarians, ctenophores, and echinoderms have modified nerve nets with rudimentary centralization (see Campbell, Figure 48.13).

Cephalization = Evolutionary trend for concentration of sensory and feeding organs on the anterior end of a moving animal; gave rise to the first brains

- Found in bilaterally symmetrical animals

Most bilaterally symmetrical animals have a peripheral nervous system and a central nervous system. In most cases, the central nervous systems consists of a brain in the head end and one or more *nerve cords*.

- Flatworms have a simple "brain" containing many large interneurons that coordinate most nervous functions. Two or more nerve trunks travel posteriorly in a ladderlike system with transverse nerves connecting the main trunks.

- Annelids and arthropods have a well-defined ventral nerve cord and a prominent brain. Often contain ganglia in each body segment to coordinate actions of that segment.

- Cephalopods have the most sophisticated invertebrate nervous system containing a large brain and giant axons.

Nervous system complexity often correlates with phylogeny, habitat, and natural history. For example, sessile animals such as clams show little or no cephalization.

B. Vertebrate nervous systems are highly centralized and cephalized

Vertebrate nervous systems are structurally and functionally diverse, however, they all have distinct central and peripheral elements and a high degree of cephalization (see Campbell, Figure 48.14).

Because vertebrate nervous systems are so complex, it is useful to group them into functional components: the peripheral nervous system and the central nervous system.

The CNS provide the basis for the complex behaviors of vertebrates by bridging the sensory and motor functions of the peripheral nervous system.

- Consists of the *spinal cord*, which is located inside the vertebral column and receives information from skin and muscles and sends out motor commands for movement; and the *brain*, which carries out complex integration for homeostasis, perception, movement, intellect and emotions.

- Both are covered with *meninges*, protective layers of connective tissue.

- In the brain, *white matter* is in the inner region and *gray matter* is in the outer region. This orientation is reversed in the spinal cord.

- *Cerebrospinal fluid* fills the *ventricles* in the brain and the *central canal* of the spinal cord; it functions in circulation of hormones, nutrients, and white blood cells and in absorption of shock, which cushions the brain.

The spinal cord integrates simple responses to certain stimuli (reflexes) and carries information to and from the brain.

- The patellar (knee-jerk) reflex is one of the simplest and involves only two neurons. A stretch receptor in the quadriceps muscle is stimulated by stretching of the patellar tendon; this activates a sensory neuron that carries the information to the spinal cord where it synapses with a motor neuron; if an action potential is generated in the motor neuron, it travels back to the quadriceps, which contracts and causes the forward knee jerk.

- Larger-scale, more complex responses result when branches of a reflex pathway carry signals to other parts of the spinal cord or to the brain.

The peripheral nervous system of mammals consists of 12 pairs of cranial nerves and 31 pairs of spinal nerves.

- Cranial nerves originate from the brain and innervate organs of the head and upper body; most contain both sensory and motor neurons, although some are sensory only (e.g., optic nerve).
- Spinal nerves innervate the entire body and contain both sensory and motor neurons.

C. The vertebrate PNS has several components differing in organization and function

The PNS consists of:

- *Sensory division*, which brings information from sensory receptors to the CNS
- *Motor division*, which carries signals from the CNS to effector cells

The two basic functions of a nervous system are to:

- Control responses to external environment
- Maintain homeostasis by coordinating internal organ functions

The sensory nervous system contributes to both functions by carrying stimuli from the external environment and monitoring the status of the internal environment.

The motor nervous system has two separate divisions associated with these functions.

1. The *somatic nervous system's* neurons carry signals to skeletal muscles in response to external stimuli; includes reflexes (automatic responses to stimuli) and is often considered "voluntary" because it is subject to conscious control.

2. The *autonomic nervous system* controls primarily "involuntary," automatic, visceral functions of smooth and cardiac muscles and organs of the gastrointestinal, excretory, cardiovascular, and endocrine systems.

   - Divided into a *parasympathetic division* that enhances activities that gain and conserve energy, and a usually antagonistic *sympathetic division* that increases energy expenditures.

IV.   Structure and Function of the Vertebrate Brain

A. The vertebrate brain develops from three anterior bulges of the spinal cord

In all vertebrates, the brain develops from the anterior region of the neural tube. The following three bilaterally symmetrical regions arise as the neural tube differentiates:

- Forebrain
- Midbrain
- Hindbrain

As vertebrates evolved, the structure of their brains became more complex. The complexity is evident in the latter stages of brain development of higher vertebrates. In 6-week old human fetuses, for example, five brain regions have formed from the three primary bulges (see Campbell, Figure 48.17b).

- Telencephalon (from forebrain)
- Diencephalon (from forebrain)
- Mesencephalon (from midbrain)
- Metencephalon (from hindbrain)
- Myelencephalon (from hindbrain)

B. The brain stem conducts data and controls automatic activities essential for life

The mesencephalon, metencephalon, and myelencephalon form the *brain stem*.

The brain stem consists of three parts: the *medulla oblongata*, the *pons,* and the midbrain.

- The *medulla oblongata* and *pons* control visceral functions including breathing, heart and blood vessel activity, swallowing, vomiting,and digestion. They also coordinate large-scale body movements such as walking.

The *superior* and *inferior colliculi* are areas of the midbrain that function in the visual and auditory systems.

C. The cerebellum controls movement and balance

The *cerebellum*, derived from part of the metencephalon, functions in balance and coordination of movement (see Campbell, Figure 48.17).

D. The thalamus and hypothalamus are prominent integrating centers of the diencephalon

The embryonic diencephalon gives rise to the *epithalamus*, the *thalamus*, and the *hypothalamus*.

The *epithalamus* includes the pineal gland (site of melatonin synthesis) and a *choroid plexus*, a cluster of capillaries that produces cerebrospinal fluid.

The *thalamus*, a prominent integrating center in the diencephalon, relays sensory information to and from the cerebrum.

- Contains many different nuclei, each one dedicated to one type of sensory information
- Sorts incoming sensory information and sends it to appropriate higher brain centers for further interpretation and integration
- Receives input from the cerebrum and from parts of the brain that regulate emotion and arousal

The *hypothalamus* is one of the most important regulators of homeostasis.

- Is the source of posterior pituitary hormones (e.g., ADH) and releasing hormones of the anterior pituitary
- Contains the body's thermostat and centers for regulating hunger and thirst
- Plays a role in sexual response and mating behavior, the fight-or-flight response, and pleasure.

1. The hypothalamus and circadian rhythms

The hypothalamus of mammals contains a pair of *suprachiasmiatic nuclei (SCN)* that function as a biological clock.

The SCN use visual information to synchronize certain bodily functions with the natural cycles of day length and darkness. This biological clock maintains daily biorhythms such as when:

- Sleep occurs
- Blood pressure is highest
- Sex drive peaks

E. The cerebrum contains the most sophisticated integrating centers

The *cerebrum*, which is derived from embryonic telencephalon, is divided into the right and left *cerebral hemispheres* (see Campbell, Figure 48.19a). Each hemisphere consists of:

- Outer covering of gray matter, the *cerebral cortex*
- Internal white matter
- Cluster of nuclei deep within the white matter, the *basal nuclei*
    ⇒ They are centers for motor coordination, relaying impulses from other motor systems.
    ⇒ They send motor impulses to the muscles.

⇒ If damaged, passivity and immobility result, because they no longer allow motor impulses to be sent to the muscles. Degeneration of cells entering the basal ganglia occurs in Parkinson's disease.

The largest, most complex part of the human brain is the cerebral cortex.

- Its highly folded convolutions result in a surface area of about 0.5 m^2.

- Bilaterally symmetrical with two hemispheres connected by a thick band of fibers (white matter) known as the *corpus callosum*.

- Each hemisphere is divided into four lobes; some functional areas within each lobe have been identified (see Campbell, Figure 48.19b).

Two functional cortical areas, the primary motor cortex and the primary somatosensory cortex, form the boundary between the frontal lobe and the parietal lobe (see Campbell, Figure 48.20).

- In response to sensory stimuli, the motor cortex sends appropriate commands to skeletal muscles.

- The somatosensory cortex receives and partially integrates signals from the body's touch, pain, pressure, and temperature receptors.

- The proportion of somatosensory or motor cortex devoted to a particular body region depends upon how important sensory or motor information is for that part.

  - For example, more brain surface is committed to sensory and motor communication with the hands than with the entire torso.

  - Impulses transmitted from receptors to specific areas of somatosensory cortex enable us to associate pain, touch, pressure, heat, or cold with specific parts of the body receiving those stimuli.

A complicated interchange of signals among receiving centers and association centers produces our sensory perceptions.

- The special senses—vision, hearing, smell and taste—are integrated by cortical regions other than the somatosensory cortex.

- Each of these functional regions, as well as, the somatosensory cortex, cooperate with an adjacent association area.

F. The human brain is a major research frontier

1. Arousal and sleep

Electrical potential between areas of the cortex can be measured and recorded with an *electroencephalogram* or *EEG* (see Campbell, Figure 48.21). Different states of sleep and arousal produce different patterns of electrical activity or brain waves, which can be classified into four types:

1. Alpha waves – Slow synchronous waves; produced in the relaxed closed-eye state of wakefulness

2. Beta waves – Faster and less synchronous than alpha waves; produced during mental alertness such as occurs during problem solving.

3. Theta waves – More irregular than beta waves; predominate in the early stages of sleep

4. Delta waves – Slow, high amplitude, highly synchronized waves; occur during deep sleep

Sleep is a dynamic process during which a person alternates between two types of sleep: nonrapid eye movement sleep (NREM) and rapid eye movement sleep (REM*)*.

- NREM sleep - In the early stages of sleep, the brain produces slow, regular theta waves; during deeper sleep, it produces high amplitude delta waves.

- REM sleep - Characterized by rapid eye movement and desynchronized EEG similar to that of wakefulness. Most dreaming occurs during REM sleep.

Sleep and arousal are controlled by several centers in the cerebrum and brain stem; the most important is the *reticular formation* (see Campbell, Figure 48.22).

- Part of the reticular formation, the reticular activating system (RAS), regulates sleep and arousal.
- RAS serves as a filter that selects what sensory information reaches the cortex.

2. Lateralization, language, and speech

Lateralization (right brain/left brain) refers to the fact that the association areas of the cerebral cortex are not bilaterally symmetrical; each side of the brain controls different functions.

- The left hemisphere controls speech, language, calculation, and the rapid serial processing of detailed information.
- The right hemisphere controls overall context, creative abilities, and spatial perception.
  - The corpus callosum transfers information between the left and right hemispheres. Severing the corpus callosum will not alter perception, but will dissociate sensory input from spoken response.

Understanding and generating language are controlled by several association areas in the left hemisphere, while the emotional content of language is processed by the right hemisphere.

- Damage to parts of the left hemisphere can cause some form of aphasia, the inability to speak coherently.
- See the Methods box to see PET scans of the different "areas" in action.

3. Emotions

Emotions depend on interactions between the cerebral cortex and the *limbic system*, a group of giant nuclei and interconnecting axon tracts in the forebrain (see Campbell, Figure 48.23).

- Includes parts of the thalamus, hypothalamus, and inner portions of the cerebral cortex, including two nuclei called the *amygdala* and *hippocampus*.
- Cerebral components of the limbic system are linked to areas of the cerebral cortex involved in complex learning, reasoning, and personality, so there is a close relationship between emotion and thought.
- Frontal lobotomies, the surgical destruction of the limbic cortex or its connection with the cerebral cortex, were once used to treat severe emotional disorders. Drug therapy is now the treatment of choice.

4. Memory and learning

Memory is the ability to store and retrieve information related to previous experiences.

Memory occurs in two stages: short-term memory and long-term memory.

- *Short-term memory* reflects immediate sensory perceptions of an object or idea and occurs before the image is stored.
- *Long-term memory* is stored information that can be recalled at a later time.
- Transfer of information from short-term to long-term memory is enhanced by rehearsal, favorable emotional state, and association of new information with previously learned and stored information.

Fact memory differs from skill memory.

- Fact memory involves conscious and specific retrieval of data from long-term memory.
- Skill memory usually involves motor activities learned by repetition, which are recalled without consciously remembering specific details.

There is no highly localized memory trace in the nervous system; instead, memories are stored in certain association areas of the cortex.

Fact memory involves a pathway in which sensory information is transmitted from the cerebral cortex to the hippocampus and amygdala, which are two parts of the limbic system.

- The amygdala may filter memory, labeling information to be saved by tying it to an event or emotion.
- In the hippocampus, certain synapses functionally change as a result of altered responsiveness by postsynaptic cells.
    - In *long-term depression (LTD),* the postsynaptic cell displays decreased responsiveness to action potentials.
    - In *long-term potentiation (LTP),* the postsynaptic cell displays increased responsiveness to action potentials.
        - Results from brief, repeated action potentials that strongly depolarize the postsynaptic membrane, so an action potential from the presynaptic cell has a much greater effect at the synapse than before.
        - Lasts for hours, days or weeks and may occur when a memory is stored or learning takes place.
    - LTP mechanism involves presynaptic release of glutamate, an excitatory neurotransmitter.
        - Glutamate binds with postsynaptic receptors and opens gated channels highly permeable to calcium ions.
        - Ca^{2+} influx triggers intracellular changes that induce LTP.
    - Postsynaptic neurons may also change to enhance LTP.
        - In a positive feedback loop, the affected postsynaptic cell may signal the presynaptic cell to release more glutamate, enhancing LTP.
        - A likely messenger for this backward signaling is the local mediator nitric oxide.

5. Consciousness

The study of consciousness has drawn considerable attention; however, its neurological basis remains a mystery.

A number of hypothesis concerning consciousness have been put forth:

- Involves simultaneous cooperation of extensive areas of the cerebral cortex.
- Cerebral neurons and functional groups of neurons are generating conscious thoughts while engaged in more specific, less complex tasks.

REFERENCES

Campbell, N., et al. *Biology*. 5th ed. Menlo Park, California: Benjamin/Cummings, 1999.

Marieb, E.N. *Human Anatomy and Physiology*. 4th ed. Menlo Park, California Benjamin/Cummings, 1997.

CHAPTER 49
SENSORY AND MOTOR
MECHANISMS

OUTLINE

I. Introduction to Sensory Reception
   A. Sensory receptors transduce stimulus energy and transmit signals to the nervous system
   B. Sensory receptors are categorized by the type of energy they transduce

II. Photoreceptors
   A. A broad array of photoreceptors has evolved among invertebrates
   B. Vertebrates have single-lens eyes
   C. The light absorbing pigment rhodopsin operates via signal transduction
   D. The retina assists the cerebral cortex in processing visual information

III. Hearing and Equilibrium
   A. The mammalian hearing organ is within the inner ear
   B. The inner ear also contains the organs of equilibrium
   C. A lateral line system and inner ear detect pressure waves in most fishes and aquatic amphibians
   D. Many invertebrates have gravity sensors and are sound-sensitive

IV. Chemoreception – Taste and Smell
   A. Perceptions of taste and smell are usually interrelated

V. Movement and Locomotion
   A. Locomotion requires energy to overcome friction and gravity
   B. Skeletons support and protect the animal body and are essential to movement
   C. Muscles move skeletal parts by contracting
   D. Interactions between myosin and actin underlie muscle contractions
   E. Calcium ions and regulatory proteins control muscle contraction
   F. Diverse body movements require variation in muscle activity

OBJECTIVES

After reading this chapter and attending lecture, the student should be able to:

1. Differentiate between sensation and perception.
2. Give the general function of a receptor cell, and explain the five processes involved in this function.
3. Explain the difference between exteroreceptors and interoreceptors.
4. List and describe the energy stimulus of the five types of receptors.

5. Using a cross-sectional diagram of human skin, identify the various receptors present, and explain the importance of having near-surface and deep-layer receptors for such stimuli as pressure.

6. Compare and contrast the structure and processing of light in eye cups of *Planaria*, compound eyes of insects, and single-lens eyes of mollusks.

7. Using a diagram of the vertebrate eye, identify and give the function of each structure.

8. Describe how the rod cells and cone cells found in the vertebrate eye function.

9. Explain how retinal signals following a horizontal pathway can enhance visual integration.

10. Using a diagram of the human ear, identify and give the function of each structure.

11. Explain how the mammalian ear functions as a hearing organ.

12. Explain how the mammalian ear functions to maintain body balance and equilibrium.

13. Compare the hearing and equilibrium systems found in non-mammalian vertebrates.

14. Describe the structure and function of statocysts.

15. Explain how the chemoreceptors involved with taste and smell function.

16. List any advantages or disadvantages associated with moving through a(n):
    a. Aquatic environment
    b. Terrestrial environment

17. Give the three functions of a skeleton.

18. Describe how hydrostatic skeletons function, and explain why they are not found in large terrestrial organisms.

19. Explain how the structure of the arthropod exoskeleton provides both strength and flexibility.

20. Explain the adaptive advantage of having different types of joints in different locations in the vertebrate skeleton.

21. Explain how the skeleton combines with an antagonistic muscle arrangement to provide a mechanism for movement.

22. Using a diagram, identify the basic components of skeletal muscle.

23. Explain how muscles contract according to the sliding filament model of contraction.

24. Describe the processes involved in excitation-contraction coupling.

25. List and explain the two mechanisms responsible for graded contraction of muscles.

26. Explain the adaptive advantage of possessing both slow and fast muscle fibers.

27. Distinguish among skeletal muscle, cardiac muscle, and smooth muscle.

KEY TERMS

sensation	sclera	outer ear	endoskeleton
perception	choroid	tympanic membrane	skeletal muscle
sensory reception	conjunctiva	middle ear	myofibrils
sensory receptor	cornea	malleus	myofilaments
exteroreceptors	iris	incus	thin filaments
interoreceptors	pupil	stapes	thick filaments
sensory transduction	retina	oval window	sarcomere
receptor potential	lens	Eustachian tube	Z lines
amplification	ciliary body	inner ear	I band
transmission	aqueous humor	cochlea	A band
integration	vitreous humor	organ of Corti	H zone
sensory adaptation	accommodation	round window	sliding-filament model

mechanoreceptors	rod cells	pitch	cross-bridge
muscle spindle	cone cells	utricle	phosphagens
hair cell	fovea	saccule	creatine phosphate
pain receptor	retinal	semicircular canals	tropomyosin
nociceptor	opsin	lateral line system	troponin complex
thermoreceptor	rhodopsin	neuromast	sarcoplasmic reticulum
chemoreceptor	photopsins	statocysts	T (transverse) tubules
gustatory receptor	bipolar cells	statoliths	tetanus
olfactory receptor	ganglion cells	taste buds	motor unit
electromagnetic receptor	horizontal cells	locomotion	recruitment
photoreceptor	amacrine cells	hydrostatic skeleton	fast muscle fiber
eye cup	lateral inhibition	peristalsis	slow muscle fiber
compound eyes	optic chiasm	exoskeleton	cardiac muscle
ommatidia	lateral geniculate nuclei	cuticle	intercalated discs
single-lens eye	primary visual cortex	chitin	smooth muscle

LECTURE NOTES

I. Introduction to Sensory Reception

An animal's interaction with its environment depends on the processing of sensory information and the generation of motor output.

Sensory receptors receive information from the environment.

- Action potentials that reach the brain via sensory neurons are termed *sensations*.
- Interpretation of a sensation leads to the *perception* of the stimulus (e.g., smells, sounds)

Motor effectors carry out the movement in response to the sensory information.

A. Sensory receptors transduce stimulus energy and transmit signals to the nervous system

Sensory reception = Ability of a cell to detect the energy of a stimulus

Sensory receptors = Structures that transmit information about changes in an animal's internal and external environment

- They are usually modified neurons or epithelial cells occurring singly or within groups in sensory organs.
  - ⇒ *Exteroreceptors* detect external stimuli, such as heat, pressure, light, and chemicals.
  - ⇒ *Interoreceptors* detect internal stimuli, such as blood pressure and body position.
- Specialized to convert stimuli energy into changes in membrane potentials and then transmit signals to the nervous system
- All receptor cells have the same four functions: transduction, amplification, transmission, and integration.

1. Sensory transduction

Sensory transduction is the conversion of stimulus energy into a change in the membrane potential of a receptor cell.

Stimulus energy changes membrane permeability of the receptor cell (via opening or closing ion channel gates, or increasing ion flow by stretching the receptor cell membrane) and results in a graded change in the membrane potential called *receptor potential*.

2. Amplification

Amplification of stimulus energy that is too weak to be carried into the nervous system often occurs.

May take place in accessory structures or be a part of the transduction process.

3. Transmission

Transmission of a sensation to the CNS occurs in two ways:

1. The receptor cell doubles as a sensory neuron (e.g., pain cells). In this case, the intensity of the receptor potential will affect the frequency of action potentials that convey sensations to the CNS.

2. The receptor cell transmits chemical signals (neurotransmitters) across a synapse to a second sensory neuron. In this case the receptor potential affects the amount of neurotransmitter that is released, which in turn influences the frequency of action potential generated by the sensory neuron.

4. Integration

Receptor signals are integrated through summation of graded potentials.

Sensory adaptation is a decrease in sensitivity during continued stimulation; a type of integration that results in selective information being sent to the CNS.

The threshold for transduction by receptor cells varies with conditions resulting in a change in receptor sensitivity.

Sensory information integration occurs at all levels in the nervous system.

B. Sensory receptors are categorized by the type of energy they transduce

Receptors can be grouped into five types depending on the type of energy they detect.

1. *Mechanoreceptors* are stimulated by physical deformation caused by pressure, touch, stretch, motion, sound—all forms of mechanical energy (see Campbell, Figure 49.1).

   - Bending of the plasma membrane increases its permeability to Na^+ and K^+ resulting in a receptor potential.

   - In human skin, Pacinian corpuscles deep in the skin respond to strong pressure, while Meissner's corpuscles and Merkel's discs, closer to the surface, detect light touch.

   - *Muscle spindles* are stretch receptors (a type of interoreceptor) that monitor the length of skeletal muscles, as in the reflex arc.

   - Hair cells detect motion.

2. *Nociceptors* are a class of naked dendrites that function as *pain receptors*

   - Different groups respond to excess heat, pressure, or specific chemicals released from damaged or inflamed tissue.

   - Prostaglandins increase pain by lowering receptor thresholds; aspirin and ibuprofen reduce pain by inhibiting prostaglandin synthesis.

3. *Thermoreceptors* respond to heat or cold and help regulate body temperature.

   - There is still debate about the identity of thermoreceptors in the mammalian skin. May be two receptors consisting of encapsulated, branched dendrites or the naked dendrites of certain sensory neurons.

   - The interothermoreceptors in the hypothalamus function as the primary temperature control of the mammalian body.

4. *Chemoreceptors* include general receptors that sense total solute concentration (e.g., osmoreceptors of the mammalian brain), receptors that respond to individual molecules, and those that respond to categories of related chemicals (e.g., *gustatory* and *olfactory receptors*).

5. *Electromagnetic receptors* respond to electromagnetic radiation such as light (*photoreceptors*), electricity, and magnetic fields (magnetoreceptors).

   • A great variety of light detectors has evolved in animals, from simple clusters of cells to complex organs.

   • Molecular evidence indicates that most, if not all, photoreceptors in animals may be homologous.

II.   Photoreceptors

A.   A broad array of photoreceptors has evolved among invertebrates

The *eye cup* of planarians is a simple light receptor that responds to light intensity and direction without forming an image (see Campbell, Figure 49.4).

   • An opening on one side of the cup permits light to enter; the opening to one cup faces left and slightly forward, the other cup opens right and slightly forward.

   • Light enters the opening and stimulates photoreceptors that contain light-absorbing pigments.

   • Planaria move away from light sources to avoid predators.

   • The proper direction is determined by the brain, which compares the rate of nerve impulses coming from the two cups; the animal turns until the impulses from each cup are equal and minimal.

Two types of image-forming eyes have evolved in invertebrates:

1. A *compound eye* contains thousands of light detectors called *ommatidia*, each with its own cornea and lens (see Campbell, Figure 49.5).

   • Found in insects, crustaceans, and some polychaete worms

   • Results in a mosaic image

   • More acute at detecting movement partly due to rapid recovery of photoreceptors

   • Superimposition eyes have ommatidia with lens that work like prisms and parabolic mirrors, focusing light entering several ommatidia onto photoreceptor (increases sensitivity to light)

2. In a *single-lens eye*, one lens focuses light onto the retina, which consists of a bilayer of photosensitive receptor cells.

   • Found in some jellies, polychaetes, spiders and many mollusks

B.   Vertebrates have single-lens eyes

The parts of the vertebrate eye are structurally and functionally diverse (see Campbell, Figure 49.6)

   • The vertebrate eye consists of a tough outer layer of connective tissue, the *sclera*, and a thin inner pigmented layer, the *choroid*. A thin layer of cells, known as the *conjunctiva*, covers the sclera and keeps the eye moist.

   • The *cornea* is located in front and is a transparent area of the sclera; it allows light to enter the eye and acts as a fixed lens.

   • The anterior choroid forms the *iris*, which regulates the amount of light entering the *pupil*. The iris is pigmented and gives the eye color; the pupil is the hole in the center of the iris.

- The *retina* is the innermost layer of the eyeball; it contains photoreceptor cells which transmit signals from the optic disc, where the optic nerve attaches to the eye.

  - The *lens* and *ciliary body* divide the eye into two chambers: a small chamber between the lens and cornea and a large chamber within the eyeball.

  - The ciliary body produces *aqueous humor* that fills the cavity between the lens and cornea.

  - *Vitreous humor* fills the cavity behind the lens and comprises most of the eye's volume.

  - Both aqueous humor and vitreous humor help to focus light onto the retina.

The lens is a transparent, protein disc that focuses an image onto the retina by changing shape (*accommodation*) (see Campbell, Figure 49.7).

- Is nearly spherical when focusing on near objects and flat when focusing at a distance

- Controlled by the ciliary muscle

The photoreceptors of the eye are *rod cells* and *cone cells* (see Campbell, Figure 45.8).

- Their relative numbers in the retina are partly correlated with whether an animal is diurnal or nocturnal.

Rod cells are sensitive to light but do not distinguish colors.

- Found in greatest density at peripheral regions of the retina; completely absent from the *fovea* (center of visual field) (see Campbell, Figure 49.6).

Cone cells are responsible for daytime color vision.

- Most dense at the fovea.

C. The light absorbing pigment rhodopsin operates via signal transduction

Cells in the retina transduce stimuli (caused by the lens focusing a light image onto the retina) into action potentials.

Each rod cell or cone cell has an outer segment with a stack of folded membranes in which visual pigments are embedded (see Campbell, Figure 49.8a).

- The visual pigments consists of light-absorbing *retinal*, which is synthesized from vitamin A, bonded to a membrane protein *opsin*.

Rods contain their own type of opsin, and when combined with retinal, makes up *rhodopsin*.

- When rhodopsin absorbs light, its retinal component changes shape. This triggers a chain of metabolic events that hyperpolarizes the photoreceptor cell membrane; thus, a decrease in chemical signal to the cells with which photoreceptors synapse serves as the message (see Campbell, Figure 49.9).

- Light-induced change in retinal is referred to as "bleaching" of rhodopsin; in bright light, rods become unresponsive and the cones take over.

- In the dark, enzymes convert retinal back to its original form.

Color vision involves more complex signal processing than the rhodopsin mechanism in rods.

- Color vision results from the presence of three subclasses of cones: red cones, green cones and blue cones, each with its own type of opsin associated with retinal for form visual pigments (*photopsins*).

D. The retina assists the cerebral cortex in processing visual information

Integration of visual information begins at the retina.

Rod and cone cell axons synapse with neurons called *bipolar cells*, which in turn synapse with *ganglion cells*.

Horizontal cells and *amacrine cells* are neurons in the retina that help integrate the information, after which ganglion cell axons convey action potentials along the optic nerve to the brain.

Rod and cone cell signals may follow vertical or lateral pathways.

- Vertical pathways involve information passing directly from receptor cells to bipolar cells to ganglion cells.
- Lateral pathways involve: horizontal cells carrying signals from one rod or cone to other receptor cells and several bipolar cells; amacrine cells spread the signals from one bipolar cell to several ganglion cells.

Horizontal cells, stimulated by rod or cone cells, stimulate nearby receptors but inhibit more distant receptors and non-illuminated bipolar cells, thus sharpening image edges and enhancing contrast (*lateral inhibition*) (see Campbell, Figure 49.10)

- Occurs at all levels of visual processing.

Optic nerves from each eye meet at the *optic chiasm* (see Campbell, Figure 49.11).

- The optic chiasm has nerve tracts arranged so that what is viewed in the left field of view is transmitted to the right side of the brain and vice versa.
- Ganglion axons usually continue through the *lateral geniculate nuclei* of the thalamus, and these neurons continue back to the *primary visual cortex* in the occipital lobe of the cerebrum.
- Additional neurons carry information to more sophisticated visual processing centers in the cortex.

III. Hearing and equilibrium

Hearing and equilibrium are related in most animals and involve mechanoreceptors.

A. The mammalian hearing organ is within the inner ear

Sound waves are collected by the outer ear (the external pinna and the auditory canal) and are channeled to the *tympanic membrane* of the *middle ear* (see Campbell, Figure 49.12).

The sound waves cause the tympanic membrane to vibrate at the same frequency; the tympanic membrane transmits the waves to three small bones—the *malleus, incus,* and *stapes*—which amplify and transmit the mechanical movements of the membrane to the *oval window*, a membrane of the *cochlea* surface.

- The middle ear opens into the *Eustachian tube*, a channel to the pharynx to aid in pressure equalization on both sides of the tympanic membrane.
- Oval window vibrations produce pressure waves in the fluid (endolymph) in the coiled cochlea of the *inner ear*.

The pressure waves vibrate the basilar membrane (forms the floor of the cochlear duct) and the attached *organ of Corti*, which contains receptor hair cells.

- The bending of the hair cells against the tectorial membrane depolarizes the hair cells and causes them to release a neurotransmitter that triggers an action potential in a sensory neuron, which then carries sensations to the brain through the auditory nerve.
- The pressure wave continues through the tympanic canal and is dissipated as it strikes the *round window*.

Volume is determined by the amplitude of the sound wave; *pitch* is a function of sound wave frequency.

- The greater the amplitude of a sound, the more vigorous the vibrations; this results in more bending of the hair cells and more action potentials.
- Different sound frequencies affect different areas of the basilar membrane, thus some receptors send more action potentials than others.

B. The inner ear also contains the organs of equilibrium

Several organs in the inner ear detect body position and balance.

- Behind the oval window is a vestibule that contains two chambers, the *utricle* and the *saccule*.
- The utricle opens into three *semicircular canals* (see Campbell, Figure 49.14a).

Hair cells in the utricle and saccule respond to changes in head position with respect to gravity and movement in one direction.

- Hair cells are arranged in clusters with their hairs projecting into a gelatinous material containing numerous otoliths (small calcium carbonate particles).
- The otoliths are heavier than endolymph in the saccule and utricle; gravity pulls them down on the hairs of receptor cells, thus causing a constant series of action potentials indicating position of the head.

Semicircular canals detect rotation of the head due to endolymph movement against the hair cells (see Campbell, Figure 49.14b and c).

C. A lateral line system and inner ear detect pressure waves in most fishes and aquatic amphibians

The inner ear of a fish has no eardrum, does not open to the outside of the body, and has no cochlea, but a saccule, utricle, and semicircular canals are present.

- Sound waves are conducted through the skeleton of the head to the inner ear. This sets otoliths in motion, stimulating the hair cells.
- Some fish have a Weberian apparatus, a series of three bones which conducts vibrations from the swim bladder to the inner ear.
- Fish can hear higher frequencies due to their inner ears.

In terrestrial amphibians, reptiles and birds, sound is conducted from the tympanic membrane to the inner ear by a single bone, the stapes.

Fishes and aquatic amphibians have a *lateral line system* running along both sides of the body (see Campbell, Figure 49.15)

- Mechanoreceptors called *neuromasts* contain hair cell clusters whose hairs are embedded in a gelatinous cap, the cupula.
- Water enters the system through numerous pores on the animal's surface and flows along the tube past the neuromasts.
- Pressure of moving water bends the cupula causing an action potential in the hair cells.
    - This provides information about the body's movement direction and velocity of water currents, and movements or vibrations caused by predators and prey.

D. Many invertebrates have gravity sensors and are sound-sensitive

Most invertebrates have mechanoreceptors called *statocysts* that function in their sense of equilibrium (see Campbell, Figure 46.16).

- Gravity causes *statoliths* (dense granules) to settle to the low point in a chamber, stimulating hair cells in that location.
- Statocysts are located along the bell fringe of many jellies and at the antennule bases in lobsters and crayfish.

Many invertebrates demonstrate a general sensitivity to sound, although specialized structures for hearing seem to be less widespread than gravity sensors.

The body hairs of many insects vibrate in response to sound waves of specific frequencies.

- Fine hairs on the antennae of male mosquitoes detect the hum produced by a female's wings.

- Vibrating body hairs of some caterpillars detect predatory wasps.

Many insects also have "ears," commonly located on their legs, consisting of a tympanic membrane stretched over an internal air chamber containing receptor cells that send nerve impulses to the brain (see Campbell, Figure 49.17).

IV.   Chemoreception—Taste and Smell

Animals rely on chemoreception for many purposes, including locating food and mates, recognizing territories, and to assist with navigation (see Campbell, Figure 49.18).

A.   Perceptions of taste and smell are usually interrelated

The perceptions of taste and smell depend on chemoreceptors that detect specific chemicals in the environment.

Insects have taste receptors within sensory hairs called sensillae on the feet and mouthparts; olfactory sensillae are usually located on antennae.

- Several chemoreceptor cells, each responding to a particular chemical, are located on each tasting hair; integrating impulses from the different receptors permits distinguishing many tastes (see Campbell, Figure 49.19).

In humans and other mammals, receptor cells for taste are organized into *taste buds* scattered in several areas of the tongue and mouth.

- Sweet, sour, salty and bitter are detected in distinct regions of the tongue.
- These tastes are associated with specific molecular shapes and charges that bind to separate receptor molecules.

In humans, olfactory receptor cells line the upper nasal cavity and send impulses along their axons directly to the olfactory bulb of the brain (see Campbell, Figure 49.20).

- Receptive ends of the cells contain cilia that extend into the coating layer of mucus lining the nasal cavity.
- Specific receptors respond to certain odorous molecules by depolarizing.
- The olfactory sense responds to airborne chemicals.
- Taste and olfaction have different receptors but interact.

V.   Movement and Locomotion

Movement is a hallmark of animals. To catch food, an animal must move through the environment or move the surrounding medium (water or air) past itself. While some animals are sessile, most are mobile and rely on *locomotion* to acquire food or to escape from becoming food and to locate mates.

A.   Locomotion requires energy to overcome friction and gravity

Different modes of transportation (running, flying, swimming) have evolved along with adaptations of animals to overcome the difficulties associated with each type of locomotion.

At the cellular level, all movements are based on the contractile systems of microfilaments and microtubules.

1.   Swimming

Swimming animals must overcome resistance; thus, many are fusiform in body shape.

Animals swim in diverse ways.

2.   Locomotion on land

A walking or running land animal must support itself and move against gravity,

- Inertia must be overcome with each step; leg muscles accelerate a leg from a standing start.

Animals that hop generate a lot of power in their hind legs by momentarily storing energy in their tendons.

Maintaining balance is also essential for running, walking, or hopping.

- Bipedal animals keep part of at least on e foot on the ground when walking.
- When running, momentum more than foot contact keeps the body upright.

Crawling animals must exert considerable effort to overcome friction.

3. Flying

Flying animals do not use a skeleton for support during motion, and must almost completely overcome gravity to become airborne.

Wings must provide enough lift to overcome gravity; the key is in the shape of the wings.

B. Skeletons support and protect the animal body and are essential to movement

Skeletons function in support, protection, and movement.

- Help maintain shape of aquatic animals.
- Hard skeletons protect soft body tissues.
- Skeletons provide a firm attachment against which muscles can work during movement.

1. Hydrostatic skeletons

Hydrostatic skeletons consist of fluid held under pressure in a closed body compartment.

Found in most cnidarians, flatworms, nematodes, and annelids

Control form and movement by using muscles to change the shape of fluid-filled compartments

Provide no protection and could not support a large land animal

The hydrostatic skeleton of earthworms and other annelids allows for the rhythmic locomotion (*peristalsis*) these animals are known for (see Campbell, Figure 49.23).

2. Exoskeletons

Exoskeleton = Hard encasement deposited on the surface of an animal

Mollusks are shed (molted) as the animals grow.

3. Endoskeletons

Endoskeleton = Hard supporting elements buried within the soft tissues of an animal

Sponges possess hard spicules of inorganic material or softer protein fibers.

Echinoderms have ossicles composed of magnesium carbonate and calcium carbonate forming hard plates beneath the skin.

Chordates have cartilage and/or bone skeletons divided into several areas.

The vertebrate frame is divided into an axial skeleton (the skull, vertebral column, and rib cage) and an appendicular skeleton (limb bones, pectoral and pelvic girdles) (see Campbell, Figure 49.24).

Bones of the vertebrate act in support and as levers when their attached muscles contract.

C. Muscles move skeletal parts by contracting

Animal movement is based on contraction of muscles working against some kind of skeleton.

- Muscles always contract actively; they can extend only passively.
- Ability to move a body part in opposite directions requires that muscles be attached to the skeleton in antagonistic pairs (see Campbell, Figure 49.25).

1. **Structure and function of vertebrate skeletal muscle**

   Skeletal muscle = Bundle of long fibers running the length of the muscle; attached to bones and responsible for their movement (see Campbell, Figure 49.26)

   - Each fiber is a single cell with many nuclei; consists of bundles of smaller *myofibrils* arranged longitudinally.
   - Two kinds of *myofilaments* are found in each myofibril.
     1. *Thin filaments* consist of two strands of actin and one strand of regulatory protein coiled together.
     2. *Thick filaments* are staggered arrays of myosin molecules.

   The unit of organization of skeletal muscle is the *sarcomere.*

   - *Z lines* are the borders of the sarcomere; aligned in adjacent myofibrils.
   - *I bands* are areas near the edge of the sarcomere containing only thin filaments.
   - *A bands* are regions where thick and thin filaments overlap and correspond to the length of the thick filaments.
   - *H zones* are areas in the center of the A bands containing only thick filaments.

D. **Interactions between myosin and actin underlie muscle contractions**

   Muscle contraction reduces the length of each sarcomere.

   This behavior is explained by the *sliding filament model* (see Campbell, Figure 49.27)

   - Thin filaments ratchet across thick filaments to pull the Z lines together and shorten the sarcomere; the myofilaments themselves do not contract.
   - Myosin molecules on thick filaments attach to actin on the thin filament to form a *cross-bridge*. It then bends inward, pulling the thin filament toward the center of the sarcomere, breaks the cross-bridge, and forms a new cross-bridge further down.
   - Energy for cross-bridge formation comes from the hydrolysis of ATP by the head region of myosin (see Campbell, Figure 49.28).
     - Muscles store only a enough ATP for a few contractions. Most of the energy is stores as *phosphagens*.
     - Creatine phosphate, the phosphagen of vertebrates, can provide a phosphate group to ADP to make the ATP as needed.

E. **Calcium ions and regulatory proteins control muscle contraction**

   Skeletal muscles contract when stimulated by motor neurons.

   - An action potential in the motor neuron innervating the muscle cell causes the axon to release a neurotransmitter (e.g., acetylcholine).
   - Binding on the neurotransmitter to the muscle cell triggers an action potential in the muscle cell
   - The action potential of the muscle cell causes contraction.

   In a muscle at rest, myosin-binding sites on the actin are blocked by the regulatory protein strand (*tropomyosin*) in the thin filament and by the *troponin complex* located at each binding site. A contraction cycle is as follows:

   - The wave of depolarization spreads rapidly in the muscle via infoldings in the muscle cell plasma membrane (*transverse* or *T tubules*).
   - The *sarcoplasmic reticulum* (the specialized endoplasmic reticulum of muscle cells) membrane becomes depolarized and releases its store of Ca^{2+}.
   - The calcium ions bind to troponin, causing the thin filament to change shape and expose the myosin-binding sites; the muscle can then contract.

- The contraction is terminated as calcium is pumped out of the cytoplasm by the sarcoplasmic reticulum; as the calcium concentration falls, the tropomyosin-troponin complex again blocks the myosin-binding sites.

F. Diverse body movements require variation in muscle activity

Graded contractions of skeletal muscles are due to summation of multiple motor unit activity (*recruitment*) and wave summation.

- Motor neurons usually deliver their stimuli rapidly, resulting in smooth contraction typical of *tetanus* (sustained contraction) rather than the jerky actions of muscle twitches.
- A *motor unit* consists of a single motor neuron and all the muscle fibers it controls; all fibers in the motor unit contract as a group when the motor neuron fires.
- As more motor neurons are recruited by the brain, tension in the muscle progressively increases.

1. Fast and slow muscle fibers

Duration of muscle contraction is controlled by how long the Ca^{2+} concentration in the cytoplasm remains elevated.

Slow muscle fibers have longer-lasting twitches because they have less sarcoplasmic reticulum; thus, Ca^{2+} remains in the cytoplasm longer.

- Have many mitochondria, a rich blood supply, and the oxygen-storing protein myoglobin
- Used to maintain posture since they can sustain long contractions

Fast muscle fibers have short duration twitches and are used in fast muscles for rapid, powerful contractions.

- Some are able to sustain long periods of repeated contractions without fatiguing.

2. Other types of muscles

Vertebrate *cardiac muscle* is found only in the heart.

- Is striated
- Muscle cells are branched, and the junction between cells contain *intercalated discs* that electrically couple all heart muscle cells, allowing coordinated action.
- Cells can also generate their own action potentials.
- Rhythmic depolarizations due to pacemaker channels in the plasma membrane trigger action potentials which last up to 20 times longer than those for skeletal muscle and have long refractory periods.

Smooth muscles lack striations and contain less myosin; the myosin is not associated with specific actin strands.

- They generate less tension, but can contract over a greater range of lengths.
- Do not have a transverse tubule system or a well developed sarcoplasmic reticulum; calcium ions must enter the cytoplasm through the plasma membrane during an action potential.
- Contractions are relatively slow but there is greater range of control.
- They are found mainly in the walls of blood vessels and digestive tract organs.

Invertebrates have muscles similar to the skeletal and smooth muscles of vertebrates, but with some interesting adaptations.

- Arthropod skeletal muscles are very similar to vertebrate skeletal muscle.

- Insect wings actually beat faster than action potentials arrive from the CNS since the flight muscles are capable of independent, rhythmic contractions.
- The thick filaments of muscles in clams which hold the shell closed contain paramyosin; this unique protein allows the muscles to stay in a fixed state of contraction for up to a month.

REFERENCES

Campbell, N., et al. *Biology*. 5th ed. Menlo Park, California: Benjamin/Cummings, 1999.

Marieb, E.N. *Human Anatomy and Physiology*. 4th ed. Menlo Park, California: Benjamin/Cummings, 1997.

CHAPTER 50
AN INTRODUCTION TO ECOLOGY AND THE BIOSPHERE

OUTLINE

I. The Scope of Ecology
   A. Ecology is the scientific study of the interactions between organisms and their environment
   B. Ecological research ranges from the adaptations of organisms to the dynamics of ecosystems
   C. Ecology provides a scientific context for evaluating environmental issues
II. Abiotic Factors of the Biosphere
   A. Climate and other abiotic factors are important determinants of the biosphere's distribution of organisms
III. Aquatic and Terrestrial Biomes
   A. Aquatic biomes occupy the largest part of the biosphere
   B. The geographical distribution of terrestrial biomes is based mainly on regional variations in climate
IV. Concepts of Organismal Ecology
   A. The costs and benefits of homeostasis affect an organism's responses to environmental variation
   B. An organism's short-term responses to environmental variations operate within a long-term evolutionary framework

OBJECTIVES

After reading this chapter and attending lecture, the student should be able to:
1. Explain why the field of ecology is a multidisciplinary science.
2. Distinguish among physiology, ecology, community ecology, and ecosystem ecology.
3. Describe the relationship between ecology and evolution.
4. Explain the importance of temperature, water, light, soil, and wind to living organisms.
5. Explain the principle of allocation.
6. Describe how environmental changes may produce behavioral, physiological, morphological, or adaptive responses in organisms.
7. Explain the concept of environmental grain and under what situation(s) a single environment may be both coarse-grained and fine-grained.
8. Describe the characteristics of the major biomes: tropical forest, savanna, desert, chaparral, temperate grassland, temperate forest, taiga, tundra.
9. Compare and contrast the types of freshwater communities.
10. Using a diagram, identify the various zones found in the marine environment.

KEY TERMS

ecology	turnover	eutrophic	benthos
abiotic components	photic zone	mesotrophic	abyssal zone
biotic components	aphotic zone	wetlands	canopy
organismal ecology	thermocline	estuary	permafrost
population	benthic zone	intertidal zone	regulator
community	benthos	neritic zone	conformer
ecosystem	detritus	oceanic zone	principle of allocation
biosphere	littoral zone	pelagic zone	acclimation
climate	limnetic zone	benthic zone	
biome	profundal zone	coral reef	
tropics	oligotrophic	oceanic pelagic biome	

LECTURE NOTES

Ecology is the scientific study of the interactions of organisms and their environments.

- A complex and critical area of biology.
- There are three key words in the definition: scientific, environment, and interactions.

I.   **The Scope of Ecology**

  A. **Ecology is the scientific study of the interactions between organisms and their environments**

  The scientific nature of ecology involves using observations and experiments to test hypothetical explanations of ecological phenomena.

  - Ecology has a long history of being a descriptive science but is young as an experimental science.
    - Very difficult to conduct experiments and control variables
    - Some scientists test hypotheses in lab experiments and by manipulating populations and communities in field experiments
    - Some scientists devise mathematical models that include important variables and hypothetical relationships; usually studied with the aid of a computer.
    - Mathematical models are also used to simulate large-scale experiments that are impossible to conduct in the field; however, the basic information on which the models are based still must be obtained through fieldwork.
  - It is a multidisciplinary field examining questions from all areas of biology as well as many physical sciences.

  The environment of an organism includes both biotic and abiotic components.

  - *Biotic components* include all other organisms that are a part of any individual organism's environment.
  - *Abiotic components* are the nonliving chemical and physical factors (e.g., temperature, light, water, nutrients) to which an organism is exposed.

  The interactions between organisms and their environments include how the environment affects an organism and how an organism can change the environment.

  - Photosynthetic bacteria began to use sunlight for energy about three billion years ago.
    - Oxygen, a by-product of photosynthesis, accumulated and resulted in an aerobic atmosphere.

- The shading of the forest floor by trees sometimes makes the floor unsuitable (due to reduced light) for their offspring to grow.

Short-term (ecological time) interactions of organisms with their environments could have long-term (evolutionary time) effects through natural selection.

- For example, predator-prey interactions may affect gene pools where individuals with protective coloration would become more prevalent.
- The current distribution and abundance of organisms are products of long-term evolutionary changes and ongoing interactions.

B. Ecological research ranges from the adaptations of organisms to the dynamics of ecosystems

Ecology can be divided into four increasingly comprehensive levels of inquiry:

1. *Organismal ecology* studies the behavioral, physiological, and morphological ways individuals meet abiotic environmental challenges.

   ⇒ The distribution of organisms is limited by their tolerance of abiotic conditions.

2. *Population ecology* studies groups of individuals of the same species living in a particular geographic area.

   ⇒ Questions in population ecology concern factors that affect population size and composition

3. *Community ecology* studies all organisms that inhabit a particular area.

   ⇒ Questions concern predation, competition, disease, and other ways in which interactions among organisms affect community structure and organization

4. *Ecosystem ecology* studies all abiotic factors as well as communities of organisms in an area.

   ⇒ Questions concern energy flow and chemical cycling among the abiotic and biotic components

Ecological study is multidisciplinary in nature, encompassing genetics, evolution, physiology, behavior, chemistry, physics, geology, and meteorology.

C. Ecology provides a scientific context for evaluating environmental issues

Although distinct, basic ecology and environmental issues have many connections.

- To properly address environmental problems, it is necessary to understand the relationships between organisms and their environments.
- Current environmental awareness began with Rachel Carson's 1962 book *Silent Spring*, which exposed the fact that widespread use of pesticides often affects nontarget organisms.
- The realization that Earth is a finite resource and that pollution, overpopulation, and habitat destruction are threatening that resource is a concern of most people.

II.   Abiotic Factors of the Biosphere

The *biosphere* is the global ecosystem—the sum of all Earth's ecosystems.

- It is a thin layer consisting of the atmosphere to an altitude of a few kilometers; the land down to and including water-bearing rocks at least 1500 meters below ground, lakes and streams, caves; and the oceans to a depth of several kilometers.

A. Climate and other abiotic factors are important determinants of the biosphere's distribution of organisms

Global and regional patterns reflect differences in climate and other abiotic factors.

- Abiotic factors such as temperature, humidity, salinity, and light influence the distribution of organisms.
- The patchiness of the global biosphere illustrates how different physical environments produce a mosaic of habitats.

1. **Major abiotic factors**

   Some of the important abiotic factors that affect distribution of species include: temperature, water, sunlight, wind, rocks and soil, and periodic disturbances.

   a. **Temperature**

   Environmental temperature affects biological processes and body temperature.

   - Most organisms are unable to regulate their body temperature precisely
   - Temperature greatly affects metabolism: few organisms have active metabolisms at temperatures close to 0°C, and temperatures above 45°C denature most essential enzymes.
   - The actual body temperature of ectotherms is affected by heat exchange with the environment.
   - Most animals maintain a body temperature only a few degrees above or below ambient temperature.
   - Even endotherms function best within the environmental temperature range to which they are adapted.

   b. **Water**

   Water is essential for life and adaptations for water balance and conservation help determine a species' habitat range.

   - Marine and freshwater animals face the problems of regulating intracellular osmolarity; terrestrial animals face the problem of desiccation.

   c. **Sunlight**

   Sunlight provides the energy that drives nearly all ecosystems although only photosynthetic organisms use it directly as an energy source.

   Light is not the most important limiting factor in terrestrial environments but may play a role as in the reduction of competition due to shading in a forest.

   In aquatic environments, the distribution of photosynthetic organisms is limited by the intensity and quality of light.

   - Water selectively reflects and absorbs certain wavelengths; therefore, most photosynthesis occurs near the water surface.

   The physiology, development, and behavior of many animals and plants are often sensitive to photoperiod.

   d. **Wind**

   Wind amplifies the effects of temperature by increasing heat loss by evaporation and convection.

   - Wind also increases the evaporation rate of animals and transpiration rate of plants, resulting in more rapid water loss.

   Mechanical pressure of wind can affect plant morphology (for example, inhibiting growth of limbs on windward side of trees).

   e. **Rocks and soil**

   The physical structure, pH, and mineral composition of soil limit distribution of plants and hence animals that feed on those plants.

   - The composition of the substrate in a stream or river greatly influences the water chemistry, which in turn influences the plants and animals.

- The type of substrate influences what animals can attach or burrow in intertidal zones.

f. Periodic disturbances

Catastrophic disturbances such as fire, hurricanes, typhoons, and volcanic eruptions can devastate biological communities.

- After the disturbance, the area is recolonized by organisms or repopulated by survivors., but the structure of the community undergoes a succession of changes.
- Those disturbances that are infrequent (volcanic eruptions) do not elicit adaptations. Adaptations do evolve to periodically recurring disturbances such as fires.

2. Climate and the distribution of organisms

Climate is the prevailing weather conditions at a locality.

- The major components of climate are temperature, water, light, and wind.
- Climate has a major impact on the distribution of organisms.

A climatograph plots temperature and rainfall in a particular region (see Campbell, Figure 50.3).

- Usually plotted in terms of annual means
- Must be careful to distinguish a correlation between climate variables and biomes from causation.
  - *Biomes* are the major types of ecosystems typical of broad geographic areas.
- Only a detailed study of a species' tolerances to water and temperature ranges could establish the direct effects of these variables.
- The fact that biomes overlap on climatographs indicate that factors other than mean temperature and rainfall play roles in determining biome location (e.g., soil composition).

3. Global climate patterns

Solar energy input and the Earth's movement in space determine the planet's global climate patterns.

About 50% of the solar energy that reaches the atmosphere's upper layers is absorbed before it reaches the surface.

- Ultraviolet and certain other wavelengths are more readily absorbed by oxygen and ozone than other wavelengths.
- Some of the solar energy that reaches the Earth's surface is reflected back into the atmosphere; large amounts are absorbed by land, water, and organisms.

The atmosphere, land, and water are heated when they absorb solar energy. This heating establishes the temperature variations, air movement cycles, and evaporation of water responsible for the latitudinal variations in climate.

- Latitudinal variation in the intensity of sunlight results from the Earth's spherical shape; seasonal variation in solar radiation in the Northern and Southern Hemispheres are due to the Earth's tilt of 23.5° relative to its plane of orbit (see Campbell, Figures 50.4 and 50.5). The tilt causes solar radiation to change daily as the Earth rotates around the sun.
- Only the *tropics* (23.5°N to 23.5°S) receive sunlight from directly overhead year round. The tropics receive the greatest annual input of solar radiation and show the least seasonal variation; only small variations in daylength and temperature occur.

- Seasonal variation in light and temperature increases steadily toward the poles; polar regions have long, cold winters with periods of continual darkness and short summers with periods of continual light.

A global circulation of air which creates precipitation and winds results from the intense solar radiation near the equator (see Campbell, Figure 50.6).

- Evaporation of surface water due to high tropical temperatures causes warm, wet air masses to rise near the equator; this rising air creates an area of light, shifting winds (doldrums) along the equator.

- As these warm air masses rise, they expand and cool; cool air can hold less water vapor so the rising air masses drop large amounts of rain in the tropics.

- The cool, dry air masses flow toward the poles at high altitudes; they continue to cool as they move farther from the equator.

- Air mass density increases as they become cooler; they begin to descend toward the surface as cool, dry air masses at about 30° latitude.

- The air masses absorb water as they descend, thus creating arid climates around 30°N and 30°S.

- Some of the descending air masses flows toward the poles at low altitudes, the rest flows toward the equator.

- The air masses flowing toward the poles are warmed and rise again at about 60°N and 60°S; as these masses of air begin to cool with increased altitude, the water vapor is lost as precipitation.

- As air from this third cell reaches the higher altitude and cools, it flows toward the poles where it descends and flows back toward the equator.

4. **Local and seasonal effects on climate**

Although global climate patterns explain the geographic distribution of major biomes, local variations due to bodies of water and topographical features create a regional patchiness in climatic conditions.

- The regional and local variations in climate and soil have a major influence on less widely distributed communities and individual species.

Proximity of large bodies of water affect local climates.

- Ocean currents are generated by the Earth's rotation and may heat or cool (depending on whether they are tropical or polar currents) air masses passing over them toward land; evaporation is also greater over the ocean than over land.

- Coastal areas are more moist than inland areas at the same latitude; the California current flows from southward along the western U.S. and helps form a cool, moist climate in this area.

- During warm summer days, air over land heats faster than that over the ocean or large inland lakes; this warmer air rises, drawing a cool breeze from the water across the land.

Topographical variations, such as mountains, also exert an influence on solar radiation, local temperature and rainfall.

- In the Northern Hemisphere, south-facing slopes receive more sunlight and are therefore warmer and drier than north-facing slopes.
  ⇒ The vegetation differs with south-facing slopes being covered by shrubby, drought resistant plants while the north-facing slopes have forests.

- Air temperature declines 6° C for each 1000m increase in elevation.

⇒ This parallels the decline in temperature associated with increasing latitude.

⇒ For this reason, mountain communities are similar to those at lower elevations farther from the equator.

- When warm, moist air moves over a mountain, its altitude is increased and it cools; cooling causes the air to release its moisture as rain on the windward side.

- The cooler, drier air then flows down the leeward side of the mountain; the cool dry air is warmed and absorbs moisture, producing the rainshadow (see Campbell, Figure 50.7).

- Deserts commonly occur on leeward sides of mountains.

The Earth's orbit around the sun causes seasonal changes in local conditions.

- The changing angle of the sun causes slight shifts in the wet and dry air masses on either side of the equator.

⇒ These shifts result in the wet and dry seasons at 20° latitude where tropical deciduous forests grow.

- Seasonal changes in wind patterns produce variations in ocean currents, sometimes causing upwellings that bring nutrient-rich, cold water from the deep ocean layer to the surface.

- Seasonal temperature changes also cause the temperature profiles that develop during the summer in temperate zone ponds and lakes.

⇒ These profiles reverse in the autumn and spring resulting in biannual mixing that brings nutrient-rich water from the bottom to the top (*turnover*; see Campbell, Figure 50.8)

Climate also varies on a smaller scale, the microclimate. Microclimate refers to small areas within a habitat that may have very different conditions than the overall area (e.g., under a rock, a forest floor).

- Cleared areas in a forest generally show greater temperature extremes than the shaded forest floor due to greater solar radiation and wind currents.

- Low-lying areas are usually moister than high ground and support different forms of vegetation.

- The area under a large stone or log is protected from extremes of temperature and moisture; a large number of small organisms usually live in such sheltered areas.

III. Aquatic and Terrestrial Biomes

The worldwide distribution of aquatic and terrestrial biomes is shown in Campbell, Figures 50.9 and 50.15.

A. Aquatic biomes occupy the largest part of the biosphere

Life arose in water and evolved there for almost three billion years before moving into terrestrial habitats.

- Aquatic biomes still account for the largest part of the biosphere.

Freshwater and marine biomes are distinguished on the basis of physical and chemical differences.

- Freshwater biomes have a salt concentration less than 1%; marine biomes average 3% salt concentration.

The oceans cover 75% of Earth's surface and contain the marine biomes.

- Evaporation from the oceans provides most of the rainfall.

- Ocean temperatures greatly effect world climate and wind patterns.

- Marine algae ad photosynthetic bacteria consume large amounts of the atmospheric carbon dioxide and produce a major portion of the world's oxygen.

Freshwater biomes are closely linked to the soils and biotic components of the terrestrial biomes in which they are located or through which they flow.

- Runoff from terrestrial habitats creates streams and rivers.
- Accumulated runoff creates ponds and lakes.
- Freshwater biomes are also influenced by the patterns and speed of water flow and the climate to which they are exposed.

1. Vertical stratification of aquatic biomes

Aquatic biomes often exhibit pronounced vertical stratification

- There is a decrease in light intensity with increasing depth as light is absorbed by the water and suspended microorganisms. This divides bodies of water into two layers:
  ⇒ The *photic zone* is the upper layer where light is sufficient for photosynthesis.
  ⇒ The lower *aphotic zone* receives little light and no photosynthesis occurs.
- Water temperature tends to be stratified, especially during winter and summer.
- Heat energy from sunlight warms the upper layers of water as far as it penetrates; the deeper waters remain cold.
  - The *thermocline* is a narrow vertical zone between the warmer and colder waters where a rapid temperature change occurs.
- At the bottom of all aquatic biomes is the *benthic zone*, which is occupied by communities of benthos organisms that consume dead organic matter.

2. Freshwater biomes

Standing bodies of water vary greatly in size, from ponds to large lakes.

Ponds and lakes usually exhibit a significant vertical stratification in light penetration and water temperature.

The distribution of plants and animals within a pond or lake also shows a stratification based on water depth and distance from the shore (see Campbell, Figure 50.10).

- The *littoral zone* is shallow, well-lit waters close to shore.
  ⇒ Characterized by the presence of rooted and floating vegetation, and a diverse attached algal community (especially diatoms)
  ⇒ There is a diverse animal fauna including suspension feeders (clams); herbivorous grazers (snails); and herbivorous and carnivorous insects, crustaceans, fishes, and amphibians.
  ⇒ Some reptiles, water fowl, and mammals also frequent this zone.
- The *limnetic zone* is the open, well-lit waters away from shore.
  ⇒ Occupants include photosynthetic phytoplankton (algae and cyanobacteria), zooplankton (rotifers and small crustaceans) that graze on phytoplankton, and small fish which feed on the zooplankton.
  ⇒ Occasional visitors to this zone are large fish, turtles, snakes, and piscivorous birds.
- The *profundal zone* is the deep, aphotic zone lying beneath the limnetic zone.
  - This is an area of decomposition where detritus (dead organic matter that drifts in from above) is broken down.

- • Water temperature is usually cold and oxygen is low due to cellular respiration of decomposers.
- • Mineral nutrients are usually plentiful due to decomposition of detritus.
- • Waters of the profundal zone usually do not mix with surface waters due to density differences related to temperature.
- Mixing of these layers usually occurs twice each year in temperate lakes and ponds; thus, oxygen enters the profundal zone and nutrients are cycled into the limnetic zone.

Lakes are often classified as oligotrophic or eutrophic, depending on the amount of organic matter produced.

- • *Oligotrophic* lakes are deep, nutrient-poor lakes in which the phytoplankton are not very productive (see Campbell, Figure 50.11a).
    - ⇒ The water is usually clear and the profundal zone has a high oxygen concentration since little detritus is produced in the limnetic zone to be decomposed.
- • *Eutrophic* lakes are usually shallow, nutrient-rich lakes with very productive phytoplankton (see Campbell, Figure 50.11b).
    - ⇒ The waters are usually murky due to large phytoplankton populations and the large amounts of detritus being decomposed may result in oxygen depletion in the profundal zone during the summer.
- • *Mesotrophic* lakes have moderate amounts of nutrients and phytoplankton productivity.

Oligotrophic lakes may develop into mesotrophic and then eutrophic lakes over a long period of time.

- • Runoff from surrounding terrestrial habitats brings in mineral nutrients and sediments.
- • Human activities increase the nutrient content of runoff due to lawn and agricultural fertilizers; municipal wastes dumped into lakes dramatically enriches the nitrogen and phosphorus concentrations, which increases phytoplankton and plant growth.
- • Algal blooms and increased plant growth results in more detritus and can lead to oxygen depletion due to increased decomposition.
    - ⇒ This cultural eutrophication usually makes the water unusable.

Streams and rivers are bodies of water that move continuously in one direction (see Campbell, Figure 50.11c).

- • There is a change in structure of these bodies of water from their headwaters (point of origin) to their mouths (where they empty into a larger body of water).
- • At the headwaters, the water is cold and clear, carries little sediment, and has few mineral nutrients.
- • The channel is narrow with a rocky substrate and the water flows swiftly.
- • Near the mouth, water moves slowly and is more turbid due to sediment entering from other streams and erosion; the nutrient content is also higher.
- • The channel is usually wider with a silty substrate that has resulted from deposition of silt.

Human activities have greatly affected many streams and rivers.

- • Channelization (increases flow rate) and damming (slows flow rate) often change associated ecosystems.

- Pollutants may be taken up by the natural flora and fauna.

3. Wetlands

A *wetland* is an area covered by water that supports aquatic vegetation (see Campbell, Figure 50.12).

- Includes a broad range of habitats from periodically flooded regions to soil that is permanently saturated during growing season
- Conditions favor hydrophytes, which are plants specially adapted to grow in water or soil that is periodically anaerobic due to being saturated with water.
  ⇒ Cattails, pond lilies, and sedges are examples of hydrophytes.
- Both the hydrology and the vegetation are important determinants in wetland classification.

A wide variety of wetlands has been recognized although they form in one of only three topographic situations:

1. Basin wetlands develop in shallow basins ranging from upland depressions to lakes and ponds that have filled in.
2. Riverine wetlands develop along shallow and periodically flooded banks of streams and rivers.
3. Fringe wetlands are found along coasts of large lakes and seas where rising lake levels or tides cause water to flow back and forth.
   ⇒ Fringe wetlands include both freshwater and marine habitats.
   ⇒ Marine coastal wetlands are closely linked to estuaries.

Wetlands are among the richest and valuable of biomes.

- A diverse invertebrate community is present which supports a wide variety of birds.
- A variety of herbivorous species consume the algae, detritus, and plants.
- They provide water storage basins that reduce the intensity of flooding.
- They improve water quality by filtering pollutants.

Although many wetlands have been destroyed for agriculture and development, a move to protect the remaining wetlands is underway.

4. Estuaries

An *estuary* is the area where a freshwater stream or river merges with the ocean.

- They are often bordered by salt marshes or intertidal mudflats
- Their salinity varies spatially within the estuary from nearly fresh water to ocean water; varies daily in these areas due to rise and fall of tides
- Estuaries are very productive due to nutrients brought in by rivers.

Because of their productivity, estuaries have a diverse flora and fauna.

- Salt marsh grasses, algae, and phytoplankton are the major producers.
- Many species of annelids, oysters, crabs, and fish are also present.
- Many marine invertebrates and fish breed in estuaries or migrate through them to freshwater habitats upstream.
- A large number of water fowl and other semiaquatic vertebrates use estuaries as feeding areas.

Human activities have had a large impact on estuaries.

- Estuaries receive the pollutants dumped into the streams and rivers that feed them.
- Residential and commercial development not only adds to pollution but eliminates some estuaries due to land filling.

- Very little undisturbed estuary habitat remains.

5. **Zonation in marine communities:** *an introduction*

Similar to freshwater communities, marine communities are distributed according to depth of the water, distance from shore, degree of light penetration, and open water versus bottom (see Campbell, Figure 50.13).

- A photic zone is present and extends to the depth at which light penetration supports photosynthesis; occupied by phytoplankton, zooplankton, and many fish species.
- The aphotic zone is below the level of effective light penetration and represents a majority of the ocean's volume.
- The *intertidal zone* is the shallow zone where terrestrial habitat meets the ocean's water.
- The *neritic zone* extends from the intertidal zone, across the shallow regions, to the edge of the continental shelf.
- The *oceanic zone* extends over deep water from one continental shelf to another; reaches great depth.
- *Pelagic zone* refers to open waters of any depth.
- *Benthic zone* refers to the seafloor.

6. **The intertidal zones**

Intertidal zones, where land and sea meet, are alternately submerged and exposed by daily tide cycles.

- Organisms in this zone are exposed to variations in temperature and availability of sea water.
- These organisms are also subjected to the mechanical forces of wave action.

Rocky intertidal zones are vertically stratified and inhabited by organisms that possess structural adaptations that allow them to remain attached in this harsh environment (see Campbell, Figure 50.14)

- The uppermost zone is submerged only by the highest tides and is occupied by relatively few species of algae, grazing mollusks, and suspension-feeding barnacles.
  - ⇒ These organisms have various adaptations to prevent dehydration.
- The middle zone is exposed at low tide and submerged at high tide.
  - ⇒ Many species of algae, sponges, sea anemones, barnacles, mussels, and other invertebrates are found in this area.
  - ⇒ The diversity is greater here due to the longer time spans this area is submerged.
- Tide pools are often found in the middle zone.
  - ⇒ These are depressions which are covered during high tide and remain as pools during low tide.
  - ⇒ Tidepool organisms face dramatic salinity increases as water evaporates at low tide.
- The low intertidal zone is exposed only during the lowest tides and shows the greatest diversity of invertebrates, fishes, and seaweeds.

Sandy intertidal zones and mudflats do not show a clear stratification.

- Wave action continually shifts sand or mud particles; few algae or plants are present.
- Predatory crustaceans and many suspension-feeding worms and clams burrow into the sand or mud and feed when the tide submerges the area.

- Predatory or scavenging crabs and shorebirds often feed on burrowing organisms in these areas.

The diversity of intertidal zones is being reduced by human impact.

- Oil pollution destroys many species.

- Polluted water, old fishing lines, and plastic debris is harmful to most species.

- Recreational use of these areas has greatly reduced the number of beach-nesting birds and turtles.

7. Coral reefs

Coral reefs are found in the neritic zone of warm tropical waters where sunlight penetrates to the ocean floor.

- Sunlight penetration permits photosynthesis and a constant supply of nutrients is provided by currents and waves.

Coral reefs are dominated by the structure of the coral. It is formed by a diverse group of cnidarians that secrete a hard, calcium carbonate external skeleton, which provides a substrate on which other corals, sponges, and algae grow (see Campbell, Figure 50.14)

- Multicellular algae that are encrusted with calcium carbonate add large amounts of limestone to the reefs, as do bryozoans.

The cnidarian coral animals feed on microscopic organisms and organic debris, and they obtain some organic molecules from the photosynthetic products of their symbiotic dinoflagellate algae.

- Coral animals can survive without the dinoflagellates, but their rate of calcium carbon deposition is much slower without them; thus, reef formation by corals depends on this symbiotic association.

Reef communities are very old and grow very slowly.

Although many coral reefs are very large, they are delicate and can be severely damaged or destroyed by pollution, human induced damage, or introduced predators (e.g., crown-of-thorns sea star).

8. The oceanic pelagic biome

The *oceanic pelagic biome* consists of the open waters far from shore.

The area is constantly mixed by circulating ocean currents.

Nutrient content is generally lower than in coastal areas because the remains of organisms sink out of the zone to the lower benthic regions.

- ⇒ In tropical waters, nutrient content of surface waters is lower than surface waters of temperate oceans because a permanent thermal stratification prevents nutrient exchange with the deep waters.

- ⇒ Temperate oceans experience periodic upwellings which carry nutrients from the bottom to the surface.

Plankton is prevalent in this zone.

- Photosynthetic phytoplankton grow and reproduce in the photic region (top 100m) of this biome.

- Zooplankton (which graze on the phytoplankton) includes protozoans, copepods, krill, jellies, and larvae of many invertebrates and fishes.

- Most species stay afloat in this zone through the aid of morphological structures like bubble-trapping spines, lipid droplets, gelatinous capsules, and air bladders.

Nekton (free-swimming animals) are also found in the oceanic pelagic biome.

Large squid, fishes, sea turtles, and marine mammals feed in this area.

Many fish are adapted to and live in the aphotic region of the pelagic zone.

⇒ Some have large eyes allowing them to see in dim light, while others have luminescent organs used to attract mates and prey.

Many pelagic bird species also feed on fish in this region.

9.   Benthos

Benthos refers to those organisms which inhabit the benthic zone, the ocean bottom below both the neritic and pelagic zones.

- The benthic zone receives nutrients in the form of detritus, which settles into the area from the waters above.
- Light and temperature decline rapidly from shallow, near-shore benthic areas to the ocean's depths.
- The benthic zone substrate may be sand or very fine sediment composed of silt and shells of dead microscopic organisms.

Neritic benthic communities are diverse and productive.

- Many bacteria, fungi, seaweeds, filamentous algae, numerous invertebrates, and fishes are found here.
- Many of the species present are burrowing forms.
- The composition of the community varies with distance from shore, water depth, and bottom composition.

Deep benthic communities living in the *abyssal zone* are exposed to very different conditions than those under the neritic and pelagic zones.

- Water temperature is continuously cold (3°C), water pressure is extremely high, there is very little (if any) light, and low nutrient concentrations are typical.
- Oxygen is usually present and a fairly diverse community of invertebrates and fishes can be found.
- The deep-sea hydrothermal vent communities are found along midocean ridges in this region (see Campbell, Figure 50.4).
   ⇒ These vent communities include chemoautotrophic bacteria as the primary producers in place of photosynthesizing organisms.
   ⇒ These bacteria obtain energy by oxidizing H_2S which forms by a reaction of the hot vent water with dissolved sulfate.
   ⇒ The bacteria are consumed by a variety of giant polychaete worms, arthropods, echinoderms, and fishes.

B. The geographical distribution of terrestrial biomes is based mainly on regional variations in climate

The climate and other abiotic factors are important in determining why a particular terrestrial biome is found in an area.

- There are latitudinal patterns of biome distribution over the Earth's surface due to the latitudinal patterns in climate.

Terrestrial biomes are often named for major physical or climatic features and for the predominant vegetation, but each is also characterized by microorganisms, fungi, and animals adapted to that particular environment.

Vertical stratification is an important feature of terrestrial biomes and the shape and sizes of plants largely defines the layering.

- In many forests, the layers are the canopy.
- Vertical stratification of a biome's vegetation provides many different habitats for animals

Biomes grade into each other without sharp boundaries and may form ecotones, transitional areas between two communities.

Species composition may vary from one location to another within a biome.

Biomes are dynamic, and disturbance rather than stability is the rule.

- Disturbance may result in patchy biomes, with several communities represented in one biome.
- Human activities have radically altered the natural patterns of periodic disturbances.

Campbell, Figure 50.16 surveys the major terrestrial biomes.

IV. Concepts of Organismal Ecology

A. The costs and benefits of homeostasis affect an organism's responses to environmental variation

Organisms survive and reproduce in areas where environmental conditions to which they are adapted are found.

- The ability to tolerate one factor may be dependent on another factor; for example, many aquatic ectotherms can tolerate reduced oxygen at low temperatures, but not at high temperatures which cause higher metabolic rates.

1. Regulators and conformers

Organisms faced with a fluctuation in an environmental variable may maintain the homeostasis of their bodies through behavioral and physiological mechanisms (regulators) or by allowing their internal conditions to vary with external conditions (conformers).

- Some species are conformers under certain conditions and become regulators under others (see Campbell, Figure 50.17)

Energy expenditure by the animal is necessary if behavioral or physiological mechanisms are used to maintain homeostasis.

- For organisms to survive and reproduce, the energy "cost" of regulation cannot exceed the benefits of homeostasis.
- Since few organisms are perfect regulators or perfect conformers, a majority of organisms represent a group of evolved strategies which permits them to live in their specific environment.

2. The principle of allocation

The *principle of allocation* is an important concept for assessing the responses of organisms to a complex environment.

- This principle holds that each organism has a limited amount of energy that can be allocated for obtaining nutrients, escaping predators, coping with environmental fluctuations, growth, and reproduction.
- Energy expended for one function reduces the amount of energy available for other functions.
- If an organism expends a large amount of energy to maintain homeostasis, less is available for growth, reproduction, and other functions.

The distribution of organisms and the homeostatic mechanisms they possess establish different priorities for energy allocation.

- Conformers living in a stable environment may have more energy available for growth and reproduction; however, their geographic distribution is restricted due to intolerance to environmental change.

- Regulators that allocate a large amount of energy to survive environmental changes have less available for other functions so they grow and reproduce less efficiently; however, they can survive and reproduce over a wider range because they can cope with changing environmental conditions.

B. An organism's short-term responses to environmental variations operate within a long-term framework

It is important to remember that behavioral, physiological, and morphological responses to environmental change have evolved over evolutionary time to their current levels through natural selection.

- What appear to be short-term adjustments are actually evolutionary adaptations to maintain homeostasis.
- Natural selection also places constraints on the distribution of populations by adapting them to localized environments.
- Organisms adapted to one type of environment may not survive if dispersed to a foreign environment or may become extinct if the local environment changes to beyond their tolerance limits.
- The existence of a species in a particular location depends on the species reaching that location and being able to survive and reproduce after getting there.

Since environments vary over space and time, the impact of the variations on a particular species depends on the scale of the variation in relation to the species' overall life history.

1. Physiological responses

Physiological responses to environmental change are generally slower than behavioral responses although some may occur very rapidly.

- An example of a slow change is when a human moves to an area of less oxygen such as at higher altitudes.
  ⇒ After several days to a few weeks the person responds with an increased number of red blood cells being produced.
- An example of a faster change would be when blood vessels in the skin constrict within seconds to reduce loss of body heat when the skin is exposed to very cold air.

Physiological adaptation is centered around regulation and homeostasis.

- Both regulators and conformers function most efficiently under certain environmental conditions which are optimal for the organism.
- Efficiency declines both above and below optimal values.
- Physiological responses to changing environments can shift tolerance limits of organisms (*acclimation*).
- Acclimation is a gradual process and is related to the range of environmental conditions experienced under natural conditions.

2. Morphological responses

Organisms can react to environmental change with responses that alter body form or internal anatomy.

May develop over the lifetime of an individual or across generations.

May be forms of acclimation since they are reversible.
  ⇒ Increase in coat fur or feather density in winter
  ⇒ Change in coat color between winter and summer

Other morphological changes are irreversible over an individual's lifespan since the environmental variation may have affected growth and differentiation patterns.

- Plants are more morphologically plastic than animals.
  - The arrowleaf plant lacks a waxy leaf cuticle when growing in water with submerged leaves (allows absorption of materials directly from the water).
  - Arrowleaf plants growing on land have a thick cuticle (reduces water loss) and an extensive root system.

3. Behavioral responses

Behavioral responses to unfavorable environmental changes can be almost instantaneous in their effects and are easily reversed.

- The quickest response of animals is to move to a new, more favorable location.
  - Desert animals return to burrows or move into the shade during the day when heat is most intense.
  - Migratory birds migrate to warmer climates to over winter.
- Some animals can modify their immediate environment by cooperative social behavior.
  - Honeybees seal the hive during cold periods to conserve heat and cool the hive on hot days by the collective beating of their wings.
  - Small mammals may huddle in burrows in cold weather which minimizes the total surface area exposed to cold air, thus reducing heat loss.

REFERENCES

Brown, James H., and Arthur C. Gibson. *Biogeography*. St. Louis, Missouri: The C.V. Mosby Company, 1983.

Campbell, N., et al. *Biology*. 5th ed. Menlo Park, California: Benjamin/Cummings, 1999.

Ricklefs, R.E. *Ecology*. 3rd ed. New York: Chiron Press, 1986.

Smith, Robert Leo. *Ecology and Field Biology*. 3rd ed. New York: Harper & Row, Publishers, 1980.

CHAPTER 51
BEHAVIORAL BIOLOGY

OUTLINE

I. Introduction to Behavior and Behavioral Ecology
   A. Behavior results from both genes and environmental factors
   B. Innate behavior is developmentally fixed
   C. Classical ethology presaged and evolutionary approach to behavioral biology
   D. Behavioral ecology emphasizes evolutionary hypotheses: *science as a process*

II. Learning
   A. Learning is experience-based modification of behavior
   B. Imprinting is learning limited to a critical time period
   C. Many animals can learn to associate one stimulus with another
   D. Practice and exercise may explain the ultimate bases of play

III. Animal Cognition
   A. The study of cognition connects nervous system function with behavior
   B. Movement from place to place often depends on internal coding of spatial relationships
   C. The study of consciousness poses a unique challenge for scientists

IV. Social Behavior and Sociobiology
   A. Sociobiology places social behavior in an evolutionary context
   B. Competitive social behaviors often represent contests for resources
   C. Mating behavior relates directly to an animal's fitness
   D. Social interactions depend on diverse modes of communication
   E. The concept of inclusive fitness can account for most altruistic behavior
   F. Sociobiology connects evolutionary theory to human culture

OBJECTIVES

After reading this chapter and attending lecture, the student should be able to:
   1. Explain the difference between innate and learned behaviors.
   2. Describe the evolutionary basis for behavioral ecology.
   3. Explain the difference between ultimate and proximate causations of behavior.
   4. Describe a fixed-action pattern and a sign stimulus.
   5. Explain the nature versus nurture controversy.
   6. Explain the effect of maturation on behavioral improvement.
   7. Define habituation.
   8. Discuss imprinting, imprinting stimulus, and critical period.

9. Define associative learning.
10. Distinguish among classical conditioning, operant conditioning, and observational learning.
11. Describe two hypotheses for the evolution of play behavior.
12. Discuss the ultimate bases of learning.
13. Describe and define kinesis, taxis, and migration.
14. Explain the differences among piloting, orientation, and navigation.
15. Compare generalist and specialist foraging strategies.
16. Explain how a search image is adaptive.
17. Describe optimal foraging strategies in terms of energetics and prey densities.
18. Describe agonistic behavior.
19. Explain what is meant by a ritual behavior, and describe the evolutionary advantage of ritual behavior.
20. Describe a dominance hierarchy, and explain the advantages to individuals in the hierarchy.
21. Explain how dominance hierarchies and territories may stabilize population densities.
22. Describe the advantages of courtship.
23. Explain how ritualized courtships may have evolved.
24. Define parental investment.
25. Discuss the ultimate bases for mate selection.
26. Compare and contrast the three main mating systems.
27. Describe the differences between polygyny and polyandry.
28. Discuss how the needs of the young influence the development of mating systems.
29. Describe how the certainty of paternity influences the development of mating systems.
30. Describe the various modes of communication.
31. Relate an animal's mode of communication with its lifestyle.
32. Discuss why altruistic behavior might evolve.
33. Define inclusive fitness and kin selection.
34. Define reciprocal altruism.
35. Define cognitive ethology.
36. Describe the premise of sociobiology.

KEY TERMS

behavior	critical period	migration	monogamous
ethology	associative learning	social behavior	polygamous
fixed-action pattern	classical conditioning	sociobiology	polygyny
sign stimulus	operant conditioning	agonistic behavior	polyandry
foraging	play	ritual	pheromones
search image	cognition	dominance hierarchy	inclusive fitness
learning	cognitive ethology	territory	coefficient of
maturation	cognitive maps	parental investment	relatedness
kin selection	habituation	kinesis	lek
imprinting	reciprocal altruism	taxis	promiscuous

LECTURE NOTES

I.   Introduction to Behavior and Behavioral Ecology

Behavior = What an animal does and how it does it

Because behavior is assumed to increase fitness, questions about *ultimate* causation, or the reasons why behaviors exist are evolutionary questions.

Ultimate causation = The evolutionary reason for the existence of a behavior

Proximate causation = The immediate cause and/or mechanism underlying a behavior

- The immediate mechanism or how a behavior is expressed is the proximate cause. Proximate causes may be internal processes or environmental stimuli
- Proximate causation limits the range of behaviors upon which natural selection can act.
- Proximate mechanisms produce behaviors that ultimately evolved because they increase fitness.

Behavioral example 1: Bluegill sunfish breeding in spring and early summer

Proximate cause: Breeding is triggered by the effect of increased day length on a fish's pineal gland.

Ultimate cause: Breeding is most successful when water temperatures and food supplies are optimal.

Behavioral example 2: Human "sweet tooth"

Proximate cause: Sweet taste buds are a proximate mechanism that increase the chances of eating high-energy foods.

Ultimate cause: Sweet, high-energy, foods were rare prior to mechanized agriculture. Increased fitness associated with consuming these foods is the ultimate reason for the natural selection of a "sweet tooth."

A.   Behavior results from both genes and environmental factors

Behavioral biologists agree that behaviors result from both genetic (nature) and environmental influences (nurture).

- The debate is about the degree to which genes and environment influence phenotypic traits, including behavior.

Like all traits, behaviors display a range of phenotypic variation depending on the environment in which the genotype is expressed (see Campbell, Figure 51.1).

The environmental factors that can affect behavior are numerous, and include the chemical environment within the cell, the hormonal and physical conditions experienced by the developing organism, and interactions with other organisms.

B.   Innate behavior is developmentally fixed

All behaviors, including those that are present at birth (innate behaviors), require an environment to be expressed.

Innate behaviors do not appear to be influenced by the vast range of environmental differences that exists among individuals.

A behavior that remains essentially the same among organisms despite environmental differences within or outside their bodies is said to be developmentally fixed.

The ultimate cause for innate behavior may be that being capable of performing some behaviors automatically, without having any specific experience, may have maximized fitness to the point that genes for variant behavior were lost.

C. Classical ethology presaged an evolutionary approach to behavioral biology

Ethology pre-dates behavioral ecology. Relying on descriptive studies, ethologists discovered many behaviors were innate.

Ethology = Descriptive science based on studies of animals in the natural environment

Konrad Lorenz, Niko Tinbergen, and Karl von Frisch shared the 1973 Nobel Prize for their work in ethology.

The most fundamental concept in classical ethology is the *fixed-action pattern (FAP)*, a highly stereotyped, innate behavior.

- A fixed-action pattern is triggered by an external *sign stimulus*.

- Once a fixed-action pattern is triggered, the behavior continues until completion even in the presence of other stimuli or if the behavior is inappropriate.

- Fixed-action patterns are adaptive responses to natural stimuli. Strange responses can be initiated by presenting unnatural situations to animals with fixed-action patterns.

Examples of sign stimuli and FAPs include:

Example 1: Niko Tinbergen noticed male three-spined stickleback fish responded aggressively to red trucks passing by their tank.

> *Fixed-action pattern:* Male sticklebacks attack other males that enter their territories.
>
> *Sign stimulus:* The red belly of the invading male; sticklebacks will attack nonfish-like models with red on the ventral surface.

Example 2: Parent/young feeding behavior in birds

> *Fixed-action pattern:* The begging behavior of newly hatched chicks (raised heads, open mouths, loud cheeps)
>
> *Sign stimulus:* Parent landing at the nest

Example 3: Greylag goose egg retrieval behavior

> *Fixed-action pattern:* Rolls the egg back to the nest using side-to-side head motions
>
> *Sign stimulus:* The appearance of an object near the nest
>
> If the goose loses the egg during the retrieval process, it stops the head motion, but continues the "pulling" motion of retrieval. It must sit down before it notices the egg at which time another retrieval FAP is initiated.
>
> If an inappropriate object (toy dog, doorknob) is placed near the nest, the goose retrieves it but may not keep or incubate it.

Example 4: Protective behavior in hen turkeys

> *Fixed-action pattern:* Mothering behavior
>
> *Sign stimulus:* Cheeping sound of chicks
>
> A deaf turkey cannot hear the sign stimulus releasing mothering behavior and kills her chicks.

Example 5: The human infant grasping response is a fixed-action pattern released by a tactile stimulus. Human babies smile when they hear certain sounds, or see a figure consisting of two dark spots on a circle (rudimentary representation of a face).

Example 6: Female digger wasps place a paralyzed cricket in the nest as food for the young wasp after it hatches.

> *Fixed-action pattern:* She places the cricket 2.5 cm from the nest
>
> *Sign stimulus:* Nest site

Fixed-action pattern: She enters the nest and inspects it

Sign stimulus: Presence of cricket 2.5 cm from nest

Fixed-action pattern: She exits the nest and retrieves the cricket

Sign stimulus: Presence of cricket 2.5 cm from nest

If the cricket is moved during the nest inspection stage, the wasp will retrieve the cricket and repeat the FAP from the first step. She cannot get past the "inspect the nest" step if the cricket is not where she left it when she tries to retrieve it.

Sign stimuli are usually simple characteristics (e.g., ultrasonic bat sounds trigger avoidance behavior in moths).

Sign stimuli may be specific choices from an array of possibilities.

Natural selection favors cues associated with the relevant behavior or object. Some randomness is probable in fixing upon one of many possible sign stimuli for an FAP.

A *supernormal stimulus* is an artificial stimulus that may elicit stronger responses than natural stimuli.

- For example, when given a choice between an egg and a volleyball, a greylag goose ignores the egg and tries to retrieve the volleyball.

Most researchers now consider the concept of FAPs as overly simplistic because animal will often display variable responses to stimuli, depending on particular circumstances.

Modern behavioral biology is more concerned with understanding the adaptive function of behavior than with defining the precise nature of behavioral sequences.

An animal's sensitivity to general stimuli and its behavior are correlated.

- For example, frogs' retinal cells are sensitive to movement. Movement is the sign stimulus that releases the tongue-shooting FAP in frogs.

- A frog starves if surrounded by motionless flies.

D. Behavioral ecology emphasizes evolutionary hypotheses: *science as a process*

Fitness is a central concept in animal behavior. Since natural selection works on genetic variation caused by mutation and recombination, organisms should have features that maximize fitness over time. Animals are expected to engage in optimal behaviors.

Optimal behavior = A behavior that maximizes individual fitness

- Optimal behavior is a valid concept because behavior is genetically influenced and subject to natural selection.

- *Behavioral ecology* is a field of study that assumes animals increase fitness through optimal behavior.

Learned behaviors are typically based upon gene created neural systems that are receptive to learning.

1. Songbird repertoires

Bird songs can be analyzed with a sound spectrograph (see Campbell, Figure 51.4).

Why has natural section favored multisong behavior? From the perspective of behavioral ecology, several hypotheses could be formulated.

Example hypothesis: A repertoire increases fitness because it makes an older, more experienced male attractive to females. To determine if the hypothesis is correct, design testable predictions.

Prediction 1: Males learn more song types as they get older, so that repertoire size is a reliable indicator of age.

Experiment 1: Determine whether there is a correlation between male age and size of song repertoire.

(As it turns out, some species of birds display this correlation and others do not.)

Prediction 2: Females prefer to mate with males having large repertoires.

Experiment 2: Determine whether females are more sexually stimulated by a large song repertoire or a small one (see Campbell, Figure 51.5).

Such an examination may lead to an evolutionary explanation: bird song repertoires result in females mating more often or earlier in the season with experienced males, which will give their offspring a greater chance of survival.

2. Cost/benefit analysis of foraging behavior

Animals feed in many ways, using various foraging behaviors that are closely linked to morphological traits.

Food habits are fundamental to an animal's niche and may be shaped by interspecies competition and evolutionary factors.

Generalists feed on many items.

- They are not efficient collectors of any single food, but take advantage of multiple options when foods are scarce.
- Generalists concentrate on abundant prey.
- Generalists develop a *search image* for a favored item. If the item becomes scarce, a new search image is developed. Search images let generalists combine efficient short-term specialization with generalist flexibility.
    - *Search image* = The ability of a generalist feeder to learn the key visual characteristics of a prey item

Specialists feed on specific items and usually have highly specific morphological and behavioral adaptations. They are extremely efficient foragers.

Natural selection should favor foraging strategies that maximize gains and minimize costs in terms of calories gained and expended. Other criteria such as nutrient gain may be equally important.

- Foraging costs include the energy needed to locate, catch and eat food; the risk of being caught by a predator while feeding; time taken from other activities such as courtship and breeding.
- Behavioral ecologists analyze tradeoffs to predict optimal foraging strategies.
- Tradeoffs include density and size of prey versus foraging distances or prey catchability.
- For example, small mouth bass eat minnows and crayfish. Since no preference is shown, each may be optimal under different conditions. Minnows have more useable energy per unit weight than crayfish, but are harder to catch. Crayfish are easier to catch, but more difficult to subdue.

Animals modify behavior to keep the ratio of energy gain to loss high.

- This ability is probably innate, although learning may be involved.

Experiment: Bluegill sunfish eat small crustaceans (see Campbell, Figure 51.6) Optimal foraging theory predicts the proportion of small to large prey will vary with the density of the prey population. At low densities, sunfish should not be selective but at high densities, they should concentrate on larger prey.

Results: Sunfish were more selective at higher prey densities, but not to the extent predicted. Young fish were less efficient than adults. Younger fish may be less able to judge size and distance due to incompletely developed neural systems.

Conclusion: Maturation and learning may result in increased foraging efficiency in adults.

II.   Learning

A.   Learning is experience-based modification of behavior

Analyzing the genetic and environmental underpinning of behavior can help scientists understand the extent to which behavior can vary among individuals of a species.

Learning = The modification of behavior by experience

1.   Learning versus maturation

Individuals may improve behaviors over time. This is often attributed to learning, but in some cases, may be due to developmental changes in neuromuscular systems as animals mature.

Maturation = Development of neuromuscular systems that allows behavioral improvement

The distinction between learning and maturation may not be obvious.

> *Example:* Herring gull chick feeding behavior. The adult lowers its head and moves its beak. The chick pecks the red spot on the beak, causing the adult to regurgitate.
>
> *Fixed-action pattern:* Pecking the red spot on the beak
>
> *Sign stimulus:* Red spot swung horizontally at the end of a long, vertical object
>
> Newly hatched chicks will indiscriminately peck at a variety of objects, but chick that are 1 to 2 weeks old show a strong response to their parents' beaks (or models).

2.   Habituation

Habituation = Learning to ignore irrelevant stimuli or stimuli that convey little or no information

Animals stop responding to stimuli that do not provide appropriate feedback.

- Gray squirrels respond to the alarm calls of other squirrels. They stop responding if the calls are not followed by an attack ("cry-wolf" effect).

B.   Imprinting is learning limited to a critical to period

Imprinting is a form of learning that is limited to a specific time period in an animal's life and that is generally irreversible.

Konrad Lorenz conducted an experiment with greylag geese to see how offspring know whom or what to follow (see Campbell, Figure 51.7).

> *Experiment:* A clutch of goose eggs was divided between the mother and an incubator.
>
> *Results:* Goslings reared by the mother behaved normally and mated with other geese. The incubator goslings spent their first hours of life with Lorenz and preferred humans for the rest of their lives. They even tried to mate with humans.
>
> *Conclusions:* Greylags have no innate sense of "mother" or "gooseness". They identify with and respond to the first object with certain characteristics they encounter. The ability or tendency to respond is innate.

The object to which the response is directed is the imprinting stimulus. For Lorenz's geese, the imprinting stimulus was movement of an object away from the young.

Imprinting stimulus = An object in the environment to which the response is directed

- For example, salmon return to the stream they were hatched in to spawn. The imprinting stimulus is the unique chemical composition (odor) of the hatching stream.

1.   Critical period

Critical period = A limited time during which imprinting can occur

Imprinting is usually thought to involve very young animals and short critical periods, but imprinting may occur at different ages with critical periods of varying durations.

- Adults require a critical period to "know" their young. Prior to imprinting on their young, adult herring gulls defend strange chicks. After the imprinting period, they kill and eat strange chicks.

Sexual imprinting (or species identity) occurs later than parental imprinting and has a longer critical period.

- In one study, male finches reared by two different finch species imprinted on the other species when they developed a sexual identity. As a result, when exposed to females of their own species, they mated reluctantly.

While irreversibility and critical period characterize imprinting, they are not always fixed.

- The cross-fostered finches eventually mated with females of their own species.

2. Song development in birds: a study of imprinting

In most songbird species, males sing complex vocalizations that have a variety of adaptive functions.

Bird songs are a complex interplay of learned and innate behaviors.

- Some species have rigid song repertoires, while others have repertoires that are more fluid, have regional dialects, and change throughout an individual bird's life.

Many studies of song development have focused on common North American species. Early experiments with male white-crowned sparrows showed:

- Birds raised in soundproof chambers developed abnormal songs.
- Birds exposed to normal song recordings at 10 to 50 days old developed normal songs.
- Birds deafened after they were exposed to the recordings, but before they began to sing, developed songs more abnormal than birds reared in isolation (see Campbell, Figure 51.8).
- White-crowned sparrows learn to sing by hearing adults and matching those songs while listening to themselves.
- 10 to 50 days was the critical period for hearing adult song. Birds who heard a recording at 50+ days never sang properly.
- Birds exposed to other species' recordings did not learn those songs. There was an innate, genetic predisposition, for their own song.
- The innate predisposition was overcome by social interaction. Sparrows 50+ days old, when in social contact with another species, learned that species' song.
- The critical period may be flexible. It was longer for a live stimulus than for a recording.

Humans also have a critical period for learning vocalizations (easier while still teens than as adults).

Song development may be more or less fixed in other species:

- Song sparrows reared in isolation develop normal songs although males have larger repertoires if they hear other birds.
- Mockingbirds have repertoires of 150+ songs. The fitness value of a large repertoire is very high for them.

There are important exceptions to the song-learning scenario based on the white-crowned sparrows.

- Males canaries sing variable numbers of songs and develop new ones each year.
  - They have a region in their forebrain that varies greatly in size according to the season and number of songs in an individual's repertoire.
  - The fitness value of learning new songs more than once in their lifetime must be critically important.

C. Many animals can learn to associate one stimulus with another

Associative learning = The ability of many animals to associate one stimulus with another

Classical conditioning is a type of associative learning in which an arbitrary stimulus is associated with a reward or punishment.

- Russian physiologist Ivan Pavlov induced dogs to salivate when they heard a bell by associating it with powdered meat.

Operant conditioning, or trial-and-error learning, is another type of associative learning, in which an animal learns to associate one of its own behaviors with a reward or punishment and then tends to repeat or avoid that behavior.

- Psychologist B.F. Skinner put lab animals in a box with a variety of levers. Test animals learned to choose only those levers which yielded food.
- English tits learned to open milk bottles left on doorsteps and drink the cream when one or more of the birds discovered its probing behavior was rewarded when directed at the bottles.

Operant conditioning is common in nature.

- Predators learn to associate certain kinds of prey with painful experiences and modify their behavior accordingly (see Campbell, Figure 51.9).
- Genes influence the outcome of operant conditioning.

D. Practice and exercise may explain the ultimate basis of play

Play has no apparent goal but uses movements closely associated with goal-directed behaviors (see Campbell, Figure 51.10).

- Young predators playfully stalk and attack each other using motions similar to those used to capture and kill prey.
- Play occurs in the absence of distracting external stimuli.

Play is potentially dangerous or costly.

- Young vervet monkeys are at higher risk of being caught and eaten by baboons when they are at play.
- In a study of young goats, 1/3 sustained play injuries that resulted in limps.

What is the selective advantage of play?

- Practice hypothesis. Play is a type of learning that allows the perfection of survival behaviors. However, play movements rarely improve after the first few practices.
- Exercise hypothesis. Play keeps the cardiovascular and muscular systems in condition and is common in young animals because they do not exert themselves in other ways while they are under parental care. (However, recent studies of beluga whales and several species of dolphins indicate that play is also common in adults in captivity.)

III. Animal Cognition

Research efforts on animal cognition seek to understand information processing at all levels, from the nervous system activities that underlie sophisticated behavior, such as problem solving, to the internal representations animals have about physical objects in their surroundings.

A. The study of cognition connects nervous system function with behavior

Cognition is the ability of an animal's nervous system to perceive, store, process, and use information gathered by sensory receptors.

The study of animal cognition is called *cognitive ethology*. Its scientists attempt to illustrate the connection between data processing by nervous systems and animal behavior.

On area of research in cognitive ethology investigates how animal brains represent physical stimuli from the environment (this is separate and distinct from questions about consciousness).

B. Movement from place to place often depends on internal coding of spatial relationships

A central hypothesis of cognitive ethology is that animals make use of *cognitive maps*, internal representations of the spatial relationships among objects in the animal's environment.

Two kinds of movement that may occur without any internal representation are kineses and taxes.

Kinesis involves a change in activity rate in response to a stimulus.

- Sowbugs are more active in dry areas and less active in humid regions. This behavior tends to keep them in moist areas.

Taxis is a semiautomatic, directed movement toward or away from a stimulus.

- Housefly larvae are negatively phototactic after feeding. Presumably this makes them less visible to predators.

- Trout are positively rheotactic. Swimming against the current keeps them from being swept downstream.

1. Migration behavior

Migration is the most commonly known type of oriented animal movement.

- Migrants generally make an annual round trip between two regions (e.g., birds, whales, some butterflies, some pelagic fish) (see Campbell, Figures 51.12).

- *Migration* = Regular movement of animals over relatively long distances

Migrating animals use one of three mechanisms or a combination of these mechanisms to find their way.

1. *Piloting* = Movement of animals from one landmark to another

   - Is used over short distances

2. *Orientation* = Movement of animals along a compass line

   - Animals that use orientation can detect compass directions and travel in a straight-line path to a destination.

3. *Navigation* = The ability of animals who can orient along compass lines to determine their location in relation to their destination

   - Migrant starlings captured in the Netherlands were released in Switzerland. Juvenile birds oriented in a straight-line to Spain. Adults navigated a new route to their wintering grounds in northern Europe.

- Many birds use celestial points for orientation and navigation. These animals need an internal clock to compensate for the movement of the sun and stars. The indigo bunting avoids the need for an internal clock by fixing on the North star.
- Some birds, bees and bacteria orient to the Earth's magnetic field. The mechanisms are poorly known, but magnetite, an iron-containing ore, has been found in animals that orient to the magnetic field.

Campbell, Figure 51.13 distinguishes between orientation and navigation.

C. The study of consciousness poses a unique challenge for scientists

The extent to which nonhuman animals are consciously aware of the themselves or their environment is difficult to determine.

- Consciousness is known only to the individual that experiences it and it is not associated with any observable behavioral or physiological change.

Donald Griffin of Princeton University believes that:

- Conscious thinking is an inherent part of animal behavior.
- Cognitive ability arises through natural selection and forms a phylogenetic continuum stretching into evolutionary history.

IV. Social Behavior and Sociobiology

A. Sociobiology places social behavior in an evolutionary context

Most sexually reproducing species must be social for part of their life cycle in order to reproduce; some species spend most of their lives in close association with conspecifics.

Social behavior = Any interaction between two or more animals, usually of the same species.

- Includes aggression, courtship, cooperation, and even deception
- Has both costs and benefits to members of species that interact extensively

Sociobiology = Study of social behavior that has evolutionary theory as its conceptual framework

- Much of the evolutionary theory underlying sociobiology stems from the work of British biologist William Hamilton. He used the concepts of fitness and the genetic basis of behavior in analyzing the evolution and maintenance of social behavior in animals.
- In 1975, E.O. Wilson's *Sociobiology: The New Synthesis* was published, which helped form sociobiology into a coherent method of analysis and interpretation.

Because members of a population share a common niche, there is potential for conflict, especially among members of species that maintain densities near what the environment can sustain.

- Sometimes social behavior involves cooperative effort, as when a group accomplishes something more efficiently than a single individual (see Campbell, Figure 51.15).
- Even when cooperation seems to be mutually beneficial, each participant usually acts to maximize its fitness, even at a cost to the other participant.

B. Competitive social behaviors often represent contests for resources

1. Agonistic behavior

Agonistic behavior = A contest of threatening and submissive behavior that determines which competitor gains access to a resource (e.g., mate, food, territory)

- Canines show agonistic behavior by trying to look larger. They bare teeth; erect ears, tail and fur; stand upright; and make eye contact. The loser submits by sleeking its fur, tucking its tail and looking away.

Ritual behaviors are prevalent, so it is rare that participants are seriously injured (see Campbell, Figure 51.16)

Natural selection favors ending a contest as soon as a winner is established because further conflict could injure the victor as well as the vanquished. Future interactions between the same animals is usually settled more quickly in favor of the original victor.

2. Dominance hierarchies

Within a *dominance hierarchy*, the top-ranked member of a social group controls the behavior of the members of the group. The second-ranked animal controls everyone except the top individual and so on down the line to the lowest-ranked animal.

- Top ranked animals are assured access to resources. Low ranked animals do not waste energy or risk harm in combat.

Wolf packs typically have a female dominance hierarchy. The top female controls mating in the pack based on food availability.

3. Territoriality

Territories are defended areas used for feeding, mating, or rearing young.

- Territory size varies with the species, territory function, and the amount of resources available (see Campbell, Figure 51.17).
- Some species defend territories during the breeding season and form social groups at other times of the year (e.g., chickadees).

Territories are not home ranges. Home ranges are areas which animals inhabit but do not defend. Territories and home ranges may overlap.

Territories are usually successfully defended by their owners.

- Owners usually win because a territory is more valuable to the owner since he is familiar with it.
- Established territory owners are likely to be older and more experienced at agonistic interactions.
- Ownership is continually proclaimed—a primary function of bird song, red squirrel chattering and the bellowing of sea lions. Others use scent marks or patrols to announce their presence (see Campbell, Figure 51.18).

Defense is usually directed only at conspecifics who are most likely to compete directly for the same resources.

Dominance hierarchies and territoriality tend to stabilize population densities by assuring enough individuals reproduce to result in relatively stable populations from year to year.

C. Mating behavior relates directly to an animal's fitness

The correlation between mating behavior and reproductive fitness is vital to behavioral ecology.

1. Courtship

Species often have a complex courtship ritual unique to that species that must occur before mating.

Courtship assures each partner that the potential mate is not a threat, is the proper species, the proper sex and in the correct physiological condition.

In some species, courtship allows one or both sexes to choose a mate form a number of candidates.

- Females are usually more discriminating than males because they normally have a greater parental investment.

- *Parental investment* = The time and resources an individual expends to produce offspring
- Eggs are usually larger and more costly to produce than sperm.
- Gametes of placental mammals are closer in size, but females invest considerable time and energy carrying young before birth.

Competition among individuals of the same sex (usually males) may determine which individuals of that sex will mate.

- Most males mate with as many females as possible. They compete with other males for mates and may try to impress females.
- Males perform more intense courtship displays than females.
- Secondary sex characteristics may be highly developed in males (e.g., deer antlers, bird colors).

There are two ultimate bases for mate selection.

1. If the other sex gives parental care, it is best to choose the most competent mate.
   - For example, male common terns bring fish to potential mates as part of the courtship ritual. This behavior may be a proximate indicator of his ability to feed the chicks.
   - Some females prefer males with the most extreme and energetic courtship displays or secondary sex characteristics. These characteristics may be proximate indicators of the male's health.
2. Genetic quality is important when males provide no parental care and sperm are their only contribution to offspring.
   - *Lek* species have a communal area where males display. Females visit the lek and choose a mate. The proximate basis for her choice is a preference for males that court the most vigorously and have the most extreme secondary sex characteristics.

It may be difficult to determine if differential mating success among males is due to male-male competition, female choice, or both.

- Three-spined stickleback courtship is based on stereotyped releasers and FAPs (see Campbell, Figure 51.19) Despite this, the female can back out of the courtship anytime.
- Female choice is probably ultimately based on the quality of the male's parental care, because only male sticklebacks give parental care.

Ritualized acts probably evolved from actions whose meaning was more direct at one time. This is evident in insects called dance flies.

- The male of some species of dance flies spin oval silk balloons which they carry while flying in a swarm. The swarm is approached by females seeking mates. A female accepts a male's balloon and then they fly off to copulate.
- In a related species, the male brings a dead insect for the female to eat while they mate.
- In another species, the insect is presented inside a silk balloon, possibly because silk helps subdue the insect or makes it look larger.
- Males of some species eat mainly nectar, so the rituals may have evolved into bringing the female something associated with food.
- It is as if, over evolutionary time, a suitor wooed a lady with diamonds, then with a box containing diamonds, and finally with an empty box.

2. **Mating systems**

Mating relationships between males and females varies greatly among species.

- *Promiscuous* = A mating system with no strong pair-bonds or lasting relationships
- *Monogamous* = A mating system in which one male mates with one female
- *Polygamous* = A mating system in which an individual of one sex mates with several of the other
  - *Polygyny* is a polygamous relationship in which one male mates with multiple females.
  - *Polyandry* is a polygamous relationship in which one female mates with multiple males.

The needs of the young are an important ultimate factor in the evolution of mating systems.

- Most birds are monogamous. Young birds often require significant parental care. A male may ultimately increase his reproductive fitness by helping a single mate rear a brood than by seeking additional mates.
- Polygyny is common in birds where the young are able to care for themselves soon after hatching. Males can maximize their fitness by seeking additional mates.

Another factor influencing mating systems and parental care is the certainty of paternity (see Campbell, Methods box on red-winged blackbirds).

- Young born or eggs laid by a female definitely contain the female's genes, but even in monogamous species, the young could have been fathered by a male other than the female's normal mate.
- The certainty of paternity is relatively low in species with internal fertilization because mating and birth (or egg laying) are separated over time. Exclusive male parental care is rare in birds or mammals.
- Certainty of paternity is higher when egg laying and mating occur together, as in external fertilization. Parental care, when present, in fishes and amphibians is as likely to be by males as by females.
- When parental care is given by males, the mating system may be polygynous with multiple females laying eggs in a nest tended by a male.

D. Social interactions depend on diverse modes of communication

Communication = The intentional transmission of information between individuals

Behavioral ecologists assume communication has occurred when an act by a "sender" produces a change in the behavior of another individual, the "receiver".

Ethologists assumed communication evolved to maximize the quantity and accuracy of information. Behavioral ecologists argue that communication evolved to maximize the fitness of communicators.

Animals lie. Mimicry often is adaptive to the sender and maladaptive for the receiver.

- Male and female *Photinus* fireflies communicate by a characteristic pattern of flashes. Females of the predatory firefly genus *Photurus* mimic the female *Photinus* flash pattern, attracting male *Photinus* fireflies which they kill and eat.
- In some mammals, a new dominant male kills young born too soon to be his offspring. Without dependent young, females ovulate sooner, allowing the new male to father their young.
- Hanuman langur females in the early stages of pregnancy solicit copulations from new dominant males. When they give birth shortly before young fathered by these males would appear, they may deceive the male into treating their young as his own.

An evolutionary consideration is the mode used to transmit information. Animals use visual, auditory, chemical, tactile and electrical signals.

The mode used to transmit information is related to an animal's lifestyle.

- Most mammals are nocturnal and use olfactory and auditory signals.
- Animals that communicate by odors emit chemical signals called *pheromones*. Pheromones are important releasers for specific courtship behaviors. They are also used by ant scouts to guide other ants to food.
- Birds are mostly diurnal and use visual and auditory signals. Diurnal humans also use visual and auditory signals. If we could detect the chemical signals of mammals, then mammal sniffing might be as popular as bird watching.

A complex communication system is found in honeybees (see Campbell, Figure 51.21).

- Pheromones maintain the social order of the hive.
- To maximize foraging efficiency, worker bees communicate the location of food sources, which change as flowers bloom and new patches are found.
- In the 1940s, Karl von Frisch studied honeybee communication. He found individual bees communicated to other bees when they returned to the hive.
  - Returning bees "dance" to indicate the location of food.
  - If the source is < 50m, the bee does the "round dance", moving rapidly sideways in tight circles and regurgitating nectar. Workers leave the hive and forage nearby.
  - If the food is farther away, the bee does a "waggle dance", a half-circle swing in one direction, followed by a straight run and then a half-circle swing in the other direction. This dance seems to indicate both direction and distance.
    1. The angle of the run in relation to the vertical surface of the hive is the same as the horizontal angle of the food in relation to the sun.
    2. Distance to the food is indicated by variations in the speed at which a bee wags its abdomen during the straight run.

E. The concept of inclusive fitness can account for most altruistic behavior

Behavior that maximizes individual reproductive success will be favored by selection, regardless of how much damage such behavior does to another individual, local population, or species.

Animals occasionally exhibit apparently unselfish or altruistic behavior.

- *Altruism* = A behavior that reduces an individual's fitness and increases the fitness of the recipient of the behavior

Natural selection favors anatomical, physiological, and behavioral traits that increase reproductive success, which in turn propagates the genes responsible for those traits.

- When parents sacrifice their well-being to produce and aid offspring, they increase their fitness because it maximizes their genetic representation in the population.
- Like parents and offspring, siblings share half their genes, so selection might favor helping one's parents produce more siblings, or even helping siblings directly.
- Selection might result in animals increasing their genetic representation in the next generation by "altruistically" helping close relatives.

Inclusive fitness is the concept which describes the total effect an individual has on proliferating its genes by producing its own offspring and by providing assistance to the reproductive efforts of close relatives

- *Coefficient of relatedness* is the proportion of genes that are identical in two individuals because of common ancestry. The higher the coefficient of relatedness, the more likely an individual is to aid a relative.
- *Kin selection* is the mechanism of increasing inclusive fitness. The contribution of kin selection to inclusive fitness varies among species. It may be rare or nonexistent in species that are not social or disperse widely.

In predicting if an individual will aid relatives, behavioral ecologists have derived a formula that combines coefficients of relatedness, costs to the altruist, and benefits to the recipient.

If kin selection explains altruism, then examples of unselfish behavior should involve close relatives.

- Belding ground squirrels give alarm calls when danger appears. These calls alert other squirrels but increase the risk to the alarm givers. Females remain near their birth sites and are usually related to other members of the group. Only females give alarm calls (see Campbell, Figure 51.22)
- Sterile worker bees labor on behalf of a single fertile queen. They sting intruders, a behavior that defends the hive, but results in the death of the worker. The queen is the mother of all the bees in the hive.
- Nesting red-cockaded woodpeckers are aided by two to four nonbreeders that assist in all aspects including incubation and feeding young. Nest helpers are older offspring of the breeding pair or siblings of one parent that have been unable to establish a breeding territory. Helpers may eventually inherit the territory.
- In naked mole rats, DNA analysis has shown that all members of a colony are closely related. The queen is a sibling, daughter, or mother of the kings and the nonreproductive rats are the queens siblings or direct descendants. By enhancing the chances of a queen or king reproducing, a nonreproductive individual increases the chances of passing genes identical to its own to the next generation.

Altruistic behavior toward nonrelatives sometimes occurs. This behavior is adaptive if there is a reasonable chance of the aid being returned in the future.

- *Reciprocal altruism* only occurs in stable social groups where individuals have many opportunities to exchange aid (rare outside of humans).

F. Sociobiology connects evolutionary theory to human culture

Wilson's *Sociobiology* presented the thesis that social behavior has an evolutionary basis.

- Behavioral characteristics are expressions of genes favored by natural selection.
- This has sparked debate about the connection between biological evolution and human culture
  - An example of the debate involves cultural taboos on incest.
    - Incest avoidance is adaptive because inbreeding may increase the frequency of genetic disorders.
    - Many species avoid incest.
    - Most human cultures have taboos forbidding incest.
    - Is there an innate aversion to incest or is this an acquired behavior?
      - The argument in favor of the "nurture" or learned behavior position says that cultural taboos are unnecessary if the behavior is innate, therefore, incest avoidance is a learned behavior and the social stigma attached to incest is based on experience.

- The argument in favor of the "nature" or genetic behavior position says that the occurrence of incest taboos in many cultures is evidence for an innate component and taboos are simply proximate mechanisms that reinforce a behavior that ultimately evolved because of its effect on fitness.

REFERENCES

Alcock, J. *Animal Behavior: An Evolutionary Approach*. 4th ed. Sunderland, Mass.: Sinauer, 1989.

Campbell, N., et al. *Biology*. 5th ed. Menlo Park, California: Benjamin/Cummings, 1999.

Krebs, J.R. and N.B. Davies (editors). *Behavioral Ecology: An Evolutionary Approach*. 2nd ed. Sunderland, Mass.: Sinauer, 1984.

Wilson, E.O. *Sociobiology: The New Synthesis*. Cambridge, Mass.: Harvard University Press, 1975.

CHAPTER 52
POPULATION ECOLOGY

OUTLINE

I. Characteristics of Populations
   A. Two important characteristics of any population are density and the spacing of individuals
   B. Demography is the study of factors that affect the growth and decline of populations
II. Life History Traits
   A. Life histories are highly diverse but exhibit patterns in their variability
   B. Limited resources mandate trade-offs between investments in reproduction and in survival
III. Population Growth Models
   A. An experimental model of population growth describes an idealized population in an unlimited environment
   B. A logistic model of population growth incorporates the concept of carrying capacity
IV. Regulation of Population Growth
   A. Density-dependent factors regulate population growth by varying with the density
   B. The occurrence and severity of density-independent factors are unrelated to population density
   C. A mix of density-dependent and density-independent factors probably limits the growth of most populations
   D. Some populations have regular boom and bust cycles
V. Human Population Growth
   A. The human population has been growing almost exponentially for centuries but will not be able to do so indefinitely

OBJECTIVES

After reading this chapter and attending lecture, the student should be able to:

1. Define the scope of population ecology.
2. Distinguish between density and dispersion.
3. Explain how ecologists measure density of a species.
4. Describe conditions which may result in clumped dispersion, random dispersion, and uniform dispersion of populations.
5. Explain how age structure, generation time, and sex structure of populations can affect population growth.

6. Describe the characteristics of populations which exhibit Type I, Type II, and Type III survivorship curves.
7. Explain how carrying capacity of the environment affects the intrinsic rate of increase of a population.
8. Explain how density-dependent factors affect population growth.
9. Describe how weather and climate can function as density-independent factors in controlling population growth.
10. Explain how density-dependent and density-independent factors may work together to control a population's growth.
11. List the three major characteristics of a life history and explain how each affects the:
    a. Number of offspring produced by an individual
    b. Population's growth
12. Explain how predation can affect life history through natural selection.
13. Distinguish between r-selected populations and K-selected populations.
14. Explain how a "stressful" environment may alter the standard r-selection and K-selection characteristics.

KEY TERMS

population	birth rate	zero population growth	opportunistic
density	fecundity	intrinsic rate of increase	populations
dispersion	death rate	exponential population	intraspecific
mark-recapture method	generation time	growth	competition
clumped	sex ratio	carrying capacity	density-dependent
grain	life table	logistic population	factor
uniform	survivorship curve	growth	density-independent
biogeography	life history	K-selected populations	factor
demography	semelparity	equilibrial populations	cohort
age structure	iteroparity	r-selected populations	

LECTURE NOTES

Population ecology is concerned with measuring changes in population size and composition and identifying the factors that cause these changes.

No population can continue to grow indefinitely.

- Many populations remain relatively stable over time with only minor increases and decreases.
- Other populations show dramatic increases followed by equally dramatic decreases.

I. Characteristics of Populations

Population = Individuals of one species simultaneously occupying the same general area, utilizing the same resources, and influenced by similar environmental factors

A. Two important characteristics of any population are density and the spacing of individuals

Every population has geographical boundaries and a population size.

Two important characteristics of populations are *density* and *dispersion*.

Population density = The number of individuals per unit area or volume

Population dispersion = The pattern of spacing among individuals within the geographical boundaries of the population

1. Measuring density

It is usually impractical or impossible to count all individuals in a population, so ecologists use a variety of sampling techniques to estimate densities and total population size.

- May count all individuals in a sample of representative plots; estimates become more accurate as sample plots increase in size or number

- May estimate by indirect indicators such as number of nests or burrows, or by droppings or tracks

- May use the mark-recapture method (see Campbell, Methods box)

  - In the *mark-recapture method*, animals are trapped within boundaries, marked in some way, and after time, retrapped.

    The number of individuals in a population (N) is estimated by the formula:

    $$N = \frac{(\text{number marked}) \times (\text{total catch the second time})}{\text{number of marked recaptures}}$$

    Assumes marked individuals have same probability of being trapped as unmarked individuals. This assumption is not always valid.

2. Patterns of dispersion

A population's geographical range is the geographic limits within which a population lives.

Local densities may vary substantially because not all areas of a range provide equally suitable habitat.

Individuals exhibit a continuum of three general patterns of dispersion in relation to other individuals: clumped, uniform, and random.

1. A *clumped* pattern is when individuals are aggregated in patches.

   May result from the environment being heterogeneous, with resources concentrated in patches; the ecological concept of *grain* relates to spatial or environmental patchiness.

   - A coarse-grained environment has large environmental patches that organisms can distinguish and choose among.

   - A fined-grained environment has small patches relative to the size and activity of an organism and they organisms may behave as though patches do not exist.

   - Temporal variation in the environment can also be coarse- or fine-grained.

   Clumping may be associated with mating or other social behavior in animals.

   May also be associated with defense against predators.

2. A *uniform* pattern is when the spacing of individuals is even.

   May result from direct, antagonistic interactions between individuals of the population. For example, competition for some resource or social interactions that set up individual territories for feeding, breeding or nesting.

3. A *random* pattern is when individual spacing varies in an unpredictable way.

   Occurs in the absence of strong attractions or repulsions among individuals.

   Not very common in nature.

While the above applies to local dispersion patterns within populations, it is important to remember that populations within a species also show dispersion patterns.

- Such populations often concentrate in clusters within the species' range.
- *Biogeography* is the study of factors that influence the distribution of a species over its range.

B. Demography is the study of factors that affect birth and death rates in a population

Demography is the study of the vital statistics affecting population size.

- Reflects relative rates of processes that add individuals to a population (birth, immigration) versus processes that eliminate individuals (death, migration).
- Birth and death rates vary among population subgroups depending on age and sex.
- A population's age structure and sex ratio are two of its most important demographic features.

1. Age structure and sex ratio

Many populations have overlapping generations where individuals of more than one generation coexist.

- Exceptions include species in which all of the adults reproduce at the same time and then die (e.g., annual plants, many insects).
- The coexistence of generations produces an age structure in most populations.
  - *Age structure* = Relative numbers of individuals of each age in a population
- Generations overlap when average life span is greater than the time it takes to mature and reproduce.

Every age group has a characteristic birth and death rate.

- *Birth rate*, or *fecundity*, is greatest for those of intermediate age.
- *Death rate* is often greatest for the very young and very old.
- In general, a population with more older, nonreproductive individuals will grow more slowly than a population with a larger percentage of young, reproductive age individuals.

Generation time is an important demographic feature related to age structure.

- *Generation time* = The average span of time between the birth of individuals and the birth of their offspring
- Strongly related to body size (see Campbell, Figure 52.3)
- A shorter generation time usually results in faster population growth, assuming birth rate is greater than death rate and all other factors being equal.

Another important demographic factor affecting population growth is the sex ratio.

- *Sex ratio* = The proportion of individuals of each sex found in a population
- The number of females is usually directly related to the expected number of births.
- The number of males may be less significant since one male may mate with several females.
- In strictly monogamous species, the number of males is more significant in affecting the birth rate than in nonmonogamous species.

2. Life tables and survivorship curves

Life tables describe how birth rates and death rates vary with age over a time period corresponding to maximum life span.

- Constructed by following the fate of a *cohort*, a group of individuals of the same age, from birth until all are dead or by using the age specific birth and death rates in a population during a specified time.

Various information about change in population size can be obtained from life tables (see Campbell, Table 52.1).

Survivorship curves plot the numbers in a cohort still alive at each age. Organisms may exhibit one of three general types of survivorship curves (see Campbell, Figure 52.4):

1. Type I curves are flat during early and middle life and drops suddenly as death rates increase among the older individuals.
   - Associated with species such as humans and other large mammals that produce few offspring that are well cared for
2. Type II curves are intermediate with mortality being more constant over the life span.
   - Seen in *Hydra*, gray squirrels, and some lizards
3. Type III curves show very high death rates for the young followed by lower death rates after individuals have survived to a certain critical age.
   - Associated with organisms, such as oysters, that produce very large numbers of offspring but provide little or no care

Many species exhibit curves between the basic types of survivorship curves and some have more complex curves.

- Great tits show a high mortality rate in young birds (Type III) but a fairly constant mortality (Type II) in adults.
- Some invertebrates show a "stair-stepped" curve with brief periods of high mortality during molts, followed by periods of lower mortality when the exoskeleton is hard (e.g., crabs).

II. Life History Traits

Natural selection, working over evolutionary time, results in traits that affect an organism's life history.

- *Life history* = An organism's schedule of reproduction and death

In many cases there are trade-offs between survival and traits such as frequency of reproduction, investment in parental care, and the number of offspring per reproductive episode.

Life histories are important in population ecology because these traits affect population growth over time.

A. Life histories are highly diverse but exhibit patterns in their variability

There is a diversity in life histories due to the varying pressures of natural selection.

- Some animal species hatch in one type of biome, migrate to another where they mature for several years, then return to the initial biome for a single massive reproductive effort, then die (e.g., salmon).
- Other animal species hatch and mature rapidly in a single habitat, then have small reproductive efforts each year for several years (e.g., lizards).
- Life history characteristics may vary significantly among populations of the same species or even among individuals within a population.

Even though life history traits vary widely, there are some basic patterns:

- Life histories often vary in parallel with environmental factors. For example, clutch sizes increase with latitude.
    - Tropical birds lay fewer eggs than those in higher latitudes, which reflects the number of offspring that can be successfully fed.
    - Since daylengths are longer at higher latitudes than the tropics during offspring-rearing season, parent birds can gather more food and feed more offspring.
    - The number of offspring per reproductive event has been found to vary in many mammals, lizards, and insects.
- Life history traits often vary in relation to one another. For example, fecundity and mortality tend to vary in close association among birds.
    - Albatrosses only average one surviving offspring every five years but adult birds have only a 5% chance of dying from one breeding season to the next.
    - Tree sparrows average six surviving offspring each year but have over a 50% chance of dying between breeding seasons.
    - Delayed maturation and high parental investment in offspring tend to be correlated with low fecundity and low mortality (see Campbell, Figure 52.5).

B. Limited resources mandate trade-offs between investments in reproduction and survival

A life history based on heritable traits that result in producing the most reproductively successful descendants will become more prevalent in a population.

A successful life history resolves the conflict between limited resources and competing functions.

- Time, energy, and nutrients used for one function are not available for other functions (see Campbell, Figure 52.6).
- The integrated life histories seen in natural populations balance the investment in the number of offspring produced against the prospects of future reproductions.
    - In general, organisms that produce fewer offspring during a reproductive effort survive longer and have more reproductive episodes.
- The life history traits exhibited by an organism are evolutionary outcomes reflected in the development and physiology of that organism.

1. Number of reproductive episodes per lifetime

Two extremes are found in life history strategies where there is a trade-off between fecundity and survival probability.

1. *Semelparity* is a type of life history in which organisms invest most of their energy in growth and development, then expend that energy in a single reproductive effort before dying.
2. *Iteroparity* is a type of life history in which organisms produce fewer offspring at a time over many reproductive seasons.

Semelparity is expected when there is a high cost to parents to stay alive between broods or if there is a trade-off between fecundity and survival.

- Seen in annual plants, salmon, and some perennial plants such as bamboo and the century plant.
- In the harsh desert climates, most plants live only a single season and put all their energy into one reproductive effort.

- Rarely found in animals or plants that live more than one or two years, although some (e.g., century plants) live for several seasons before investing all their energy into a single reproduction (see Campbell, Figure 52.7).
  - Century plants live in arid climates with unpredictable rainfall.
  - They grow vegetatively and store nutrients for several years.
  - The stored nutrients are used to produce seeds when an unusually wet year occurs that favors offspring survival.

Iteroparity is expected when established individuals are likely to survive but mortality is high in immature individuals.

- Iteroparous plants are more common in the tropics where competition and predation hinder seedling establishment, but mature plants survive for many seasons.
- More resources are invested in survival (e.g., root system, resistant buds) to prepare for multiple breeding episodes than is allocated by single season plants.

2. Number of offspring per reproductive episode

Organisms with a low probability of surviving to the next reproductive season usually invest more energy into producing a large number of offspring.

Organisms with a high probability of survival invest less energy and produce fewer offspring.

Clutch, litter, or seed crop size may vary seasonally within a single population in some species (see Campbell, Figure 52.8).

There is usually a trade-off between the number and quality of offspring produced.

- In general, organisms that produce many offspring produce small ones (see Campbell, Figure 52.9).
  - Each offspring thus starts with a limited amount of energy.
  - A large number of offspring and small young are typical of organisms with a Type III survivorship curve.
    - Typical of organisms that colonize disturbed or harsh environments or are subject to high levels of predation.
  - Organisms that produce small clutches, litters, or seed crops generally have larger offspring.
    - Offspring have a longer initial amount of energy which improves their chance of survival.
    - This is typical of organisms with Type I and II survivorship curves.
    - Parental investment is higher but is offset by increased survival of offspring.

a. Age at first reproduction

The timing of the first reproduction greatly influences the female's lifetime reproductive output in organisms that have several reproductive episodes during their lifespan.

- Balances the cost between current reproduction and survival plus future reproduction.
- Reproduction at a younger-than-average age may reduce a female's reproductive potential by reducing the amount of energy available for growth and maintenance.

- Female's that delay reproduction tend to be larger due to energy used for growth and maintenance; older (larger) females produce larger clutches and appear to maximize their reproductive output by delaying.

III. Population Growth Models

Indefinite increases in population size do not occur.

- A population may increase rapidly from a low level under favorable environmental conditions, but this increase in numbers will eventually approach the level where resources cannot support continued increases.
- The combination of limited resources and other factors will stop the growth of the population.

A. An exponential model of population growth describes an idealized population in an unlimited environment

A combination of observation, experimentation, and mathematical modeling is used to determine answers to many ecological questions.

- Birth rates and death rates can be quantified, by measurement or observation, in many populations and used to predict changes in the population size.
- Laboratory studies on small animals can determine how various factors affect population growth rates and a few natural populations can be manipulated experimentally to answer a range of questions.
- Mathematical models can test hypotheses about the effects of various factors on population growth in organisms that are difficult or impossible to study experimentally.

A population consisting of a few individuals that live in an environment without limiting factors (no restrictions on available energy, growth, or reproduction) will increase over time in proportion to the birth and death rates:

$$\text{Change in population size during time interval} = \text{Births during time interval} - \text{Deaths during time interval}$$

In a mathematical form, this equation becomes:

$$\Delta N / \Delta t = B - D$$

where N = population size, t = time, B = absolute number of births during the time interval, and D = absolute number of deaths during the time interval, ΔN = change in population size, Δt = the time interval (lifespan or generation time).

Since populations differ in size, the basic model must be altered in order to apply it to *any* population. This alteration involves a conversion from absolute birth and death rates to average numbers of births and deaths per individual during the specified time interval.

- If b = the annual per capita birth rate, its value would equal 0.034 in a population of 1000 individuals where 34 births occurred (34/1000 = 0.034) in a year.
- If d = per capita death rate, its value would equal 0.016 in a population of 1000 individuals where 16 deaths occurred (16/1000 = 0.016) in a year.

Including average (or per capita) birth and death rates alters the previous equation so that expected numbers of births and deaths can be predicted for a population of any size:

$$\Delta N / \Delta t = bN - dN$$

- The change in size of our 1000 individual population would be:

$$\Delta N / \Delta t = (0.034)(1000) - (0.016)(1000)$$

$\Delta N/\Delta t = 34 - 16$

$\Delta N/\Delta t = 18$

- If another population of the same species (same b and d values) contained 1500 individuals, then:

$\Delta N/\Delta t = (0.034)(1500) - (0.016)(1500)$

$\Delta N/\Delta t = 51 - 24$

$\Delta N/\Delta t = 27$

Population ecologists study overall changes in population sizes and use r to represent the difference between per capita birth rates and per capita death rates:

$r = b - d$ thus $\Delta N/\Delta t = rN$

- The value r is thus the per capita population growth rate and can be used to determine if a population is growing or declining.

- *Zero population growth (ZPG)* occurs where per capita birth and death rates are equal, thus $r = 0$. (Note that births and deaths still occur, but are equal in number.)

- A population is increasing in size if $r > 0$ (more births than deaths); it is decreasing in size if $r < 0$ (fewer births than deaths).

Many population ecologists use a slightly different equation based on differential calculus to express population growth in terms of instantaneous growth rates:

$dN/dt = rN$

- This expresses the population growth over very short time intervals.

A population living under ideal conditions will increase at the fastest rate possible; nutrients are abundant and only the physiological capacity of the individuals limits reproduction.

- The maximum population growth rate is called the *intrinsic rate of increase* and is symbolized by r_{max}.

Exponential population growth is the population increase under ideal conditions due to intrinsic rate of increase. It is expressed as:

$dN/dt = r_{max}N$

- The size of the population increases rapidly due to ideal conditions of unlimited resources.

- Produces a J-shaped growth curve (see Campbell, Figure 52.11)

- Although the intrinsic rate of increase is constant, more new individuals accumulate when the population is large than when it is small; this is due to the fact that N gets larger.

A population with a higher intrinsic rate of increase will grow faster than one with a lower rate of increase (see Campbell, Figure 52.11).

- A population's r_{max} value is influenced by its life history features (age of first reproduction, clutch size, offspring survival rate).

- Generation time and r_{max} are usually inversely related over a range of species (see Campbell, Figure 52.12).

- Exponential growth is characteristic of populations introduced into a new or unfilled environment or whose numbers have been decimated by a catastrophe and are rebounding.

B. A logistic model of population growth incorporates the concept of carrying capacity

No population can grow exponentially for an indefinite period of time. As a population increases in size, the higher density may influence the ability of individuals to obtain resources necessary for maintenance, growth and reproduction.

- Populations inhabiting environments which contain a finite amount of available resources reach a size where further increases in number reduces the share of resources available to each individual.

The *carrying capacity (K)* of a habitat is the maximum stable population size that the particular environment can support over a relatively long time period.

- Carrying capacity is an environmental property that varies over space and time with the abundance of limiting resources.
- Carrying capacities can be determined by many factors:
    - Energy limitations (food resources) are the most common determinant of *K*.
    - Other factors include the availability of specialized nesting sites required by some birds; roosting sites as for some bats; shelters and refuges from potential predators.

Crowding and resource limitation can greatly effect the population growth rate.

- Insufficient resources may reduce per capita birth rates in a population (lower *b*).
- If enough energy cannot be obtained for maintenance, per capita death rates increase (higher *d*).
- A decrease in *b* and/or an increase in *d* results in a lower overall population growth rate (smaller *r*).

1. The logistic growth equation

A *logistic population growth* model assumes the rate of population growth (*r*) slows as the population size (*N*) approaches the carry capacity (*K*) of the environment.

A mathematical model for logistic population growth incorporates the effect of population density on *r*, allowing it to vary from r_{max} when resources are plentiful to zero when the carrying capacity is reached.

The equation for logistic population growth is:

$$dN/dt = r_{max}N \left(\frac{K-N}{K} \right.$$

- *K* = carrying capacity; the maximum sustainable population.
- *K − N* = the number of new individuals the environment can accommodate.
- *(K − N)/K* = percentage of *K* available for population growth.
- Multiplying r_{max} by *(K − N)/K* reduces the value of *r* as *N* increases (see Campbell, Table 52.2)
- The actual growth rate of a population of any size is r_{max} *N(K − N)/K*.

The implications of the logistic growth equation at varying population sizes for a growing population are:

- When *N* is low, *(K − N)/K* is large and *r* is only slightly changed from r_{max}.
- When N is large and resources are limiting, (K − N)/K is small; this reduces *r* substantially from r_{max}.
- When *N = K*, *(K − N)/N* is 0 and *r* = 0; this means the number of births is equal to the number of deaths and zero population growth occurs.

The logistic model of population growth produces a sigmoid (S-shaped) growth curve (see Campbell, Figure 52.14).

- Intermediate population sizes add new individuals most rapidly since the breeding population is of a substantial size and the habitat contains plentiful amounts of resources and available space.
- As N approaches K, the population growth rate slows due to limitations in available resources.
- The logistic model is density-dependent since the rate at which a population grows changes as the density of the population changes.

2. How well does the logistic model fit the growth of real populations?

Laboratory populations of paramecia and *Daphnia*, as well as other animals, show population growth rates which fit the predicted S-shaped curve fairly well (see Campbell, Figure 52.15).

- These represent relatively unnatural situations of idealized conditions without predators and other species.
  - Even under these conditions, some populations show deviations from smooth, sigmoid curves and do not stabilize at a clear carrying capacity.
- Studies of wild populations that have been introduced into new habitats and populations rebounding after near elimination by disease or hunting, provide general support for logistic population growth (see Campbell, Figure 52.15c).

Some assumptions of the logistic model do not hold true for all populations. For example, the logistic model assumes that:

1. Even at low levels, each individual added may have the same negative effect on population growth rate; any increase in N reduces $(K - N)/N$.
  - The Allee effect points out that individuals may benefit by population increase which improves the chances for survival or reproduction.
    - For example, a plant standing alone would suffer from excessive wind, but be protected from the wind in a clump of individuals.
    - Solitary animals like rhinoceros have a greater chance of locating a mate during breeding season if populations are higher.
  - When a population is at low levels, there is a greater possibility that chance events will eliminate all individuals or inbreeding will lead to reduction in fitness.
2. Populations approach carrying capacity smoothly.
  - Often see a lag time before the negative effects of an increasing population are realized, this causes the population to overshoot carrying capacity.
    - Eventually, deaths exceed births and population size drops below carrying capacity.
    - Many populations thus appear to oscillate about a general carrying capacity.

Populations do not necessarily remain at, or even reach, levels where population density is an important factor. In these cases, the idea of carrying capacity does not really apply.

- Seen in short-lived, quickly reproducing insects and other small organisms sensitive to environmental fluctuations

Although the logistic model fits few if any real populations, it incorporates ideas that apply to many.

3.   Population growth models and life histories

The logistic population growth model predicts that there will be different growth rates for populations with high and low densities in relation to the environmental carrying capacity.

- Each individual has fewer resources available at high densities and the population is growing slowly.
- Abundant resources are available to individuals at low densities and the population is growing rapidly.

The concept that different life history adaptations would be favored under high densities and low densities was introduced by ecologist Martin Cody in the late 1960s. For example,

- Under high population densities, selection would favor adaptations that enhanced survival and reproduction with few resources.
  - Competitive ability and maximum efficiency of resource utilization would be favored in a population that was maintained at or near the carrying capacity.
- Under low population densities, selection would favor adaptations that enhanced rapid and high rates of reproduction regardless of efficiency.
  - Increased fecundity and reaching maturity quickly would be favored.

The development of Cody's concept lead to designations of *K-selected* (also called *equilibrium populations*) and *r-selected* (also called *opportunistic populations*) traits of life history strategies.

- The designations *r* and *K* refer to variables in the logistic growth model equation.
- *K*-selected populations are those living at a density near the limits of their resources (*K*).
- *r*-selected populations are more likely to be found in variable environments where population densities fluctuate or in open or disturbed habitats where individuals have little competition.
- Most populations show a mixture of *r*-selected and *K*-selected traits since life history evolves in a natural setting of complex factors (see Campbell, Table 52.3).

IV.   Population Limiting Factors

Populations are regulated by density-dependent factors and density-independent factors, either separately or in combination.

- The relative importance of these factors differs among species and their specific circumstances.

A.   Density-dependent factors regulate population growth by varying with the density

The prime implication of the logistic growth model is that increasing population density reduces resource availability and resource limitations ultimately limits population growth.

- This model can thus be applied to *intraspecific competition*: the reliance of two or more individuals of the same species on the same limited resource.
- Competition becomes more intense as population size increases, and *r* is reduced in proportion to the intensity of competition.

In restricting population growth, a *density-dependent factor* intensifies as the population size increases, affecting each individual more strongly. They also affect a greater percentage of individuals in a population as the number of individuals increases.

- Population growth declines because death rate increases, birth rate decreases or both.
- Resource limitation is one such factor.
  - A reduction in available food often limits reproductive output as each individual produces fewer eggs or seeds (see Campbell, Figure 52.16).
- Resources other than nutrients may also limit populations.
  - Territoriality (the defense of a well-bounded physical space) is a behavioral mechanism to reduce intraspecific competition since each individual protects resources only within their own territory.
  - Competition still occurs when individuals compete for space to establish their territories.
- Health and survivorship also decrease as crowding results in smaller, less robust individuals (see Campbell, Figure 52.17).
- Many predators concentrate on a particular prey when its population density is high, taking a greater percentage than usual.
  - Predators may switch to other more dense prey populations if energy expenditure to capture prey increases.
- The accumulation of toxic metabolic wastes may also limit a population.

Intrinsic factors may also play a role in regulating population size.

- Population growth rate decreases may occur even when food and shelter are abundant.
- High densities may cause stress syndromes resulting in hormonal changes that delay sexual maturation or otherwise inhibit reproduction.
- High densities have also been shown to produce a stress syndrome which suppresses the immune system.
- High densities can thus reduce birth rates and increase death rates.

B. The occurrence and severity of density-independent factors are unrelated to population density

Density-independent factors are unrelated to population size and affect the same percentage of individuals regardless of the size of the population.

- Weather, climate and natural disasters such as freezes, seasonal changes, hurricanes and fires are examples.
  - The severity and time of occurrence is the determining factor on what proportion of the population is affected.
  - In some natural populations, these effects routinely control population size before density-dependent factors become important.

C. A mix of density-dependent and density-independent factors probably limits the growth of most populations

Over the long term, species' populations exhibit varied dynamics.

Many populations remain fairly stable in size, close to the carrying capacity determined by density-dependent factors.

- Although exhibiting stability, they often display short-term fluctuations due to density-independent factors.
- Long-term population size may remain the same, masking short-term effects of some factor such as an extremely cold winter.

Density-dependent and density-independent factors sometimes work together to regulate a population, although the relative importance of each may vary seasonally.

- A severe winter may greatly reduce a population due to cold temperatures (density-independent) and intraspecific competition for limited food (density-dependent).
- This reduction in population size may benefit the surviving adults by reducing competition for food in the following spring.

D. Some populations have regular boom and bust cycles

Some bird, mammal and insect populations show a regular fluctuation in density. Among mammals:

- Small herbivores (e.g., lemmings) show 3- to 5- year cycles.
- Larger herbivores (e.g., snowshoe hares) show 9- to 11-year cycles.

Several hypotheses have been proposed to explain population cycles:

Crowding may regulate cyclical population by affecting the organisms endocrine systems.

- Stress from high density may alter hormone balance and reduce fertility, increase aggressiveness, and induce mass emigrations.

Population cycles may result from a time lag in the response to density-dependent factors. This lag causes the population to overshoot and undershoot the carrying capacity.

- A high density of snowshoe hares may cause a deterioration of food quality which makes it unfit for consumption.
- Predation may also play a role if predators take enough prey to cause a decrease in prey population density.

Much longer cycles are also known. The periodical cicadas have population cycles of 13 or 17 years.

- This may be an adaptation to reduce predation.
- Cicada populations are controlled in some local regions by a fungus whose spores can survive in soil for the years between cicada outbreaks.

V.   Human Population Growth

A. The human population has been growing almost exponentially for centuries but will not be able to do so indefinitely

The exponential growth of the human population has caused severe environmental degradation.

- Until 1650, the human population increased slowly.
- The population doubled by 1850 (200 years), doubled again by 1920 (80 years), and doubled yet again by 1975 (45 years) (see Campbell, Figure 52.21).

Human population growth is affected by the same parameters (birth and death rates) that affect other plant and animal populations.

- The advent of agriculture 10,000 years ago increased birth rates and decreased death rates.
- Since the Industrial Revolution, virtually exponential growth has resulted mainly from a drop in deaths, especially infant mortality.
  - The decrease was due to improved nutrition, better medical care, and improved sanitation.
- The population growth rates in most developing countries are actually increasing due to relatively high birth rates coupled with decreasing mortality.

Population ecologists cannot agree on Earth's carrying capacity for the human population or what factor will eventually limit the population.

- Limited nutrients is the usual factor in restraining populations of other animals, but agricultural technology has advanced to the point that food supplies have maintained an equivalent pace to population increases. Malnutrition and famine result from unequal distribution as opposed to inadequate production.
- Space limitations may play a role. As the human population continues to increase, conflicts over land use will intensify.
- Resources other than nutrients and space may play a role.
  - How will the disappearance of nonrenewable fossil fuels and metals affect the human population?
  - Will the accumulation of pollutants and other forms of wastes eventually reach toxic levels?

Because of the interrelatedness of food production, land use, and energy consumed in producing food, it is likely that the carrying capacity for humans will be determined by several interacting factors.

Worldwide population growth is a mosaic with some countries having near zero population growth while others have relatively high growth rates.

- Although variation in growth rates is found, the human population as a whole continues to grow.

Age structure within each country appears to be a major factor in the variation of population growth rates (see Campbell, Figure 52.23)

- A relatively uniform age distribution, where individuals of reproductive age or younger are not disproportionately represented, lends itself to a stable population size.
- A population which contains a large proportion of reproductive age or younger individuals will face a sudden increase in the rate of population growth in the future.

A unique feature of human reproduction is the ability to be consciously controlled by voluntary contraception or government-sponsored family planning.

- Social change, such as women delaying reproduction in favor of employment and advanced education, can also result in decreases in population growth rates.

Human cultural evolution has had an impact on Earth's carrying capacity.

- The development of agriculture and industrial technology has twice increased the Earth's carrying capacity for humans.
- What Earth's carrying capacity is and how we approach it are questions generating concern and debate.

The human population will eventually stop growing due to social changes, individual choice, government intervention, or increased mortality due to environmental limitations.

- Hopefully, if the human population reaches K, it will approach it smoothly and level off.
  - This will occur when birth rates and death rates are equal.
- If the population fluctuates around K, periods of increase will be followed by times of mass death due to famine, disease, or conflict.

REFERENCES

Campbell, N., et al. *Biology*. 5th ed. Menlo Park, California: Benjamin/Cummings, 1999.

Krebs, C.J. *Ecology: The experimental analysis of distribution and abundance*, 4th ed. New York, NY: HarperCollins College Publishers, 1994.

Ricklefs, R.E. *Ecology*. 3rd ed. New York: Chiron Press, 1990.

OUTLINE

I.   Early Hypotheses of Community Structure

    A. The interactive and individualistic hypotheses pose alternative explanations of community structure: *science as a process*

II. Interactions between Populations of Different Species

    A. Intraspecific interactions can be strong selection factors in evolution

    B. Interspecific interactions may have positive, negative, or neutral effects on a population's density: *an overview*

    C. Predation and parasitism are (+ / −) interactions: *a closer look*

    D. Interspecific competitions are (− / −) interactions: *a closer look*

    E. Commensalism and mutualism are (+ / 0) and (+ / +) interactions, respectively: *a closer look*

III. Interspecific Interactions and Community Structure

    A. Predators can alter community structure by moderating competition among prey species

    B. Mutualism and parasitism can have community-wide effects

    C. Interspecific competition influences populations of many species and can affect community structure

    D. A complex interplay of interspecific interactions and environmental variability characterizes community structure

IV. Disturbance and Nonequilibrium

    A. Nonequilibrium resulting from disturbance is a prominent feature of most communities

    B. Humans are the most widespread agents of disturbance

    C. Succession is a process of change that results from disturbance in communities

    D. The nonequilibrial model views communities as mosaics of patches at different stages of succession

V.   Community Ecology and Biogeography

    A. Dispersal and survivability in ecological and evolutionary time account for the geographical ranges of species

    B. Species diversity on some islands tends to reach a dynamic equilibrium in ecological time

OBJECTIVES

After reading this chapter and attending lecture, the student should be able to:

1. Compare and contrast the individualistic hypothesis of H.A. Gleason and the interactive hypothesis of F.E. Clements with respect to communities.

2. Explain the relationship between species richness, relative abundance, and diversity.

3. List four properties of a community, and explain the importance of each.

4. Explain how interspecific competition may affect community structure.

5. Describe the competitive exclusion principle, and explain how competitive exclusion may affect community structure.

6. Distinguish between an organism's fundamental niche and realized niche.

7. Explain how resource partitioning can affect species diversity.

8. Describe the defense mechanisms evolved by plants to reduce predation by herbivores.

9. Explain how cryptic coloration and aposematic coloration aid an animal in avoiding predators.

10. Distinguish between Batesian mimicry and Mullerian mimicry.

11. Describe how predators use mimicry to obtain prey.

12. Explain the role of predators in community structure.

13. Distinguish among parasitism, mutualism, and commensalism.

14. Explain why it is difficult to determine what factor is most important in structuring a community.

15. Distinguish between primary succession and secondary succession.

16. Explain how inhibition and facilitation may be involved in succession.

17. Describe how natural and human disturbances can affect community succession.

18. Explain how the intensity of disturbances can affect equilibrium and species diversity.

19. List the factors involved in limiting a species to a particular range.

20. Describe the mechanisms which contribute to the global clines in diversity.

21. Explain the factors which determine what species eventually inhabit islands.

KEY TERMS

species richness	cryptic coloration	principle	stability
relative abundance	aposematic coloration	ecological niche	disturbances
species diversity	mimicry	fundamental niche	ecological
predation	predator	prey	succession
individualistic hypothesis	Batesian mimicry	realized niche	primary
interactive hypothesis	Mullerian mimicry	resource partitioning	succession
secondary succession	parasite	character displacement	recruitment
interspecific interactions	host	symbiosis	dynamic
coevolution	endoparasites	symbiont	equilibrium
parasitism	ectoparasites	parasitism	hypothesis
parasitoidism	interspecific competition	commensalism	intermediate
herbivory	interference competition	mutualism	disturbance
community	exploitative competition	keystone species	hypothesis
biogeography	competitive exclusion	exotic species	

LECTURE NOTES

A *community* consists of all the organisms inhabiting a particular area; an assemblage of populations of different species living close enough together for potential interaction.

I.   Early Hypotheses of Community Structure

Communities, even those that appear similar, vary with regard to their species diversity.

- *Species diversity* = The number and relative abundance of species in a biological community
  - Based on both species richness and relative abundance
    - *Species richness* is the number of species in a community.
    - *Relative abundance* is a measure of the proportion of a species in the community as a whole.

A community with a certain species richness with equal relative abundance of each species would have a greater diversity than a second community with the same species richness with a few common species and many rare ones.

A. The interactive and individualistic hypotheses pose alternative explanations of community structure: *science as a process*

Among the pioneers of community ecology, there were two divergent views on why certain combinations of species are found together as members of a community: the individualistic hypothesis and the interactive hypothesis.

The *individualistic hypothesis* was proposed by H.A. Gleason. This hypothesis depicted a community as a chance assemblage of species found in an area because they have similar abiotic requirements.

- Emphasizes studying single species
- Predicts each species will have an independent distribution along an environmental gradient and there are no discrete geographical boundaries between communities (see Campbell, Figure 53.1)
- In most cases, the composition of communities would change continuously along some environmental gradient due to addition or loss of particular species

The *interactive hypothesis* proposed by F.E. Clements saw each community as an assemblage of closely linked species having mandatory biotic interactions that cause the community to function as an integrated unit.

- Based on the observation that certain plant species are consistently found together
- Emphasizes entire assemblages of species as the essential units for understanding the interrelationships and distributions of organisms
- Predicts species should be clustered, with discrete boundaries between communities. The presence or absence of one species is governed by the presence or absence of other species with which it interacts (see Campbell, Figure 53.2).

Animals in a community are often linked more rigidly to other organisms.

- Some animals feed primarily on certain food items so their distribution is linked to distribution of their prey.
- Other animals feed on a variety of food items and tend to be distributed in a variety of communities.

Distribution of almost all organisms is affected by both abiotic gradients and interactions with other species.

- Among the most significant abiotic factors are disturbances (e.g,. floods, fire, storms) that destabilize existing relationships among organisms.

II. Interactions between Populations of Different Species

Interspecific interaction are those that occur between populations of different species living together within a community.

A. Interspecific interactions can be strong selection factors in evolution

Interactions between species are as important to adaptation by natural selection as the physical and chemical features of the environment.

Coevolution is a change in one species that acts as a selective force on another species. Counteradaptation of the second species, in turn, affects selection of individuals in the first species.

- Studied most extensively in predator-prey relationships, in mutualism, and in parasite-host relationships

The association between passionflower vines (*Passiflora*) and the butterfly *Heliconius* is believed to be an example of coevolution.

- The vines produce toxic chemicals to reduce damage to young shoots and leaves by herbivorous insects.
- Butterfly larvae of *Heliconius* can tolerate these chemicals due to digestive enzymes that break down the toxic chemicals (a counteradaptation) (see Campbell, Figure 53.3).
- Females of some *Heliconius* species avoid laying eggs (which are bright yellow) on leaves where other yellow egg clusters have been laid.
  - This may reduce intraspecific competition on individual leaves.
- Some species of passionflowers develop large, yellow nectaries which resemble *Heliconius* eggs; an adaptation that may divert egg-laying butterflies to other plants.
- These nectaries, as well as smaller ones, also attract ants and wasps which prey on butterfly eggs and larvae.
- Thus, what appears to be coevolution may actually result from interactions with many species (not just the obvious two).

Biologists agree that adaptation of organisms to other species in a community is a fundamental characteristic of life.

B. Interspecific interactions may have positive, negative, or neutral effects on a population's density: *an overview*

While adaptations to abiotic factors largely determine the geographic distributions of many species, all organisms are influenced by biotic interactions with other individuals.

Interspecific interactions may take a variety of forms (see Campbell, Table 53.2).

The two species involved in the interaction may have a positive (+), negative (–), or neutral (0) effect on their population densities.

Both population densities may increase (+ / +), one may increase while the other decreases (+ / –), one may increase while the other is not affected (+ / 0), or both may decrease
(– / –).

C. Predation and parasitism are (+ / –) interactions: *a closer look*

There are several (+ / –) interactions, some are obvious, some are not.

- *Predation* involves a *predator* that eats its *prey*.
- *Parasitism* involves predators that live in or on their *hosts* and seldom involves outright host death.
- *Parasitoidism* involves insects, such as small wasps, who lay their eggs on living hosts. After hatching, the larvae feed within the host's body and eventually cause its death.
- *Herbivory* occurs when animals eat plants. May only cause damage to a portion of the plant (grazing) or may kill an entire plant (seed eating).

1. Predation

Predation = A community interaction where one species, the predator, eats another, the prey. Includes both animal-animal interactions and animal-plant interactions.

Adaptations of predators to this interaction are obvious and familiar. Predators have:

- Acute senses that are used to locate and identify prey items (e.g., heat-sensing pits of rattlesnakes, chemical sensors of herbivorous insects)
- Structures such as claws, teeth, fangs, stingers, and poisons that function to catch, subdue, or chew the prey item
- Speed and agility to pursue prey or camouflage that permits them to ambush prey

Various defensive adaptations have evolved in prey species as a result of repeated encounters with predators over evolutionary time.

a. Plant defenses against herbivores

Since plants cannot escape herbivores, they have evolved several ways to protect themselves from predation.

Plants have mechanical and chemical defenses against herbivores.

- Thorns or microscopic hooks, spines, and crystals may be present to discourage herbivores.
- Plants may produce chemicals (formed as byproducts of major metabolic pathways) that make vegetation distasteful or harmful.
- Plants may produce analogues of insect hormones that cause abnormal development in insects that prey on them.

Plant defenses may act as selective factors leading to counteradaptations in herbivores that can then nullify those defenses and consume the plant.

b. Animal defenses against predators

Animal defenses against predators may be passive (hiding), active (escaping and physical defense), mechanical, or chemical (toxins).

Active defenses include fleeing, active self-defense, alarm calls, mobbing (see Campbell, Figure 53.4), direct attack, and distraction.

Passive forms of defense involving adaptive coloration have evolved repeatedly among animals.

- *Cryptic coloration* (coloration making prey difficult to spot against its surroundings) is common and only requires the animal to remain still to avoid detection (see Campbell, Figure 53.5).
- The shape of an animal can also help camouflage it.
- Deceptive marking such as large, fake eyes can startle predators, allowing prey to escape, or cause predators to strike a nonvital area (see Campbell, Figure 53.6).
- *Aposematic coloration* (bright coloration that acts as a warning of effective physical or chemical defense) appears to be adaptive since predators quickly learn to avoid prey with this coloration (see Campbell, Figure 53.7).

Some mechanical and chemical defenses actively discourage predators (e.g., porcupine quills and skunk spray).

- Some insects acquire chemical defense passively by accumulating toxins from plants they eat; these make them distasteful to predators.

c. Mimicry

Mimicry is a phenomenon in which a mimic bears a superficial resemblance to another species, the model.

- Occurs in both predatory and prey species
- Defensive mimicry in prey usually involves aposematic models

Batesian mimicry = A palatable species mimics an unpalatable or harmful model (see Campbell, Figure 53.8)

- Mimic must be much less abundant than the model to be effective since predators must learn the coloration indicates a bad or harmful food item.

Mullerian mimicry = Two or more unpalatable, aposematically colored species resemble each other

- Each gains additional advantage since predators learn more quickly to avoid prey with this coloration.

Predators also use mimicry to lure prey.

- The tongue of snapping turtles resembles a wriggling worm which attracts small fish into capture range.

2. Parasitism

Parasitism is a (+ / −) interaction in which one organism, the *parasite*, derives its nourishment from another organism, the *host*, which is harmed in some way.

- *Endoparasites* live within the host tissues or body cavities.
- *Ectoparasites* attach to or briefly feed on the external surface of the host.

Natural selection favors parasites that are best able to locate and feed on a host.

- Infection by a parasite (locating a host) may be passive as when an endoparasite's egg is swallowed by a host.
- Active location of a host may involve thermal or chemical cues that help the parasite identify the host.

Natural selection has also favored the evolution of defensive capabilities in potential hosts.

- Secondary plant products toxic to herbivores may also be toxic to parasitic fungi and bacteria.
- The vertebrate immune system provides defense against endoparasites (see Campbell, Figure 53.9).
- Many parasites are adapted to specific hosts, often a single species.
- Coevolution generally results in stable relationships in which the host is not killed (host death would also kill the parasite).

Some forms of parasitism do not involve the consumption of the host, but are the exploitation of the host's behavior by the parasite.

- Some species of birds (cowbirds, European cuckoos) lay their eggs in the nests of other species. Often a newly hatched brood parasite will move other eggs out of the nest. Host parents invest their energy in feeding the brood parasite instead of their own offspring – a (+ / −) interaction.
  - An evolutionary adaptation in many host species is the ability to detect parasite eggs which are removed or the nest is abandoned.
  - A counteradaptation to this host defense is egg mimicry in some brood parasites.

D. Interspecific competitions are (– / –) interactions: *a closer look*

Interspecific competition occurs when two or more species in a community rely on similar limiting resources.

- May take the form of *interference competition* (actual fighting) or *exploitative competition* (consumption or use).
- As the population density of one species increases, it may limit the density of the competing species as well as its own.

1. The competitive exclusion principle

The *competitive exclusion principle* predicts that two species competing for the same limiting resources cannot coexist in the same community. One will use resources more efficiently, thus reproducing more rapidly and eliminating the inferior competitor.

- Derived independently by A.J. Lotka and V. Volterra, each of whom modified the logistic model of population growth to incorporate interspecific competition.
- Experiments by G.F. Gause confirmed competitive exclusion among species of *Paramecium* (see Campbell, Figure 53.10).
- Laboratory experiments have confirmed the principle for other organisms; however, natural communities are much more complex and require field experiments and observations to determine the importance of competition.

2. Ecological niches

An *ecological niche* is the sum total of an organism's use of biotic and abiotic resources in its environment; how it "fits into" an ecosystem.

A *fundamental niche* is the resources a population is theoretically capable of using under ideal circumstances.

Biological constraints (competition, predation, resource limitations) restrict organisms to their *realized niche*: the resources a population actually uses.

Two species cannot coexist in a community if their niches are identical.

Ecologically similar species can coexist in a community if there are one or more significant differences in their niches.

3. Evidence for competition in nature

If competition is as important as indicated by the competitive exclusion principle, it should be rare in natural communities, since only two outcomes are possible:

- The weaker competitor will become extinct.
- One of the species will evolve to the point of using a different set of resources.

Several lines of evidence, including resource partitioning and character displacement, indicate that competition in the past has had a major role in shaping current ecological relationships.

Resource partitioning is well documented among animals.

- Sympatric species consume slightly different foods or use other resources in different ways.
  - For example, in the Dominican Republic, several species of *Anolis*, small arboreal lizards, are sympatric. They feed on small arthropods that land within their territories. Each species uses a characteristic perching site, which presumably reduces competition (see Campbell, Figure 53.11).

- Natural selection appears to have favored perch site selection since each species has morphological characteristics (body size, leg length) which are adaptive to their microhabitats.

Character displacement is the tendency for characteristics to be more divergent in sympatric populations of two species than in allopatric populations of the same two species.

- Often occurs in areas where populations of closely related species which are normally allopatric become sympatric.
- Allopatric populations of such species are similar in structure and use similar resources.
- Sympatric populations show morphological differences and use different resources.
- Two closely related species of Galapagos finches (*Geospiza fuliginosa* and *G. fortis*) have beaks of similar size when the populations are allopatric; however, on an island where they are sympatric, a significant difference in beak depth has evolved (see Campbell, Figure 53.12).
  - Indicates they feed on seeds of different sizes when sympatric.

Controlled field experiments also provide evidence for the influence of one species on the density and distribution of another species. J. H. Connell manipulated the densities of two species of barnacles that compete for attachment space in the rocky intertidal zone (see Campbell, Figure 53.13).

- *Balanus balanoides* is usually concentrated in the lower portions of the rocky intertidal zone due to a lower tolerance to desiccation.
- *Chthamalus stellatus* is usually found in the upper portions of the zone.
- After Connell removed *Balanus* from the lower strata, *Chthamalus* was able to grow there.
- This showed that one species is able to exclude another from the area where their fundamental niches overlap. That is, in the lower part of the zone, *Balanus* outcompeted *Chthamalus* for attachment sites, thus limiting it to the upper part of the rocks.

E. Commensalism and mutualism are (+ / 0) and (+ / +) interactions, respectively: *a closer look*

Symbiosis is a form of interspecific interaction in which a *host species* and a *symbiont* maintain a close association.

- May be parasitism, commensalism, or mutualism
- All are important determinants of community structure

1. Commensalism

Commensalism is a (+ / 0) interaction in which the symbiont benefits and the host is unaffected.

- Few absolute examples exist since it is unusual for one species not to be affected in some way by a close association with another species.
- Since only one species in a commensalistic association benefits, any evolutionary change in the relationship will likely occur in the beneficiary.

Cowbirds and cattle egrets and grazing herbivores appear to have a commensal association (see Campbell, Figure 53.14).

- Cattle egrets feed on insects and other small animals.

- The birds concentrate their feeding activity near grazing herbivores whose movements flush prey items from the vegetation.
- The egrets benefit by an increase in feeding rates by following the herbivores.
- Most of the time, the herbivores are not affected by the activity of the egrets. In fact, they may derive some benefit when the birds remove and eat ticks or other ectoparasites from the herbivores. However, a possible negative effect may occur if the birds somehow make an herbivore more susceptible to predators.

Commensalism, parasitism, or mutualism are best described as when two associating populations are together, they exhibit a particular symbiotic relationship most of the time.

2. **Mutualism**

Mutualism is a (+ / +) interaction requiring the evolution of adaptations in both species.

- A change in either species is likely to affect the survival and reproduction of the other.

> Examples:
> - Nitrogen fixation by bacteria in the root nodules of legumes
> - Cellulose digestion by microorganisms in the alimentary canals of termites and ruminant mammals
> - Photosynthesis by algae in the tissues of corals

Mutualistic relationships may have evolved from predator-prey or host-parasite relationships.

- Some angiosperms have adaptations that attract animals for pollination or seed dispersal.
  - May represent counteradaptations to herbivores feeding on seeds or pollen.
  - Pollen is spared when the pollinator can feed on nectar and seeds are dispersed when animals eat fruits.
- Plants that could derive a benefit from sacrificing tissues other than pollen or seeds would increase their reproductive success.

III. Interspecific Interactions and Community Structure

The relationships of organisms within a community are tied together in a complex web of community structure and processes.

One way to examine the community's web of interspecific interactions is to study the various feeding (trophic) relationships, or what eats what.

- The functional groupings approach involves placing species in groups with similar trophic positions.
  - All plants are grouped together as producers; herbivores, regardless of size, are primary consumers; organisms that feed on primary consumers are secondary consumers, and so on.
  - This method provides a broad overview of the community, but it may mask some important aspects of the population interactions.
- Food web analysis emphasizes the trophic connections among community members.
  - Includes species-level information about the community and emphasizes species-species connections

- Can be very complex
- Presents a detailed knowledge of feeding relationships within the community structure

A. Predators can alter community structure by moderating competition among prey species

Predation plays a complex role in helping shape community structure and may actually help maintain diversity.

Predators do not always reduce diversity.

The most important effect of a predator on a community structure is to moderate competition among its prey species.

- In a predator-free environment, it has been shown that the species richness of the community declines markedly (see Campbell, Figure 53.16).
- A *keystone species* has an impact on a community that is disproportionately large relative to its own abundance.
- A *keystone* predator may maintain a higher community species diversity by reducing the densities of strong competitors, thus preventing competitive exclusion of poorer competitors.

B. Mutualism and parasitism can have community-wide effects

The interactions that characterize mutualistic and parasitic relationships play important roles in communities.

- Keystone mutualists (e.g., mycorrhizal fungi) help maintain processes that impact all other species in communities.
- Parasitic diseases that reduce populations of one species also impact other species
- Parasites also can modify host behavior (the trypanosome/tsetse relationship)

C. Interspecific competition influences populations of many species and can affect community structure

Competition is recognized as an important factor in determining community structure but whether it is the major factor is open to debate.

- Many studies have documented interspecific competition, but it does not always result in competitive exclusion: competing species sometimes coexist at reduced densities.
  - *Exotic species* (species introduced into new communities by humans) often outcompete native species and alter the community structure.
  - The competition for space between barnacles discussed earlier resulted in exclusion of *Chthamalus* from the lower intertidal zone but it remained in the upper zone.
- It is difficult to demonstrate that two species are currently competing and even more difficult to assess what has happened in their evolutionary past.
- It is generally recognized that competition becomes more important as population sizes approach carrying capacities and resources are limiting.

Determining the effects of interspecific factors on species diversity and structure in communities is an important goal in ecological research, as is determining the relative significance of environmental patchiness in community structure.

D. A complex interplay of interspecific interactions and environmental variability characterizes community structure

The more heterogeneous the habitat, the more ecological niches available to organisms, and the more diverse the community.

Vegetation greatly influences the types of animals found in a community. Animal populations can more easily partition a structurally complex system than one of less complexity.

Environmental patchiness is a spatial and temporal characteristic of all ecosystems.

- For example, the chemical composition of rocks influences the mineral content of the soil derived from their erosion. Soil moisture varies with topography.
- Environmental patchiness can increase community diversity by facilitating resource partitioning among potential competitors adapted to different local conditions.

Community diversity is also affected by the temporal use of habitats.

- Nocturnal animals utilize a habitat at night and are replaced by diurnal animals when it becomes daytime.
- A community will contain a variety of plants which bloom during different seasons.

Evaluating which factors are important in developing the characteristics of a community is difficult since interspecific interactions and abiotic factors that cause patchiness all have significant impact.

- Communities are structured by multiple interactions of organisms with the biotic and abiotic factors in their environment.
- Which types of interactions are most important varies from one type of community to another, and even among different components of the same community.

IV.   Disturbance and Nonequilibrium

Communities vary in their responses to both natural and human-induced disturbances.

- The *stability* of a community is its tendency to reach and maintain an equilibrium (or relatively constant condition) in response to disturbance.
- Responses depend on the type of community and nature of the disturbance.

A.   Nonequilibrium resulting from disturbance is a prominent feature of most communities

Disturbances are events that disrupt communities.

- They change resource availability and create opportunities for new species to become established.
- The magnitude of the disturbance's impact depends on the size, frequency, and severity of the disturbance.
- Natural disturbances include storms, fire, floods, freezes, and drought.
- Animals, including humans, can also cause disturbances by overgrazing, damaging communities, removing organisms within them, or altering resource availability.

Increasing evidence indicates that some amount of disturbance and nonequilibrium is normal for communities. Some ecologists suggest that a given community is usually in some state of recovery from a disturbance.

B.   Humans are the most widespread agents of disturbance

Human disturbances have the greatest impact on communities.

- Logging and clearing for farmland have reduced and disconnected forests.
- Agricultural development has disrupted grasslands.
- After a community disturbance, the early stages of succession, characterized by weedy and shrubby vegetation, may persist for many years.

- Centuries of overgrazing and agricultural disturbance have contributed to current famine in parts of Africa by turning grasslands into barren areas.
- Human disturbances usually reduce species diversity in communities.

It is possible for disturbances to have *positive* impact on a community.

- Small-scale natural disturbances may help maintain species diversity in a community (see Campbell, Figure 53.18).
  - When patches are formed during different successional stages, habitat heterogeneity is increased.
- Frequent small-scale disturbances can prevent large-scale disturbances of greater negative impact.
  - Small fires prevent the accumulation of flammable materials that would feed a larger, more damaging fire.
  - Decades-long fire suppression efforts in Yellowstone NP provided a great accumulation of fuel that fed a large-scale fire in 1988 (see Campbell, Figure 53.19).

C. Succession is a process of change that results from disturbance in communities

Ecological succession is the transition in species composition over ecological time.

- Called *primary succession* if it begins in areas essentially barren of life due to lack of formed soil (e.g., volcanic formations) or on rubble (e.g., retreat of glaciers)
- Called *secondary succession* if an existing community has been cleared by some disturbance that leaves the soil intact (e.g., abandoned agricultural fields)

A variety of interrelated factors determines the course of succession.

Individual species compete for available resources and replace each other.

- Different species compete better at different successional stages since resource availability changes over the course of succession.
- Early stages are typically characterized by *r*-selected species that are good colonizers.
  - They have a high fecundity and excellent dispersal mechanisms.
  - Others may be fugitive species that do not compete well except in newly disturbed areas.

The species composition during early successional stages is affected by the abiotic conditions in a barren area.

- *K*-selected species may colonize an area but fail to become abundant because of environmental conditions that are outside of their tolerance limits.
- Variation in growth rates and maturation times of colonizers impact the community.
  - For example, herbaceous species are more prevalent in early successional stages than trees because they grow faster and mature quicker.

Changes in community structure during succession may be induced by the organisms themselves.

- Direct biotic interactions include *inhibition* of some species by others through exploitative competition, interference competition, or both.
- Organisms also affect the abiotic environment by modifying local conditions.

- In facilitation, the group of organisms representing one successional stage "paves the way" for species typical of the next stage (e.g., alder leaves lower soil pH as they decompose, and lower soil pH facilitates the movement of spruce and hemlock into the area).
  - Sometimes facilitation for other species makes the environment unsuitable for the very species responsible for the changes.
- Inhibition and facilitation may be involved throughout the stages of succession.

D. The nonequilibrial model views communities as mosaics of patches at different stages of succession

A traditional view of ecological succession held that the community passed through a series of predictable transitional stages to reach a relatively stable state, the *climax community*.

- The climax was reached when the web of interactions was so intricate that the community was saturated.
- No new species could enter unless a localized extinction occurred.

Ecologists now consider the concept of stable communities flawed and overly simplistic.

- Many communities are also routinely disturbed by outside factors during the course of succession.
- Communities that appear stable actually change over long periods of time.

The prevailing view is the nonequilibrial model, which sees communities as being in continual flux.

- It emphasizes the importance of less predictable factors (dispersal, disturbance) in shaping community composition and structure.
- The succession pathway depends on the size, frequency, and severity of disturbance.
- The course of succession may vary with the identity of colonizing species.
- Disturbances may prevent the community from ever reaching equilibrium.
- A mature community is an unpredictable mosaic of patches at different successional stages.
- Local environmental heterogeneity contributes to the mosaic structure as different species inhabit different patches.

What are the effects of disturbance in a community?

Disturbances result in minor changes, such as regrowth and migration from adjacent areas; usually, species originally in the area refill it.

Disturbances may also trigger major changes in community structure that result in colonization of disturbed patches by *recruitment*.

- *Recruitment* = Colonization by species from distant areas not directly associated with the disturbed patch or its immediate vicinity

Several hypotheses have been offered to explain why some communities are more diverse than others.

1. The *dynamic equilibrium hypothesis* holds that diversity depends mainly on the effect of disturbance on the competitive interactions of populations.
2. The *intermediate disturbance hypothesis* states that species diversity is greatest where disturbances are moderate in frequency and severity because organisms typical of several successional stages will be present.
   - Supported by studies of clearings formed by trees falling in tropical rain forests.

- In these disturbed areas, immigrations and extinctions occur rapidly and species from different successional stages coexist.
3. The nonequilibrial model places interspecific interactions in secondary roles and suggests that high species diversity results mainly from abiotic disturbance-induced environmental patchiness.

V.   Community Ecology and Biogeography

Biogeography is the study of the past and present distribution of individual species and entire communities.

- Provides a different approach to understanding community properties by analyzing global and local phenomena, mostly from a historical perspective.
- Traditionally concerned with the identities of species comprising particular communities.
- Currently also applies the principles of community ecology to the analysis of geographical distribution.

Terrestrial life can be divided into biogeographical realms having boundaries associated with continental drift patterns (see Campbell, Figure 53.20).

- Thus species distribution reflects past history as well as present interactions of organisms with their environment.

A.   Dispersal and survivability in ecological and evolutionary time account for the geographical ranges of species

Limitation of a species to a particular range may be due to:

1. Failure of the species to disperse beyond its current range
2. Failure to survive of pioneer individuals that spread beyond the observed range
3. The species having retracted from a once larger range to its current boundaries

The first two limitations can be distinguished by transplant experiments in which specimens of a species are moved to similar environments outside of their range.

- If the transplants survive, they probably have not dispersed to suitable locations outside the existing range. If they die, it indicates an intolerance to different abiotic conditions or an inability to compete with resident species.
- Most species fail to survive outside their range, but there are notable exceptions.

The third limitation is well supported by data from paleontology and historical biogeography.

- The fossil record includes evidence that close relatives of some living animals with limited ranges once had much wider ranges.
- For example, fossils of close relatives of elephants and camels have been found in North America.

B.   Species diversity on some islands tends to reach a dynamic equilibrium in ecological time

Islands provide opportunities to study factors affecting species diversity of communities due to their isolation and limited size.

- An island is any area surrounded by an environment not suitable for the "island" species.
- Islands can be oceanic islands or habitat islands on land (lakes or mountain peaks separated by lowlands).

Ecologists Robert MacArthur and E.O. Wilson developed a general hypothesis of island biogeography that predicts species diversity on an island will reach an equilibrium directly proportional to island size and inversely proportional to the island's distance from a mainland.

- The number of species to inhabit islands is determined by immigration and extinction rates, which are themselves affected by island size and distance from the mainland.
- The smaller the island and the greater the distance from mainland, the lower the immigration rate, due to the difficulty in finding the island and the distance colonizers must travel.
- Smaller islands also have higher extinction rates since they have fewer resources and less diverse habitat to be partitioned; these factors increase the probability of competitive exclusion.
- Immigration and extinction rates also are affected by the number of species already present.
    - As species numbers increase, immigration rate of new species decreases since new arrivals are less likely to represent a new species.
    - Extinction rate increases due to increased chance of competitive exclusion.
- Eventually, an equilibrium will be reached where immigration rates match extinction rates.
    - The number of species present at the equilibrium point is correlated to island size and distance from the mainland (see Campbell, Figure 53.21).
- The species composition may change since equilibrium is dynamic with immigrations and extinctions continuing.
- Experimental research by E.O. Wilson and D. Simberloff, in the late 1960s, on the small mangrove islands off the tip of South Florida, supported the hypothesis.

Over the past decades, the island biogeography hypothesis has come under considerable fire.

- May only apply in a limited number of cases and over short time periods, where colonization is the most important process determining species composition.
- Over longer periods, the species composition and community structure will be affected by speciation and evolutionary change in the island species.

REFERENCES

Campbell, N. *Biology*. 5th ed. Menlo Park, California: Benjamin/Cummings, 1999.

Krebs, C.J. *Ecology: The experimental analysis of distribution and abundance*, 4th ed. New York, NY: HarperCollins College Publishers, 1994.

MacArthur, Robert H., and Edward O. Wilson. *The Theory of Island Biogeography*. Princeton, New Jersey: Princeton University Press, 1969.

Ricklefs, R.E. *Ecology*. 3rd ed. New York: Chiron Press, 1990.

CHAPTER 54
ECOSYSTEMS

OUTLINE

OBJECTIVES

After reading this chapter and attending lecture, the student should be able to:

1. Explain the importance of autotrophic organisms with respect to energy flow and nutrient cycling in ecosystems.
2. List and describe the importance of the four consumer levels found in an ecosystem.
3. Explain how gross primary productivity is allocated by the plants in an ecosystem.
4. List the factors that can limit productivity of an ecosystem.
5. Explain why productivity declines at each trophic level.
6. Distinguish between energy pyramids and biomass pyramids.
7. Describe the hydrologic (water) cycle.

8. Describe the carbon cycle, and explain why it is said to result from the reciprocal processes of photosynthesis and cellular respiration.

9. Describe the nitrogen cycle, and explain the importance of nitrogen fixation to all living organisms.

10. Explain how phosphorus is recycled locally in most ecosystems.

11. Explain why the soil in tropical forests contains lower levels of nutrients than soil in temperate forests.

12. Describe how agricultural practices can interfere with nitrogen cycling.

13. Describe how deforestation can affect nutrient cycling within an ecosystem.

14. Describe how the carbon cycle differs in terrestrial and aquatic systems.

15. Explain how "cultural eutrophication" can alter freshwater ecosystems.

16. Explain why toxic compounds usually have the greatest effect on top-level carnivores.

17. Describe how increased atmospheric concentrations of carbon dioxide could affect the Earth.

18. Describe how human interference might alter the biosphere.

KEY TERMS

ecosystem	food chain	biomass	biogeochemical cycle
trophic structure	food web	standing crop biomass	nitrogen fixation
trophic level	production	limiting nutrient	nitrification
primary producers	consumption	secondary productivity	ammonification
primary consumers	decomposition	ecological efficiency	long-term ecological
secondary consumers	primary productivity	pyramid of productivity	research (LTER)
tertiary consumers	gross primary	biomass pyramid	biological
detritivores	productivity	turnover time	magnification
detritus	net primary productivity	pyramid of numbers	greenhouse effect
denitrification			

LECTURE NOTES

Ecosystem = All organisms living in a given area along with the abiotic factors with which they interact

The boundaries of ecosystems are not usually discrete.

This is the most inclusive level of biological organization.

Ecosystems involve two processes that cannot be described at lower levels: energy flow and chemical cycling.

- Energy flows through ecosystems and matter cycles within them.

I. **Trophic Relationships in Ecosystems**

 Each ecosystem has a *trophic structure* of feeding relationships that determine the paths of energy flow and chemical cycling.

 - Ecologists divide the species in a community or ecosystem into different *trophic levels* based on their main source of nutrition.

 A. **Trophic relationships determine an ecosystem's routes of energy flow and chemical cycling**

 The five trophic levels typically recognized include:

1. *Primary producers* = Autotrophs (usually photosynthetic) that support all other trophic levels either directly or indirectly by synthesizing sugars and other organic molecules using light or chemical energy
2. *Primary consumers* = Herbivores that consume primary producers (plants and algae)
3. *Secondary consumers* = Carnivores that eat herbivores
4. *Tertiary consumers* = Carnivores that eat other carnivores
5. *Detritivores (decomposers)* = Consumers that derive energy from *detritus* (organic wastes) and dead organisms from other trophic levels. Detritivore often form a major link between primary producers and the consumers in an ecosystem.

An ecosystem's trophic structure determines the routes of energy flow and chemical cycling.

Food chain is the pathway along which food is transferred from trophic level to trophic level, beginning with primary producers (see Campbell, Figure 54.1).

- Rarely are unbranched since several different primary consumers may feed on the same plant species and a primary consumer may eat several species of plants
- Feeding relationships are usually woven into elaborate *food webs* within an ecosystem (see Campbell, Figure 54.2)

It is important to distinguish between ecosystem structure (trophic levels) and ecosystem processes (production, consumption, decomposition). All organisms carry out each of the ecosystem processes to some extent.

- *Production* refers to the rate of incorporation of energy and materials into the bodies of organisms.
  - In this sense, all organisms are producers; however, primary producers are often referred to as "producers" because their production supports all other organisms.
- *Consumption* refers to the metabolic use of assimilated organic molecules for organismal growth and reproduction.
- *Decomposition* is the breakdown of organic molecules into inorganic molecules.

B. Primary producers include plants, algae, and many species of bacteria

The main primary producers will vary depending on the ecosystem.

Plants are the main producers in most terrestrial ecosystems.

- Debris falling from terrestrial plants that reaches streams (directly or through runoff) is a major source of organic material.
- Phytoplankton (algae and bacteria) are the most important autotrophs in the limnetic zone of lakes and in the open ocean.
- Multicellular algae and aquatic plants are often more important primary producers in the shallow, near-shore areas of freshwater and marine ecosystems.
- The aphotic zone of the deep sea receives energy and nutrients (dead plankton, detritus) from the overlying photic zone.

Organisms in communities surrounding the hot water vents on the deep-sea floor depend more on chemical energy than solar energy.

- The main producers are chemoautotrophic bacteria that derive energy from the oxidation of hydrogen sulfide.

C. Many primary and higher-order consumers are opportunistic feeders

Consumers also vary with the type of ecosystem.

Primary consumers in terrestrial ecosystems are mostly insects, snails, plant parasites, grazing mammals, and seed-eating and fruit-eating birds and mammals.

- Primary consumers are considered opportunistic because they supplement their diet of autotrophs with heterotrophic material if it is available. Many consumers that mainly eat live organisms also scavenge dead organic material.

In aquatic ecosystems, the primary consumers are the zooplankton (heterotrophic protists, small invertebrates, numerous larval stages) and some fish. As with their terrestrial counterparts, aquatic consumers also are opportunistic

Secondary consumers in terrestrial ecosystems are spiders, frogs, insect-eating birds, carnivorous mammals, and animal parasites.

Secondary consumers in aquatic ecosystems are fish and benthic forms such as sea stars and other carnivorous invertebrates.

D. Decomposition interconnects all trophic levels

Organic matter in that composes living organisms in ecosystems is eventually recycled, decomposed, and returned to the abiotic environment in a form that can be used by autotrophs.

The most important decomposers are bacteria and fungi, which digest materials externally and then absorb the products.

Decomposition links all trophic levels.

II.   Energy Flow in Ecosystems

Energy for growth, maintenance, and reproduction is required by all organisms; some species also require energy for locomotion.

A. An ecosystem's energy budget depends on primary productivity

Light energy is used by most primary producers to synthesize organic molecules (photosynthesis), which are later broken down to produce ATP (cellular respiration).

- Since only primary producers can directly utilize solar energy, an ecosystem's entire energy budget is determined by the photosynthetic activity of the system.

Consumers obtain energy in the form of organic molecules that were produced at the previous trophic level. Thus, energy flows to higher trophic levels through food webs.

1. The global energy budget

Earth receives an estimated 10^{22} joules (J) of solar radiation each day.

The amount of solar radiation striking the Earth's surface shows dramatic regional variation that limits the photosynthetic output of ecosystems in different places.

- The intensity of solar radiation also varies with latitude resulting in the tropics receiving the most input.

- Most of the solar radiation is reflected, absorbed, or scattered by the atmosphere, clouds, and dust particles in the air; this amount varies over different regions.

Only a fraction of the solar radiation which reaches the biosphere strikes plants, photosynthetic bacteria, and algae (much hits bare ground or is absorbed or reflected by water) and these primary producers can only use some wavelengths for photosynthesis.

- Only about 1% to 2% of the visible light reaching photosynthetic organisms is converted to chemical energy by photosynthesis.

  - The photosynthetic efficiency also varies with the type of plant, light levels, and other factors.

Even with all the variations mentioned above, primary production of Earth collectively creates about 170 billion tons of organic material each year.

2. **Primary productivity**

Primary productivity is the amount of light energy converted to chemical energy by autotrophs of an ecosystem.

- The total is known as *gross primary productivity (GPP)*, which may be determined by measuring the total oxygen produced by photosynthesis.

Net primary productivity (NPP) = GPP - Rs (energy used by producers for respiration)

- NPP accounts for the organic mass of plants (growth) and represents storage of chemical energy available to consumers.
- The NPP:GPP ratio is generally smaller for large producers with elaborate nonphotosynthetic structures (such as trees) which support large metabolically active stem and root systems.

Primary productivity can be expressed as *biomass* (expressed as dry weight since water contains no usable energy) added to an ecosystem per unit area per unit time ($g/m^2/yr$) or as energy per unit time ($J/m^2/yr$).

- Primary productivity should not be confused with standing crop biomass.
    - Primary productivity is the rate at which new biomass is synthesized by photosynthetic organisms.
    - *Standing crop biomass* is the total biomass of photosynthetic autotrophs present at a given time, which may have accumulated over several growing seasons.

Primary productivity varies among ecosystems, and an ecosystem's size affects its contribution to the Earth's total productivity (see Campbell, Figure 54.3).

- Tropical rain forests are very productive and contribute a large proportion to the planet's overall productivity since they cover a large portion of the Earth's surface.
- Estuaries and coral reefs are also very productive but make only a small contribution to planetary productivity since they do not cover an extensive area.
- The open ocean has a relatively low productivity but makes the largest contribution to overall productivity of any ecosystem due to its very large size.
- Deserts and tundra also have low productivity.

Factors important in limiting productivity depend on the type of ecosystem and temporal changes such as seasons.

Generally, precipitation, temperature, and light intensity are factors limiting productivity in terrestrial ecosystems.

- Productivity increases as latitudes approach the equator because availability of water, heat, and light increases in the tropics.
- Productivity in terrestrial ecosystems may also be limited by availability of inorganic nutrients.
    - Plants require a variety of nutrients, some in large quantities and some in small quantities.
    - Primary productivity sometimes removes nutrients from the system faster than they can be replenished.
    - If a nutrient is removed in such quantities that sufficient amounts are no longer available, it becomes the *limiting nutrient*.

- Adding the limiting nutrient will stimulate the system to resume growth until another nutrient or it becomes limiting (usually nitrogen or phosphorus).
- Carbon dioxide availability sometimes limits productivity.

An aquatic ecosystem's productivity is usually determined by light intensity, water temperature, and availability of inorganic nutrients.

- Productivity is greatest in shallow waters near continents and along coral reefs due to abundant nutrients and sunlight.
- Light intensity and temperature affect primary productivity of phytoplankton in the open oceans; productivity is highest near the surface and decreases with depth.
- Inorganic nutrients are limiting at the surface of open ocean waters with nitrogen and phosphorus in especially short supply.
  - This is a primary reason for the relatively low productivity of open oceans.
- Marine phytoplankton is most productive where upwellings bring nutrient-rich waters to the surface.
  - These areas (usually in Antarctic seas) are more productive than tropical seas.
  - Thermal vent communities are also very productive though they are not very widespread and contribute little to marine productivity.
- Freshwater ecosystem productivity also varies from the surface to the depths in relation to light intensity.
  - Availability of inorganic nutrients is sometimes limiting, but biannual turnovers bring nutrients to the surface waters.

B. As energy flows through an ecosystem, much is lost at each trophic level

The transfer of energy from one trophic level to another is not 100%.

The amount of energy available to each trophic level is determined by NPP and the efficiencies with which food energy is converted to biomass in each link of the food chain.

1. Secondary productivity

Secondary productivity is the rate at which consumers convert the chemical energy in the food they eat into their own biomass.

Consider that herbivores consume only a small fraction of available plant material and they cannot digest all of the organic compounds in what they do ingest (see Campbell, Figure 54.4)

- About 1/6 of the calories is used for growth which adds biomass to the trophic level.
- The remaining organic material consumed is used for cellular respiration or is passed out of the body as feces.
- The energy in the feces stays in the system and is consumed by decomposers.
- The energy used in cellular respiration is lost from the system.
- Carnivores are more efficient at converting food into biomass but more is used for cellular respiration, further decreasing energy available to the next trophic level.

Consequently, energy flows through an ecosystem, it does not cycle within the ecosystem.

2. Ecological efficiency and ecological pyramids

Ecological efficiency is the ratio of net productivity at one trophic level compared to net productivity at the level below, or the percentage of energy transferred from one trophic level to the next.

- Efficiencies can vary greatly depending on the organisms involved, but usually range from 5% to 20%.
- This means that 80% to 95% of the energy available at one trophic level never transfers to the next.

Loss of energy in a food chain can be represented diagrammatically in several ways:

1. A *pyramid of productivity* has trophic levels stacked in blocks proportional in size to the productivity of each level.
   - Usually bottom heavy since ecological efficiencies are low (see Campbell, Figure 54.5)

2. A *biomass pyramid* has tiers that each symbolize the total dry weight of all organisms (standing crop biomass) in a trophic level (see Campbell, Figure 54.6)
   - Most narrow sharply from producers at the base to top-level carnivores at the apex because of energy transfers between trophic levels are so inefficient.
   - Some aquatic ecosystems are inverted because producers have a short *turnover time*. They grow rapidly but are consumed rapidly, leaving little standing crop biomass.

3. A *pyramid of numbers* is comprised of blocks which are proportional in size to the numbers of individuals present at each trophic level (see Campbell, Figure 54.7).
   - Biomass of top-level carnivores is usually small compared to the total biomass of producers and lower-level consumers.
   - Only about 1/1000 of the chemical energy fixed by photosynthesis flows through a food web to a tertiary consumer.
   - Only 3 to 5 trophic levels can be supported since the energy in the webs is insufficient to support another trophic level.
   - Predators (top-level consumers) are highly susceptible to extinction when their ecosystem is disturbed due to their small population and wide spacing within the habitat.

For humans, eating meat is a relatively inefficient way of tapping photosynthetic productivity—eating grains directly as a primary consumer provides far more calories.

III. Cycling of Chemical Elements in Ecosystems

Despite an inexhaustible influx of energy in the form of sunlight, continuation of life depends on recycling of essential chemical elements.

- These elements are continually cycled between the environment and living organisms as nutrients are absorbed and wastes released.
- Decomposition of wastes and the remains of dead organisms replenishes the pool of inorganic nutrients available to autotrophs.

Biogeochemical cycles = Nutrient circuits involving both biotic and abiotic components of ecosystems

A. Biological and geological processes move nutrients among organic and inorganic compartments

There are two general categories of biogeochemical cycles:

1. Elements such as carbon, oxygen, sulfur, and nitrogen have gaseous forms, thus, their cycles are global in character and the atmosphere serves as a reservoir.

2. Elements less mobile in the environment like phosphorus, potassium, calcium and trace elements generally cycle on a more localized scale over the short term. The soil serves as the main reservoir for these elements.

A general scheme of nutrient cycling includes the four main reservoirs of elements and the processes that transfer elements between reservoirs (see Campbell, Figure 54.8).

Reservoirs are defined by two characteristics: whether they contain organic or inorganic materials; and whether or not the materials are directly available for use by organisms.

- The available organic reservoir contains the living organisms and detritus.
    - The nutrients are readily available when consumers feed on one another and when detritivores eat nonliving organic matter..
- The unavailable organic reservoir is comprised of coal, oil, and peat which formed from organisms that died and were buried millions of years ago.
    - These nutrients cannot be directly assimilated.
- The available inorganic reservoir includes all matter (elements and compounds) present in the soil or air and those dissolved in water.
    - Organisms can directly assimilate these nutrients from the soil, air, or water.
- The unavailable inorganic reservoir contains nutrients tied up in limestone and minerals of other rocks.
    - These nutrients cannot be assimilated until released by weathering or erosion.

Various processes are involved in the transfer of nutrients between the four reservoirs which form the basis for biogeochemical cycling. The general schemes were determined by adding small amounts of radioactive tracers to systems in order to follow the movement of elements.

- Weathering and erosion are the primary processes which move nutrients from the unavailable inorganic reservoir to the available inorganic reservoir.
- Erosion is also important, along with the burning of fossil fuels, in moving nutrients from the unavailable organic reservoir to the available inorganic reservoir.
- Nutrients are transferred from the available organic reservoir to the unavailable organic reservoir only by the covering of detritus by sediments and its eventual fossilization to oil, coal, or peat.
- Sedimentary rock formation is the process which moves nutrients from the available inorganic reservoir to the unavailable inorganic reservoir.
- Nutrients enter the available organic reservoir from the available inorganic reservoir through photosynthesis and assimilation by living organisms.
- Nutrients are transferred from the available organic reservoir to the available inorganic reservoir by respiration, decomposition, excretion, and leaching.

The cycling of materials through an ecosystem depends on both biological and geological processes.

1. **The water cycle**

    The essential nature of water to living organisms has many facets:
    - It is essential to maintaining homeostasis in every organism.
    - It contributes to the fitness of the environment.
    - Its movement within and between ecosystems transfers other materials in several biogeochemical cycles.

Most of the water cycle occurs between the oceans and the atmosphere (see Campbell, Figure 54.9).

- Solar energy results in evaporation from the oceans.
- Water vapor rises, cools, and falls as precipitation.
- Over the oceans, evaporation exceeds precipitation; the excess water vapor is moved onto land by winds.
- Precipitation exceeds evaporation and transpiration over land; runoff and ground water balance the net flow of water vapor to land.

The water cycle differs from other cycles in that it occurs primarily due to physical processes, not chemical processes.

2. The carbon cycle

In the carbon cycle, photosynthesis and cellular respiration form a link between the atmosphere and terrestrial environments (see Campbell, Figure 54.10).

- During the carbon cycle, autotrophs acquire carbon dioxide (CO_2) from the atmosphere by diffusion through leaf stomata, incorporating it into their biomass. Some of this becomes a carbon source for consumers, and respiration returns CO_2 to the atmosphere.

Carbon recycles relatively quickly. Plants have a high demand for CO_2, yet CO_2 is present in the atmosphere at a low concentration (0.03%).

- Carbon loss by photosynthesis is balanced by carbon release during respiration.

Some carbon is diverted from cycling for longer periods of time, as when it accumulates in wood or other durable organic material.

- Decomposition eventually recycles this carbon to the atmosphere.
- Can be diverted for millions of years, such as in the formation of coal and petroleum.

The amount of atmospheric CO_2 decreases in the Northern Hemisphere in summer due to increased photosynthetic activity.

- Amounts increase in the winter when respiration exceeds photosynthesis.

Atmospheric CO_2 is increased by combustion of fossil fuels by humans, disturbing the balance.

In aquatic environments photosynthesis and respiration are also important but carbon cycling is more complex due to interaction of CO_2 with water and limestone.

- Dissolved CO_2 reacts with water to form carbonic acid, which reacts with limestone to form bicarbonates and carbonate ions.
- As CO_2 is used in photosynthesis, bicarbonates convert back to CO_2; thus bicarbonates serve as a CO_2 reservoir and some aquatic autotrophs can use dissolved bicarbonates directly as a carbon source.
- The ocean contains about 50 times the amount of carbon (in various inorganic forms) as is available in the atmosphere. The ocean may act as a buffer to absorb excess CO_2.

3. The nitrogen cycle

Nitrogen is a key chemical in ecosystems as it is found in all amino acids which comprise the proteins of organisms.

Although the Earth's atmosphere is almost 80% N_2, it is unavailable to plants since they cannot assimilate this form.

- Nitrogen is available to plants in only two forms: ammonium (NH_4^+) and nitrate (NO_3)

Nitrogen enters ecosystems by either atmospheric deposition or nitrogen fixation.

Atmospheric deposition accounts for only 5 to 10% of the usable nitrogen that enters an ecosystem.

- NH_4^+ and NO_3 are added to the soil by being dissolved in rain or by settling as part of fine dust or other particulates.
- Some plants (epiphytic bromeliads) in the canopy of tropical rain forests have aerial roots that can take up NH_4^+ and NO_3^- from the atmosphere.

Nitrogen fixation is the reduction of atmospheric nitrogen (N_2) to ammonia (NH_3), which can be used to synthesize nitrogenous organic compounds such as amino acids.

- Only certain prokaryotes can fix nitrogen (see Campbell, Figure 54.11).
  - In terrestrial ecosystems some nonsymbiotic soil bacteria and some symbiotic (*Rhizobium*) bacteria fix nitrogen.
  - Cyanobacteria fix nitrogen in aquatic ecosystems.
  - Nitrogen fixing prokaryotes are fulfilling their own metabolic needs, but other organisms benefit since excess ammonia is released into the soil or water.
  - Industrial fixation in the form of fertilizer makes significant contributions to the nitrogen pool in agricultural regions.
- The slightly acidic nature of soil results in NH_3 being protonated to ammonium (NH_4^+).
  - NH_3 is a gas and can evaporate quickly to the atmosphere.
  - NH_4^+ can be used directly by plants.

The nitrogen cycle involves three processes in addition to nitrogen fixation: nitrification, denitrification, and ammonification (see Campbell, Figure 54.11).

1. *Nitrification* is a metabolic process by which certain aerobic soil bacteria use ammonium (NH_4^+) as an energy source by first oxidizing it to nitrite (NO_2^-) and then to nitrate (NO_3^-).
   - While plants can use NH_4^+ directly, the nitrifying bacteria use most of the available NH_4^+ as an energy source.
   - Plants assimilate the NO_3^- released from these bacteria and convert it to organic forms, such as amino acids and proteins.
   - Animals can only assimilate organic nitrogen which they obtain by eating plants and other animals.
2. *Denitrification* occurs when bacteria obtain the oxygen necessary for their metabolism from NO_3^- rather than O_2 under anaerobic conditions. This process returns nitrogen to the atmosphere by converting NO_3^- to N_2.
3. *Ammonification* is the decomposition of organic nitrogen back into ammonium.
   - Carried out mainly by decomposer bacteria and fungi
   - Process is especially important because it recycles large amounts of nitrogen to the soil

Some important aspects of the nitrogen cycle to remember include:

- Prokaryotes serve as vital links at several points in the cycle.
- Most of the nitrogen cycling involves nitrogenous compounds in the soil and water.

- While atmospheric nitrogen is plentiful, nitrogen fixation contributes only a small fraction of the nitrogen assimilated by plants; however, many species of plants depend on symbiotic, nitrogen-fixing bacteria in their root nodules as a source of nitrogen in a form that can be assimilated.
- Denitrification returns only a small amount of N_2 to the atmosphere.
- Most assimilated nitrogen comes from nitrate, which is efficiently recycled from organic forms by ammonification and nitrification.
- The majority of nitrogen in most ecosystems is recycled locally by decomposition and reassimilation, although nitrogen exchange between the soil and atmosphere are of long-term importance.

4. **The phosphorus cycle**

Phosphorus is a major component of nucleic acids, phospholipids, ATP, and a mineral in bones and teeth.

The phosphorus cycle is relatively simple since it does not have a gaseous form and it occurs in only one important inorganic form, phosphate.

Phosphorus cycles locally as follows (see Campbell, Figure 54.12):

- Weathering of rock adds phosphate to the soil.
- Producers absorb the soil phosphate and incorporate it into molecules.
- Phosphorus is transferred to consumers in organic form.
- Phosphorus is added back to the soil by excretion by animals and by decomposition of detritus by decomposers.
- Phosphorus cycling is localized since humus and soil particles bind phosphate.
  - Some leaching does occur and phosphate is lost to the oceans through the water table.
  - Weathering of rocks keeps pace so terrestrial systems are not depleted.
  - Phosphate that reaches the oceans accumulates in sediments and becomes incorporated into rocks which may eventually be exposed to weathering.

Phosphorus may limit algal productivity in aquatic habitats.

- Production in these habitats is stimulated by the introduction of phosphorus in the form of sewage or runoff from fertilized agricultural areas.

B. **Decomposition rates largely determine the rates of nutrient cycling**

The rate of decomposition has a great impact on the timetable for nutrient cycling.

- The rate of decomposition (and thus nutrient cycling) is affected by water availability, oxygen, and temperature.
- Decomposition of organic material in the tropical forests usually occurs in a few months to a few years.
- It takes an average of four to six years for decomposition to occur in temperate forests.
- Decomposition in the tundra may take 50 years.
- In aquatic ecosystems, where most decomposition occurs in anaerobic bottom muds, decomposition may occur even more slowly than in the tundra.

Soil chemistry and the frequency of fires also influence nutrient cycling times.

Some key nutrients are present in the soil of tropical rain forests in levels much lower than those found in temperate forests. Several conditions influence this paradox:

- There is rapid decomposition in tropical areas due to warm temperature and abundant water.

- The large biomass of tropical rain forests creates a high demand for nutrients, which are absorbed as soon as they become available through the action of decomposers.
  - About 10% of the nutrients are in the soil; 75% are present in the woody parts of trees.
- Relatively little organic material accumulates as litter due to the rapid decomposition.
- The low nutrient content of the soil results from the rapid cycling time.

The soil in temperate forests may contain 50% of all organic material in the ecosystem.

- The rate of decomposition is slow.
- The nutrients present in detritus and soil may stay there for long periods before being assimilated.

The sediments of aquatic systems form a nutrient sink and there must be an interchange between the bottom layers of water and the surface for the ecosystem to be productive.

- The rate of decomposition in the sediments is very slow.
- Algae and aquatic plants usually assimilate their nutrients directly from the water.

C. Field experiments reveal how vegetation regulates chemical cycling: *science as a process*

Long-term ecological research (LTER) is being used to examine the dynamics of many natural ecosystems over relatively long periods of time.

Since 1963, scientists have been studying nutrient cycling in a forest ecosystem under natural conditions and after vegetation is removed. The study site is the Hubbard Brook Experiment Forest in New Hampshire.

- The team first determined mineral budgets of six valleys by measuring inflow and outflow of several key nutrients.
- Rainwater was collected to measure amounts of water and dissolved minerals added to the ecosystem.
- Water and mineral loss were monitored by using small concrete dams with a V-shaped spillway across the creek at the bottom of each valley (see Campbell, Figure 54.14).
- Scientists found that 60% of the water added by rainfall exits through streams and 40% is lost by plant transpiration and evaporation from soil.
- They also found that mineral inflow and outflow were nearly balanced and were small compared to minerals being recycled within the forest ecosystem.
  - Only about 0.3% more Ca^{++} exited a valley through its creek than was added by rainwater. Net mineral losses were probably replaced by chemical decomposition of bedrock.
  - During most years, some net gains of a few mineral nutrients occurred.

In 1966, after logging an experimental area and preventing reforestation, comparisons were made over a three-year period.

- Water runoff increased by 30 to 40% (no plants were left to absorb and transpire water).
- Net losses of minerals were very large:
  - Nitrate loss increased 60-fold (water nitrate levels made the water unsafe for drinking).
  - Calcium loss increased 400%
  - Potassium loss increased 1500%.

The study demonstrated the importance of plants in retaining nutrients within an ecosystem and the effects of human intrusion into a system.

- None of the watersheds was undisturbed by human activity even when the study began. Acid precipitation has leached most of the Ca^{2+} from forest soil, resulting in increased levels of Ca^{2+} in stream water. By the 1990s, the forest plants stopped adding new growth, apparently due to the lack of Ca^{2+}.

IV. Human Impacts on Ecosystems

The ever increasing human population has intruded into the dynamics of most ecosystems through human activities or technology.

- Some natural systems are totally destroyed while others have had major components (trophic structure, energy flow, chemical cycling) disrupted.
- Most effects are local or regional, while others are global in scale (e.g., acid rain).

A. The human population is disrupting chemical cycles throughout the biosphere

Human activity often removes nutrients from one part of the biosphere and adds them to another.

- May deplete one area of key nutrients while creating an excess in another area
- These occurrences disrupt the natural equilibrium of chemical cycles in both areas.

Farming exhausts the natural store of nutrients as crop biomass is removed from an area, this greatly reduces the amount of nutrients recycled. Supplements must then be added in the form of fertilizer.

- Nutrients in crops soon appear in human and livestock wastes, and then turn up in lakes and streams through sewage discharge and field run-off.
- Once in aquatic systems, these nutrients may stimulate excessive algal growth which degrades the system.
- Consequently, disruptions can flow from one system to another.

1. Agricultural effects on nutrient cycling

As the human population has continued to grow, greater demands for production of food has resulted in natural habitats being converted to agricultural use. This has resulted in:

- Intrusions into the cycling of nutrients
- Overharvesting of natural populations of food species
- Introductions of toxic compounds into ecosystems in the form of pesticides

After natural vegetation is cleared from an area, the time period during which no additional nutrients need to be added to new agricultural ecosystems varies greatly.

- Nutrient reserves in the soil will support crops for some time after the natural vegetation has been removed.
- These nutrients are not recycled locally since they are removed from the system as crop biomass.
- Some new farmlands in the tropics can be farmed for only one or two years.
  - Remember, in the tropical rain forests only about 10% of the nutrients are in the soil.
- In temperate areas, crops may be grown for many years due to the nutrients present in the soil.
- When nutrients are added, it is normally in the form of industrially synthesized fertilizers.

The nitrogen cycle of an area is greatly impacted by agriculture.

- Breaking up and mixing the soil increase the rate of decomposition of organic matter.
- This releases usable nitrogen, which is taken up by the crop and exported from the system at harvest.
- Nitrates remaining in the soil are quickly leached out of the system.
- Fertilizers are applied to replace the lost nitrogen.
  - Human activities have approximately doubled the Earth's supply of fixed nitrogen.
  - Excess nitrogen in fertilizers leaches into the water table.
  - Increased nitrogen fixation is also associated with a greater release of nitrogen compounds into the air by denitrifiers.
    - Nitrogen oxides can contribute to atmospheric warming, to the depletion of atmospheric ozone, or to acid precipitation.
  - Excess algal and bacterial growth typically results from an overabundance of nitrogen entering surface waters.

2. Accelerated eutrophication of lakes

Lakes are classified on a scale of increasing nutrient availability as oligotrophic, mesotrophic, or eutrophic.

- Oligotrophic lakes have low primary productivity because mineral nutrient levels will not support large phytoplankton populations.
- In other lakes, basic and watershed characteristics cause the addition of more nutrients that are captured by the primary producers and continuously recycled through the lake's food webs.
- Overall productivity is higher in mesotrophic lakes and highest in eutrophic ones.

Sewage, factory wastes, livestock runoff, and fertilizer leaching increases inorganic nutrient levels in waters and results in cultural eutrophication.

- This enrichment often results in explosive growth of photosynthetic organisms.
  - Large algal blooms occur; shallow areas become choked with weeds.
- As these producers die, metabolism of detritivores consumes all the oxygen in the water and many species die.

B. Toxins can become concentrated in successive levels of food webs

A variety of toxic chemicals, including unnatural synthetics, are dumped into ecosystems.

- Many cannot be degraded by microbes and persist for years or decades.
- Some are harmless when released but are converted to toxic poisons by reactions with other substances or by the metabolism of microbes (e.g., conversion of mercury to methyl mercury).

Organisms acquire toxic substances along with nutrients or water, some are metabolized and excreted wile others accumulate in their tissues (e.g., DDT, PCBs).

Biological magnification = Process by toxins become more concentrated with each successive trophic level of a food web; results from biomass at each trophic level being produced from a much larger biomass ingested from the level below.

- Top level carnivores are usually most severely affected by toxic compounds released into the environment.

The pesticide DDT is a well known example of biological magnification (see Campbell, Figure 54.16).

- It is used to control mosquitoes and agricultural pests.
- DDT persists in the environment and is transported by water to areas away from the point of application.
- It is lipid-soluble and collects in fatty tissues of animals.
- One of the first signs that DDT was a serious environmental problem was the decline in bird populations that feed at the top of food chains.
  - Reproductive rates declined dramatically because DDT interfered with the deposition of calcium in eggshells and the weight of nesting birds broke the weakened shells.
- DDT use was banned in the United States in 1971 and the affected bird populations have recovered.
- The use of DDT still continues in other parts of the world.

C. Human activities are causing fundamental changes in the composition of the atmosphere

Human activities have resulted in the release of many gaseous waste products into the atmosphere.

One problem directly related to nutrient cycling is the rising levels of carbon dioxide.

1. Carbon dioxide emissions and the greenhouse effect

Carbon dioxide emissions have caused atmospheric CO_2 concentrations to increase 14% since 1958. This increase is due to combustion of fossil fuels and burning of wood removed by deforestation.

Some effects of increased carbon dioxide levels might appear to be beneficial while others are definitely detrimental.

- Increased productivity by vegetation would occur with increased CO_2.
  - C_3 plants are more limited than C_4 plants by CO_2, so spread of C_3 species into habitats previously favoring C_4 species may have important natural and agricultural implications.
- Temperature increases with increased CO_2 concentration since CO_2 and water vapor absorb infrared radiation and slows its escape from Earth.
  - Called the greenhouse effect
  - A number of studies predict a doubling of CO_2 by the end of the 21st century and an associated average temperature increase of about 2°C above that in 1990.
  - Scientists are predicting a variety of scenarios based on the global warming trend.
    - Some predict warming near the poles will result in melting of polar ice and flooding of current coastal areas.
    - A warming trend would probably alter geographical distribution of precipitation, which could have major agricultural implications.
  - Ecologists are studying the records of pollen cores to determine how past temperature changes have affected vegetation.

2. Depletion of atmospheric ozone

Depletion of atmospheric ozone weakens a protective layer in the stratosphere that absorbs ultraviolet radiation.

- Much of the ultraviolet radiation is absorbed by an ozone layer 17 to 25 km above the Earth's surface.

- Destruction of the ozone layer is largely due to accumulation of chlorofluorocarbons used as aerosol propellants and in refrigeration.
- Breakdown products of chlorofluorocarbons include chlorine which rises to the stratosphere where it reacts with ozone (O_3) and reduces it to atmospheric oxygen (O_2).
  ⇒ The chlorine is released in other reactions and reacts with additional ozone molecules.
- Ozone depletion is best documented over Antarctica but levels in the middle latitudes have decreased 2% – 10% in the last 20 years.

Ozone depletion could have serious consequences.

- Increases are expected in lethal and nonlethal forms of skin cancer and cataracts among humans.
- Unpredictable effects on crops and natural communities (especially phytoplankton) are expected.

D. The exploding human population is altering habitats and reducing biodiversity worldwide

The growth of human populations, human activities, and our technological capabilities have disrupted the trophic structure, energy flow, and chemical cycling of ecosystems in most areas of the world.

Some effects are local while others affect the biosphere's distribution and diversity of organisms.

The destruction of natural systems due to human encroachment has resulted in only a small proportion of natural, undisturbed habitat remaining in existence.

- Only 15% of the original primary forest and just 1% of original tallgrass prairie remain in the United States.
- Tropical rainforests are being cut at a rate of 500,000 km^2 per year and will be eliminated in a couple of decades.
- Human activities that disrupt entire systems include development, logging, war, and oil spills.

One result of the destruction of natural habitat will be the loss of biodiversity.

REFERENCES

Campbell, N., et al. *Biology*. 5th ed. Menlo Park, California: Benjamin/Cummings, 1999.

Cohen, Joel E. *Food Webs and Niche Space*. Princeton, New Jersey: Princeton University Press, 1978.

Krebs, C.J. *Ecology: The experimental analysis of distribution and abundance*, 4th ed. New York, NY: HarperCollins College Publishers, 1994.

Odum, H.T. *Systems Ecology: And Introduction*. New York: Wiley, 1983.

Ricklefs, R.E. *Ecology*. 3rd ed. New York: Chiron Press, 1990.

CHAPTER 55
CONSERVATION BIOLOGY

OUTLINE

I. The Biodiversity Crisis: An Overview
   A. Numerous examples indicate that estimates of extinction rates are on track
   B. The major threats to biodiversity are habitat destruction, over-exploitation, and competition by exotic species
   C. Biodiversity is vital to human welfare
   D. Change in ecological and evolutionary time is the focus of conservation biology

II. The Geographic Distribution of Biodiversity
   A. Gradual variation in biodiversity correlates with geographical gradients
   B. Biodiversity hot spots have high concentrations of endemic species
   C. Migratory species present special problems in conservation

III. Conservation at the Population and Species Levels
   A. Sustaining genetic diversity and the environmental arena for evolution is an ultimate goal
   B. The dynamics of subdivided populations apply to problems caused by habitat fragmentation
   C. Population viability analyses examine the chances of a species persisting or becoming extinct in the habitats available to it
   D. Analyzing the viability of selected species may help sustain other species: *science as a process*
   E. Conserving species involves weighing conflicting demands

IV. Conservation at the Community, Ecosystem and Landscape Levels
   A. Edges and corridors can strongly influence landscape biodiversity
   B. Nature reserves must be functional parts of landscapes
   C. Restoring degraded areas is an increasingly important conservation effort
   D. Sustaining development goals are reorienting ecological research and will require changing some human values

OBJECTIVES

After reading the text and attending lecture, the student should be able to do the following:
1. List the major threats to biodiversity and give example of each.
2. Describe why biodiversity is important to humans.
3. Describe the three concepts upon which the field of biodiverstiy emerged.
4. Describe the goal of conservation biology.
5. Describe how biodiversity is distributed.
6. Define the term, "biodiversity hot spot."

7. Describe the problems presented to conservation by migratory species

8. Describe how the biodiversity crisis extends throughout the hierarchy of biological organization.

9. Describe how habitat fragmentation affects population dynamics.

10. Define "source habitat" and "sink habitat" and discuss how these terms relation to conservation efforts.

11. Describe how population viability analysis as well as estimates of minimum viability size and effective population size are used to evaluate the chances of a species persisting or becoming extinct.

12. Give examples of how predictive models are being used in conservation efforts.

13. Describe the conflicting demands that arise in conservation management plans.

14. Describe how edges and corridors influence landscape biodiversity.

15. Discuss why nature reserves are important to preserving biodiversity and why conservation efforts will involve working in landscapes dominated by humans.

16. Describe why restoring degraded areas is an important part of conservation biology an how bioremediation and augmentation play a role in restoration efforts.

17. Describe how sustainable development goals are reorienting ecological research and will require changes some human values.

KEY TERMS

conservation biology	endangered species	population size	restoration ecology
biodiversity crisis	threatened species	minimum dynamic area	bioremediation
source habitat	metapopulation	effective population size	sustainable development
sink habitat	population viability	landscape ecology	Sustainable Biosphere
biodiversity hot spot	analysis	movement corridor	Initiative
endemic species	minimum viable	zoned reserve systems	

LECTURE NOTES

The planet is populated with a vast richness of living organism—so vast that we have only cataloged a minority of the species that exist. Unfortunately, we are altering ecosystems and ecosystem process to an extent that we are accelerating the extinction of species, creating a *biodiversity crisis*. *Conservation biology*, a recently conceived subdiscipline of biology, is dedicated to countering the biodiversity crisis.

I.   **The Biodiversity Crisis: An Overview**

Extinction is a natural phenomenon, but it is the current *rate* of extinction that underlies the biodiversity crisis.

The high rate of ecosystem degradation is being causes by one species—*Homo sapiens*.

A. **Numerous examples indicate that estimates of extinction rates are on track**

Extinction rates are usually expressed as the number or percentage of species expected to become extinct in an area in a unit of time.

Estimates are difficult at best.

Since birds are among the most studied animals, the extinction rates for less well-known, nonavian taxa are sometimes based on the rate of loss of bird species.

Most often, extinction rates are estimated from the concept of species-area relations in which the number of species in an area is directly related to the size of the area.

- This rule predicts that, on average, about 50% of the total number of species will be lost in an area where 90% of the habitat is lost.

The absence of clear documentation of the rate of extinction has led some to argue that there is no reason to worry at this time. Population census data, however, indicate that extinction of known organisms is occurring at an alarming rate.

- 11% of the 9040 known species of bird are endangered
- 680 of the ca. 20,000 known plant species are in danger of becoming extinct by the year 2000
- Approximately 20% of the known species of freshwater fish have become extinct or have become threatened during recorded history.

In order to know for certain that a given species is extinct, we must know its exact distribution and habits. However, we do not have a complete catalogue of biodiversity and knowledge of the geographic distribution and ecological roles of Earth's species, thus, our understanding remains incomplete.

B. The major threats to biodiversity are habitat destruction, over-exploitation, and competition by exotic species

The most significant threat to biodiversity is human alteration of habitat. Human activities which disrupt entire systems include development, logging, war, and oil spills.

- 73% of the IUCN's designations of extinct, endangered, vulnerable, and rare species are related to destruction of natural habitats.
- Marine biodiversity also is threatened by human activity. About 93 % of the coral reefs (reefs are estimated to support about 1/3 of the known species of fish) have been damaged; if the current rate of destruction continues, 40% to 50 % of the reefs could be lost within 30 to 40 years.

Overexploitation of wildlife by humans is another source of threat.

- Species threatened by excessive commercial harvest or sport hunting include whales, American bison, Galapagos tortoises, and numerous fishes.
- The often illegal trade of rare animals and animal products also jeopardizes many species.

The introduction of exotic species can cause a variety of problems. Although most transplanted species fail to survive, there are notable exceptions.

- The introduction of Nile perch into Lake Victoria in east Africa has resulted in the loss of 200 of 300 species of chichlids (see Campbell, Figure 55.1).
- Fire ants, which were accidentally introduced into the southern United States from Brazil in 1918, have continued to spread northward (see Campbell, Figure 55.2).
- Displacement by introduced species is considered at least partially responsible for 68% of the IUCN's listings of extinct, endangered, vulnerable, and rare species.

C. Biodiversity is vital to human welfare

Why should we care about the loss of biodiversity?

Answers to this question range from general to specific:

- *Biophilia*, the human sense of connection to nature and other forms of life, is centered around aesthetics and ethics.
- Biodiversity is a crucial natural resource and threatened species could provide crops, fibers, and medicines.
  - 25% of the prescriptions dispensed from pharmacies in the United States contain substances derived from plants.

- In the 1970s, alkaloids isolated from the rose periwinkle of Madagascar were found to inhibit cancer cell growth and result in remission of childhood leukemia and Hodgkin's disease.

  - The loss of species results in the loss of genes and all the genetic potential.

  - Humans are dependent on ecosystems and other species. By allowing the extinction of species and degradation of habitats to continue, we are taking a risk with our own species survival.

In an effort to influence policy-making, ecologists and economists have estimated the cost of replacing ecosystem "services" as a measure of the services' value at US$33 trillion.

D. Change in ecological and evolutionary time is the focus of conservation biology

Globally, an area half the size of the U.S. has been protected as natural areas (about 3% of the plant's land surface).

Three concepts form the roots of conservation biology

1. *Preservation* is the practice of setting aside select areas to remain natural and undeveloped.

2. *Resource conservation* is a management scheme aimed at balancing "multiple uses" of natural resources (e.g., agricultural, industrial, preservation, recreation)

3. *Evolutionary/ecological view* recognizes that natural systems are the result of millions of years of evolution and that ecosystems processes are critical for the maintenance and proper function of the biosphere.

The goal of conservation biology is preserve individuals species *and* to sustain ecosystems, where natural selection can continue to function and to maintain the genetic variability upon which it acts.

- Conservation biology follows the ecological tenets of nonequilibrium discussed in Chapter 53 and recognizes that disturbance is a natural force.

- Consideration of human presence is vital in conservation biology because no ecosystems are unaffected by humans. Conservation biology seeks to foster human activities that sustain ecosystems and reduce the current rate of environmental degradation.

II. The Geographic Distribution of Biodiversity

Biodiversity is not evenly distributed and there are recognizable patterns of distribution, including clines (gradual variation), hot spots, concentrations of diversity, and ranges of migratory species.

A. A gradual variation in biodiversity correlates with geographical gradients

Biogeographers have long recognized the existence of clines in species diversity in the form of major geographical gradients.

- The number of terrestrial bird species in North and Central America increases steadily from the Arctic to the tropics (see Campbell, Figure 55.4).

- The number of marine benthic species increases with depth.

Four hypotheses have been developed to explain clines, and more generally, the factors that influence patterns of diversity in all communities.

1. *Energy availability.* Holds that because solar radiation is greatest in the tropics, the resource base is greatest there.

2. *Habitat heterogeneity.* Holds that tropical regions experience more local disturbances that contribute to greater environmental patchiness; the greater patchiness allows a greater diversity of plants species to form a resource base for diverse communities of animals.

3. *Niche specialization.* Holds that the stability and predictability of tropical climate may allow organisms to specialize on a narrower range of resources; smaller niches would reduce competition and contribute to greater species diversity.

4. *Population interactions.* Holds that diversity is self-propagating because population interactions coevolve, and the resulting predator-prey and symbiotic interactions in a diverse community prevent any populations from becoming dominant.

Many ecologist believe that a complex combination of factors is responsible for clines.

B. Biodiversity hot spots have high concentrations of endemic species

Biodiversity hot spot = Relatively small areas with exceptional concentrations of species

Endemic species = A species found nowhere else

Biogeographers have identified 18 vascular plant hot spots (see Campbell, Figure 55.5).

- These hot spots contain about 20% of the known vascular plant species and 7% of all land vertebrate species.

- Six of the 18 hot spots have lost close to 90% of their original habitats to human development; as a result, the biodiversity hot spots also are hot spots of extinction.

Islands are hot spots of bird extinction.

- 30% of all bird species are endemic.

- Approximately 90% of the 104 species of birds lost in the last 400 years were endemic on islands

- Today, all of the areas where over 10% of the bird species are threatened with extinction are islands (e.g., Hawaii, Philippines, New Zealand).

- It seems likely that most of the non-avian threatened species also are endemic on islands.

In the U.S., most of the endangered species are found in the areas with the most endemic species: Hawaii, southern California, southern Appalachians, southeastern coastal states. Most of these species are threatened because of loss of habitat due to human population growth and agriculture.

Studies of biodiversity and recent extinctions show that many threatened, endangered, and potentially endangered species are concentrated in biodiversity hot spots.

- This pattern suggests that with appropriate measures, many species could be saved in relatively small areas.

- The biodiversity crisis is a global problem and focus on hot spots should not detract form efforts to preserve biodiversity in other areas.

C. Migratory species present special problems in conservation

The preservation of migratory species is complicated by a life history that involves residence in multiple jurisdictions.

- Monarch butterflies, for example, migrate from Canada to Mexico (see Campbell, Figure 55.6). Human intrusion is making the migration an "endangered phenomenon."

- Similar situations exist for migratory songbirds, sea turtles, and marine mammals.

Successful conservation efforts for such species require international cooperation and the careful preservation of habitat in both parts of the species' range.

III.   Conservation at the Population and Species Level

Much of the attention of the biodiversity crisis has focused on species.

Endangered species are species that are in danger of extinction in all or a significant portion of its range

Threatened species are species that are likely to become endangered in the foreseeable future throughout all or a significant portion of its range.

A.   Sustaining genetic diversity and the environmental arena for evolution is an ultimate goal

Species are only one component of earth's biodiversity. Other components include:
- Genetic variability within populations of species
- Myriad biotic and abiotic factors that provide the arena for evolution

Modern conservation science attempts to concentrate more on sustaining ecosystem processes and the evolutionary lineages that species represent than on conserving individual species.

Alteration of ecosystems by human activities already makes it impractical to conserve all the genetic diversity within most species.

Conservation biology has focused on understanding the dynamics of small populations, diagnosing declines, assessing the factors responsible for a population's decline, and determining how to revise declines and sustain small, often fragmented populations.

- Currently, conservation efforts lag far behind the rate of decline and loss of species.
- Many species are at critically low numbers and the strategy has mainly been to reverse the trend.
- Conservation biologists also use some features of crisis management and apply some untested hypotheses and concepts.

B.   The dynamics of subdivided populations apply to problems caused by habitat fragmentation

Degradation of habitats often leads reduction in the area of suitable habitat as well as to fragmentation of the remaining area (see Campbell, Figure 55.7).

- Some ecologists have likened fragmentation to islands surrounded by areas of human activity (see Campbell, Figure 53.21).
- The island model may be overly simplistic, and concepts developed by studying subdivided populations may prove to be more useful to conservation efforts.

Metapopulation = A subdivided population or a network of subpopulations of a species

- Vary greatly, depending on size, quality, spatial arrangement, and persistence of habitat patches

The subpopulations of a metpopulation are separated into habitat patches that vary in quality.

- Patches with abundant, high-quality resources tend to have persistent subpopulations that produce more offspring.
- Low-quality patches may be populated only when new individuals reared in high-quality patches disperse to them.
- Dispersal is essential to maintaining genetic variability within subpopulations; a subpopulation that is cut off from others may eventually become genetically extinct.

Human activity may impact population structure.

- Fragmentation may result in the conversion a population to a metapopulation with reduced genetic variability.

- Human encroachment on a metapopulation (e.g., loss or reduction of habitat patches; restriction of dispersal) may decrease the number, size, and or genetic variability of subpopulations.

1. The source and sink dynamics in metapopulations

Reproductive rates of metapopulations vary widely among habitat patches

- A *source habitat* is one where a population's reproductive success exceeds mortality.
- A *sink habitat* is one where a population's mortality exceeds reproductive success.

Distinguishing sources from sinks requires:

- Detailed analysis of birth rates and death rates
- Identification of factors that affect dispersal
- Identification of habitat factors that are critical to a species

Sustaining metapopulations created or altered by human habitat fragmentation requires the identification and protection of source habitats.

- For example, efforts to sustain the peregrine falcon population in California focused on stocking the wild subpopulation in southern California with captive-reared birds until it was discovered that southern California was a sink habitat. Efforts are now directed at the source habitat in northern California.

Understanding source and sink dynamics is also essential to designing the most effective nature reserves.

C. Population viability analyses examine the chances of a species persisting or becoming extinct in the habitats available to it

Population viability analysis (PVA) is a method of predicting whether or not a particular given will persist in a specific environment.

- It is generated by computer simulation.
- It integrates information about a population's genetic variability and life history (e.g., sex ratio, fecundity)
- It considers information of a population's response to environmental factors (e.g., predation, competition)
- It also considers potential effects of planned human activities (e.g., logging, mining).

A PVA usually predicts long-term viability.

- Periodic natural disasters are factored in.
- However, because threatened populations are small, their survival may depend on chance events.

The development of PVAs may be a multistep process

- An initial PVA may be developed to predict the viability of a population for a specific number of generations.
- More extensive PVAs may be developed following the initial prediction by further modeling and additional research.

1. Estimating minimum viable population size (MVP)

Most PVAs are designed to predict a species' *minimum viable population size (MVP)*, the smallest number of individuals needed to propagate a population, subpopulation, or species.

- Prediction usually indicates a percent chance of survival and time span.
- MVP size varies widely.

- Some populations of rare birds have remained viable with only 10 breeding pairs.
  - Estimates of MVP aid in predictions about the *minimum dynamic area*, the amount of suitable habitat needed to sustain a viable population.

2. Estimating effective population size

Meaningful MVP estimates require the determination of the *effective population size* (N_e).

- N_e is based on the number of adults that successfully breed (contribute gametes to the next generation).
- N_e is calculated by the following formula:

Ne = (4Nm x Nf) / (Nm + Nf)

where N_m and N_f are the numbers of males and female, respectively, that successfully breed.

- Numerous life history traits can affect N_e; other formulas factor in family size, age at maturation, genetic relatedness among population members, the effects of gene flow between geographically separated populations, and population fluctuations.

3. The effect of genetic diversity on survivability

The management goal of sustaining effective population sizes, N_e, is based on the concern that populations should possess sufficient genetic diversity to be evolutionarily adaptable.

- Populations with low N_e are prone to inbreeding, reduced heterozygosity, and the random effects of genetic drift and bottlenecking.
- For many species (e.g., especially those that reproduce slowly, such as the cheetah and grizzly bear), however, low genetic variability appear normal.

Low genetic variability does not always lead to permanently small populations.

- After intense hunting reduced the number of northern elephant seals to about 20 individuals, the population has rebounded to about 30,000.

Reduced genetic variability by itself may not be critical the survival of wild populations.

D. Analyzing the viability of selected species may help sustain other species: *science as a process*

Modeling requires extensive background research and time, thus only populations of relatively few threatened or endangered species will be systematically analyzed.

What we learn from the viability studies of one species may help us develop strategies to sustain other species.

Patrick Nantel's PVA of two species of edible herbaceous plants, American ginseng and wild leek, in Canada, exemplifies how predictive models can be used in planning conservation strategy (see Campbell, Figure 55.11).

- His data included trends in the numbers of individuals capable of reproducing in several populations for two-, three-, and four-year periods.
- Computer simulations projected the likely effect of environmental influences on these populations.
- The PVA indicated that most populations of these two species in Canada are too small to persist if harvested.

One of the first PVAs was part of a long-term study of grizzly bears in Yellowstone National Park (see Campbell, Figure 55.12).

- Grizzly bears require very large habitat sizes (e.g., in western Canada, a population of 50 individual requires 5 million hectares)

- In the U.S., the grizzly bear population is fragmented into 6 isolated subpopulations; most of the subpopulations have fewer than 100 individuals (the Yellowstone subpopulation numbers about 200).
- PVA models predicted that a grizzly bear population of 70 to 90 individuals in a suitable habitat have about a 95% chance of surviving for 100 years.
- The N_e of grizzly bear populations was estimated to be 25% of the total subpopulation size (e.g., 50 for the Yellowstone subpopulation).

E. Conserving species involves weighing conflicting demands

The determination of viable population numbers and habitat needs make up only part of the effort to save species.

Often, it is necessary to weight a species' needs against other conflicting demands.

- In the Pacific Northwest, for example, an ongoing debate exists over saving habitat for populations of spotted owl, timber wolf, grizzly bear, and bull trout versus demands for jobs in timber, mining, and other resource extraction industries.
- Habitat use is almost always at issue.

The magnitude of the biodiversity crisis raises practical considerations. Since we will not be able to save every endangered species, which ones are most important?

- Keystone species have disproportionately large impacts relative to their numbers. Identifying them and finding ways to sustain their populations can ensure the continuance of numerous other species and can be central to the survival of whole communities.

IV. Conservation at the Community, Ecosystem, and Landscape Levels

Today, conservation efforts are aimed at sustaining the biodiversity of entire communities and ecosystems. Some efforts are even directed at the broader level of landscapes, regional assemblages of interacting ecosystems (e.g., forest or forest patches, adjacent open fields, wetlands, streams)

Landscape ecology is the application of ecological principles to the study of land use patterns.

A. Edges and corridors can strongly influence landscape biodiversity

The boundaries or edges between and within ecosystems are the defining features of landscapes (see Campbell, Figure 55.13).

Edges have their own communities of organisms in association with their physical features; some organisms thrive in edge communities because they need resources from each of the adjacent ecosystems.

The proliferation of edge species can have either positive or negative effects on the biodiversity of the community.

- In the tropical rain forest of Cameroon, edge areas are important regions of speciation.
- In communities in which edges have expanded because of human alteration, however, there is often a reduction of biodiversity as a result of abundance of edge-adapted species (e.g., brown-headed cowbird takes over the nests of other birds).

Another important feature of landscapes, particularly in habitats that have been severely fragmented is a *movement corridor*, a strip or series of clumps of quality habitat that connect patches (e.g., streamside habitats often function as corridors, and in some areas where human alteration is significant, artificial corridors have been constructed).

- Corridors promote dispersal and help sustain metapopulations.

- In some instances, corridors can be detrimental (e.g., avenues for the spread of disease).

B. Nature reserves must be functional parts of landscapes

Terrestrial and aquatic parks as well as other wilderness and protected areas are important to the efforts of conservation biology.

- In some regions, protected sites are the sole habitat of endangered or threatened species.
- Protected reserves are subject to outside influences.

A preserve should be planned so as to be self-regulating and to allow natural disruption; policies to the contrary are counterproductive (e.g., preventing the spread of wildfires).

Consideration of patch dynamics, metapopulation dynamics, edges, and corridor effects in the design of management plans for protected areas is necessary because of ever-increasing human-induced disturbance and fragmentation. Still, there are questions:

- *Is one large habitat better than a group of smaller ones?* The ultimate answers may be governed more by human land use patterns than on ecological considerations.
- Some countries use a *zoned reserve system* for landscape management (see Campbell, Figure 55.14). In such a system, the core preserve is surrounded by successive layers that increase in human alteration as distances increase from the core.

Projections estimate that less than 10% of the biosphere will ever be protected as nature reserves.

- Conservation measures must, therefore, be put to work in landscapes dominated by humans.

C. Restoring degraded areas is an increasingly important conservation effort

Some areas are damaged so badly by human activity that they are abandoned.

- Soils of tropical lands become unproductive quickly and are often abandoned with five years of being cleared for farming.
- Mining is a protracted human activity that can render an area unusable by wildlife for many years, even following cessation of the activity.
- Many ecosystems are damaged accidentally (e.g., oil spills).
- Because the natural rates of recovery are slower than the rate of human-induced degradation, the area of degraded habitats and ecosystems is increasing.

Restoration ecology applies ecological principles to find ways to restore degraded ecosystems as close to their original state as possible.

Two key approaches to restoration ecology are bioremediation and augmentation of ecosystem processes.

1. *Bioremediation* makes use of living organisms (e.g., prokaryotes, fungi) to detoxify a polluted ecosystem. For example, the bacterium *Pseudomonas* has been used to clean up oil spills on beaches.
2. Augmenting ecosystem processes is accomplished by supplying the critical components (e.g., nutrients) that have been identified to restrict the rate of recovery. For example, encouraging the growth of plants that perform well in nutrient-deficient soils can speed up succession and ultimately lead to the recovery of a damaged habitat.

D. Sustainable development goals are reorienting ecological research and will require changing some human values

In order to make informed decisions about how best to conserve the Earth's resources, it is necessary to understand the complex interactions of the biosphere.

Many countries have adopted the concept of *sustainable development*, a plan than provides for the long-term prosperity of human societies and the ecosystems that support them.

The goal of the *Sustainable Biosphere Initiative* is to acquire the basic ecological information needed for intelligent development, management, and conservation of the Earth's resources. This will include studies on:

- Global change, including interactions between climate and ecological processes
- Biological diversity and its role in maintaining ecological processes
- The ways in which productivity of natural and artificial ecosystems can be sustained

The nature of ecological research will have to be reoriented but the importance cannot be overstated due to the current state of the biosphere.

REFERENCES

Campbell, N., et al. *Biology*. 5th ed. Menlo Park, California: Benjamin/Cummings, 1999.

Krebs, C.J. *Ecology: The experimental analysis of distribution and abundance*, 4th ed. New York, NY: HarperCollins College Publishers, 1994.

Meffe, G.K. and Carroll, C.R. Principles of Conservation Biology, 2nd ed. Sunderland, MA: Sinauer, 1997.

Ricklefs, R.E. *Ecology*. 3rd ed. New York: Chiron Press, 1990.